Atomic Masses of the Elements

Name	Symbol	Atomic Number	Atomic Mass[a]	Name	Symbol	Atomic Number	Atomic Mass[a]
Actinium	Ac	89	(227)	Mendelevium	Md	101	(258)
Aluminum	Al	13	26.98	Mercury	Hg	80	200.6
Americium	Am	95	(243)	Molybdenum	Mo	42	95.94
Antimony	Sb	51	121.8	Moscovium	Mc	115	(289)
Argon	Ar	18	39.95	Neodymium	Nd	60	144.2
Arsenic	As	33	74.92	Neon	Ne	10	20.18
Astatine	At	85	(210)	Neptunium	Np	93	(237)
Barium	Ba	56	137.3	Nickel	Ni	28	58.69
Berkelium	Bk	97	(247)	Nihonium	Nh	113	(286)
Beryllium	Be	4	9.012	Niobium	Nb	41	92.91
Bismuth	Bi	83	208.9	Nitrogen	N	7	14.01
Bohrium	Bh	107	(264)	Nobelium	No	102	(259)
Boron	B	5	10.81	Oganesson	Og	118	(294)
Bromine	Br	35	79.90	Osmium	Os	76	190.2
Cadmium	Cd	48	112.4	Oxygen	O	8	16.00
Calcium	Ca	20	40.08	Palladium	Pd	46	106.4
Californium	Cf	98	(251)	Phosphorus	P	15	30.97
Carbon	C	6	12.01	Platinum	Pt	78	195.1
Cerium	Ce	58	140.1	Plutonium	Pu	94	(244)
Cesium	Cs	55	132.9	Polonium	Po	84	(209)
Chlorine	Cl	17	35.45	Potassium	K	19	39.10
Chromium	Cr	24	52.00	Praseodymium	Pr	59	140.9
Cobalt	Co	27	58.93	Promethium	Pm	61	(145)
Copernicium	Cn	112	(285)	Protactinium	Pa	91	231.0
Copper	Cu	29	63.55	Radium	Ra	88	(226)
Curium	Cm	96	(247)	Radon	Rn	86	(222)
Darmstadtium	Ds	110	(281)	Rhenium	Re	75	186.2
Dubnium	Db	105	(262)	Rhodium	Rh	45	102.9
Dysprosium	Dy	66	162.5	Roentgenium	Rg	111	(272)
Einsteinium	Es	99	(252)	Rubidium	Rb	37	85.47
Erbium	Er	68	167.26	Ruthenium	Ru	44	101.1
Europium	Eu	63	152.0	Rutherfordium	Rf	104	(267)
Fermium	Fm	100	(257)	Samarium	Sm	62	150.4
Flerovium	Fl	114	(289)	Scandium	Sc	21	44.96
Fluorine	F	9	19.00	Seaborgium	Sg	106	(266)
Francium	Fr	87	(223)	Selenium	Se	34	78.96
Gadolinium	Gd	64	157.3	Silicon	Si	14	28.09
Gallium	Ga	31	69.72	Silver	Ag	47	107.9
Germanium	Ge	32	72.63	Sodium	Na	11	22.99
Gold	Au	79	197.0	Strontium	Sr	38	87.62
Hafnium	Hf	72	178.5	Sulfur	S	16	32.07
Hassium	Hs	108	(277)	Tantalum	Ta	73	180.9
Helium	He	2	4.003	Technetium	Tc	43	(98)
Holmium	Ho	67	164.9	Tellurium	Te	52	127.6
Hydrogen	H	1	1.008	Tennessine	Ts	117	(294)
Indium	In	49	114.8	Terbium	Tb	65	158.9
Iodine	I	53	126.9	Thallium	Tl	81	204.4
Iridium	Ir	77	192.2	Thorium	Th	90	232.0
Iron	Fe	26	55.85	Thulium	Tm	69	168.9
Krypton	Kr	36	83.80	Tin	Sn	50	118.7
Lanthanum	La	57	138.9	Titanium	Ti	22	47.87
Lawrencium	Lr	103	(262)	Tungsten	W	74	183.8
Lead	Pb	82	207.2	Uranium	U	92	238.0
Lithium	Li	3	6.941	Vanadium	V	23	50.94
Livermorium	Lv	116	(293)	Xenon	Xe	54	131.3
Lutetium	Lu	71	175.0	Ytterbium	Yb	70	173.0
Magnesium	Mg	12	24.31	Yttrium	Y	39	88.91
Manganese	Mn	25	54.94	Zinc	Zn	30	65.38
Meitnerium	Mt	109	(268)	Zirconium	Zr	40	91.22

[a]Values in parentheses are the mass number of the most stable isotope.

CONCISE | PRACTICAL | INTEGRATED

GENERAL, ORGANIC, AND BIOLOGICAL
CHEMISTRY
FOURTH EDITION

Laura Frost
Florida Gulf Coast University

Todd Deal
Georgia Southern University

Pearson

Senior Courseware Portfolio Manager: Jessica Moro
Director of Portfolio Management: Jeanne Zalesky
Managing Producer: Kristen Flathman
Courseware Senior Analyst: Mary Ann Murray
Courseware Director, Content Development: Barbara Yien
Courseware Editorial Assistant: Amanda Davis
Content Producer: Susan McNally
Rich Media Content Producer: Paula Iborra
Full-Service Vendor: CSC
Full-Service Project Manager: Mary Tindle, SPi-Global
Copyeditor: Joanna Dinsmore
Compositor: SPi-Global
Art Coordinator: Lachina Creative, Inc.
Design Manager: Mark Ong
Interior Designer: Tamara Newnam
Cover Designer: Tamara Newnam
Rights & Permissions Project Manager: Eric Schrader
Rights & Permissions Manager: Ben Ferrini
Photo Researcher: SPi-Global
Manufacturing Buyer: Stacey Weinberger
Director of Field Marketing: Tim Galligan
Director of Product Marketing: Allison Rona
Field Marketing Manager: Chris Barker
Product Marketing Manager: Elizabeth Bell
Director MasteringChemistry Content Development: Amir Said
MasteringChemistry Senior Content Producer: Margaret Trombley
MasteringChemistry Content Producer: Meaghan Fallano

Cover Photo Credit: utima/123RF (front); Anna Kucherova/123RF (back cover)

Library of Congress Cataloging-in-Publication Data
Names: Frost, Laura D., author. | Deal, S. Todd, author.
Title: General, organic, and biological chemistry / Laura Frost (Florida Gulf
    Coast University), Todd Deal (Georgia Southern University).
Description: Fourth edition. | Hoboken, NJ : Pearson, [2020] | Includes index.
Identifiers: LCCN 2018055566| ISBN 9780134988696 (hardcover) | ISBN
    9780134999500 (loose leaf)
Subjects: LCSH: Chemistry--Textbooks. | Biochemistry--Textbooks.
Classification: LCC QD251.3 .F76 2020 | DDC 540--dc23 LC record available at https://lccn.loc.gov/2018055566

www.pearson.com

ISBN 10: 0-13-498869-8;
ISBN 13: 978-0-13-498869-6
(Student edition)

About the **Authors**

LAURA FROST is a Professor of Chemistry at Florida Gulf Coast University and Director of the Whitaker Center for Science, Technology, Engineering, and Mathematics (STEM) Education. She received a Bachelor's degree in chemistry from Kutztown University and her Ph.D. in chemistry with a biophysical focus from the University of Pennsylvania. She has been teaching chemistry in higher education for over 20 years and continues to teach chemistry to students in the health professions.

Professor Frost is actively engaged in teaching and learning in all STEM subjects, particularly chemistry, and uses a guided inquiry approach in her classes. She is very involved in the scholarship of teaching and learning and has demonstrated that the use of inquiry-based activities increases student learning in her one-semester chemistry course for health professions.

Dr. Frost is a member of the American Chemical Society and its Chemical Education division. A member of the University System of Georgia—Faculty Hall of Fame, Dr. Frost has been honored on more than one occasion for her teaching excellence. Her writing has also been recognized, and in 2014, she was elected to the governing council of the Textbook and Academic Authors Association.

Dr. Frost is engaged in STEM education reform through evidence-based teaching and learning at all levels (K–16), as well as through community outreach. She was principal investigator of a successful NSF grant that provides professional development for faculty in evidence-based practices in STEM. She has spoken and attended numerous conferences and workshops on this topic.

DEDICATION

This book is dedicated to my family, Baxter, Iris, and Baxter, whose constant support sustains me, and to the students and faculty who continue to inspire me. **—Laura Frost**

TODD DEAL received his B.S. degree in chemistry in 1986 from Georgia Southern College (now University) in Statesboro, Georgia, and his Ph.D. in chemistry in 1990 from The Ohio State University. He returned to his undergraduate alma mater in 1992 and has served as a faculty member and administrative leader for more than 25 years.

An award-winning educator, Dr. Deal has received numerous awards for excellence in teaching and in service, and was named Professor of the Year by the students of Georgia Southern. Professor Deal has also been recognized by Project Kaleidoscope for his innovative teaching in the sciences.

Dr. Deal is a member of American Chemical Society and its Chemical Education division. He is also a member of the Omicron Delta Kappa National Leadership Honor Society, the International Leadership Association, and the Association of Leadership Educators. Dr. Deal's teaching interests span from chemistry and the hard sciences to the soft, human sciences of leadership and leadership development.

DEDICATION

I dedicate this book to my loving wife, Karen, and to our daughters, Abbie and Anna. Thank you for believing in me. And to my students who inspired me to help them learn; this book is written for you. **—Todd Deal**

Brief Contents

Contents

5 Chemical Reactions 181

6 Carbohydrates—Life's Sweet Molecules 219

9 Acids, Bases, and Buffers in the Body 352

10 Proteins—Workers of the Cell 392

11 Nucleic Acids—Big Molecules with a Big Role 439

Preface

To the Student

How does the body regulate carbon dioxide levels? Why are some pharmaceuticals injected and others taken orally? The key to understanding the answers to these questions starts with chemistry.

General, Organic, and Biological Chemistry was written especially for students interested in pursuing a health science career like nursing, nutrition, dental hygiene, or respiratory therapy. Yet this textbook has applications for all students interested in discovering the concepts of chemistry in everyday situations. Throughout the text, you will find that we have integrated the concepts of general, organic, and biological chemistry to create a seamless framework to help you relate chemistry to your life.

One of our goals in writing this book is to help you become better problem solvers so that you can critically assess situations at your workplace, in the news, and in your world. In this edition, we have kept the problem-solving strategies while increasing their depth to encourage greater understanding.

As you explore the pages of this book, you will encounter materials that

- apply chemistry to your life
- apply chemistry to health careers that interest you
- encourage you to develop problem-solving skills
- help you to work with and learn from your fellow students
- demonstrate how to be successful in this chemistry course and other courses.

As you read this book, you will notice that the language is less formal. Wherever possible, we relate the chemical concepts to objects in everyday life to help you understand chemistry. We also provide several study strategies with this edition, including materials for you to engage with before, during, and after class. Cognitive research in learning tells us that new ideas stick with us better if they are related to things that we already know and if we practice retrieving this information from our memories.

New to This Edition

The Fourth Edition continues to strengthen our strategy of integrating concepts from general, organic, and biological chemistry to give students a focused introduction to the fundamental and relevant connections between chemistry and life. With an emphasis on developing problem-solving skills, guiding the students' reading through Inquiry Questions, and helping students retain information through iterative retrieval practice, this text empowers students to solve problems in applied contexts relating to health and biochemistry.

- Each chapter now begins with a Learning Tip relevant to the chapter content. As students acquire the Learning

Tips throughout the book, they can become more independent learners. The Learning Tips are rooted in the cognitive science literature and are supported with a list of references in the Credits section.

- While the order of chapters remains the same, some sections within the chapters have been reorganized for the better flow of concepts. In Chapter 1, we moved Section 1.6 (How Matter Changes) up to follow Section 1.2, before we begin significant figures and unit conversion. Chapter 7 now begins with Gases and Gas Laws prior to discussing attractive forces and the physical properties of liquids and solids.

- This edition includes an increased number of problems stamped with the health icon 🕇 to highlight health applications of the chemistry content, making the book more relevant for students in the health professions.

- The biochemistry applications offer even more depth than the Third Edition, providing new content on drug solubility and delivery, peptides in celiac disease, common viral diseases, and CRISPR.

- We created new Practicing the Concepts videos for this edition. Each chapter now has two supporting videos. The videos, which run from 3 to 5 minutes, feature author Todd Deal. In the videos, he reviews a big idea or concept from the chapter, then helps students deepen their knowledge and develop their skills. Carefully developed visuals portray concepts vividly, and a pause-and-predict stopping point gives the student a chance for a meaningful concept check.

- Every chapter has been revised, including the sample problems and practice problems. To support areas of chapters with expanded coverage, we added new practice problems.

To the Instructor
Actively Engaging Your Students with Discovering the Concepts

From Laura Frost

Each chapter in *General, Organic, and Biological Chemistry* contains at least two guided inquiry activities, called Discovering the Concepts, at the beginning of some sections. These activities are offered to engage students in groups during class as they construct an understanding of the content in cooperation with their peers. Active learning strategies that include Discovering the Concepts have been shown to increase engagement, learning, and retention (see Freeman et al., 2014). All the information necessary to answer the questions in the activity is included at the beginning of the activity, so students should not need to use other parts of their textbook. An outline of some key points regarding the use of these activities follows.

Facilitation   Faculty can facilitate the use of these activities by managing class time, guiding students to the correct answers instead of giving them the correct answers, and interjecting information during the group work where appropriate to guide student learning. It is also important for the instructor to be familiar with the activity in advance, to anticipate where students might struggle with the questioning.

Organizing Groups   Faculty new to active learning often have questions about organizing student groups: Should I assign students to particular groups? Should I let students pick their own groups? Should I rotate the groups during the term? Because this course is an introductory course with few (if any) prerequisites, random assignment on the first day of class can suffice. However, I encourage diversification by gender, ethnicity, and problem-solving strategies whenever possible. With few exceptions, I have found that students become comfortable with their group members and almost insist on staying in the same group. That being said, I have met many colleagues who use active learning strategies who do rotate groups. This too can be a successful approach, with the caveat that instructors must inform students of their intention to rotate groups well in advance. Some faculty members rotate groups after an exam, some rotate them more frequently, and others keep the same groups for the entire semester. More information on group work can be found in CBE Life Sciences Evidence-Based Teaching Guide on Group Work (Wilson et al., 2018).

Skill Development   Encourage students to review and extend their understanding by completing problems outside of the classroom. The practice problems at the end of each textbook section, and the additional problems and challenge problems at the end of the chapters, are ideal for this purpose.

Group Accountability   To develop skill in written communication, it is essential that a record be kept of the students' activities during the class period by a student acting as group recorder. Other roles such as group manager, presenter, and technician should also be considered to keep the groups on task. An instructor may choose to grade some, all, or none of the activity, or assign participation points during the class period. However you choose to do it, group accountability should be incorporated into your grading scheme in some form.

Individual Accountability   Some form of individual understanding of the activity should also be a part of the assessment. This can be done through quizzing, which also encourages skill development and retrieval practice. I give a short quiz at the beginning of a class based on the activity from the previous class. Quizzing can be done online, as a clicker quiz (works well in large classes), or on paper. The quiz should be reviewed immediately after students complete it, offering some teachable moments for student understanding.

To be efficient, the quizzing process should take no more than 10 minutes of the class period. Quizzing also has the effect of increasing student attendance and decreasing student tardiness if done at the beginning of the class. Individual accountability can also be assessed through monitoring homework, either online or on paper.

Metacognition   To bring their learning full circle, students must reflect on their own learning. I provide groups with a feedback sheet that is to be completed at the end of class by one of the members of the group. The purpose of this reporting is to have students assess what they did and did not learn during the class period. The wording can be general, or changed to reflect the contents of the activity, but should provide the instructor with feedback regarding the effectiveness of the activity. The following general items are suggested for inclusion in the feedback sheet:

- List the primary topic(s) that you learned during today's activity.
- Is there anything that is still not clear regarding today's activity? Please be specific.

Many instructors are concerned with covering the required material in a course, and wonder whether an active classroom will slow the pace of coverage. I offer two thoughts related to this very real concern:

1. Because the activities help students develop problem-solving and critical-thinking skills, students are better able to apply the learned knowledge to related topics not explicitly taught in class. For example, identifying functional groups does not mean that the instructor must show the students all common functional groups in class (as may be done in a lecture). Once the concept of organic families and functional groups is introduced, students should be able to use a table to identify other functional groups, thereby maximizing instructional time in the classroom.
2. Just because a lecturer covers a topic in the classroom does not mean that a student learned it. In fact, faculty members are regularly amazed at the lack of knowledge retention from one course to the next.

I invite you to try Discovering the Concepts. If you find that these activities are helpful in engaging and reaching students, you may want to explore the full set of activities at pearsoncustomlibrary.com for use with the textbook.

I have been field-testing and revising these activities with students since 2006. My own course-based research has shown that students who use these activities perform better on the final exam and have a deeper understanding of the material (as measured by the learning level of the questions they are capable of answering) than students in my courses prior to my use of guided inquiry activities (Frost, 2010). I welcome your comments and questions at lfrost@fgcu.edu regarding their use.

Guiding Student Reading Using the Inquiry Questions

You may have noticed the Inquiry Question ? at the beginning of each section, which encapsulates the main learning goal for that section of text. Although these questions may seem obvious to the instructor, they can provide powerful guidance for the novice learner if you point them out as study strategy. Students can use these questions to guide their reading. I recommend suggesting that as they read, they jot down notes that might help to answer the Inquiry Question by the end of that section. This form of self-explanation is very useful for student retention and understanding. Research shows that when students who are less structured in their studies and think more concretely use embedded questions to guide their reading, they can retain information longer than students who don't use embedded questions (Callendar & McDaniel, 2007).

Chapter Organization and Revision

Throughout the text, we integrated general, organic, and biological chemistry topics using relevant examples and applications to solidify concepts. This text intentionally contains only 12 chapters, allowing all chapters to be covered in a single semester. Each chapter builds upon conceptual understanding and skills learned from previous chapters, providing students with an efficient path through the content and a clear context for how all of the topics connect to one another.

In this edition, at least a quarter of the chapter problems are new or have been modified. We also made extensive chapter revisions as discussed below.

1 Chemistry Basics— Matter and Measurement

- Former Section 1.6, How Matter Changes, has been moved to appear earlier in the chapter, since physical changes and chemical reactions naturally flow from understanding the nature of matter.
- The periodic table has been updated to include all elements to 118.
- The significant figures discussion in Section 1.4 now uses readings on a syringe and a digital thermometer in the discussion of measuring.
- The content on nutrition labels has been updated to reflect recent changes to nutrition labeling.

2 Atoms and Radioactivity

- The more common unit *millirem* is used to describe the biological effects of radiation.
- The Integrating Chemistry feature on "Radioisotopes and Radiation in Cancer Treatment" is expanded to include proton therapy and neutron capture therapy.

3 Compounds—How Elements Combine

- A new Solving a Problem feature, "Drawing Lewis Structures," has been added to Section 3.4.
- Table 3.7 has been updated with examples of molecules showing the preferred bonding patterns for carbon.

4 Introduction to Organic Compounds

- The chapter has been updated throughout to add a focus on the structural aspects of organic chemistry.
- The Integrating Chemistry feature "Pharmaceuticals Are Organic Compounds," has been completely rewritten with a focus on opioids.

- The Integrating Chemistry feature "Fatty Acids in Our Diets" has been updated to include recent research findings.
- The discussion of chiral molecules has been rewritten to focus on limonene.

5 Chemical Reactions

- Expanded thermodynamics coverage in Section 5.1 now includes the terms enthalpy, entropy, as well as Gibbs free energy.
- Section 5.5 now includes content on hydrolyzable and nonhydrolyzable lipids.
- The analogy used to explain equilibrium reactions has been updated for clarity.

6 Carbohydrates—Life's Sweet Molecules

- New Table 6.1 provides Fischer projections and names for D-monosaccharides containing three to six carbons, with the highest numbered chiral center highlighted.
- New Figure 6.16 consolidates the storage and structural polysaccharides into a single figure, so students can more readily compare and contrast the structures.

7 States of Matter and Their Attractive Forces: Gas Laws, Solubility, and Applications to the Cell Membrane

- The revised chapter title highlights the new conceptual balance of the chapter.
- Gas laws have moved to the beginning of the chapter.
- Section 7.1 has been expanded to include the kinetic molecular theory of gases.
- In Section 7.2, the introduction of attractive forces has been streamlined, using boiling point and vapor pressure as concrete examples.
- In Section 7.3, the content on solubility has been updated to include the terms *hydrophobic* and *hydrophilic* when referring to aqueous solutions.

8 Solution Chemistry—Sugar and Water Do Mix

- The new chapter title asks the student to look to the chapter for answers about solutions.
- The "Unique Behavior of Water" section has been updated to include coverage of specific heat.
- A reference to osmolarity in Section 8.6 connects the new concentration units to ones familiar to the student.

9 Acids, Bases, and Buffers in the Body

- Chapter 9 now has more pharmacologically relevant problems.
- Section 9.7 now discusses the relationship between pH and pK_a in terms of drug solubility and diffusion.

- A new Integrating Chemistry feature focuses on the role the kidneys and liver play in regulating CO_2 levels.
- The Henderson-Hasselbalch equation has been added to Section 9.7 as a tool for determining the ratio of conjugate base to acid.

10 Proteins—Workers of the Cell

- The discussion of amino acids as weak acids is positioned in Section 10.1 to reinforce the relationship between pH and pK_a from Chapter 9.
- A new Integrating Chemistry feature describes the role of peptides in celiac disease.

11 Nucleic Acids—Big Molecules with a Big Role

- In new Figure 11.11, the genetic code is presented using a genetic wheel instead of a table.
- A new Integrating Chemistry feature takes a closer look at common viruses.
- A new discussion of CRISPR has been added to Section 11.8.

12 Food as Fuel—An Overview of Metabolism

- New problems shift the focus toward problem solving.
- Updated art better represents the protein structures in the electron transport chain.
- Section 12.8 includes updated information on protein metabolism.

Acknowledgments

We have learned much since the first edition was published. Faculty and reviewer feedback has allowed us to enhance the Fourth Edition with some much-needed coverage while keeping the book length reasonable for a one-semester course.

The editorial staff at Pearson has been exceptional. We are extremely grateful for the assistance of Mary Ann Murray, Senior Analyst, Content Development, whose fresh eyes on the content allowed for clarity from the student perspective. Her years of textbook development were apparent. We also want to welcome back Jessica Moro to the project, Senior Courseware Portfolio Analyst, whose patience and understanding are much appreciated. We greatly appreciate the efforts of Susan McNally, Content Producer, and Mary Tindle, Project Manager, who have gone through much of the material with a fine-tooth comb, making sure that author comments were interpreted correctly by production. Thanks also to Eric Schrader, Senior Manager, Rights and Permissions, and Ben Ferini, Manager, Rights and Permissions, for their efforts on the book. We want to thank other members of the production team, including Joanna Stein, Project Manager, and Joanna Dinsmore, copyeditor. They have been very patient with us throughout the production process.

Laura Frost would like to also thank Adam Jaworski, Senior Vice President, Portfolio Management–Science, for his continued support of this project, his shared vision that an actively engaged classroom can enhance student understanding of chemistry, and his support for the inclusion of the inquiry activities. She would also like to thank Jeanne Zalesky, Director of Portfolio Management, whose strong effort on the second edition continues to allow this textbook to thrive. She also recognizes the continued mentorship and friendship of Karen Timberlake in the area of GOB chemistry writing and publishing.

Todd Deal would like, once again, to express a special appreciation to Jim Smith, our original editor, whose enthusiasm for the integrated strategy used in this project and belief in us as authors provided the foundation upon which this text is built.

This text reflects the contributions of many professors who took the time to review and edit the manuscript and provided outstanding comments, help, and suggestions. We are grateful for your contributions.

In addition, this project could not have been completed without the support of several exceptional colleagues in the Department of Chemistry at Georgia Southern University and Florida Gulf Coast University, who have taught using this text and reviewed materials, offering many comments and corrections.

If you would like to share your experience using this textbook, as either a student or faculty member, or if you have questions regarding its content, we would love to hear from you.

Laura Frost   **Todd Deal**
lfrost@fgcu.edu   stdeal@georgiasouthern.edu

Accuracy Reviewer

Melody Jewell, *South Dakota State University*

Reviewers

Fourth Edition

Cindy Ault, *University of Jamestown*
Allison Babij, *Ivy Tech Community College*
Karen Glover, *Clarke University*
Sarah Pierce, *Cumberland University*
Ria Ramoutar, *Georgia Southern University*
Duane Smith, *Nicholls State University*
Dennis Viernes, *University of Mary*

Previous Edition Reviewers

Anthony Amaro, *University of Jamestown*
Sara Egbert, *Walla Walla Community College*
Lisa Sharpe Elles, *Washburn University*
Tom Huxford, *San Diego State University*
Ria Ramoutar, *Georgia Southern University*
Tanea Reed, *Eastern Kentucky University*
John Singer, *Jackson College*
James Stickler, *Allegany College of Maryland*
Amy Taketomo, *Hartnell College*

Resources in Print and Online

Supplement	Available in Print	Available Online	Instructor or Student Supplement	Description
Mastering Chemistry (ISBN: 0-13-517528-3 / 978-0-13-517528-6)		X	Instructor and Student Supplement	The Mastering platform delivers engaging, dynamic learning opportunities—focused on your course objectives and responsive to each student's progress—that are proven to help students absorb course material and understand difficult concepts. Practicing the Concepts videos with pause-and-predict quizzes in Mastering Chemistry bring chemistry to life. These 3- to 5-minute videos feature coauthor Todd Deal introducing key topics in general, organic, and biological chemistry that students find difficult. Students are asked to solve a problem while they watch the video content. Mastering also offers Learning Catalytics questions that directly relate to the content of the text. Learning Catalytics is a "bring your own device" student engagement, assessment, and classroom intelligence system.
Instructor Solutions Manual (ISBN: 0-13-559332-8 / 978-0-13-559332-5)		X	Instructor Supplement	This Solutions Manual provides detailed solutions to all in-chapter as well as end-of-chapter exercises in the text.
Test Bank		X	Instructor Supplement	This Test Bank contains over 600 multiple-choice, true/false, and matching questions. It is available in the TestGen program, in Word format, and is included in the item library of Mastering Chemistry.
Instructor Resources (ISBN: 0-13-559331-X / 978-0-13-559331-8)		X	Instructor Supplement	This provides an integrated collection of online resources to help instructors make efficient and effective use of their time. Includes all artwork from the text, including figures and tables in PDF format for high-resolution printing, as well as four pre-built PowerPoint™ presentations. The first presentation contains the images embedded within PowerPoint slides. The second includes a complete lecture outline that is modifiable by the user. Also available are PowerPoint slides of the parent text "in-chapter" sample exercises. Also includes electronic files of the Instructor's Resource Manual, as well as the Test Bank. Access resources through http://www.pearsonhighered.com/.
Study Guide (ISBN: 0-13-416051-7 / 978-0-13-416051-1)	X		Student Supplement	This manual for students contains complete solutions to the odd-numbered end-of-chapter problems in the text.
Laboratory Manual (ISBN: 0-32-181925-X / 978-0-32-181925-3)	X		Student Supplement	Written by T. Deal, the lab manual continues the strategy of integration of concepts to help students understand chemistry. Several of the experiments included in the lab manual are original works developed by Professor Deal and his students in support of the integrated strategy. Most experiments are designed around a question, which is intended to engage students and to demonstrate the applicability of chemistry concepts to real-world problems. Many of the experiments highlight concepts from multiple chapters of the text, once again building on the strategy of integration.
Guided Inquiry Activities (ISBN: 0-13-559337-9 / 978-0-13-559337-0)	X		Instructor and Student Supplement	Guided Inquiry Activities, authored by L. Frost, are available through the Mastering Instructor Resource Area. These activities are designed for in-class use by groups of students with facilitation by the instructor. Students explore information, develop chemical concepts, and apply the concepts to further examples.

An integrated and applied approach to *General, Organic, and Biological Chemistry*

With the **Fourth Edition** of *General, Organic, and Biological Chemistry*, authors Laura Frost and Todd Deal apply their knowledge and experience in the science of learning to this focused, concise text. Practical connections and applications are highlighted, showing both allied-health and non-science majors how to use their understanding of chemistry in future health professions and in their everyday lives. Enhanced digital tools in **Mastering Chemistry** and embedded in the Pearson eText guide students through all stages of the course, providing support when and where students need it.

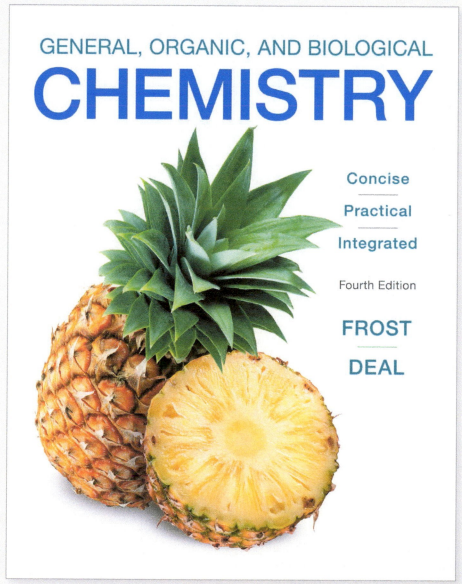

GENERAL, ORGANIC, AND BIOLOGICAL

CHEMISTRY

Concise

Practical

Integrated

Fourth Edition

FROST

DEAL

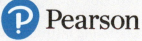

Apply the science of learning to the way students learn

DISCOVERING THE CONCEPTS

? INQUIRY ACTIVITY—Reaction Energy Diagrams

Information

The diagrams shown are called reaction energy diagrams and graphically show the progress of two different chemical reactions on the x-axis and the amount of energy required as the reaction moves forward on the y-axis. The *activation energy* is the amount of energy required to get the reactants into position for collision with enough energy so that they react. Reactions with larger activation energies occur more slowly than reactions with smaller activation energies.

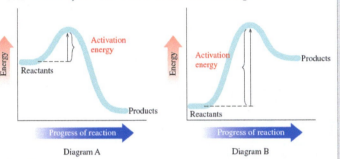

Diagram A   Diagram B

Questions

1. Which reaction has the larger activation energy?
2. Based on the diagrams, which reaction can form products more quickly?
3. A *catalyst* speeds up a chemical reaction by lowering the activation energy. Sketch diagram A and draw a second line on the same diagram for the reaction when a

p. 189

UPDATED! Discovering the Concepts can be used to engage students in groups during class as they construct an understanding of the content in cooperation with their peers.

Activity: Balancing Chemical Equations

Learning Objectives

Balance a simple chemical equation

Information

Scientists use chemical formulas as a short-hand to represent substances made from the elements. Similarly, a chemical equation is used to represent how substances change during a chemical reaction.

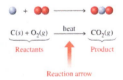

$C(s) + O_2(g) \xrightarrow{\text{heat}} CO_2(g)$

Reactants   Product

Reaction arrow

Scheme 1. Basic parts of a chemical equation.

Questions

1. a. Fill in the following table for Scheme 1.

Element	Number in Reactants	Number in Product
Carbon		
Oxygen		

   b. How is the number of each element in the reactants of a chemical equation related to the number of each element in the product?

Guided Inquiry Activities, authored by Laura Frost, engage students with more of the topics in the textbook through exploration, concept development, and application. These Activities are available in the Mastering Instructor Resources.

Relate chemistry to students' future careers

INTEGRATING Chemistry

Find out how ▶ the fatty acids in coconut oil are different from those in other oils.

Fatty Acids in Our Diets

Fats are important in a balanced diet because they play important roles as insulators and protective coverings for internal organs and nerve fibers. Mono- and polyunsaturated fats are part of a healthy diet. The Food and Drug Administration (FDA) recommends that a maximum of 30% of the calories in a normal diet come from such fatty acid-containing compounds. The FDA also recommends that the majority of our fat intake come from foods containing a higher percentage of mono- and polyunsaturated fatty acids. **TABLE 4.6** shows the fatty acid composition of some common fats. Highly saturated oils like coconut and palm oils have found uses as natural substitutes for hydrogenated oils, which are chemically saturated. More about hydrogenated oils is found in Chapter 5.

TABLE 4.6 Fatty Acid Composition of Common Dietary Fats

Coconut oil
Butter
Palm oil
Lard
Soybean oil
Olive oil
Sunflower oil
Corn oil
Canola oil

■ Saturated
■ Monounsaturated
■ Polyunsaturated

Percent by weight of total fatty acid

0   20   40   60   80   100

p. 148

Health-Related Problems

are integrated throughout each chapter and are tied to real-life applications from allied-health fields, helping to promote critical thinking skills and to connect the chemistry learned with their future professions.

Additionally, **clinical examples** throughout the text pay particular attention to topics such as acid-base and biochemistry.

Practice Problems

9.53 The antihypertensive medication alprenolol is shown as an acid below. The pK_a for the acid is 9.6.

a. Which form is charged: acid, conjugate base, or both?

b. Which form (acid, conjugate base, or both) will predominate in a stomach with a pH between 1 and 3?

c. Which form, acid or conjugate base, will be able to more easily diffuse through a cell membrane?

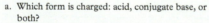

Alprenolol, an antihypertensive,
$pK_a = 9.6$

9.54 The anticonvulsant medication valproic acid is shown as an acid below. The pK_a for the acid is 4.8.

a. Which form is charged: acid, conjugate base, or both?

b. Which form (acid, conjugate base, or both) will predominate in the first part of the small intestine (jejunum), where the pH is between 6 and 7?

c. Which form, acid or conjugate base, will be able to more easily diffuse through a cell membrane?

9.55 Consider the vitamin niacin, which has a pK_a value of 4.85.

a. Will the acid or the conjugate base predominate at the following pH values: 3.00, 4.85, and 7.40?

b. At each pH, is that form charged or uncharged?

c. Calculate the ratio of [c. base]/[acid] present at each pH.

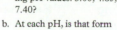

Niacin, $pK_a = 4.85$

9.56 Consider the neurotransmitter dopamine, which has a pK_a value of 8.90.

a. Will the acid or the conjugate base predominate at the following pH values: 7.45, 8.90, 13.40?

b. At each pH, is that form charged or uncharged?

c. Calculate the ratio of [c. base]/[acid] present at each pH.

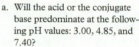

Dopamine, acid form, $pK_a = 8.90$

9.57 Procaine and lidocaine are used to numb the gums during dental procedures. Procaine has a pK_a of 9.1, whereas lidocaine has a pK_a of 7.9. Which do you think will relieve pain faster in the gums? Explain.

p. 379

Reach every student . . .

EXPANDED! 11 new Practicing the Concepts videos explain major concepts that students struggle with in the chapter. The videos are 3–5 minutes long and are narrated by author Todd Deal, with video assessment questions written for use in Learning Catalytics, the Pearson eText, and assignable in Mastering Chemistry.

The Chemistry Primer in Mastering Chemistry helps students remediate their chemistry math skills and prepare for their first college chemistry course. Scaled to students' needs, remediation is only suggested to students that perform poorly on an initial assessment. Remediation includes tutorials, wrong-answer specific feedback, video instruction, and stepwise scaffolding to build students' abilities.

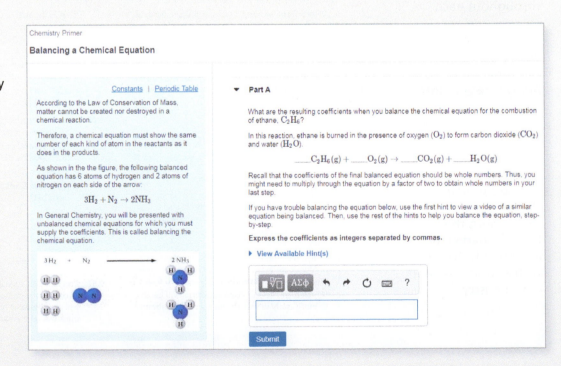

with Mastering Chemistry

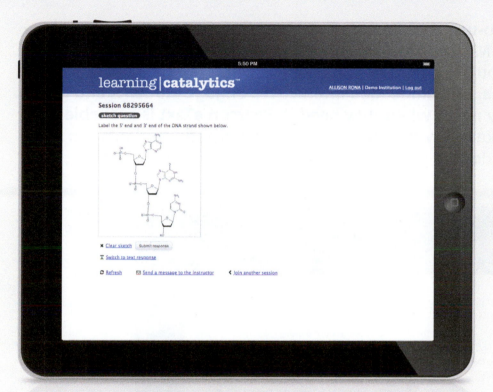

With Learning Catalytics, you'll hear from every student when it matters most. You pose a variety of questions that help students recall ideas, apply concepts, and develop critical-thinking skills. Your students respond using their own smartphones, tablets, or laptops. You can monitor responses with real-time analytics and find out what your students do—and don't—understand, to help students stay motivated and engaged.

ENHANCED! End-of-chapter questions with answer-specific feedback use data gathered from all of the students using the Mastering Chemistry to offer wrong-answer feedback that is specific to each student, where and when they need it. Rather than simply providing feedback of the "right/wrong/try again" variety, Mastering guides students towards the correct final answer without giving the answer away.

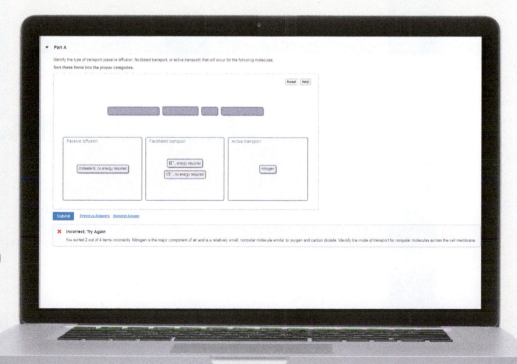

Give students anytime, anywhere access with Pearson eText

Pearson eText is a simple-to-use, mobile-optimized, personalized reading experience available within Mastering. It allows students to easily highlight, take notes, and review key vocabulary all in one place—even when offline. Seamlessly integrated videos, rich media, and embedded interactives engage students and give them access to the help they need, when they need it. Pearson eText is available within Mastering when packaged with a new book, as an upgrade students can purchase online, or can be adopted separately as your main course material.

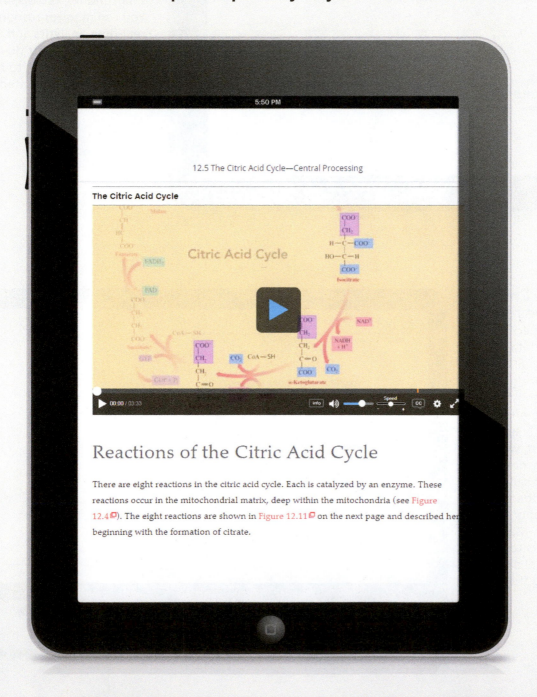

Improve learning with Dynamic Study Modules

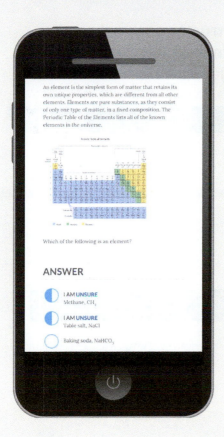

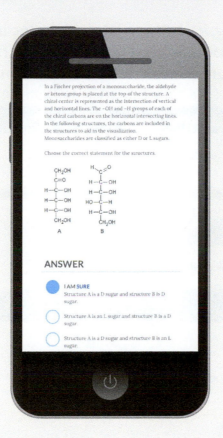

Dynamic Study Modules in Mastering Chemistry help students study effectively—and at their own pace—by keeping them motivated and engaged. The assignable modules rely on the latest research in cognitive science, using methods—such as adaptivity, gamification, and intermittent rewards—to stimulate learning and improve retention.

Each module poses a series of questions about a course topic. These question sets adapt to each student's performance and offer personalized, targeted feedback to help them master key concepts. With **Dynamic Study Modules**, students build the confidence they need to deepen their understanding, participate meaningfully, and perform better—in and out of class.

Instructor support you can rely on

General, Organic, and Biological Chemistry includes instructor support materials in the Instructor Resources area in Mastering Chemistry. Resources include customizable PowerPoint lecture presentations and all images in JPEG format.

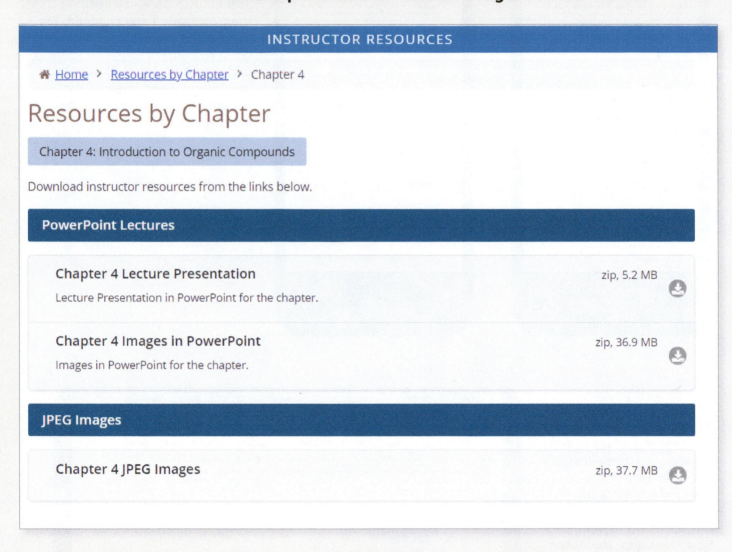

Chemistry Basics—Matter and Measurement

We are constantly measuring things: our body height and weight, the air temperature, spaces for furniture. Even medications must be measured exactly before a dose can be given. Chemists measure matter and its changes. To introduce you to the study of chemistry and measurements, we begin with Chapter 1.

LEARNING TIP   Guide your Reading

As you read, take notes in your own words on information that answers the inquiry question that appears in the margin at the beginning of each section. Research shows that using these questions as a prompt to explain ideas in your own words helps with remembering the information later when you need it on a test.

DID YOU KNOW that everything you do every day involves chemistry? Yes, everything. From the water and shampoo in your shower, to the food you ate for breakfast, to the electronics that power your cell phone, to the therapeutic drugs for treating diseases, to the sunscreen lotion that protects your skin, and even the clothes that you wear—all of these somehow involve chemistry. Learning chemistry is really learning about how chemistry impacts our everyday lives and even provides life itself. Chemistry helps us understand concepts as diverse as how our bodies function to all of the modern conveniences that make our lives easier. So, come explore with us. It will be challenging, but fun. We promise!

All of the "stuff" that we just mentioned is composed of something that chemists call matter. **Matter** can be defined as anything that takes up space (scientists call this *volume*) and weighs something (scientists call this *mass*). From the smallest tablet dispensed by a pharmacist to the shampoo in a bottle to the air in a balloon, each of these has mass and takes up some amount of space and is a form of matter.

DISCOVERING THE CONCEPTS

❓ INQUIRY ACTIVITY—Classifying Matter

Information

Chemistry is the study of matter and its changes. There are two main types of matter: *pure substances* and *mixtures* of substances. A mixture that is mostly water is called an *aqueous solution*. Matter can exist in several different *states*, the three most common being solid, liquid, or gas.

Types of Matter Flow Chart with Examples, Their Chemical Formulas, and States

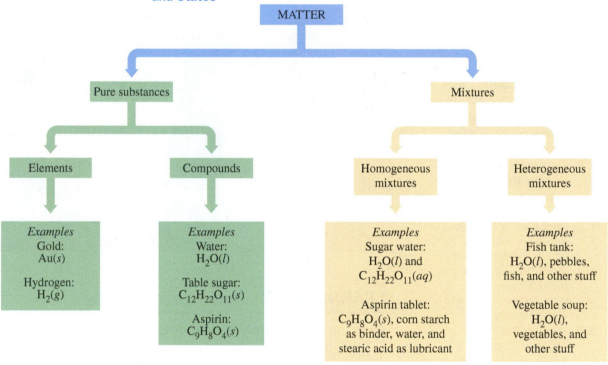

Elements can exist as individual *atoms* (for example, neon [Ne]), or as pairs of atoms (for example, hydrogen, [H_2]), forming *molecules*. Atoms of different elements can also combine forming compounds.

Questions

1. Consider the examples on the flow chart. How does the formula of an element differ from that of a compound?
2. Can an element be a pure substance? Can a compound be a pure substance? How did you decide?
3. Using the information given, how might you define a pure substance? How does a pure substance differ from a mixture?
4. Based on the information in the data set, complete the following table.

Matter	Element or Compound	Atom or Molecule
He		
O_2		
CH_2O (formaldehyde)		
CH_3COOH (vinegar)		

5. As a group, devise a definition for a compound.
6. Describe the difference between a homogeneous and a heterogeneous mixture.
7. Would you classify the following matter as element, compound, or mixture? If you classify it as a mixture, classify it as homogeneous or heterogeneous.
   a. table salt (NaCl)   b. nickel (Ni)
   c. chocolate chip cookie dough   d. air
8. What do you think the labels (s), (l), (g), and (aq) on the formulas in the figure mean?

1.1 Classifying Matter: Pure Substance or Mixture

Homo sapiens is the genus and species name for the modern day human. Biologists classify all living organisms by assigning a genus and species name. In chemistry, we classify matter into different types based on both physical and chemical characteristics.

The flow chart shown in **FIGURE 1.1** gives you an idea of how we classify matter. This flow chart is your guide to classifying matter as chemists do.

As two examples of matter, consider blood and a diamond. If you are a blood donor, you may have heard blood discussed in terms of whole blood, plasma, white blood cells, and red blood cells. These terms give you a hint that blood is a mixture containing many components. In contrast, if you have ever bought a diamond, you know that one of the measures used to grade these gemstones is clarity, which is a measure of purity. The clearer (more pure) a diamond is, the more it costs. So, how do the complexity of a sample of blood and the purity of a diamond contribute to how they are classified?

Looking at the flow chart in Figure 1.1 from the top, we see that all matter is divided into two large categories or classifications—pure substances and mixtures. Let's consider these one at a time.

Pure Substances

Diamonds are composed only of carbon. Because a diamond is composed of a single substance, carbon, it is not a mixture but rather a pure substance. A **pure substance** is matter that is made up of only one type of substance and can be represented with one chemical formula.

Note from the flow chart that pure substances can be one of two types—elements or compounds. Substances containing only one type of atom are considered elements, with

What's an Inquiry Question?

? Inquiry Questions are designed to focus your reading on the main concepts by section. As you read each section, see if you can answer its Inquiry Question in your own words.

? **1.1 Inquiry Question**

How is matter classified?

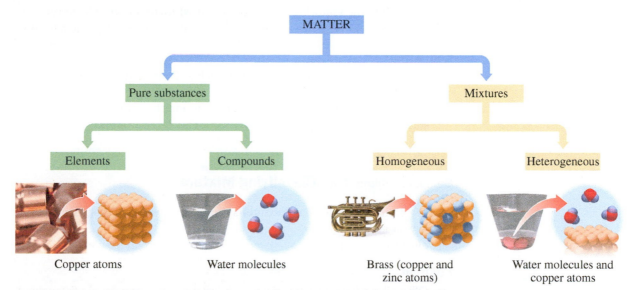

FIGURE 1.1 Flow chart used for classifying matter. Matter can be broadly classified as a pure substance or a mixture. Pure substances can be further classified as elements or compounds, and mixtures can be further classified as homogeneous or heterogeneous. The red balls on the water molecules represent oxygen, and the gray balls represent hydrogen.

Blood is a mixture of many components, whereas a diamond is composed of a single substance, the element carbon. A specialist in blood banking separates and stores blood.

diamonds being one such example. An **element** is the simplest type of matter because it is made up of only one type of atom. An **atom** is the smallest unit of matter that can exist and keep its chemically unique characteristics. Substances that contain more than one type of atom, but contain these elements in a fixed composition, are also pure substances. Pure water, for example, contains only water and has the chemical formula H_2O. Water that is distilled contains only the substance water, H_2O, and is therefore a pure substance. Most water on the planet, however, is not pure water (H_2O), but has other substances dissolved in it so is a mixture. Sea water is very salty; spring water and even purified water contain dissolved solids that make the water taste better to us.

Because pure water has more than one type of atom—it contains both hydrogen and oxygen—it is not an element but is instead a compound. A **compound** is a pure substance that is made of two or more elements that are chemically joined together.

Mixtures

As noted earlier, blood contains red blood cells, white blood cells, plasma (which is mostly water), and several other components. Blood is an example of a **mixture**, a combination of two or more substances. One of the defining characteristics of a mixture is that it can be separated into its different components. A specialist in blood banking is trained to separate blood into red blood cells and plasma for storage and use in different medical procedures at a blood bank.

Next, we consider different types of mixtures. Air contains nitrogen, oxygen, carbon dioxide, argon, and small amounts of other gases. Every time you take a breath, you inhale the same substances in the same ratio as another person breathing the same air. Air is classifed as a **homogeneous mixture** because its composition is the same throughout. In contrast, trail mix is a **heterogenous mixture** because its composition is not uniform, but varies throughout. One handful of trail mix may contain raisins, chocolate chips, and a peanut while another handful contains sunflower seeds, almonds, and dried banana. No two handfuls contain the same substances in the same amount.

Pure substances are easier to characterize, but chemists do work with and study mixtures. The air we breathe, the food we eat, the concrete we walk on are all mixtures. Most matter we encounter every day is a mixture. However, due to their varied composition, mixtures are not easily characterized.

Sample Problem 1.1   Mixture Versus Pure Substance

Classify each of the following substances as a mixture or a pure substance:
  a. cake batter   b. the helium gas inside a balloon

Solution
  a. Cake batter is a mixture of substances such as flour, butter, sugar, and other ingredients.
  b. The helium gas in the balloon is a pure substance because it contains only one substance, the element He.

Sample Problem 1.2   Classifying Mixtures

Classify each of the following mixtures as homogeneous or heterogeneous. Briefly justify your answer.
  a. olive oil   b. rocky road ice cream

Solution
  a. Olive oil is a homogeneous mixture. It has the same composition throughout.
  b. Rocky road ice cream is a heterogeneous mixture. Each spoonful contains different amounts of nuts, marshmallow, and ice cream.

Practice Problems

Find the answers to the odd-numbered Practice Problems at the end of the chapter.

1.1 Classify each of the following substances as a pure substance or a mixture:
a. iron
b. gelato
c. cinnamon
d. spring water

1.2 Classify each of the following substances as a pure substance or a mixture:
a. well water
b. ocean water
c. cement
d. a tropical smoothie

1.3 Classify each of the following mixtures as homogeneous or heterogeneous. Briefly justify your answer.
a. a bowl of vegetable soup
b. mouthwash
c. an unopened can of cola
d. a dinner salad

1.4 Classify each of the following mixtures as homogeneous or heterogeneous. Briefly justify your answer.
a. a bottle of sports drink
b. a blueberry pancake
c. gasoline
d. a box of raisin bran

DISCOVERING THE CONCEPTS

? INQUIRY ACTIVITY—The Periodic Table

All *elements* are listed individually on the *periodic table of the elements*. There is a periodic table on the inside front cover of your textbook and an alphabetical listing of all the elements on the facing page. Refer to this as you answer the questions in this activity.

The horizontal rows on the periodic table are referred to as *periods*, and the vertical columns are referred to as *groups*. The groups are numbered across the top of the periodic table, and the periods are numbered down the left side of the table. The table is organized with metals on the left and nonmetals on the right, with a staircase dividing line between the two. Find this staircase in the periodic table on the inside front cover.

Water's chemical formula is H_2O, meaning that it contains two hydrogen atoms and one oxygen atom. Chemical formulas show the type and number of each element present in a compound.

Questions

1. In which group on the periodic table are the following elements found?
   a. sodium   b. oxygen   c. calcium   d. carbon

2. In which period are the following elements found?
   a. hydrogen   b. nitrogen   c. sulfur   d. phosphorus

3. Provide names for the following elements and identify them as metals or nonmetals:
   a. Cu   b. Na   c. Cl   d. C   e. K   f. P

4. Look at the periodic table of the elements. About how many elements are there?

5. In each of the chemical formulas below, identify the number of atoms of each element and provide the name of each element:
   a. $C_6H_{12}O_6$ (dextrose)   b. NaOH (lye, found in drain cleaners)
   c. $NaHCO_3$ (baking soda)   d. $C_{15}H_{21}NO_2$ (Demerol, a painkiller)

6. Are most of the elements on the periodic table metals or nonmetals? The Earth's biomass is made up mostly of carbon, hydrogen, and oxygen. Are these elements metals or nonmetals?

1.2 Elements, Compounds, and the Periodic Table

If a pure substance can be an element or a compound, how do we distinguish which is which? The tool to guide us is the **periodic table of the elements**. At its most basic level, the periodic table is a listing of all the elements found on Earth.

You may have previously encountered the periodic table, most likely hanging on the wall in a science classroom. But have you ever tried to decipher it? This strangely shaped chart contains an amazing collection of data, both in the way that it is organized and in the actual information listed. We will make extensive use of the periodic table and the information it contains. Take a look at the periodic table in **FIGURE 1.2**.

The periodic table consists of many small blocks in a simple layout. Each has a letter or two in its center and numbers above and below these letters. The letters are known as the **chemical symbol** and represent the name of each element. For many of the elements, the symbols are derived directly from the name of the element—for example, lithium = Li, carbon = C, hydrogen = H, oxygen = O, and so on. A few of the elements have symbols that do not match the first few letters of their name—for example, sodium = Na and gold = Au. These elemental symbols are derived from the Latin names for the element, natrium and aurum, respectively. Because the elements form the

Periodic Table of the Elements

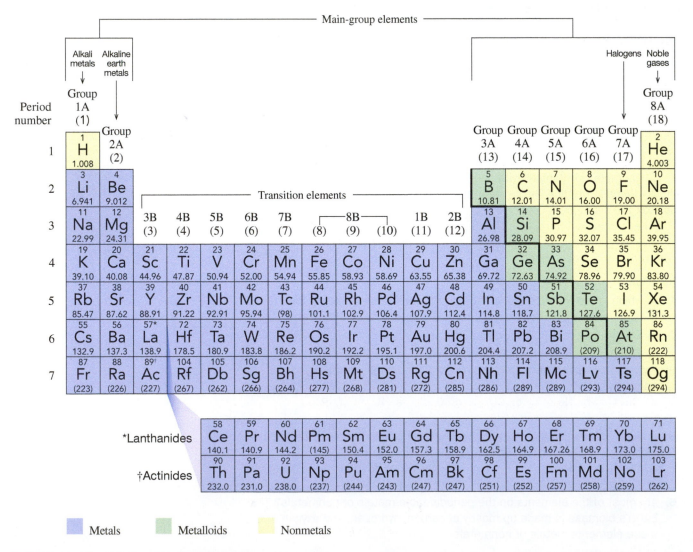

FIGURE 1.2 The periodic table of the elements.

basic vocabulary of chemistry, it is important to know the names and symbols of the more common elements, especially those that are found in living things (carbon, hydrogen, oxygen, nitrogen [N], phosphorus [P], and sulfur [S]).

The blocks on the periodic table are organized in specific arrangements. Each block holds a different element. A vertical column of blocks is known as a **group** of elements. (Some chemists refer to the vertical columns as families; we will use the term groups in this textbook.) The elements in a group have similar chemical behaviors. Each group has a number and letter designation. The groups with A designations (1A–8A) are known as the main-group elements, and the groups with B designations are the transition elements. This system of numbering the groups is most common in North America and will be used throughout this textbook. The system using the numbers 1 through 18 for the columns has been recommended by the International Union of Pure and Applied Chemistry (IUPAC) and is also in common use. Several of the groups have special names to designate the members of the group as shown at the top of Figure 1.2.

A horizontal row of blocks is known as a **period**. The periods are numbered from 1 to 7, with sections of Periods 6 and 7 set apart at the bottom of the periodic table. The bold, descending staircase beginning at the element boron (#5) and running diagonally down and to the right separates the metals from the nonmetals. Elements bordered by the line, with the exception of aluminum (Al), are metalloids. Metalloids contain some properties of metals and some of nonmetals.

Metals and nonmetals have different properties. Aluminum can be made into sheets or blocks and is a metal. Carbon is more brittle, and we find it as a powder or as the "lead" in a pencil. Carbon is a nonmetal.

Elements in Nutrition

Which elements are found in the food we eat? The elements found most commonly in food are the nonmetals carbon, hydrogen, oxygen, and nitrogen, or the CHON elements. These four elements also make up most of the biological molecules on Earth. Along with the CHON elements, the U.S. Food and Nutrition Board at the National Academies has also identified several other elements on the periodic table as being essential for human health. These elements are classified as macronutrients if required in quantities greater than 100 milligrams per day or as micronutrients if required in trace quantities, less than 100 milligrams per day.

Macronutrients include several elements—sodium, magnesium, potassium, calcium, and chlorine—that are found in the body as charged particles and are essential for the transport of electrical signals through cells. Micronutrients are often found concentrated in particular organs. For example, iodine is found in the thyroid, fluorine in the teeth, and zinc mainly in muscle and bone. Red blood cells contain the protein hemoglobin, which contains iron. **TABLE 1.1** summarizes the uses and locations of the micronutrients.

Chromium, once considered a micronutrient, has recently been suggested to be nonessential because removal of this element from the diet has been shown to have no adverse health effects and no known molecule in the body uses chromium.

INTEGRATING Chemistry

◄ Find out which elements the body uses.

H																	
													C	N	O	F	
Na	Mg												P	S	Cl		
K	Ca				Mn	Fe		Cu	Zn				Se				
			Mo										I				

▢ Most common elements in living things

▢ Macronutrients

▢ Micronutrients

—Continued next page

Continued—

TABLE 1.1 Micronutrients

Element	Use in Body	Main Location in Body
Copper (Cu)	Used in several enzymes that are essential for human health and development	Liver
Fluorine (F)	Mineralizes teeth and bones	Teeth and bone
Iodine (I)	Found in thyroid hormones	Thyroid
Iron (Fe)	Found in hemoglobin, a protein that transfers oxygen from lungs to tissues	Red blood cells
Manganese (Mn)	Found in enzymes that act as antioxidants, enzymes involved in bone development and wound healing	Pituitary gland, bones, and connective tissue
Molybdenum (Mo)	Used with enzymes that control oxidation processes	Liver and kidney
Selenium (Se)	Used in proteins that regulate reproduction, thyroid, DNA synthesis, and oxidative damage	Skeletal muscle, thyroid, immune system
Zinc (Zn)	Used in cell growth and division, and in immune system function; in males, zinc is vital to fertility	Throughout the body, particularly in bones and muscle

---

Sample Problem 1.3   Using the Periodic Table

Use the periodic table to answer the following questions:
   a. What is the elemental symbol for oxygen?
   b. To what group does oxygen belong?
   c. What period is oxygen in?

Solution

   a. Oxygen's symbol is O.
   b. Oxygen is located in Group 6A.
   c. Oxygen is in Period 2.

Compounds

How do we classify substances like water (H_2O) or table salt (NaCl) that contain more than one element from the periodic table? Remember that we said a pure substance containing two or more elements chemically combined is classified as a compound. Therefore, both water and table salt are compounds. Compounds combine elements in specific ratios. How and why elements combine in these ratios are discussed in Chapter 3.

Sample Problem 1.4   Classifying Pure Substances

Classify each of the following as an element or compound and explain your classification:
   a. carbon dioxide (the gas we breathe out)
   b. xenon (a gas sometimes used as an anesthetic)

Solution

a. Carbon dioxide is a compound. From its name, you can tell that this substance contains carbon and "something else" (even if you did not know that oxide is a name often used for oxygen when it is in compounds). Because carbon dioxide contains two different elements, it is a compound.

b. Xenon is an element. Look for xenon on the periodic table. You will find it in group 8A (18). This gas is becoming more common as an anesthetic agent because unlike other anesthetics—such as nitrous oxide, which depletes the ozone layer of the atmosphere—it has minimal effects on the environment.

Before we continue, let's take a quick look at the representations we use for compounds. Why does the representation for water (H_2O) contain the subscript 2 after hydrogen, but the one we use for table salt (NaCl) does not have a subscript? These representations, known as **chemical formulas**, show that water is composed of two atoms of hydrogen and one atom of oxygen and table salt is composed of one sodium atom and one chlorine atom. The subscript tells us how many atoms of the preceding element are present in the compound. The absence of a subscript on oxygen, sodium, and chlorine in the formulas for water and sodium chloride is understood to mean *one* of each element. A compound's chemical formula identifies which element and how many atoms of that element are present in a compound.

Sample Problem 1.5   Elements in Compounds

Identify and give the number of atoms of each element in the following compounds:
a. $C_2H_4O_2$, acetic acid, a component of vinegar
b. H_3PO_4, phosphoric acid

Solution

a. Acetic acid contains 2 atoms of carbon, 4 atoms of hydrogen, and 2 atoms of oxygen.
b. Phosphoric acid contains 3 atoms of hydrogen, 1 atom of phosphorus, and 4 atoms of oxygen.

Practice Problems

1.5 Classify each of the following as an element or compound and explain your classification:

a. aluminum, used in soft drink cans and foil

b. sodium hypochlorite, found in bleach

c. hydrogen, fuel of the sun

d. potassium chloride, a table-salt substitute

1.6 Classify each of the following as an element or a compound:

  a. Fe   b. $CaCl_2$ c. Si   d. KI

1.7 Use the periodic table to supply the missing information in the following chart:

Name	Elemental Symbol	Group	Period	Metal or Nonmetal
Fluorine				
		1A	2	
	Cl			
		4A	2	Nonmetal

—Continued next page

Continued—

1.8 Use the periodic table to supply the missing information in the following chart:

Name	Elemental Symbol	Group	Period	Metal or Nonmetal
Sodium				
		2A	4	
		P		
		5A	2	Nonmetal

1.9 Identify and give the number of atoms of each element in the following compounds:

a. $LiCO_3$, lithium carbonate, used in the treatment of bipolar disorder

b. H_2O_2, hydrogen peroxide, a disinfectant

c. $C_8H_8O_3$, vanillin, the primary flavor compound in vanilla beans

1.10 Identify and give the number of atoms of each element in the following compounds:

a. $MgSO_4$, magnesium sulfate, commonly called Epsom salts

b. $NaHCO_3$, sodium bicarbonate, commonly called baking soda

c. $C_{14}H_{18}N_2O_5$, aspartame, an artificial sweetener

1.3 How Matter Changes

1.3 Inquiry Question

How are physical and chemical changes in matter different, and how can we represent them?

On a hot summer day, you add several ice cubes to your glass of tea and come back later to find that the ice has disappeared. Now your glass seems to have two liquids—the tea at the bottom and a new, clear liquid on top. You may have experienced the popular experiment where a clear, vinegar-smelling liquid is poured into a papier-mâché volcano and then a foamy "lava" flows out. Each of these instances represents a change in matter, but each is a different type of change: One is a physical change, and the other is a chemical reaction.

Physical Change

The ice in a glass of tea is water in the solid state. As it melts, the ice changes from a solid to a liquid (forming a clear liquid on top of your tea). The identity of the water did not change—both ice and liquid water are still H_2O (see **FIGURE 1.3**). In fact, if you add enough heat to the liquid water, it will eventually become steam—a gas—but it is still water (H_2O). A change in the state of matter—from a solid to a liquid or a liquid to a gas, for example—represents a physical change of matter. In a **physical change**, the form of the matter is changed, but its identity or chemical formula remains the same.

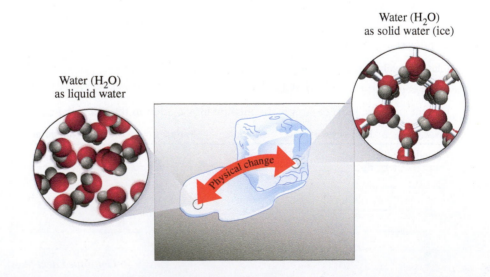

Water (H_2O) as solid water (ice)

Water (H_2O) as liquid water

FIGURE 1.3 Physical change of water from solid to liquid. As water changes from solid to liquid, the arrangement of the molecules becomes less orderly and their motion increases. The red balls represent oxygen atoms and the gray balls represent hydrogen atoms in the water molecules.

Chemical Reaction

The volcano in our example was most likely filled with baking soda before the vinegar was poured into it. When the vinegar and baking soda combine, water and carbon dioxide bubbles form. Chemically, neither vinegar nor baking soda is the same as water or carbon dioxide. So when these two substances combined, new substances were formed. Unlike a physical change, a chemical change results in a change in the chemical identity of the substance (or substances) involved. When a substance (or substances) undergoes such a change, we more commonly refer to it as a **chemical reaction**. **TABLE 1.2** gives some examples of physical changes and chemical reactions.

Sample Problem 1.6   Physical Change or Chemical Reaction

Determine whether each of the following is a physical change or a chemical reaction:
a. a plant using carbon dioxide and water to make sugar
b. water vapor condensing on a cool evening to form dew on the grass

Solution

a. This is a chemical reaction. Carbon dioxide and water are chemically different substances from sugar.
b. This is a physical change. Water vapor is a gas that condenses to form dew, which is simply liquid water.

TABLE 1.2 Comparing Physical Changes and Chemical Reactions
Physical Change
Sawing a log in half
An ice cube melting
Molten iron solidifying to form a nail
Mixing oil, water, and eggs to make mayonnaise
Mixing cooked spaghetti and sauce to make a meal
Chemical Reaction
Burning a log in your fireplace
Hydrogen and oxygen combining to form water
A nail rusting
Mayonnaise becoming spoiled
A spaghetti supper being metabolized to provide energy for a morning run

Chemical Equations

Have you ever wondered how the big pile of charcoal that you start with in an outdoor grill is reduced to a small pile of gray dust when the fire finally dies out? When charcoal burns with oxygen, or combusts, a chemical reaction is occurring. To show what happens to the element carbon (the main component of charcoal) as it burns, we can write a **chemical equation** like the one shown here.

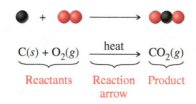

$$\underbrace{C(s) + O_2(g)}_{\text{Reactants}} \xrightarrow{\text{heat}} \underbrace{CO_2(g)}_{\text{Product}}$$
Reaction arrow

The number of atoms of each element in the reactants must equal the number in the products.

The carbon in charcoal burns with oxygen, forming carbon dioxide.

A chemical equation is the chemist's way of writing a sentence about what happens in a chemical reaction. It is meant to be read from left to right. This chemical equation explains that solid carbon (from charcoal) and oxygen gas (from air) react to form carbon dioxide gas when heated. In this example carbon and oxygen are the **reactants**, and carbon dioxide is the **product**. Reactants are found on the left side of a chemical equation and products are found on the right side. The reaction arrow means "react to form."

Special reaction conditions are often written as words or symbols above the reaction arrow. Typical examples include heat, light, or the name of a catalyst or other substance required for the reaction to occur.

The labels in parentheses after each substance (called state labels) indicate its physical state—(s)olid, (l)iquid, or (g)as. A fourth label, (aq), meaning aqueous or dissolved in water, will be discussed later. So, when solid charcoal is burned, it combines with oxygen in the air and forms carbon dioxide, a gas.

Balancing Chemical Equations

The reaction of carbon with oxygen clearly illustrates the *parts* of a chemical reaction. Notice something else. There is one atom of carbon and there are two atoms of oxygen on the reactant side and the same number of each on the product side even though a chemical change took place. The fact that the number of atoms of each element in the reactants equals the number of atoms of each element in the products illustrates the **law of conservation of mass**. Matter can neither be created nor destroyed. Matter only changes form, so the amounts of matter on the reactant side and the product side must be equal.

The number of atoms of each element in the reactants must equal the number in the products.

For any chemical equation that we write, the number of atoms of each element must be the same on both sides of the equation, or balanced. We can balance chemical equations when necessary by adding a number, called a **coefficient**, in front of the chemical formula for a substance in the chemical equation. This is easiest to understand by working some examples.

Solving a Problem

Balancing a Chemical Equation

Balance the chemical equation for the combustion of natural gas, also called methane. This occurs when oxygen (O_2) reacts with methane (CH_4), producing carbon dioxide and water. The unbalanced equation is written as follows:

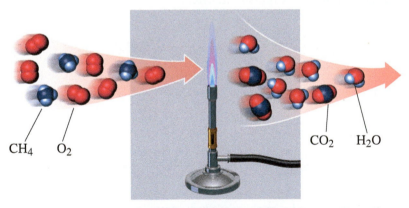

CH_4 O_2 CO_2 H_2O

$$CH_4(g) + O_2(g) \rightarrow CO_2(g) + H_2O(g)$$

Once the chemical equation is written, follow steps 1 to 3.

STEP 1 **Examine the original equation.** Is it balanced? If not, proceed to step 2.

Element	Number in Reactants	Number in Products
H	4	2
C	1	1
O	2	3

There are different numbers of atoms in the reactants and products, so no, it is not balanced.

STEP 2 Balance the equation one element at a time by adding coefficients. Balance elements that appear only once on a side first. For our example, we save oxygen for last because it appears in both CO_2 and H_2O in the products. The number of carbon is balanced, but the number of H is not. Writing the coefficient 2 in front of H_2O will give us 4 Hs ($2 \times H_2O$) in the products. This addition of 2 in front of H_2O likewise changes the total number of oxygen atoms on the product side to 4. Two oxygens in $2H_2O$ and 2 in CO_2. Then, writing a 2 in front of the O_2 in the reactants gives 4 oxygen atoms ($2 \times O_2$) on the reactant side.

$$CH_4(g) + 2O_2(g) \rightarrow CO_2(g) + 2H_2O(g)$$

STEP 3 Check to see if the equation is balanced. The coefficients should represent the smallest possible set of numbers (not all divisible by another number).

Element	Number in Reactants	Number in Products
H	4	4
C	1	1
O	4	4

Now the number of atoms of each element in the reactants equals that in the products, so the equation is balanced.

■

Sample Problem 1.7   Balancing Chemical Equations

Add coefficients to balance the chemical equations shown.
  a. $H_2(g) + O_2(g) \rightarrow H_2O(g)$ b. $C(s) + SO_2(g) \rightarrow CS_2(l) + CO(g)$

Solution

a.
STEP 1 Examine.

Element	Number in Reactants	Number in Products
H	2	2
O	2	1

STEP 2 Balance. The element not balanced is oxygen. If we write a 2 in front of the H_2O in the product, we balance the oxygen, but now the hydrogen is doubled to 4. The hydrogen can then be balanced by writing a 2 in front of the hydrogen in the reactants.

$$2H_2(g) + O_2(g) \rightarrow 2H_2O(g)$$

STEP 3 Check.

Element	Number in Reactants	Number in Products
H	4	4
O	2	2

The equation written in step 2 is balanced.

b.
STEP 1 Examine.

Element	Number in Reactants	Number in Products
C	1	2
S	1	2
O	2	1

—*Continued next page*

Continued—

STEP 2 Balance. None of the elements are balanced. If we start by looking at sulfur, which appears only once on both sides, and we write a 2 in front of the SO_2 in the reactant to balance the sulfur, we get a total of 4 oxygens on the reactant side. This can be balanced by writing a 4 in front of the CO in the products. Now we have 5 carbons in the product, which can be balanced by writing a 5 in front of the C on the reactant side.

$$5C(s) + 2SO_2(g) \rightarrow CS_2(l) + 4CO(g)$$

STEP 3 Check.

Element	Number in Reactants	Number in Products
C	5	5
S	2	2
O	4	4

The equation written in step 2 is balanced.

Practice Problems

1.11 Determine whether each of the following is a physical change or a chemical reaction:

a. A copper penny turns green.

b. Sugar is melted to make candy.

c. An antacid tablet is dropped into water, and bubbles of carbon dioxide form.

1.12 Determine whether each of the following is a physical change or a chemical reaction:

a. Cream, sugar, and flavorings are mixed and frozen to make ice cream.

b. Milk spoils and becomes sour.

c. A puddle of spilled nail polish remover "disappears."

1.13 Balance each of the following reactions by adding coefficients in the blanks. For the purpose of this exercise, place a 1 where appropriate. Usually, if no coefficient appears, "1" is implied.

a.  ___ ●● + ___ ●● → ___ ●●

b.  ___ ●● + ___ ●● → ___ ●●●

1.14 Balance each of the following reactions by adding coefficients in the blanks. For the purpose of this exercise, place a 1 where appropriate. Usually, if no coefficient appears, "1" is implied.

a.  ___ ●● + ___ ● → ___ ●●●

b.  ___ ●● + ___ ●● → ___ ●● + ___ ●●

1.15 Add coefficients to balance the chemical equations shown.

a. $Al(s) + Cl_2(g) \rightarrow AlCl_3(s)$

b. $HCl(aq) + Zn(s) \rightarrow ZnCl_2(aq) + H_2(g)$

c. $SiO_2(s) + C(s) \rightarrow SiC(s) + CO(g)$

1.16 Add coefficients to balance the chemical equations shown.

a. $NO_2(g) + H_2O(l) \rightarrow HNO_3(aq) + NO(g)$

b. $NH_3(g) + O_2(g) \rightarrow N_2(g) + H_2O(l)$

c. $C_7H_{16}(g) + O_2(g) \rightarrow CO_2(g) + H_2O(g)$

1.4 Math Counts

The proper tools can make a job easier. Mathematics gives us some tools that can make understanding chemistry easier. This section covers some basic concepts from mathematics that are applicable to the field of chemistry. As you move through the textbook, you will encounter these math concepts again. You can always come back to this section to review them.

? 1.4 Inquiry Question

What basic math concepts are important to chemistry?

Units, Prefixes, and Conversion Factors

In science, health care, and business, we often make a measurement or report a value. For measurements to be consistently and easily compared, a defined set of standards or a measurement system is required. The *Système International d'Unités* (**SI**) or International System of Units used throughout the world by scientists and health care professionals alike is the modern day version of the **metric system.** Because of this, SI units and therefore metric units will be used throughout this text, so it is important that you are familiar with them.

The SI system is based on a series of standard units for each type of measurable quantity like mass and volume. The standard SI unit for mass is the **kilogram (kg)** while the standard metric unit for mass is the **gram (g),** which is a thousand times smaller than the kilogram. The standard SI unit for volume is the **liter (L),** and the standard SI unit for length is the **meter (m).** Although these measures may seem unfamiliar now, a kilogram is about 2.2 pounds, a gram is about the mass of a paper clip, a liter is a little more than a quart, and a meter is slightly more than 3 feet.

Imagine how many paper clips it would take to equal your body mass. This would be a huge number and would be awkward to work with. Or think of trying to get the mass of one human hair by comparing it to a paper clip. It would take a very large number of hairs to equal the mass of a paper clip.

To deal efficiently with quantities that are much larger or much smaller than one another, the SI system employs a set of prefixes that can be applied to the base unit, changing the meaning of the unit by a power of ten. The power-of-ten concept is similar to our currency in that 10 pennies equal 1 dime and 10 dimes equal 1 dollar. **TABLE 1.3** lists the common SI and metric prefixes, as well as the abbreviation and value for each.

For example, because the average person has a mass much greater than a paper clip, we apply the prefix kilo to the base unit of mass, the gram. "Kilo" means "1000 times greater than," so a *kilo*gram is 1000 g. Thus, an average person's mass would be expressed as 70 kg, which is a much more manageable number than 70,000 g. Note that while the kilogram is the standard SI unit of mass, it is 1000 times larger than the base unit (the gram), which has no prefix.

TABLE 1.3 SI and Metric Prefixes

Prefix	Abbreviation	Relationship to Base Unit
giga	G	× 1,000,000,000
mega	M	× 1,000,000
kilo	k	× 1000
base unit (has no prefix)		× 1 (gram, liter, meter)
deci	d	÷ 10
centi	c	÷ 100
milli	m	÷ 1000
micro	μ or mc	÷ 1,000,000
nano	n	÷ 1,000,000,000

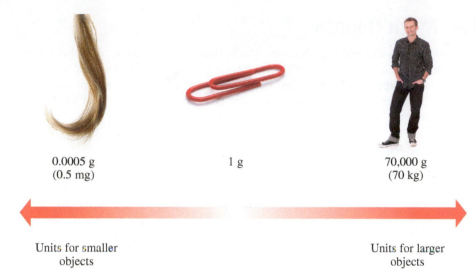

0.0005 g   1 g   70,000 g
(0.5 mg)   (70 kg)

Units for smaller   Units for larger
objects   objects

Prefixes in the SI and metric system help us to scale objects.

Mastering Video
Practicing the Concepts
Conversion Factors Are
Relationships

We can use Table 1.3 to obtain equivalency relationships between different units. The numbers in the right-hand column provide the relationship between any unit and the base unit. For example, when moving from deciliters to liters, the last column says we should divide the base unit by 10; therefore, the equivalency is 1 dL = 0.1 L. Quantities that can be related to each other by an equal sign are called **equivalent units**. By multiplying both sides by 10, we can avoid working with decimals and say that 10 dL = 1 L.

Such equivalencies can be used as **conversion factors** to convert one unit to another using one or more of these factors. For example, when the equivalency stated previously, 10 dL = 1 L, is written as a conversion factor, it can take the following forms:

$$\frac{10 \text{ dL}}{1 \text{ L}} \quad \text{or} \quad \frac{1 \text{ L}}{10 \text{ dL}}$$

These conversion factors are read as, "There are 10 deciliters in 1 liter" and "1 liter contains 10 deciliters," respectively. Conversion factors allow you to convert a quantity from one unit to the equivalent quantity in a different unit. This use of converting units to an equivalent unit is also called **dimensional analysis**. Let's try an example.

Solving
a Problem   *Converting Units*

How many grams of vitamin C are in a tablet containing 100 mg?

STEP 1 **Determine the unit on your final answer.** In this case, the unit will be grams.

STEP 2 **Establish the given information.** In this case, you are given the value of 100 mg in the problem. You also have an equivalency relationship (see Table 1.3) showing that 1 mg = 0.001 g or 1000 mg = 1 g, which gives the following conversion factors:

$$\frac{1000 \text{ mg}}{1 \text{ g}} \quad \text{or} \quad \frac{1 \text{ g}}{1000 \text{ mg}}$$

STEP 3 **Decide how to set up the problem.** Which conversion factor should we use? The given units must cancel out (appearing in both a numerator and denominator), leaving the desired unit in the answer. This problem gives a number of milligrams and asks for the number of grams. We must *convert* from milligrams to grams. To do this, set up an equation using a conversion factor with the desired unit in the numerator. The form of this equation is as follows:

$$\text{Given unit} \times \underbrace{\frac{\text{desired unit}}{\text{given unit}}}_{} \quad \longleftarrow \text{ Units of answer in numerator}$$

This is the conversion factor.

For our problem, the equation is as follows:

$$100 \text{ mg} \times \frac{1 \text{ g}}{1000 \text{ mg}} =$$

STEP 4 Solve the problem. When the calculation is carried out, the given unit in the denominator of the conversion factor will cancel the given unit from the problem.

$$100 \text{ m\!g} \times \frac{1 \text{ g}}{1000 \text{ m\!g}} = \frac{100}{1000} = 0.1 \text{ g}$$

Sample Problem 1.8 Unit Conversions

Using the appropriate conversion factor, convert each of the following quantities to the indicated unit:

a. 10,000 cg = _____ g

b. 5 g/mL = _____ mg/dL

Solution

a.

STEP 1 Determine the unit on your final answer. In this case, the unit will be grams.

STEP 2 Establish the given information. In this case, you have a value of 10,000 cg given. You also have an equivalency relationship (Table 1.3) showing that 1 g = 100 cg.

$$\frac{100 \text{ cg}}{1 \text{ g}} \quad \text{or} \quad \frac{1 \text{ g}}{100 \text{ cg}}$$

STEP 3 & 4 Decide how to set up the problem and then solve the problem. The answer must be in grams, so grams must be in the numerator of our conversion factor. The equation is set up and solved like this:

$$10,000 \text{ c\!g} \times \frac{1 \text{ g}}{100 \text{ c\!g}} = 100 \text{ g}$$

b.

STEP 1 Determine the unit on your final answer. In this case, the unit will be mg/dL.

STEP 2 Establish the given information. In this case, you have a value of 5 g/mL given. You have equivalency relationships (see Table 1.3) showing that 1 L = 1000 mL, 1 L = 10 dL, and 1 g = 1000 mg.

STEP 3 Decide how to set up the problem. In this problem you will need to convert from g/mL to mg/dL. This will require two conversion factors. You will want to set up the equivalencies determined in step 2 as conversion factors so that the mg unit is in the numerator and the dL unit is in the denominator, or

$$\frac{1000 \text{ mg}}{1 \text{ g}}, \quad \frac{1 \text{ L}}{10 \text{ dL}}, \quad \frac{1000 \text{ mL}}{1 \text{ L}}$$

—Continued next page

Mastering Video
Practicing the Concepts
Using Multiple Conversion Factors

Continued—

STEP 4 Solve the problem. All the units that are not part of the answer should be organized so that the unwanted units cancel each other out, leaving only the desired unit(s) for the answer. The equation can be set up like this:

$$\frac{5\ \cancel{g}}{1\ \cancel{mL}} \times \frac{1000\ mg}{1\ \cancel{g}} \times \frac{1\ \cancel{L}}{10\ dL} \times \frac{1000\ \cancel{mL}}{1\ \cancel{L}} = \frac{5{,}000{,}000\ mg}{10\ dL} = 500{,}000\ \frac{mg}{dL}$$

Significant Figures

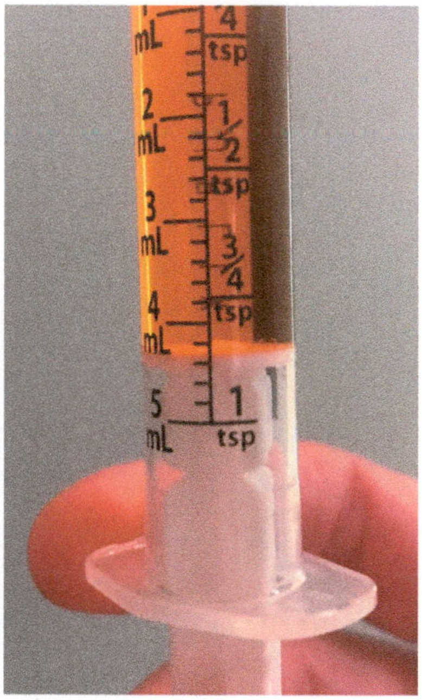

Before proceeding further with any calculations, it is important that we understand that measuring matter relies on the precision of the instruments that we use to measure it. For example, a bathroom scale only measures to the closest half-pound, so we wouldn't consider recording a person's weight as 150.2356 lb! It is important that we report calculated answers reasonably, or as scientists refer, to the correct number of significant figures.

Suppose you are asked to read the number of mLs in the syringe shown at the right. Measurements on a syringe are made by reading across the top ring of the plunger (closest to the liquid). What would you report? To how many decimals?

Because the syringe shows marks (calibrations) every two-tenths (0.2) mL, it looks like the there is definitely more than 4.2 mL but perhaps not quite 4.4 mL. You might report your measure as 4.35 mL since the plunger looks like it is sitting closer to 4.4 than to 4.2. By reporting a value of 4.35, we are reporting that we are sure that there is between 4.3 and 4.4 mL in the syringe.

We report this measured value to three numbers, or three significant figures. **Significant figures** are the digits in a value known with certainty *plus* an estimated digit. When reading a measurement from a nondigital device like a syringe, there is always a level of uncertainty in the measurement. Significant figures allow us to report the certainty of a measurement. Digital devices like a digital thermometer automatically show us the number of significant figures in the digital readout. In any measurement, all nonzero numbers are considered significant.

We can report the measure of this syringe as 4.35 mL, estimating the last digit, 5.

This digital thermometer shows three significant figures.

Zeros tend to be the "troublemakers" because their significance depends on their position in a number. For example, when a person's weight is reported as 70 kg, the number has *only one* significant figure, so all we know about the person's weight is that it is between 65 grams and 75 kg. However, if a person's weight is reported as 70.0 kg, the last zero *is* significant and this measurement tell us the person weighs between 69.9 and 70.1 kg. When reading measurements from nondigital devices, one should always estimate the last reported number just as we did with the syringe example.

If a zero in a number without a decimal point is significant, its significance can still be shown. To indicate that the zero in a measurement of 10 inches is significant, simply put in a decimal point after the zero; 10. inches means that both digits are significant. Otherwise, to indicate significant zeros in large numbers, draw a line above the last significant zero; 10,0$\overline{0}$0 has four significant figures.

It should be noted that numbers used in defined conversion factors and counted items have more certainty. For example, if you are counting drops per minute (gtt/min) and you counted 20 of these, you counted exactly 20. Your certainty is infinite (the drop rate was exactly 20) and there is not a limited number of significant figures in such a case. Significant figures tend to not be considered for these numbers in calculations. Only measured numbers determine the number of significant figures given in a final answer.

The rules for counting significant figures in measurements are summarized in **TABLE 1.4**.

TABLE 1.4   Counting Significant Figures in Measurements

Rule	Measurement	Number of Significant Figures
1. A digit is significant if it is		
a. not a zero	41 g	2
	15.3 m	3
b. a zero between nonzero digits	101 L	3
	6.071 kg	4
c. a zero at the end of a number with a decimal point	20. g	2
	9.800 °C	4
2. A zero is not significant if it is		
a. at the beginning of a number with a decimal point	0.03 L	1
	0.00024 g	2
b. in a large number without a decimal point	12,000 km	2
	3,450,000 m	3

Sample Problem 1.9   Determining Significant Figures

Give the number of significant figures in the following measurements:
a. A digital scale reads 56.7 kg.
b. A pharmacist dispenses 30 pills.
c. A newspaper reports that about 500 people attended the event.
d. A baby's temperature is measured to be 100.° F.

Solution

a. Three significant figures; see rule 1a in Table 1.4.
b. Infinite number of significant figures; the pharmacist counted the number of pills, so the measurement is exact.
c. One significant figure; see rule 2b in Table 1.4.
d. Three significant figures; the presence of the decimal point indicates that the zeros are not just placeholders but are a part of the measurement. See rule 1c in Table 1.4.

Calculating Numbers and Rounding

Next, let's look at how we consider significant figures in calculations. Suppose you ask the person next to you their age, and they tell you that they are 27. How many days old is the person? To calculate the number of days old we could multiply 27 by 365.24 days (the precise number of days in a year) on our calculator, giving us an answer of 9,861.48 days.

Consider the answer from the calculator. We know the person's age to the nearest year, but by multiplying that value by the precise number of days in a year, we now know the age to within a tenth of a day! Somehow that doesn't seem logical. Manipulating measurements (which have some built-in uncertainty) with arithmetic cannot *increase* the certainty of the measurement. A calculator does not know the certainty in your measurement and will often make an attempt at more certainty than makes sense, so the person doing the calculation must be prepared to round the answer correctly.

How many of the resulting numbers from our previous calculation of the person's age are certain and, therefore, meaningful? Keeping all of the numbers results in an answer with six significant figures, yet we started with a number (27) that has only two.

In general, your answer can be no more certain than the least certain number that you started with. The rules for determining the number of significant figures in a calculated answer depend on the arithmetic operation being used.

Rules for Significant Figures in Calculations

- **Addition and Subtraction.** Answers should match the least number of decimal places in the measured numbers.

$$
\begin{array}{lll}
1.002 & \longleftarrow & \text{three decimal places} \\
2.5 & \longleftarrow & \text{one decimal place} \\
+10.16 & \longleftarrow & \text{two decimal places} \\
\hline
13.662 & & \text{report as 13.7}
\end{array}
$$

Because one of the numbers contains only one decimal place, the final answer must be rounded to one decimal place.

- **Multiplication and Division.** Answers should match the least number of significant figures in the measured numbers. For our age-in-days calculation, the answer can have only two significant figures because we were given the person's age as 27.

$$
27 \text{ years old} \times \frac{365.24 \text{ days}}{1 \text{ year}} = 9{,}861.48 \text{ days old; report as 9,900 days old}
$$

Rules for Rounding Numbers

- If the leftmost digit to be dropped (the one following your last retained significant figure) is 4 or less, simply remove it and the remaining digits.
- If the leftmost digit to be dropped is 5 or greater, increase the last retained digit by 1 and remove all other digits.
- If rounding a large number with no decimal point, substitute zeros for numbers that are not significant.
- When conducting multiple-step calculations, do not round answers until the end of the calculation. *Rounding at each step introduces rounding errors and produces inaccurate answers.*

Sample Problem 1.10   **Rounding Calculated Numbers**

Complete each of the following calculations and round the answer to the correct number of significant figures:

a. $75 - 21.3 =$
b. $8.24 \times 3.0 =$
c. $0.005 + 1.23 =$
d. $12.4 \div 0.06 =$

Solution

a. Answer is 54. The calculator result of 53.7 is rounded so that the answer has no decimal places because 75 has no numbers after the decimal point.
b. Answer is 25. The calculator result of 24.72 is rounded to two significant figures because 3.0 has only two significant figures.
c. Answer is 1.24. The calculator result of 1.235 is rounded to the least number of decimal places, which is two.
d. Answer is 200. The calculator result of 206.66666 is rounded to one significant figure as 200, which has the same number of significant figures as 0.06.

Scientific Notation

In chemistry, we sometimes use extremely large and extremely small numbers in the same calculation. Suppose that we wanted to figure out how many sheets of paper we could stack in a cabinet that is 3 m high. Each sheet of paper has a thickness of one-tenth of a millimeter or 0.0001 m. The number of sheets of paper would be

$$\frac{1 \text{ sheet of paper}}{0.0001 \text{ m}} \times \frac{3 \text{ m}}{1 \text{ cabinet}} = \frac{30{,}000 \text{ sheets of paper}}{\text{cabinet}}$$

This is a lot of pieces of paper! We can represent this with this large number. Yet it is more easily written using scientific notation. To explore scientific notation, consider that 30,000 pieces of paper is $3 \times 10 \times 10 \times 10 \times 10$ (try it on your calculator). In scientific notation, we can express the four times we multiplied by 10 as an exponent, showing it as 10^4. The scientific notation for 30,000 is written as 3×10^4.

A number smaller than 1, like the 0.0001 m for each sheet of paper, can be determined by dividing 1 by $10 \times 10 \times 10 \times 10$ or $1/10^4$. Dividing by a number is equivalent to multiplying by its reciprocal. So, in exponential form, $1/10^4$ can be written as 1×10^{-4}. The general form for scientific notation is

$$C \times 10^n$$

where C is the **coefficient** and is a number that is at least 1 but less than 10, and n is the exponent telling us the number of tens places that apply.

A positive exponent tells us that the actual number is greater than 1, and a negative exponent tells us the number is between 0 and 1. In scientific notation, only significant figures are shown in the coefficient (any place-holding zeros are not part of the notation). For example, the scientific notation of the number 0.00650 is 6.50×10^{-3} because the zeros between the decimal point and the 6 are placeholders and the zero after the 5 is significant.

We could rewrite an equivalent calculation in scientific notation as

$$\frac{1 \text{ sheet of paper}}{1 \times 10^{-4} \text{ m}} \times \frac{3 \text{ m}}{1 \text{ cabinet}} = \frac{3 \times 10^4 \text{ sheets of paper}}{\text{cabinet}}$$

1 piece of paper = 0.0001 m

TABLE 1.5 shows how numbers and scientific notation are related for a wide range of numbers.

Scientific notation can be displayed on your calculator as long as you know which buttons to press. Most scientific calculators have a button labeled EE, EXP, or SCI, which displays the coefficient (C) and the power-of-ten exponent (*n*). It is worth exploring your calculator to be sure that you can input scientific notation correctly. To input 3×10^4, try pressing 3 EE or EXP 4. To input 1×10^{-4}, try pressing 1 EE or EXP +/− 4. With a scientific calculator (which requires a reverse input), press 1 EE or EXP 4 +/− .

TABLE 1.5 Relating Numbers to Scientific Notation

Number	Meaning	Scientific Notation
1,000,000	$10 \times 10 \times 10 \times 10 \times 10 \times 10$	1×10^6
100,000	$10 \times 10 \times 10 \times 10 \times 10$	1×10^5
10,000	$10 \times 10 \times 10 \times 10$	1×10^4
1000	$10 \times 10 \times 10$	1×10^3
100	10×10	1×10^2
10	10	1×10^1
1	1	1×10^0
0.1	1/10	1×10^{-1}
0.01	$1/(10 \times 10)$	1×10^{-2}
0.001	$1/(10 \times 10 \times 10)$	1×10^{-3}
0.0001	$1/(10 \times 10 \times 10 \times 10)$	1×10^{-4}
0.00001	$1/(10 \times 10 \times 10 \times 10 \times 10)$	1×10^{-5}
0.000001	$1/(10 \times 10 \times 10 \times 10 \times 10 \times 10)$	1×10^{-6}

Sample Problem 1.11 Expressing Numbers in Scientific Notation

Express the following numbers in scientific notation:

a. 820,000   b. 0.00096

Solution

a. To determine the correct scientific notation for a number greater than 1, count the number of places to the right of the first number, 8. In this case, there are 5 places to the right of 8. This becomes the exponent on the number 10 in scientific notation. Any other nonzero numbers go after a decimal point. The answer is 8.2×10^5.

b. To determine the correct scientific notation for a number less than 1, count the number of places to the left of the first number, 9, and add 1. In this case, there are 3 places to the left of the 9, and adding 1 would give a total of 4. The negative of this number becomes the exponent on the number 10 in scientific notation. Any other nonzero numbers go after the decimal point. The answer is 9.6×10^{-4}.

Sample Problem 1.12 Changing Numbers from Scientific Notation to Decimal Form

Express the following numbers in decimal form:

a. 6.5×10^4 b. 8.34×10^{-6}

Solution

a. The number in the exponent tells you how many places are beyond the decimal place in the scientific notation when converting from scientific notation to decimal form. Notice that the number of significant figures does not change when the format of the number changes. In this problem, the number 5 will occupy one of the places, so 3 more zeros after the 5 are necessary. The answer is 65,000.

b. Negative exponents indicate that this number is smaller than 1 but greater than zero. The number in the exponent tells you how many places should appear to the left of the decimal place when converting from scientific to decimal form. The number 8 will occupy one of these places, so 5 more zeros should appear in the final answer to the right of the decimal place. All digits to the right of the decimal place in the scientific notation should appear in the final answer. The answer is 0.00000834.

Percent

If your chemistry class has 40 students enrolled and 20 students got an A on the first test, you might comment that one-half, or 50%, of the class got an A. Percent, represented by the symbol %, means the part out of 100 total, or hundredths. So, 50% means 50 out of a total of 100 or 50 hundredths (0.50), which reduces to $^1/_2$, the same fraction as our original example using grades. Percent allows us to directly compare two sets of numbers that have different total sizes.

$$\text{Percent (\%)} = \frac{\text{part}}{\text{whole}} \times 100$$

A fraction can be converted to a percent by dividing the numerator by the denominator, multiplying by 100, and adding a percent sign. A decimal number can be converted to a percent by multiplying by 100 and adding a percent sign.

Sample Problem 1.13   **Expressing Fractions and Decimals as a Percent**

a. Express 1/5 as a percent.   b. Express 0.35 as a percent.

Solution

a. For a fraction, divide the numbers of the fraction, multiply by 100, and add a percent sign.

$$1/5 = 0.20 \times 100 = 20\%$$

b. For a decimal, multiply by 100 and add the percent sign.

$$0.35 \times 100 = 35\%$$

Sample Problem 1.14   **Expressing Percents as Fractions and Decimals**

a. Express 25% as a decimal.   b. Express 25% as a fraction.

Solution

a. To express 25% as a decimal, divide the value by 100.

$$\frac{25}{100} = 0.25$$

b. Percent means parts out of a total of 100 parts, so the fraction is

$$\frac{25}{100}, \text{which reduces to } \frac{1}{4}$$

Sample Problem 1.15   **Determining the Percent of a Number**

a. What is 40.0% of 275?   b. What is 80.0% of 65?

Solution

To determine the percent of a number, change the percent to a decimal and multiply by the number.

a. 40.0% = 0.400   b. 80.0% = 0.800
   0.400 × 275 = 110.   0.800 × 65 = 52

Sample Problem 1.16   **Determining the Percent That One Number Is of Another Number**

a. What percent of 1500 is 30?   b. Thirty-five is what percent of 140?

Solution

To determine what percent one number is of a second number, a fraction is created to represent parts out of the whole. This gives a decimal answer that can then be converted to a percent by multiplying by 100.

a. $\dfrac{30}{1500} = 0.02$ b. $\dfrac{35}{140} = 0.25$

   0.02 × 100 = 2%   0.25 × 100 = 25%

Solving a Problem

Calculating Percent

Suppose you go out to a nice lunch with three other friends and the final bill is $125.86. If you decide to split the bill evenly and leave an 18% tip, how much would each person contribute?

STEP 1 **Determine the units on your final answer.** In this case, the final answer is in dollars.

STEP 2 **Establish the given information.** We know the total bill, $125.86. We also know the percent we are going to tip, 18%.

STEP 3 **Decide how to set up the problem.** To determine 18% of the total bill, you express the percent as a decimal and multiply by the total bill. The total amount that will be spent is the sum of the bill plus the 18% tip.

$$18\% = 0.18$$
$$0.18 \times \$125.86 \text{ total bill} = \$22.65$$
$$\text{Total bill} = \$125.86 + \$22.65 = \$148.51$$

STEP 4 **Solve the problem.** Since the total bill is $148.51, you divide this number by the four of you.

$$\$148.51/4 = \$37.13$$

Note that you must round to the nearest penny since that is your smallest unit of measure.

Practice Problems

1.17 Give the number of significant figures in each of the following measurements:

    a. 25 min   b. 44.30 °C

    c. 0.037010 m   d. 800. L

1.18 Give the number of significant figures in each of the following measurements:

    a. 0.000068 g   b. 100 °F

    c. 9,237,200 years   d. 25.00 m

1.19 Complete each of the following calculations and give your answer to the correct number of significant figures:

    a. $100 \times 23 =$ b. $97.5 - 43.02 =$

    c. $8064/0.0360 =$ d. $54.00 + 78 =$

1.20 Complete each of the following calculations and give your answer to the correct number of significant figures:

    a. $340/96 =$ b. $0.305 + 43.0 =$

    c. $0.0065 \times 11 =$ d. $19.029 - 0.00801 =$

1.21 A typical aspirin tablet contains 325 mg of the active ingredient acetylsalicylic acid. Convert to grams.

1.22 On average, an adult's lung volume is 5 L. Convert to mL.

1.23 Write scientific notation for the following numbers:

    a. 203,000,000   b. 12.4   c. 0.0000000278

1.24 Write scientific notation for the following numbers:

    a. 1,400,000   b. 0.079   c. 0.00000354

1.25 Express the following numbers in a decimal form:

    a. 1.56×10^{-5} b. 2.8×10^{5} c. 9.0×10^{-2}

1.26 Express the following numbers in a decimal form:

    a. 7.4×10^{-2} b. 3.75×10^{3} c. 1.19×10^{-8}

1.27 Write scientific notation for the following numbers:

    a. a platelet dose of 500,000,000,000

    b. a red blood cell count of 5,500,000

1.28 Express the following numbers in decimal form:

    a. an asthma medication of 1.00×10^{-8} g

    b. vitamin D dose of 5.0×10^{-6} g

1.29 Express the following numbers as a percent. Report two significant figures in your answer.

    a. 1/4   b. 3/8   c. 66/100

1.30 Express the following numbers as a percent. Report two significant figures in your answer.
  a. 0.58   b. 0.36   c. 0.125

1.31 Express the following numbers in decimal form:
  a. 4.5%   b. 13.0%
  c. 66%   d. 78%

1.32 Express the following numbers as a fraction:
  a. 20%   b. 75%
  c. 40%   d. 12%

1.33 Determine the number from the percent given:
  a. What is 25% of 80?   b. What is 0.9% of 1000?
  c. What is 5.0% of 750?   d. What is 66.6% of 200.?

1.34 Determine the percent from the numbers given here.
  a. Fifty is what percent of 125?
  b. Six is what percent of 600?
  c. What percent of 300 is 15?
  d. What percent of 400 is 30?

1.5 Matter: The "Stuff" of Chemistry

Chemists study matter that we cannot see, even with a high-powered microscope. Even though we cannot see this matter, we can see and measure its properties in bulk. Here we discuss some of those properties and characteristics of matter.

1.5 Inquiry Question
What properties of matter can be measured?

Mass

Consider that anything that takes up space can also be placed on a scale and weighed—that is, it has mass. So, matter can be more completely defined as anything that takes up space and has mass. **Mass** is a measure of the amount of material in an object. As we saw in Section 1.4, a common unit used to measure the mass of a substance is the gram (g).

A raisin or a paper clip has a mass of about 1 g. The term *weight* may be more familiar to you than the term *mass*. These terms are related, but they do not mean the same thing. The **weight** of an object is determined by the pull of gravity on the object, and that force changes depending on location. For instance, astronauts weigh less on the moon than on Earth. This is because the gravitational pull of the moon is less than that of Earth. The astronauts' mass did not change, but their weight did. The SI unit for weight, the force on an object of a given mass due to gravity, is the newton (N).

When you step on the bathroom scale, the scale does not read your weight in newtons. Are you determining your mass or your weight? How are mass and weight related to each other?

A scale or balance is used to measure mass. When balances are manufactured, they are set at the factory so that a 1-kg (1000-g) object has the same mass on all of the balances. This is known as "calibrating" the balance and accounts for the gravitational pull of Earth. After calibrating, the mass and the weight of the 1-kg object are the same. If the object and balance were transported to the moon, the balance would have to be re-calibrated to account for the lesser gravitational pull of the moon if it were to read the mass on the moon as 1 kg.

As long as an object is weighed in roughly the same location on Earth's surface, its mass and weight will have the same measured value. To avoid confusion, this book will, from this point forward, refer to weighable quantities using the term *mass*.

Measuring mass in the lab

Measuring mass at home

Volume

Another way of saying that matter "takes up space" is to say that it has **volume**. Large soft drink bottles are often called "2-liter" bottles. The volume of the soft drink in the bottle is 2 L (1 L = 1.057 quart). Similarly, as you blow up a balloon, it gets bigger. The volume of air in the balloon increases. So, volume is a three-dimensional measure of the amount of space occupied by matter.

FIGURE 1.4 The relationship between volume and length. One cubic centimeter (abbreviated cc or cm³) equals 1 mL of volume. For water, this volume weighs 1 g.

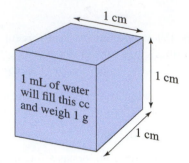

When cooking or baking at home, you might use a measuring cup or a measuring spoon to measure volume. In the lab, volumes are routinely measured with a graduated cylinder or a pipet. The unit of volume typically used in the lab is the milliliter (mL). In a clinical setting, volumes are often measured with calibrated syringes. The typical unit in the clinical setting is the cubic centimeter (cc, or cm³). One milliliter equals one cubic centimeter (see **FIGURE 1.4**). A teaspoon of cough syrup contains about 5 mL or 5 cc.

Measuring spoons

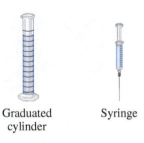

Measuring cup

Measuring volume at home

Graduated cylinder   Syringe

Measuring volume in the lab

Density

Regardless of its mass or volume, most types of wood will float on water while a piece of metal will sink. This observation implies that the mass of wood that fits into a certain space is less than that for metal. This property of matter is called density. **Density** (*d*) is a comparison (also called a ratio) of a substance's mass (*m*) to its volume (*V*). This can be displayed mathematically as

$$d = \frac{m}{V}$$

One gram of water has a volume of one milliliter, so the density of water is 1.00 g/mL. A piece of wood will float on water because it is less dense than water, while a piece of metal will sink because it is more dense than water (see **FIGURE 1.5**).

We can think of density as a measure of the packing of matter. As another example, the density of our bones can change even though the size of our bones doesn't change. As we age, minerals are lost from our bones faster than they are replenished. The bones' strength and integrity can become compromised. In this case, the mass of the bone lessens while the size or volume stays the same, making bones less dense. This can lead to the condition known as *osteoporosis*.

FIGURE 1.5 Visualizing density. Objects that are less dense than water will float while objects more dense than water will sink.

> ### Sample Problem 1.17 Calculating Density
>
> A brick stamped "14-karat gold" has a mass of 1100 g and a volume of 85 mL. Calculate its density. Will the brick float or sink when placed in water?
>
> #### Solution
>
> You are given the mass and volume and asked to calculate the density in g/mL. Insert the given values into the density equation.
>
> $$d = \frac{m}{V}$$
>
> $$d = \frac{1100 \text{ g}}{85 \text{ mL}} = 13 \text{ g/mL}$$
>
> The final answer is reported to two significant figures because the minimum number of significant figures measured in the problem is two. The gold brick is more dense than water (which has a density of 1.0 g/mL), so it will sink.

Because the density of a substance is constant at a given temperature, we can use density values as conversion factors to determine either the mass or the volume of a substance.

Sample Problem 1.18 **Problem Solving with Density**

Table sugar has a density of 1.29 g/mL. Calculate the mass in grams of 1.00 teaspoon of sugar (1.00 teaspoon has a volume of 4.93 mL).

Solution

These problems are worked similarly to the unit-conversion problems solved in the previous section.

STEP 1 Determine the units on your final answer. In this case, the units will be the mass in grams.

STEP 2 Establish the given information. In this case, the density is given and you will solve the density equation for mass. You are also reminded of the volume equivalency that 1.00 tsp = 4.93 mL.

STEP 3 Decide how to set up the problem. You are given the density and the volume. To solve for mass, the density equation must be rearranged algebraically.

$$d = \frac{m}{V}$$

To rearrange the density equation, multiply both sides by V so the mass is now the variable that can be solved for.

$$d \times V = \frac{m}{V} \times V$$
$$m = d \times V$$

STEP 4 Solve the problem. The values and conversion factor are plugged in and the mL units cancel out, leaving g as the unit on the final answer.

$$m = \frac{1.29 \text{ g}}{1 \text{ mL}} \times \frac{4.93 \text{ mL}}{1.00 \text{ tsp}} \times 1.00 \text{ tsp} = 6.36 \text{ g}$$

Because the density and mass contain three significant figures, the answer must be given to three significant figures.

Specific Gravity

Which is more dense, water or honey? Both contain water, but if you pour honey into water (or tea, which is mostly water), it drops to the bottom. From this experiment, we can observe that honey must be more dense than liquid water. Honey, although a liquid, possesses the property of density because it has mass and takes up space.

For liquids, density often is measured with respect to water. The density of water is 1.00 g/mL at 4 °C and close to that value at body temperature. The ratio of the density of a substance to the density of water is called **specific gravity** (sp gr).

$$\text{Specific gravity} = \frac{\text{density of sample}}{\text{density of water}}$$

To calculate the specific gravity of a liquid, the density of the liquid and the density of water must have the same units. Specific gravity is a unitless quantity because it is a ratio of two densities.

The specific gravity of urine is routinely measured as an indicator of kidney function. If a person's kidneys are functioning normally, the density of urine will be similar to that of water. Normal specific gravity values are between 1.005 and 1.030. These urine values also indicate whether a person is over or underhydrated. Specific gravity values closer to 1.000 indicate overhydration, while values above 1.025 can signify dehydration. For liquids, specific gravity is measured with a simple instrument called a *refractometer* (see **FIGURE 1.6**).

FIGURE 1.6 A refractometer measures specific gravity. Measuring the specific gravity of urine using a refractometer (in lab tech's hand) can indicate kidney malfunction and hydration level.

Temperature

We measure the **temperature** of a substance to determine its hotness or coldness. This is often done using a thermometer or an electronic temperature probe. In different parts of the world, temperature is measured using different scales. In the United States, we continue to use the Fahrenheit scale, while the rest of the world uses the Celsius scale. However, scientists use yet another scale called the absolute, or Kelvin, scale where the temperature unit is the kelvin. Kelvin is the SI unit for temperature.

The most straightforward way to compare the temperature scales is to compare temperatures that we are most familiar with and observe their values on each scale. Consider the thermometers in **FIGURE 1.7**.

Pay particular attention to the freezing point and boiling point of water on the three scales. Notice that the Celsius and Kelvin scales have degrees of the same size, but the two scales are offset by 273 units. Mathematically, this relationship is expressed as:

$$\text{Kelvin (K)} = \text{Celsius (°C)} + 273\ °C$$

The relationship between the Fahrenheit scale and the others is not as straightforward. Notice that 100 degrees (100°) on the Celsius scale (the difference between water freezing and water boiling) is equal to 180 degrees (180°) on the Fahrenheit scale, or 100 °C = 180 °F. When reduced to the lowest whole numbers, this relation provides a conversion factor of $\dfrac{5\ °C}{9\ °F}$, which reduces further to $\dfrac{1\ °C}{1.8\ °F}$. Also in Figure 1.7 notice that the "zero points" between Fahrenheit and Celsius are offset by 32 degrees. The relationship between °F and °C is shown:

$$°C = (°F - 32\ °F) \times \frac{1\ °C}{1.8\ °F}$$

$$°F = \left(\frac{1.8\ °F}{1\ °C} \times °C\right) + 32\ °F$$

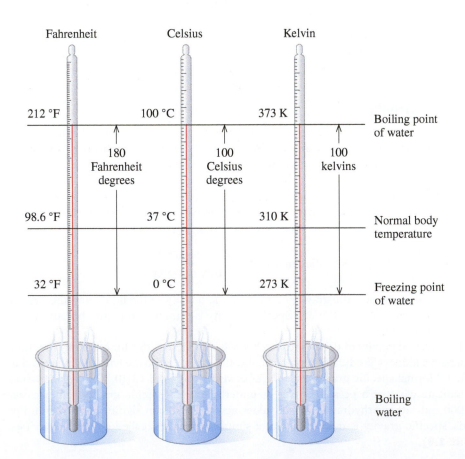

FIGURE 1.7 Comparison of tempera-ture scales. The zero point (freezing of water) is different for each of the scales. Comparing the boiling point to the freezing point of water on each scale, we see that the "size" of a degree is the same on the Celsius and Kelvin scales, but smaller on the Fahrenheit scale. This provides us with the conversion factor $\dfrac{1\ °C}{1.8\ °F}$.

After converting a Fahrenheit temperature to degrees Celsius, you can convert the Celsius temperature to Kelvin using the Celsius-to-Kelvin scale equation. A separate equation to convert Fahrenheit to Kelvin is not needed.

Body Temperature

Normal body temperature is considered to be 98.6 °F or 37.0 °C. However, body temperature varies from person to person and changes throughout the day. Body temperature also changes based on level of exertion. Our bodies expend a lot of energy maintaining a constant body temperature within a narrow range. If human body temperature rises above 40.0 °C (104 °F), a condition known as *hyperthermia* exists. This condition can be caused by heatstroke, where sweat production ceases. At body temperatures this high, a person can go into convulsions or coma and experience permanent brain damage. One immediate remedy for this situation is to immerse the person in ice-cold water.

If body temperature drops below 35 °C (95 °F), *hypothermia* occurs. A person in this condition feels cold, has an irregular heartbeat, and a slow breathing rate. Because these bodily functions are operating slower than normal, the person may lapse into unconsciousness. Treatments may include increasing blood volume, administering oxygen, or, in extreme cases, injecting warm (37 °C) fluid.

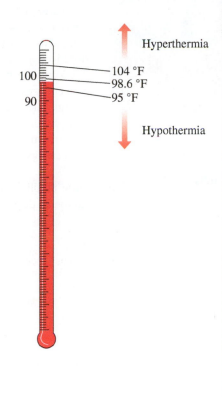

Hyperthermia

104 °F
98.6 °F
95 °F

Hypothermia

INTEGRATING
Chemistry

◀ Find out what happens if body temperature goes to extremes.

Sample Problem 1.19 **Temperature Conversion**

On a hot summer day in the southern United States, the temperature often rises as high as 102 °F. What is this temperature in °C and Kelvin?

Solution

Use the Fahrenheit-to-Celsius equation and substitute the appropriate number:

$$°C = (102 \, °F - 32 °F) \times \frac{1 \, °C}{1.8 \, °F}$$

$$°C = (70 \, °F) \times \frac{1 \, °C}{1.8 \, °F}$$

$$°C = 39 \, °C$$

Now that we have the temperature in °C, we can convert it to the Kelvin scale using the Celsius-to-Kelvin equation.

$$K = 39 \, °C + 273$$

$$K = 312$$

FIGURE 1.8 **Potential and kinetic energy.** The runner contains potential energy in her muscles while waiting for the start of the race. Her body movements during the race transform the potential energy into kinetic energy as she runs.

Energy

After you eat, your stomach breaks down the food, allowing the nutrients to be absorbed in the digestive system. We often say that food is converted to energy by the body. We use the energy produced by food to move muscles, maintain our heartbeat, and perform other bodily functions. But what is the nature of this energy?

Energy is defined as the capacity to do work or supply heat. Food contains nutrients like carbohydrates, proteins, and fats that hold energy that our bodies can transform to do work. The stored chemical energy is called **potential energy**. When you flex a muscle you are using chemical energy to produce motion. The energy of motion is called **kinetic energy** (see **FIGURE 1.8**). Energy takes various forms, but it is never created or destroyed. This concept is called the **law of conservation of energy**. All matter contains some amount of energy in one form or another. This energy is measured in **joules (J),** the SI unit for energy.

While joules is the SI unit of measurement, you may be more familiar with the energy unit called the **calorie (cal).** The calorie is defined as the amount of energy required to raise the

Energies of Various Processes and Occurrences in Joules

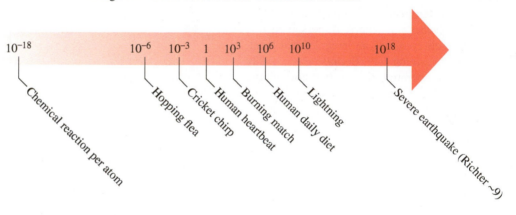

temperature of one gram of water one degree Celsius, but is now typically defined in terms of a joule.

$$1 \text{ calorie} = 4.184 \text{ joule}$$

We can convert between the two units by applying a conversion factor, as follows:

$$\frac{1 \text{ calorie}}{4.184 \text{ joules}} \quad \text{or} \quad \frac{4.184 \text{ joules}}{1 \text{ calorie}}$$

A nutritional Calorie, which is the unit on nutrition labels, is termed a **Calorie (Cal)** and is 1000 times larger than a calorie. (Notice that the nutritional Calorie uses a capital C in its unit, and the calorie uses a lowercase c.)

$$1 \text{ Calorie} = 1000 \text{ calories}$$

The two units can be converted using a conversion factor, as follows:

$$\frac{1000 \text{ cal}}{1 \text{ Cal}} \quad \text{or} \quad \frac{1 \text{ Cal}}{1000 \text{ cal}}$$

Sample Problem 1.20   **Converting Energy Units**

The nutritional label on a granola bar says it contains 220 Calories.
  a. Calculate the number of calories (lowercase) in the granola bar.
  b. Calculate the number of joules in the granola bar.

Solution

  a. There are 1000 calories in 1 nutritional Calorie.

$$220 \text{ Calories} \times \frac{1000 \text{ calories}}{1 \text{ Calorie}} = 220,000 \text{ calories}$$

  b. One calorie is equivalent to 4.184 joules. Converting from Calories to joules requires two conversion factors, as follows:

$$220 \text{ Calories} \times \frac{1000 \text{ calories}}{1 \text{ Calorie}} \times \frac{4.184 \text{ joules}}{1 \text{ calorie}} = 920,000 \text{ joules}$$

The final answer is reported to two significant figures because the measured value in the problem, 220 Calories, contains two significant figures.

Heat and Specific Heat

When you get into a parked car on a sunny day, the car feels warm. If you get into the same car on a cold winter's day, the car is cold. In a warm car, because you are not as hot as the car, the excess warmth of the car is being transferred to your body. In a cold car, your body is warmer than the car and the warmth from your body is transferred to the car, so you perceive the car as cold. This example describes the phenomenon we know as heat. **Heat** is kinetic energy flowing from a warmer body to a colder one. Note that heat is a form of energy.

  When you cook food and then store the leftovers in the refrigerator, heat is being transferred into the food during cooking and out of the food in the cool air of the fridge. Every substance has this ability to absorb or lose heat as the temperature changes. Each substance, though, is different and these values can be measured. The **specific heat capacity** or specific heat of a substance is the amount of heat needed to raise the temperature of exactly 1 g of that substance by exactly 1 °C. To represent a change in temperature, we include the Greek capital delta, Δ, with the symbol for temperature, T, as ΔT, where the change represents the final temperature minus the initial temperature. The specific heat of a substance is represented mathematically as

$$\text{Specific heat } (SH) = \frac{\text{heat}}{\text{grams} \times \Delta T}$$

  TABLE 1.6 shows the specific heat values for several different substances. Notice that metals have low specific heat values and water has a very high specific heat. We enjoy this property of water on a hot summer day when we cool off in a swimming pool. Because it takes a lot of heat to raise the temperature of water, the water does not warm up as fast as the surrounding air and refreshes us when we are hot. We will discuss specific heat and other unique properties of water further in Chapter 8.

Sample Problem 1.21   **Specific Heat**

Based on Table 1.6, would a silver or stainless steel spoon heat up faster, potentially removing more heat from your coffee while you stir it? Support your answer.

Solution

The silver spoon gets hotter faster because of its low heat capacity. Stainless steel has a higher heat capacity than silver, so it takes more heat energy to raise its temperature. It will warm more slowly, giving you more time to stir your coffee.

In warm weather, water stays cooler than air due to its high specific heat.

TABLE 1.6 Specific Heats of Various Substances

Substance	Specific Heat (cal/g °C)
Water (liquid)	1.00
Human body	0.83
Paraffin wax	0.60
Wood, soft	0.34
Wood, hard	0.29
Air	0.24
Aluminum	0.21
Table salt	0.21
Brick	0.20
Stainless steel	0.12
Iron	0.11
Copper	0.092
Silver	0.056
Gold	0.031

A snowflake has a six-sided (hexagonal) structure because of the way molecules of water arrange themselves in the solid state.

States of Matter

Did you know that every snowflake (made from frozen water) is unique, yet every snowflake has a six-sided structure? What is it about water, H_2O, that causes it to freeze into a structure with six sides? The behavior and composition of matter are defined for every substance at the submicroscopic level, in its particles. In the case of water, the H_2O particles are molecules in which the elements oxygen and hydrogen are chemically bonded together (we will discuss molecules in greater detail in Chapter 3). When water molecules freeze (as snowflakes or ice), they organize into a hexagonal structure giving a single snowflake its six-sided structure. This structure allows optimal contact between the water molecules. The properties of individual molecules are responsible for the properties that we can actually measure in substances.

Ice, liquid water, and steam (water vapor) represent different forms of matter. As long as it is kept frozen, ice has a definite, unchanging shape and volume, while liquid water has a definite volume, even as its shape changes depending on its container. It is hard to see steam, but we can witness the changes that it causes. When a kettle whistles, the water vapor building up inside escapes into the room, filling up the new, larger space it is given. Each of these examples represents a different state of matter. A **state of matter** is the physical form in which the matter exists. The three most common states of matter are solid, liquid, and gas, which for water would be ice, liquid water, and water vapor, respectively.

Let's look at the organization of the molecules of water that make up ice, liquid water, and water vapor as shown in **FIGURE 1.9**.

The molecules in ice, a solid, are in an orderly arrangement and are tightly packed together. They are moving, but only very slightly, not enough to affect the shape of the solid. Like ice, **solids** have a definite shape and a define volume.

The molecules in liquid water, are somewhat less orderly and are only loosely associated with each other. They are moving freely, colliding with and moving past each other. The behaviors of these molecules show that a **liquid** has a definite volume, but it takes the shape of its container.

The molecules in steam, the gaseous form of water, have no orderly arrangement and are far apart from each other. They move at high rates of speed and collide with each other and with the walls of their container. A **gas** has no definite shape or volume. Instead, it expands to fill the container in which it is placed.

A substance's properties, such as shape, volume, and kinetic energy, are greatly influenced by its molecular makeup. As we will explore in later chapters, the characteristics of a substance's small molecular building blocks influence the measurable properties of matter. Some of the basic properties of matter are summarized in **TABLE 1.7**.

Solid	Liquid	Gas
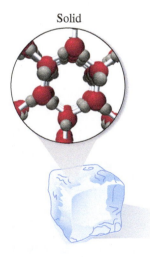		
The particles in a solid are tightly packed and barely moving.	The particles in a liquid are less orderly than those in a solid and are freely moving.	The particles in a gas are disordered and rapidly moving.

FIGURE 1.9 Comparing the particles of a solid, liquid, and gas.

TABLE 1.7 Properties of Solids, Liquids, and Gases

Property of a Substance	Solid	Liquid	Gas
Shape	Definite shape	Adopts shape of container	Adopts shape of container
Volume	Definite volume	Definite volume	Fills volume of container
Kinetic energy	Lowest of the three states	More than solid, less than gas	Highest of the three states
Positioning of particles	Closely packed and fixed	Loosely packed, but random	Far apart and random
Attraction between particles	Very strong	Strong	Practically none

Sample Problem 1.22 States of Matter

Based on the properties of matter in Table 1.7, in which state of matter are the particles moving the most, meaning that they possess the most kinetic energy?

Solution

Energy of motion is kinetic energy. Because the kinetic energy present in gases is the highest of the three states of matter, the particles in a gas are moving the most. In fact, they are in constant motion.

Practice Problems

1.35 Based on your experience, is each of the following more or less dense than liquid water?

  a. vegetable oil   b. piece of gold jewelry   c. cork

1.36 Based on your experience, is each of the following more or less dense than liquid water?

  a. paper clip   b. table sugar   c. ice

1.37 Calculate the grams of sugar present in a 355-mL can of soda (assume mostly sugar and water) if the water in the can weighs 355 g and the density of the soda is 1.11 g/mL.

1.38 Calculate the density in g/mL of 2.0 L of gasoline that weighs 1.32 kg.

1.39 A person's urine has a specific gravity of 1.002. Is this person more likely dehydrated or overhydrated?

1.40 The specific gravity of the ocean is 1.025. Is this more or less dense than tap water?

1.41 A family visiting Europe goes to the hospital because their child is sick. The child has a temperature of 40.3 °C. Calculate the temperature in °F.

1.42 Fetal cord blood is stored at −112 °F. Calculate this temperature in Kelvin.

1.43 Discuss the differences in the type of energy present in a snack before, while, and after you eat it.

1.44 Discuss the differences in the type of energy present as a child climbs to the top of a slide, stands at the top, and slides down.

1.45 A defibrillator delivers about 360 joules per shock. Calculate the number of calories.

1.46 A moderate walk burns 225 Calories per hour. Calculate the number of joules burned by this activity.

1.47 Two warehouses have space available for storage. One is made of brick and the other of pine, which is a soft wood. Which one will keep its contents cooler on a 90 °F day? Support your answer.

1.48 Which has a higher specific heat, water or saltwater? Support your answer.

1.49 Indicate if each of the following describes a solid, liquid, or gas:

  a. This state of matter has no definite volume or shape.

  b. The particles in this state of matter are rigidly fixed in place.

1.50 Indicate if each of the following describes a solid, liquid, or gas:

  a. This state of matter has a definite volume, but takes the shape of its container.

  b. The particles in this state of matter are so far apart they do not interact with each other.

1.6 Measuring Matter

In this section, we apply the math concepts in Section 1.4 with the properties of matter from Section 1.5 to work some practical problems in health and wellness.

Accuracy and Precision

Accuracy and precision are often equated in everyday language, but from a scientist's perspective, they have different meanings. To illustrate the difference, have you ever known someone who constantly runs late even after setting watches and clocks ahead 20 minutes? All the timepieces are off from the true time by 20 minutes, but they are all *exactly* 20 minutes ahead. In this case, none of the clocks are displaying the true time, so none of the clocks are accurate. Yet because the clocks all read exactly the same time, the clocks are considered precise.

When you read a watch or make a measurement, measuring with **accuracy** means that you are taking measurements close to the actual or true value. Measuring with **precision** means that your repeat measurements are similar in value—that is, reproducible—even if they are not close to the actual value. In taking measurements, it is best to measure with both accuracy and precision. Accurate measurements require that instruments are calibrated regularly so they reflect true values. Taking precise measurements requires reproducibility of results and good technique. Taking several measurements and calculating the standard deviation of those measurements provides an indication of precision. Lower standard deviations indicate more precise measurements. Another example illustrating the difference between accuracy and precision using a football kicker is shown in **FIGURE 1.10**.

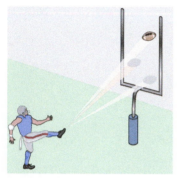

The kicker is accurate because the ball goes through the goalpost 3 out of 3 times.

The kicker is precise because the ball hits near the same spot 3 out of 3 times. However, he is not accurate.

The kicker is both accurate and precise because the ball goes through the goalposts at the same spot 3 out of 3 times.

FIGURE 1.10 Accuracy and precision have different meanings in science. A kicker is shown kicking a football three times in each panel to illustrate the difference between accuracy and precision.

Sample Problem 1.23 **Distinguishing Accuracy from Precision**

Identify the following as examples of good accuracy, good precision, or both:
   a. Adjusting a balance 5 pounds lighter and reporting weights of 125, 125, and 124.5 when a person steps on the scale three times in a row.
   b. Throwing four baseballs outside the upper right corner of the strike zone.
   c. Using medical syringes that deliver 5.00 cc (mL) of medicine to patients requiring 5 cc of medicine.

Solution

   a. Good precision, but not good accuracy. If the balance is offset by 5 pounds, it is not measuring the true value.
   b. Good precision, but not good accuracy. The pitcher is aiming for the strike zone (the true value), but throws all the balls to the same off-center spot (precise).
   c. Both good precision and good accuracy. By the number of significant figures in the measurement, the device measures to the nearest 1/100 of a cc. This measurement is both precise and accurate because each patient will get an exact dose of the correct amount of medicine from a syringe.

Measurement in Health: Units and Dosing

Imagine that a thermometer for measuring body temperature is offset by 2 °F. It would not be a very accurate instrument. Measurements in health care require great accuracy and a level of precision that leads to correct diagnoses. Next, we take a look at common units outside the SI system that are used in measurement.

Comparing Metric and U.S. Customary Systems

Health care professionals use SI or metric units like the cc and the mg in many applications, but must also be familiar with the U.S. customary system of measurement. Similar to the metric system, the U.S. customary system (formerly called the English system) has a set of basic units including the pound (lb) and ounce (oz) for mass and the foot (ft) and inch (in.) for length.

The metric and SI systems are "power of 10" systems allowing us easily to move from smaller to larger or larger to smaller units simply by moving the decimal. In contrast,

TABLE 1.8 Equivalent Units in SI and U.S. Customary Systems			
Property	**U.S Customary Unit**	**SI or Metric Equivalent**	**U.S. Customary Equivalent Unit**
Mass	Pound (lb)	2.205 lb = 1 kg	1 lb = 16 oz
Volume	Quart (qt)	1.057 qt = 1 L	1 qt = 4 cups
	Fluid ounce (fl oz)	1 fl oz = 29.6 mL	1 cup = 8 fl oz
	Teaspoon (tsp)	1 tsp = 4.93 mL	1 fl oz = 6 tsp
Length	Mile (mi)	1 mi = 1.6 km	1 mi = 5280 ft
	Inch (in.)	1 in. = 2.54 cm	1 ft = 12 in.

the U.S. system has a more complex set of equivalencies. Perhaps the most complex of these are the equivalencies for volume. U.S. units of volume include teaspoons, tablespoons, cups, fluid ounces, pints, quarts, and gallons. The equivalencies for conversions between units of volume are not intuitive and may require memorization.

TABLE 1.8 shows some U.S. customary units for mass, volume, and length, with equivalent metric units. Because only one of the units is exactly defined, significant figures should be taken into account with these conversion units. For example, when using the equivalent unit 1 kg = 2.205 lb, 1 kg is exact and 2.205 lb has 4 significant figures. U.S. customary equivalent units in Table 1.8 show exact units as in Table 1.5 for the metric units.

Recall from Section 1.4 that the metric units can be adjusted to larger or smaller units by using prefixes such as *kilo* (1000 × larger) or *milli* (1000 × smaller). The metric prefixes allow us to work with smaller, more manageable numbers instead of forcing us to use numbers with lots of zeros at the beginning or end.

Reading Lab Reports

Blood can be tested for any number of substances including chemical substances, cells, and cholesterols. Often this type of analysis is called "blood work." Depending upon the test, the report can be long and complex. Patient results are given along with normal values. If the patient's result is out of the normal range, that value is usually highlighted. As an introduction, **TABLE 1.9** shows a portion of a sample blood chemistry report. It is called blood chemistry because these substances are simple chemicals (versus cells or cholesterols) found in the liquid portion of the blood. Does this sample patient have normal blood chemistry?

INTEGRATING Chemistry

◀ Find out how to read and interpret the results of a patient's blood work.

TABLE 1.9 Sample Blood Chemistry Lab Results			
Patient Name: Jane Patient		**Age: 40**	
Test	**Result**	**Normal Range**	**Units**
Blood Chemistry			
Sodium	137	135–145	mmole/L or mEq/L
Potassium	3.9	3.5–5.2	mmole/L or mEq/L
Chloride	100	97–108	mmole/L or mEq/L
Glucose	88	65–99	mg/dL
Urea nitrogen	14	5–26	mg/dL
Calcium	9.1	8.5–10.6	mg/dL
Phosphorus	3.5	2.5–4.5	mg/dL
Protein, total	7.1	6.0–8.5	g/dL
Iron	113	35–155	μg/dL

—Continued next page

Continued—

It is easier to read numbers that are greater than 1 to a patient, so all the values are listed this way, but this makes the units on the numbers vary widely. Nurse practitioners and other health care professionals *must* be well acquainted with units of measure to read and interpret these reports. Notice that most of the units are metric. Often the deciliter (dL) is used in clinical lab reports as the volume measure. One deciliter is equal to a tenth of a liter, or 100 mL. One of the units, mmole (millimole), is equivalent to one-thousandth of a mole. A mole is a unit used to count the particles in matter. A similar unit to the millimole is the milliequivalent (mEq). For the electrolytes using mmole/L shown in Table 1.9, mEq/L is an equivalent unit used for these substances. This unit will be described in detail in Chapter 8.

In the United States, body weight is usually measured in units of pounds. Because pharmaceuticals are often dispensed by body weight in kilograms, it is important for health care professionals to be able to convert between the two systems. Let's try a problem calculating dosage.

Solving a Problem

Calculating a Dosage

A doctor's prescription for acetaminophen is 7.5 mg per kg of body weight per dose. For a child who weighs 42 lb, calculate the number of milligrams to be given in a dose.

STEP 1 **Determine the unit for the final answer.** You are being asked to calculate the number of milligrams in one dose, or mg. This unit will appear in the final answer.

STEP 2 **Determine the given information.** You have the following information given:

$$\frac{7.5 \text{ mg acetaminophen}}{1 \text{ kg body weight}}, \quad 42 \text{ lb body weight}$$

STEP 3 **Determine the conversion factors needed to cancel units.** Since the final units of the medicine are in milligrams, you must convert and cancel units for the mass of the child through a conversion factor (see Table 1.8).

$$\frac{1 \text{ kg}}{2.205 \text{ lb}}$$

STEP 4 **Set up the equation with the given information and conversion factors so that all units cancel except the final answer unit.** In this case, the final unit is mg. This unit should appear in the numerator, and all other units should cancel out, as shown below.

$$\frac{7.5 \text{ mg acetaminophen}}{1 \text{ kg body weight}} \times \frac{1 \text{ kg}}{2.205 \text{ lb}} \times 42 \text{ lb body weight} = 140 \text{ mg acetaminophen}$$

The final answer is reported to two significant figures because the minimum number of significant figures measured in the problem is two.

Drop Units

Because many medicines are introduced through an IV, such delivery is often measured through drops per milliliter, abbreviated gtt/mL. The abbreviation *gtt* is short for the Latin *gutta*, which means drop. This value, called the **drop factor**, can be used to determine drip rates when a prescribed volume of medicine is required in a given time period. Drop factors vary with the diameter of the IV tubing.

Sample Problem 1.24 Calculating Drop Factor

How many hours would it take to deliver 1.0 L of fluid through an IV with a drop factor of 60 gtt/mL if the drip rate is 100 gtt/min?

Using the process outlined previously:

STEP 1 Determine the unit for the final answer. You are being asked to calculate the amount of time it will take in hours. Hours, abbreviated h, will appear in the numerator of the final answer.

STEP 2 Determine the given information. The volume to be administered, the drop factor, and the drip rate are given.

$$1.0 \text{ L}, \quad \frac{60 \text{ gtt}}{1 \text{ mL}}, \quad \frac{100 \text{ gtt}}{1 \text{ min}}$$

STEP 3 Determine the conversion factors needed to cancel units. Volume units of L and mL are given, and a conversion factor will be necessary to convert them. The final answer must be given in hours, so a conversion factor for hours to minutes is also required.

$$\frac{1000 \text{ mL}}{1 \text{L}}, \quad \frac{1 \text{ h}}{60 \text{ min}}$$

STEP 4 Set up the equation with the given information and conversion factors so that all units cancel except the final answer unit. The final unit of hours must appear in the numerator and all other units must cancel out.

$$\frac{1 \text{ min}}{100 \text{ gtt}} \times \frac{1 \text{ h}}{60 \text{ min}} \times \frac{60 \text{ gtt}}{1 \text{ mL}} \times \frac{1000 \text{ mL}}{1 \text{ L}} \times 1.0 \text{ L} = 10. \text{ h}$$

Note that because drops are items that are counted, the number of significant figures is determined by the volume, 1.0 L, so the final answer has two significant figures.

Percents in Health

Percent values are prevalent in health and nutrition. Here are some applications.

Percent Active Ingredient

Because of the high potency of many medicines, these are administered in very small doses, in the range of micrograms (mcg) to milligrams (mg). However, such a small amount of material would be extremely difficult to handle, so inert binders such as dextrose, starch, or microcrystalline cellulose are often added to increase the size of a pill.

Sample Problem 1.25 Percent Active Ingredient

An aspirin tablet weighs 0.670 g. It contains 325 mg of the active ingredient acetylsalicylic acid. Calculate the percent acetylsalicylic acid in the tablet.

Solution

Recall from Section 1.4 that percent is a fraction comparing the part to the whole. In this case, the total mass is the tablet mass, which is 0.670 g. The acetylsalicylic acid is a part of that mass, 325 mg. Units must be the same to calculate percent, so first convert 0.670 g to mg:

$$0.670 \text{ g} \times \frac{1000 \text{ mg}}{1 \text{ g}} = 670. \text{ mg}$$

Then calculate the percent acetylsalicylic acid in an aspirin tablet, as follows:

$$\frac{325 \text{ mg}}{670. \text{ mg}} \times 100 = 48.5\%$$

The final answer is reported to three significant figures because both measured values contain three significant figures.

Percent of an Adult Dose

Because children weigh less than adults, they are sometimes administered a percent of the dose required for an adult.

Sample Problem 1.26 **Percent of an Adult Dose**

The adult dose for sodium thiopental, an anesthetic, is 280 mg. A 19-lb child should only receive 17% of the adult dose. Calculate this dose.

Solution

Remember that percent means part divided by the whole. Here, 17% means that the child should receive 17 mg for each 100 mg of the adult dose. Mathematically, this is shown as

$$\frac{17 \text{ mg}}{100 \text{ mg}}$$

Because we want to know what 17% of 280 mg is, we can solve for the missing dose by changing the percent into a decimal and multiplying by the total amount.

$$17\% = 0.17$$
$$0.17 \times 280 \text{ mg} = 48 \text{ mg}$$

Alternatively, we can solve for the missing dose using a ratio:

$$\frac{17 \text{ mg}}{100 \text{ mg}} = \frac{? \text{ mg}}{280 \text{ mg}}$$

$$\frac{(17 \text{ mg})(280 \text{ mg})}{100 \text{ mg}} = ? \text{ mg}$$

$$= 48 \text{ mg}$$

The final answer is reported to two significant figures because both given values contain two significant figures.

Percent in Nutrition Labeling

Nutrition labels give us a set of nutrition facts about foods that we eat. They list the amount of carbohydrate, protein, and fat found in the food per serving along with other important nutrients. The labels also indicate the Percent Daily Value (%DV), showing the contribution of a single serving to the (suggested) daily dietary requirement for a specific nutrient. This helps consumers easily determine whether they are getting enough of a nutrient by reading the nutrition label.

A Nutrition Facts label. Percent daily value for protein is not required on nutrition labels.

Sample Problem 1.27   Percent in Nutrition

A package of six cheese crackers weighs 39 g. The crackers contain 11 g of fat, 3 g protein, 22 g carbohydrate, and 3 g other ingredients. Calculate the percent fat, protein, and carbohydrate present in the crackers.

Solution

Percent is part over the whole times 100. So for each nutrient, we can calculate the percent in the package.

$$\text{Fat: } \frac{11 \text{ g}}{39 \text{ g}} \times 100 = 28\%$$

$$\text{Protein: } \frac{3 \text{ g}}{39 \text{ g}} \times 100 = 8\%$$

$$\text{Carbohydrate: } \frac{22 \text{ g}}{39 \text{ g}} \times 100 = 56\%$$

The final answer for percent protein can only be reported to one significant figure. Note that because there are other ingredients in the crackers, the nutrient percent shown does not add up to 100%.

Practice Problems

1.51 Identify the following as good accuracy, good precision, both, or neither:

a. Tossing three horseshoes, none of which lands near the center pole. The horseshoes are scattered on either side.

b. Tossing three horseshoes and all three are ringers on the center pole.

c. Tossing three horseshoes and all three land 3 feet in front of the center pole on top of each other.

1.52 Consider the following measurements determined for a known volume of 10.0 mL:

Volumes Measured with a Graduated Cylinder	Volumes Measured with a Beaker
10 mL	12 mL
9.5 mL	8 mL
10.5 mL	9 mL

a. Which set of measurements is more precise?

b. Which set of measurements is more accurate?

c. Which measuring device for volume is more accurate?

1.53 A low dose of aspirin is often recommended for patients who are at risk for coronary artery disease. Low-dose aspirin contains 81 mg of acetylsalicylic acid per tablet. How many grams of acetylsalicylic acid will a patient take over the course of 5.0 years if they take one aspirin tablet daily?

1.54 A mother is to give her child 10 cc's of medicine every 6 h. Unfortunately, she lost the syringe provided with the liquid medicine. She calls your clinic for help. How much medicine in teaspoons can you instruct her to give to the child per dose? Per day?

1.55 Give "Drug X" 5 mg/kg per day in two divided doses. The patient weighs 44 lb. How many milligrams should be given per dose? Per day?

1.56 A 38-lb child is prescribed acyclovir for chicken pox in an amount of 80 mg/kg body weight per day to be divided into four doses. Each tablet contains 700 mg of the medicine. How many tablets should be given per day? Per dose?

1.57 A child is prescribed morphine sulfate at 50.0 mcg/kg/dose. The child weighs 85 lb and the solution is available as a 5.0-mg/mL vial. How many milliliters should be injected?

1.58 A child must take 10 mL of cephalexin, an antibiotic, by mouth 2 times a day for 10 days. The solution is provided as a 50 mg/mL suspension. How many milligrams will the child have taken when the course of medication is over?

1.59 A patient gets 2.0 L of fluid over 18 h through an IV. The drop factor is 20 gtt/mL. Calculate the drip rate in drops per minute (gtt/min).

1.60 How long would it take in hours to administer exactly 500 mL of fluid through an IV with a drop factor of 30 gtt/mL if the drip rate is 60 gtt/min?

1.61 A tablet of Benadryl™, an antihistamine, has a mass of 0.50 g. It contains 25 mg of the active ingredient, diphenhydramine HCl. What percent of the tablet is active ingredient?

1.62 A medium-sized carrot weighs 61 g and contains 6 g of carbohydrate. What percent of the carrot is carbohydrate?

1

CHAPTER REVIEW

The study guide will help you check your understanding of the main concepts in Chapter 1. You should be able to answer problems for each learning outcome in this list. To check your mastery, try the problems listed after each.

STUDY GUIDE	CHAPTER GUIDE

1.1 Classifying Matter: Mixture or Pure Substance

Classify the basic forms of matter.

- Classify matter as a pure substance or a mixture. (Try 1.1, 1.63)
- Classify mixtures as homogeneous or heterogeneous. (Try 1.3, 1.67)
- Classify pure substances as elements or compounds. (Try 1.5, 1.65)

Inquiry Question

How is matter classified?

Besides classifying matter as solid, liquid, or gas, chemists use a broader classification of mixture or pure substance. Mixtures can be separated into their component parts and are further classified as homogeneous (evenly mixed throughout) or heterogeneous (unevenly mixed). Pure substances are made up of a single component and are classified as elements or compounds. Compounds contain more than one element, while elements contain a single type of atom. Atoms are the smallest unit of matter with unique chemical characteristics.

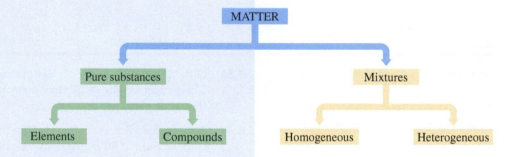

1.2 Elements, Compounds, and the Periodic Table

Examine the periodic table and its organization.

- Distinguish between groups and periods. (Try 1.7, 1.69)
- Locate metals and nonmetals on the periodic table. (Try 1.7, 1.69)
- Identify the number of atoms of each element in a chemical formula. (Try 1.9, 1.71)

Inquiry Question

How is the periodic table organized?

The periodic table of the elements is a useful catalogue of all of the elements. Each block on the table contains the symbol of a single element along with other useful information about that element. The blocks are arranged in columns known as groups and in rows known as periods. The elements in Groups 1A through 8A are known as the main-group elements, and those in the B groups are known as the transition elements. A staircase line on the right side of the periodic table separates the elements that are metals from nonmetals. Compounds are chemical combinations of elements represented by chemical formulas, which provide the identity and number of atoms of each element in the compound.

1.3 How Matter Changes

Represent changes in matter with a chemical formula.

- Distinguish between physical changes and chemical reactions. (Try 1.11, 1.73)
- Balance a given chemical equation. (Try 1.15, 1.75)

Inquiry Question

How are physical and chemical changes in matter different, and how can we represent them?

Matter can undergo two types of changes: a physical change and a chemical change called a chemical reaction. In a physical change, the substance changes states, but its identity remains the same. In a chemical reaction, the identity of the reacting substance(s) is changed. A chemical reaction is represented using a chemical equation, which identifies the reacting substance(s) and the product(s), the physical state of all substances in the reaction, and any conditions necessary for the reaction to occur.

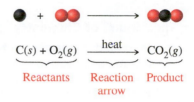

$$\underbrace{C(s) + O_2(g)}_{\text{Reactants}} \xrightarrow[]{\text{heat}} \underbrace{CO_2(g)}_{\text{Product}}$$

Reactants   Reaction   Product
arrow

The law of conservation of mass dictates that a balanced chemical equation must have equal numbers of each atom in both the reactants and products. An outline for balancing chemical equations is given as three steps: (1) Examine. (2) Balance. (3) Check.

1.4 Math Counts

Gain familiarity with math concepts central to chemistry.

- Convert between metric units. (See Sample Problem 1.8. Try 1.79, 1.83a)
- Apply the appropriate number of significant figures to a measurement or calculation. (Try 1.17, 1.91)
- Convert numbers to scientific notation and scientific notation to standard numerical notation. (Try 1.23, 1.25)
- Convert numbers and fractions to percent and percent to numbers and fractions. (Try 1.29, 1.31, 1.35)

Inquiry Question

What basic math concepts are important to chemistry?

Some basic mathematical concepts apply to chemistry. The SI system of units is based on powers of ten and includes the metric system. In both, the prefixes reflect the powers of ten. Conversion factors are used to convert SI units through dimensional analysis. Significant figures allow us to designate the certainty of any measurement. Rounding answers to no more certainty than measured ensures meaningful answers. Scientific notation is used to write very large and very small numbers economically. This notation also allows a direct comparison of extremely large and very small numbers. Direct comparison of different sample sizes can also be examined using percent.

$$\text{Percent (\%)} = \frac{\text{part}}{\text{whole}} \times 100$$

TABLE 1.4 Counting Significant Figures in Measurements

Rule	Measurement	Number of Significant Figures
1. A digit is significant if it is		
a. not a zero	41 g	2
	15.3 m	3
b. a zero between nonzero digits	101 L	3
	6.071 kg	4
c. a zero at the end of a number with a decimal point	20. g	2
	9.800 °C	4
2. A zero is not significant if it is		
a. at the beginning of a number with a decimal point	0.03 L	1
	0.00024 g	2
b. in a large number without a decimal point	12,000 km	2
	3,450,000 m	3

—Continued next page

Continued—

1.5 Matter: The "Stuff" of Chemistry

Apply math concepts to matter measurements.

- Define mass and units for mass. (See Section 1.5)
- Define volume and units for volume. (See Section 1.5)
- Calculate and solve problems using density and specific gravity. (Try 1.37, 1.39, 1.95)
- Convert temperatures among Celsius, Kelvin, and Fahrenheit. (Try 1.41, 1.99)
- Distinguish between kinetic and potential energy. (Try 1.43, 1.101)
- Convert between energy units. (Try 1.45, 1.103, 1.105)
- Define specific heat and compare values of various materials. (Try 1.47, 1.107)
- Contrast the properties of solids, liquids, and gases. (Try 1.49, 1.109)

Inquiry Question

What properties of matter can be measured?

Matter is anything that takes up space and has mass. Chemists measure properties such as mass, volume, density, specific gravity, temperature, energy, heat, and specific heat. Mass measures the amount of matter and can be measured on a balance. Volume is a three-dimensional measure of the space that matter occupies. Density is a physical property of matter and is a ratio of a substance's mass to volume. Measuring the temperature of matter is useful because it indicates the amount of energy present. Temperature units include Fahrenheit, Celsius, and Kelvin. The Kelvin and Celsius degree are the same size unit but offset by 273. The Fahrenheit degree is smaller. There are nine Fahrenheit degrees for every five degrees Celsius. Energy can be either potential (stored) or kinetic (moving). Energy is neither created nor destroyed but simply changes form. Heat is kinetic energy that flows from a warmer body to a colder one. The specific heat of a particular material measures how much heat energy it takes to raise its temperature. Most matter exists in one of three different states: solid, liquid, or gas. The three states of matter differ in the motion, kinetic energy, and positioning of their particles.

1.6 Measuring Matter

Apply matter measurements to health measurements.

- Distinguish between accuracy and precision. (Try 1.51, 1.111)
- Convert between SI or metric and U.S. units. (Try 1.55 and 1.83b)
- Apply conversion factors, units, drop units, and percent to measurements in health. (Try 1.61, 1.86, 1.89, 1.116)

Inquiry Question

How are the properties and measurement of matter applied to problems in health?

This section applies conversion factors and the units for measuring matter from Sections 1.4 and 1.5 to solve dimensional analysis problems in health. The U.S. system of units is introduced and compared to the SI system. Practical dosing calculations show how conversion factors, units, drop factors, and percent can be applied to solve problems in health care.

Mastering Videos

PRACTICING THE CONCEPTS

The following videos can be accessed through the Pearson eText or your Mastering Chemistry course.

- Conversion Factors Are Relationships
- Using Multiple Conversion Factors

Important Equations

Percent

$$\text{Percent (\%)} = \frac{\text{part}}{\text{whole}} \times 100$$

Temperature Conversions

$$\text{Kelvin (K)} = \text{Celsius (°C)} + 273°C$$

$$°C = (°F\text{-}32°F) \times \frac{1°C}{1.8°F}$$

$$°F = \left(\frac{1.8°F}{1°C} \times °C \right) + 32°F$$

Density

$$d = \frac{m}{V}$$

Specific Gravity

$$\text{Specific gravity} = \frac{\text{density of sample}}{\text{density of water}}$$

Additional Problems

1.63 Classify each of the following as a mixture or a pure substance:

a. ink in a pen   **b.** wood   **c.** blood

d. ammonia (NH_3)   **e.** silicon

1.64 For each of the substances that you classified as a mixture in Problem 1.63, determine whether it is a homogeneous or heterogeneous mixture.

1.65 For each of the substances that you classified as a pure substance in Problem 1.63, determine whether it is an element or a compound.

1.66 Classify each of the following as a mixture or a pure substance:

a. paint   **b.** $CaCO_3$ (calcium carbonate)

c. tin   **d.** milk

e. table sugar ($C_{12}H_{22}O_{11}$)

1.67 For each of the substances that you classified as a mixture in Problem 1.66, determine whether it is a homogeneous or heterogeneous mixture.

1.68 For each of the substances that you classified as a pure substance in Problem 1.66, determine whether it is an element or a compound.

1.69 Use the periodic table to supply the missing information in the following chart:

Name	Elemental Symbol	Group	Period	Metal or Nonmetal
Nitrogen				
		3A	3	
	S			
	P	5A		Nonmetal

1.70 Use the periodic table to supply the missing information in the following chart:

Name	Elemental Symbol	Group	Period	Metal or Nonmetal
Oxygen				
		2A	3	
	Si			
		1A	4	Metal

1.71 Give the name and number of atoms of each element in the following compounds:

a. N_2O, nitrous oxide, laughing gas

b. NH_3, ammonia, a household cleaner

c. TiO_2, titanium dioxide, white pigment used in paint

d. $C_8H_{10}N_4O_2$, caffeine, a stimulant

e. $C_{19}H_{25}BN_4O_4$, bortezomib, an anti-cancer drug

1.72 Give the name and number of atoms of each element in the following compounds:

a. $CaCO_3$, calcium carbonate, found in limestone

b. N_2H_4, hydrazine, found in rocket fuel

c. $C_7H_5BiO_4$, found in Pepto-Bismol

d. $C_{12}H_{22}O_{11}$, table sugar

e. $C_{12}H_{18}O$, propofol, a powerful sedative

1.73 Identify each of the following as a physical change or a chemical reaction:

a. bleaching your hair

b. cutting your finger

c. a cut healing

d. nitric acid dissolving copper metal to form copper nitrate

e. striking a match

1.74 Identify each of the following as a physical change or a chemical reaction:

   a. water in a cloud forming into snow
   b. gasoline burning in a car's engine
   c. chewing a stick of gum
   d. a hand grenade exploding
   e. breaking a china plate

1.75 Add coefficients to balance the chemical equations shown.

   a. $LiOH(s) + CO_2(g) \rightarrow LiHCO_3(s)$
   b. $HCl(aq) + Mg(s) \rightarrow MgCl_2(aq) + H_2(g)$
   c. $Fe(s) + S(s) \rightarrow Fe_2S_3(s)$

1.76 Add coefficients to balance the chemical equations shown.

   a. $NO_2(g) + H_2O(l) \rightarrow HNO_3(aq) + NO(g)$
   b. $P(s) + O_2(g) \rightarrow P_2O_5(s)$
   c. $C_2H_2(g) + O_2(g) \rightarrow CO_2(g) + H_2O(g)$

1.77 If red spheres represent oxygen atoms and blue spheres represent nitrogen atoms:

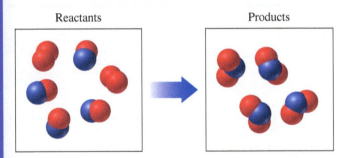

Reactants   Products

Write a balanced chemical equation for the reaction.

1.78 If red spheres represent oxygen atoms and green spheres represent chlorine atoms:

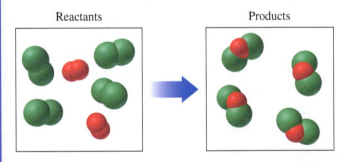

Reactants   Products

Write a balanced chemical equation for the reaction.

1.79 When completing each of the following conversions, would you expect the number to get larger or smaller?

   a. 50 mg to g   **b.** 0.23 g to μg
   c. 5.2 m to km   **d.** 800 kL to dL

1.80 When completing each of the following conversions, would you expect the number to get larger or smaller?

   a. 789 cm to m   **b.** 0.45 kg to mg
   c. 694 mL to μL   **d.** 800 dm to m

1.81 Supply the missing information in each of the following conversion factors:

   a. $\dfrac{? \text{ mL}}{1 \text{ L}}$ **b.** $\dfrac{? \text{ kg}}{1000 \text{ g}}$ **c.** $\dfrac{100 \text{ cm}}{? \text{ m}}$

   d. $\dfrac{1 \text{ L}}{? \text{ dL}}$ **e.** $\dfrac{? \text{ } \mu\text{g}}{1 \text{ g}}$

1.82 Supply the missing information in each of the following conversion factors:

   a. $\dfrac{? \text{ g}}{1000 \text{ mg}}$ **b.** $\dfrac{10 \text{ dg}}{? \text{ g}}$ **c.** $\dfrac{? \text{ g}}{1,000,000 \text{ } \mu\text{g}}$

   d. $\dfrac{? \text{ m}}{1 \text{ km}}$ **e.** $\dfrac{1 \text{ L}}{? \text{ cL}}$

1.83 A runner completes a 50-K (50-kilometer) ultramarathon.

   a. If a typical runner's stride is about 1 m, how many strides did he take, assuming that all strides were the same length?
   b. How many miles did the runner race?

1.84 A soft drink is sold in a 1.0-L container.

   a. How many milliliters are in the container?
   b. How many fluid ounces are in the container?

1.85 A raisin has a mass of approximately 1 g.

   a. Write a conversion factor relating number of raisins and number of grams.
   b. If a box contains 4500 cg of raisins, how many 1-g raisins are present in the box?

1.86 A soft gel capsule contains 300 mcg of vitamin A.

   a. How many grams of vitamin A are present?
   b. If a capsule weighs 1.1 g, what percent of the capsule is vitamin A?

1.87 A prescription calls for 70 mcg per day of levothyroxine (Synthroid) to combat hypothyroidism. If your pharmacist only has the 137-mcg tablets available, how many tablets should you take a day?

1.88 If a drop of blood is 0.05 mL, how many drops of blood are in a blood collection tube that holds 2 mL?

1.89 If a prescription calls for 1.0 g of acetylsalicylic acid, the active ingredient in aspirin, per dose, and each tablet contains 325 mg of acetylsalicylic acid, how many aspirin tablets will a patient take in one dose?

1.90 A vial of amoxicillin contains 500. mg in 2.0 mL of solution. What volume of the solution contains 180 mg of the drug?

1.91 Round the following numbers to three significant figures:

   a. 651,457   **b.** 0.004500
   c. 6.665   **d.** 2000

1.92 Round the following numbers to two significant figures:

   a. 0.33333   **b.** 300,000
   c. 155,555   **d.** 0.02105

1.93 Consider the following data set for three bags of jelly beans:

Bag Number	Number of Black Jelly Beans	Total Number of Jelly Beans
1	25	320
2	50	532
3	60	680

Which bag of jelly beans contains the highest percent of black jelly beans?

1.94 It is flu season. Professor F has a class with 50 students enrolled, and on Monday, 12 students are absent. Professor D has a class with 30 students enrolled, and on Monday, 9 students are absent. Calculate the percent of students absent from each class. Which class has the higher percent of students attending on Monday?

1.95 Calculate the volume in liters that 5.00×10^2 g of air will occupy if the density of air is 1.30 g/L.

1.96 An adult human femur weighs about 225 g and has a volume of 118 cm³. Calculate the density of this bone.

1.97 A simple syrup has a specific gravity of 1.25. Will this syrup drop to the bottom or stay on the top of a glass of water when poured?

1.98 A person's urine has a specific gravity of 1.04. Is this person likely overhydrated or dehydrated?

1.99 To keep a room comfortable, the air is heated or cooled to remain at approximately 75 °F. This is often referred to as ambient or room temperature. What is room temperature in °C?

1.100 On the Kelvin scale, the lower limit of the temperature scale is zero. Zero kelvin is known as absolute zero. What is this temperature in °C? In °F?

1.101 Indicate whether each of the following contains more kinetic or potential energy:
   a. firewood before starting a fire
   b. exercising
   c. standing at the starting line ready to sprint
   d. a phone charging bank

1.102 Indicate whether each of the following contains more kinetic or potential energy:
   a. riding a longboard
   b. decorating a cake
   c. chocolate cake
   d. diesel fuel

1.103 A cup of yogurt contains 130 Calories. Calculate the number of joules present.

1.104 A person can burn 550 Calories by running at a moderate pace for 1 h. Calculate the number of Calories that would be burned if the person only ran for 22 min. Calculate the number of joules burned in 22 min.

1.105 A single heartbeat requires about 0.5 J of energy. Calculate the number of calories.

1.106 A banana contains 420,000 J of energy. Calculate the number of nutritional Calories.

1.107 If you are shopping for a new set of cookware and one set is made of aluminum, one is made of stainless steel, and one is made of copper, which one would use the least heat energy from your stove during cooking? Support your answer.

1.108 Softwood trees have a higher specific heat than hardwood trees. Provide a reasonable explanation for this.

1.109 Contrast the kinetic energy of particles in a liquid with those in a gas.

1.110 Contrast the kinetic energy of particles in a liquid with those in a solid.

1.111 A student is weighing a standard 5.00-g weight four times on two different balances to check them for accuracy and precision. The following table shows the data:

Trial	Balance A	Balance B
1	4.2 g	5.1 g
2	4.1 g	4.9 g
3	4.3 g	4.7 g
4	4.2 g	5.2 g

Which of the balances is more accurate? Which is more precise?

1.112 GE Healthcare sells three varieties of vaporizers for anesthesia delivery. The accuracy specs for each are listed below for the same flow rate of vapor. Which unit is the most accurate?

Unit	Accuracy Spec
Tec 6 Plus	±0.5% of delivered agent
Tec 7 Vaporizer	±0.25% of delivered agent
Aladdin Cassette	±0.15% of delivered agent

1.113 A patient undergoing open heart surgery is to receive 175 units/kg body weight. If the medicine contains 5000 units/mL, how many milliliters should be injected into a 250-lb patient?

1.114 Your patient gets a prescription for 62.5 mcg (micrograms, μ g) of digoxin in liquid form. The label reads 0.0250 mg/mL. How many milliliters of digoxin should you give?

1.115 A mother calls you to ask about a proper dosage of cough medicine for her 2-year-old. The bottle says to administer 5.0 mg/kg three times a day for children under 3 years of age. The cough syrup is supplied in a liquid that contains 50.0 mg/mL. How many teaspoons of cough syrup should she give the child, who weighs 28 lb, in each dose?

1.116 A prescription for amoxicillin comes in an oral suspension that is 250 mg/5.0 mL. A patient is prescribed 500 mg every 6 h. How many milliliters of amoxicillin should the patient take per dose? How many teaspoons? How many teaspoons per day?

1.117 A patient needs exactly 1000 mL of a fluid over a 6 h period. The drop factor is 15 gtt/mL. What is the drip rate in gtt/min?

1.118 A patient has an order for a drip at 7.0 mg of medicine per hour. The IV drip is provided from a pharmacy as exactly 1000 mg in 500 mL of dextrose 5% in water. Calculate the flow rate of the pump in milliliters per hour.

1.119 How does the arrangement of particles in a liquid differ from that in a solid?

1.120 The brakes in your car work on a hydraulic system. When you depress the brake pedal, fluid in your brake line is compressed. In turn, this fluid presses against the brake pads, which then press against the wheels, causing them to stop. If air gets into your brake lines, the brakes will work but will require you to step on the brake pedal harder to make your car stop. Using your knowledge of the difference in the molecular arrangement in liquids and gases and your understanding of gas behavior, explain why air in your brake lines causes your brakes to work less efficiently.

Challenge Problems

1.121 To donate blood, your blood must have a density greater than 1.053 g/mL, indicating that adequate levels of hemoglobin and red blood cells are present. Before donating blood, a simple test is done to determine eligibility by placing a drop of the donor's blood into a tube containing a copper sulfate solution having a density of 1.053 g/mL. How does the technician decide whether or not the person can donate blood?

1.122 The following equation shows the reaction of magnesium metal and chlorine to form magnesium chloride, which is used as a deicer for roads and airport runways:

$$Mg(s) + Cl_2(g) \rightarrow MgCl_2(s)$$

a. Is magnesium metal an element or a compound? How do you know?

b. Is chlorine an element or a compound? How do you know?

c. What do the labels (s) and (g) represent?

d. What do you notice about the number of atoms on either side of the chemical equation?

e. Is this a physical change or a chemical reaction?

1.123 A physician orders a Heparin drip at 8.0 units per kg body weight per hour via an IV pump. The patient weighs 212 lb. The IV is available at 25,000 units Heparin in exactly 500 mL of IV fluid. Calculate the flow rate in mL/h that should be set for the IV pump.

1.124 A child who weighs 66 lb is receiving a dose of 125 mg of a drug three times a day. The safe range is 10–20 mg/kg/day. Is the child receiving a safe dose?

1.125 Consider the syringe shown below. Report the volume in this syringe to the correct number of significant figures. _____ cc

1.126 According to the U.S. Census Bureau's American Community Survey, the size of an American family household is 3.34. Why does this number have three significant figures when a third of a person is impossible?

1.127 We know high-density lipoprotein (HDL) and low-density lipoprotein (LDL) as *good* and *bad* cholesterol, respectively. HDL and LDL aren't actually cholesterol molecules. They are aggregations of cholesterol, other dietary fats, and protein that circulate in the bloodstream. If an LDL particle weighs about half that of an HDL particle and has 20% the density of the HDL, which one takes up more volume?

Answers to Odd-Numbered Problems

Practice Problems

1.1   **a.** pure substance   **b.** mixture
          c. mixture   **d.** mixture

1.3   **a.** heterogeneous   **b.** homogeneous
          c. homogeneous   **d.** heterogeneous

1.5   **a.** element   **b.** compound
          c. element   **d.** compound

1.7

Name	Elemental Symbol	Group	Period	Metal or Nonmetal
Fluorine	F	7A	2	Nonmetal
Lithium	Li	1A	2	Metal
Chlorine	Cl	7A	3	Nonmetal
Carbon	C	4A	2	Nonmetal

1.9   **a.** 1 lithium, 1 carbon, 3 oxygen
          b. 2 hydrogen, 2 oxygen
          c. 8 carbon, 8 hydrogen, 3 oxygen

1.11 **a.** chemical reaction
          b. physical change
          c. chemical reaction

1.13 **a.** $\underline{1}$ ⚬⚬ + $\underline{1}$ ●● → $\underline{2}$ ⚬●

          b. $\underline{2}$ ⚬⚬ + $\underline{1}$ ●● → $\underline{2}$ ⚬●⚬

1.15 **a.** $2Al(s) + 3Cl_2(g) \rightarrow 2AlCl_3(s)$
          b. $2HCl(aq) + Zn(s) \rightarrow ZnCl_2(aq) + H_2(g)$
          c. $SiO_2(s) + 3C(s) \rightarrow SiC(s) + 2CO(g)$

1.17 **a.** 2   **b.** 4   **c.** 5   **d.** 3

1.19 **a.** 2000   **b.** 54.5   **c.** 224,000   **d.** 132

1.21 **a.** 0.325 g

1.23 **a.** 2.03×10^8 **b.** 1.24×10^1 **c.** 2.78×10^{-8}

1.25 **a.** 0.0000156 **b.** 280,000   **c.** 0.090

1.27 **a.** 5×10^{11} **b.** 5.5×10^6

1.29 **a.** 25%   **b.** 38%   **c.** 66%

1.31 **a.** 0.045   **b.** 0.130   **c.** 0.66   **d.** 0.78

1.33 **a.** 20   **b.** 9   **c.** 38   **d.** 133

1.35 **a.** less   **b.** more   **c.** less

1.37 39 g

1.39 overhydrated

1.41 104.5 °F

1.43 Before eating, the food is mostly potential energy, during eating it is a mixture of potential and kinetic energy as it is broken down, and after eating the food gets transformed into mostly kinetic energy to carry out the functions of the body.

1.45 86 cal

1.47 The warehouse made of pine. The pine has a higher specific heat, so it will take more heat energy to raise the temperature of the wood versus the brick. This will result in a slower transfer of the heat to the contents of the warehouse.

1.49 **a.** gas   **b.** solid

1.51 **a.** neither   **b.** both   **c.** good precision

1.53 150 g

1.55 50 mg/dose; 100 mg/day

1.57 0.39 mL

1.59 37 gtt/min

1.61 5.0%

Additional Problems

1.63 **a.** mixture   **b.** mixture   **c.** mixture
          d. pure substance **e.** pure substance

1.65 **d.** compound   **e.** element

1.67 **a.** homogeneous   **d.** homogeneous

1.69

Name	Elemental Symbol	Group	Period	Metal or Nonmetal
Nitrogen	N	5A	2	Nonmetal
Aluminum	Al	3A	3	Metal
Sulfur	S	6A	3	Nonmetal
Phosphorus	P	5A	3	Nonmetal

1.71
          a. 2 nitrogen, 1 oxygen
          b. 1 nitrogen, 3 hydrogen
          c. 1 titanium, 2 oxygen
          d. 8 carbon, 10 hydrogen, 4 nitrogen, 2 oxygen
          e. 19 carbon, 25 hydrogen, 1 boron, 4 nitrogen, 4 oxygen

1.73 **a.** chemical reaction   **b.** physical change
          c. chemical reaction   **d.** chemical reaction
          e. chemical reaction

1.75 **a.** $LiOH(s) + CO_2(g) \rightarrow LiHCO_3(s)$ (balanced)
          b. $2HCl(aq) + Mg(s) \rightarrow MgCl_2(aq) + H_2(g)$
          c. $2Fe(s) + 3S(s) \rightarrow Fe_2S_3(s)$

1.77 $O_2(g) + 2NO(g) \rightarrow 2NO_2(g)$

1.79 **a.** smaller   **b.** larger
          c. smaller   **d.** larger

1.81 **a.** 1000   **b.** 1   **c.** 1   **d.** 10
          e. 1,000,000

1.83 **a.** 50,000 strides   **b.** 31 miles

1.85 **a.** $\dfrac{1 \text{ raisin}}{1 \text{ g}}$ **b.** 45 raisins

1.87 0.5 tablets or half a tablet

1.89 The answer rounds to 3 tablets

1.91 **a.** 651,000   **b.** 0.00450

      c. 6.67   **d.** $20\overline{0}0$ or 2.00×10^3

1.93 Bag 2

1.95 385 L

1.97 drop to the bottom

1.99 24 °C

1.101 **a.** potential   **b.** kinetic

      c. potential   **d.** potential

1.103 5.4×10^5 J or 540,000 J

1.105 0.1 cal

1.107 The copper cookware. Copper has the lowest specific heat of the three metals. It will take the copper cookware the shortest time to heat up, which will result in a shorter overall cooking time, reducing the amount of time the stove must be on to warm up the pan.

1.109 The gas particles contain more kinetic energy than the liquid particles. In a gas, the particles are not touching each other and can move more freely and quickly.

1.111 Balance B is more accurate. Balance A is more precise.

1.113 4 mL

1.115 Approximately 1.27 mL equals $^1\!/_4$ teaspoon in each dose.

1.117 42 gtt/min

1.119 In a solid, the particles are packed together more rigidly than in a liquid; their movement is more limited in a solid.

Challenge Problems

1.121 Because the density of healthy blood is greater than the density of the copper sulfate solution, healthy blood will drop to the bottom of the tube. This happens within a few seconds.

1.123 15 mL/h

1.125 2.30 cc

1.127 The LDL takes up 2.5x more volume.

2

Atoms and Radioactivity

Radiation is used in medicine to identify and treat diseases. Radiation can be introduced into the body as x-rays, radioisotopes, protons, and positrons. To find out more about the small particles that make up atoms, isotopes, and radioisotopes and how they are used in medicine, begin reading Chapter 2.

CHAPTER OUTLINE

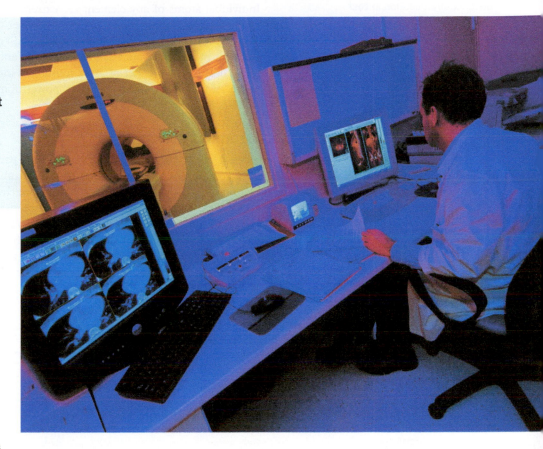

LEARNING TIP   Take Notes as You Read

Highlighting every word in the textbook won't allow you to remember key ideas later. Instead, try to read a paragraph or two and write in the margin a key word or words focusing on the main point of the paragraph. If you can distinguish the key points in the paragraphs as you take notes, this will deepen your understanding more than blindly highlighting entire paragraphs.

TODAY, cancer and disorders of the heart and brain are often diagnosed through the use of a positron emission tomography, or PET scan. What is a positron? How does the scanner work? To understand the answers to these questions, first we have to take a closer look at the particles that make up matter in its simplest form, the atom. In the first part of this chapter, we introduce the properties of atoms so that we can then begin to explore radioactive atoms. Radioactivity and radioactive elements used in medicine are the focus of the remainder of the chapter. Radioactive elements emit energy as radiation. In medicine, radiation and radioactive elements can be used to both diagnose and treat diseases.

What's an Inquiry Question?

Inquiry Questions are designed to focus your reading on the main concepts by section. As you read each section, see if you can answer each Inquiry Question in your own words.

? **2.1 Inquiry Question**

What are the subatomic particles that make up an atom and where are they located?

2.1 Atoms and Their Components

Imagine the number of fine grains of sand it would take to cover the entire east and west coastlines of the mainland United States with beaches as wide as a football field and 1 m deep. This is about the same number of atoms of carbon that are found in a half-carat diamond (see **FIGURE 2.1a**). That is a lot of grains of sand and a lot of carbon atoms.

Individual atoms of any element are extremely small, so huge numbers of them fit into a small space. Carbon atoms (element number 6 on the periodic table) are found in combination with other elements in living systems. The element carbon can also be found in pure form in nonliving materials like diamond and graphite (see **FIGURE 2.1a** and **b**). If we were able to crush either of these substances into a very fine powder, we could view the outline of individual atoms through a powerful microscope called a scanning tunneling microscope (see **FIGURE 2.1c**).

Subatomic Particles

If we could crush our half-carat diamond into particles even smaller than the atoms themselves, the atoms would no longer be recognizable as carbon. Instead, we would have a collection of small parts called **subatomic particles** that organize to form all atoms. The three basic subatomic particles are the proton, neutron, and electron. **Protons** and **electrons** are charged particles; that is, they have electrical properties. **Neutrons,** as their name implies, are neutral, or uncharged. Protons have a positive (+) charge, and electrons have an opposite negative (−) charge. Overall, atoms have *no charge* because the number of protons is *equal* to the number of electrons (see **TABLE 2.1**).

TABLE 2.1 Properties of Particles in an Atom

Subatomic Particle	Symbol	Electrical Charge	Relative Mass	Location in Atom
Electron	e^-	1−	0.0005 (1/2000)	Outside nucleus
Proton	p or p^+	1+	1	Nucleus
Neutron	n or n^0	0	1	Nucleus

Structure of an Atom

If an atom were the size of a hot-air balloon, the protons and neutrons would be clustered together in an area the size of a grain of salt in the center of the balloon, in a location known as the **nucleus** of an atom. The electrons would be dispersed throughout the balloon away from the nucleus. The space occupied by the electrons is called the **electron cloud** since the electrons are constantly moving in this space and it is difficult to pinpoint their exact location (see **FIGURE 2.2**). The rest of the balloon's interior would represent empty space. In fact, most of an atom consists of empty space.

Even though atoms are small and their subatomic particles are even smaller, both still qualify as matter (they have mass and take up space). Most of the mass of an atom lies in the nucleus. The electrons contribute very little to the mass of an atom because their mass is about 2000 times less than that of a proton or neutron. For this reason, we will ignore the electron's contribution to the mass of an atom.

If an atom were the size of a hot-air balloon, the nucleus would be the size of a grain of salt in the center.

FIGURE 2.1 **Carbon atoms.**
Photographs of (a) a half-carat diamond ring and (b) a pencil containing carbon in the form of graphite in the center. These are both forms of pure carbon. If we could see the atoms in graphite, a form of pure carbon, we would see many sheets like the one shown in (c) stacked on top of each other. One atom of carbon is found at the corner of each hexagon in the image.

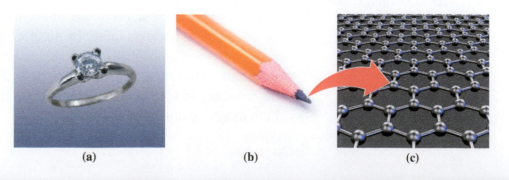

(a)   (b)   (c)

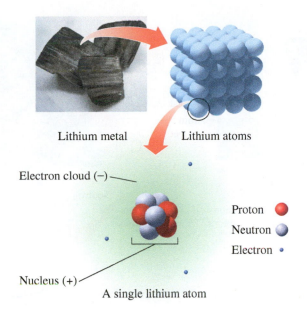

Lithium metal   Lithium atoms

Electron cloud (−)

Proton 🔴
Neutron ⚪
Electron ·

Nucleus (+)

A single lithium atom

FIGURE 2.2 The location of subatomic particles in an atom. Using the element lithium (Li) as an example, in an atom the protons and neutrons are packed into the center (nucleus), and the electrons are in motion outside of this nucleus in an area called the electron cloud. An atom is mainly empty space between the nucleus and the electron cloud.

Because the mass of an atom is so tiny, chemists use a unit called the **atomic mass unit (amu)** when discussing the mass of atoms. An amu is defined as one-twelfth of the mass of a carbon atom containing six protons and six neutrons. Therefore, the mass of a carbon atom is 12 amu. Because a proton and a neutron have roughly the same mass, each is defined as weighing about 1 amu.

Sample Problem 2.1   Characterizing Subatomic Particles

Name the subatomic particles that make up all atoms, and the charge and relative mass associated with each.

Solution

The subatomic particles found in all atoms are the proton, electron, and neutron. Protons have a positive charge, electrons have an opposite negative charge, and neutrons have no charge (neutral). The proton and neutron have a relative mass of 1 amu each, and the electron is about 1/2000 of an amu and does not contribute significantly to the mass.

Practice Problems

Find the answers to the odd-numbered Practice Problems at the end of the chapter.

2.1   Where are the subatomic particles located in an atom?

2.2   How does the mass of a proton compare to the mass of a neutron?

2.3   How does the mass of an electron compare to the mass of a proton?

2.4   What units are used for describing the mass of a proton or neutron?

2.2 Atomic Number and Mass Number

All atoms contain protons, neutrons, and electrons, but they contain different numbers of each, making them different elements. Atoms of the same element always have the same number of protons. This feature distinguishes atoms of one element from the atoms of a different element. In this section, we examine some basic characteristics of atoms that distinguish one element from another: atomic number and mass number.

 2.2 Inquiry Question

What do the atomic number and mass number indicate?

The atomic number from the periodic table indicates the number of protons and electrons in an atom.

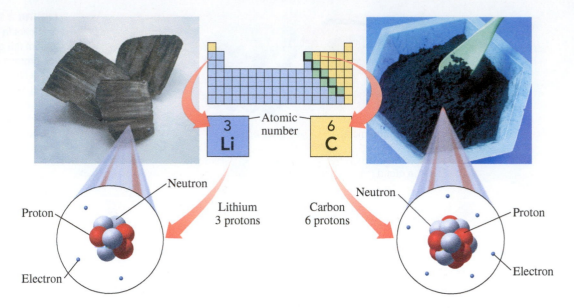

Lithium
3 protons

Carbon
6 protons

Atomic Number

The number of protons in an atom of any given element can be determined from the periodic table. Refer to the periodic table on the inside front cover of this text. Notice that every element has a number above the element's symbol. This number is the atom's number, so it is called the **atomic number.** The atomic number indicates the *number of protons* present in an atom of each element. The number of protons gives an atom its unique properties. For example, a carbon atom, atomic number 6, contains six protons. *All* atoms of carbon have six protons. Lithium, atomic number 3, contains three protons. *All* atoms of lithium have three protons. Because atoms are neutral (no charge), the number of electrons in an atom is equal to the number of protons. To be considered neutral, carbon must contain six electrons and lithium must contain three.

Mass Number

How many neutrons are present in an atom of carbon? Remember that almost all of an atom's mass is in the protons and neutrons and both particles weigh about the same. Suppose that the mass of a carbon atom is given as 12. The number of neutrons present must be six because the number of protons in a carbon atom is six. The number of neutrons in an atom can be found from an atom's **mass number,** which is simply the number of protons plus the number of neutrons. This information is summarized in **TABLE 2.2**.

$$\text{Mass number} = \text{number of protons} + \text{number of neutrons}$$

Once the atomic number and the mass number for a given atom are known, you can determine the number of subatomic particles present. We can represent an atom with its mass number and atomic number in *symbolic notation:*

$$\text{Mass number} \longrightarrow {}^{12}_{6}\text{C} \longleftarrow \text{Atomic symbol}$$
$$\text{Atomic number} \longrightarrow$$

TABLE 2.2 Determining the Number of Subatomic Particles in an Atom	
Subatomic Particle	**Number of Subatomic Particles in Atom**
Protons	Atomic number *(found above element in periodic table)*
Electrons	Equal to number of protons in an atom
Neutrons	Mass number minus number of protons *(mass number designated for a given atom)*

Sample Problem 2.2	**Determining the Number of Subatomic Particles and Symbolic Notation**

Determine the number of protons, electrons, and neutrons present in a chlorine (Cl) atom that has a mass number of 35. Write the element in symbolic notation.

Solution

STEP 1 Find the element on the periodic table. In this case, chlorine is element 17.

STEP 2 Use the atomic number to find the number of protons and electrons. Because the atomic number is 17, this tells us chlorine has 17 protons. The number of electrons = the number of protons for an atom, so there are 17 electrons.

STEP 3 Use the mass number to find the number of neutrons. The number of neutrons = mass number − number of protons. The mass number is given as 35. Therefore, the number of neutrons = 35 − 17, which is 18.

STEP 4 Write the element in symbolic notation. When writing symbolic notation, the mass number appears as a superscript to the left of the symbol, and the atomic number appears as a subscript to the left of the symbol:

$$^{35}_{17}\text{Cl}$$

Practice Problems

2.5 How can you determine the following?

a. the number of protons present in an atom

b. the number of neutrons present in an atom

c. the number of electrons present in an atom

2.6 What can be determined from the following?

a. the mass number

b. the atomic number

c. the mass number minus the atomic number

2.7 Provide the name and atomic symbol of the element that has the following atomic number:

a. 7   b. 16   c. 50   d. 30   e. 27

2.8 Provide the name and atomic symbol of the element that has the following atomic number:

a. 79   b. 19   c. 82   d. 33   e. 11

2.9 What is the mass number for an oxygen atom that contains 8 neutrons?

2.10 What is the mass number for a fluorine atom that contains 10 neutrons?

2.11 Determine the number of protons, neutrons, and electrons in the following atoms:

a. an iodine atom that has a mass number of 131

b. a potassium atom that has a mass number of 39

2.12 Determine the number of protons, neutrons, and electrons in the following atoms:

a. a helium atom that has a mass number of 4

b. a carbon that has a mass number of 14

2.13 Write the symbolic notation for atoms with

a. 5 protons and 6 neutrons.

b. 35 protons and 46 neutrons.

c. 14 protons and 14 neutrons.

d. 54 protons and 70 neutrons.

2.14 Write the symbolic notation for atoms with

a. 3 protons and 4 neutrons.

b. 7 protons and 10 neutrons.

c. 15 protons and 16 neutrons.

d. 46 protons and 60 neutrons.

2.15 Determine the number of protons, neutrons, and electrons for each of the following atoms:

a. $^{18}_{8}\text{O}$ b. $^{40}_{20}\text{Ca}$ c. $^{108}_{47}\text{Ag}$ d. $^{207}_{82}\text{Pb}$

2.16 Determine the number of protons, neutrons, and electrons for each of the following atoms:

a. $^{2}_{1}\text{H}$ b. $^{27}_{13}\text{Al}$ c. $^{23}_{11}\text{Na}$ d. $^{79}_{35}\text{Br}$

DISCOVERING THE CONCEPTS

? INQUIRY ACTIVITY—Isotopes

Information

Symbolic Notation for Isotopes

Isotopes of Calcium and the Number of Particles in Each

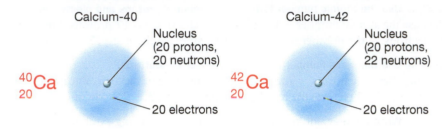

Calcium-40

$^{40}_{20}Ca$

Nucleus
(20 protons,
20 neutrons)

20 electrons

Calcium-42

$^{42}_{20}Ca$

Nucleus
(20 protons,
22 neutrons)

20 electrons

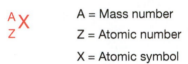

$^A_Z X$

A = Mass number

Z = Atomic number

X = Atomic symbol

Questions

1. How many protons are found in calcium-40 and calcium-42?
2. How many neutrons are found in calcium-40 and calcium-42?
3. What is the relationship between the number of protons and the number of electrons in an atom?
4. Based on the information, what is common to all calcium atoms?
5. What could the atomic number Z indicate?
6. How is the mass number A found?
7. From the information given, how would you define *isotope?*
8. Where are the protons, neutrons, and electrons located in an atom?
9. What do all calcium isotopes have in common? How do they differ?
10. Where is most of the mass of an atom?
11. For the following isotopes, determine the number of protons, neutrons, and electrons.

   $^{18}_9F$

   Protons = _____

   Neutrons = _____

   Electrons = _____

   $^{19}_9F$

   Protons = _____

   Neutrons = _____

   Electrons = _____

2.3 Isotopes and Atomic Mass

? 2.3 Inquiry Question

What is the difference between the mass number for an isotope and the atomic mass of an element?

Among the carbon atoms found in nature, most have a mass number of 12. Yet a few have a mass number of 13. Fewer still have a mass number of 14. These atoms must have the same number of protons to be considered carbon atoms, so what is different about them? It must be the number of neutrons. Atoms of the same element can have different numbers of neutrons, so not all atoms of the same element have the same mass number. Atoms of the same element with different mass numbers are called **isotopes.** Isotopes have the same number of protons (therefore, the same atomic number), but different numbers of neutrons (different mass numbers). We can show isotopes in two ways. Using symbolic notation, we can write the isotopes of carbon with mass numbers 12, 13, and 14 as

$$^{12}_6C, \ ^{13}_6C, \text{ and } ^{14}_6C$$

Or, using words, we can state the mass numbers after the name of the element—carbon-12, carbon-13, or carbon-14—since carbon atoms always have six protons.

Sample Problem 2.3 Subatomic Particles in Isotopes

Determine the number of protons, electrons, and neutrons present in the following isotopes of nitrogen:

a. $^{15}_{7}N$ b. $^{14}_{7}N$ c. $^{16}_{7}N$

Solution

a.

STEP 1 Find the element on the periodic table. In this case, nitrogen is element 7. Its atomic number is 7.

STEP 2 Use the atomic number to find the number of protons and electrons. Because the atomic number is 7, this tells us nitrogen has 7 protons. The number of electrons = the number of protons for an atom so there are 7 electrons.

STEP 3 Use the mass number to find the number of neutrons. The number of neutrons = mass number − number of protons. The mass number is given in the upper left of the symbolic notation as 15. Therefore, the number of neutrons = 15 − 7, which is 8.

Using the same step-by-step procedure as in part a, we can arrive at the answers below for parts b and c.

b. 7 protons, 7 electrons, 7 neutrons

c. 7 protons, 7 electrons, 9 neutrons

Atomic Mass

Next, we explore atomic mass. Consider the carbon atoms present in the half-carat diamond discussed earlier. All the atoms will have six protons and six electrons, but the number of neutrons will differ for different isotopes. Carbon has three natural isotopes, carbon-12, carbon-13, and carbon-14, which have six, seven, and eight neutrons, respectively.

Are all the isotopes found in equal amounts? The periodic table gives us a clue. The number below each element on the periodic table shows the *average* atomic mass for that element. Carbon's average atomic mass is 12.01 amu. If the most common isotopes were the heavier isotopes, we would expect the average mass of carbon given in the periodic table to be closer to 13 or 14 amu. The atomic mass of carbon depends on the proportion of each isotope in a sample of carbon. Therefore, carbon's atomic mass of 12.01 amu indicates that the most abundant isotope is carbon-12. So, the atomic number tells us that the three carbon isotopes are not found in equal amounts.

In our half-carat diamond, there are mostly carbon-12 atoms. However, there are a few carbon-13 atoms (1 in every 100) and even fewer carbon-14 atoms (1 in every 1,000,000,000,000; that is, *one trillion*). Because the carbon-13 and carbon-14 isotopes have more neutrons, the *average* atomic mass of the carbon atoms in a sample of pure carbon such as diamond is 12.01, slightly more than 12.00. The **atomic mass** shown below each of the elements on the periodic table is the average atomic mass weighted for all the isotopes of that element found naturally (see **FIGURE 2.3**).

Remember, the atomic mass is an average value. It is no different from calculating averages for other things in everyday life, like the average age of a population or the average price of gas across the country.

(a)

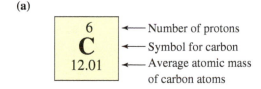

6	← Number of protons
C	← Symbol for carbon
12.01	← Average atomic mass of carbon atoms

Isotope	^{12}C	^{13}C	^{14}C
Atomic number	6	6	6
Protons	6	6	6
Neutrons	6	7	8
Abundance	Most	1/100	1/1,000,000,000,000

(b)

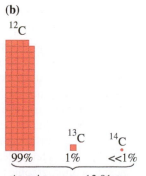

^{12}C

99% ^{13}C 1% ^{14}C <<1%

Atomic mass = 12.01 amu

FIGURE 2.3 The periodic table and atomic mass. (a) Each symbol on the periodic table gives information regarding the number of protons and the average atomic mass of the element. (b) Because the atomic mass of carbon depends on the proportion of each isotope in a sample of carbon, the atomic mass of 12.01 amu is closest to that of the most abundant isotope ^{12}C.

Mastering Video
Practicing the Concepts
*Distinguishing Mass Number
and Atomic Number*

Sample Problem 2.4 **Isotopes and Atomic Mass**

Copper has two naturally occurring isotopes: copper-63 and copper-65. Based on the atomic mass from the periodic table, which of these is more abundant?

Solution

The atomic mass of copper (element number 29) is 63.55 amu. Because the average atomic mass is closer to 63 than 65, there must be more copper-63 atoms present in the natural world than copper-65 atoms.

Practice Problems

2.17 The mass of the isotope carbon-12 is exactly 12 amu. The atomic mass for carbon (as seen on the periodic table) is 12.01 amu. Explain this difference.

2.18 How are atomic mass and mass number similar? How are they different?

2.19 There are three naturally occurring isotopes of magnesium: magnesium-24, magnesium-25, and magnesium-26.

    a. How many neutrons are present in each isotope?

    b. Write complete symbolic notation for each isotope.

    c. Based on the average atomic mass given in the periodic table, which isotope of magnesium is the most abundant?

2.20 Some of the naturally occurring isotopes of tin are tin-118, tin-119, tin-120, and tin-124.

    a. How many neutrons are present in each isotope?

    b. Write complete symbolic notation for each isotope.

2.4 Radioactivity and Radioisotopes

? 2.4 Inquiry Question

How does a radioisotope emit radiation?

Do you recognize the symbol at the left?

    This symbol is called the *trefoil* and is the international symbol for radiation. What is radiation? Energy given off spontaneously from the nucleus of an atom is called **nuclear radiation.** We are encouraged to avoid nuclear radiation in everyday life because it can sometimes be life-threatening. Elements that emit radiation are said to be radioactive. What is radioactivity? What elements possess this property and why? If radiation is so dangerous, why do we see these stickers in different areas in a hospital? These questions are answered in the next few sections.

    Radiation is a form of energy that we get from both natural and human-made sources. Light and heat from the sun are both natural forms of radiation. Microwaves used in ovens and radio waves used for communication are examples of human-made radiation. Similarly, the nuclei of some unstable atoms can also "radiate," which gives off energy. Radiating nuclei were first detected in the late 1800s with materials that are still used today in professional film photography.

    In 1896, Henri Becquerel noticed that when a mineral containing uranium was put next to a photographic plate in the dark, an image developed without light (see **FIGURE 2.4**). In those days, they used photographic plates, not film or digital cameras to take pictures. Becquerel thought that the uranium in the mineral was acting like light or was emitting light. Elements that spontaneously emit radiation from their nucleus are said to be **radioactive.** In 1903, Becquerel shared the Nobel Prize with Marie and Pierre Curie for discovering this new property of matter, radioactivity.

Are all atomic nuclei radioactive? No. Most naturally occurring isotopes have a stable nucleus, and these isotopes are therefore *not* radioactive. Isotopes that are not stable can become stable by spontaneously emitting radiation from their nuclei. This process is called **radioactive decay.** Elements like uranium that emit radiation through radioactive decay are called radioactive isotopes, or **radioisotopes.** Several naturally occurring radioisotopes can be mined from the earth. Many more are prepared in scientific laboratories. All the isotopes of elements with atomic number 83 and higher are radioactive. Some of the lower elements also have radioisotopes.

The same symbolic notation used to distinguish isotopes (Section 2.2) is also used to distinguish radioisotopes. For example, the most abundant isotope of gallium is $^{69}_{31}\text{Ga}$, which is not radioactive. However, the radioisotope $^{67}_{31}\text{Ga}$ is, and is used in medicine to image certain types of cancer, such as lymphoma. Some radioisotopes used in medicine are shown in **TABLE 2.3**.

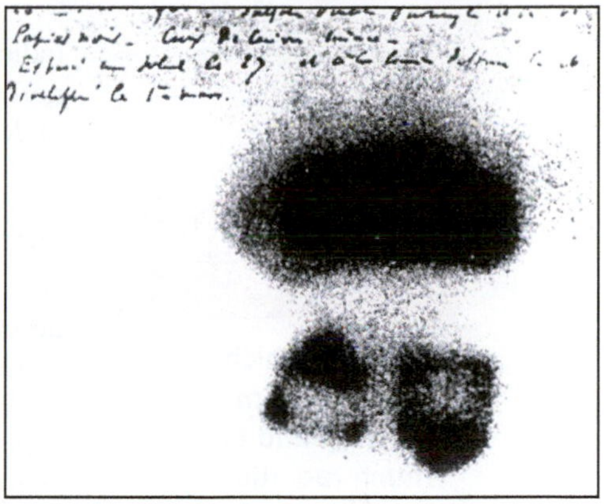

FIGURE 2.4 Radioactive decay. When Henri Becquerel placed uranium, a radioactive element, on a photographic plate in the dark, the uranium exposed the plate, causing an image to appear. Shown is Becquerel's first image of radioactivity.

TABLE 2.3 Some Radioisotopes and Their Use in Diagnosing and Treating Disease

Radioisotope	Clinical Use
Cesium-131	Radioisotope implant (brachytherapy) in brain cancer
Fluorine-18	Cancer metabolism and tumor imaging
Iodine-123, iodine-131	Thyroid imaging and uptake
Phosphorus-32	Bone marrow therapy, liver cancer therapy
Technetium-99m	Bone, kidney, breast tumor, heart, and lung-perfusion imaging
Xenon-133	Lung function imaging

Source: https://isotopes.gov/outreach/med_isotopes.html

Forms of Radiation

The first three forms of nuclear radiation that were discovered are the positively charged alpha (α) particle, the negatively charged beta (β) particle, and the neutral gamma (γ) ray. Two other forms are the positron (+ charge) and the subatomic particle, the neutron (no charge). The properties of these five types are summarized in **TABLE 2.4**.

An **alpha particle** is represented by the Greek letter α or as $^{4}_{2}\text{He}$, a helium nucleus containing two protons and two neutrons. Because it is a helium atom without electrons, it carries a 2+ charge from its two protons.

In some atoms with an unstable nucleus, one of the neutrons may eject a high-energy electron. This electron is called a **beta particle.** The neutron that emits the beta particle becomes a proton as a result. A beta particle is a high-energy electron and carries a charge of 1−. It is represented by the Greek letter β, or symbolically as $^{0}_{-1}\text{e}$ since e⁻ is the symbol for an electron.

Gamma rays are high-energy radiation emitted during radioactive decay, but they have no mass and so are rays instead of particles. Gamma rays are released when an unstable nucleus rearranges to a more stable state. Neither the charge nor the mass of the radioisotope will change, so the gamma ray is represented simply by the Greek letter γ.

Although not emitted from a radioactive nucleus, we mention **X-rays** here because they are commonly used in the clinical setting and share many properties with gamma rays. X-rays are generated by high-energy electron processes. While X-rays are also high-energy rays, they are not quite as high energy as gamma rays.

A **positron** has the same mass as a beta particle but is positively charged. It is represented as $^{0}_{1}\text{e}$. Some radioisotopes with an unstable nucleus will change a proton to a neutron and emit a positron. During the decay process, the positron collides with an electron (equal in mass and opposite in charge), ultimately emitting energy in the form of gamma rays. Positron decay events, commonly called *emissions,* are used in the medical imaging technique called positron emission tomography, or PET (see Section 2.7).

TABLE 2.4 Forms and Properties of Nuclear Radiation

Emission	Symbol	Charge
Alpha	α or $^{4}_{2}\text{He}$	2+
Beta	β or $^{0}_{-1}\text{e}$	1−
Gamma	$^{0}_{0}\gamma$	0
Positron	$^{0}_{1}\text{e}$	1+
Neutron	$^{1}_{0}\text{n}$	0

A neutron has no charge and is represented as $_0^1$n. Neutrons can be used to cause radioactive decay if they are aimed at a radioisotope and added to the nucleus. This is done in neutron capture therapy, causing boron-10 to become boron-11, which then undergoes radioactive decay.

INTEGRATING Chemistry

Find out which ▶ radiation is most harmful and how much radiation it takes to make you sick.

Biological Effects of Radiation

Why is radiation dangerous to living organisms? Radioactive emissions like the ones just described contain a lot of energy. When emitted, they will interact with any atoms they come into contact with, whether or not in a living organism. Alpha and beta particles, neutrons, positrons, gamma rays, and X-rays are also called **ionizing radiation.** When they interact with another atom, they have the effect of ejecting one of that atom's electrons. Remember, electrons are on the outside of the atom, far from the nucleus. This ejection makes the atom more reactive and less stable.

The loss of too many electrons from too many atoms over long periods of time in living cells can affect a cell's chemistry and genetic material. In humans, this can cause a number of health problems, the most common of which is cancer.

Not all ionizing radiation has the same amount of energy. Radiation of higher energy can penetrate farther into a tissue, affecting cells located deeper in the body. The penetrating power for different forms of ionizing radiation is shown in **FIGURE 2.5**. Persons who work with radioactive materials—for example, radiologists—must take special precautions to protect themselves from radiation exposure by wearing a heavy lab coat, lab glasses, and gloves. Depending on the type of radiation with which they are working, they may have to stand behind a plastic or lead shield (see **TABLE 2.5**). People who

FIGURE 2.5 Penetration of radiation. Different forms of ionizing radiation have different energies and therefore penetrate the body to different extents. Alpha particles are stopped at the skin surface, while beta particles penetrate about a centimeter into the skin, and gamma rays penetrate deeply into tissue. X-rays penetrate skin but not bone. Depending on the ionizing radiation, different materials like Al, Pb, or concrete are used for shielding.

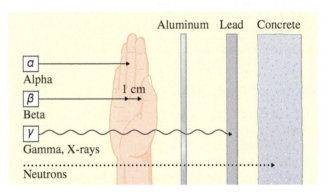

TABLE 2.5 Properties of Common Ionizing Radiation

	Travel Distance Through Air	Tissue Penetration	Protective Shielding
Alpha (α)	A few centimeters	Stops at the skin surface; dangerous only if inhaled or eaten	Paper, clothing
Beta (β)	A few meters	Will not penetrate past skin layer	Heavy clothing, plastic, aluminum foil, gloves
Gamma (γ)	Several hundred meters	Fully penetrates body	Thick lead, concrete, water layer
X-ray	Several meters	Penetrates tissues but not bone	Lead apron, concrete barrier
Neutron (n)	Hundreds to thousands of meters	Fully penetrates body	Concrete barrier, water layer

routinely work with radioactive materials or X-rays usually wear a film badge to monitor their total exposure to radiation over a given time period. Health care workers who handle beta or gamma radiation often wear smaller monitoring rings (see **FIGURE 2.6**).

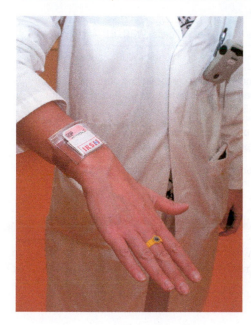

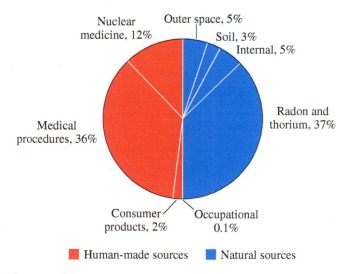

FIGURE 2.6 Film badge. A film badge can be used to monitor exposure to nuclear radiation or X-rays over a given time period. A ring dosimeter is worn by those whose hands will touch radiation-emitting substances.

How much radiation exposure is too much? The **sievert (Sv)** is the SI unit used to measure the effect of biological damage. A more common unit used to measure biological damage is the **millirem (mrem).** One sievert is equal to 100,000 millirems, or 100 rem. These units take into account the activity (or amount) and type of radiation (alpha, beta, gamma) absorbed per kilogram of body tissue. According to the U.S. Nuclear Regulatory Commission, the average annual dosage from natural and human-made sources of radiation is about 0.0062 sieverts, or 620 millirems. Generally, this annual dose from all radiation sources is not harmful to humans.

We are exposed to low levels of ionizing radiation every day in the form of cosmic rays from outer space, radon gas, and other naturally radioactive elements in the earth. This radiation is called natural background radiation. We are also exposed to human-made sources of radiation during medical and dental treatments, during air travel, and from consumer products such as televisions and smoke detectors. One transcontinental flight exposes a person to an additional 2–3 mrem, while one X-ray scan at the airport delivers at most 0.1 mrem, or 1 μSv. Human exposure to radiation has increased 40% since the 1980s mainly due to increased use of radiation in medical procedures.

Nuclear medicine, 12%　Outer space, 5%　Soil, 3%　Internal, 5%　Radon and thorium, 37%　Medical procedures, 36%　Consumer products, 2%　Occupational 0.1%

■ Human-made sources　■ Natural sources

Sources of human exposure to radiation

Radiation Sickness

Although X-ray radiation is not caused by a radioactive event (disintegration), X-ray energies are similar to gamma-ray energies and are considered ionizing radiation. A typical chest X-ray provides a 10 mrem dose, which is minimal. X-ray technicians stand behind a plastic screen to stop scattered X-rays from hitting them because they administer a number of X-rays daily. Clinically, no effects of radiation exposure can be noted below doses of 20,000 mrem.

The effects of radiation exposure are dependent on both time and dose. Long-term exposure to low levels of radiation is not as life-threatening as a short-term intense dosage

TABLE 2.6 Radiation Sickness

Exposure in mrem	Clinical Effect
5,000–20,000	Cannot be detected.
20,000–50,000	Temporary decrease in white blood cells (WBCs).
50,000–100,000	Mild radiation sickness. Large decrease in WBC, nausea, vomiting.
100,000–200,000	Light radiation poisoning. Delayed appearance of appetite loss; nausea; diarrhea; hair loss; spontaneous abortion in pregnant women; recovery in 3 months.
200,000–300,000	Moderate radiation poisoning. Hair loss; fatigue; general illness; permanent sterility possible; recovery takes several months.
300,000–400,000	LD_{50}: Lethal dose for half the population after 30 days.

of radiation. An intense short-term dosage will swiftly destroy tissue in the exposed area. Some of the symptoms of such an acute dose are given in **TABLE 2.6**. Cell types that are rapidly dividing, such as white blood cells (WBCs) and cancer cells, are killed off more quickly than others. Therefore, this intense radiation is very useful in the treatment of some cancers.

Sample Problem 2.5 **Properties of Radiation**

Based on its penetrating power into tissue, which form of ionizing radiation is the least dangerous?

Solution

An examination of Table 2.5 indicates that alpha radiation is the least penetrating and does not pass through the skin.

Practice Problems

2.21 If the symbol for an alpha particle is $^4_2$He, explain how an alpha particle differs from a helium atom.

2.22 What is the charge on a beta particle? An alpha particle? A positron?

2.23 Which form of ionizing radiation has the greatest penetrating power?

2.24 Which form of ionizing radiation is most similar to X-ray radiation? How is it different?

2.25 How much radiation is the average person exposed to annually from all sources?

2.26 If high doses of radiation can be harmful, what is the rationale behind radiation therapy for people with cancer?

2.27 A routine dental exam often includes four bite-wing X-rays, exposing a patient to a total of 5 mrem of radiation. Would this cause radiation sickness in the patient? If so, what would be the effects?

2.28 How many simultaneous chest X-rays would a person have to endure to have an exposure of 20,000 mrem, causing a temporary decrease in white blood cells? Based on this example, is one chest X-ray harmful?

DISCOVERING THE CONCEPTS

 INQUIRY ACTIVITY—Radioactivity

Information

Types of Ionizing Radiation Produced in Nuclear Reactions

Emission	Symbol	Charge	Mass Number
Alpha	α or $_2^4\text{He}$	2+	4
Beta	β or $_{-1}^{0}\text{e}$	1−	0
Gamma	$_0^0\gamma$	0	0
Positron	$_1^0\text{e}$	1+	0
Neutron	$_0^1\text{n}$	0	1

Nuclear Equation

$$_{92}^{238}\text{U} \longrightarrow {}_{90}^{234}\text{Th} + {}_2^4\text{He}$$
$$\text{Reactant} \qquad \text{Products}$$

Questions

1. Using the table provided, what type of ionizing radiation (radioactivity) is produced in the nuclear equation shown?
2. Using the nuclear equation shown, compare the following in the reactants versus the products. Answer as the same, more in products, or more in reactants.
   a. Total number of protons
   b. Total number of neutrons
   c. Total mass number
3. Balance the mass numbers and the atomic numbers, and write the correct atomic symbol to complete the following equation by filling in the blanks.

$$_{53}^{131}\text{I} \rightarrow \underline{\quad\quad} + {}_{-1}^{0}\text{e}$$

4. Write a nuclear equation for the following statement. Iron-59 emits a beta particle to form cobalt-59.

2.5 Nuclear Equations and Radioactive Decay

Uranium-238 is a radioactive isotope that emits an alpha particle when it undergoes radioactive decay. We can represent this process with a special type of chemical equation, called a nuclear decay equation, as follows:

$$_{92}^{238}\text{U} \longrightarrow {}_{90}^{234}\text{Th} + {}_2^4\text{He}$$
$$\text{Reactant} \qquad \text{Products}$$

Notice that the mass number of the reactant (238 for uranium) is equal to the total mass number of the products (234 + 4: thorium and an alpha particle) and the atomic number of the reactant (92) is equal to the sum of the atomic numbers of the products (90 + 2). The symbol changed from uranium to thorium because the number of protons present changed (92 to 90). Atoms with different numbers of protons are different elements.

The general form for a **nuclear decay equation** is

Radioactive nucleus undergoing decay → new nucleus formed + radiation emitted

? 2.5 Inquiry Question

How is a nuclear decay equation written?

Remember that the number of protons plus neutrons on both sides of the nuclear decay equation will be the same whether they are all in the same atom or divided up due to the decay. Let's practice writing nuclear decay equations for ionizing radiation.

Alpha Decay

In alpha decay, an alpha particle, consisting of two protons and two neutrons, is emitted. The mass number of the isotope produced decreases by 4 and the atomic number decreases by 2.

Solving a Problem

Writing a Nuclear Decay Equation for Alpha Decay

Write a nuclear decay equation for radium-224 undergoing alpha decay.

STEP 1 Write the symbolic notation for the radioisotope undergoing decay. Radium-224 has an atomic number of 88, so the symbolic notation is

$$^{224}_{88}\text{Ra} \rightarrow$$

STEP 2 Place the ionizing radiation on the product side of the equation. An alpha particle is ^4_2He.

$$^{224}_{88}\text{Ra} \rightarrow \boxed{} + {}^4_2\text{He}$$

STEP 3 Determine the missing product radioisotope. Remember that the mass and atomic numbers must be equal on both sides of the equation. If radium (Ra) has a mass number of 224 and it loses an alpha particle, which has a mass of 4, the resulting atom will have a mass number of 220 (224 minus 4). During alpha decay, 2 protons are lost (in the alpha particle), so the resulting atom will have two fewer protons than Ra, or 86 (88 minus 2). The new symbol for the element will be the element that has an atomic number of 86, radon. The symbol of the resulting atom is $^{220}_{86}\text{Rn}$.

$$^{224}_{88}\text{Ra} \longrightarrow {}^{220}_{86}\text{Rn} + {}^4_2\text{He}$$

Sample Problem 2.6   Writing Nuclear Decay Equations

Write a nuclear decay equation for americium-241 undergoing alpha decay.

Solution

STEP 1 Write the symbolic notation for the radioisotope undergoing decay.
Americium is element number 95, so the reactant is written as

$$^{241}_{95}\text{Am} \longrightarrow$$

STEP 2 Place the ionizing radiation on the product side of the equation. An alpha particle is ^4_2He.

$$^{241}_{95}\text{Am} \longrightarrow \boxed{} + {}^4_2\text{He}$$

STEP 3 Determine the missing product radioisotope. If americium (Am) has a mass number of 241 and it loses an alpha particle, which has a mass of 4 with two of these being protons, the resulting atom will have a mass number of 237 and an atomic number of 93. This element is neptunium, $^{237}_{93}\text{Np}$. The nuclear decay equation is

$$^{241}_{95}\text{Am} \longrightarrow {}^{237}_{93}\text{Np} + {}^4_2\text{He}$$

Beta Decay and Positron Emission

In beta decay, a high-energy electron is emitted from the nucleus and, in the process, a neutron becomes a proton. A radioisotope that produces a positron is also considered a beta emitter because, in both, the mass number of the isotope does not change. Although the two forms of ionizing radiation are identical except for their charge, typically beta decay refers to the emission of an electron.

Electron   Positron

$_{-1}^{0}e$ $_{1}^{0}e$

Electrons and positrons are both beta emitters.

Writing a Nuclear Decay Equation for Beta Decay

Solving a Problem

Write a nuclear equation for manganese-56 undergoing beta decay.

STEP 1 Write the symbolic notation for the radioisotope undergoing decay. Manganese is element 25, so the reactant is written as

$$_{25}^{56}\text{Mn} \longrightarrow$$

STEP 2 Place the ionizing radiation on the product side of the equation. A beta particle is $_{-1}^{0}e$.

$$_{25}^{56}\text{Mn} \longrightarrow \square + _{-1}^{0}e$$

STEP 3 Determine the missing product radioisotope. Remember: The total mass (56) does not change, but one more proton is present. The number of protons increases on the right side by one, so the resulting isotope has a total of 26 protons. The symbol for the element that has 26 protons is Fe, iron. The symbol of the resulting atom is $_{26}^{56}\text{Fe}$. The nuclear decay equation is

$$_{25}^{56}\text{Mn} \longrightarrow _{26}^{56}\text{Fe} + _{-1}^{0}e$$

Sample Problem 2.7 Writing Nuclear Decay Equations

Write a nuclear decay equation for the positron emission of fluorine-18.

Solution

STEP 1 Write the symbolic notation for the radioisotope undergoing decay. Fluorine is element 9, so the reactant is written as

$$_{9}^{18}\text{F} \longrightarrow$$

STEP 2 Place the ionizing radiation on the product side of the equation. A positron is $_{1}^{0}e$.

$$_{9}^{18}\text{F} \longrightarrow \square + _{1}^{0}e$$

STEP 3 Determine the missing product radioisotope. Remember: The total mass (18) does not change, and one less proton is present in the resulting isotope if a positron is emitted, meaning 8 protons. The symbol for the element that has 8 protons is O, oxygen. The symbol of the resulting atom is $_{8}^{18}\text{O}$. The nuclear decay equation is

$$_{9}^{18}\text{F} \longrightarrow _{8}^{18}\text{O} + _{1}^{0}e$$

Gamma Decay

Because gamma rays are energy only, an isotope that is a pure gamma emitter will not change its atomic number or mass number upon decay. Radioactive technetium (atomic symbol Tc) is one of the few pure gamma emitters. It is used extensively in medical imaging of many organs because most (~96%) of the radioactivity decays from the metastable

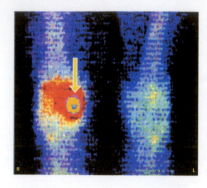

Technetium-99m is commonly used in bone scans to detect breaks and osteoarthritis. Do you see the darkened area on the knee to the left, where the patient was indicating pain?

(temporarily stable) state containing the gamma ray within 24 hours. Short decay times mean that there is radioactivity in the patient for a shorter time period, which is a desirable feature for medical radioisotopes. This radioactive isotope is designated in its metastable state with the symbol m, technetium-99m ($^{99m}_{43}$Tc). By emitting a gamma ray, the technetium becomes more stable. The equation for such a gamma decay is

$$^{99m}_{43}\text{Tc} \longrightarrow {}^{99}_{43}\text{Tc} + {}^{0}_{0}\gamma$$

Producing Radioactive Isotopes

Although some radioisotopes occur in nature, many more are prepared in chemical laboratories. Radioisotopes can be prepared by bombarding stable isotopes with fast-moving alpha particles, protons, or neutrons. The source for technetium-99m is the radioisotope molybdenum-99, which is produced in a nuclear reactor by bombarding the stable isotope molybdenum-98 with neutrons.

$$^{98}_{42}\text{Mo} + {}^{1}_{0}\text{n} \longrightarrow {}^{99}_{42}\text{Mo}$$

Here, the ionizing radiation appears on the reactant side of the equation because a radioisotope is being produced. The nuclear equation is still balanced. The sums of the mass numbers and the atomic numbers on both sides of the equation are equal.

Because technetium-99m is used so much in nuclear medicine, many labs keep a supply of molybdenum-99 on hand, which decays by beta emission, producing technetium-99m.

$$^{99}_{42}\text{Mo} \longrightarrow {}^{99m}_{43}\text{Tc} + {}^{0}_{-1}\text{e}$$

Sample Problem 2.8 Producing Radioisotopes

Cobalt-60 is produced by bombarding a naturally occurring cobalt-59 with a neutron. Write the nuclear equation for this reaction.

Solution

The symbol for a neutron is $^{1}_{0}$n (Table 2.4). A neutron is added to cobalt-59, and cobalt-60 is produced:

$$^{59}_{27}\text{Co} + {}^{1}_{0}\text{n} \longrightarrow {}^{60}_{27}\text{Co}$$

Practice Problems

2.29 Write a balanced nuclear equation for the decay of each of the following:

   a. carbon-14 undergoing beta decay
   b. polonium-212 undergoing alpha decay
   c. copper-66 undergoing beta decay
   d. carbon-11 undergoing positron emission

2.30 Write a balanced nuclear equation for the decay of each of the following:

   a. thorium-232 undergoing alpha decay
   b. strontium-92 undergoing beta decay
   c. nitrogen-13 undergoing positron emission
   d. californium-251 undergoing alpha decay

2.31 Write a balanced nuclear equation for the beta decay of cobalt-60, used in brachytherapy.

2.32 Write a balanced nuclear equation for the production of iodine-131, used in the treatment of some cancers, from the beta decay of tellurium-131.

2.6 Radiation Units and Half-Lives

Radioactivity Units

Earlier, we discussed the units for measuring the biological effects of radiation, the millirem and sievert. Here we discuss the units for radiation emitted from a radioisotope.

When ionizing radiation leaves the nucleus of a radioactive atom, this event is called a disintegration. The amount of radioactivity present in a given isotope is measured as the number of radioactive emissions, or disintegrations, that occur in a second. The unit for measuring disintegrations per second is called the **curie (Ci).** The curie was named for the Polish scientist Marie Curie, who studied radioactivity in France at the turn of the twentieth century along with Henri Becquerel. The SI unit for measuring disintegrations is called the **becquerel (Bq)** after Becquerel.

The *activity* of a radioactive isotope defines how quickly (or slowly) radiation is emitted. A curie is a unit of activity equal to 3.7×10^{10} Bq or disintegrations per second. One curie is a dangerously high level of radiation, so a fraction of a curie, like a millicurie (mCi—one-thousandth of a curie) or a microcurie (μCi—one-millionth of a curie), is often used in medical applications (see **TABLE 2.7**).

2.6 Inquiry Question

How can the amount of radiation be determined from radiation units and half-life?

TABLE 2.7 Units for Radiation Activity

Common Unit	Relationship to Other Units
becquerel (Bq)	1 Bq = 1 disintegration per second
curie (Ci)	1 Ci = 3.7×10^{10} disintegrations per second
millicurie (mCi)	1 Ci = 1000 mCi
microcurie (μCi)	1 Ci = 1,000,000 μCi

Sample Problem 2.9 Determining Doses of Radioisotopes

A 1.0-mL sample of phosphorus-32 for tumor treatment contains 5.0 mCi.
  a. Calculate the number of becquerels present in the sample.
  b. If a patient is to receive a 12- mCi dose, how many milliliters should be injected?

Solution

  a. Use the relationships defined in Table 2.7 to convert millicuries to becquerels.

$$\frac{5.0 \ \cancel{mCi}}{\text{sample}} \times \frac{1 \ \cancel{Ci}}{1000 \ \cancel{mCi}} \times \frac{3.7 \times 10^{10} \ Bq}{1 \ \cancel{Ci}} = 185{,}000{,}000 \text{ Bq per sample}$$

There are two significant figures in the problem, so the answer should be reported as 190,000,000 Bq or in scientific notation as 1.9×10^8 Bq per sample.
  b. In this type of dosage conversion, the unit on the final answer is mL. This unit should appear in the numerator of the problem.

$$\frac{12 \ \cancel{mCi}}{\text{injection}} \times \frac{1.0 \text{ mL}}{5.0 \ \cancel{mCi}} = 2.4 \text{ mL/injection}$$

Half-Life

Every radioactive isotope emits its radiation at different rates. In other words, some isotopes are more unstable than others and emit radiation more rapidly. How quickly this decay occurs, or the rate of decay, is measured as the **half-life.** This is the time it takes for one-half (50%) of the atoms in a radioactive sample to decay (emit radiation). This half-life is more specifically referred to as the **physical half-life.**

FIGURE 2.7 A Geiger counter. The Geiger counter records radioactive disintegrations per second.

Two processes affect the rate of clearance of a radioisotope from the body: the physical half-life and a second rate of elimination from the body through bodily processes called the **biological half-life.** The biological half-life shortens the physical half-life due to increased elimination, establishing an **effective half-life** for medical radioisotopes.

The amount of radiation emitted is measured on an instrument called a Geiger counter, which counts the number of disintegrations (decay events) over a period of time in curies or becquerels (**FIGURE 2.7**). Radioisotopes that are more unstable emit radiation more rapidly and have shorter half-lives.

Naturally occurring radioisotopes tend to have long half-lives, disintegrating slowly over a number of years. Radioisotopes used in medicine tend to have much shorter half-lives, decaying rapidly (in hours or days). Their shorter half-lives allow radioactivity to be eliminated quickly from the body. The physical half-lives of several radioactive elements and effective half-lives of radioisotopes used in medicine are shown in **TABLE 2.8**.

TABLE 2.8 Half-Lives of Some Radioisotopes

Radioisotope	Symbol	Physical Half-Life	Effective Half-Life
Naturally occurring radioisotopes			
Hydrogen-3 (tritium)	^{3}H	12.3 years	
Carbon-14	^{14}C	5730 years	
Radium-226	^{226}Ra	1600 years	
Uranium-238	^{238}U	4.5 billion years	
Radioisotopes used in medicine			
Chromium-51	^{51}Cr	28 days	27 days
Fluorine-18	^{18}F	110 minutes	85 minutes
Iron-59	^{59}Fe	45 days	42 days
Phosphorus-32	^{32}P	14.3 days	13.5 days
Technetium-99m	^{99m}Tc	6.0 hours	4.8 hours
Iodine-123	^{123}I	13.2 hours	12 hours
Iodine-131	^{131}I	8 days	7.6 days
Indium-111	^{111}In	2.8 days	2.5 days

Knowing the half-life allows us to determine how much radioactivity is left after a given amount of time has passed. Let's look at an example.

Solving a Problem

Determining Half-Life

Radioactive iodine-131 has an effective half-life of 7.6 days. If a dose with an activity of 400 µCi is given to a patient today, about how much of the radioactivity will still remain after 38 days?

STEP 1 Determine the total number of half-lives. How many half-lives are in 38 days? Since 1 half-life is 7.6 days, there are 5 half-lives in 38 days. Mathematically, you could arrive at this answer in the following way:

$$38 \text{ days} \times \frac{1 \text{ half-life}}{7.6 \text{ days}} = 5 \text{ half-lives}$$

STEP 2 Determine the amount of isotope remaining. Halve the original dose (400 μCi) dividing by four successive times, since five half-lives were determined in step 1. This can be diagrammed as

$$400 \ \mu\text{Ci} \xrightarrow[\text{1 half-life}]{\text{7.6 days}} 200 \ \mu\text{Ci} \xrightarrow[\text{2 half-lives}]{\text{15.2 days}} 100 \ \mu\text{Ci} \xrightarrow[\text{3 half-lives}]{\text{22.8 days}} 50 \ \mu\text{Ci} \xrightarrow[\text{4 half-lives}]{\text{30.4 days}} 25 \ \mu\text{Ci} \xrightarrow[\text{5 half-lives}]{\text{38 days}} 12.5 \ \mu\text{Ci}$$

Another way this can be solved is by using the equation

$$\text{Isotope remaining} = \left(\frac{1}{2}\right)^n \times \text{starting amount}$$

where n = the number of half-lives determined in Step 1.

$$\text{Isotope remaining} = \left(\frac{1}{2}\right)^5 \times 400 \ \mu\text{Ci} = \left(\frac{1}{32}\right) \times 400 \ \mu\text{Ci} = 12.5 \ \mu\text{Ci}$$

This second method is especially useful when the number of half-lives is not a whole number.

Mastering Video
Practicing the Concepts
Radioactive Decay: Half-Life Calculations

■

Sample Problem 2.10   Half-Life

Iron-59 has a physical half-life of 45 days. If 96 g of a radioactive iron sample (^{59}Fe) is received in the lab today, how many grams of the iron will still be radioactive after 135 days?

Solution

Notice that in this example we are using the physical half-life because this sample is not going into the body.

STEP 1 Determine how many half-lives will pass in 135 days.

$$135 \ \text{days} \times \frac{1 \ \text{half-life}}{45 \ \text{days}} = 3 \ \text{half-lives}$$

STEP 2 Determine the amount of isotope remaining. Diagram the solution using the number of half-lives you determined in step 1:

$$96 \ \text{g} \xrightarrow[\text{1 half-title}]{\text{45 days}} 48 \ \text{g} \xrightarrow[\text{2 half-lives}]{\text{90 days}} 24 \ \text{g} \xrightarrow[\text{3 half-lives}]{\text{135 days}} 12 \ \text{g}$$

Alternatively,

$$\text{isotope remaining} = \left(\frac{1}{2}\right)^3 \times 96 \ \text{g} = \left(\frac{1}{8}\right) \times 96 \ \text{g} = 12 \ \text{g}$$

Practice Problems

2.33 What does the unit mrem measure?

2.34 What does the unit curie measure?

2.35 How do the radioisotopes used in medical imaging differ from naturally occurring radioisotopes?

2.36 Define half-life in your own words.

2.37 A brain scan uses the radioisotope oxygen-15. The recommended dosage is 50 mCi. A supply of 250 mCi in 20 mL arrives at the lab. How many milliliters will be injected into a patient?

2.38 Xenon-133 is used for imaging the lung. The recommended dose is 15 mCi. If xenon-133 arrives in the laboratory in a package containing 120 mCi, how many patients can be treated with the contents of this package?

—*Continued next page*

Continued—

2.39 Radioactive indium-111 has an effective half-life of 2.5 days. If a dose with an activity of 3.0 mCi is injected in a patient to detect blood clots, how much of the ^{111}In will be active 10 days after the injection is given?

2.40 Technetium-99m is very useful in diagnostic imaging since it has a short effective half-life of 4.8 hours. If a patient receives a dose with an activity of 50.0 mCi of technetium-99m for cardiac imaging, how much radioactivity will be left in the patient's body 48 hours after injection?

2.7 Medical Applications for Radioisotopes

 2.7 Inquiry Question

How does the medical field use radioisotopes?

Nuclear radiation has many forms. It can be high-energy particles or high-energy rays. Like the uranium rock that left a pattern on Becquerel's photographic plate, some radioisotopes of elements are useful in medical imaging. Certain radioisotopes concentrate in particular tissues, providing images of specific body tissues. For example, iodine is used in the body only by the thyroid gland, so any iodine—radioactive or nonradioactive—put into the body will accumulate in the thyroid (see **FIGURE 2.8**). The emitted radiation from those locations can create an image on a photographic plate or be detected by scanning sections of the body.

Because it is important to expose patients to the smallest possible dose of radiation for the shortest time period, radioisotopes with short half-lives are selected for use in nuclear medicine. Radioisotopes used in medicine are prepared in a laboratory (unlike naturally occurring isotopes). Several radioisotopes and their medical uses were shown earlier in Table 2.3.

The two main uses of medical radioisotopes are in (1) diagnosing diseased states and (2) therapeutically treating diseased tissues.

When diagnosing a diseased state, a minimum amount of radioisotope is administered because the isotope is used for detection only and thus should have minimal effect on body tissue. A radioisotope used in this way is called a **tracer.** As an example, technetium-99m (Tc-99m) can be used to diagnose proper blood flow through the lungs. It is a gamma emitter and has an effective half-life of 4.8 h. Gamma emitters are useful for diagnosis because gamma radiation is highly penetrating and can more easily exit the body. After inhalation, the radioisotope can be used to take an image of the surface of the lungs. After injection, the radioisotope begins circulating (perfusing) through the bloodstream. The ionizing radiation (gamma rays) emitted from the Tc-99m is detected by a scanner that produces images of the lungs. If the lungs are functioning normally, the radioisotope will be evenly distributed. This can be seen in the ventilation scan in **FIGURE 2.9a**, where normal breathing is seen. If there is a nonfunctioning area in the lungs, the tracer will not be distributed in this area, and it shows up as a "cold" spot (absence of tracer) as seen in the perfusion scan in **FIGURE 2.9b**. Unusually high areas of activity, like an area where the blood is being retained too long in the lungs, would show up as a "hot" spot (more tracer than normal). Areas with rapidly dividing cancer cells will also retain extra tracer and can also show up as hot spots.

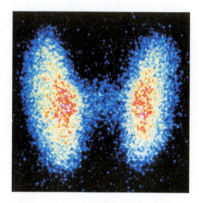

FIGURE 2.8 Normal thyroid image taken using a radioactive element. Radioactive elements like iodine-123 can be used to specifically image the thyroid since it is one of the few organs where the element iodine accumulates.

FIGURE 2.9 Ventilation and perfusion scans of the lungs. Poor blood flow to the lungs or a tumor can be detected with radioisotopes. (a) Normal breathing in the lungs shows a uniform dispersion of radioactive xenon-133 in a ventilation scan, but does not show areas of poor blood flow. (b) Areas with poor blood flow to the lungs show up as "cold spots" when the patient is injected and imaged with technetium-99m.

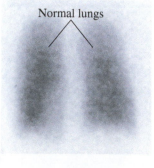

Normal lungs

(a)

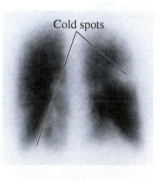

Cold spots

(b)

Radioisotopes can be used therapeutically to destroy diseased or cancerous tissues. In the case of thyroid cancer, iodine-131, a beta emitter, is administered in a dose approximately 1000 times higher than that of an iodine-123 tracer used to detect the disease. The radioactive iodine will be absorbed only by the thyroid, and the beta emissions will destroy cells in that specific location. Cells that are rapidly dividing (cancer cells) are more susceptible to ionizing radiation damage because their genetic material is exposed more often during cell division. These cells are destroyed at a higher rate.

Sample Problem 2.11   Radioisotopes in Medicine

A patient suspected of having thyroid cancer can be diagnosed and, if necessary, treated with radioactive iodine. Compare the dose and radioisotope that would be given for (a) diagnosis and (b) treatment.

Solution

a. For diagnosis, a patient would be receiving only a low dose or trace amount (typically in the μCi range) of radioactive iodine to capture a scan of the thyroid gland. A gamma emitter is preferred because gamma radiation travels farther. Iodine-123, a gamma emitter, works well for this purpose.

b. For treating cancer, a higher dose (typically in the mCi range) and a less penetrating radioisotope is more suitable. Iodine-131, a beta emitter, works well for treatment.

In brachytherapy, small titanium seeds containing a radioisotope like Pd-103 or I-125 are implanted at the tumor site.

Radioisotopes and Radiation in Cancer Treatment

Today there are many options for the therapeutic treatment of cancer using radioisotopes or radiation. Both can be applied either externally or internally without delivering the radioisotope through the bloodstream. In *external beam radiation therapy*, gamma radiation generated from the radioisotope cobalt-60 can be aimed directly at a tumor area, destroying the tissue. In *brachytherapy*, small titanium capsules, or "seeds," containing radioisotopes are implanted at a tumor site. Brachytherapy is useful for cancers of the lung, prostate, and breast.

In *proton therapy*, protons are generated from nonradioactive hydrogen-1 atoms and accelerated to penetrate precisely to the depth of a particular tumor, sparing more of the healthy tissue than gamma rays or X-rays, which are more penetrating. Proton therapy can be especially useful for tumors that are close to other important organs.

A final radiation treatment, which is used in brain and other head and neck tumors, is called *neutron capture therapy*. In this therapy, the patient is injected with a nonradioactive isotope like boron-10 at the site of the tumor. Once the patient is irradiated with neutrons, boron-10 picks up the neutron, becoming radioactive boron-11, which then emits gamma rays and an alpha particle when it decays to lithium-7.

In these ways, a high dose of radiation can be localized to a cancer tumor while minimizing damage to surrounding tissue.

INTEGRATING Chemistry

◀ Find out how radiation is used to treat cancer.

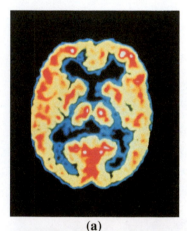

Vizamyl contains the radioisotope fluorine-18 and is used in brain scans for Alzheimer's disease. Do you see the radioisotope ^{18}F on the chemical structure of Vizamyl? We will look at more chemical structures of pharmaceuticals in Chapter 4.

Positron Emission Tomography

Another scan that uses radioisotopes is positron emission tomography, or PET. PET scans are used to identify functional abnormalities in organs and tissues. One of the tracers used in PET is Vizamyl, containing fluorine-18, which has a short half-life of 110 min. This fluorine isotope emits a positron as it decays to form oxygen-18. During the emission, the positron comes into contact with an electron (its opposite), and gamma radiation is produced and detected by the scanner. This type of scan is commonly used on the brain. It has been used successfully in the diagnosis of Alzheimer's disease because areas of the brain covered with plaques do not scan as normal brain tissue (see **FIGURE 2.10**).

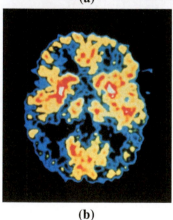

(a)

(b)

FIGURE 2.10 PET scans can reveal abnormalities in tissue function. A cross section of a brain with normal activity (a) is compared with the brain of a person with Alzheimer's (b). Can you see a difference?

Sample Problem 2.12   **Medical Isotopes and Half-Life**

Gold-198 is a beta emitter used in the treatment of leukemia. It has a physical half-life of 2.7 days. How long would it take for a dose with an activity of 96 mCi to decay to an activity of 3.0 mCi?

Solution

This problem is easily done by diagramming the problem to determine the number of half-lives that will have passed. After the passing of each half-life, one-half of the previous amount is still present.

$$96 \text{ mCi} \xrightarrow[\text{1 half-life}]{\text{2.7 days}} 48 \text{ mCi} \xrightarrow[\text{2 half-lives}]{\text{5.4 days}} 24 \text{ mCi} \xrightarrow[\text{3 half-lives}]{\text{8.1 days}} 12 \text{ mCi} \xrightarrow[\text{4 half-lives}]{\text{10.8 days}} 6.0 \text{ mCi} \xrightarrow[\text{5 half-lives}]{\text{13.5 days}} 3.0 \text{ mCi}$$

Therefore, it would take about 14 days.

Practice Problems

2.41 Because calcium is an element in bone, why do you think it might be useful to use the radioisotope calcium-47 in the diagnosis and treatment of bone diseases?

2.42 What is a diagnostic tracer? What are the preferable characteristics of a diagnostic tracer?

2.43 Chromium-51 is used in imaging red blood cells and has an effective half-life of 27 days. If a dose with an activity of 40 μCi is given to a patient, how long will it take for the patient to have less than 5 μCi present?

2.44 Radioactive iodine-123 has an effective half-life of 12 hours. If a dose with an activity of 1.50 mCi of ^{123}I is given to a patient for a thyroid test, how much of the ^{123}I will still be active 36 hours later?

CHAPTER REVIEW

The study guide will help you check your understanding of the main concepts in Chapter 2. You should be able to answer problems for each learning outcome in this list. To check your mastery, try the practice problem listed after each.

STUDY GUIDE

2.1 Atoms and Their Components

Characterize the subatomic particles that make up the atom.
- Name the kind of subatomic particles that make up an atom. (See Sample Problem 2.1. Try 2.45)
- Locate the subatomic particles in an atom. (Try 2.1, 2.46b)

2.2 Atomic Number and Mass Number

Relate the atomic number and mass number to the parts of an atom.
- Distinguish atomic number and mass number. (Try 2.5)
- Predict the mass number of an atom given the subatomic particles. (Try 2.9, 2.51)
- Determine the number of protons, neutrons, and electrons for an element given the mass number or symbolic notation. (Try 2.11, 2.15, 2.47, 2.57)
- Write symbolic notation for atoms when the number of protons and neutrons is given. (Try 2.13, 2.55)

2.3 Isotopes and Atomic Mass

Distinguish mass number of an isotope and atomic mass of an element.
- Define isotope. (See Section 2.3.)
- Distinguish between mass number and atomic mass. (Try 2.17, 2.51, 2.57)
- Predict the most common isotope based on the atomic mass of an element. (See Sample Problem 2.4. Try 2.19)

CHAPTER GUIDE

Inquiry Question

What are the subatomic particles that make up an atom and where are they located?

An atom consists of three subatomic particles: protons, neutrons, and electrons. Protons have a positive charge, neutrons have no charge, and electrons have a negative charge. Most of the mass of an atom comes from the protons and neutrons located in the center, or nucleus, of an atom. The unit for the mass of an atom is the atomic mass unit (amu); each proton and neutron present in an atom weighs approximately 1 amu.

Properties of Particles in an Atom				
Subatomic Particle	Symbol	Electrical Charge	Relative Mass	Location in Atom
Electron	e^-	$1-$	0.0005 (1/2000)	Outside nucleus
Proton	p or p^+	$1+$	1	Nucleus
Neutron	n or n^0	0	1	Nucleus

Inquiry Question

What do the atomic number and mass number indicate?

The atomic number of an atom indicates the number of protons in an atom. Because atoms are uncharged, the atomic number also indicates the number of electrons in an atom. All atoms of a given element have the same number of protons. The mass number of an atom is the total number of protons and neutrons present in a given atom of an element.

$$\text{Mass number} \longrightarrow \,^{12}_{6}\text{C} \longleftarrow \text{Atomic symbol}$$
$$\text{Atomic number} \longrightarrow$$

Inquiry Question

What is the difference between the mass number for an isotope and the atomic mass of an element?

The mass number is the sum of protons and neutrons for a *given* isotope. For example, nitrogen-14 has seven protons and seven neutrons. The atomic mass is the average atomic mass for *all* the isotopes of an element found in nature. This number is found on the periodic table, often below the element symbol.

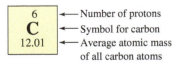

6 ← Number of protons
C ← Symbol for carbon
12.01 ← Average atomic mass of all carbon atoms

—Continued next page

Continued—

2.4 Radioactivity and Radioisotopes

Distinguish and characterize types of ionizing radiation.
- Define radioactivity. (See Section 2.4.)
- Distinguish the forms of ionizing radiation. (Try 2.24, 2.53)
- Differentiate the penetrating power of the forms of ionizing radiation. (Try 2.23, 2.54)
- Use biological exposure to radiation (mrem units) values to determine whether treatments are harmful. (Try 2.25, 2.27, 2.59)

Inquiry Question

How does a radioisotope emit radiation?

Some atomic isotopes emit radiation (a form of energy) sponta-neously from their nucleus in a process called radioactive decay. Isotopes that undergo radioactive decay are called radioisotopes, and the high-energy particles given off in this process are referred to as ionizing radiation, or radioactivity. Three common forms of radioactivity are alpha (α) and beta (β) particles and gamma (γ)rays. Biological damage to tissue by radiation is measured by the units sievert and millirem. An X-ray is also a form of ionizing radiation, although it is not caused by a radioactive decay event. Different forms of ionizing radiation penetrate the body differ-ently, producing different biological effects. The biological effects of high levels of radiation include low white blood cell counts, nausea, hair loss, and ultimately death.

2.5 Nuclear Equations and Radioactive Decay

Write a nuclear decay equation.
- Write a balanced nuclear decay equation for alpha, beta, gamma, and positron emissions. (Try 2.29, 2.31, 2.61, 2.63a–c)
- Write a balanced nuclear equation for the production of radioisotopes. (Try 2.32, 2.63d, 2.65)

Inquiry Question

How is a nuclear decay equation written?

The radioactive decay of a radioisotope can be represented sym-bolically in the form of a nuclear decay equation. The number of protons and the mass number found in the reactant (the decaying radioisotope) is equal to the sum of the number of protons and the mass numbers found in the products (the stable isotope and the radioactive particle).

$$\underset{\substack{\text{Decaying isotope}\\\text{(Reactant)}}}{^{238}_{92}\text{U}} \longrightarrow \underset{\text{Decay product}}{^{234}_{90}\text{Th}} + \overset{\text{Ionizing radiation}}{^{4}_{2}\text{He}}$$

2.6 Radiation Units and Half-Lives

Determine the amount of radiation present from radia-tion units and half-lives.
- Distinguish the units for radioactive emission and biological damage to tissue. (Try 2.33, 2.67)
- Perform dosing calculations using radiation activity units. (Try 2.37, 2.69)
- Determine the remaining dose of a radioactive isotope given the half-life. (Try 2.39, 2.71, 2.73)

Inquiry Question

How can the amount of radiation be determined from radiation units and half-life?

Radioactive decay is measured as the number of decay events, or disintegrations, that occur in 1 second. The common unit for measuring radioactive decay is the curie (Ci), and the SI unit is the becquerel (Bq). For the smaller quantities in medical applications, the millicurie (mCi) and microcurie (μCi) are often used. A becquerel is equal to one disintegration per second. A curie is equivalent to 3.7×10^{10} becquerels. The half-life of a radioisotope is the amount of time it takes for one-half of the radiation in a given sample to decay. Most radioisotopes used in medicine have short half-lives and use other biological pro-cesses in addition to radioactive decay for elimination. Medical isotopes have a measured effective half-life value that is shorter than their physical half-life, which is commonly referred to as just half-life.

2.7 Medical Applications for Radioisotopes

Apply radioactivity to medical diagnosis and treatment.

- Apply the use of radioisotopes to the diagnosis and treatment of disease. (Try 2.41, 2.75)

Inquiry Question

How does the medical field use radioisotopes?

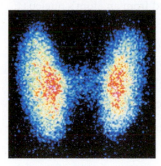

Certain elements concentrate in particular organs of the body. If a radioisotope of this element can be made, this area of the body can be imaged using that radioisotope. A patient can be injected with a trace amount of a radioisotope to diagnose a diseased state. Radioisotopes can also be used to treat diseases. Radioisotopes can be applied externally (external beam radiation therapy) or internally (brachytherapy, neutron capture therapy, and proton therapy) by applying radiation directly at the tumor site in high doses, destroying cancerous cells. Positron emission tomography (PET) uses a radioisotope to image tissues that are not functioning normally.

 Mastering Videos

PRACTICING THE CONCEPTS

The following videos can be accessed through the Pearson eText or your Mastering Chemistry course.

- Distinguishing Mass Number and Atomic Number
- Radioactive Decay: Half-Life Calculations

Additional Problems

2.45 Complete the following statements:

a. A _____ is a subatomic particle that has a neutral charge.

b. The atomic number on the periodic table equals the number of _____ in the atom.

c. Two atoms that have the same number of protons and electrons but a different number of neutrons are called _____.

2.46 Complete the following statements:

a. The mass number is the sum of the number of _____ and _____.

b. Most of the mass of an atom is found in the _____.

c. In an atom, the number of _____ equals the number of _____.

2.47 Provide the number of protons and electrons in the following atoms:

a. magnesium   b. Ag

c. element number 28

2.48 Provide the number of protons and electrons in the following atoms:

a. chlorine   b. element number 11   c. S

2.49 Determine the number of protons, neutrons, and electrons in the following atoms:

a. $^{55}_{26}Fe$ b. $^{15}_{7}N$ c. $^{52}_{24}Cr$ d. $^{137}_{56}Ba$

2.50 Determine the number of protons, neutrons, and electrons in the following atoms:

a. $^{39}_{19}K$ b. $^{16}_{8}O$ c. $^{81}_{35}Br$ d. $^{4}_{2}He$

2.51 Complete the following table:

Symbol	Number of Protons	Number of Neutrons	Number of Electrons	Mass Number	Name
$^{1}_{1}H$					
		12		12	24
$^{9}_{4}Be$			5		

2.52 Complete the following table:

Symbol	Number of Protons	Number of Neutrons	Number of Electrons	Mass Number	Name
$^{20}_{10}Ne$				20	
		35	44	35	
$^{202}_{80}Hg$				80	

2.53 Identify the radiation associated with the following:

   a. particle that has the same number of protons and neutrons as a helium nucleus

   b. an electron that is emitted from an atom's nucleus

   c. has no mass

2.54 Identify the radiation associated with the following:

   a. can be stopped by paper

   b. the highest energy (most penetrating) emission

   c. can be stopped by plastic or aluminum foil

2.55 Write the symbolic notation for the following radioactive isotopes:

   a. phosphorus-32

   b. cobalt with 33 neutrons

   c. element number 24 containing 27 neutrons

2.56 Write the symbolic notation for the following radioactive isotopes:

   a. uranium-238

   b. xenon with 89 neutrons

   c. element number 53 containing 78 neutrons

2.57 Complete the following table:

Isotope Name	Symbolic Notation	Number of Protons	Number of Neutrons	Mass Number	Medical Use
Thallium-201					Tumor imaging
	$^{123}_{53}\text{I}$				Thyroid imaging
			54	133	Lung function imaging
Fluorine-18					Positron emission tomography (PET) imaging

2.58 Complete the following table:

Isotope Name	Symbolic Notation	Number of Protons	Number of Neutrons	Mass Number	Medical Use
Carbon-14					Used in breast tumor detection
	$^{103}_{46}\text{Pd}$				Prostate cancer treatment
Ytterbium-169			70		Gastrointestinal tract diagnoses
Xenon-127	$^{127}_{54}\text{Xe}$				Brain imaging for mental disorders

2.59 If a single airport X-ray scan has a biological radiation effect of 0.1 mrem, how many of these X-ray scans would a person have to have before any radiation sickness were detected at 20 rem?

2.60 If a normal dose CT scan delivers 700 mrem of radiation, how many CT scans would a patient have to have to feel mild radiation sickness (50 rem)?

2.61 Complete the following nuclear equations:

   a. $^{15}_{8}\text{O} \rightarrow {}^{15}_{7}\text{N} + ?$
   b. $^{46}_{23}\text{V} \rightarrow ? + {}^{0}_{-1}\text{e}$

   c. $^{234}_{92}\text{U} \rightarrow ? + {}^{4}_{2}\text{He}$
   d. $^{8}_{4}\text{Be} \rightarrow ? + {}^{4}_{2}\text{He}$

2.62 Complete the following nuclear equations:

   a. $^{214}_{82}\text{Pb} \rightarrow ? + {}^{0}_{-1}\text{e}$
   b. $^{18}_{8}\text{O} + {}^{0}_{-1}\text{e} \rightarrow ? + {}^{1}_{0}\text{n}$

   c. $^{218}_{84}\text{Po} \rightarrow {}^{214}_{82}\text{Pb} + ?$
   d. $^{60}_{27}\text{Co} \rightarrow ? + {}^{0}_{0}\gamma$

2.63 Write balanced nuclear decay equations for each of the following emitters:

   a. copper-66 (β)

   b. platinum-192 (α)

   c. tin-126 (β)

   d. gallium-72 produces germanium-72

2.64 Write balanced nuclear decay equations for each of the following emitters:

   a. thorium-225 (α)

   b. bismuth-210 (α)

   c. cesium-137 (β)

   d. sulfur-35 (β)

2.65 When a boron-10 is bombarded with an alpha particle, nitrogen-12 and neutrons are produced. Write a nuclear equation for this reaction.

2.66 Iron-52 is a positron emitter. Write a nuclear equation for this reaction.

2.67 What is measured by the bequerel unit?

2.68 What is measured by the unit millirem?

2.69 A 25-mL sample of chromium-51 contains 1.00 mCi. If a patient is to receive a 50-μCi dose to undergo a red blood cell count, how many milliliters should be injected?

2.70 Thallium-201 is a radioisotope used in brain scans. If the recommended dose is 3.0 mCi and a vial contains 60 mCi in 50 mL, how many milliliters should be injected?

2.71 Fluorine-18, which has a physical half-life of 110 min, is used in PET scans. If 1.00 g of fluorine-18 is shipped at 5:00 P.M., how many milligrams of the radioisotope are still active if the sample arrives at the nuclear medicine laboratory at 9:30 A.M. the next morning?

2.72 A 400-mg sample of technetium-99m is used for a diagnostic test. If technetium-99m has an effective half-life of 4.8 h, how much of the technetium-99m remains 48 h after the test?

2.73 A patient receives an 80-mCi dose of gold-198 for the treatment as part of his cancer therapy. If after 8 days the patient has 10 mCi in the body, what is the effective half-life of gold-198?

2.74 If the amount of radioactive phosphorus-32 in a sample prepared for treatment of leukemia decreases from 1.2 g to 0.30 g in 28.6 days, what is the half-life of phosphorus-32?

2.75 What is the importance of a cold spot when using a radioactive tracer? A hot spot?

2.76 How are neutron capture therapy and brachytherapy the same? How are they different?

Challenge Problems

2.77 Iron-59 has a physical half-life of 45 days. If 168 g of a radioactive iron (^{59}Fe) is received in the lab today, what percentage of the original sample is left after 270 days?

2.78 On an archaeological dig, a wooden canoe is unearthed and analyzed for carbon-14. About 25% of the carbon-14 that was initially present remains. What is the approximate age of the canoe? The half-life of carbon-14 is 5730 years.

2.79 PET scans are useful for imaging areas of high activity in the brain. One of the compounds commonly used

is an fluorine-18-labeled isotope of the sugar glucose called fludeoxyglucose, or FDG.

a. Fluorine-18 is produced by reacting oxygen-18 with a proton, ^1_1p. The products of this reaction are fluorine-18, a neutron, and a gamma ray. Write a nuclear equation for the production of fluorine-18.

b. Glucose is used in the brain for energy. How do you think a PET scan of the brain of a patient with higher than normal activity would compare to that of a person with normal activity? Support your answer.

Answers to Odd-Numbered Problems

Practice Problems

2.1 Protons and neutrons are located in the nucleus (center) of an atom, and the electrons are found in constant motion around the nucleus, in a space called the electron cloud.

2.3 The mass of an electron is about 2000 times smaller than that of a proton.

2.5 **a.** The number of protons (or electrons) is the atomic number.

   b. The number of neutrons is the mass number minus the atomic number.

   c. The number of electrons is the same as the number of protons in an atom.

2.7 **a.** nitrogen, N   **b.** sulfur, S

   c. tin, Sb   **d.** zinc, Zn

   e. cobalt, Co

2.9 16

2.11 **a.** 53 protons, 78 neutrons, 53 electrons

   b. 19 protons, 20 neutrons, 19 electrons

2.13 **a.** $^{11}_5B$ **b.** $^{81}_{35}Br$ **c.** $^{28}_{14}Si$ **d.** $^{124}_{54}Xe$

2.15 **a.** 8 protons, 10 neutrons, 8 electrons

   b. 20 protons, 20 neutrons, 20 electrons

   c. 47 protons, 61 neutrons, 47 electrons

   d. 82 protons, 125 neutrons, 82 electrons

2.17 The isotope carbon-12 contains exactly 6 protons and 6 neutrons. The atomic mass as seen on the periodic table is an average mass, taking into consideration the abundance of all the carbon isotopes. Because about 1% of the carbon isotopes are carbon-13, the atomic mass is slightly higher than 12 amu.

2.19 **a.** Magnesium-24 has 12 neutrons, magnesium-25 has 13 neutrons, and magnesium-26 has 14 neutrons.

   b. $^{24}_{12}Mg$, $^{25}_{12}Mg$, $^{26}_{12}Mg$

   c. magnesium-24

2.21 An alpha particle is a helium nucleus. There are no electrons in an alpha particle.

2.23 neutrons

2.25 620 mrem, or 6.2 mSv

2.27 No; you would have to receive 4000 sets of X-rays before observing a clinical effect (which starts to occur at 20,000 mrem). This is more than a dental technician would likely take in a day.

2.29 **a.** $^{14}_6C \longrightarrow {}^{14}_7N + {}^{0}_{-1}e$

   b. $^{212}_{84}Po \longrightarrow {}^{208}_{82}Pb + {}^4_2He$

   c. $^{66}_{29}Cu \longrightarrow {}^{66}_{30}Zn + {}^{0}_{-1}e$

   d. $^{11}_6C \longrightarrow {}^{11}_5B + {}^{0}_{1}e$

2.31 $^{60}_{27}Co \longrightarrow {}^{60}_{28}Ni + {}^{0}_{-1}e$

2.33 Millirems measure the biological damage caused by radiation.

2.35 The half-lives are shorter, and they are prepared in the lab.

2.37 4 mL

2.39 0.19 mCi or 190 μCi

2.41 Ca-47 will concentrate in bone.

2.43 After three half-lives (81 days), the patient will have less than 5 μCi present.

Additional Problems

2.45 **a.** neutron   **b.** protons (and electrons)

    c. isotopes

2.47 **a.** 12 protons, 12 electrons

    b. 47 protons, 47 electrons

    c. 28 protons, 28 electrons

2.49 **a.** 26 protons, 29 neutrons, 26 electrons

    b. 7 protons, 8 neutrons, 7 electrons

    c. 24 protons, 28 neutrons, 24 electrons

    d. 56 protons, 81 neutrons, 56 electrons

2.51

Symbol	Number of Protons	Number of Neutrons	Number of Electrons	Mass Number	Name
$^{1}_{1}H$	1	0	1	1	Hydrogen-1
$^{24}_{12}Mg$	12	12	12	24	Magnesium-24
$^{9}_{4}Be$	4	5	4	9	Beryllium-9

2.53 **a.** alpha particle   **b.** beta particle

    c. gamma ray

2.55 **a.** $^{32}_{15}P$ **b.** $^{60}_{27}Co$ **c.** $^{51}_{24}Cr$

2.57

Isotope Name	Symbolic Notation	Number of Protons	Number of Neutrons	Mass Number	Medical Use
Thallium-201	$^{201}_{81}Tl$	81	120	201	Tumor imaging
Iodine-123	$^{123}_{53}I$	53	70	123	Thyroid imaging
Xenon-133	$^{133}_{54}Xe$	54	79	133	Lung function imaging
Fluorine-18	$^{18}_{9}F$	9	9	18	Positron emission tomography (PET) imaging

2.59 200,000 airport X-ray scans

2.61 **a.** $^{15}_{8}O \longrightarrow ^{15}_{7}N + ^{0}_{1}e$

    b. $^{46}_{23}V \longrightarrow ^{46}_{24}Cr + ^{0}_{-1}e$

    c. $^{234}_{92}U \longrightarrow ^{230}_{90}Th + ^{4}_{2}He$

    d. $^{8}_{4}Be \longrightarrow ^{4}_{2}He + ^{4}_{2}He$

2.63 **a.** $^{66}_{29}Cu \longrightarrow ^{66}_{30}Zn + ^{0}_{-1}e$

    b. $^{192}_{78}Pt \longrightarrow ^{188}_{76}Os + ^{4}_{2}He$

    c. $^{126}_{50}Sn \longrightarrow ^{126}_{51}Sb + ^{0}_{-1}e$

    d. $^{72}_{31}Ga \longrightarrow ^{72}_{32}Ge + ^{0}_{-1}e$

2.65 $^{10}_{5}B + ^{4}_{2}He \longrightarrow ^{12}_{7}N + 2^{1}_{0}n$

2.67 The bequerel is the SI unit for measuring disintegrations per second from a radioactive material.

2.69 1.3 mL

2.71 1.95 mg

2.73 64 hours (2.7 days)

2.75 A cold spot indicates a diseased area of an organ, and a hot spot indicates an area of the organ where the cells are rapidly dividing or a cancerous growth.

Challenge Problems

2.77 1.56%

2.79 **a.** $^{18}_{8}O + ^{1}_{1}p \longrightarrow ^{18}_{9}F + ^{1}_{0}n + ^{0}_{0}\gamma$

    b. Because a person with a more active brain uses more glucose, more FDG will be present in the brain. The PET image would be brighter in more areas than for the normal person.

Compounds—How Elements Combine

Why is the formula for water H_2O and not H_3O or H_4O? How do elements combine to form compounds? How can we use the periodic table as a guide? Read on to explore compound formation in Chapter 3.

CHAPTER OUTLINE

LEARNING TIP   Try Practice Problems Instead of Rereading the Textbook

Rereading does not help you master a topic; it only makes you more familiar with it. To make sure you understand a topic, be sure to do some practice problems.

IN THE PREVIOUS two chapters, you were introduced to matter and its component parts—atoms. In nature, most elements do not exist as individual atoms. Instead, atoms combine with other atoms to form compounds. For example, the formula for water is H_2O, which means that water is the unique combination of two hydrogen atoms with one oxygen atom.

What advantage does an atom gain through combining with other atoms? What is special or unique about the Group 8A elements that do exist as single atoms?

As we explore how elements form compounds, we will see that the periodic table helps predict how elements combine. We will also see that metals and nonmetals combine differently from the way nonmetals combine with other nonmetals.

What's an
Inquiry Question?

Inquiry Questions are designed to focus your reading on the main concepts by section. As you read each section, see if you can answer each Inquiry Question in your own words.

3.1 Inquiry Question

How are electrons arranged in an atom?

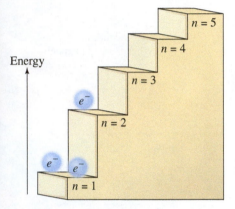

FIGURE 3.1 Electron arrangements.
In atoms, electrons occupy the lower energy levels first. Electrons are only found at distinct energy levels (steps), not between them. The number of the energy level is represented by n. Energy level $n = 1$ is closest to the nucleus. Which element is represented by the three electrons shown? Refer to Table 3.1 to help you answer this question.

3.1 Electron Arrangements and the Octet Rule

In Chapter 2, we saw that an atom's nucleus can be reactive (radioactive). In this chapter we turn our focus to the electrons and the electron cloud surrounding the nucleus. One of the reasons that elements form compounds is that the electron arrangements in the atoms are more stable when elements combine. In Chapter 2, we saw that the electrons occupy a large space outside the nucleus of an atom. The exact location of any given electron outside the nucleus is difficult to determine, but most of them lie within a space called the electron cloud. Because electrons are charged and are in constant motion, they possess energy. The electrons in an atom are arranged by energy level.

You can think of the electrons in an atom as occupying different steps on an uneven staircase where each step is a different energy level. The height of the steps in the staircase decreases as you ascend.

In atoms, the electrons can only occupy certain energy levels (steps on the staircase), not energies in between (see **FIGURE 3.1**). In general, electrons occupy the lowest energy level first. This lowest energy level is closest to the nucleus. Unlike a real staircase, the energy levels get closer together the farther you get from the nucleus. The maximum number of electrons that can be found in any given energy level can be calculated by the formula $2n^2$, where n is the number of the energy level. In the first energy level ($n = 1$), the maximum number of electrons present is two; in the second energy level ($n = 2$), the maximum number of electrons is eight; in the third energy level ($n = 3$), the maximum number of electrons is 18; and so on. Remember, when using exponents, do the calculation with the exponent before the multiplication.

TABLE 3.1 shows the electron arrangements for the first 20 elements. Look at potassium (K) and calcium (Ca) at the bottom of the table. They have electrons in the fourth energy level even though they do not have the maximum 18 electrons filling up the third level. Once there are eight electrons in $n = 3$, the next two electrons are placed in the higher energy shell, $n = 4$. This trend continues as the energy levels increase.

TABLE 3.1 Energy-Level Arrangements of Electrons for the First 20 Elements (Valence Electrons in Red)

			Number of Electrons in Energy Level			
Element	Group Number	Total Number of Electrons	$n = 1$	$n = 2$	$n = 3$	$n = 4$
H	1A	1	1			
He	8A	2	2			
Li	1A	3	2	1		
Be	2A	4	2	2		
B	3A	5	2	3		
C	4A	6	2	4		
N	5A	7	2	5		
O	6A	8	2	6		
F	7A	9	2	7		
Ne	8A	10	2	8		
Na	1A	11	2	8	1	
Mg	2A	12	2	8	2	
Al	3A	13	2	8	3	
Si	4A	14	2	8	4	
P	5A	15	2	8	5	
S	6A	16	2	8	6	
Cl	7A	17	2	8	7	
Ar	8A	18	2	8	8	
K	1A	19	2	8	8	1
Ca	2A	20	2	8	8	2

Being able to locate the number of electrons present in each energy level provides insight into how the elements in a group or period in the periodic table are related to each other. For example, look at boron (B) and aluminum (Al) in Table 3.1. Both of these elements are in Group 3A and have three electrons in their highest energy level. Elements in the same group in the periodic table have the same number of electrons in their highest energy level.

For any atom, the highest energy level that contains electrons is called the **valence shell,** or valence level, and the electrons residing in that energy level are called **valence electrons.** The valence electrons are farthest from the nucleus and are important because they are the electrons that can combine when elements make compounds.

The group (column) number in the periodic table represents the number of valence electrons for the main-group elements, those in Groups 1A–8A. Group 1A elements have one valence electron, Group 2A elements have two, and so on, across the periodic table. The period (row) represents the outermost energy level or shell containing electrons. The electrons in the elements in Period 1, hydrogen (H) and helium (He), are located in energy level 1, which can hold a maximum of only two electrons. The elements in Period 2, lithium (Li) through neon (Ne), have their valence electrons in energy level 2, which can hold a maximum of eight electrons. Valence electrons for the main-group elements in Period 3 and higher are similarly found in energy level 3 and higher. Our focus will be on the electron arrangements in the main-group elements.

As we will see throughout this chapter and the remainder of this book, the reactivity of an element is determined by the arrangement of the electrons in the atom, specifically the valence electrons, in the electron cloud. For example, carbon is atomic number 6 on the periodic table and has six electrons. In its lowest energy state, the electrons in carbon are arranged with two in the first energy level, $n = 1$, and four in the second energy level, $n = 2$, meaning that it has four valence electrons in a shell that can hold up to eight. Carbon will seek to complete its valence shell by reacting with other elements.

Groups indicate the number of valence electrons in atoms for the main-group elements.

Periods indicate the valence energy level for atoms for the main-group elements.

A portion of the main-group elements is shown in the expanded area to highlight the group and period numbers. Note that the main-group elements (Groups 1A–8A) are outlined in red on the table.

Sample Problem 3.1 **Valence Electrons**

Use the periodic table to determine the group number and number of valence electrons in

a. O   b. Cl

Solution

The number of valence electrons can be determined by the group number for the main-group elements.

a. Oxygen is in Group 6A, so the number of valence electrons is six.
b. Chlorine is in Group 7A, so the number of valence electrons is seven.

The Octet Rule

As you go across the periodic table, you will notice that the valence electrons in the main-group elements in the same period increase by 1 until the number 8 is reached in Group 8A. The Group 8A elements are called the **noble gases** or inert gases, and they are chemically unreactive under all but extreme conditions. Elements that are unreactive are stable and exist as atoms.

The properties of stability and reactivity are *inversely related.* What makes the noble gas elements so much more stable and less reactive than other atoms? The answer is found in the valence electrons of the noble gases. The valence electrons are the electrons in atoms that combine when forming compounds. The noble gases (8A) have eight valence electrons (except for helium, which has only two valence electrons).

Experimentally, it has been observed that possessing eight valence electrons is a highly stable state for an atom. Most of the atoms we will see in this book will react with other atoms to achieve a total of eight electrons, an octet, in their valence shell. This is known as the **octet rule.**

To achieve a valence octet, some atoms give away electrons, others will accept or take electrons, and still others will share electrons with other atoms. These different modes of achieving a valence octet lead to different kinds of compounds.

Practice Problems

Find the answers to the odd-numbered Practice Problems at the end of the chapter.

3.1 How many electrons are in each energy level of the following elements?
   a. He   b. C   c. Na   d. Ne

3.2 How many electrons are in each energy level of the following elements?
   a. H   b. F   c. Ar   d. K

3.3 How many valence electrons are present in the following atoms?
   a. N   b. B   c. Si   d. K

3.4 How many valence electrons are present in the following atoms?
   a. C   b. Al   c. Li   d. F

3.5 Which of the following elements are stable as atoms?
   a. H   b. N   c. Ne   d. Na

3.6 Which of the following elements are stable as atoms?
   a. Be   b. Ar   c. F   d. Xe

3.2 In Search of an Octet, Part 1: Ion Formation

3.2 Inquiry Question

How do ions form?

Recall that any atom of a given element has an equal number of protons and electrons, which means that the atom is electrically neutral—it has no charge. One of the ways that an atom can achieve a valence octet is by gaining or losing electrons, meaning that there will be a corresponding change in net charge, with the gain or loss of electrons.

Consider chlorine, element number 17 in the periodic table. As an element in Group 7A, a chlorine atom has seven valence electrons. To achieve an octet in its valence shell, the chlorine atom can gain one electron. By gaining that one electron, the chlorine atom would then have an extra electron in its electron cloud, which means that it has more negative charges than positive charges and, therefore, a net negative charge. If an atom gains or loses electrons to achieve an octet, this newly formed particle will have an *unequal* number of protons and electrons and, therefore, will have a net charge. We refer to these charged atoms as **ions.**

When chlorine gains the one electron necessary to complete its octet, the newly formed ion has a net charge of $1-$. Ions formed when atoms gain electrons and become negatively charged are called **anions.** As **FIGURE 3.2** shows, adding one electron to the chlorine atom means that it now has 18 electrons but still only 17 protons, resulting in the negative charge.

Notice that when a chlorine atom gains an electron to complete the octet in its valence shell, thus becoming an ion, it is more stable than the chlorine atom, having satisfied the octet rule. Notice also that in gaining an electron, the chlorine ion achieves not only the same number of valence electrons as argon (eight) but also has the same total number of electrons. The ion formed by chlorine is **isoelectronic** (*iso*, a prefix meaning "same") with the argon atom, meaning that it has the same number of electrons as argon.

Next, consider sodium, element number 11 in the periodic table. The sodium atom is in Group 1A and has one valence electron. The sodium atom would have to gain seven more

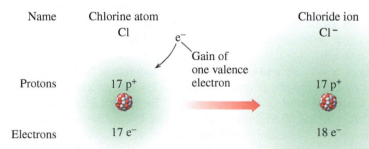

FIGURE 3.2 Chlorine forming an anion. A chlorine atom gains an electron, achieving a stable eight valence electrons.

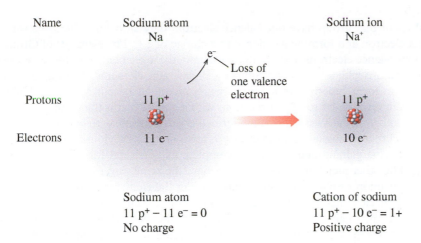

FIGURE 3.3 Sodium forming a cation.
A sodium atom gives up an electron, achieving a stable eight valence electrons.

electrons to complete its octet. This would take more energy than the atom has available for attracting electrons and, therefore, is not favorable. However, look at the possibility shown in **FIGURE 3.3**. If the sodium atom loses one electron, it becomes isoelectronic with neon. This means that the ion formed by the loss of that one electron would have a complete octet in its valence shell—exactly like neon. After losing the electron, the ion would have only 10 electrons but still 11 protons, resulting in a net charge of 1+ (11 protons − 10 electrons = 1+). Ions formed when atoms lose electrons and become positively charged are called **cations.**

Trends in Ion Formation

Which atoms gain electrons to form anions, and which atoms lose electrons to form cations? The periodic table is a useful tool to help answer these questions.

Recall from Chapter 1 that we can use the periodic table to distinguish the elements that are nonmetals from those that are metals (see **FIGURE 3.4**). As a rule, when elements form ions, those that are metals lose electrons to form cations, whereas elements that are nonmetals gain electrons to form anions. Among the main-group elements, those in Groups 1A, 2A, and 3A form cations, and the nonmetallic elements in Groups 5A, 6A, and 7A form anions. (The elements of Group 4A tend not to form ions but gain their stability in ways we will discover later in this chapter.)

How is the charge on an ion determined? The periodic table is our tool for determining this as well. Previously, we saw that a sodium atom loses one electron to form a cation with a 1+ charge. The same is true of all the elements of Group 1A. Because all

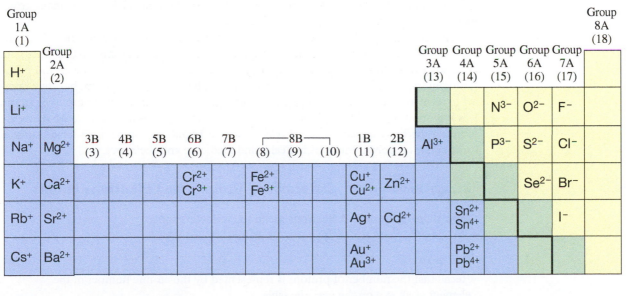

FIGURE 3.4 Trends in ion formation. Positive ions are produced from metals, and negative ions are formed from nonmetals. Common ions formed are shown.

members of that group have one valence electron, they all behave the same way, giving up that electron and forming a cation with a charge of 1+. The elements of Group 2A all have two valence electrons. During ion formation, these elements lose those two electrons to form cations with a 2+ charge. Similarly, the metallic elements of Group 3A lose three electrons to form cations with a 3+ charge. Notice that there is a pattern. The metallic elements in Groups 1A to 3A form cations with a charge that is the same as the element's group number. In each case, the cation formed is isoelectronic with one of the noble gases.

Recall that nonmetals gain electrons to form anions. Previously, we used the example of chlorine gaining one electron to complete its octet and form an anion with a 1− charge. The other members of Group 7A behave the same way, each gaining one electron to form an anion with a 1− charge. Next, consider the elements in Group 6A. Recall that each of these elements has six valence electrons. For an element of Group 6A to complete its octet, it must gain *two* electrons. In so doing, the element forms an anion with two more electrons than the atom from which it originated and, therefore, has a charge of 2−. Following this trend, the nonmetallic elements of Group 5A (nitrogen and phosphorus) form anions with a charge of 3−. The charge on the anions formed by nonmetallic main-group elements is equal to the group number of the element minus eight.

The rules for predicting charges on ions formed by the main-group elements can be summarized as follows:

> ### *Predicting Charge on an Ion Formed by a Main-Group Element*
> - **Cations**: Charge = group number;
>   for example, K = 1+, Mg = 2+, Al = 3+
> - **Anions**: Charge = group number − 8;
>   for example, Br (7 − 8) = 1−, O (6 − 8) = 2−, N (5 − 8) = 3−

For elements other than the main-group elements, the charge on the ions they form is not as simple to predict. These elements do not follow the octet rule. In fact, some of the transition metals form more than one ion. For example, both copper and iron can each form more than one cation. Copper forms the ions Cu^+ and Cu^{2+}, while iron forms Fe^{2+} and Fe^{3+}. Others, like silver (Ag) and zinc (Zn), do not vary their charge and form only Ag^+ and Zn^{2+} ions, respectively. Unlike the Group 1A–3A metals, it is not possible to determine the charge of the ions formed by the transition metals from their group number. Later, we will show how it is possible to predict the charges on these metal ions, but for now, common charges on some of the transition metal ions and main-group metals in Periods 5 and 6 are shown in Figure 3.4.

Another group of common ions whose charges cannot be determined from the periodic table are the **polyatomic ions.** These are groups of atoms that interact with each other to form an ion. Two common examples are HCO_3^- (bicarbonate) and NH_4^+ (ammonium). Notice that each ion contains more than one type of atom, but they interact to form an ion with a single charge. The charge shown is for the entire group of atoms. It is important to be able to recognize these polyatomic groups as ions because many are found in the body or have medical relevance. See **TABLE 3.2** on page 83 for a listing of some common polyatomic ions in health and medicine and their uses.

Sample Problem 3.2 Determining Protons and Electrons in Ions

How many protons and electrons are in the following ions?
  a. Mg^{2+} b. O^{2-}

Solution
Recall that the number of protons is represented by the atomic number of the element as shown on the periodic table.
  a. Protons = 12. An ion with a 2+ charge has two fewer electrons than protons or, in this example, 10 electrons.
  b. Protons = 8. An ion with a 2− charge has two more electrons than protons or, in this example, 10 electrons.

TABLE 3.2 Common Polyatomic Ion Names, Formulas, and Uses

Main Element	Formula of Ion[a]	Name of Ion	Common Uses
Hydrogen	OH^-	Hydroxide	Found in antacids
	H_3O^+	Hydronium	Measure of acidity (pH)
Nitrogen	NH_4^+	Ammonium	High levels indicate liver and kidney malfunction
	$\boxed{NO_3^-}$	**Nitrate**	Used as a preservative; large amounts have been linked to cancer
	NO_2^-	Nitrite	Added to cured meats to prevent botulism; large amounts have been linked to cancer
Chlorine	$\boxed{ClO_3^-}$	**Chlorate**	Used as a disinfectant
	ClO_2^-	Chlorite	Used as a disinfectant
Carbon	$\boxed{CO_3^{2-}}$	**Carbonate**	Lithium carbonate, used to treat depression; calcium carbonate, used in antacids
	HCO_3^-	Hydrogen carbonate (or bicarbonate)	Found in some antacids; also used to control the acid–base balance in blood
	CN^-	Cyanide	Poison that disrupts cellular respiration
	$C_2H_3O_2^-$	Acetate	Calcium acetate, used in patients with liver failure
	$C_3H_5O_3^-$	Lactate	Used in electrolyte fluids; also a by-product of anaerobic respiration
Sulfur	$\boxed{SO_4^{2-}}$	**Sulfate**	Magnesium sulfate (found in Epsom salts), used to reduce inflammation
	HSO_4^-	Hydrogen sulfate (or bisulfate)	Used in cleaning products
	SO_3^{2-}	Sulfite	Used as a preservative in wines and dried fruits
	HSO_3^-	Hydrogen sulfite (or bisulfite)	Used as a preservative in wines and dried fruits
Phosphorus	$\boxed{PO_4^{3-}}$	**Phosphate**	Referred to as inorganic phosphate, or P_i, used by the body for energy transfer
	HPO_4^{2-}	Hydrogen phosphate	Used to control acid–base balance in cells
	$H_2PO_4^-$	Dihydrogen phosphate	Used to control acid–base balance in cells
	PO_3^{3-}	Phosphite	Applied to crops as fungicide
Manganese	MnO_4^-	Permanganate	Used in care of wounds

[a]Boxed formulas with bold ion names are the most common polyatomic ion for that element.

Naming Ions

To distinguish an ion and the atom from which it was formed, we give the ion a different name. For metal ions, this simply involves adding the word *ion* to the name of the metal, so Na^+ is called sodium *ion*. Transition metals that form more than one ion use a Roman numeral in parentheses following the name of the metal to indicate the charge on the ion. So, Fe^{2+} is the iron(II) ion, and the Fe^{3+} ion is the iron(III) ion.

For the nonmetals, the suffix *-ide* replaces the last few letters of the name of the element. For example, the anion formed from fluorine is fluor*ide* and that of oxygen is ox*ide*. In most cases, the first syllable of the element is kept and the *-ide* suffix is substituted for the dropped letters.

Most of the common polyatomic ions end in *-ate* (see Table 3.2). The *-ite* ending is used for the names of related ions that have one fewer oxygen atom. By recognizing these endings, you can identify when a polyatomic ion is present in a compound. The hydroxide (OH^-), hydronium (H_3O^+), ammonium (NH_4^+), and cyanide ions (CN^-) are exceptions to this naming pattern. Just as with vocabulary words, you may have to commit to memory the formula and charges associated with the polyatomic ions. By memorizing the formulas in the boxes, the other related ions can be derived. For example, the sulfate ion is SO_4^{2-}; the formula for the sulfite ion with one fewer oxygen atom is therefore SO_3^{2-}.

Name the following ions:

a. Ba^{2+}

b. S^{2-}

c. Fe^{2+}

d. ClO_3^-

Solution

a. Barium is a main-group metal, so the name will be the metal name plus the word *ion*. The name is *barium ion*.

b. Sulfur is a nonmetal, so the ending *-ide* will replace the current ending. The name is *sulfide*.

c. Iron is a transition metal, so it is necessary to identify the charge using a Roman numeral after the metal name. The name is *iron(II) ion*.

d. This is a polyatomic ion. The name is *chlorate*.

INTEGRATING Chemistry

Find out which ▶ cations and anions are important in your body.

Sports drinks contain water and important ions called electrolytes necessary during prolonged exercise.

Important Ions in the Body

A number of ions found in the fluids and cells of the human body perform important functions. The main cations found in the body include Na^+, K^+, Ca^{2+}, and Mg^{2+}, which are important in maintaining solution concentrations inside and outside of cells. The main anions are Cl^-, HCO_3^-, and HPO_4^{2-}. These ions help maintain the charge neutrality between the blood and fluids inside cells, and are often referred to as electrolytes, with sodium being the main electrolyte. During periods of extreme exercise (three or more hours of continuous exercise), it is important to consume electrolytes in addition to rehydrating with water.

Another important ion in the body is the phosphate ion, also called inorganic phosphate and abbreviated P_i. It is involved in cellular energy transfer during chemical reactions.

In contrast to ions in fluids, ions can also form hard minerals. Tooth enamel contains the mineral hydroxyapatite, consisting of calcium (Ca^{2+}), phosphate (PO_4^{3-}), and hydroxide (OH^-) ions. **TABLE 3.3** lists some of the main ions in bodily fluids and their functions.

TABLE 3.3 Biologically Important Ions

Ion	Function	Sources
Cations		
Na^+	Regulates fluids outside cells	Table salt, seafood
K^+	Maintains ion concentration in cells; induces heartbeat	Dairy, bananas, meat
Ca^{2+}	Found outside cells; involved in muscle contraction, formation of bones and teeth; regulates heartbeat	Dairy, whole grains, leafy vegetables
Mg^{2+}	Found inside cells; involved in transmission of nerve impulses	Nuts, seafood, leafy vegetables
Fe^{2+}	Found in the protein hemoglobin, which is responsible for oxygen transport from lungs to tissue	Liver, red meat, leafy vegetables
Anions		
Cl^-	Found in gastric juice and outside cells; involved in fluid balance in cells	Table salt, seafood
HCO_3^-	Controls acid–base balance in blood	Body produces own supply through breathing and breakdown of foods
HPO_4^{2-}	Controls acid–base balance in cells	Fish, poultry, dairy

Practice Problems

3.7 What is the difference between a metal atom and its cation?

3.8 What is the difference between a nonmetal atom and its anion?

3.9 How are the names of a nonmetal atom and its anion different?

3.10 How are the names of a transition metal atom and its cation different?

3.11 Provide the charge when an ion is formed from each of the following:

    a. calcium   b. oxygen

    c. nitrogen   d. potassium

3.12 Provide the charge when an ion is formed from each of the following:

    a. sodium   c. magnesium

    b. phosphorus   d. fluorine

3.13 How many protons and electrons are present in the following ions?

    a. Mg^{2+} b. O^{2-} c. Cl^{-} d. Cu^{2+}

3.14 How many protons and electrons are present in the following ions?

    a. Al^{3+} b. Br^{-} c. Hg^{+} d. N^{3-}

3.15 Name the ions in Problem 3.13.

3.16 Name the ions in Problem 3.14.

3.17 Give the name and symbol of the ion with the following number of protons and electrons:

    a. 35 protons, 36 electrons

    b. 27 protons, 25 electrons

3.18 Give the name and symbol of the ion with the following number of protons and electrons:

    a. 12 protons, 10 electrons

    b. 26 protons, 23 electrons

3.19 Name the following ions:

    a. NH_4^{+} b. $C_2H_3O_2^{-}$ c. CN^{-}

3.20 Name the following ions:

    a. Cu^{2+} b. SO_4^{2-} c. HPO_4^{2-}

DISCOVERING THE CONCEPTS

❓ INQUIRY ACTIVITY—Ionic Compounds

Part 1. Information

Ionic compounds form between elements that can give or take electrons (exist as cations or anions) to yield a full valence shell. In an ionic compound, the element forming the cation gives its electron(s) to the element forming the anion and a strong attraction of opposite charge is formed between the two. This type of chemical bond is called an *ionic bond*.

    Examine the following data set. Notice that none of the compounds have a net charge (+ and − charges add up to zero).

Data Set

Formula	Name
NaI	Sodium iodide
K_2S	Potassium sulfide
CaF_2	Calcium fluoride
Mg_3N_2	Magnesium nitride
$CuBr_2$	Copper(II) bromide
Fe_2O_3	Iron(III) oxide
FeO	Iron(II) oxide
AlP	Aluminum phosphide

—Continued next page

Continued—

Questions

1. In naming ionic compounds, which goes first in the name, the metal or the nonmetal? The anion or the cation?
2. How are anions and cations the same? How are they different?
3. a. In potassium sulfide, the charge on the potassium ion is _____ and the charge on the sulfide is _____.
   b. Why does the formula for potassium sulfide contain two potassium ions and one sulfide ion?
4. Some of the names of ionic compounds have Roman numerals after the metals and others do not. Consider their position on the periodic table. How do these two groups of metals differ?
5. a. The charge on one Fe in Fe_2O_3 is _____.
   b. The charge on the Fe in FeO is _____.
6. a. Based on your answer to question 5, what does the Roman numeral in the names in the data set represent?
   b. Based on your answer to question 4, when is the Roman numeral used in the name of an ionic compound?
7. How does the name of the nonmetal (anionic) part of the compound differ from that of its original element?

Part 2. Information

Polyatomic ions are groups of nonmetal atoms that together act like an anion or cation. When combined with a cation or anion, respectively, they form ionic compounds.

Formula	Name	Formula	Name
NH_4^+	Ammonium	OH^-	Hydroxide
$C_2H_3O_2^-$	Acetate	HPO_4^{2-}	Hydrogen phosphate
HCO_3^-	Bicarbonate	NO_3^-	Nitrate
CO_3^{2-}	Carbonate	SO_4^{2-}	Sulfate

8. Complete the following table:

Formula	Name
KCl	
	Sodium sulfate
$Fe(OH)_3$	
$NaHCO_3$	
	Ammonium nitrate
	Silver acetate
$CaHPO_4$	
	Mercury(II) chloride

3.3 Ionic Compounds—Electron Give and Take

? 3.3 Inquiry Question

How do ionic compounds form?

Previously, we saw that main-group metal atoms lose electrons, forming cations to achieve an octet, whereas nonmetal atoms gain electrons, forming anions for the same reason. Where do the electrons go, or where do they come from? Ion formation does not occur in isolation. In other words, cations do not form unless anions are also formed and vice versa. When a metal and a nonmetal combine, a compound is formed and electrons are transferred between the atoms, forming oppositely charged ions as shown in **FIGURE 3.5**.

FIGURE 3.6 demonstrates this process for the formation of common table salt, sodium chloride. Because opposite charges attract, the newly formed cation and anion are strongly attracted to each other. This attraction of cation and anion is called an **ionic bond.** The resulting combination of cation and anion held together by this strong attraction is called an **ionic compound.**

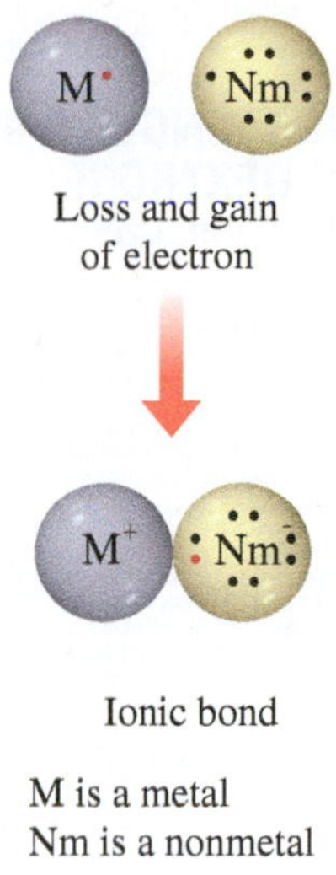

Electron transfer   Ions formed   Ionic bond

Sodium and   Sodium and   Sodium chloride, NaCl
chlorine atoms   chloride ions

FIGURE 3.6 **Formation of the ionic compound sodium chloride.** Both sodium and chlorine are able to complete their octet by giving and taking electrons, respectively.

Loss and gain
of electron

Ionic bond

M is a metal
Nm is a nonmetal

FIGURE 3.5 **Ion formation.** When an ion forms, the metal transfers its electron(s) to the nonmetal.

Formulas of Ionic Compounds

In the example shown in Figure 3.6, a sodium atom gives up one electron to a chlorine atom to form a sodium ion (charge = 1+) and a chloride ion (charge = 1−). The ions now have completed octets of valence electrons through this give and take of the electron, and because they are oppositely charged, the ions are attracted to each other, forming an ionic bond. When ions combine, they do so to form compounds with an overall charge of zero, as shown at left.

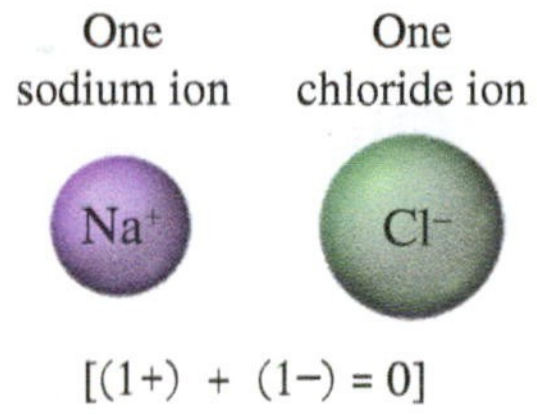

Sodium chloride, NaCl, is an ionic compound because it contains a metal combined with a nonmetal, but the formula does not indicate the charges on the ions. The formula is written with no charges shown. For sodium chloride, only one sodium ion (Na^+) and one chloride ion ($Cl^−$) combine to form a neutral compound. When writing formulas for chemical compounds, we represent the number of each particle (ions in this case) in the compound with a subscript. The "1" is understood in the formula for sodium chloride and is written as NaCl instead of Na_1Cl_1.

How would we write a formula if one of the ions has a greater and still opposite charge to the other? Let's look at how calcium ions and chloride might combine. Remember that cations and anions combine so that the resulting compound has a net charge of zero. No charges appear in the compound's formula.

We saw that calcium forms ions with a 2+ charge, and chlorine forms ions with a 1− charge. When calcium and chlorine form an ionic compound, calcium must give away two electrons, but chlorine can accept only one. Therefore, a calcium atom forms ionic bonds with *two* chlorine atoms, giving up one electron to each of the chlorine atoms to form a calcium ion (Ca^{2+}) and two chloride ions ($Cl^−$). The resulting compound, calcium chloride, is used as road salt to deice roads. **FIGURE 3.7** demonstrates the formation of the ionic compound from calcium and chlorine.

One   One
sodium ion   chloride ion

Na^+ Cl^-

$[(1+) + (1−) = 0]$

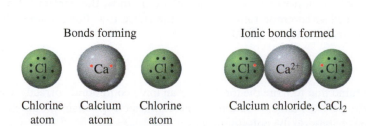

Bonds forming   Ionic bonds formed

Chlorine Calcium Chlorine   Calcium chloride, $CaCl_2$
atom   atom   atom

FIGURE 3.7 **The formation of an ionic compound from calcium and chlorine.** Calcium gives up two electrons, and each chlorine takes one of these electrons, forming ions.

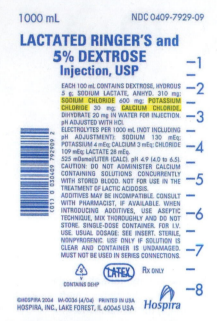

Electrolyte solutions used in IV drips like Lactated Ringer's contain many ionic compounds.

Once again the charges on the ions that were formed and the resulting ionic bond formation result in a compound whose net (overall) charge is zero as shown:

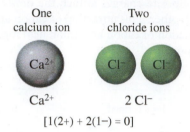

One calcium ion

Two chloride ions

Ca^{2+}

Cl^- Cl^-

Ca^{2+}

$2 \, Cl^-$

$[1(2+) + 2(1-) = 0]$

Because the neutral ionic compound has one calcium ion and two chloride ions, the correct formula is $CaCl_2$.

If we examine the formulas more closely, we see that the subscript on each ion is the magnitude of the charge (number without the sign) on the other ion. Using $CaCl_2$ as an example, the calcium ion has a charge magnitude of 2. Notice that the subscript on Cl in the formula is 2. Similarly, chloride has a charge magnitude of 1, and the subscript on Ca in the formula is also 1.

In many cases, this pattern can be used as a check that you have determined the correct formula for an ionic compound.

Solving a Problem

Checking a Chemical Formula

Determine the correct formula for the ionic compound formed from aluminum ions and oxide ions.

STEP 1 **Determine the charge of the ions from the periodic table if they are main-group elements.** In this case, aluminum is in Group 3A and its ion has a 3+ charge, and oxygen is in Group 6A, and the oxide ion has a 2− charge.

STEP 2 **Combine the ions in a formula so that the total charge present sums to zero.** In this case, two aluminum ions would provide a total 6+ charge, and three oxide ions would provide a total 6− charge.

$(2 \times (3+)) + (3 \times (2-)) = 0$. The formula is written as Al_2O_3.

STEP 3 **Check your formula by examining the charges on the ions and subscripts in the formula.**

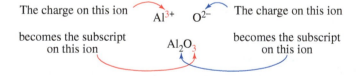

The charge on this ion Al^{3+} O^{2-} The charge on this ion

becomes the subscript on this ion Al_2O_3 becomes the subscript on this ion

When writing ionic formulas, in many cases, the charge on the anion becomes the subscript on the cation and vice versa. For ionic compounds, the correct formula always contains the smallest common ratio of ions in the subscripts. For example, the formula for aluminum oxide is Al_2O_3, not Al_4O_6.

Keep in mind that to write the formula for an ionic compound correctly, you must be able to use the periodic table to determine the charges on the main-group elements and remember the charges on the polyatomic ions.

Sample Problem 3.4	**Writing Formulas for Ionic Compounds of Main-Group Elements**

Determine the correct formula of the ionic compound formed from the following combination of ions:

  a. sodium ions and sulfide ions   b. magnesium ions and oxide ions

  c. calcium ions and hydroxide ions

Solution

a.

STEP 1 Determine the charge. Sodium ions have a 1+ charge, Na^+, and sulfide ions have a 2− charge, S^{2-}.

STEP 2 Combine the ions in a formula. Two sodium ions (total charge 2+) combine with one sulfide (total charge 2−) to form the neutral ionic compound sodium sulfide. The two sodium ions are represented in the formula by the subscript 2 after the symbol for sodium, Na. The formula is written Na_2S.

STEP 3. Check your formula.

$$\left(Na^+ \quad S^{2-} \right) = Na_2S_1 \longrightarrow Na_2S$$

b.

STEP 1 Determine the charge. Magnesium ions have a charge of 2+, Mg^{2+}, and oxide ions have a charge of 2−, O^{2-}.

STEP 2 Combine the ions in a formula. Because they have the same charge, one magnesium ion (total charge 2+) will combine with one oxide (total charge 2−) to form the neutral ionic compound magnesium oxide. The formula is written MgO.

STEP 3 Check your formula.

$$\left(Mg^{2+} \quad O^{2-} \right) = Mg_2O_2 \longrightarrow MgO$$

This formula reminds us of an important exception. When the charges on the ions are equal but opposite, the correct formula should contain the smallest common ratio (here 1:1). The correct formula for this compound combines one Mg^{2+} ion and one O^{2-} ion in a neutral compound with the formula MgO.

c.

STEP 1 Determine the charge. Calcium ions have a 2+ charge, Ca^{2+}, and the polyatomic ion, hydroxide, has a 1− charge, OH^-.

STEP 2 Combine the ions in a formula. One calcium ion (total charge 2+) will combine with two hydroxides (total charge 2−) to form the neutral ionic compound calcium hydroxide. In this case, parentheses are used around the hydroxide to indicate that the compound contains multiple polyatomic ions. The formula is written $Ca(OH)_2$.

STEP 3 Check your formula.

$$\left(Ca^{2+} \quad OH^- \right) = Ca_1(OH)_2 \longrightarrow Ca(OH)_2$$

So far, we have written formulas for ionic compounds containing main-group elements. Next, we address transition metals and metals residing in groups 4A and 5A, where the charges often vary. If given the formula, can we predict the charge on a transition metal in a compound? To demonstrate, let's work through an example in the following Solving a Problem feature.

Solving a Problem

Predicting Number and Charge in an Ionic Compound

Predict the number and charge on the ions present in the compound CuO.

We can determine the transition metal charge in an ionic compound by remembering a few rules.

- **RULE 1.** The formula of an ionic compound has no net charge (+ and − charges add up to zero).
- **RULE 2.** The charge of the anion is known.

Apply rule 1 and rule 2. In the case of CuO, an oxygen anion (Group 6A) has a 2− charge. The formula tells us that the compound contains one copper ion and one oxide. Because the + charge of the copper ion must equal the negative charge of the oxide for the compound to have no net charge, the charge of the copper is 2+.

Because the number of charges on the anion and cation are the same (2) and the lowest ratio is 1:1, in this case, the charges are *not* the subscripts on the formula.

Sample Problem 3.5   Predicting Charges of Ions

Determine the number and charge of each ion present in the following formulas:
  a. KBr, used to treat epilepsy in dogs
  b. $Al_2(SO_4)_3$, used in water purification
  c. $CoCl_2$, used in the synthesis of vitamin B_{12}

Solution

  a. For the main-group elements, the charge can be determined from the periodic table. Potassium is in Group 1A, so its charge as an ion is 1+. Bromine is in Group 7A, so its charge as an ion is 1−. The ionic compound of potassium and bromide contains one potassium ion (K+) and one bromide (Br−).
  b. Aluminum is in Group 3A, so its charge as an ion is 3+. This compound contains sulfate, a polyatomic ion (see Section 3.2). This group of atoms in sulfate functions together as a single ion and has a 2− charge as an anion. The formula for the ionic compound contains two aluminum ions and three sulfate ions.
  c. Cobalt is a transition metal, so we must also determine its charge by applying the rules previously discussed. Chlorine is an element in Group 7A, so its charge is 1−. Because the compound combines two chlorides (total charge 2−) with one cobalt ion, the cobalt must have a 2+ charge for the compound to have a net charge of zero. The compound contains one cobalt(II) ion (Co^{2+}) and two chloride ions (Cl^-).

Naming Ionic Compounds

Now that we can write formulas for ionic compounds, the next step is to name them. To do this, put the names of the two ions together, always remembering to list the cation first. So, MgO is magnesium oxide and NaCl is sodium chloride. Notice that you do not include the word *ion*—sodium ion, for example—in the name. For transition metals, recall that a Roman numeral is used to designate the charge on the transition metal because it can vary depending on the compound. The example compound CuO would be written as copper(II) oxide because the charge on the copper ion is 2+.

If a polyatomic ion is present, the name of the ion remains unchanged in the name of the compound. For example, $Ca_3(PO_4)_2$ is named calcium phosphate.

Sample Problem 3.6	**Naming Ionic Compounds**

Name the ionic compounds represented by the formulas given.
a. $BaSO_4$, used to image intestines
b. MnO_2, used in disposable batteries
c. SnF_2, used to strengthen teeth
d. NH_4Cl, used to treat electrolyte imbalances

Solution

a. Barium is in Group 2A, so it has a 2+ charge. The polyatomic anion sulfate has a 2− charge. The two ions combine one-to-one. The compound is barium sulfate.
b. This compound contains the transition metal manganese. Before naming this compound, we must determine the charge on the manganese ion. Because an oxygen ion (oxide) always has a 2− charge, and two oxides (total 4− charge) combine with one manganese ion, the manganese ion has a 4+ charge. (This gives the compound no net charge.) This compound is manganese(IV) oxide.
c. Tin (Sn) is a metal in group 4A, which requires the use of a Roman numeral. Fluoride has a 1− charge. Because two fluorides (total charge 2−) combine with one tin ion, the tin ion must have a 2+ charge. The compound is tin(II) fluoride.
d. The cation is a polyatomic ion, NH_4^+, the ammonium ion, and the anion is Cl^-, the chloride ion. The name of the compound is ammonium chloride. Note that this is an example of an ionic compound that does not contain a metal combined with a nonmetal. Ammonium compounds are ionic compounds that do not contain a metal ion.

Practice Problems

3.21 Give the formula and name for the ionic compounds formed from the following pairs of ions:

a. gold(III) and chloride, an antimicrobial
b. calcium and sulfate, used in plaster casts
c. magnesium and hydroxide, the main compound in milk of magnesia

3.22 Give the formula and name for the ionic compounds formed from the following pairs of elements:

a. potassium and chloride, the main ingredients in salt substitutes used by people on low-sodium diets
b. zinc and oxide, used in some sunscreens *(Note: The charge on zinc does not vary; see Figure 3.4.)*
c. copper(I) and sulfide, a common copper mineral mined for copper metal

3.23 Give the number of each type of ion present in each of the ionic compounds in Problem 3.21.

3.24 Give the number of each type of ion present in each of the ionic compounds in Problem 3.22.

3.25 Give the formula and name for the ionic compound formed from the combination of the indicated metal and carbonate, CO_3^{2-}, a polyatomic ion.

a. Na   b. Fe(II)   c. Al

3.26 Repeat Problem 3.25 using the acetate, $C_2H_3O_2^-$, ion.

3.27 Give the number of each type of ion present in each of the ionic compounds in Problem 3.25.

3.28 Give the number of each type of ion present in each of the ionic compounds in Problem 3.26.

3.4 In Search of an Octet, Part 2: Covalent Bond Formation

In Section 3.2, we saw that an atom can achieve a stable octet of electrons by forming an ion. Cations are usually formed from metals, and anions are formed from nonmetals. While such ionic compounds are abundant in nature in nonliving things like rocks and minerals, compounds containing mostly nonmetal components are most abundant in living things.

Consider carbon dioxide, CO_2. This compound is a product of cellular respiration and is used by plants to manufacture their food source. Carbon dioxide is everywhere. Notice that CO_2 contains one atom of carbon (a nonmetal) and two atoms of oxygen

3.4 Inquiry Question

How do covalent compounds form?

Nonliving minerals are mostly composed of ionic and metallic compounds.

Living organisms are mostly composed of nonmetal compounds.

(also a nonmetal). When nonmetals like carbon and oxygen combine, they do *not* form ionic compounds.

Earlier in this chapter, we noted that the Group 4A elements like carbon do not form ions. But carbon *does* form compounds with other elements. How does it form compounds without forming ions?

Nonmetals combine by *sharing* valence electrons to achieve an octet. This sharing of electrons results in the formation of a **covalent bond.** Unlike an ionic bond, no electrons are transferred between the atoms in a covalent bond; instead, electrons are shared, making both atoms more stable. In covalent bonding, the valence electrons in the bond belong to both of the atoms.

When atoms share electrons to form covalent bonds, the resulting new compound is called a **covalent compound.** The smallest (or fundamental) unit of a covalent compound is a **molecule.** So, carbon dioxide, CO_2, is a molecule and contains atoms held together by covalent bonds.

Sample Problem 3.7   **Distinguishing Covalent and Ionic Compounds**

Determine whether each of the following is a covalent or ionic compound:
a. $AgNO_3$, used to treat warts
b. CS_2, used as an insecticide
c. NH_4CO_3, used in smelling salts

Solution

a. $AgNO_3$ contains a metal (silver) and a nonmetal group (nitrate). Compounds formed from the combination of a metal and a nonmetal are ionic.
b. CS_2 is made up of carbon (a nonmetal) and sulfur (also a nonmetal). Compounds formed from the combination of nonmetals are covalent compounds.
c. Initial inspection of the formula NH_4CO_3 shows that all the atoms involved are nonmetals (N, H, C, and O). However, in Section 3.2, we saw that both NH_4^+ and CO_3^{2-} are polyatomic ions. Even though this compound is composed entirely of nonmetals, it is an ionic compound because it contains polyatomic ions.

Covalent Bond Formation

Atoms share electrons with other atoms, forming covalent bonds to complete an octet, but how many covalent bonds does an atom form? A simple rule of thumb is this: *The number of covalent bonds that an atom will form equals the number of electrons necessary to complete its octet.*

We can see how this works with atoms by using dots to represent valence electrons. The use of electron-dot symbols was first developed by chemist G. N. Lewis to show how atoms share electrons in covalent bonding. The electron-dot symbol for any atom consists of the elemental symbol plus a dot for each valence electron. The electron-dot symbol for chlorine is as follows:

We know that chlorine, a member of Group 7A, has seven valence electrons. We construct its electron-dot symbol by placing one electron on each side of the elemental symbol (top, bottom, left, and right) and then adding the remaining electrons to form pairs until all the valence electrons are represented. After the first four electrons are in place, each of the remaining electrons can be placed with any one of the existing electrons to form pairs. So, these two electron-dot symbols for chlorine are equivalent:

$$:\ddot{C}l\cdot \quad \text{or} \quad \cdot\ddot{C}l:$$

To complete its octet through a covalent bond, chlorine must share its unpaired electron with another nonmetal atom, here a chlorine atom. A shared pair of electrons, known as a **bonding pair,** or simply a *bond,* is represented as a dash or line connecting the two electron-dot symbols (see **FIGURE 3.8**). Note that each of the atoms in the newly formed

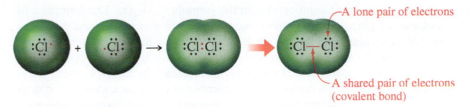

A lone pair of electrons

A shared pair of electrons
(covalent bond)

FIGURE 3.8 Formation of a covalent bond between two chlorine atoms.

molecule now has an octet—three **lone pairs** of electrons, which are not shared, and the shared bonding pair.

A glance at the properly constructed electron-dot symbols of the main-group non-metals in the second period of the periodic table helps to determine the preferred covalent bonding patterns for these elements.

Ignoring carbon for the moment, from the electron-dot symbols, we see that nitrogen forms three covalent bonds, oxygen forms two covalent bonds, and fluorine, like its group member chlorine, forms one covalent bond. Note that hydrogen is an exception to the octet rule. As a member of Period 1, hydrogen requires only two electrons to complete its valence shell; therefore, hydrogen forms just one covalent bond.

Now let's focus on carbon, the element on which life on Earth is based. With its four unpaired valence electrons, carbon must share its four electrons and form four covalent bonds to complete its octet. To accomplish this, carbon can share one electron each with four other atoms to form four **single bonds**. Or it can form **double bonds** by sharing two of its electrons with one atom. Carbon can also share three of its electrons with one atom to form a **triple bond**. The most common covalent bonding patterns observed for carbon are shown in **FIGURE 3.9**. Note that in each case carbon has four bonding pairs—a complete octet—surrounding it.

TABLE 3.4 shows the common bonding patterns for the majority of the main-group nonmetals that are encountered in living systems.

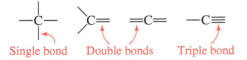

Single bond   Double bonds   Triple bond

FIGURE 3.9 Preferred bonding patterns used by carbon to complete its octet when forming bonds in a molecule.

Formulas and Lewis Structures of Covalent Compounds

Like the chemical formulas for ionic compounds, the **molecular formula** for a covalent compound completely identifies *all* the components in a molecule. Glucose, the sugar

TABLE 3.4 Preferred Covalent Bonding Patterns for Main-Group Elements

Group 1A	Group 4A	Group 5A	Group 6A	Group 7A
H—	—C̈—	—N̈—	—Ö—	:F̈—
	>C=	—N̈=	Ö=	
	=C=	:N≡		
	—C≡			
			—S̈—	:C̈l—
				:B̈r—
				:Ï—

we use for energy, is a covalent compound with the formula $C_6H_{12}O_6$. The formula tells us that a molecule of glucose has 6 carbon atoms, 12 hydrogen atoms, and 6 oxygen atoms. Note that molecular formulas do *not* reduce to the smallest whole-number ratio like the chemical formulas for ionic compounds, and so $C_6H_{12}O_6$ would not be reduced to CH_2O—they are different formulas, for glucose and formaldehyde, respectively.

While the molecular formula does tell us the number of atoms in a molecule, it does not tell us *how* the atoms are joined together. The formula does not show the structure. The electron-dot symbols help us visualize the bonding in the molecule.

Consider natural gas, also known as methane, with a molecular formula of CH_4. This is a covalent compound (all nonmetals). Its molecular formula only tells us that methane has one carbon atom and four hydrogen atoms. How are the atoms connected? Start by drawing the electron-dot symbols for all the atoms in the molecule. Because carbon has four unpaired valence electrons, it must make four bonds to complete its octet. Each of the hydrogen atoms has one valence electron and requires two to complete its valence shell. Therefore, each hydrogen atom makes only one bond.

So, carbon forms four bonds, and each of the *four* hydrogen atoms forms one bond. Connecting the atoms as shown fulfills these requirements. Recall that a line connecting two atoms represents a bonding pair of electrons,

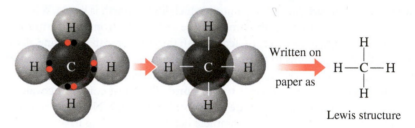

The drawing to the far right shows the number and type of each atom in the molecule and their connectivity. This representation is called a **Lewis structure.**

Solving a Problem

Drawing Lewis Structures for Covalent Compounds

Draw the correct Lewis structure for NH_3.

STEP 1 Draw the correct electron-dot symbols. For each of the atoms in the formula, draw its electron-dot symbol.

$$\cdot \overset{\cdot \cdot}{\underset{\cdot}{N}} \cdot \qquad \cdot H \qquad \cdot H \qquad \cdot H$$

STEP 2 Line up the atoms. Place the atoms on the page so that unpaired electrons on adjacent atoms are next to each other, much as you line up the pieces of a puzzle to fit them together.

$$H \! : \! \overset{\cdot \cdot}{\underset{\cdot \cdot}{N}} \! : \! H$$
$$\underset{H}{}$$

STEP 3 Draw the Lewis structure. Replace the pairs of dots between atoms with lines to represent the bonds connecting the atoms.

$$H \! - \! \overset{\cdot \cdot}{N} \! - \! H$$
$$\underset{|}{}$$
$$H$$

STEP 4 Confirm valence shells are complete. Make sure that each atom has a complete octet except hydrogen, which requires only two electrons to complete its valence shell. If any atoms lack a complete valence shell, make multiple bonds as necessary. Here, nitrogen has three bonds ($3 \times 2 \text{ e}^- = 6 \text{ e}^-$) and one lone pair ($2 \text{ e}^-$), for a total of eight electrons, and so has a complete octet. Each hydrogen makes one bond to complete its valence shell with two electrons.

STEP 5 Check your structure. Compare your structure to Table 3.4 to confirm that it is one of the preferred bonding patterns. The table shows that a structure with three bonds and one lone pair is a preferred bonding pattern for nitrogen.

■

Sample Problem 3.8   Drawing Lewis Structures for Covalent Compounds

Draw the correct Lewis structure for each of the following covalent compounds:

a. C_2H_4 b. CCl_4

Solution

a.

STEP 1 Draw the correct electron-dot symbols.

$$\cdot \ddot{C} \cdot \quad \cdot \ddot{C} \cdot \quad \cdot H \quad \cdot H \quad \cdot H \quad \cdot H$$

STEP 2 Line up the atoms. In this case, two carbon atoms are present, and since each hydrogen atom can form only one bond, the two carbons must bond to each other. If you were to arrange the atoms using just single bonds, you would get

Two unpaired
carbon electrons

$$\begin{array}{c} \text{H} \overset{\displaystyle \cdot}{\underset{\displaystyle \cdot \cdot}{\text{C}}} \overset{\displaystyle \cdot}{\underset{\displaystyle \cdot \cdot}{\text{C}}} \text{H} \\ \text{H} \quad \text{H} \end{array}$$

STEP 3 Draw the Lewis structure. This would give a hypothetical Lewis structure with each carbon having three bonds and an unpaired electron.

STEP 4 Confirm valence shells are complete. In this structure, each carbon only has seven valence electrons, not eight.

$$\text{H} - \overset{\displaystyle \cdot}{\underset{\displaystyle |}{\text{C}}} - \overset{\displaystyle \cdot}{\underset{\displaystyle |}{\text{C}}} - \text{H}$$
$$\quad\quad \text{H} \quad \text{H}$$

This means that each of the carbon atoms in this structure still has an unpaired electron *and* each has only seven valence electrons (three bonding pairs = 6 electrons + 1 unpaired electron = 7 electrons). In this book, carbon will always form four bonds to complete its octet. So, to complete their octets in this case, the carbons share their unpaired electrons to give

$$\begin{array}{cc} \text{H} & \quad \text{H} \\ \quad \diagdown & \diagup \\ \quad \text{C} = \text{C} \\ \quad \diagup & \diagdown \\ \text{H} & \quad \text{H} \end{array}$$

STEP 5 Check your structure. In Table 3.4, we find that the arrangement with two single bonds and one double bond is one of carbon's preferred bonding patterns.

b. Following the steps as outlined, the correct structure is

$$\begin{array}{c} :\ddot{C}l: \\ | \\ :\ddot{C}l - C - \ddot{C}l: \\ | \\ :\ddot{C}l: \end{array}$$

Naming Covalent Compounds

Naming ionic compounds is a relatively straightforward process—name the cation and then name the anion. Naming covalent compounds is a little different. Here we focus on naming the simplest covalent compounds and look at a few common exceptions.

Covalent compounds composed of only two elements are known as **binary compounds**, and can be named using a three-step procedure as follows.

Solving a Problem

Naming Covalent Compounds

Name the covalent compounds SO₂ and SO₃.

STEP 1 Name the first element in the formula. In both compounds, the first element is sulfur.

STEP 2 Name the second element in the formula and change the ending to *-ide*. In both compounds, the second element is oxygen, so its name would be oxide.

STEP 3 Designate the number of atoms of each element present using one of the Greek prefixes shown in TABLE 3.5. In the first compound, there are two oxides, and in the second compound, there are three. The two compounds are named as follows:

$$SO_2: \text{Sulfur dioxide}$$
$$SO_3: \text{Sulfur trioxide}$$

TABLE 3.5 Greek Prefixes Used When Naming Binary Covalent Compounds

Prefix	Meaning
mono-	1
di-	2
tri-	3
tetra-	4
penta-	5
hexa-	6
hepta-	7

It is important to state the number of atoms of each element present when naming covalent compounds because nonmetals can combine with each other in multiple ways. Notice that in both cases, the *mono*-prefix for sulfur was understood and was not included in the name. This exception (for *mono*) holds true for only the first element in the name. The compound CO is named carbon *mono*xide.

Some binary covalent compounds are so prevalent in our world that their long-standing traditional names have never been replaced by the rules-derived names. One of the most notable exceptions is H_2O. According to the naming steps, this compound is dihydrogen monoxide, but of course, we know it as water. Ammonia—a component of smelling salts, many household cleaners, and agricultural fertilizers—has a molecular formula of NH_3 and is also an exception. N_2O is an important binary covalent compound used as an inhalation anesthetic. This compound is called nitrous oxide or, commonly, laughing gas, instead of its rules-derived name, dinitrogen monoxide.

Sample Problem 3.9 Naming Binary Covalent Compounds

Give the correct name for each of the following binary covalent compounds:
a. CS_2, used in manufacturing artificial silk
b. NI_3, a highly sensitive contact explosive
c. $SiCl_4$, used in the manufacture of optical fibers

Solution

a. carbon disulfide   b. nitrogen triiodide   c. silicon tetrachloride

Sample Problem 3.10 Writing Formulas of Binary Covalent Compounds

Give the correct formula for each of the following binary covalent compounds:
a. dichlorine monoxide, a degradation product found in the ozone layer
b. phosphorus trichloride, used in the manufacture of flame retardants
c. dinitrogen pentoxide, a greenhouse gas

Solution

a. Cl_2O b. PCl_3 c. N_2O_5

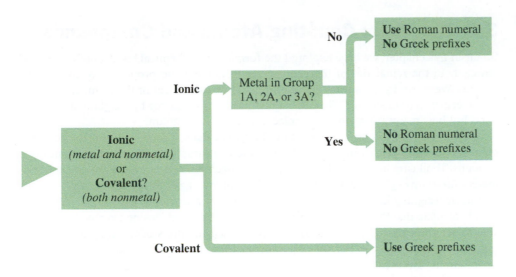

FIGURE 3.10 Flow chart for naming ionic and binary covalent compounds.

Although we have introduced the naming of ionic and binary covalent compounds as separate topics, in practice, before naming any compound, we must first determine whether the compound is ionic (metal with nonmetal) or covalent (all nonmetals). The flow chart in **FIGURE 3.10** offers guidance for the correct identification and naming of compounds.

Practice Problems

3.29 Draw the correct Lewis structure for each of the following covalent compounds:

a. PCl_3, used in the manufacture of herbicides

b. H_2Se, used in production of semiconductors

c. CS_2, an insecticide

d. C_2H_6, ethane, a minor component of natural gas

3.30 Draw the correct Lewis structure for each of the following covalent compounds:

a. H_2O

b. C_3H_8, propane, a fuel used for heating

c. N_2, nitrogen gas

d. SiF_4, used in making integrated circuits

3.31 Draw the correct Lewis structure for each of the following covalent compounds:

a. SiO_2, quartz

b. C_2H_4, ethylene, a plant-ripening hormone

c. O_2, molecular oxygen, the form we breathe

d. CH_2O, formaldehyde

3.32 Draw the correct Lewis structure for each of the following covalent compounds:

a. $CHCl_3$, chloroform, used in the production of Teflon

b. $AsCl_3$, a highly volatile poison

c. Cl_2, chlorine gas

d. C_2H_2, acetylene, used as a fuel in welding

3.33 Determine whether each of the following is a covalent or ionic compound:

a. NF_3 b. Fe_2O_3 c. Cs_2CO_3 d. PBr_3

e. C_5H_{12} f. NH_4OH g. SeO_2 h. BaS

i. CBr_4 j. OF_2

3.34 Determine whether each of the following is a covalent or ionic compound:

a. K_2O b. CO_2 c. GaF_3 d. C_2H_6

e. $NiCl_2$ f. $BaSO_4$ g. SO_2 h. N_2O

i. NH_4Cl j. SiH_4

3.35 Complete the table by supplying the missing name or formula for each covalent compound.

Formula	Name
	Nitrogen dioxide
PCl_3	
	Silicon tetrafluoride
H_2O	
	Nitrogen trichloride

3.36 Complete the table by supplying the missing name or formula for each covalent compound.

Formula	Name
N_2O_4	
	Carbon monoxide
P_2O_5	
	Carbon dioxide
NH_3	

3.5 The Mole: Counting Atoms and Compounds

720 g = 12 eggs = 1 dozen

So far in this chapter, we have explored the formation of chemical bonds, both ionic and covalent. In the remainder of the chapter, we examine some properties of these compounds. We begin by answering the question, "How can we count the number of atoms or molecules in a substance?" We can get the mass of a substance by weighing it on a balance, but how many atoms, ions, or molecules are in that amount?

Just as a *dozen* is a unit for counting things like eggs, a mole is a unit for counting atoms. The **mole** relates the mass of an element in grams to the number of atoms it contains. To illustrate, the number of eggs in 1 dozen is 12. A large egg weighs approximately 60 grams (g). We could determine how many eggs are present in a closed egg carton by weighing it. For example, if the contents of a carton of eggs weigh 720 g, we could calculate that there were 12 large eggs, or 1 dozen eggs, inside even without being able to see the eggs inside the carton. We have used the mass of the eggs to count the number of eggs. Similarly, exactly 12 g of the isotope carbon-12 contains a certain number of atoms; this is called a *mole*. By measuring a mass of a substance, we can count the number of atoms.

We cannot see atoms, but can relate the atomic mass from the periodic table to the number of atoms using the mole unit. For example, on the periodic table, carbon has an atomic mass of 12.01 amu. One mole of any element has a **molar mass** in grams numerically equal to the atomic mass of that element. Therefore, we can say the molar mass of carbon is 12.01 g per one mole of carbon atoms (12.01 g/mole). This is analogous to our earlier example with the eggs: a dozen eggs has a mass of 720 g (720 g/doz),

Avogadro's Number

The molar mass tells us the mass of one mole of a substance and corresponds to the mass of a single atom in amu. Just as we know that there are 12 objects in the unit dozen, the number of atoms present in the unit called the mole is experimentally found to be about 602,000,000,000,000,000,000,000 atoms. Wow! That is a large number. Huge numbers like this are awkward to handle. It is more convenient to express such a large number in scientific notation. The number of atoms in a mole can be represented in scientific notation as 6.02×10^{23} atoms.

The large number of atoms defined as one mole is known as **Avogadro's number (N)** in honor of the Italian physicist Amedeo Avogadro.

$$6.02 \times 10^{23} \text{ atoms} = 1 \text{ mole of atoms}$$

In general, for any substance, Avogadro's number is

$$N = 6.02 \times 10^{23} \text{ particles/mole}$$

We use the mass in grams for the carbon atoms to count the number of carbon atoms.

$$12.01 \text{ g C} = 1 \text{ mole of C} = 6.02 \times 10^{23} \text{ atoms of C}$$

Just as a dozen eggs, a dozen paper clips, and a dozen bowling balls have different masses, a mole of silver atoms and a mole of carbon atoms will not have the same mass. Since elements have different atomic masses, the same number of atoms of two different elements have different masses.

Let's solve a problem to see how atoms and moles are related to each other. Two quantities—like atoms and moles—that can be related to each other with an equal sign are referred to as **equivalent units**. From our definition of Avogadro's number, we know the equivalent unit relating atoms and moles is

$$6.02 \times 10^{23} \text{ atoms} = 1 \text{ mole of atoms}$$

This equivalent unit can be rearranged to

$$\frac{6.02 \times 10^{23} \text{ atoms}}{1 \text{ mole}} \quad \text{or} \quad \frac{1 \text{ mole}}{6.02 \times 10^{23} \text{ atoms}}$$

and can be used as a conversion factor when we solve a problem.

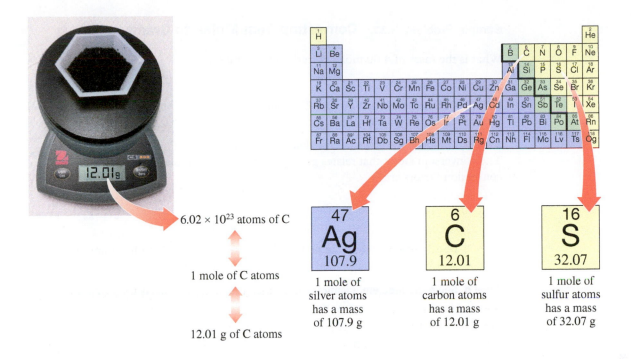

6.02 × 10²³ atoms of C

$\updownarrow$

1 mole of C atoms

$\updownarrow$

12.01 g of C atoms

47		6		16
Ag		**C**		**S**
107.9		12.01		32.07

1 mole of silver atoms has a mass of 107.9 g

1 mole of carbon atoms has a mass of 12.01 g

1 mole of sulfur atoms has a mass of 32.07 g

Converting Between Units

Solving a Problem

How many atoms are present in 2.00 moles of carbon?

To determine the answer to a problem like this, the problem can be set up using the strategy developed in Chapter 1 for converting units:

STEP 1 Determine the unit on your final answer. The unit will be atoms.

STEP 2 Establish the given information. You are given moles and need to find atoms. The conversion factor that relates these two is Avogadro's number. Two possible conversion factors are

$$\frac{6.02 \times 10^{23} \text{ atoms C}}{1 \text{ mole C}} \quad or \quad \frac{1 \text{ mole C}}{6.02 \times 10^{23} \text{ atoms C}}$$

STEP 3 Decide how to set up the problem. Because you are looking for atoms, atoms must appear in the numerator, and the other unit (mole) must cancel out.

STEP 4 Solve the problem.

The equation can be set up as:

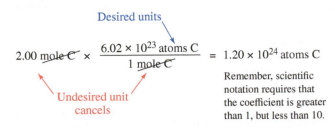

Desired units

$$2.00 \text{ mole C} \times \frac{6.02 \times 10^{23} \text{ atoms C}}{1 \text{ mole C}} = 1.20 \times 10^{24} \text{ atoms C}$$

Undesired unit cancels

Remember, scientific notation requires that the coefficient is greater than 1, but less than 10.

STEP 5 Check your answer. When solving a problem for the number of atoms, your answer should be a large number having a large positive exponent.

Sample Problem 3.11 Converting from Moles to Grams

What is the mass of 4.00 moles of carbon in grams?

Solution

STEP 1 Determine the unit on your final answer. Because the problem asks for mass, the unit on the final answer will be grams.

STEP 2 Establish the given information. You are given moles and asked to find grams. The conversion factor that relates grams and moles is the molar mass. Two possible conversion factors are

$$\frac{12.01 \text{ g C}}{1 \text{ mole C}} \quad \text{or} \quad \frac{1 \text{ mole C}}{12.01 \text{ g C}}$$

STEP 3 Decide how to set up the problem. Because you are looking for grams, grams must appear in the numerator, and the other unit (mole) must cancel out.

STEP 4 Solve the problem. Using the conversion factor that cancels the given unit (mole), we can set up the problem as

$$4.00 \text{ mole C} \times \frac{12.01 \text{ g C}}{1 \text{ mole C}} = 48.0 \text{ g C}$$

STEP 5 Check your answer. If 1 mole of carbon weighs about 12 g, then 4 moles should weigh four times that amount.

Sample Problem 3.12 Using Avogadro's Number to Find the Number of Atoms

a. How many atoms of aluminum are present in 1.00 g of aluminum?
b. How does this compare to the number of atoms in 1.00 g of lead?

Solution

STEP 1 Determine the unit on your final answer. In this case, the final unit will be atoms.

STEP 2 Establish the given information. To solve these two problems, we use two conversion factors, the molar mass for the element from the periodic table (located beneath the atomic symbol with units of g/mole) and the conversion factor for Avogadro's number to relate the mole to the number of atoms. The unit of Avogadro's number will be atoms per mole.

STEP 3 Decide how to set up the problem. Because you are looking for atoms, atoms must appear in the numerator, and the other units (moles and grams) must cancel out.

STEP 4 Solve the problem.

a. How many atoms of aluminum?

$$\underset{\substack{\uparrow \\ \text{Information} \\ \text{given}}}{1.00 \text{ g Al}} \times \underset{\substack{\uparrow \\ \text{Molar mass} \\ \text{(conversion factor)}}}{\frac{1 \text{ mole Al}}{26.98 \text{ g Al}}} \times \underset{\substack{\uparrow \\ \text{Avogadro's number} \\ \text{(conversion factor)}}}{\frac{6.02 \times 10^{23} \text{ atoms of Al}}{1 \text{ mole Al}}} = 2.23 \times 10^{22} \text{ atoms of Al}$$

b. How many atoms of lead?

$$1.00 \text{ g Pb} \times \frac{1 \text{ mole Pb}}{207.2 \text{ g Pb}} \times \frac{6.02 \times 10^{23} \text{ atoms of Pb}}{1 \text{ mole Pb}} = 2.91 \times 10^{21} \text{ atoms of Pb}$$

$\uparrow$ $\uparrow$ $\uparrow$

Information   Molar mass   Avogadro's number
given   (conversion factor)   (conversion factor)

STEP 5 Check your answer. The exponents in the two answers show that there are about 10 times more atoms in 1.00 g of aluminum than in 1.00 g of lead. Because aluminum atoms are lighter, it takes more aluminum atoms to obtain the same mass.

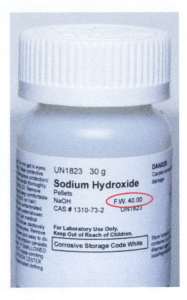

Molar Mass and Formula Weight

So far, all our calculations have been with atoms. We can also use Avogadro's number to determine the number of molecules or particles present in a sample of a compound. To do this, we must first determine the molar mass of the compound. Just as the atomic mass for one atom of an element is expressed in atomic mass units, the formula weight for a compound is the sum of the atomic masses. The molar mass of a compound is numerically equal to the formula weight with units of grams/mole, and this can be used to calculate the number of molecules or atoms in a given sample. Let's look at some examples, starting with calculating molar mass.

Chemical labels often display the formula weight of the compound.

Calculating Molar Mass for a Compound

Solving a Problem

Calculate the molar mass for (a) NaCl and (b) H_2O.

(a) The molar mass for NaCl is the sum of the atomic masses of the elements.

22.99	Atomic mass for Na
+35.45	Atomic mass for Cl
58.44 g/mole	Molar mass of NaCl

(b) The molar mass for H_2O is the sum of the atomic masses for the elements.

2 × 1.01	Atomic mass for two H
+16.00	Atomic mass for O
18.02 g/mole	Molar mass of H_2O

Sample Problem 3.13 Calculating Molar Mass for a Compound

Determine the molar mass for the compounds shown.
a. KOH   b. NH_3 c. $NaHCO_3$

Solution

a.

39.10	Atomic mass for K
16.00	Atomic mass for O
+1.01	Atomic mass for H
56.11 g/mole	Molar mass of KOH

b.

14.01	Atomic mass for N
+3 × 1.01	Atomic mass for three H
17.04 g/mole	Molar mass of NH_3

—Continued next page

Continued—

c.	22.99	Atomic mass for Na
	1.01	Atomic mass for H
	12.01	Atomic mass for C
	$+3 \times 16.00$	Atomic mass for three O
	84.01 g/mole	Molar mass of $NaHCO_3$

Note that because the numbers are added, the answers are reported to two decimal places. For a review, see significant figures in Section 1.3.

Now that we have calculated the molar mass of a compound, let's see how we can use this information to calculate the number of molecules in a specific amount of a compound. We will use water as our example.

Solving a Problem

Finding the Number of Molecules in a Sample

Determine the number of molecules of H_2O present in an ice cube of pure water that weighs 4.00 g.

PRELIMINARY STEP **Calculate the molar mass of the compound.** For this problem, before you use the steps outlined for solving a problem using conversion factors, you must determine the molar mass of the compound. As calculated in the previous Solving a Problem feature, the molar mass of H_2O is 18.02 g/mole.

STEP 1 **Determine the unit on your final answer.** In this case, the final unit will be molecules.

STEP 2 **Establish the given information.** We have the molar mass of water (grams/mole) and the conversion factor for Avogadro's number to relate the mole to the number of molecules. The unit of Avogadro's number will be molecules per mole.

STEP 3 **Decide how to set up the problem.** Because you are looking for molecules, molecules must appear in the numerator, and the other units (moles and grams) must cancel out.

STEP 4 **Solve the problem.**

$$4.00 \text{ g} \times \frac{1 \text{ mole}}{18.02 \text{ g}} \times \frac{6.02 \times 10^{23} \text{ molecules}}{1 \text{ mole}} = 1.34 \times 10^{23} \text{ molecules}$$

↑	↑	↑
Information given	Molar mass (conversion factor)	Avogadro's number (conversion factor)

A Unit from Biochemistry: The Dalton

We briefly bring up the dalton because it is commonly used in biochemistry as a unit to measure mass of biological molecules. The **dalton** (Da) is the same as the amu (Chapter 2). It is similar to molar mass in that it is a measure of the mass of a compound. It is mainly used in biochemistry to measure the weight of large molecules like proteins. It is defined as one-twelfth the mass of a carbon-12 atom, which, in large molar masses, gives approximately the same numerical value as molar mass.

Practice Problems

3.37 Compare (a) the number of atoms and (b) the number of grams present in 1.00 mole of silver (Ag) and 1.00 mole of gold (Au).

3.38 Compare (a) the number of atoms and (b) the number of grams present in 1.00 mole of helium and 1.00 mole of argon.

3.39 Calculate the following:

  a. the number of Na atoms in 0.250 mole of Na

  b. the number of moles in 4.0 g of Pb

  c. the number of atoms in 12.0 g of Si

3.40 Calculate the following:

a. the number of S atoms in 0.660 mole of S

b. the number of moles in 15.0 g of Fe

c. the number of atoms in 66.0 g of Cu

3.41 Determine the molar mass for the following compounds:

a. calcium sulfate, used in calcium supplements, $CaSO_4$

b. the carbohydrate glucose, $C_6H_{12}O_6$

c. hydrogen peroxide, a germicidal also used to whiten teeth, HOOH

3.42 Determine the molar mass for the following compounds:

a. ethylene, a ripening hormone, C_2H_4

b. citric acid, the sour taste found in citrus fruits, $C_6H_8O_7$

c. sodium bicarbonate, baking soda, $NaHCO_3$

3.43 Determine the number of molecules in a 325-mg dose of aspirin, molecular formula $C_9H_8O_4$.

3.44 Determine the number of molecules in a 200-mg dose of ibuprofen, molecular formula $C_{13}H_{18}O_2$.

DISCOVERING THE CONCEPTS

? INQUIRY ACTIVITY—Molecular Shape

Information

Electron pairs joining atoms (whether in single, double, or triple bonds) will get as *far away from other electron pairs as possible* when forming bonds in a molecule. Electrons in bonds will also repel lone pair electrons, and lone pairs will repel each other more than they repel electrons involved in bonds. This repulsion between electrons is the basis for the valence-shell electron-pair repulsion theory, or *VSEPR*. VSEPR allows us to predict the three-dimensional shape (also called molecular geometry) that a molecule will adopt.

Activity

Using toothpicks for bonds, a half toothpick for lone pairs on the central atom(s), large marshmallows for C, N, and O, and small marshmallows for H, build the five structures a–e. You will need 24 toothpicks, 8 large marshmallows, and 13 small marshmallows. Keep in mind the bonding patterns and Lewis structures outlined in Section 3.4.

a. Make a model of CO_2, being sure to get all bonds (toothpicks) as far away from each other as possible. What is the bond angle between O—C—O?

b. Make a model of H_2CCH_2, being sure to get all bonds (toothpicks) as far away from each other as possible. What is the bond angle between H—C—H?

c. Make a model of CH_4, being sure to get all bonds (toothpicks) as far away from each other as possible. Do you think the bond angle between H—C—H less than, equal to, or greater than 90°?

d. Make a model of NH_3, being sure to get all bonds and lone pairs (toothpicks) as far away from each other as possible. Do you think the H—N—H bond angle is greater or less than the H—C—H bond angle in CH_4? Support your answer.

e. Make a model of H_2O, being sure to get all bonds and lone pairs (toothpicks) as far away from each other as possible. Do you think the H—O—H bond angle is greater or less than the H—C—H bond angle in CH_4? Support your answer.

Keep your structures for reference.

—Continued next page

Continued—

Questions

You built the following molecular shapes using toothpicks and marshmallows: tetrahedral, trigonal planar, linear, pyramidal, and bent.

1. Based on the names given and the molecules that you built, assign each of the following molecules a shape:
   a. CO_2 _____
   b. H_2CCH_2 _____
   c. CH_4 _____
   d. NH_3 _____
   e. H_2O _____

2. Complete the following table:

Molecular Formula	Lewis Structure	Molecular Shape (Geometry) [from question 1]	Number of Atoms Bonded to Carbon or Central Atom	Number of Lone Pairs on Carbon or Central Atom
CO_2				
H_2CCH_2				
CH_4				
NH_3				
H_2O				

3. Complete the following table:

Molecular Formula	Lewis Structure	Molecular Shape (Geometry)	Number of Atoms Bonded to Carbon or Central Atom	Number of Lone Pairs on Carbon or Central Atom
CH_3F				
HCCH				
CH_2CHCl				
NH_4^+				
CH_3CH_3				

4. Based on the tables in questions 2 and 3, does the number of atoms attached to a central carbon atom affect its shape? List the shapes that carbon atoms can adopt and the number of atoms attached in each case.

3.6 Getting Covalent Compounds into Shape

3.6 Inquiry Question

How is the shape of a covalent compound determined?

Just as a road map allows us to see how towns and cities are connected by roads but gives very little information about the topography, the Lewis structures discussed in Section 3.4 tell us how atoms are connected in molecules but do not tell us anything about a molecule's three-dimensional shape. Our world is three-dimensional, so to understand molecules, it is important to identify the three-dimensional shapes they can adopt.

For example, the Lewis structure for methane appears flat, or two-dimensional, on paper. If this were how methane was arranged in three dimensions, the angle between two adjacent pairs of electrons would be about 90°.

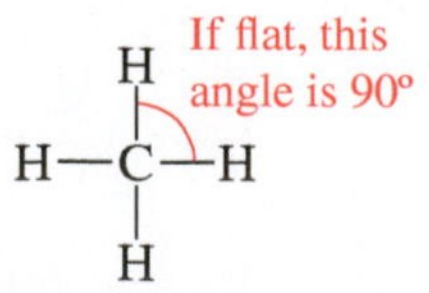

To understand how methane looks in three dimensions, remember first that electrons are negatively charged and that like charges repel each other. So, the four bonding pairs of electrons around the carbon atom in CH_4 form four clouds of negative charge that want to get *as far apart as possible*.

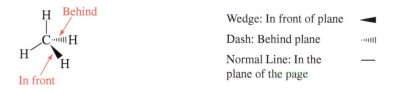

If the electrons are allowed to rearrange in three-dimensional space to get as far apart as possible, the result is an arrangement of the atoms of CH_4 that achieves an angle of 109.5° between the negatively charged clouds of electrons (that is, greater than 90°). The molecule is no longer two dimensional. Instead, it has adopted a three-dimensional shape called **tetrahedral,** as shown in **FIGURE 3.11**.

Because it is difficult to draw the tetrahedral shape shown in Figure 3.11, another way to draw a three-dimensional shape on a flat plane like this page is to use wedges and dashes to indicate in front of the plane and behind the plane, respectively. If you drew the methane shown in Figure 3.11, it would look like this:

	Wedge: In front of plane ◀
	Dash: Behind plane ·‧‧‧‧‧
	Normal Line: In the plane of the page —

The wedges and dashes are a simpler way to represent a three-dimensional structure.

When carbon forms four bonds to four atoms, the bonded pairs of electrons repel each other equally, forming the tetrahedral shape. This reasoning describes the **valence-shell electron-pair repulsion model.** VSEPR is used to predict the shape of a molecule based on the number of electron **charge clouds** on a given atom. According to VSEPR, in the valence shell of an atom the charge clouds formed by groups of electrons will arrange themselves to be as far away from each other as possible to reduce repulsions between electron pairs.

Determining the Shape of a Molecule

To determine the shape (also called geometry) of a molecule using VSEPR, a molecule is assigned a VSEPR form. The VSEPR form is a tool used to relate the number and type of charge clouds on an atom to the shape around that atom. To determine the VSEPR form and, from that, the shape of a molecule, first determine the atom around which you want to know the shape. For small molecules, this is the central atom. (We discuss larger molecules later.) The central atom is represented with a capital A in the VSEPR form.

Next, determine the number of charge clouds around the central atom. When counting charge clouds, keep in mind that a single bond, a double bond, and a triple bond are each counted as one charge cloud. Charge clouds of bonds are called *bonding clouds* and are represented with a capital B in the VSEPR form.

A lone pair of electrons is also counted as one charge cloud, but as we will see, these charge clouds affect the shape of the molecule in a slightly different way. Charge clouds from lone pairs of electrons are called *nonbonding clouds* and are represented with a capital N in the VSEPR form.

Lewis structures are like road maps, showing connections between atoms the way maps depict connections between cities. However, no information about the three-dimensional shape is given.

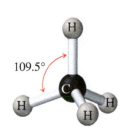

FIGURE 3.11 The tetrahedral shape of methane, CH_4. The atoms are represented by balls, and the bonds are represented with sticks.

Here are some examples to guide you as you learn to count charge clouds to determine a molecule's VSEPR form.

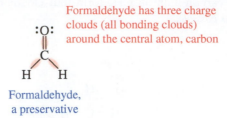

Formaldehyde has three charge clouds (all bonding clouds) around the central atom, carbon

Formaldehyde,
a preservative

The VSEPR form for formaldehyde is AB_3—a central atom surrounded by three bonding charge clouds.

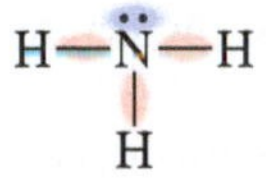

Ammonia

Ammonia has three bonding charge clouds and one nonbonding charge cloud (its lone pair) around the central atom, nitrogen

The VSEPR form for ammonia is AB_3N—a central atom surrounded by three bonding charge clouds and one nonbonding charge cloud.

The molecular shapes associated with the VSEPR forms are shown in **TABLE 3.6** for the common forms containing the elements C, N, and O found in living things.

TABLE 3.6 Predicting Molecular Shape Using VSEPR

VSEPR Form	Molecular Shape	Bond Angle	Ball-and-Stick Example	Wedge-Dash Example
AB_4	Tetrahedral	109.5°	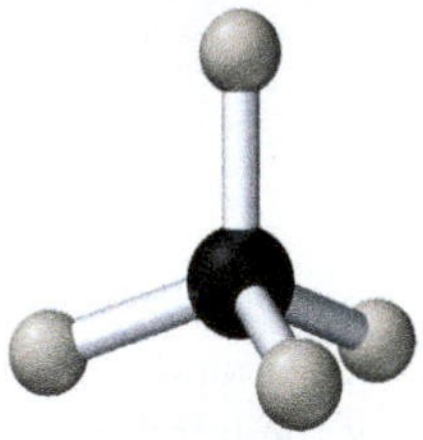	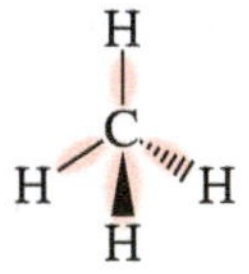
AB_3N	Pyramidal	<109.5°		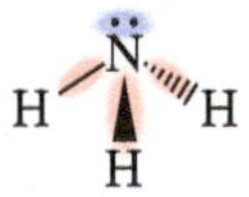
AB_2N_2	Bent	<109.5°		
AB_3	Trigonal planar	120°	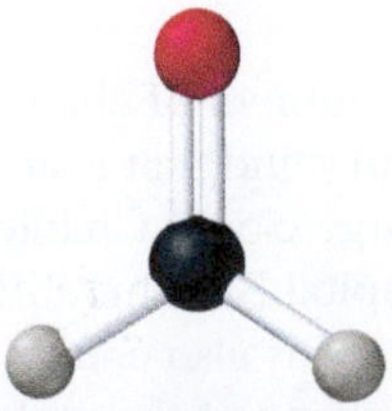	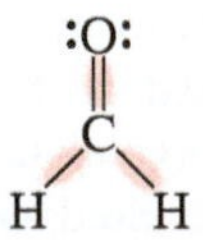
AB_2	Linear	180°		

Sample Problem 3.14   Determining a Molecule's Shape

From the following Lewis structures, assign the VSEPR form for each molecule and, using Table 3.6, determine the shape of each molecule:

a. $\ddot{O}{=}Si{=}\ddot{O}$

b. $H{-}\ddot{S}{-}H$

c. $:\!\ddot{B}r{-}\overset{\displaystyle |}{P}{-}\ddot{B}r\!:$
   $\qquad\quad\;\;\;:\!\ddot{B}r\!:$

Solution

a. The central atom in the molecule is silicon. Silicon is participating in two double bonds, indicating it has two bonding charge clouds in its valence shell. So this molecule has the VSEPR form AB_2. This VSEPR form corresponds to the molecular shape of linear (Table 3.6). Note that we did not concern ourselves with the oxygen atoms in this molecule. Because oxygen is not the central atom, its arrangement of charge clouds is not included when we assign the VSEPR form. The molecular shape is linear.

b. Sulfur is the central atom in this molecule. Careful examination shows that sulfur is participating in two single bonds and has two nonbonding pairs (also called lone pairs) of electrons. Therefore, the sulfur is surrounded by four charge clouds with a VSEPR form of AB_2N_2. This VSEPR form corresponds to the molecular shape of bent (Table 3.6). Note that the Lewis structures of SiO_2 and H_2S look fairly similar—a central atom connected to two other atoms. Both Lewis structures can be drawn with all the atoms in a straight line, but only SiO_2 actually has a linear shape—the molecular shape of H_2S is bent. It is important to remember that Lewis structures show only the *connectivity* of the atoms in a molecule, not the molecule's *shape*.

c. Phosphorus is the central atom in this molecule; it is participating in three single bonds and has one nonbonding pair of electrons. The VSEPR form of this molecule is AB_3N. This VSEPR form corresponds to the molecular shape of pyramidal (Table 3.6).

TABLE 3.7 Predicting Molecular Shapes for Carbon Compounds	
Molecular Shape	**Preferred Bonding Patterns for Carbon**
Tetrahedral	$-\overset{\displaystyle |}{\underset{\displaystyle |}{C}}-$
Trigonal planar	$\overset{\diagdown}{\underset{\diagup}{}}C{=}$
Linear	${=}C{=}$
	$-C{\equiv}$

Carbon atoms adopt one of three molecular shapes determined by the number of atoms bonded directly to the carbon. If carbon is bonded to four atoms, the shape is tetrahedral; if it is bonded to three atoms, the shape is trigonal planar; and if it is bonded to two atoms, the shape is linear (see **TABLE 3.7**).

Nonbonding Electrons and Their Effect on Molecular Shape

The first three entries in Table 3.6 have the VSEPR forms AB_4, AB_3N, and AB_2N_2. Each has a total of four charge clouds around a central atom, but each has a different shape. How can this be? Isn't there one optimal arrangement of four charge clouds around a central atom? To explore this, let's look more closely at the molecules methane (CH_4), ammonia (NH_3), and water (H_2O), which represent these three VSEPR forms.

Nonbonding electrons do affect the shape of a molecule. The tetrahedral shape of methane, which has four atoms around a central atom, changes to pyramidal for ammonia because a nonbonding pair occupies the fourth position (see **FIGURE 3.12**). Nonbonding pairs of electrons take up space but are invisible in the molecule's shape. A molecule's shape is determined by the relative positions of the *atoms* in the molecule.

Nonbonded pairs have a subtle effect on the shape of a molecule. In the Bond Angle column of Table 3.6, notice that the presence of nonbonded pairs on the central atom changes the bond angle, making it smaller. A nonbonded pair of electrons forces the bonded pairs closer together. But why?

A bonding pair of electrons is confined to the space directly between the two atoms being held. A nonbonded pair of electrons has no such restriction, so the nonbonded pair can move around more in the valence shell. A lone pair, therefore, takes up more space in the valence shell than does a bonding pair. So, using our example from Figure 3.12,

Tetrahedral
(methane)

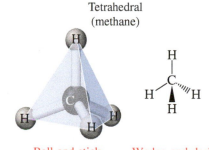

Ball-and-stick   Wedge-and-dash

Pyramidal
(ammonia)

Lone pair

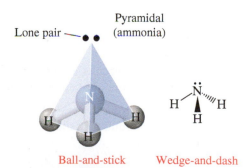

Ball-and-stick   Wedge-and-dash

FIGURE 3.12 Methane and ammonia represented with ball-and-stick and wedge-and-dash examples. A non-bonded pair changes a tetrahedral shape to a pyramidal shape.

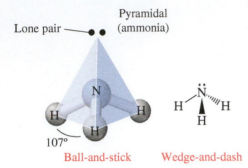

Lone pair — Pyramidal (ammonia)

107°

Ball-and-stick   Wedge-and-dash

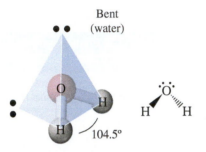

Bent (water)

104.5°

Ball-and-stick   Wedge-and-dash

FIGURE 3.13 Ammonia and water represented with ball-and-stick and wedge-and-dash examples. The bonding electron clouds are pushed closer together by the extra space requirements of the nonbonded electrons, making the bond angles different.

methane, which has four bonding pairs, has a bond angle of 109.5°, whereas the nonbonded (lone) pair in ammonia repels the bonded pairs more than the bonded pairs repel each other, resulting in a smaller angle between the bonded pairs. In another example, notice that the two nonbonded pairs in water push the bonded electron clouds closer to each other than the one nonbonded pair in ammonia, resulting in an even smaller bond angle (see **FIGURE 3.13**).

Molecular Shape of Larger Molecules

The majority of molecules you will see throughout the remainder of this text will not be as structurally simple as CH_4 or NH_3. Most molecules contain many carbon atoms with several charge clouds. Can we determine the shape of a molecule that does not have a single clearly identifiable central atom?

Using VSEPR, the shape around any single atom that is bonded to at least two other atoms can be determined. The Lewis structure of a molecule of ethanol is shown below. Ethanol is found in wine, beer, and other spirits as a product of fermentation, and, more recently, has found new uses as a fuel additive. We can determine the shape around each of the carbons and the oxygen in ethanol, but VSEPR does not allow us to determine an overall shape for the molecule. Typically, we choose one or more atoms of interest in the molecule and determine the shape around those atoms.

Tetrahedral   Tetrahedral   Bent

VSEPR allows us to determine the shape around each central atom but not the overall shape of the molecule.

Practice Problems

3.45 For the molecules shown, indicate whether the orange-colored atoms are in front of, behind, or in the plane of this book.

a.

b.

3.46 For the molecules shown, indicate whether the orange-colored atoms are in front of, behind, or in the plane of this book.

a.

b.

3.47 For the molecules in Problem 3.45, determine the shape around the central carbon.

3.48 For the molecules in Problem 3.46, determine the shape around the central carbon.

3.49 Determine the shape around the orange-colored atom (or atoms) in each of the following Lewis structures:

a. dimethylamine, an insect pheromone

b. acrylonitrile, found in plastics

3.50 Determine the shape around the orange-colored atom (or atoms) in each of the following Lewis structures:

a. dimethyl ether, found in wart treatments

b. ethylene, a plant-ripening hormone

DISCOVERING THE CONCEPTS

 INQUIRY ACTIVITY—Bond Polarity

Information

A *covalent bond* is formed when two atoms share electrons between them. If the two atoms are different elements, they will not share the electrons equally, and a *polar covalent bond* is formed. In this case, one of the atoms will have the electrons more of the time and be partially negatively charged (symbol δ^-) and the other atoms will have the electrons less of the time and be partially positively charged (symbol δ^+). We say that the bond has a dipole, meaning two poles. The amount of charge present can be measured as a quantity called a *dipole moment*. This is represented by an arrow with a hash mark crossing the tail ⊢——→. The head of the arrow points toward the more negative atom.

Electronegativity

Fluorine is the most electronegative element. The closer an element is to fluorine on the periodic table, the more electronegative it is. For example, oxygen is more electronegative than nitrogen, which in turn is more electronegative than carbon.

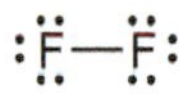

:F̈—F̈:   In a nonpolar covalent bond, the electrons are shared equally.
This occurs if the atoms in the bond are identical.
One exception: C—H bonds are also nonpolar covalent.

FIGURE 1A A nonpolar covalent bond.

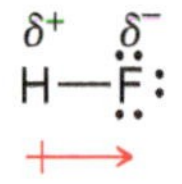

δ^+ δ^-
H—F̈:   In a polar covalent bond, the electrons are not shared equally.
⊢——→   The more electronegative element is the negative pole of the dipole. This polarity can be represented with delta symbols (above) or the dipole moment arrow (below).

FIGURE 2A A polar covalent bond.

Questions

Bond Polarity

1. Based on their ability to share electrons equally or unequally, are the following bonds considered polar covalent or nonpolar covalent? Explain your reasoning.
   a. C—N   b. C—Cl   c. O—H
   d. H—H   e. C—H   f. N—H

2. For the polar covalent bonds in question 1, indicate the polarity using the partial symbols (δ^+/δ^-) and the dipole moment arrow.

Molecule Polarity

3. Draw the Lewis structure for ammonia, NH_3. Show the polarity of the bonds between N and H on the structure you drew using the dipole moment arrow. What is the molecular shape (geometry) around the N in ammonia? Draw the overall dipole using the dipole moment arrow with the head pointing toward the negative side of your Lewis structure and the tail positioned near the positive side. Draw this to the right of your molecule.

—Continued next page

Continued—

4. Draw the Lewis structure for H_2O. Show the polarity of the bonds on the structure you drew using the dipole moment arrow. Considering that the shape of H_2O is bent, do you think that H_2O has a dipole? (In other words, is there a side of the molecule that is more negative and a side that is more positive?) If so, indicate this to the right of your structure using the dipole moment arrow.

5. Draw the Lewis structure for carbon dioxide. What is the molecular shape (geometry) around the carbon in carbon dioxide? Show the polarity of the bonds between C and O on the structure you drew using the dipole moment arrow. Unlike NH_3 and H_2O, CO_2 overall has no dipole. Although CO_2 has polar bonds, it is a nonpolar molecule. Provide an explanation for this. [Hint: Consider its shape.]

6. Draw the Lewis structure for carbon tetrachloride, CCl_4. What is the molecular shape (geometry) around the carbon in carbon tetrachloride? Show the polarity of the bonds between C and Cl on the structure you drew using the dipole moment arrow. What is the overall (net) dipole for this molecule? Explain.

7. Why do some molecules have polar bonds and yet are nonpolar molecules, whereas other molecules like water and ammonia have polar bonds and are polar molecules?

8. Using a dipole moment arrow (⊢——→), indicate the overall molecular polarity for the following molecules (you may have to draw out Lewis structures to do this). If the molecule is nonpolar, state no dipole.
   a. HBr   b. OCS   c. CH_3F
   d.   e. CH_2Br_2

   $$\begin{array}{c} H \diagdown \qquad \diagup F \\ C = C \\ F \diagup \qquad \diagdown H \end{array}$$

3.7 Electronegativity and Molecular Polarity

3.7 Inquiry Question

How is the polarity of a molecule determined?

Electrons are shared in covalent bonds, but often the electrons are not shared equally. In this section, we explore a fundamental property of atoms, called *electronegativity*, that causes this inequality to occur.

Electronegativity

How can we tell which atoms in a covalent bond will attract the shared electrons more? The ability of an atom to attract the bonding electrons of a covalent bond to itself is known as **electronegativity. FIGURE 3.14** shows many of the main-group elements from the periodic table with the electronegativity values for each element below the symbol. Notice that the element with the greatest electronegativity on the periodic table is fluorine. Also note that the electronegativity increases as you move closer to fluorine. For example, oxygen is more electronegative than carbon, and chlorine is more electronegative than iodine.

An inequality occurs when two *different* atoms are involved in a covalent bond; for example, in H—Cl, the sharing of electrons is not equal. In the HCl molecule, the chlorine atom attracts the shared electrons more than the hydrogen atom, which means that the shared electrons in the bond between H and Cl spend more time near the chlorine atom. In other words, the electrons are shared unequally. When two identical atoms share electrons to form a covalent bond, as in H_2, the electrons are equally attracted to each atom and are shared equally. A covalent bond in which the electrons are not shared

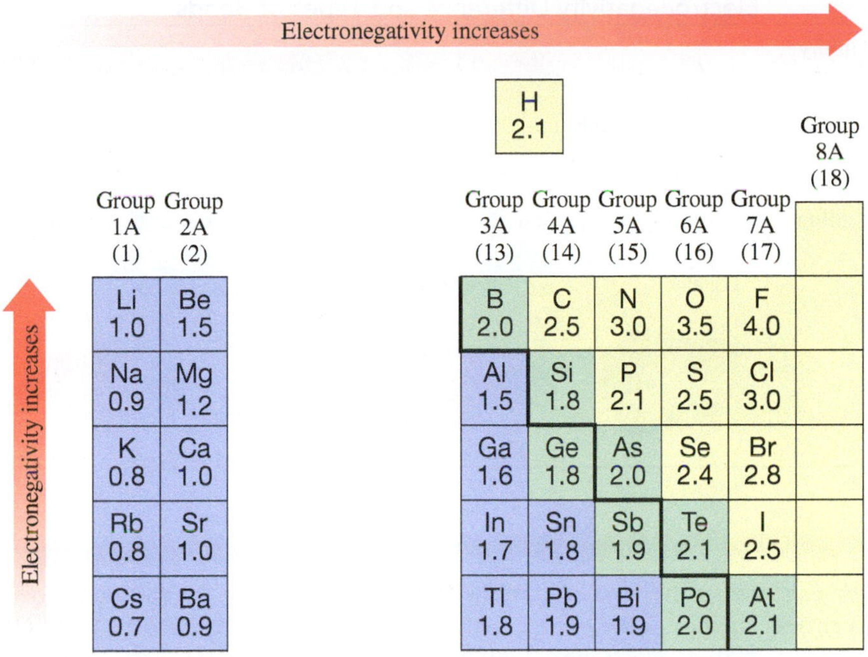

Electronegativity increases →

	H 2.1	

Group 1A (1)	Group 2A (2)		Group 3A (13)	Group 4A (14)	Group 5A (15)	Group 6A (16)	Group 7A (17)	
Li 1.0	Be 1.5		B 2.0	C 2.5	N 3.0	O 3.5	F 4.0	
Na 0.9	Mg 1.2		Al 1.5	Si 1.8	P 2.1	S 2.5	Cl 3.0	
K 0.8	Ca 1.0		Ga 1.6	Ge 1.8	As 2.0	Se 2.4	Br 2.8	
Rb 0.8	Sr 1.0		In 1.7	Sn 1.8	Sb 1.9	Te 2.1	I 2.5	
Cs 0.7	Ba 0.9		Tl 1.8	Pb 1.9	Bi 1.9	Po 2.0	At 2.1	

Electronegativity increases ↑

FIGURE 3.14 Electronegativities of some of the main-group elements. Fluorine is the most electronegative element. The noble gases are not considered because they are not very reactive.

equally is called a **polar covalent bond**. When the electrons in a covalent bond are shared equally, the bond is called a **nonpolar covalent bond**.

We can denote the uneven sharing or distribution of electrons in a polar covalent bond using the symbol δ (lowercase Greek delta, meaning "partial"). **FIGURE 3.15** demonstrates how this notation is used in the case of HCl. We can also use an arrow with a hash mark (⊢→), referred to by chemists as a dipole moment arrow, to represent bond polarity. Both notations are shown in Figure 3.15. Note that the arrow points toward the more electronegative element and the hash mark on the tail of the arrow is at the partial positive end.

Like the opposite poles on a magnet, a covalent bond that does not share electrons equally has two distinct poles or ends—one that is partially negative and one that is partially positive. We refer to these bonds as *polar*. Of course, a bond that is "nonpolar" (sharing equally) has no ends or poles.

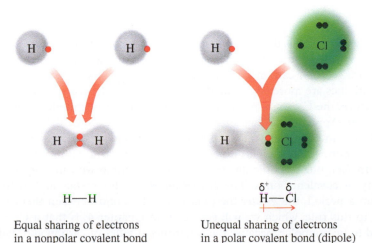

H—H

Equal sharing of electrons
in a nonpolar covalent bond

$\overset{\delta^+}{H}\!\!-\!\!\overset{\delta^-}{Cl}$
⊢→

Unequal sharing of electrons
in a polar covalent bond (dipole)

FIGURE 3.15 Covalent bonding. In the nonpolar covalent bond of H_2, electrons are shared equally. In the polar covalent bond of HCl, electrons are shared unequally.

FIGURE 3.16 Electronegativity difference and types of bonds. The greater the electronegativity difference between two atoms, the more polar the bond.

Electronegativity Difference and Types of Bonds

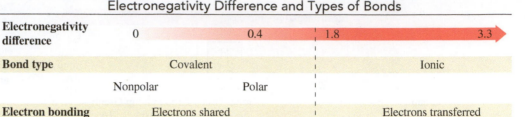

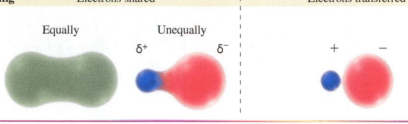

We can predict the type of bond likely to form by taking the difference between the electronegativities of the two elements, by using the table in **FIGURE 3.16**. Generally, elements with an electronegativity difference of 1.8 or more form an ionic bond resulting from the *transfer* of electrons from the less electronegative element to the more electronegative element. For example, NaCl, which we know as an ionic compound, has an electronegativity difference of $2.1 [3.0(Cl) - 0.9(Na)]$ between chlorine and sodium.

In turn, two elements with an electronegativity difference of less than 1.8 most likely *share* electrons to form a covalent bond. When two atoms are involved in a covalent bond, the greater the difference in electronegativity, the more polar the bond.

While it is possible to use the information in Figures 3.14 and 3.16 to calculate an electronegativity difference for any bond and determine where along the bond-type continuum it lies, without doing any calculations we can distinguish ionic bonds from covalent bonds by looking at the types of elements present (metals versus nonmetals).

Let's see how we can distinguish between nonpolar bonds, slightly polar bonds, and strongly polar bonds for covalent compounds. We can do this without numbers by remembering two facts:

1. Fluorine is the most electronegative element.
2. Electronegativities increase as you move toward fluorine on the periodic table.

With these rules in mind, let's consider the following carbon-containing bonds. Which of these bonds is the most polar?

$$C—N \qquad C—O \qquad C—F$$

Because each of the three bonds contains a carbon atom, we can compare the electronegativity of each of the other atoms relative to carbon. Looking at the periodic table, we see that carbon and nitrogen are side by side, which means that their electronegativities are more alike than those of carbon and oxygen or carbon and fluorine. Therefore, the C—N bond, with the smallest difference in electronegativities, is the least polar. Applying the same reasoning, we can determine that the C—O bond would fall next in the order and the C—F bond would be the most polar. When we want to determine relative polarities of bonds, we can do it without electronegativity numbers. In fact, our example illustrates a rule of thumb we can use for comparing the polarity of covalent bonds: The *farther apart* two nonmetals are on the periodic table within a period, the *greater* the polarity of the bond between them. One major exception to this rule, which we will see again in Chapter 4, is that a C—H bond is considered to be nonpolar, because the electronegativities of carbon and hydrogen are not that different.

Sample Problem 3.15   Polarity of Covalent Bonds

For each of the following covalent bonds, label the less electronegative atom with a δ^+ and the more electronegative atom with a δ^-. Then label the bond using the dipole moment arrow ($\longmapsto$).

   a. N—C   b. N—O   c. P—F

Solution

   a. $\overset{\delta^-}{\text{N}} \rightleftharpoons \overset{\delta^+}{\text{C}}$ On the periodic table, nitrogen is closer to fluorine than carbon, so nitrogen is more electronegative and attracts the bonding electrons more strongly, giving it a partial negative charge.

   b. $\overset{\delta^+}{\text{N}} \rightleftharpoons \overset{\delta^-}{\text{O}}$ Oxygen is closer to fluorine than carbon, so oxygen is more electronegative and attracts the bonding electrons more strongly, giving it a partial negative charge.

   c. $\overset{\delta^+}{\text{P}} \rightleftharpoons \overset{\delta^-}{\text{F}}$ Fluorine is the most electronegative of all elements, so it attracts the electrons in this bond more strongly, giving it a partial negative charge.

Sample Problem 3.16   Comparing Covalent Bonds

In each of the following bond pairs, circle the pair with the more polar bond:
   a. C—Br and C—Cl
   b. N—O and N—S
   c. Si—F and Si—O

Solution

In general, the more polar bond will have two elements with a greater electronegativity difference.
   a. C—Cl is the more polar bond.
   b. N—O is the more polar bond.
   c. Si—F is the more polar bond.

Molecular Polarity

Molecules, like the bonds that hold them together, can be polar or nonpolar. A **polar molecule** is a molecule in which one end or area of the molecule is more negatively charged than the rest of the molecule. In other words, the electrons in a polar molecule are unevenly distributed over the molecule. In contrast, a **nonpolar molecule** has an even distribution of electrons over the entire molecule.

    For simple molecules containing only two atoms connected by a covalent bond, we can determine the polarity of the molecule simply by determining the polarity of the bond. In our previous examples, the bond between the hydrogens in H_2 is nonpolar, so H_2 is a nonpolar molecule. Likewise, since the bond in HCl is a polar bond, HCl is a polar molecule.

    What happens when a molecule is composed of three or more atoms connected by several bonds? How do we determine the polarity of a molecule if it has more than one bond dipole? We *can* determine the polarity of larger molecules, but when we do, we must consider both the electronegativity of the atoms involved *and* the shape of the molecule.

    As a first example, consider carbon dioxide and water. Their Lewis structures are shown here.

$$:\ddot{\text{O}}\!=\!\text{C}\!=\!\ddot{\text{O}}: \qquad \text{H}\!-\!\ddot{\text{O}}\!-\!\text{H}$$

Mastering Videos
Practicing the Concepts
Determining if a Molecule Is Polar or Nonpolar

First, determine the bond polarity for each bond in each molecule:

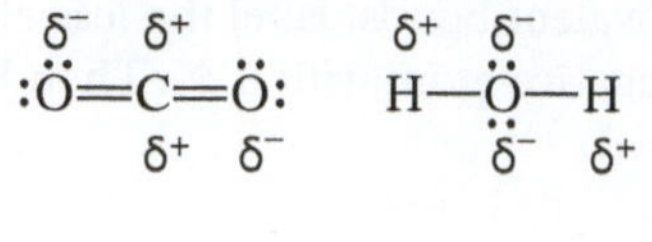

$$\overset{\delta^-\quad\delta^+}{:\ddot{O}=C=\ddot{O}:}\qquad \overset{\delta^+\quad\delta^-}{H-\ddot{O}-H}$$
$$\underset{\delta^+\quad\delta^-}{}\qquad\qquad \underset{\delta^-\quad\delta^+}{}$$

Each molecule has two polar bonds. Despite this apparent similarity, we know from experimental evidence that carbon dioxide is a nonpolar molecule whereas water is a polar molecule. How can two molecules that seem so similar actually be so different? Remember, these two molecules have different shapes.

Applying the principles of VSEPR from Section 3.6, we know that carbon dioxide is a linear molecule, but water is bent. To see how shape affects the polarities of these molecules, let's use the dipole moment arrow to represent the bond dipoles and draw the molecules to more closely represent their true shape.

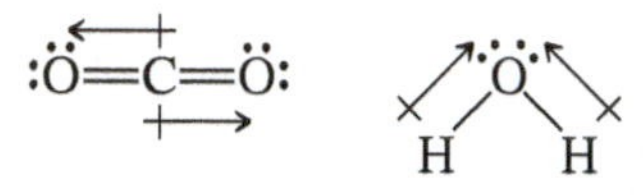

Notice that in CO_2 the bonding electrons are being pulled equally toward each of the oxygen atoms, but in opposite directions. So, like a tug of war that is equally matched, the electrons in the molecule are being pulled by one of the oxygen atoms, but the other oxygen is pulling equally in the opposite direction. This results in an even distribution of the electrons. In other words, the bond dipoles cancel each other.

Because of the bent shape of the water molecule, instead of canceling each other as they do in carbon dioxide, the bond dipoles add together to give an uneven distribution of electrons. The area around the oxygen atom where the lone pairs of electrons are positioned has more negative charge than the rest of the molecule. This results in an overall molecular dipole for water, shown as an orange arrow.

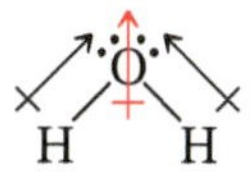

Both the electronegativity and molecular shape must be considered when determining whether a molecule is polar or nonpolar.

Sample Problem 3.17 **Determining the Polarity of Molecules**

Determine whether each of the following molecules is polar or nonpolar. If polar, show the direction of the molecular dipole using a dipole moment arrow.

a. CCl_4 b. NH_3 c. CH_3F

Solution

To see all the bonds and the shape, we draw the molecules using a wedge-dash formula.

a.

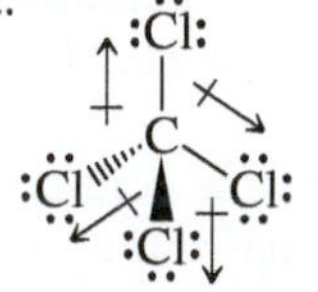

No overall dipole

Each $C-Cl$ bond is polar because Cl is more electronegative than C. As in our example above using CO_2, the Cl's are pulling electrons equally and oppositely in all directions, so the charge is evenly distributed in this tetrahedral molecule. The individual bond dipoles are shown in black. Despite the many nonbonded pairs on the chlorines in this molecule, the electrons are evenly distributed and this molecule has no overall dipole and is *nonpolar*.

b.

↕ Overall dipole

Each N—H bond is polar and N has a lone pair of electrons on it. The nitrogen pulls electrons towards itself, resulting in an uneven distribution of electrons in this pyramidal molecule. This molecule is *polar*. The individual bond dipoles are shown in black. The overall dipole (orange) points toward the lone pair of electrons on the central nitrogen atom.

c.

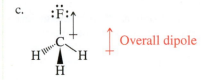

↑ Overall dipole

Recall that C—H bonds are nonpolar. The only polar bond in this tetrahedral molecule is the C—F bond, and its dipole is shown (black). The fluorine pulls the electrons toward itself more, causing an uneven distribution of charge, so the molecule is *polar*. The overall dipole (orange) mirrors the bond dipole.

Practice Problems

3.51 For each of the following molecules, (1) draw the correct Lewis structure; (2) label each polar covalent bond with a dipole moment arrow; and (3) determine if the molecule is polar or nonpolar and draw the dipole moment arrow for the molecule:

a. HCl

b. COS (C is the central atom)

c. H₂S

d. CH₂Br₂ (C is the central atom)

3.52 For each of the following molecules, (1) draw the correct Lewis structure; (2) label each polar covalent bond with a dipole moment arrow; and (3) determine if the molecule is polar or nonpolar and draw the dipole moment arrow for the molecule:

a. CH₃Br (C is the central atom)

b. HF

c. PBr₃

d. OF₂

CHAPTER

3

CHAPTER REVIEW

The study guide will help you check your understanding of the main concepts in Chapter 3. You should be able to answer problems for each learning outcome in this list. To check your mastery, try the problems listed after each.

STUDY GUIDE	CHAPTER GUIDE

3.1 Electron Arrangements and the Octet Rule

Arrange electrons in shells surrounding an atom.

- Predict the number of valence electrons and energy level for the main-group elements in the first four periods. (Try 3.1, 3.3, 3.53, 3.55)
- Explain and apply the octet rule. (Try 3.5, 3.57)

Inquiry Question

How are electrons arranged in an atom?

In atoms, electrons are distributed only at distinct energy levels in an electron cloud. These energy levels can be designated as n, where $n = 1, 2, 3$, and so on. The electrons in the outermost energy level or shell are called the valence electrons. For the main-group elements, the number of valence electrons is the same as the group number. The noble gases (Group 8A) are stable atoms, and each has an octet of electrons in its valence shell. An octet is eight electrons for all elements except hydrogen and helium, where a full valence shell is two electrons. Atoms, other than the noble gases, attain a stable valence octet by combining with each other.

3.2 In Search of an Octet, Part 1: Ion Formation

Apply the octet rule to ion formation.

- Predict the ionic charge of a main-group element using the periodic table. (Try 3.11, 3.59)
- Name ions given their symbol. (Try 3.15, 3.63, 3.67)
- Write symbols for ions given their name or the number of protons and electrons present. (Try 3.13, 3.17, 3.61, 3.65)
- Gain familiarity with polyatomic ions and their charges. (Try 3.19, 3.69)

Inquiry Question

How do ions form?

Ions are formed by gaining or losing electrons; this is one way that atoms can attain a valence octet. An atom that gains electrons becomes a negatively charged anion, and an atom that loses electrons becomes a positively charged cation. Nonmetal atoms form anions and metal atoms form cations. During ion formation, atoms gain or lose electrons to attain the same number of electrons as the noble gas with the nearest atomic number. For the main-group elements, the atom's position on the periodic table determines its charge when it forms an ion. Cations have the same name as the element with the word *ion* added to the end, but the name of an anion is derived by changing the ending of the elemental name to *ide*. Polyatomic ions are groups of atoms that together have an ionic charge. Their naming is different from single-element ions. In naming transition metals and metals in Groups 4A and 5A that form more than one cation, a Roman numeral is used in parentheses after the name of the metal to designate the charge.

3.3 Ionic Compounds—Electron Give and Take

Combine ions to form ionic compounds.

- Name ionic compounds given the formula. (Try 3.25, 3.75)
- Give the formula and name of an ionic compound given the ions. (Try 3.21, 3.25, 3.71)
- Write the formula for ionic compounds given the name. (Try 3.73, 3.79)
- Predict the ionic charges present in an ionic compound. (Try 3.23, 3.77)
- Predict the ionic charge of a transition metal using the compound's formula and the anionic charge. (Try 3.23a, 3.27b, 3.75c, 3.75e)

Inquiry Question

How do ionic compounds form?

Ionic bond

An ionic compound is formed when an ionic bond is created between a metal and nonmetal. Ionic bonds are formed by the attraction between an anion and a cation. Ionic compounds are neutral. The total charge of the cations must equal the total charge of the anions in the compound. To name an ionic compound, first name the cation (dropping the word *ion*) and then name the anion. If a transition metal with varying charge is present, a Roman numeral follows the name of the metal to designate its charge. In writing ionic formulas, the cation always goes first.

3.4 In Search of an Octet, Part 2: Covalent Bond Formation

Combine atoms to form covalent compounds.

- Distinguish between ionic and covalent compounds. (Try 3.33, 3.83)
- Establish the relationship between the number of valence electrons present in nonmetals in Periods 1–3 and the number of bonds that they typically make in a molecule. (Try 3.81)
- Draw Lewis structures for covalent compounds containing C, O, N, H, and the halogens (Group 7A). (Try 3.29, 3.31, 3.89)
- Name binary covalent compounds given the formula. (Try 3.35, 3.93)
- Write the formula for a binary covalent compound given the name. (Try 3.35)

Inquiry Question

How do covalent compounds form?

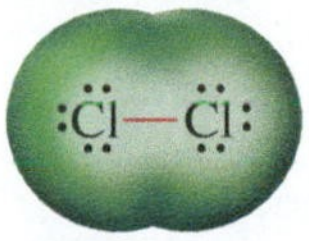

Covalent compounds are formed when non-metals share electrons to achieve valence octets. Covalent bonds are formed by the sharing of electrons between nonmetals. The number of covalent bonds that an atom will form is determined by the number of electrons necessary to complete its valence octet and leads to the atom's preferred covalent bonding pattern. The molecular formula of a covalent compound gives the number of each type of atom in a molecule, and the Lewis structure shows how those atoms are bonded. Covalent compounds containing only two elements can be named by naming the first element, naming the second element, and changing the ending to *-ide*. Greek prefixes indicate the number of atoms of each in the compound.

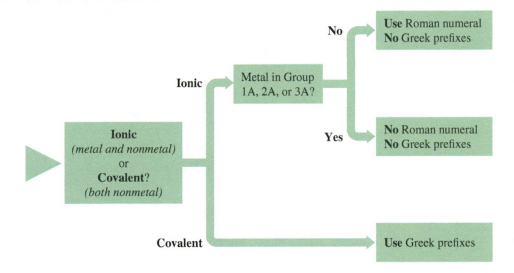

3.5 The Mole: Counting Atoms and Compounds

Calculate the number of moles and number of atoms in a substance.

- Describe the mole and Avogadro's number. (See Sample Problem 3.12. Try 3.37a)
- Calculate molar mass for a compound. (Try 3.41, 3.97)
- Convert among the units of mole, number of particles, and gram. (Try 3.37b, 3.39, 3.43, 3.99)

Inquiry Question

How is the molar mass used to determine the number of moles or particles present in a known mass of a compound?

The molar mass relates the mass of a substance to a counting unit called the mole. Avogadro's number is a measure of the number of particles (atoms or molecules) in a substance: 6.02×10^{23} per one mole. It is possible to calculate a substance's molar mass, which is equivalent to the atomic mass for an atom or the formula weight for a compound. By using Avogadro's number and molar mass as conversion factors, the number of moles or particles present in the mass of a compound can be determined.

—*Continued next page*

Continued—

3.6 Getting Covalent Compounds into Shape

Predict the shape of covalent compounds.

- Predict the shape of covalent compounds using VSEPR. (Try 3.47)
- Identify the location of a wedge and dash bonded atom in three-dimensional space. (Try 3.45)
- Predict the molecular shapes around atoms in Lewis structures. (Try 3.49, 3.103, 3.107)

Inquiry Question

How is the shape of a covalent compound determined?

Tetrahedral

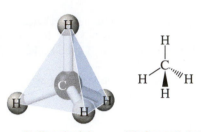

Ball-and-stick   Wedge-and-dash

The arrangement of the electrons in a molecule determines the three-dimensional shape of covalent compounds. A Lewis structure does not imply the three-dimensional shape of a molecule, yet shapes can be represented using wedges or dashes when drawing the structure. Valence-shell electron-pair repulsion theory (VSEPR) explains that the pairs of electrons around any given atom arrange themselves to get as far away from each other as possible. For carbon atoms, this leads to shapes such as tetrahedral, trigonal planar, and linear. The presence of nonbonding pairs on an atom reduces the bond angles around that atom and alters the shape of the molecule.

3.7 Electronegativity and Molecular Polarity

Predict the polarity of covalent compounds.

- Predict covalent bond polarity based on electronegativity. (See Sample Problems 3.15 and 3.16. Try 3.109)
- Predict molecular polarity from bond polarities and molecular shape. (Try 3.51, 3.111, 3.117)

Inquiry Question

How is the polarity of a molecule determined?

A molecule's polarity is based on the individual bond polarities in the molecule and molecular shape. The bond polarities are determined by comparing the electronegativities of the atoms in the bond. Nonpolar covalent bonds are formed when atoms share electrons equally (electronegativities are equal), and polar covalent bonds are formed when the sharing of electrons is unequal. Polarity can be represented by the delta (δ) symbol or the dipole moment arrow ($\longmapsto$).

 Mastering Videos

PRACTICING THE CONCEPTS

The following videos can be accessed through the Pearson eText or your Mastering Chemistry course.

- Drawing Lewis Structures
- Determining if a Molecule Is Polar or Nonpolar

Additional Problems

3.53 Complete the following table.

	Name	Symbol	Group Number	Number of Valence Electrons	Charge when an Ion
a.	Calcium			2	2+
b.		F			
c.		K			1+
d.	Sulfur				
e.		Al			

3.54 Complete the following table.

	Name	Symbol	Group Number	Number of Valence Electrons	Charge when an Ion
a.		Mg		2	
b.	Bromine				1−
c.		N			
d.		Na			
e.	Oxygen				

3.55 How many valence electrons are present in the following atoms?
- **a.** C
- **b.** Cl
- **c.** K
- **d.** Al

3.56 How many valence electrons are present in the following atoms?
- **a.** N
- **b.** Mg
- **c.** S
- **d.** Si

3.57 For each of the atoms in Problem 3.55, determine the minimum number of electrons the atom must gain, lose, or share to become stable (isoelectronic with a noble gas).

3.58 For each of the atoms in Problem 3.56, determine the minimum number of electrons the atom must gain, lose, or share to become stable (isoelectronic with a noble gas).

3.59 Complete the following statements:
- **a.** A cation/anion (*circle one*) has fewer protons than electrons.
- **b.** The valence electrons are found in the _____ of the atom.
- **c.** Ions formed from elements of Groups 6A and 7A have a positive/negative (*circle one*) charge.

3.60 Complete the following statements:
- **a.** An ion is charged because the numbers of _____ and protons are not the same.
- **b.** The ion formed from an iodine atom is named _____.
- **c.** A cation/anion (*circle one*) has more protons than electrons.

3.61 Determine the number of protons and electrons in the following ions:
- **a.** chloride
- **b.** Fe^{3+}
- **c.** Cr^{6+}
- **d.** N^{3-}
- **e.** sodium ion
- **f.** H^+

3.62 Determine the number of protons and electrons in the following ions:
- **a.** oxide
- **b.** Ag^+
- **c.** copper(I) ion
- **d.** H^-
- **e.** zinc(II) ion
- **f.** I^-

3.63 Give the name of each of the unnamed ions in Problem 3.61.

3.64 Give the name of each of the unnamed ions in Problem 3.62.

3.65 Give the name and symbol of the ion described in each of the following statements:
- **a.** A group 7A ion that is isoelectronic with Ar
- **b.** An ion with 7 protons and 10 electrons
- **c.** An ion with 11 protons that is isoelectronic with Ne

3.66 Give the name and symbol of the ion described in each of the following statements:
- **a.** A group 6A ion that is isoelectronic with Kr
- **b.** An ion with 16 protons and 18 electrons
- **c.** An ion with 3 protons that is isoelectronic with He

3.67 Each of the following ions is isoelectronic with a noble gas. Name the ion and give the symbol of the noble gas.
- **a.** Rb^+
- **b.** F^-
- **c.** S^{2-}
- **d.** Ca^{2+}

3.68 Each of the following ions is isoelectronic with a noble gas. Name the ion and give the symbol of the noble gas.
- **a.** Li^+
- **b.** Mg^{2+}
- **c.** Al^{3+}
- **d.** O^{2-}

3.69 Complete the table of polyatomic ions by supplying the missing information.

Name	Formula
a. acetate	
b. bicarbonate	
c.	NO_3^-
d.	CN^-

3.70 Complete the table of polyatomic ions by supplying the missing information.

	Formula
a.	MnO_4^-
b.	HSO_4^-
c. carbonate	
d. lactate	

3.71 Give the formula for the ionic compound formed by the ammonium ion and each of the following anions:
- **a.** sulfide
- **b.** chloride
- **c.** sulfate
- **d.** hydroxide

3.72 Give the formula for the ionic compound formed by the phosphate ion and each of the following cations:
- **a.** sodium
- **b.** aluminum
- **c.** magnesium
- **d.** iron(III)

3.73 Give the formula for each of the following ionic compounds:
- **a.** sodium bicarbonate, used as an antacid
- **b.** copper(II) sulfate, used as an algicide in swimming pools
- **c.** barium sulfate, used for X-rays

3.74 Give the formula for each of the following ionic compounds:

    a. magnesium oxide, used as a mineral supplement

    b. calcium sulfate, plaster of Paris used in casts

    c. iron(III) chloride, used in wastewater treatment

3.75 Name the following ionic compounds:

    a. Na_2O **b.** $BaSO_4$

    c. $CuCl_2$ **d.** $Mg(NO_3)_2$

    e. Fe_2O_3 **f.** KF

3.76 Name the following ionic compounds:

    a. $AlCl_3$ **b.** $NH_4C_2H_3O_2$

    c. $AgBr$ **d.** K_3PO_4

    e. CsF **f.** MnO

3.77 Magnesium salicylate is sold as an over-the-counter drug to treat pain and inflammation. If there are two salicylate ions for each magnesium ion in the compound, what is the charge on the salicylate ion?

3.78 Two common additives to bottled water are the compounds calcium chloride and potassium bicarbonate. Write the formulas for these ionic compounds.

3.79 Calcium sorbate is used as a preservative in baked goods. The formula of the sorbate ion is $C_6H_7O_2^-$. Give the formula for calcium sorbate.

3.80 Stannous, or tin(II) fluoride, is a compound found in many toothpastes. Write the formula for this compound.

3.81 How many covalent bonds does each of the following atoms tend to form?

    a. C   **b.** F

    c. P   **d.** O

3.82 How many covalent bonds does each of the following atoms tend to form?

    a. Cl   **b.** Si

    c. S   **d.** N

3.83 For each of the following compounds, indicate whether it is covalent or ionic and give its correct name:

    a. CBr_4 **b.** SiO_2 **c.** $MgBr_2$

    d. NCl_3 **e.** $CrCl_3$

3.84 For each of the following compounds, indicate whether it is covalent or ionic and give its correct name:

    a. Cu_2O **b.** $SiCl_4$ **c.** H_2O

    d. BaS **e.** CS_2

3.85 Explain the difference between an ionic formula and a molecular formula.

3.86 Explain the difference between a Lewis structure and a molecular formula.

3.87 Draw the electron-dot symbol for each of the following atoms:

    a. C   **b.** N

    c. Cl   **d.** O

3.88 Draw the electon-dot symbol for each of the following atoms:

    a. H   **b.** S

    c. P   **d.** Br

3.89 Draw a Lewis structure for each of the following covalent compounds:

    a. HCN   **b.** CS_2

    c. PCl_3 **d.** CH_5N

3.90 Draw a Lewis structure for each of the following covalent compounds:

    a. CH_4O **b.** N_2H_4

    c. Br_2 **d.** C_2H_3Cl

3.91 Give the molecular formula for each compound whose Lewis structure is shown. *(Note: In the molecular formula, list C and H first, then all other elements in alphabetical order.)*

3.92 Give the molecular formula for each compound whose Lewis structure is shown. *(Note: In the molecular formula, list C and H first, then all other elements in alphabetical order.)*

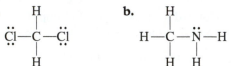

3.93 Give the name of each of the following covalent compounds:

    a. SeO_2 **b.** SiF_4

    c. P_4S_3 **d.** OF_2

3.94 Give the name of each of the following covalent compounds:

    a. SO_2 **b.** CI_4

    c. SiS_2 **d.** SCl_2

3.95 Explain the difference between an ionic bond and a covalent bond.

3.96 What are the units of Avogadro's number?

3.97 Determine the molar mass of the following compounds.

    a. KNO_3, potassium nitrate, used in some toothpastes for sensitive teeth

    b. $C_3H_6O_3$, lactic acid, a by-product of anaerobic respiration

    c. C_2H_6O, ethanol, the alcohol found in beer, wine, and spirits

3.98 Determine the molar mass of the following compounds.

a. $NaC_2H_3O_2$, sodium acetate, used in infant heel warmers

b. $KMnO_4$, potassium permanganate, used in wound healing

c. $C_2HBrClF_3$, halothane, an inhalation anesthetic

3.99 What is the mass of 4.00 moles of the following?

a. He   **b.** Li

c. KNO_3 **d.** C_2H_6O

3.100 How many atoms or molecules are in 5.0 moles of the following?

a. O   **b.** N

c. $MgCl_2$ **d.** C_2H_3NO

3.101 A pencil mark (made with graphite, a form of carbon) on paper has a mass of about 0.000005 g (5×10^{-6} g). How many atoms of carbon are present in the pencil mark?

3.102 The recommended daily allowance of niacin (vitamin B_3) is 14 mg per day for women. The molecular formula for niacin is $C_6H_5NO_2$. Calculate the number of molecules of niacin in 14 mg.

3.103 Determine the shape around the orange-colored atom (or atoms) in each of the following Lewis structures:

a. ethyl acetate, component of nail polish remover

b. methyl carbamate, used in the manufacture of textiles

3.104 Determine the shape around the orange-colored atom (or atoms) in each of the following Lewis structures:

a. acetylene, a fuel used in welding

$$H—C≡C—H$$

b. formic acid, used as a preservative in livestock feed

3.105 Aspartic acid, a naturally occurring amino acid and used in the brain as a neurotransmitter, is also a component of aspartame (Nutrasweet™), an artificial sweetener widely used in diet soft drinks.

a. Determine the shape around each orange-colored atom in the Lewis structure of aspartic acid.

b. Give the bond angle around each of the orange-colored atoms.

3.106 Cyanoacrylic acid is one of the compounds used to make Super Glue®. Determine the shape of the molecule around each orange-colored atom in its Lewis structure.

a. Determine the shape around each orange-colored atom in the Lewis structure of cyanoacrylic acid.

b. Give the bond angle of each of the orange-colored atoms.

3.107 Methyl isothiocyanate is used as an antifungal in agricultural applications. Determine the shape around the orange-colored atoms in the Lewis structure of methyl isothiocyanate.

3.108 Fumaric acid is an intermediate in the metabolism of carbohydrates. Determine the shape around the orange-colored atoms in the Lewis structure of fumaric acid.

3.109 Identify the more electronegative atom in each of the following pairs:

a. S and Se   **b.** P and N

c. C and Cl   **d.** N and I

3.110 Identify the more electronegative atom in each of the following pairs:

a. H and Cl   **b.** N and O

c. C and N   **d.** F and Br

3.111 Using the Lewis structures provided,

a. label each polar covalent bond with a dipole moment arrow.

b. determine if the molecule is polar or nonpolar and draw the molecular dipole.

3.112 Using the Lewis structures provided,

a. label each polar covalent bond with a dipole moment arrow.

b. determine if the molecule is polar or nonpolar and draw the molecular dipole.

Challenge Problems

3.113 Tooth enamel contains a hard mineral called hydroxyapatite, composed of calcium, phosphate, and hydroxide ions. If the formula for the compound contains six phosphate ions and two hydroxide ions, provide the complete formula for the ionic compound.

3.114 Two elements, E and X, are combined and react to form a compound containing E^{2+} and X^{2-} ions. Answer the following questions about E and X based on this information.

a. Which of the elements, E or X, is more likely to be a metal?

b. Which of the elements, E or X, is more likely to be a nonmetal?

c. What is the formula of the compound formed from E and X?

d. To which main group on the periodic table does each element belong?

3.115 Precious metals are commonly measured out in troy ounces. One troy ounce of platinum is equivalent to 31.10 g of platinum. How many atoms of platinum are in 1.00 troy ounce of platinum?

3.116 Vinyl chloride, C_2H_3Cl, is used in the production of polyvinyl chloride (PVC), a plastic used to make tubing and pipes for a variety of medical applications.

a. Draw the Lewis structure of vinyl chloride.

b. Determine the shape and the bond angle around each of the carbon atoms.

c. Label each of the polar covalent bonds with a dipole moment arrow.

d. Determine if this compound is polar or nonpolar.

3.117 One of the most common compounds used in commercial dry cleaning is tetrachloroethylene, which has the molecular formula C_2Cl_4.

a. Draw the Lewis structure of tetrachloroethylene.

b. Determine the shape and the bond angle around each of the carbon atoms.

c. Label each of the polar covalent bonds with a dipole moment arrow.

d. Determine if this compound is polar or nonpolar.

Answers to Odd-Numbered Problems

Practice Problems

3.1
a. 2 e⁻ in first energy level

b. 2 e⁻ in first energy level, 4 e⁻ in second energy level

c. 2 e⁻ in first energy level, 8 e⁻ in second energy level, 1 e⁻ in third energy level

d. 2 e⁻ in first energy level, 8 e⁻ in second energy level

3.3 a. 5 b. 3 c. 4 d. 1

3.5 c. Ne

3.7 In atoms, the number of protons and electrons are the same and there is no net charge. In a metal cation, there are more protons than electrons present, giving the cation a positive charge.

3.9 In naming an anion, the last several letters of the nonmetal element name are dropped and the suffix -ide is applied.

3.11 a. 2+ b. 2− c. 3− d. 1+

3.13 a. 12 p, 10 e⁻ b. 8 p, 10 e⁻
c. 17 p, 18 e⁻ d. 29 p, 27 e⁻

3.15 a. magnesium ion b. oxide
c. chloride d. copper(II) ion

3.17 a. bromide, Br⁻ b. cobalt(II) ion, Co^{2+}

3.19 a. ammonium b. acetate c. cyanide

3.21 a. gold(III) chloride, $AuCl_3$

b. calcium sulfate, $CaSO_4$

c. magnesium hydroxide, $Mg(OH)_2$

3.23 a. one Au^{3+}, three Cl⁻

b. one Ca^{2+}, one SO_4^{2-}

c. one Mg^{2+}, two OH⁻

3.25 a. sodium carbonate, Na_2CO_3

b. iron(II) carbonate, $FeCO_3$

c. aluminum carbonate, $Al_2(CO_3)_3$

3.27 a. two Na^+, one CO_3^{2-} **b.** one Fe^{2+}, one CO_3^{2-}

c. two Al^{3+}, three CO_3^{2-}

3.29 a. 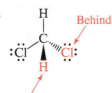   **b.** H—S̈e—H

c. :S̈=C=S̈:   **d.**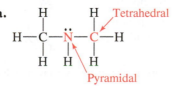

3.31 a. Ö=Si=Ö   **b.** (structure of ethene)

c. Ö=Ö   **d.** (structure of formaldehyde)

3.33 a. covalent   **b.** ionic   **c.** ionic   **d.** covalent

e. covalent   **f.** ionic   **g.** covalent   **h.** ionic

i. covalent   **j.** covalent

3.35 NO_2, phosphorus trichloride, SiF_4, water, NCl_3

3.37 a. They are equal, 6.02×10^{23} atoms.

b. Gold weighs more—197.0 g versus 107.9 g for silver.

3.39 a. 1.51×10^{23} atoms   **b.** 0.019 mole

c. 2.57×10^{23} atoms

3.41 a. 136.15 g/mole   **b.** 180.16 g/mole

c. 34.02 g/mole

3.43 a. 1.09×10^{21} molecules

3.45 a. 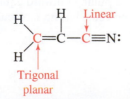   **b.** (structure of water, H and Ö labeled Behind, In front)

3.47 a. tetrahedral   **b.** bent

3.49 a. (structure H—C—N—C—H labeled Tetrahedral and Pyramidal)

b. (structure labeled Linear, Trigonal planar)

3.51 a. polar   H—C̈l:
Polar

Because this molecule contains one bond, the bond and molecular dipoles are the same.

b. polar   S̈=C=Ö
Polar

C—O bonds are polar and C—S bonds are not. The molecular dipole is shown in orange.

c. polar
Polar

S—H bonds are polar. This molecule has a bent shape. The lone pairs of electrons are on one side of the molecule in three dimensions. The molecular dipole is shown in orange.

d. polar   (tetrahedral structure with C, H, Br)
Polar

C—Br bonds are polar. Considering the tetrahedral shape, the bromine side of the molecule is the negative side. The molecular dipole is shown in orange.

Additional Problems

3.53 a. Ca, 2A

b. fluorine, 7A, 7, 1−

c. potassium, 1A, 1

d. S, 6A, 6, 2−

e. aluminum, 3A, 3, 3+

3.55 a. 4   **b.** 7   **c.** 1   **d.** 3

3.57 a. 4   **b.** 1   **c.** 1   **d.** 3

3.59 a. anion   **b.** outer shell   **c.** negative

3.61 a. 17 p, 18 e^- **b.** 26 p, 23 e^-

c. 24 p, 18 e^- **d.** 7 p, 10 e^-

e. 11 p, 10 e^- **f.** 1 p, 0 e^-

3.63 a. iron(III)   **b.** chromium(VI)

c. nitride   **d.** hydrogen ion, (proton)

3.65 a. chloride, Cl^- **b.** nitride, N^{3-}

c. sodium ion, Na^+

3.67 a. rubidium ion, Kr   **b.** fluoride, Ne

c. sulfide, Ar   **d.** calcium ion, Ar

3.69 a. $C_2H_3O_2^-$ **b.** HCO_3^-

c. nitrate   **d.** cyanide

3.71 a. $(NH_4)_2S$ **b.** NH_4Cl

c. $(NH_4)_2SO_4$ **d.** NH_4OH

3.73 **a.** NaHCO$_3$ **b.** CuSO$_4$ **c.** BaSO$_4$

3.75 **a.** sodium oxide   **b.** barium sulfate

c. copper(II) chloride   **d.** magnesium nitrate

e. iron(III) oxide   **f.** potassium fluoride

3.77 1$^-$

3.79 Ca(C$_6$H$_7$O$_2$)$_2$

3.81 **a.** 4   **b.** 1   **c.** 3   **d.** 2

3.83 **a.** covalent, carbon tetrabromide

b. covalent, silicon dioxide

c. ionic, magnesium bromide

d. covalent, nitrogen trichloride

e. ionic, chromium(III) chloride

3.85 Ionic formulas are written for ionic compounds and use the smallest ratio of ions; molecular formulas are written for covalent compounds and indicate the exact number of atoms in the molecule.

3.87 **a.** ·Ċ·   **b.** ·N̈·

c. :C̈l·   **d.** :Ö·

3.89 **a.** H—C≡N: **b.** S̈=C=S̈

c. 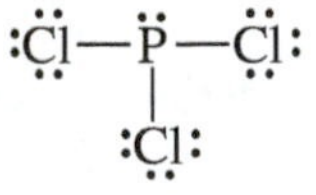 :C̈l—P̈—C̈l: **d.**

:C̈l:

H
|
H—C—N̈—H
| |
H   H

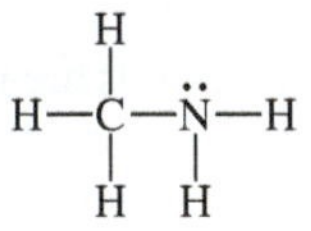

3.91 **a.** CH$_2$Cl$_2$ **b.** CH$_5$N

3.93 **a.** selenium dioxide

b. silicon tetrafluoride

c. tetraphosphorus trisulfide

d. oxygen difluoride

3.95 A covalent bond results when two atoms share one or more pairs of electrons. An ionic bond is the attraction between two or more ions that are formed by the loss and gain of electrons.

3.97 **a.** 101.10 g/mole   **b.** 90.09 g/mole

c. 46.08 g/mole

3.99 **a.** 16.0 g   **b.** 27.8 g

c. 404 g   **d.** 184 g

3.101 3 × 10^{17} atoms of carbon

3.103 **a.**

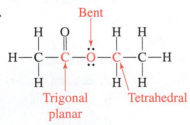

b.

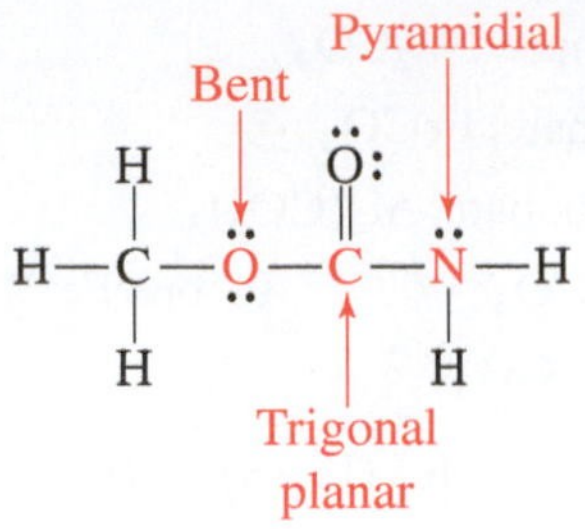

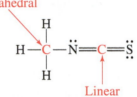

3.105

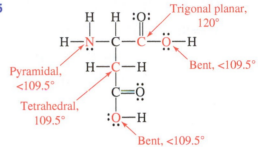

3.107 Tetrahedral

H
|
H—C—N̈=C=S̈
|
H

Linear

3.109 **a.** S   **b.** N   **c.** Cl   **d.** N

3.111 **a.** All bonds are nonpolar

b. Nonpolar

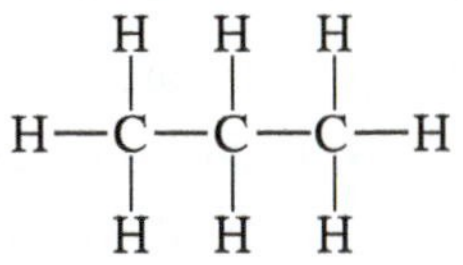

H H H
| | |
H—C—C—C—H
| | |
H H H

a. C—N bonds and N—H bond are polar

b. Polar, molecular dipole in orange

H   H
| |
H—C⇄N̈⇆C—H
| ↑ |
H   H   H

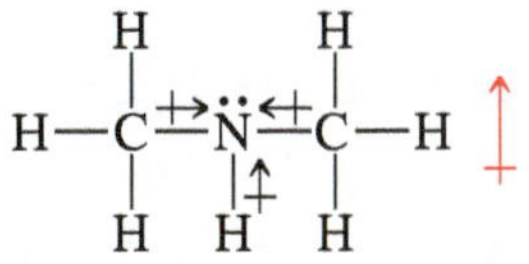

a. C=O bond is polar

b. Polar, molecular dipole in orange

H   Ö
| ‖ ↑
H—C—C—H
|
H

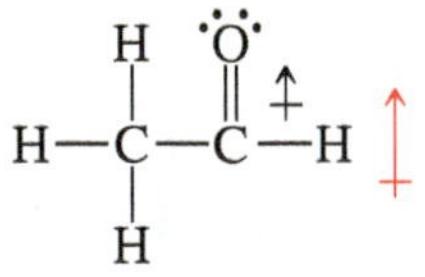

3.113 Ca$_{10}$(PO$_4$)$_6$(OH)$_2$

3.115 9.60 × 10^{22} atoms of Pt

3.117 **a.** Lewis structure shown at left.

b. Both carbons are trigonal planar and the bond angle is 120 degrees.

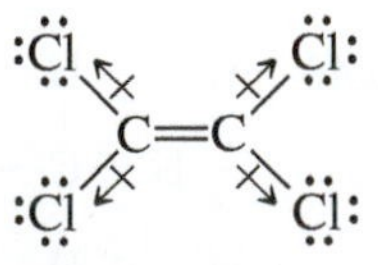

c. C—Cl bonds are polar, dipoles shown at left.

d. The polar bonds are pulling outward equally and oppositely so the molecule has no molecular dipole and is nonpolar.

Introduction to Organic Compounds

The word "organic" may bring to mind many things—food grown without pesticides or a consumer product that is all-natural. In chemistry, organic is a term we use to describe carbon-containing molecules. Most medicines are organic compounds. In Chapter 4, you will begin to understand the meaning behind the line drawings of the molecules shown on the prescription insert here. We will see that the structures represented with these drawings are essential in helping us understand the properties and behaviors of these medicines and all organic compounds.

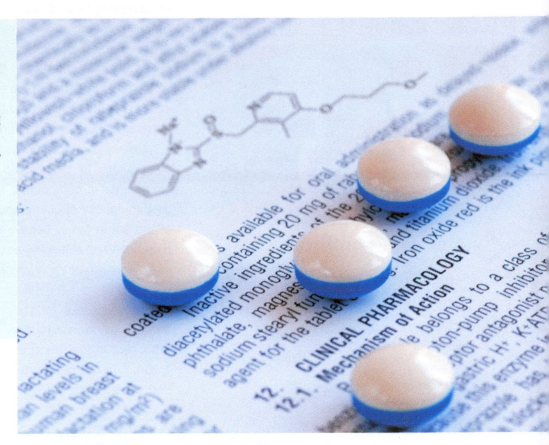

LEARNING TIP   Draw Pictures in Your Notes

We learn more from text with illustrations than from text alone. That is why there are so many illustrations in this textbook! Try making line drawings as you take notes in class or while you study.

"ORGANIC" CAN DESCRIBE a set of environmentally friendly farming practices or imply that a product is in some way more natural, but chemists use the word *organic* to identify covalent compounds containing carbon. **Organic compounds** are composed mainly of carbon and hydrogen but may also include oxygen, nitrogen, sulfur, phosphorus, and other elements. The molecules of life, or **biomolecules**—such as proteins, carbohydrates, lipids, and DNA—are all organic compounds. This chapter combines the information about nonmetal elements from Chapter 2, and how they merge to form molecules from Chapter 3, to begin the study of the organic compounds that make up living things and the chemistry of life. Organic compounds, whether naturally occurring or synthesized in the laboratory, are found in familiar substances such as gasoline, cotton, plastics, vitamins, medicines, and cosmetics.

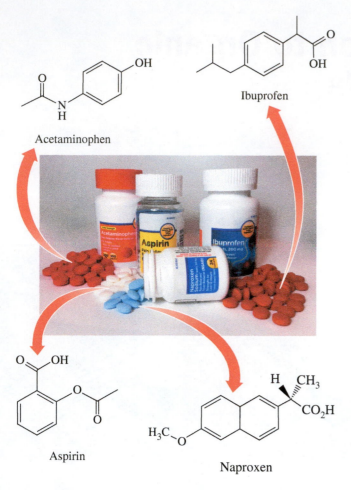

Acetaminophen

Ibuprofen

Aspirin

Naproxen

Skeletal structures of the common over-the-counter analgesics.

Why carbon? Carbon is unique in its ability to form bonds with other atoms of carbon and can create chains and rings of various sizes and shapes. In fact, most of the over 60 million known chemical compounds are organic.

Carbon's ability to form such a tremendous variety of compounds and our ability to manipulate and react these compounds in the laboratory contribute to an ever-increasing interest in the chemistry subdiscipline called **organic chemistry**. Organic chemistry is dedicated to the study of the structure, properties, and reactivity of carbon-containing compounds. Compounds that do not contain carbon and hydrogen are called **inorganic compounds**.

Luckily, it is not necessary to study millions of organic compounds to understand organic chemistry. Chemists group organic compounds into distinct families based on their molecular structure and composition, where members of the same family behave similarly in chemical reactions. By recognizing a few key structural features, you can gain a basic understanding of organic structure and reactivity.

Organic molecules can be very large, so drawing Lewis structures for these molecules can be cumbersome. We start Chapter 4 by exploring two forms of representational shorthand chemists use when drawing organic molecules containing many carbons: condensed and skeletal structures.

DISCOVERING THE CONCEPTS

? INQUIRY ACTIVITY—Representing Molecules on Paper

Information

The following structures are representations of two covalent compounds that are *structural isomers:* isopropyl alcohol (rubbing alcohol) and *n*-propyl alcohol.

Isopropyl alcohol

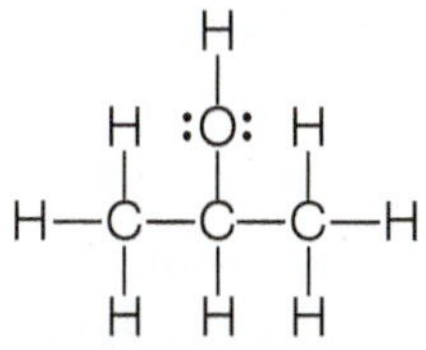

Lewis structure

$CH_3CH(OH)CH_3$

Condensed structure

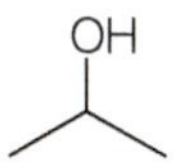

OH

Skeletal
(or line) structure

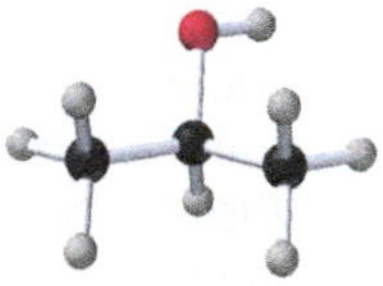

Ball-and-stick model

n-Propyl alcohol

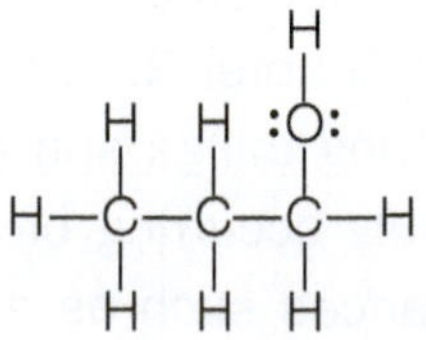

Lewis structure

$CH_3CH_2CH_2OH$

Condensed structure

OH

Skeletal
(or line) structure

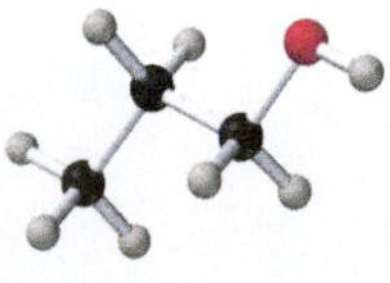

Ball-and-stick model

Questions

1. How are *n*-propyl alcohol and isopropyl alcohol the same? How are they different?
2. How do a molecular formula and a condensed structure differ?
3. How do a Lewis structure and a condensed structure differ?
4. What does the OH in parentheses in the condensed structure of isopropyl alcohol signify?
5. a. What atoms do the black spheres in the ball-and-stick models represent?
   b. What two atoms do the red sphere and the small gray sphere to the right of it represent?
6. What atom is always at the intersection of two line segments in the skeletal structure?
7. Provide a definition for *structural isomer*. (*Hint*: What is the same about the *n*-propyl alcohol and isopropyl alcohol molecules? What is different?)
8. Provide the corresponding condensed and skeletal structures for the Lewis structures of isobutane and *n*-butane.

Isobutane   *n*-Butane

9. Are *n*-butane and isobutane structural isomers? How do you know?
10. Provide the corresponding skeletal structure for each of the following hydrocarbons:

$$CH_3CH_2CH_2\overset{\underset{\displaystyle CH_3}{|}}{C}HCH_2CH_2CH_3$$

$$CH_3\overset{\underset{\displaystyle CH_3}{|}}{C}HCH_2CH_2\overset{\underset{\displaystyle |}{\overset{\displaystyle CH_2CH_2CH_3}{|}}}{C}HCH_2CH_3$$

11. Explain how the number of hydrogens written after each carbon in the condensed structures in question 10 was determined.
12. Provide the condensed structure for the following skeletal structures.

4.1 Representing the Structures of Organic Compounds

In Chapter 3, molecules were drawn using Lewis structures. Lewis structures show the connectivity of all the atoms in a molecule by showing all atoms, bonds, and lone pairs of electrons.

All the information provided by a Lewis structure is important, but as molecules get larger, chemists simplify the representations by still showing the whole molecule but omitting some of the details. For example, because we have established that an uncharged oxygen will always have two bonds and two lone pairs of electrons in a covalent compound, the lone pairs of electrons are often omitted (yet implied) when drawing oxygen in an organic compound.

What's an
 Inquiry Question?

Inquiry Questions are designed to focus your reading on the main concepts by section. As you read each section, see if you can answer each Inquiry Question in your own words.

? 4.1 Inquiry Question

How are organic compounds represented?

Condensed Structural Formulas

The first representation we will consider is the **condensed structural formula,** also called the condensed structure. Like Lewis structures, these structures show *all the atoms* in a molecule, but condensed structures show *as few bonds as possible.* Condensed structures may or may not show lone pairs.

When drawing or interpreting condensed structures, it is helpful to remember the preferred covalent bonding patterns shown in Chapter 3 (see Table 3.4). For example, carbon always makes four bonds to other atoms. Because few bonds are shown between atoms in condensed structures, knowing these patterns will help you quickly interpret condensed structural formulas.

The condensed structural formula for the organic molecule propane, C_3H_8, is shown between its molecular formula and its Lewis structure in the figure below. Note the differences. The molecular formula shows only the number of each atom in the molecule, while the Lewis structure shows the molecule's complete connectivity—all atoms and all bonds. With regard to the amount of information provided, the condensed structural formula is somewhere between the two. The arrangement of the carbons and hydrogens is expanded in the condensed structure as compared to the molecular formula, but bonds are not shown as in the Lewis structure. Notice that the hydrogens are shown after the carbon to which they are bonded in propane's condensed structure.

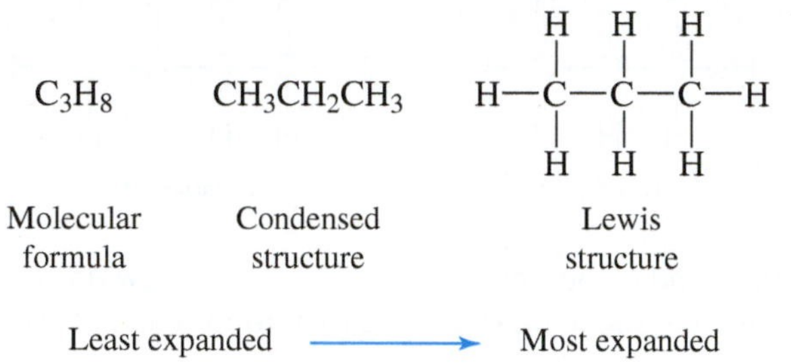

C_3H_8	$CH_3CH_2CH_3$	(Lewis structure of propane)
Molecular formula	Condensed structure	Lewis structure

Least expanded ⟶ Most expanded

In the condensed structural formula, the atoms are listed from left to right in the order in which they are connected in the molecule. It is understood that the octet of each atom is completed according to its preferred pattern of covalent bonding.

Notice that a carbon bonded to three hydrogen atoms in the Lewis structure is written as CH_3 (or sometimes H_3C) in the condensed structure and a carbon bonded to two hydrogen atoms is written as CH_2. In each case, the carbon atom's preferred bonding pattern of four bonds is completed by bonding to other carbon atoms.

Sample Problem 4.1   Drawing Condensed Structural Formulas

Draw a condensed structural formula for each of the Lewis structures shown:

a. (Lewis structure, five-carbon chain)

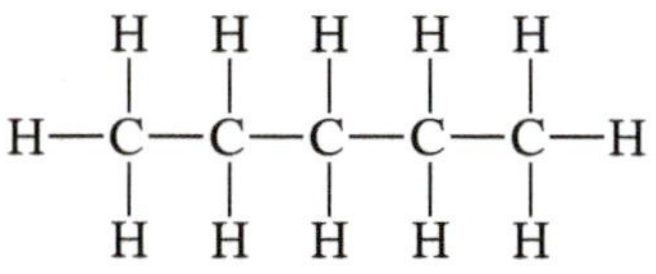

b. (Lewis structure, two-carbon chain)

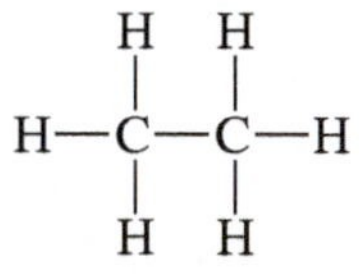

Solution

The condensed structural formula shows all atoms but omits the bonds between atoms. To write a condensed formula, begin with the carbon atom farthest to the left in the structure. List that carbon and then the hydrogens bonded to it, using a subscript to indicate the number of hydrogens. Proceed in the same manner for each carbon atom in the Lewis structure.

a. The correct condensed structure is $CH_3CH_2CH_2CH_2CH_3$. To further condense this structure, it could be written as $CH_3(CH_2)_3CH_3$.

b. The correct condensed structure is CH_3CH_3.

Sample Problem 4.2	**Converting Condensed Structures to Lewis Structures**

Draw a Lewis structure for each of the condensed structures shown:

a. $CH_3CH_2CH_2CH_3$
b. $CH_3CH=CHCH_2CH_2CH_2CH_3$

Solution

Reviewing the preferred bonding patterns from Table 3.4, notice that hydrogen can make only one bond, while carbon makes four. The H following the C typically implies that those hydrogens are bonded to the preceding carbon. Bond the hydrogens to their respective carbons to draw the Lewis structures.

a.

$$\begin{array}{ccccccc} & H & H & H & H \\ & | & | & | & | \\ H- & C & - & C & - & C & - & C & -H \\ & | & | & | & | \\ & H & H & H & H \end{array}$$

b. Notice the double bond between the second and third carbons in this condensed structure. Remember that when drawing Lewis structures, only the connectivity of the atoms is shown, not the shape of the molecule.

Skeletal Structures

The other representation that we will consider is commonly used on pharmaceutical inserts and on the labels of pesticides and herbicides. This type of structure, called a **skeletal structure,** is truly a "bare-bones" structure because it shows the bonds between carbon atoms as lines. Carbon and hydrogen atoms are not shown explicitly. The carbon atoms in skeletal structures are understood to exist at a corner where two lines meet or at the end of a line segment (bond) if no other atom is shown.

The skeletal structure of propane is shown in **FIGURE 4.1** along with other representations. Propane contains three carbons, so it is drawn with just two lines. All bonds to carbon are drawn in a skeletal structure, *except* those to hydrogen. In propane, the middle carbon atom shows two bonds to other carbons, implying that two hydrogens are also bonded to it. The end carbons show one bond. Therefore, they would have three hydrogens bonded to them. Bonds to hydrogen are not shown, but bonds between carbon and other atoms are shown, as you can see in the structure of lansoprazole, the active ingredient in Prevacid®, in the figure in the margin. Because of these rules for skeletal structures, it is not possible to draw a skeletal

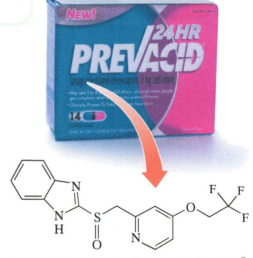

Lansoprazole, the active ingredient in Prevacid®

The package insert of a pharmaceutical often shows the active ingredient in skeletal structure.

Molecular formula	Condensed structure	Lewis structure	Skeletal structure	Ball-and-stick model
C_3H_8	$CH_3CH_2CH_3$			

FIGURE 4.1 Representations of propane. Notice how the skeletal structure compares to the others. In the skeletal structure, the carbon and hydrogen atoms are not shown.

structure for CH_4, methane, because it has only one carbon. In fact, we typically do not draw skeletal structures for compounds with fewer than three carbon atoms.

Drawing skeletal structures takes some practice but can be accomplished by keeping in mind a few simple rules as you draw the structures.

Mastering Video
Practicing the Concepts
Drawing Skeletal Structures of Organic Molecules

Rules for Drawing Skeletal Structures

- Bonds to carbon are shown.
- Bonds between carbon and hydrogen are not shown but are implied.
- Other elements bonded to carbon are drawn at the end of the bond using their symbol.
  - If these atoms have hydrogens bonded to them, these hydrogens are shown.
  - Lone pairs of electrons are not shown.

Solving a Problem

Drawing Skeletal Structures

Draw a skeletal structure for the following given their Lewis structure.

a.
$$H-\underset{\underset{H}{|}}{\overset{\overset{H}{|}}{C}}-\underset{\underset{H}{|}}{\overset{\overset{H}{|}}{C}}-\underset{\underset{H}{|}}{\overset{\overset{H}{|}}{C}}-\underset{\underset{H}{|}}{\overset{\overset{H}{|}}{C}}-H$$

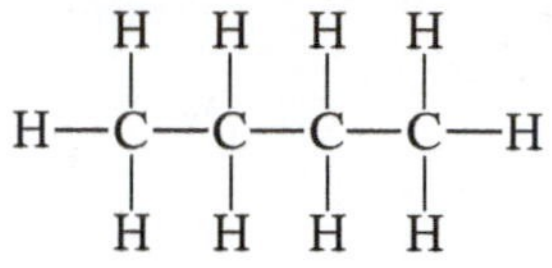

Butane,
used as a fuel in lighters

b.
$$H-\underset{\underset{H}{|}}{\overset{}{C}}-\underset{\underset{H}{|}}{\overset{\overset{H}{|}}{\overset{:O:}{C}}}-\underset{\underset{H}{|}}{\overset{}{C}}-H$$

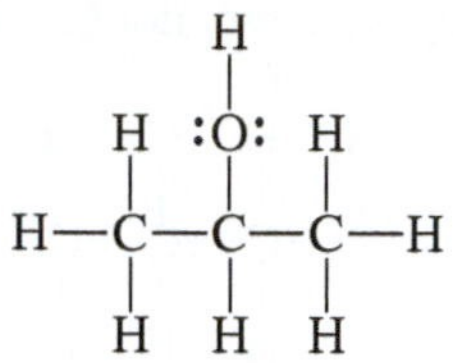

Isopropyl alcohol,
rubbing alcohol

STEP 1 **Determine the number of carbons connected end to end.**

a. In butane, there are four carbons.
b. In isopropyl alcohol, there are three carbons.

STEP 2 **Draw the bonds between the carbons (the carbon skeleton).** These are circled in the molecules that follow. Zigzag the lines so that you can see where one bond ends and the next begins. Sometimes, it is useful to put a dot at the end of the bond so that you can see the position of the carbons.

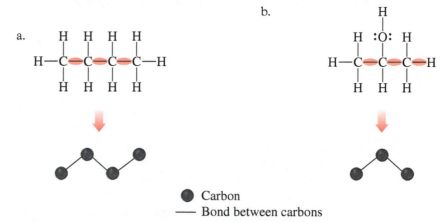

● Carbon
— Bond between carbons

STEP 3 **Draw bonds to noncarbon atoms.** Butane only contains carbon and hydrogen. Isopropyl alcohol contains an oxygen with a hydrogen bonded to it. This is represented in skeletal structure, and while not shown, the lone pair of electrons on the oxygen is implied in the skeletal structure.

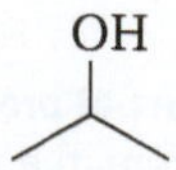

Sample Problem 4.3   Drawing Skeletal Structures

Draw a skeletal structure for each of the following compounds:

a. $CH_3CH_2CH_2CH_2CH_2CH_3$

b. NH_2
    $|$
    CH_3CHCH_3

Hexane, used in the production of glue

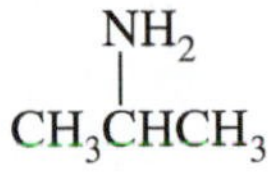

Isopropylamine, used in the manufacture of herbicides

Solution

STEP 1 Determine the number of carbons connected end to end.

a. In hexane there are six carbons.
b. In isopropylamine there are three carbons.

STEP 2 Draw the carbon skeleton. Zigzag the six carbons of hexane. Sometimes it is useful to draw dots to keep track of the carbon atoms while assembling the skeletal structure. For isopropylamine, zigzag the three carbons.

a.

b.

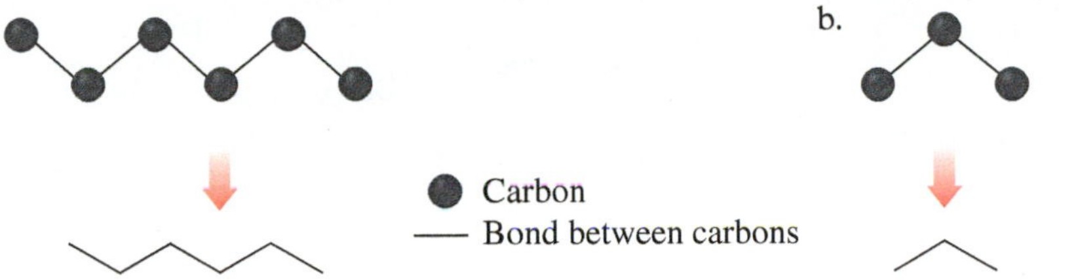

● Carbon
— Bond between carbons

Hexane

Isopropylamine
(carbon skeleton)

STEP 3 Draw bonds to noncarbon atoms. Because hexane has only carbon and hydrogen, its skeletal structure is complete. Isopropylamine has a nitrogen with two hydrogens bonded to it. These are added to the structure as shown below.

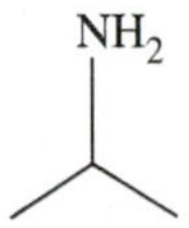

NH_2

Isopropylamine

Practice Problems

Find the answers to the odd-numbered Practice Problems at the end of the chapter.

4.1 Describe the difference between a Lewis structure and a condensed structure in terms of atoms and bonds shown in the structures.

4.2 Describe the difference between a Lewis structure and a skeletal structure in terms of atoms and bonds shown in the structures.

4.3 Explain why it is not possible to draw a skeletal structure for methane.

4.4 Explain what a condensed structure shows that a molecular formula does not.

4.5 Draw a skeletal structure for each of the following compounds:

a. 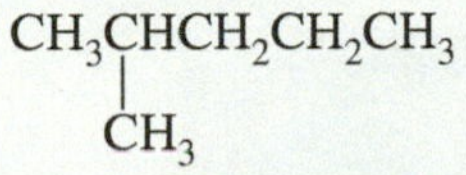 $CH_3CHCH_2CH_2CH_3$
        $|$
        CH_3

b.

H   H   H
$|$ $|$ $|$
H—C—C—C=C
$|$ $|$ $|$ $|$
H   H   H   H

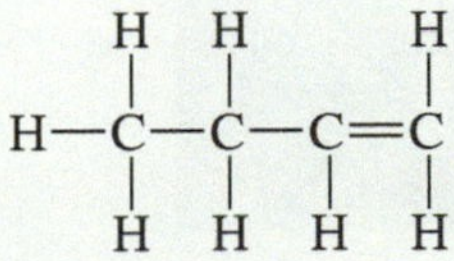

—Continued next page

Continued—

4.6 Draw a skeletal structure for each of the following compounds:

a. 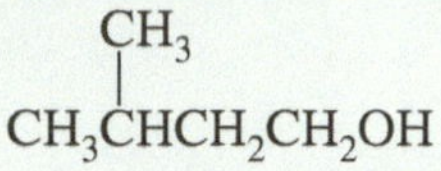 CH₃
   CH₃CHCH₂CH₂OH

b. 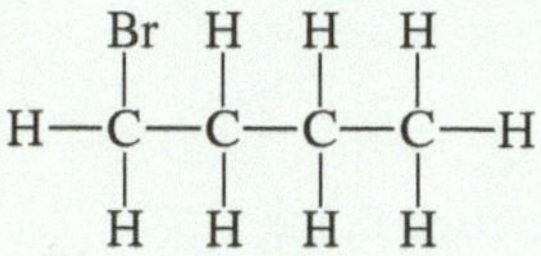 Br H H H
   H—C—C—C—C—H
   H H H H

4.7 Draw a Lewis structure for each of the following compounds:

a. 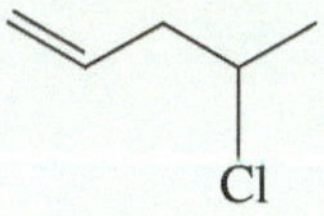 [structure with Cl]

b. 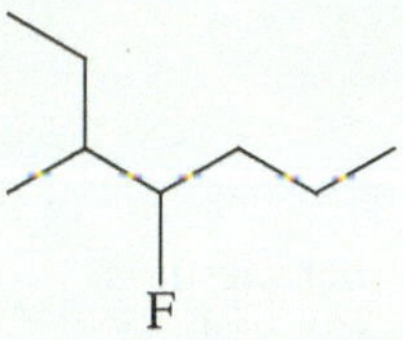 [structure with F]

4.8 Draw a Lewis structure for each of the following compounds:

a. 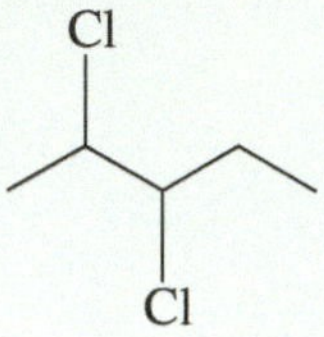 Cl [structure with Cl]

b. 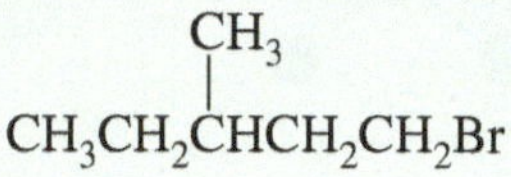 CH₃
   CH₃CH₂CHCH₂CH₂Br

4.9 Draw a condensed structure for each of the following compounds:

a. 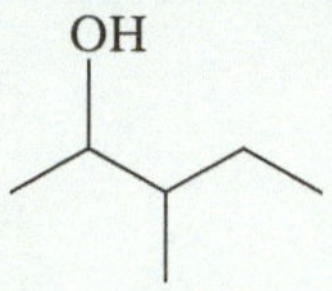 OH [structure]

b. 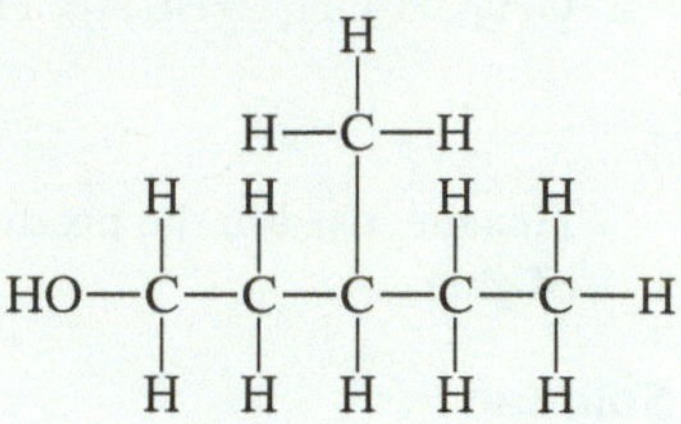 H
   H—C—H
   H H   H H
   HO—C—C—C—C—C—H
   H H   H H H

4.10 Draw a condensed structure for each of the following compounds:

a. 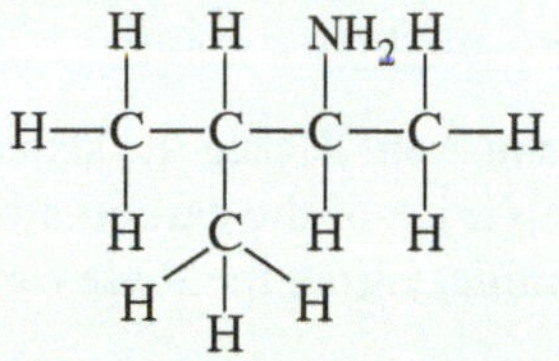 H H NH₂ H
   H—C—C—C—C—H
   H C H H
   H   H

b. 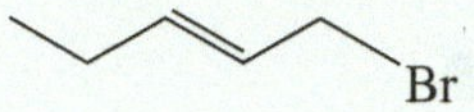 [structure with Br]

4.2 Alkanes: The Simplest Organic Compounds

4.2 Inquiry Question

How are alkanes and cycloalkanes represented?

Now that we can draw organic molecules using condensed and skeletal structures, let's look at some of the families of organic molecules and their properties. We begin with the simplest group, those containing only single-bonded carbon and hydrogen. This family of compounds is known as the **alkanes.** The alkanes are typically referred to as **saturated hydrocarbons.** The term *hydrocarbon* indicates that alkanes are made up entirely of hydrogen and carbon, and *saturated* indicates that these compounds contain only single bonds. Each carbon atom, in addition to being bonded to other carbon atoms, is bonded to the maximum number of hydrogen atoms (which saturate it). Maybe you haven't heard the word *alkane* before, but it is likely you have used some of these molecules in your everyday life. Many of our fossil fuels used for heating, transportation, and generating electricity belong to the alkane family.

Propane representations

C₃H₈
Molecular formula

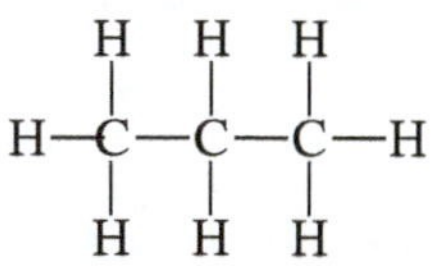

H H H
H—C—C—C—H
H H H
Lewis structure

CH₃CH₂CH₃
Condensed structure

[skeletal structure]
Skeletal structure

Straight-Chain Alkanes

A portable form of heating fuel often used in gas grills, recreational vehicles, and remote locations is the alkane called propane. Compounds like propane belong to the simplest group of alkanes, known as **straight-chain alkanes,** and are made up of carbon atoms joined to one another to form continuous, unbranched chains of varying length. To distinguish individual alkanes from one another, each compound is given a name that is based on the

Propane is one of the straight-chain alkanes. Note in the ball-and-stick model that the chain is not literally "straight." The carbons have a tetrahedral geometry that leads to the zigzag lines we use in skeletal structures.

TABLE 4.1 Names and Structures of the First Ten Straight-Chain Alkanes

Number of Carbon Atoms	Prefix	Name of Alkane	Molecular Formula	Condensed Structure	Skeletal Structure
1	Meth-	Methane	CH_4	CH_4	
2	Eth-	Ethane	C_2H_6	CH_3CH_3	
3	Prop-	Propane	C_3H_8	$CH_3CH_2CH_3$	
4	But-	Butane	C_4H_{10}	$CH_3CH_2CH_2CH_3$	
5	Pent-	Pentane	C_5H_{12}	$CH_3CH_2CH_2CH_2CH_3$	
6	Hex-	Hexane	C_6H_{14}	$CH_3CH_2CH_2CH_2CH_2CH_3$	
7	Hept-	Heptane	C_7H_{16}	$CH_3CH_2CH_2CH_2CH_2CH_2CH_3$	
8	Oct-	Octane	C_8H_{18}	$CH_3CH_2CH_2CH_2CH_2CH_2CH_2CH_3$	
9	Non-	Nonane	C_9H_{20}	$CH_3CH_2CH_2CH_2CH_2CH_2CH_2CH_2CH_3$	
10	Dec-	Decane	$C_{10}H_{22}$	$CH_3CH_2CH_2CH_2CH_2CH_2CH_2CH_2CH_2CH_3$	

number of carbon atoms in its chain. For example, propane is the name for an alkane containing three carbon atoms connected in a straight chain and saturated with hydrogen.

The names, formulas, and condensed structures of the first ten alkanes are shown in **TABLE 4.1**. Skeletal structures are shown for alkanes containing three carbons and higher. The names of the first four alkanes may seem odd, but they are historical names still used today. The names of the straight-chain alkanes with five or more carbons are derived from a Greek numerical prefix (for example, *pent* = 5) followed by the ending *ane* (indicating the alkane family). It is important to recognize the names of these first ten alkanes because the names of many organic compounds are derived from them.

As you look down the column of molecular formulas in Table 4.1, notice that the formulas systematically increase. For every carbon present in an alkane, the number of hydrogens is two times the number of carbons plus two. For propane, this means that for its three carbons, there are $(2 \times 3) + 2$ or a total of eight hydrogens, giving a molecular formula of C_3H_8. We can write an alkane general molecular formula as C_nH_{2n+2} where n is the number of carbons in the alkane.

Cycloalkanes

Besides forming long chains, carbon can also form rings. These compounds are called **cycloalkanes** (*cyclo* implies circle). Because they are saturated hydrocarbons, they are still alkanes. The names of the cycloalkanes are formed by adding the prefix *cyclo* to the alkane name for the compound containing the same number of carbon atoms. For example, the simplest cycloalkane is cyclopropane. Notice that the name indicates that this compound is a ring compound (cyclo) of three carbon atoms (propane).

Cyclopropane representations

Skeletal structure

Lewis structure

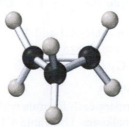

Ball-and-stick model

TABLE 4.2 Cycloalkanes: Names and Common Structures

Name

Cyclopropane	Cyclobutane	Cyclopentane	Cyclohexane

Molecular Formula

C_3H_6	C_4H_8	C_5H_{10}	C_6H_{12}

Ball-and-Stick Models

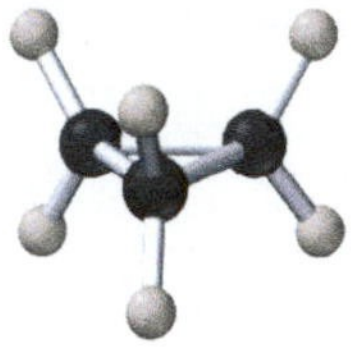

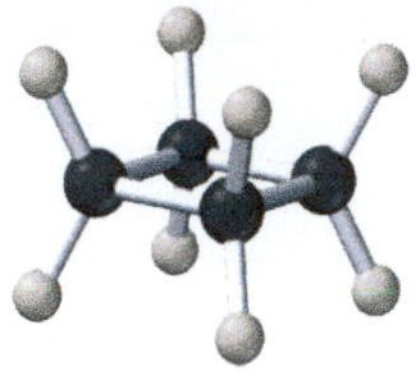

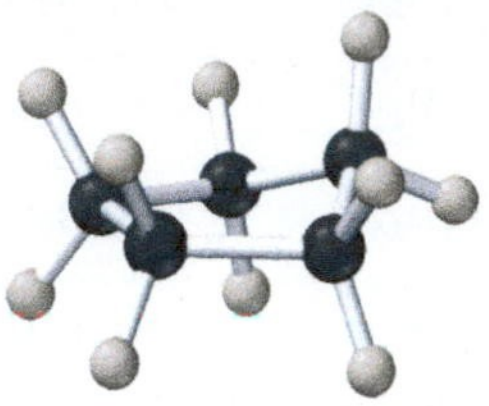

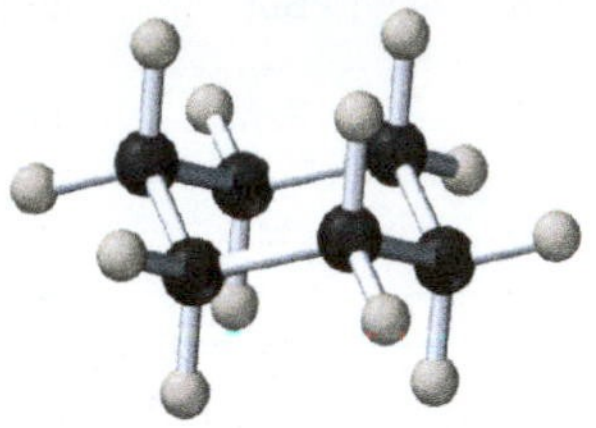

Skeletal Structure

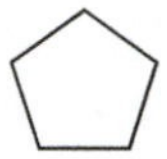

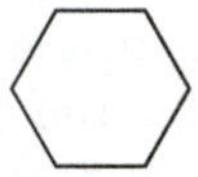

In theory, cycloalkanes (and similar ring-containing organic compounds) can exist in many different ring sizes. However, rings of five and six carbon atoms are the most common in nature; we will focus on these more common rings.

Skeletal structures are particularly useful for representing the cycloalkanes. Lewis structures are rarely used to represent cycloalkanes. When these molecules are represented using skeletal structures, they appear as familiar geometric figures. The three-carbon ring of cyclopropane looks like a triangle, the four-carbon ring of cyclobutane looks like a square, the five-carbon ring of cyclopentane looks like a pentagon, and the six-carbon ring of cyclohexane looks like a hexagon. **TABLE 4.2** gives the names and structures of cycloalkanes with three to six carbons.

Sample Problem 4.4   The Formulas for Cycloalkanes

The general molecular formula for a straight-chain alkane is C_nH_{2n+2}. What is the general molecular formula for a cycloalkane? Explain.

Solution

Inspection of Table 4.2 shows the molecular formulas C_3H_6, C_4H_8, C_5H_{10}, and C_6H_{12} for cyclopropane, cyclobutane, cyclopentane, and cyclohexane, respectively. Because the carbons in a cycloalkane connect as a ring, two fewer hydrogens are present as opposed to the corresponding straight-chain alkane. The general molecular formula is therefore C_nH_{2n}.

Sample Problem 4.5   Distinguishing Alkanes and Cycloalkanes

Using Tables 4.1 and 4.2, name the alkanes or cycloalkanes with the molecular formulas shown:

   a. C_2H_6 b. C_6H_{12} c. C_9H_{20}

Solution

Compare each formula with the general formulas for straight-chain and cycloalkanes. Use Table 4.1 or 4.2 to find the name of the appropriate alkane.

   a. ethane   b. cyclohexane   c. nonane

Alkanes Are Nonpolar Compounds

In Chapter 3, the polarity of molecules was determined by examining both the electronegativity of the atoms in the bonds and the overall shape of the molecule. Nonpolar molecules have an even distribution of electrons over their entire surface, while polar molecules have a partially negative part and a partially positive part (poles) due to an uneven distribution of electrons in the molecule.

The electronegativities of carbon and hydrogen are so similar that when these two elements form covalent bonds, the electrons are shared equally and the bond is nonpolar. Alkanes are composed solely of carbon and hydrogen, so regardless of their shape, alkanes are nonpolar. The nonpolar nature of alkanes affects their behavior in aqueous systems. (Remember that water is polar.)

Alkanes as Fuel Sources

Because they are nonpolar, alkanes are not considered reactive molecules. However, they readily undergo a reaction with oxygen, called *combustion*. Alkanes therefore make excellent fuel sources. When hydrocarbons are completely combusted, or burned cleanly, the products are carbon dioxide and water. The chemical reaction of combustion is explored further in Chapter 5. Several hydrocarbon fuel sources and their chemical structures are shown in **TABLE 4.3**.

The incomplete combustion of hydrocarbons from fossil fuels results in carbon soot and the greenhouse gas carbon monoxide as products. Natural gas, methane, is one of the cleanest burning alkanes, but exists as a gas at room temperature and is difficult to contain. We continue to search for cleaner, cheaper hydrocarbon fuels to replace fossil fuels isolated from oil deposits. One alternative is biodiesel. Biodiesel can be produced from the hydrocarbons present in vegetable oils such as soybean, canola, and rapeseed. These crop plants provide a readily available source for making biodiesel. Much of the biodiesel today is dispensed as "B20" containing 20% biodiesel and 80% petroleum diesel. A report issued by the U.S. Environmental Protection Agency concluded that biodiesel decreases the tailpipe emission of particulate matter (carbon soot), carbon monoxide, and hydrocarbons when compared to petroleum-derived diesel.

INTEGRATING Chemistry

◀ Find out why alkanes make good fuels.

TABLE 4.3 Some Alkane Fuel Sources

Alkane Name, Formula	Where Commonly Found	Chemical Structure
Methane, CH_4	Natural gas	CH_4
Propane, C_3H_8	Propane gas	
Isooctane, C_8H_{18}	Gasoline	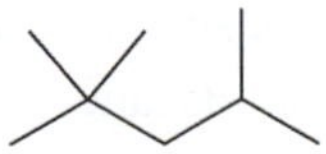
Hexadecane, $C_{16}H_{34}$	Diesel fuel	
Hentriacontane, $C_{31}H_{64}$	Candle wax (paraffin)	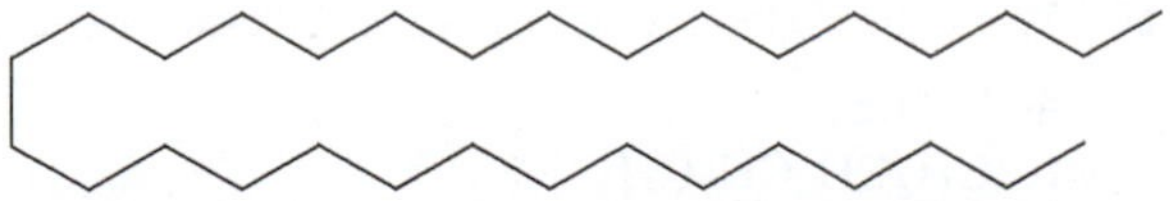

—*Continued next page*

Continued—

Notice in the following figure that the structure of one of the main components of bio-diesel, methyl stearate (19C), is very close in hydrocarbon length to hexadecane (16C) found in diesel fuel, except that the biodiesel molecule contains two oxygen atoms to facilitate its combustion.

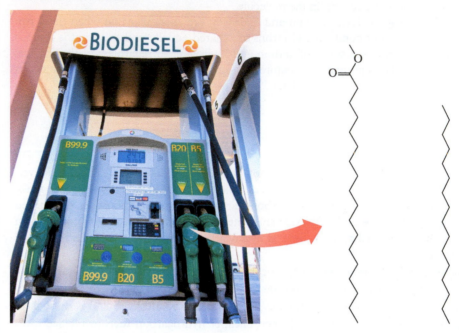

Methyl stearate   Hexadecane
Found in biodiesel   Found in diesel

Methyl stearate, $C_{19}H_{38}O_2$, (shown above) is one of the hydrocarbons found in biodiesel. Do you see the structural similarity to hexadecane found in diesel from fossil fuels?

Practice Problems

Use Tables 4.1 and 4.2 to help you answer these practice problems.

4.11 Name the straight-chain alkanes or cycloalkanes whose structure or formula is shown:

a. $CH_3CH_2CH_2CH_2CH_3$ b. C_6H_{12} c.

4.12 Name the straight-chain alkanes or cycloalkanes whose structure or formula is shown:

a. C_2H_6 b.   c. C_4H_{10}

4.13 Write the condensed structure for the straight-chain alkanes shown:

a. heptane   b. methane   c. decane

4.14 Write the condensed structure for the straight-chain alkanes shown:

a. butane   b. octane   c. hexane

4.15 Write the skeletal structure for the alkane or cycloalkane shown:

a. C_6H_{12}

b. $CH_3CH_2CH_2CH_3$

c. $CH_3CH_2CH_2CH_2CH_2CH_2CH_2CH_3$

4.16 Write the skeletal structure for the alkane or cycloalkane shown:

a. C_4H_8 b. $CH_3CH_2CH_2CH_2CH_2CH_3$ c. $C_{10}H_{22}$

DISCOVERING THE CONCEPTS

? INQUIRY ACTIVITY—Meet the Hydrocarbons

Information
Organic Compounds

- Contain carbon
- Form covalent bonds
- Are grouped into families
  - The families, with the exception of the alkanes, are classified by the *functional group(s)* present.
  - Functional groups are the reactive parts of molecules. Polar groups are more reactive than nonpolar groups.

The Hydrocarbon Family of Organic Compounds	
Family Member	**Functional Group**
Alkane	None; contain only C and H single bonds
Alkene	C=C double bond
Alkyne	C≡C triple bond
Aromatic	Planar, ring structures based on benzene, shown here. Can contain *heteroatoms* such as N, O, or S. 

Hydrocarbon Molecule Set

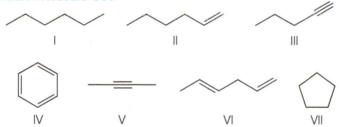

Questions

1. Decide whether the following molecular formulas represent organic or inorganic compounds.
   a. KCl   b. C_4H_{10} c. CH_3CH_2OH d. H_2SO_4
2. Consider the hydrocarbon molecule set shown. How are all the molecules the same? What do all hydrocarbons have in common?
3. Which molecules in the molecule set are polar? Which are nonpolar?
4. Based on the information given above, are hydrocarbons very reactive molecules?
5. Identify the molecules in the molecule set that belong to the following families.

Family Member	Molecule Number
Alkane	
Alkene	
Aromatic	
Alkyne	

6. What is the molecular shape (geometry) of the carbon atoms in molecule IV in the molecule set?
7. Distinguish between an organic family of compounds and its functional group.

4.3 Families of Organic Compounds— Functional Groups

? **4.3 Inquiry Question**

How are carbon compounds organized?

There are innumerable organic compounds, which means there must be a way of organizing all of them into manageable groups or families. So far, we have met the saturated hydrocarbon family of organic compounds called alkanes. Yet organic compounds can also contain oxygen, nitrogen, sulfur, phosphorus, and other elements. Elements other than carbon and hydrogen that are present in an organic compound are called **heteroatoms.**

There are distinct patterns to bonding in organic compounds. A group of atoms bonded in a particular way is called a **functional group.** Organic compounds are classified into families based on common functional groups (their bonding patterns). Each functional group has specific properties and chemical reactivity. Organic compounds that contain the same functional group behave alike. By identifying the properties and reactivity of fewer than 20 functional groups, we can understand a lot about the properties and reactivity of many of the millions of organic compounds.

Some common functional groups are listed in **TABLE 4.4**. The first four groups are the hydrocarbons, families of compounds that contain just carbon and hydrogen. Notice, however, that most of the functional groups listed in the table contain heteroatoms—oxygen, nitrogen, and sulfur. Also notice that many of the functional groups contain a carbon double bonded to an oxygen. The $C=O$ group found in several families is called a **carbonyl** (pronounced carbon-NEEL). We will encounter many of the functional groups listed here in later chapters as they appear in the structures of biomolecules.

The functional group is the reactive part of an organic molecule. To keep the focus on the functional group, an *R* is often used to represent the *R*est of the molecule. The R part is usually bonded to the functional group through a carbon bond. For example, many molecules fit into the family of **carboxylic acids.** This family's functional group contains a carbonyl bonded to an —OH group. Acetic acid, the main component of vinegar, and palmitic acid, a fatty acid discussed later in this section, are both part of the carboxylic acid family because both contain this functional group, even though the rest of the molecule, R, is different.

Representative structure of
a carboxylic acid

Acetic acid,
a carboxylic acid

Palmitic acid,
a carboxylic acid

Both acetic acid and palmitic acid contain the carboxylic acid functional group (orange rectangle) and are in the same family of organic molecules (R group highlighted in blue).

The use of R allows us to simplify a structure and highlight just the functional group of interest. Remember that the abbreviation R can represent anything from one carbon to a complex group containing many carbons.

TABLE 4.4 Families of Organic Compounds: Common Functional Groups

Family Name	Representative Structure of the Functional Group	Example of Compound Containing Functional Group
Alkane (saturated hydrocarbon)	All C—C bonds are single	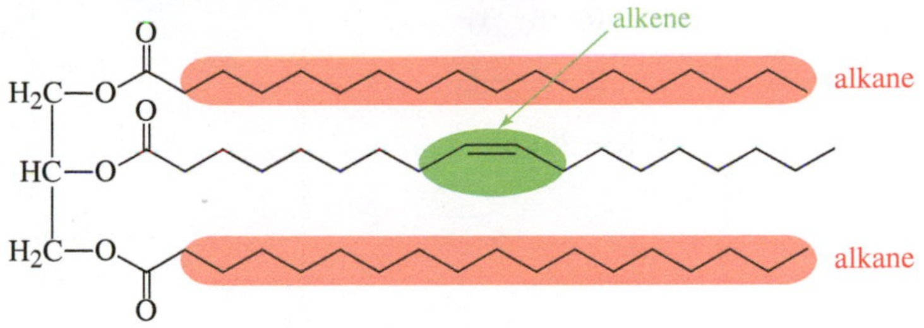 A dietary fat called a triglyceride
Alkene (unsaturated hydrocarbon)	Contain C=C double bonds	
Alkyne (unsaturated hydrocarbon)	Contain C≡C triple bonds	
Aromatic	Planar, ring structures, based on benzene. Can contain heteroatoms. 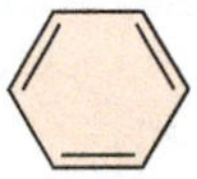	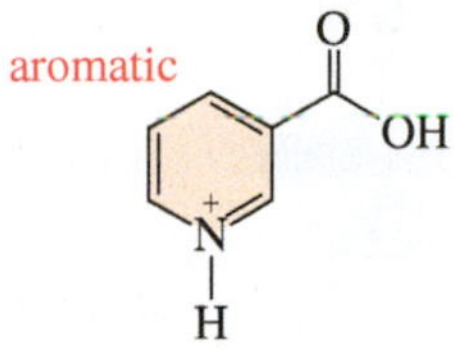 Niacin, vitamin B$_3$
Alcohol	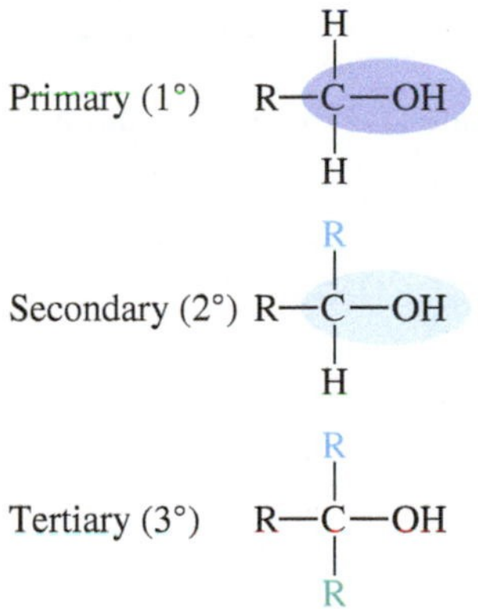 Primary (1°) R—C—OH Secondary (2°) R—C—OH Tertiary (3°) R—C—OH	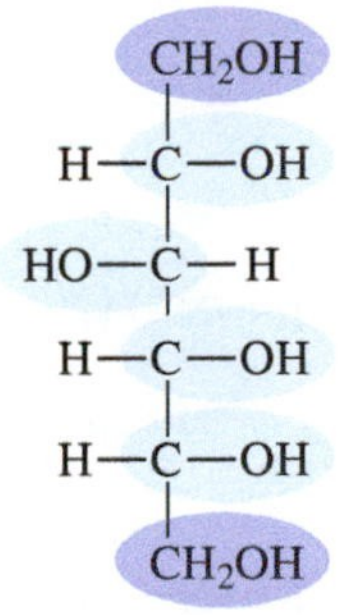 Sorbitol, a low-calorie sweetner
Phenol (aromatic alcohol)	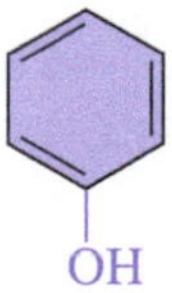 OH	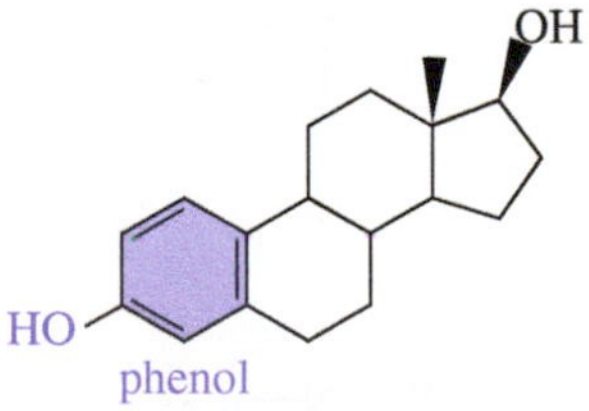 Estradiol, primary female sex hormone
Ether	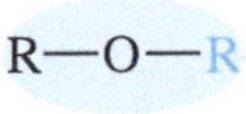 R—O—R	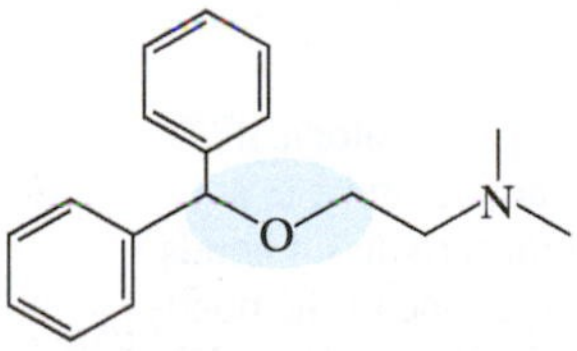 Diphenhydramine, the active ingredient in the antihistamine Benadryl®

Note: R represents the "Rest" of the molecule bonded through carbon. See text.

—Continued next page

TABLE 4.4 Families of Organic Compounds: Common Functional Groups (*continued*)

Family Name	Representative Structure of the Functional Group	Example of Compound Containing Functional Group
Thiol Sulfide (thioether) Disulfide	$R-S-H$ $R-S-R$ $R-S-S-R$	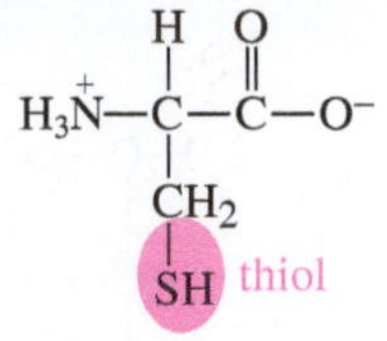 The amino acid cysteine 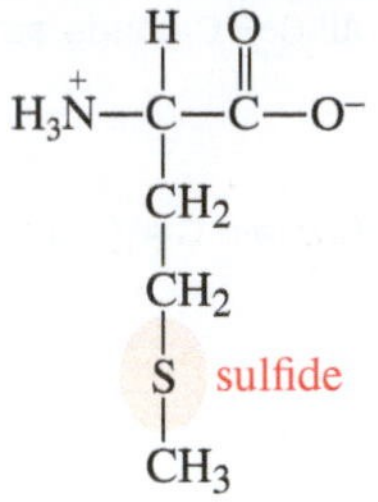 The amino acid methionine
Phosphate	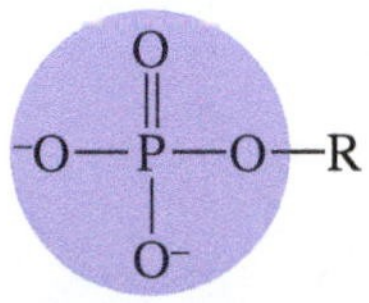	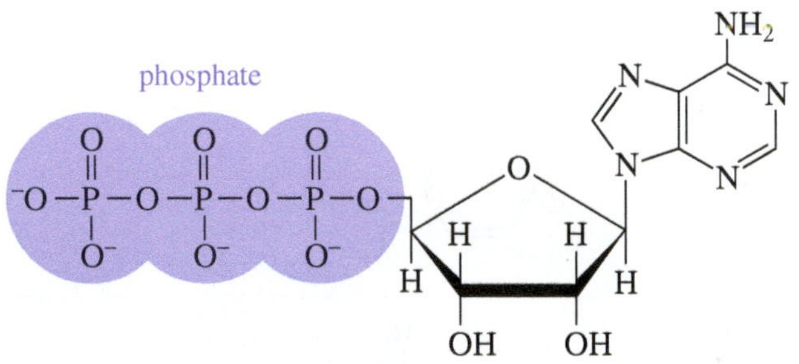 Adenosine triphosphate (ATP) contains three phosphates
Amines and protonated amines	Primary (1°) $R-N-H$ with H below Secondary (2°) $R-N-H$ with R below Tertiary (3°) $R-N-R$ with R below Quaternary (4°) $R-\overset{+}{N}-R$ with R above and below	 The stimulant amphetamine The neurotransmitter acetylcholine
	Protonated Under normal physiological conditions, amines exist in a protonated form (N has 4 bonds and a + charge; one of the bonds must be to H). Can be 1° — RNH_3^+, 2° — $R_2NH_2^+$ or 3° — R_3NH^+.	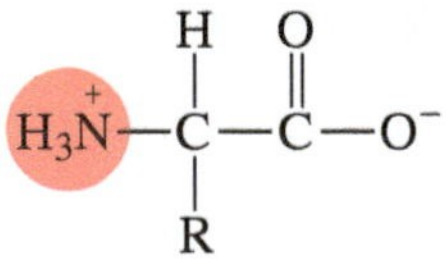 Amino acids contain a protonated amine

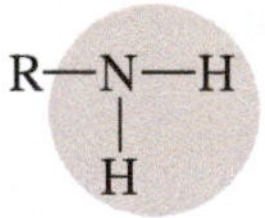

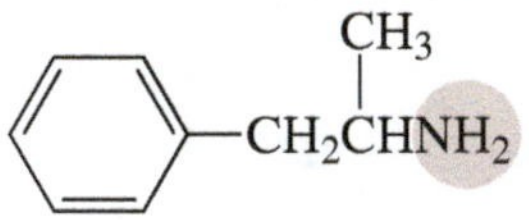

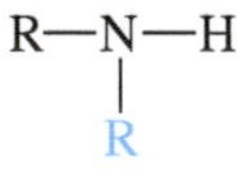

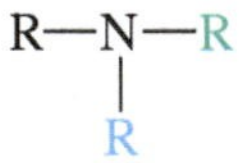

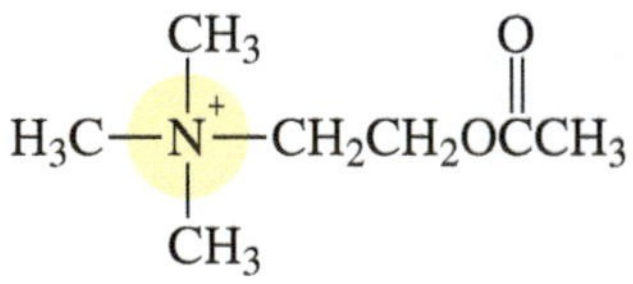

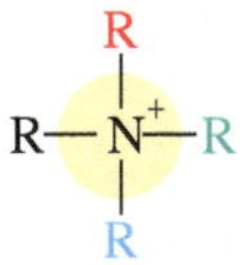

Note: R represents the "Rest" of the molecule bonded through carbon. See text.

TABLE 4.4 Families of Organic Compounds: Common Functional Groups (*continued*)

Family Name	Representative Structure of the Functional Group	Example of Compound Containing Functional Group
Carbonyl	All the remaining functional groups contain the carbonyl group, $C{=}O$.	

Acetyl

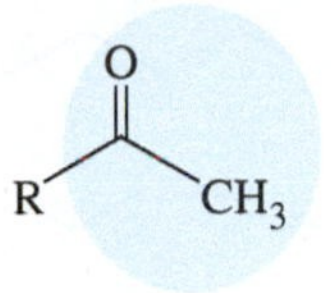

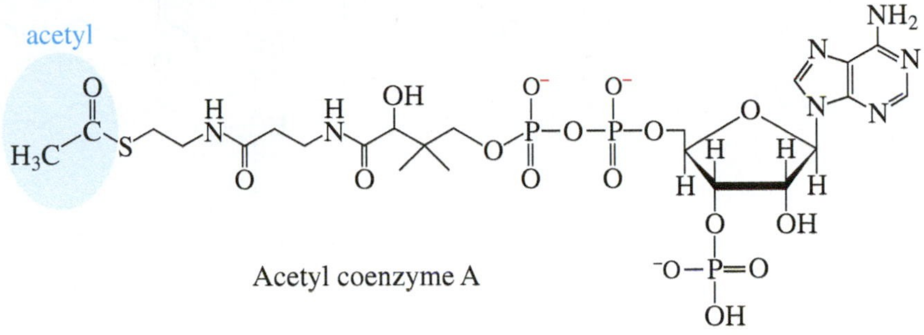

acetyl

Acetyl coenzyme A

Aldehyde

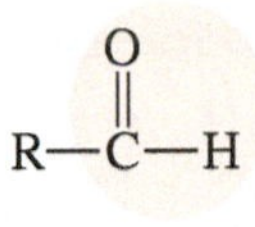

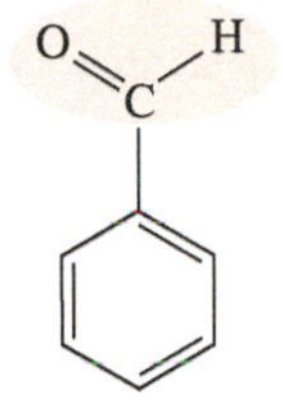

$$R{-}\overset{\overset{\displaystyle O}{\|}}{C}{-}H$$

aldehyde

Benzaldehyde, almond flavoring

Ketone

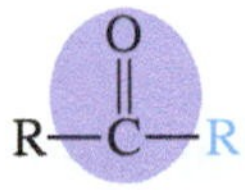

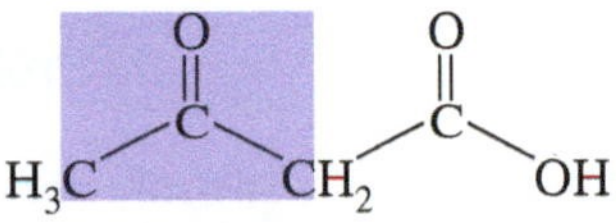

$$R{-}\overset{\overset{\displaystyle O}{\|}}{C}{-}R$$

ketone

Acetoacetic acid, by-product of fatty acid metabolism

Carboxylic acid

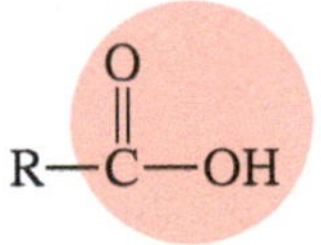

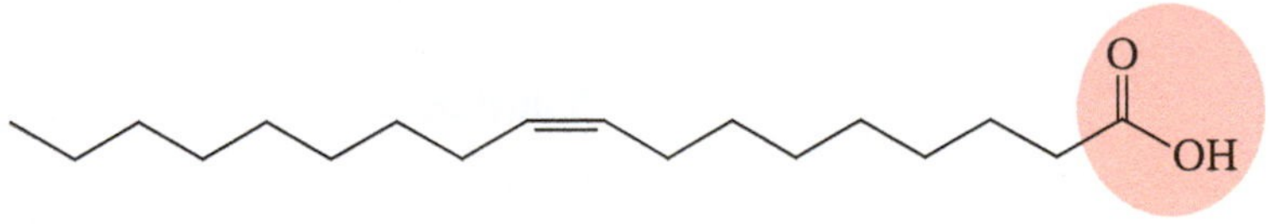

$$R{-}\overset{\overset{\displaystyle O}{\|}}{C}{-}OH$$

The monounsaturated fatty acid, oleic acid   carboxylic acid

Carboxylate

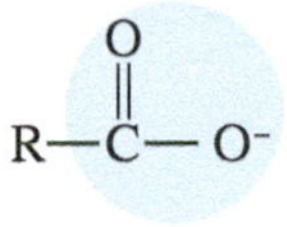

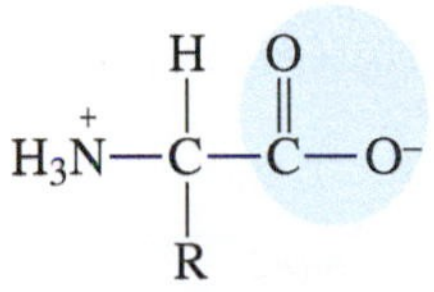

$$R{-}\overset{\overset{\displaystyle O}{\|}}{C}{-}O^-$$

carboxylate

Amino acids have a carboxylate

Note: R represents the "Rest" of the molecule bonded through carbon. See text.

—*Continued next page*

TABLE 4.4 Families of Organic Compounds: Common Functional Groups (*continued*)

Family Name	Representative Structure of the Functional Group	Example of Compound Containing Functional Group
Ester	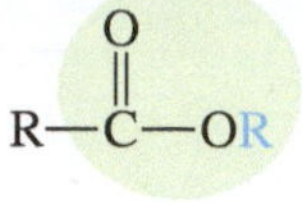	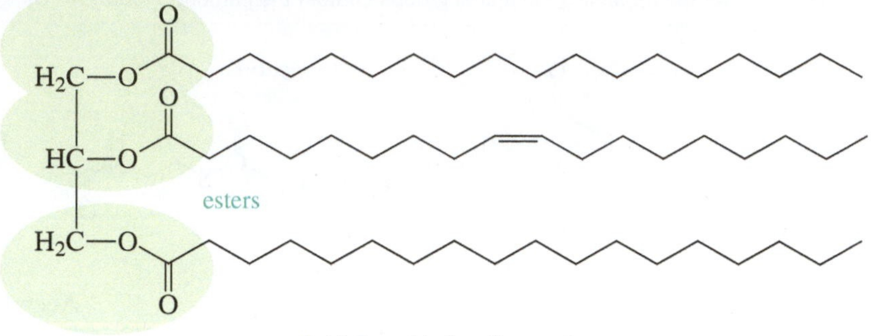 A triglyceride has three esters
Amide	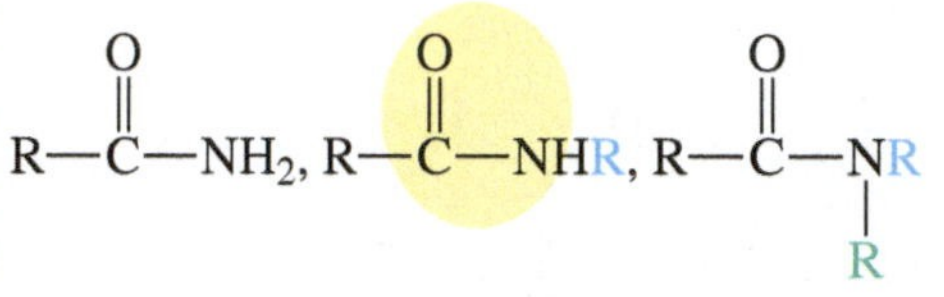	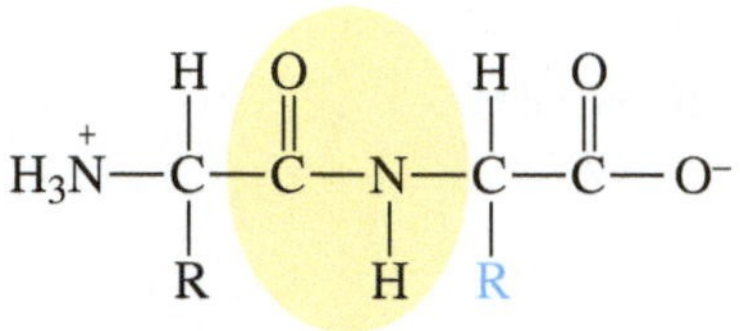 Amino acids join, forming an amides

Note: R represents the "Rest" of the molecule bonded through carbon. See text.

Sample Problem 4.6 Identifying Functional Groups

Identify the functional group in the molecules shown:

a.

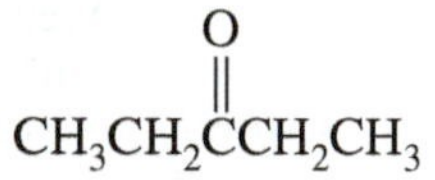

$$CH_3CH_2\overset{\overset{\displaystyle O}{\|}}{C}CH_2CH_3$$

b. $CH_3CH_2NH_2$

c.

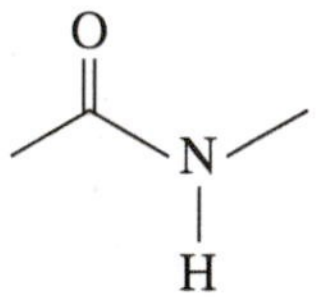

Solution

Each of these molecules has one functional group present. The rest of molecule (R) is alkane. Use Table 4.4 to identify the functional group.

a.

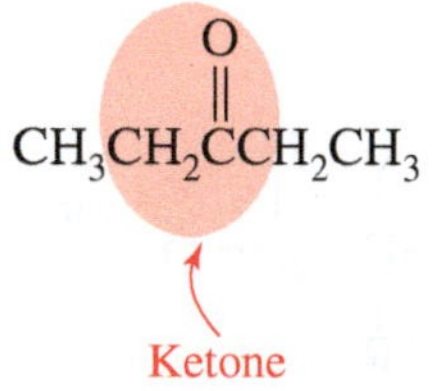

$$CH_3CH_2\overset{\overset{\displaystyle O}{\|}}{C}CH_2CH_3$$
Ketone

b. 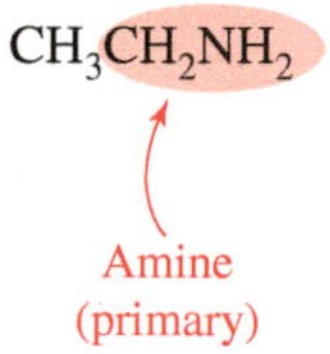 $CH_3CH_2NH_2$
Amine (primary)

c.
Amide

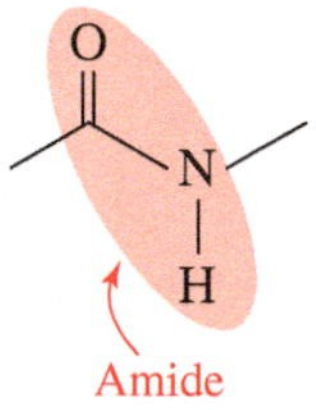

Unsaturated Hydrocarbons—Alkenes, Alkynes, and Aromatics

In Section 4.2, we took a first look at alkanes, the simplest hydrocarbons. Here we look at the remaining hydrocarbon-only families and their functional groups. Organic compounds that contain at least one multiple bond are broadly characterized as unsaturated. Three families of organic compounds contain unsaturated hydrocarbons: alkenes, alkynes, and aromatics.

Alkenes

As tomatoes, bananas, and other fruits ripen on the vine or tree, the plant hormone ethene (also called ethylene) controls the ripening process. An ethene molecule contains a carbon–carbon double bond. A carbon–carbon double bond is an unsaturated hydrocarbon functional group that is the defining characteristic of the **alkene** family. Alkenes are considered **unsaturated hydrocarbons** because they have more than one bond between two carbon atoms, which means those carbons are no longer saturated with hydrogen atoms. In the skeletal structure, the double bond is represented with a second line between the two carbons.

Carbon atoms joined by a double bond are held closer together and more strongly than those joined by just a single bond. So, double bonds between carbon atoms are shorter and stronger than single bonds. However, the second bond of a double bond is not as strong as the first bond. When alkenes react in a reaction called *addition*, the second bond of the double bond is broken and the carbon atoms remain joined by the single bond. This means that alkenes are more reactive than alkanes. We explore addition reactions in Chapter 5.

Alkenes are also found in a set of naturally occurring compounds called **terpenes.** These compounds contain at least one carbon-carbon double bond and are a multiple of five carbons (5, 10, 15, 20, and so on). Beta-carotene, the alkene responsible for the orange pigment in carrots and sweet potatoes, is a common terpene used by our bodies to produce vitamin A. Essential oils in plants also are terpenes. D-Limonene and α-pinene are essential oils found in citrus fruits and pine trees, respectively. These oils give the plant their characteristic smell. Cholesterol and the hormones testosterone and estrogen also contain the alkene functional group and are synthesized from terpenes.

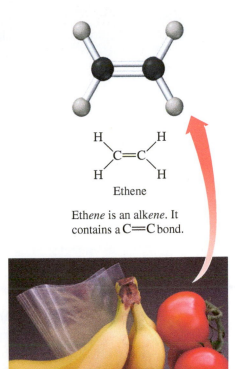

Eth*ene* is an alk*ene*. It contains a C=C bond.

Ethene (ethylene) is a ripening hormone in plants and is also the starting material for many plastics, such as the polyethylene used in plastic bags and milk jugs.

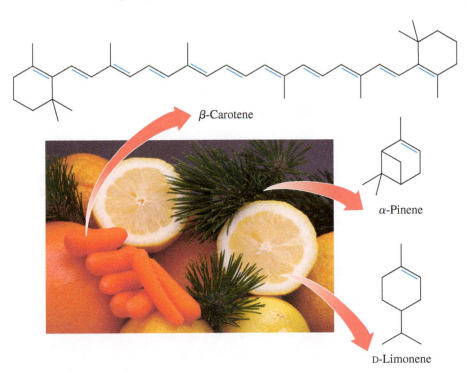

β-Carotene

α-Pinene

D-Limonene

β-Carotene, α-pinene, and D-limonene are terpenes and all contain one or more alkene functional group (blue).

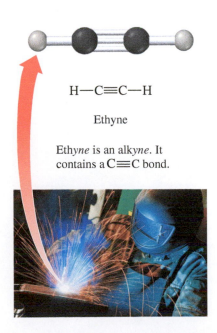

H—C≡C—H

Ethyne

Eth*yne* is an alk*yne*. It contains a C≡C bond.

Alkynes

The fuel used in high-temperature welding torches to melt iron and steel is called ethyne or, more commonly, acetylene. Compounds like acetylene that contain one or more carbon–carbon triple bonds are members of the **alkyne** family. Alkynes are also unsaturated hydrocarbons because they have more than one bond between two carbon atoms.

As you might expect, the alkyne's triple bond is even shorter and stronger than the alkene's double bond. However, the two additional bonds of the triple bond mean that alkynes are even more reactive (and, therefore, less stable) than alkenes. Because they are so reactive, alkynes are rare in nature, occurring mainly in short-lived compounds.

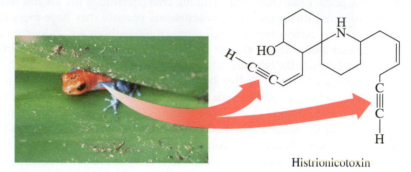

Histrionicotoxin

The poison-dart frog secretes a poison that contains two alkyne functional groups.

Aromatics

The aromatic family of unsaturated hydrocarbons got their name because many of the first compounds discovered had pleasant aromas. Oil of spearmint and peppermint are in the aromatic family. **Aromatic compounds** have a cyclic structure like benzene. With so many double bonds, aromatics might seem to be highly reactive like alkenes. However, aromatics are unusually stable and not very reactive at all. Let's explore why.

Many members of the aromatic family contain the unsaturated six-membered ring that is called a benzene ring. When the benzene ring is part of a larger molecule, it is called a phenyl group. Benzene itself has the structure shown in **FIGURE 4.2**. Notice that benzene appears to have a series of alternating carbon–carbon single and double bonds. From the previous discussion of the lengths of double versus single bonds, we would expect benzene to have three longer single bonds and three shorter double bonds in an alternating pattern. We might also expect such a structure to have reactivity like an alkene.

However, experimental evidence shows that all the carbon–carbon bonds of benzene are the same length and that benzene is an exceptionally stable compound. In fact, benzene resists reactions that would break any (or all) double bonds. Unsaturated cyclic aromatic compounds like benzene, which are unusually and unexpectedly stable (more stable than normal alkenes), are said to exhibit **aromaticity.** The term aromaticity describes the unexpected stability of these compounds and has nothing to do with their smell.

The properties and behavior of benzene and other aromatic compounds arise because the double bonds between adjacent carbon atoms are not static. In other words, the electrons shared to create those double bonds could be in either of the two arrangements shown in **FIGURE 4.3**.

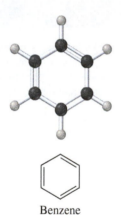

Benzene

FIGURE 4.2 Benzene is aromatic. Many aromatic compounds contain the six-membered ring seen in benzene.

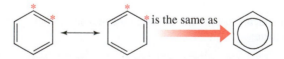

is the same as

FIGURE 4.3 Benzene exhibits resonance. The electrons in the double bonds in benzene are distributed evenly among the six carbons in the ring. The starred carbon atoms are equivalent in both structures.

In fact, the electrons in the double bonds are shared evenly by all six carbons. This is why benzene and benzene rings are often represented by the structure on the far right in the figure. This representation is called the **resonance hybrid** form and implies equal sharing of electrons in benzene instead of distinct double and single bonds.

Because these electrons are free to move among the six carbons, they are much less likely to react with other molecules. This decreased reactivity translates to the enhanced stability exhibited by aromatic compounds.

Aromatic compounds containing two or more benzene rings are called polycyclic aromatic hydrocarbons, or PAHs. Phenanthrene and benzo[*a*]pyrene are two such molecules found in tobacco smoke. Polycyclic aromatic hydrocarbons are often formed when organic matter, such as the tobacco leaf, is burned. Some of these PAHs have been shown to be cancer-causing or carcinogenic. Aromatic compounds are also components of many plastics and pharmaceuticals.

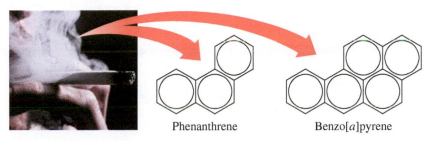

Phenanthrene   Benzo[*a*]pyrene

Tobacco smoke produces many polycyclic aromatic hydrocarbons.

Sample Problem 4.7   **Identifying Unsaturated Hydrocarbons**

Identify the unsaturated hydrocarbon functional group seen in the molecules shown.

a. $CH_3C \equiv CH$ b.    c.

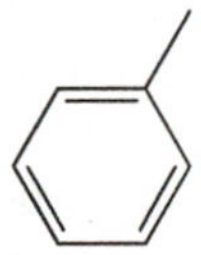

Solution

Recall the definitions of the three unsaturated hydrocarbon functional groups introduced in the section or refer to Table 4.4.

a. $CH_3C \equiv CH$ b. 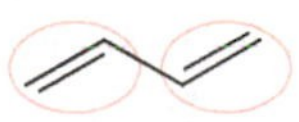   c.

Alkyne   Alkene

Aromatic

Pharmaceuticals Are Organic Compounds

Open the package of any prescription medication and you will find an insert that includes prescribing information, a description of the medication, and its chemical structure. Many pharmaceuticals are large organic molecules that contain many atoms and more than one functional group. As we saw in Chapter 3, we can predict the shape of simple molecules from the arrangement of the bonds around a central atom. In complex molecules, the shapes around the atoms taken as a whole give rise to the overall structure of the compound.

INTEGRATING
Chemistry

◀ Find out why the functional groups and overall shapes of medicines matter.

—*Continued next page*

Continued—

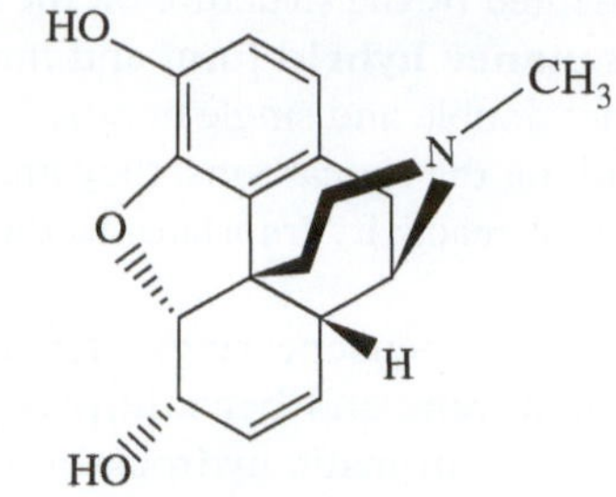

Morphine

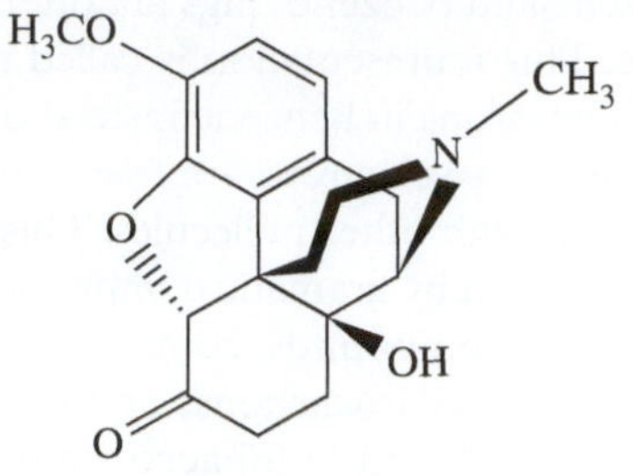

Oxycodone

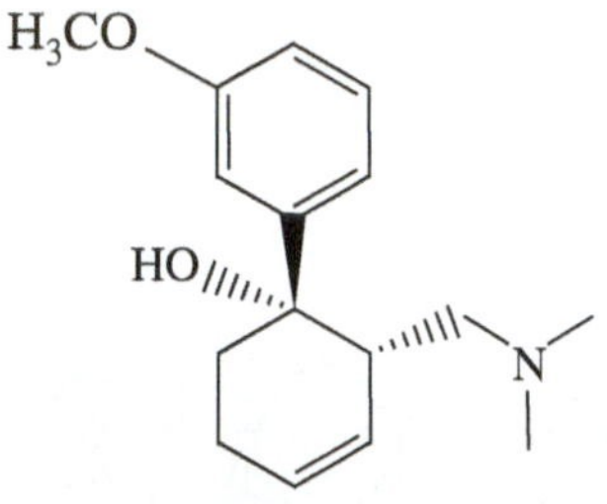

Tramadol

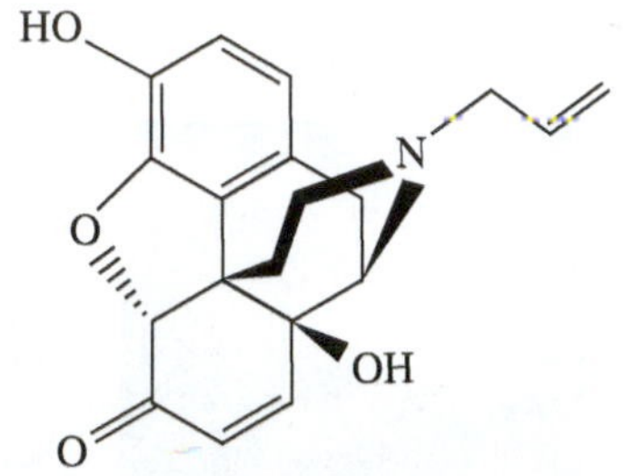

Naloxone

The compounds shown above are a class of powerful pain relievers known as opioids. Morphine, which is isolated from the opium poppy, is one of the most common opioids used to treat moderate to severe pain. Opiods act by binding to specific receptors in the brain and central nervous system. When a molecule binds to a receptor, the shape of the molecule and its functional groups are key to the binding. Compare morphine to oxycodone, another powerful opioid. While these structures may look very complicated, notice that the rings, thick black wedges, dashes, and functional groups are located in similar places. These similarities mean that oxycodone will interact with the opioid receptors and induce a similar therapeutic effect.

When chemists create new pharmaceuticals, they try to mimic the characteristics (functional groups and structure) of the compounds that are known to produce the desired effects. Look above at the structure of tramadol, a pain reliever that is synthesized in the lab. While it is much simpler than morphine or oxycodone, tramadol has similar functional groups and a shape that modestly resembles the others. These similarities result in an effective pain reliever, but one that is less powerful than the other opioids.

Treatments for opioid overdose also exploit molecular similarities. Naloxone, commonly called Narcan®, has a similar structure and some of the same functional groups as the opioids. It binds to the same receptors as the opioid pain relievers; however, it binds much more strongly, effectively blocking the receptors and the effect of the opioids.

Lipids Are Hydrocarbons—Fatty Acids

The media are filled with stories about health issues related to diets high in *saturated* fats. We are encouraged to include monounsaturated and polyunsaturated fats as part of a healthy diet. Earlier, alkanes were described as saturated hydrocarbons, and alkenes, aromatics, and alkynes as unsaturated hydrocarbons. When describing hydrocarbons or fats, the terms *saturated* and *unsaturated* have the same meaning: carbon–carbon single bonds (saturated) or carbon–carbon multiple bonds (unsaturated). Fats with one double bond are called **monounsaturated,** and those with two or more double bonds are called **polyunsaturated.** Here we take a brief look at the main component of fats, saturated fatty acids.

Fatty acids belong to a class of biomolecules called **lipids.** Although lipids are part of a structurally diverse class, they all share the common property of being mainly nonpolar biomolecules. This means most of a lipid molecule is hydrocarbon. Fatty acids also contain a carboxylic acid functional group. We saw carboxylic acids in Table 4.4. See **FIGURE 4.4** for more details on this important functional group.

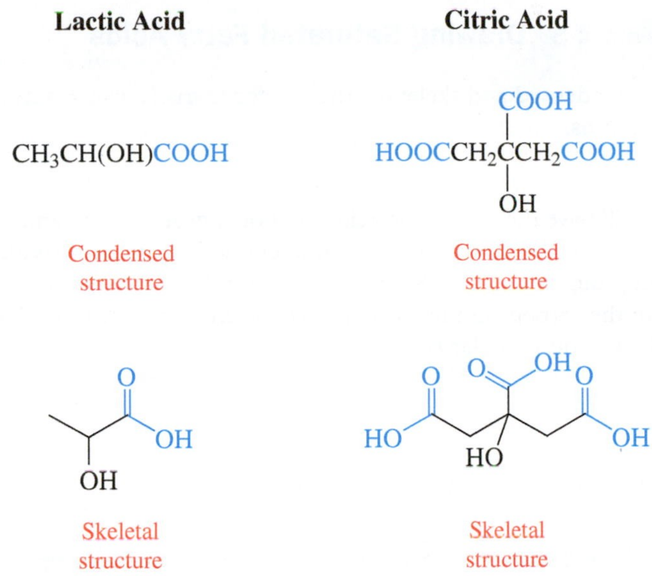

Lactic Acid

$$CH_3CH(OH)COOH$$

Condensed
structure

Skeletal
structure

Citric Acid

$$HOOCCH_2\overset{\displaystyle COOH}{\underset{\displaystyle OH}{C}}CH_2COOH$$

Condensed
structure

Skeletal
structure

FIGURE 4.4 The carboxylic acid functional group. Like the fatty acids, other organic acids such as citric acid, found in fruits, and lactic acid, a by-product of strenuous muscular activity, also contain the carboxylic acid functional group (blue). This group is often represented as COOH in a condensed formula.

Saturated fatty acids are long, straight-chain alkane-like compounds that do not contain double bonds and have a carboxylic acid functional group at one end. The most common and biologically most important fatty acids are compounds containing 12 to 22 carbon atoms (see **TABLE 4.5**). Most naturally occurring fatty acids have even numbers of carbon atoms.

Compare the structures of the compounds in Table 4.5 with those of the straight-chain alkanes in Table 4.1. Notice that the structures of alkanes and saturated fatty acids both contain straight chains of carbon atoms, but the fatty acids contain the carboxylic acid at one end. Organic acids that have the word *acid* as part of their name often contain the carboxylic acid functional group. We will explore unsaturated fatty acids in more detail in Section 4.5.

TABLE 4.5 Saturated Fatty Acids: Names and Structures

Name	Carbon Atoms	Source	Structures
Lauric acid	12	Coconut	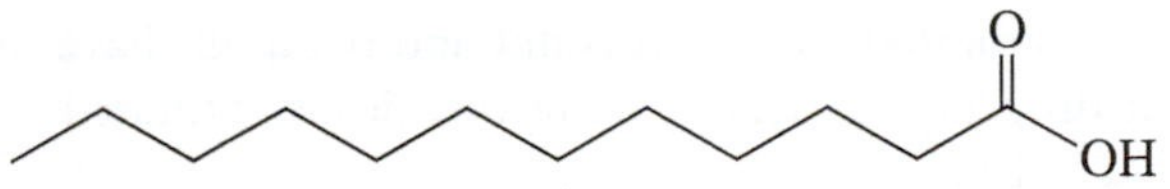
Myristic acid	14	Nutmeg	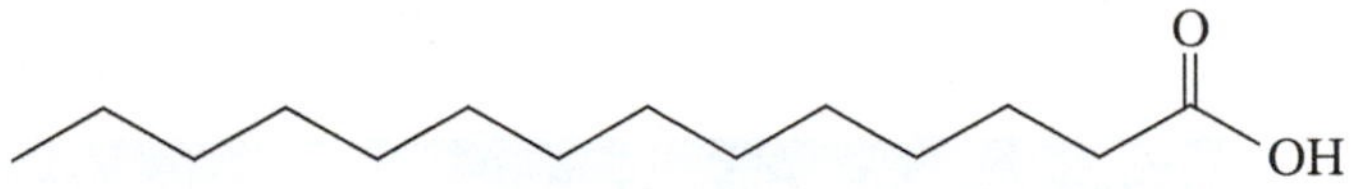
Palmitic acid	16	Palm	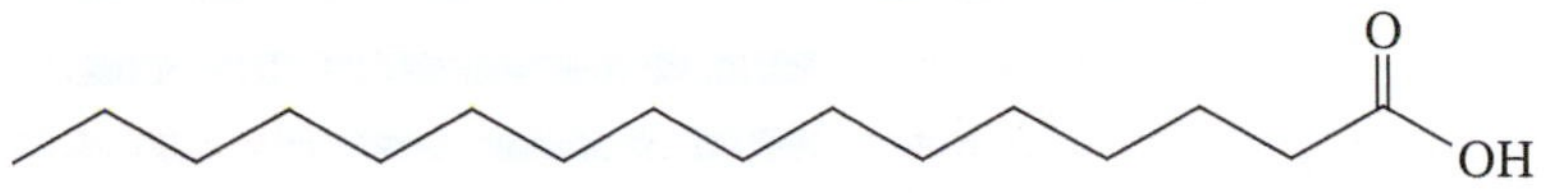
Stearic acid	18	Animal fat	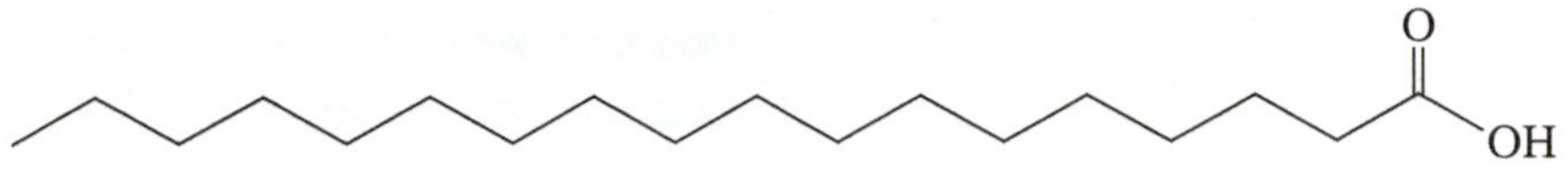
Arachidic acid	20	Peanut	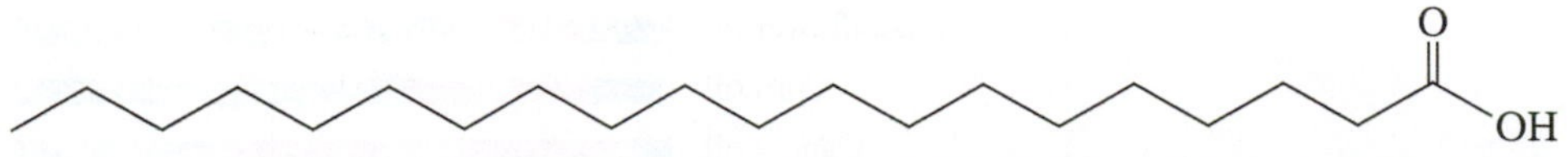
Behenic acid	22	Canola	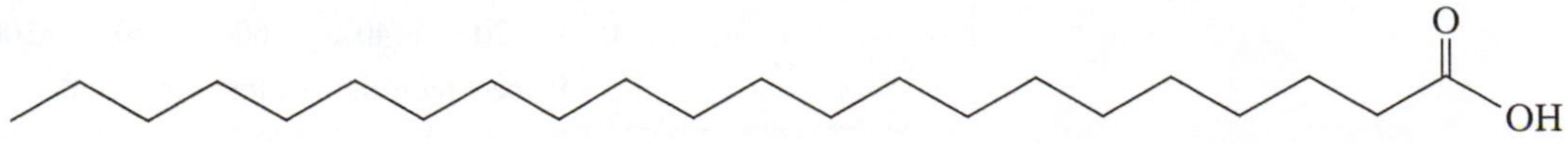

Sample Problem 4.8 Drawing Saturated Fatty Acids

Draw both the condensed and skeletal structure for myristic acid, a saturated fatty acid with 14 carbons.

Solution

All fatty acids will have the carboxylic acid functional group at one end.

For myristic acid, 14 carbons are drawn in condensed structure with a carboxylic acid group at one end. Keep in mind that the carboxylic acid group contains one of the carbons in the 14. The carbons are numbered in the skeletal structure in the solution for clarity.

$$CH_3CH_2CH_2CH_2CH_2CH_2CH_2CH_2CH_2CH_2CH_2CH_2CH_2\overset{\displaystyle O}{\overset{\|}{C}}OH$$

Condensed structure

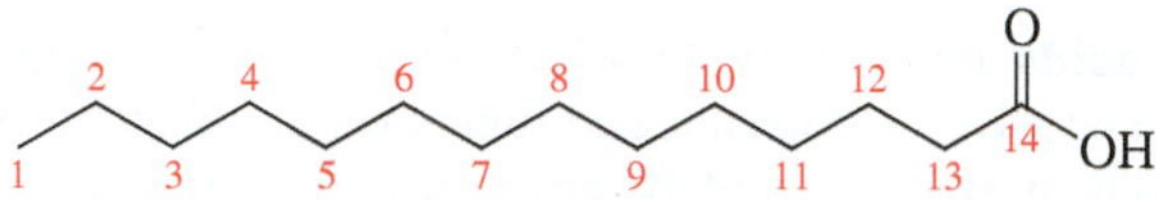

Skeletal structure

INTEGRATING Chemistry

Find out how ▶ the fatty acids in coconut oil are different from those in other oils.

Fatty Acids in Our Diets

Fats are important in a balanced diet because they play important roles as insulators and protective coverings for internal organs and nerve fibers. Mono- and polyunsaturated fats are part of a healthy diet. The Food and Drug Administration (FDA) recommends that a maximum of 30% of the calories in a normal diet come from such fatty acid-containing compounds. The FDA also recommends that the majority of our fat intake come from foods containing a higher percentage of mono- and polyunsaturated fatty acids. **TABLE 4.6** shows the fatty acid composition of some common fats. Highly saturated oils like coconut and palm oils have found uses as natural substitutes for hydrogenated oils, which are chemically saturated. More about hydrogenated oils is found in Chapter 5.

TABLE 4.6 Fatty Acid Composition of Common Dietary Fats

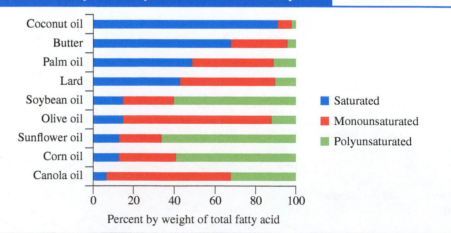

Coconut oil has become increasingly popular in recent years, but its purported benefits are still not well understood. The fatty acids in coconut oil are more than 90% saturated; however, most of the fatty acid chains are between 8 and 12 carbons, much shorter than most dietary sources. While the medium-chain fatty acids found in coconut oil (predominantly the 12-carbon lauric acid) are more easily absorbed and digested by the body, both the U.S. FDA and the American Medical Association still recommend that coconut oil be avoided due to its high saturated fat content. Because most of the fatty acids do not contain double bonds, they are not very reactive, so coconut oil has a longer shelf life than other common oils, butter, or lard. As with many health fads, being an educated and informed consumer is the key to making wise choices. For fats in our diets, the research is clear—unsaturated fats, in moderation, are the best choice.

Coconut oil has high amounts of saturated fats.

Practice Problems

4.17 Identify the family of hydrocarbon present in the following:

a. $H_3CC{\equiv}CH$ b. ⌇⌇⌇   c. ⬡

4.18 Identify the family of hydrocarbon present in the following:

a. $CH_3CH_2CH{=}CH_2$

b.

Naphthalene, makes up moth balls

c.
$$CH_3\overset{\overset{\displaystyle CH_3}{|}}{C}CH_2\underset{\underset{\displaystyle CH_3}{|}}{C}HCH_3$$

Isooctane, a component in gasoline

4.19 Identify all the functional groups present in the following:

a.

Benzocain, active ingredient in Orajel®

b. ⌇⌇SH

Allyl mercaptan, found in chives

4.20 Identify all the functional groups present in the following:

a.

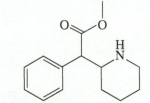

Aspartame, an artificial sweetner

b.

Methylphenidate, the active ingredient in Ritalin®

4.21 The most prevalent fatty acid in coconut oil is lauric acid, a saturated fatty acid containing 12 carbons. Draw lauric acid in skeletal structure.

4.22 The most common fatty acid found in animals is stearic acid, a saturated fatty acid containing 18 carbons. Draw stearic acid in skeletal structure.

4.4 Nomenclature of Simple Alkanes

Much as a limb on a tree branches off a tree's trunk, carbon atoms can branch off the straight chain of carbon atoms in an alkane. Alkanes that do not have all their carbon atoms connected in a single continuous chain are called **branched-chain alkanes.** Straight-chain alkanes and branched-chain alkanes are given different names to avoid confusion.

 4.4 Inquiry Question

How are alkanes named?

Here we explore the systematic naming (also called *nomenclature*) developed to precisely name organic compounds. This naming system was developed by the International Union of Pure and Applied Chemistry (IUPAC) and is used by chemists worldwide. The IUPAC nomenclature rules provide a unique name for any organic compound and ensure that every compound, like two different octanes, has just one correct systematic name.

The names of organic compounds have three basic parts.

Substituents	+	Parent Name	+	Suffix
(Attachments)		*(Number of carbons in the longest continuous chain)*		*(Family name)*

The naming of any organic compound follows this outline. Here, we name some simple alkanes to give you a working understanding of organic nomenclature.

Branched-Chain Alkanes

In Section 4.2, the names of the first ten straight-chain alkanes were introduced. These names form the basis upon which the names of the branched-chain alkanes are built.

Solving a Problem

Naming Branched-Chain Alkanes

Provide the IUPAC name for the compound shown.

$$
\begin{array}{ccccc}
 & H & H & H & \\
 & | & | & | & \\
H- & C & -C- & C & -H \\
 & | & | & | & \\
 & H & | & H & \\
 & & H-C-H & & \\
 & & | & & \\
 & & H & &
\end{array}
$$

STEP 1 Find the longest continuous chain of carbon atoms. This is called the parent chain. To find the longest chain, place your finger on a carbon at one end of the molecule and trace to an end carbon without picking up your finger, counting each carbon along the way. Do this for each end carbon. The longest chain will be the one with the most carbons. Name the parent chain according to the alkane names given in Table 4.1. (*Hint*: The longest chain may not be a straight, horizontal carbon chain.)

$$
\begin{array}{ccccc}
 & H & H & H & \\
 & | & | & | & \\
H- & C & -C- & C & -H \\
 & | & | & | & \\
 & H & | & H & \\
 & & H-C-H & & \\
 & & | & & \\
 & & H & &
\end{array}
$$

In this example, the name of the parent chain is *propane* since the longest chain has three carbons (circled).

STEP 2 Identify the groups bonded to the main chain but not included in the main chain (ignore hydrogen). These groups are known as **substituents,** and, in the case of branched-chain alkanes, they are called **alkyl groups.** The name of each alkyl group is derived from the alkane with the same number of carbon atoms by changing the *ane* ending of the name of the corresponding alkane to *yl*. For example, a one-carbon alkyl group is named as a *methyl* group. [Methane (CH_4) minus *ane* = meth, and meth plus *yl* = methyl ($—CH_3$).] Similarly, a two-carbon alkyl group is an ethyl group, a three-carbon alkyl group is propyl, and so forth (see **TABLE 4.7**). So, in our example, the substituent is a methyl group (circled).

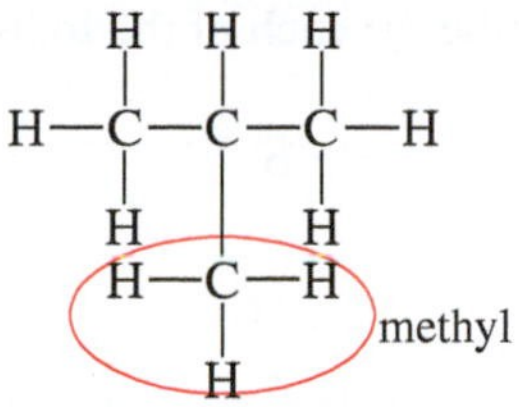

STEP 3 **Number the carbons of the parent chain starting at the end nearer to a substituent.** In this example, the substituent is on carbon number 2 no matter from which end you start numbering.

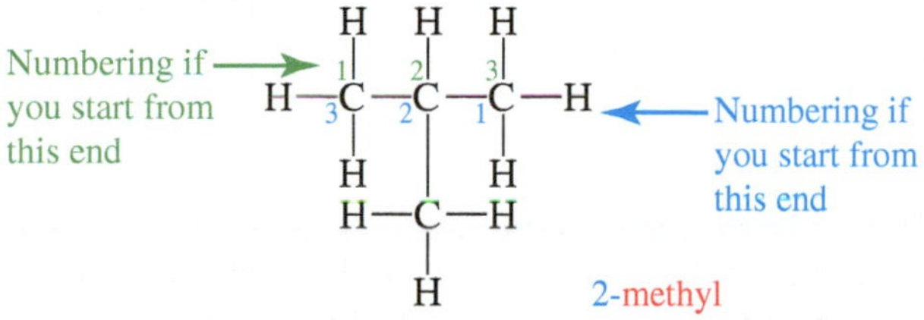

STEP 4 **Assign a number to each substituent based on location, listing the substituents in alphabetical order at the beginning of the name.** Separate numbers and words in the name by a dash.

The IUPAC name of this compound is 2-methylpropane.

(*Note*: If more than one of the same type of substituent is present in the compound, you would indicate this using the Greek prefixes *di-*, *tri-*, and *tetra-* but ignore these prefixes when alphabetizing. For example, two methyl groups branching from a parent chain would be named *di*methyl.)

TABLE 4.7 Four Simplest Alkyl Substituents

	Methyl	Ethyl	Propyl	Isopropyl
Lewis structure	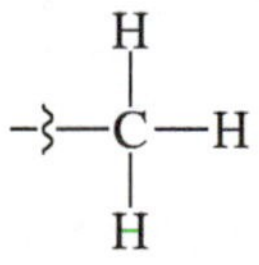	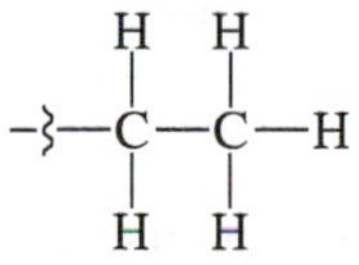	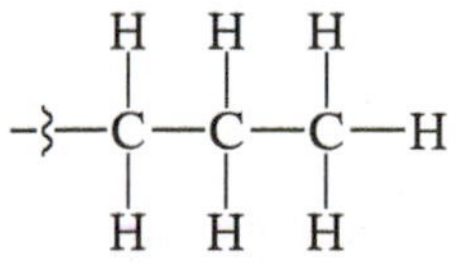	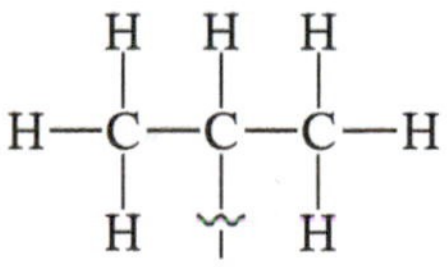
Condensed structure	$-\}-CH_3$	$-\}-CH_2CH_3$	$-\}-CH_2CH_2CH_3$	CH_3CHCH_3
Line structure	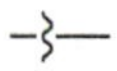	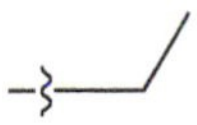		
Ball-and-stick model	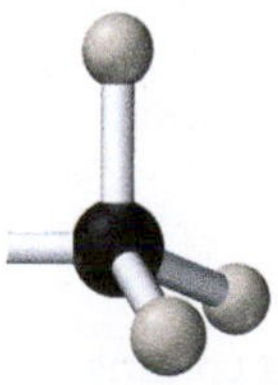	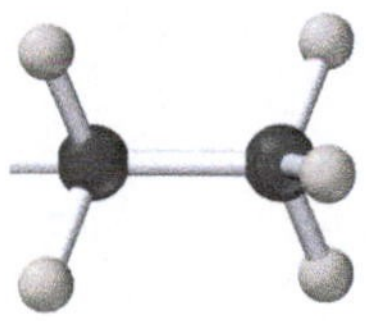	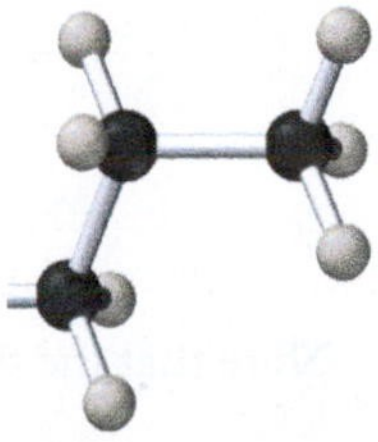	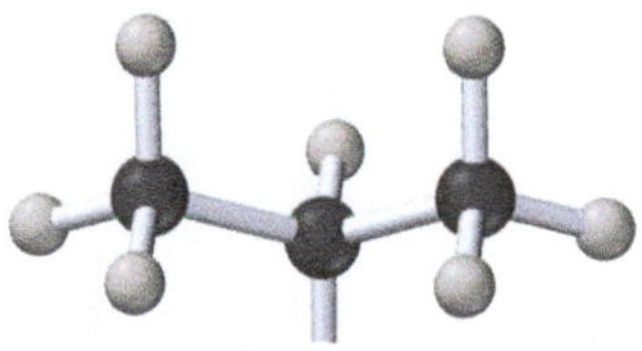

} represents the point where the group is bonded to the parent chain.

Sample Problem 4.9 Naming Branched-Chain Alkanes

Determine the correct IUPAC name for each of the following branched-chain alkanes:

a.

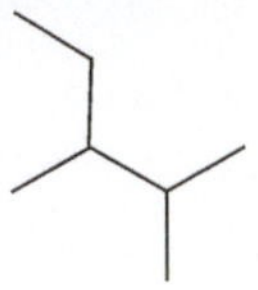

b.

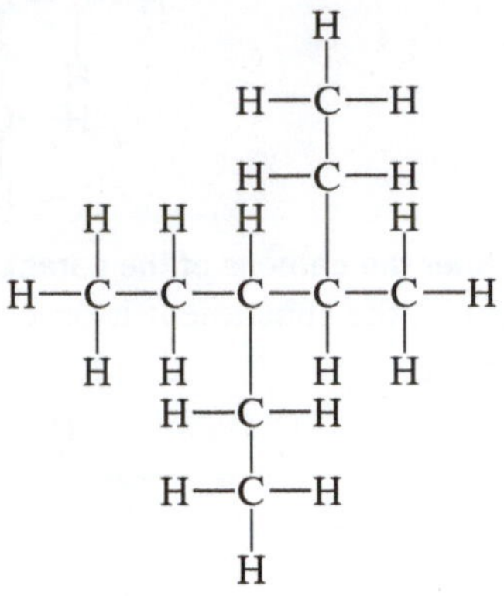

$$
\begin{array}{c}
H \\
| \\
H{-}C{-}H \\
| \\
H{-}C{-}H \\
\qquad\quad | \qquad\qquad H \\
H \;\; H \;\; H \qquad\quad | \\
| \;\;\; | \;\;\; | \qquad\quad H \\
H{-}C{-}C{-}C{-}C{-}C{-}H \\
| \;\;\; | \qquad | \;\;\; | \\
H \;\; H \qquad H \;\; H \\
H{-}C{-}H \\
| \\
H{-}C{-}H \\
| \\
H
\end{array}
$$

Solution

Carefully follow steps 1 to 4 to determine the correct name for each compound.

a.

STEP 1 Find the longest continuous chain of carbons.

pentane

STEP 2 Identify the substituents.

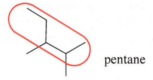

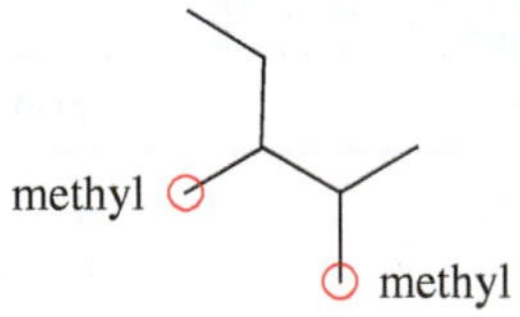

methyl

methyl

STEP 3 Number the carbons of the parent chain. Numbering from right to left gives the methyl substituents the lower number combination (2,3), whereas numbering from left to right would give a higher number combination (3,4).

STEP 4 Assign numbers to the substituents and alphabetize them in the name.
2,3-dimethylpentane

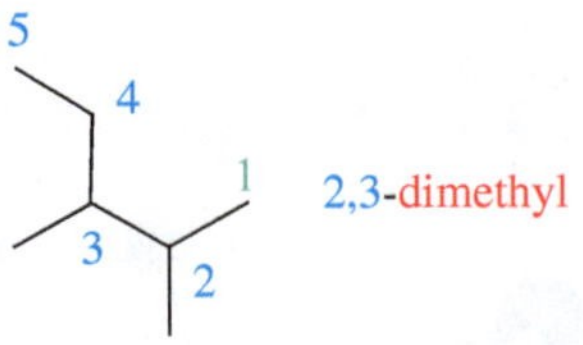

5
4
1   2,3-dimethyl
3   2

Note that the two numbers are separated by a comma.

b.

STEP 1 **Find the longest continuous chain of carbons.** Remember, the longest chain may not be a straight, horizontal one.

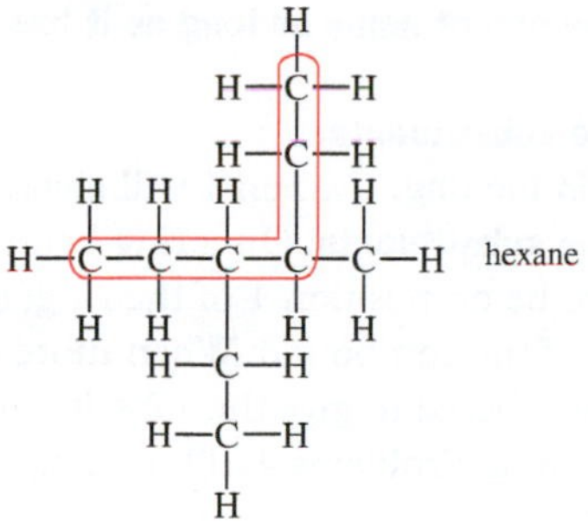

hexane

STEP 2 **Identify the substituents.**

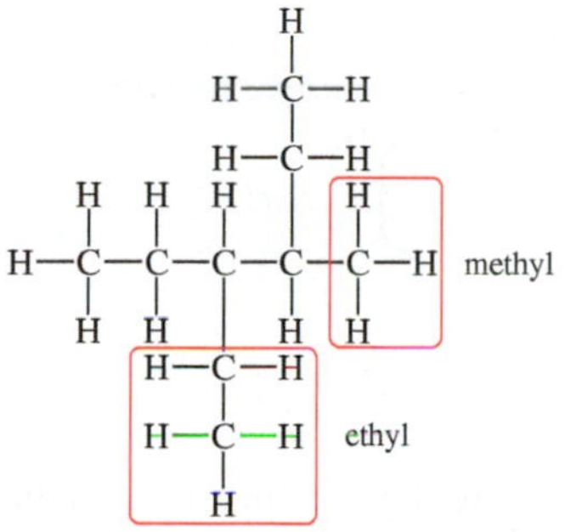

methyl

ethyl

STEP 3 **Number the carbons of the parent chain.** Begin numbering from the end closest to a substituent. When you have a choice in numbering the chain (as in this case, where the substituents are on carbons 3 and 4 regardless of which direction you count), the lower number goes to the group that is first alphabetically.

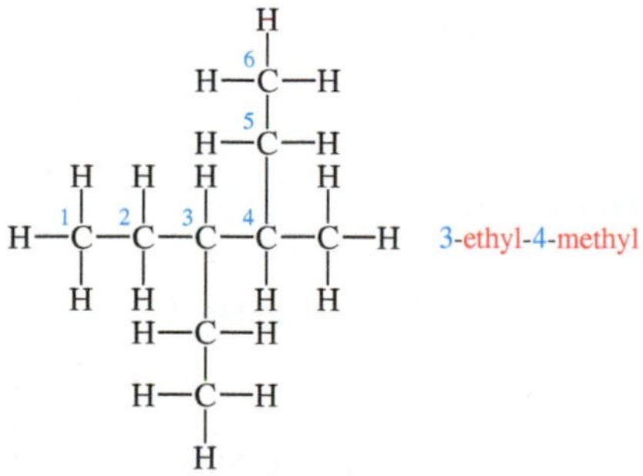

3-ethyl-4-methyl

STEP 4 **Assign numbers to the substituents and alphabetize them in the name.**
3-ethyl-4-methylhexane

Haloalkanes

The members of Group 7A on the periodic table, collectively known as the **halogens**, are common substituents on alkane chains. Alkanes with a halogen substituent in place of a hydrogen atom are known as **haloalkanes** or alkyl halides. When a halogen is a substituent, the name is changed by replacing the *ine* ending of the name of the element with an *o*. The substituent names for the halogens become *fluoro, chloro, bromo,* and *iodo*. The rules for naming haloalkanes are the same as those for naming branched-chain alkanes with the halogen being the substituent.

Cycloalkanes

The rules for determining the IUPAC names for cycloalkanes are much the same as those for the branched-chain alkanes with a few modifications to the previous steps:

STEP 1 **The ring serves as the parent name as long as it has more carbons than any substituent.**

STEP 2 **As before, identify the substituents.**

STEP 3 **Number the carbons in the ring.** Carbon 1 will always have a substituent.

STEP 4 **Assign numbers to the substituents.** On a ring bearing only one substituent, that substituent is assumed to be on position 1 of the ring; the 1 is implied and need not be included in the name of the compound. When more than one substituent is present, the ring should be numbered to give the substituents the lowest possible combination of numbers. (See Sample Problems 4.10b and c.)

Sample Problem 4.10 **Naming Haloalkanes and Cycloalkanes**

Determine the correct IUPAC name for each of the following compounds:

a. Cl   b. CH₃   c.

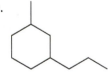

Solution

For haloalkanes, follow steps 1 to 4 given for the branched-chain alkanes to determine the correct name for each compound.

a.

STEP 1 **Find the longest continuous chain of carbons.**

Cl

pentane

STEP 2 **Identify the substituents.**

Cl   chloro

          methyl

STEP 3 **Number the parent chain.**

Cl   2-chloro

1 2 3 4 5

          3-methyl

STEP 4 **Assign numbers to the substituents and alphabetize them in the name.**

2-chloro-3-methylpentane

Often, alkyl substituents on cycloalkanes are drawn in condensed structure. In parts b and c, we provide an example where substituents are drawn in condensed structure (part b) and drawn in skeletal structure (part c). Follow steps 1 to 4 as modified for cycloalkanes to determine the correct name for each compound.

b.

STEP 1 Determine the number of carbons in the ring.

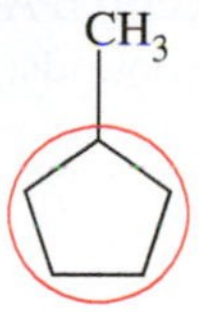

cyclopentane

STEP 2 Identify the substituents.

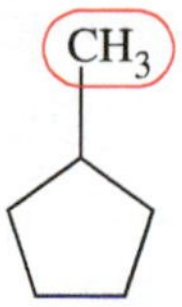

methyl

STEP 3 Number the carbons in the ring.

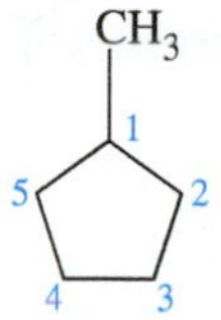

Single substituent on a cycloalkane will always be on carbon #1.
methyl

STEP 4 Name the cycloalkane. methylcyclopentane

c.

STEP 1 Determine the number of carbons in the ring.

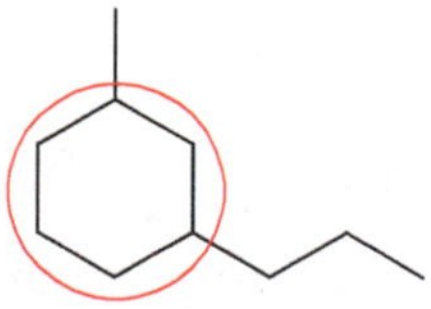

cyclohexane

STEP 2 Identify the substituents.

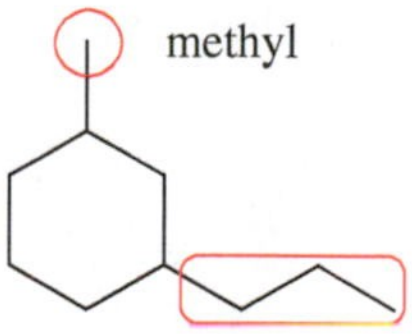

methyl

propyl

STEP 3 Number the carbons in the ring.

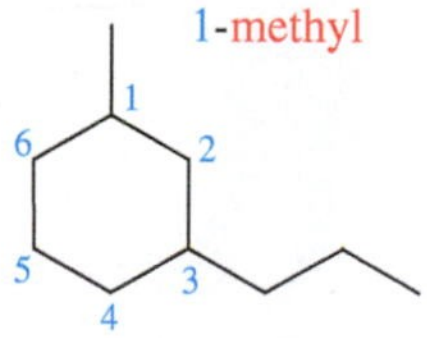

1-methyl

3-propyl

STEP 4 Assign numbers to the substituents and alphabetize them in the name.
1-methyl-3-propylcyclohexane

Notice that the numbering of the ring could have started at either the methyl or the propyl substituent to get the combination of 1 and 3. In such a case, give the lower number to the substituent that comes first alphabetically.

Practice Problems

4.23   Draw the condensed structural formula for each of the following alkyl groups:

    a. propyl   b. methyl

4.24   Give the correct name for each of the following substituents:

    a. CH_3CH_2—

    b. $CH_3CH_2CH_2CH_2$—

    c. I—

4.25   Draw the skeletal structure for each of the following compounds:

    a. 2,3-dimethylpentane

    b. 2-ethyl-1,4-dimethylcyclohexane

    c. 1,2-dichlorohexane

4.26   Draw the skeletal structure for each of the following compounds:

    a. 3-ethylhexane

    b. 1-bromo-3-iodocyclopentane

    c. propylcyclopentane

4.27   Give the correct IUPAC name for each of the following compounds:

    a.

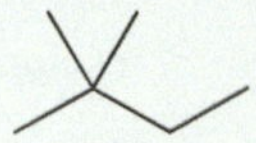

    b. 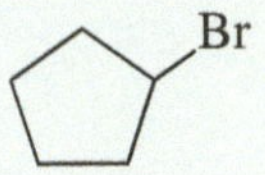   Br

    c. 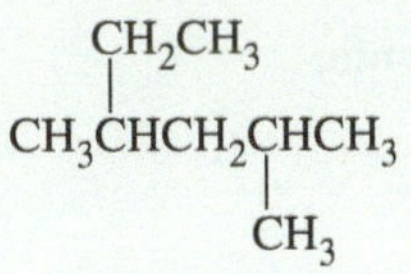 CH_2CH_3

       $CH_3CHCH_2CHCH_3$

              CH_3

4.28   Give the correct IUPAC name for each of the following compounds:

    a.

    b. 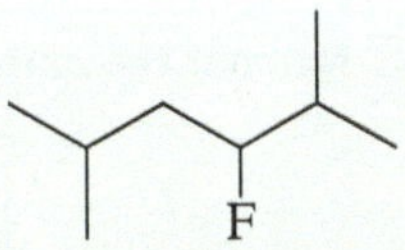   F

    c.

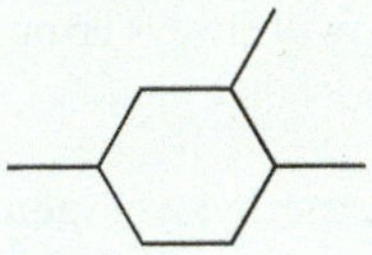

4.5 Isomerism in Organic Compounds

? 4.5 Inquiry Question

How are isomers the same, and how are they different?

Consider a cooking contest in which the chefs are given the same basket of ingredients and asked to prepare their best dish from just those ingredients. Because each chef is unique, each chef will likely prepare a different dish from those same ingredients. Some of the dishes may differ greatly from each other, while others may be more alike.

Just as chefs can cook different dishes from the same ingredients, isomers have the same atoms but are arranged differently.

Much like the cooking contest, organic compounds with the same number of carbon atoms can have the carbons connected in many different ways. Molecules with the same molecular formula but different connectivity or arrangements of the atoms are called **isomers,** from the Greek *iso* meaning "same" and *meros* meaning "part." In this section, we compare and contrast different types of isomers possible in organic compounds.

Structural Isomers and Conformational Isomers

When a hydrocarbon has four or more carbons, the carbons can be connected in more than one way. Let's look at the following two compounds, both of which have the molecular formula C_4H_{10}:

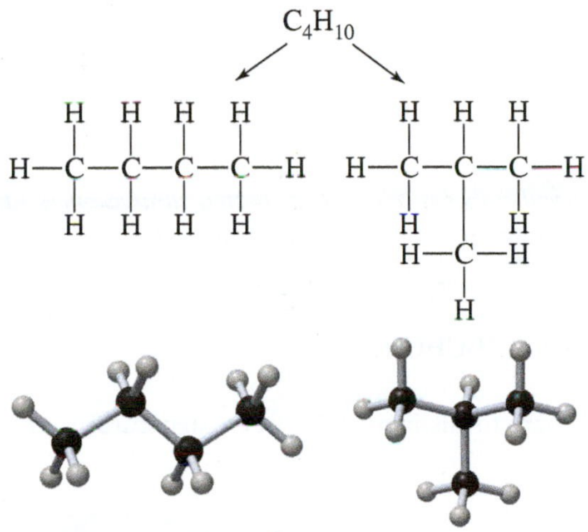

Butane (left) and isobutane (right) are structural isomers.

The compound on the left is the straight-chain alkane butane. The compound on the right is a branched-chain alkane and has the IUPAC name 2-methylpropane, commonly called isobutane. The compounds are isomers, but the carbons are not bonded in the same way. Butane (left) has four carbons bonded in a row, while 2-methylpropane (right) has three carbons bonded in a row with one carbon branching off the center carbon. When two molecules have the same molecular formula but a different connectivity of atoms, the molecules are related to each other as **structural isomers.**

Next, look at the following two structures shown in Lewis and skeletal structure. The two structures on the left are the straight-chain butane that you've seen before. The structures on the right look different. Are these representations of the same molecule, or are they structural isomers?

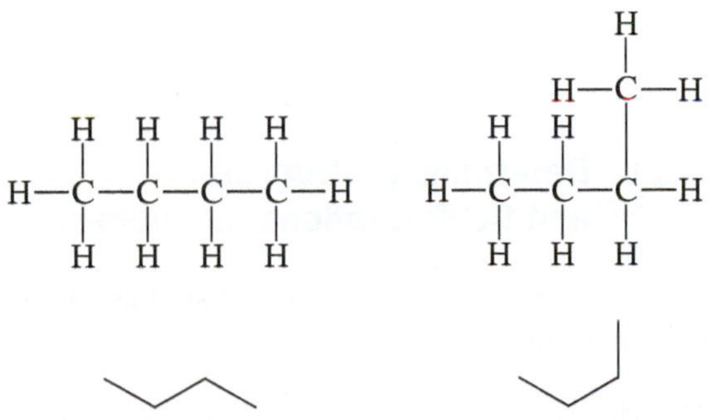

Because the connectivity of the carbons is the same, these are representations of the same molecule. You can trace the chain from the first to the fourth carbon on both structures without picking up your finger. This means that these compounds have four carbon atoms connected in a continuous chain. Although these two Lewis structures and skeletal structures may look different, the structures represent the same compound—butane—because the carbons are connected in the same way in both cases.

Another way you can convince yourself that these are the same molecule is by naming each structure. The representations have the same name, butane, because the longest continuous chain is four carbons and there are no substituents.

Both structures have the same four-carbon chain, but the atoms in the chain are not arranged the same on the page. These structures represent a type of isomer called a **conformational isomer.** These isomers, also called **conformers,** are not different compounds like structural isomers but are different arrangements of the same compound. Single bonds in a noncyclic compound can rotate and make molecules look different, but they are the same molecule.

Solving a Problem

Distinguishing Structural and Conformational Isomers

Indicate if the pair of molecules shown are related as structural isomers (different molecules), related as conformational isomers (the same molecule), or not related as isomers.

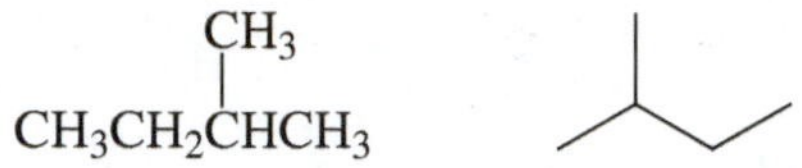

$$\overset{\displaystyle CH_3}{\underset{\displaystyle |}{CH_3CH_2CHCH_3}}$$

Use the following flow chart to distinguish between structural isomer and conformational isomer:

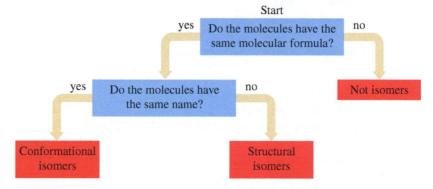

Flow chart for distinguishing structural isomers and conformational isomers.

In this case, the two molecules have the same molecular formula, C_5H_{12}. When correctly named, both compounds are 2-methylbutane, so the molecules are related as conformational isomers (conformers).

Sample Problem 4.11 **Determining Structural and Conformational Isomers**

For each of the following pairs, indicate if they are related as structural isomers, related as conformational isomers, or not related.

a.

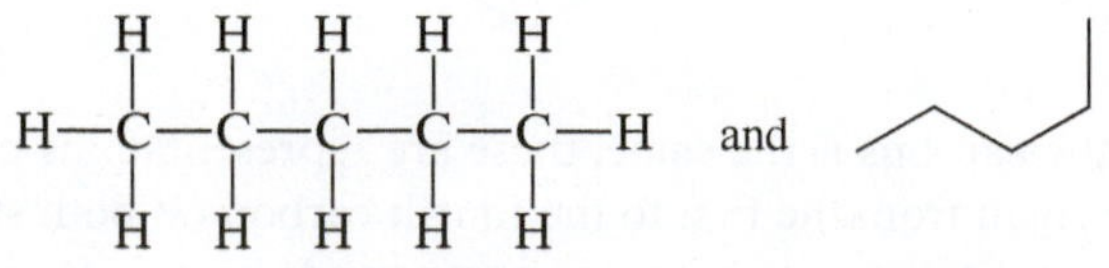

$$H-\overset{\displaystyle H}{\underset{\displaystyle H}{C}}-\overset{\displaystyle H}{\underset{\displaystyle H}{C}}-\overset{\displaystyle H}{\underset{\displaystyle H}{C}}-\overset{\displaystyle H}{\underset{\displaystyle H}{C}}-\overset{\displaystyle H}{\underset{\displaystyle H}{C}}-H$$

and

b.

and

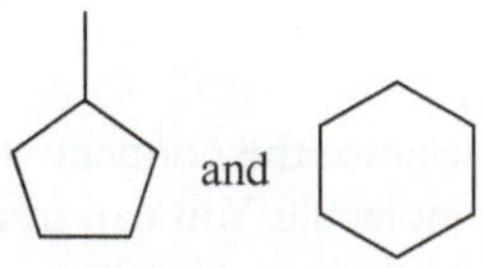

Solution

Use the flow chart given previously in the Solving a Problem feature to establish the relationship between the molecules shown.

    a. The molecular formula of each of these compounds is C_5H_{12}, so they are isomers. Notice that both structures have a continuous chain of five carbon atoms and are named pentane. These are conformational isomers.

    b. The molecular formula of each of these cycloalkanes is C_6H_{12}, so they are isomers. The name of the molecule on the left is methylcyclopentane, and the name of the molecule on the right is cyclohexane. The carbons are bonded differently. These are structural isomers.

Cis–Trans Stereoisomers in Cycloalkanes and Alkenes

Look at the two molecules represented in **FIGURE 4.5**. These two compounds have the same molecular formula and the *same* connectivity of their atoms. So, these two compounds are *not* structural isomers. Are they conformers? If they were conformers, we could change one into the other without breaking any bonds. This is not the case. The ring of carbons makes it impossible for one of the molecules to rotate and produce the other.

If we look at the ball-and-stick models, we see that they are different, with the positioning of the methyl groups on the ring being different in the two compounds. The skeletal structure shows the compound on the left with a wedge-shaped bond (◄) and a dashed bond (⑈⑈⑈), while the compound on the right has two dashed bonds.

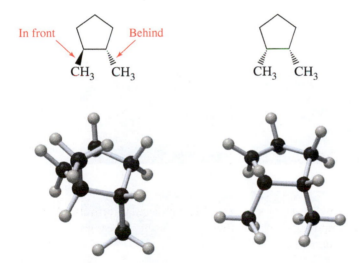

FIGURE 4.5 Stereoisomers of 1,2-dimethylcyclopentane. Two methyl groups can be arranged on a cyclopentane in two ways. The ball-and-stick models show two spatial arrangements. In the skeletal structures, this spatial arrangement is represented by wedges (in front of the plane of the page) and dashes (behind the plane of the page).

Unlike straight-chain alkanes, in which bonds freely rotate, the cycloalkanes are limited in their bond rotations and flexibility. Because of this restricted rotation about the carbon–carbon bonds in cycloalkane rings, these compounds have two distinct sides or faces. To view this better, we can look at the cyclopentane ring across one edge in a projection where the point at the top of the structure is behind the plane of the page and the bottom edge is in front of the plane of the page. Using this representation, we can see that the ring has a "top" face and a "bottom" face.

With this in mind, look back at Figure 4.5 and compare the skeletal structures of the two substituted cyclopentane compounds with the ball-and-stick models. Notice how the methyl groups on the wedge bonds are on the top face of the ring, while those on the dashed bonds are on the bottom face of the ring. The wedges and dashes allow us to show the three-dimensional nature of a molecule in the two-dimensional space of a page.

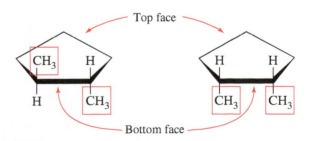

The methyl groups are connected to the same carbons but arranged differently on the two molecules.

Naming the molecules using the IUPAC rules we've learned so far would give us the name 1,2-dimethylcyclopentane for both. Yet in the representations shown, the two molecules are clearly different. How can we distinguish the two by name?

When two molecules have the same molecular formula and the same attachments to the carbon skeleton but a different spatial arrangement, they are related to each other as **stereoisomers.** The two cyclopentane compounds shown in Figure 4.5 belong to a class of stereoisomers called **cis–trans stereoisomers.**

In this example, the stereoisomer with the methyls on the same face of the cyclopentane ring (both on bottom in this case) is called the cis stereoisomer. Use this memory device to remember cis: *same* *s*ide is *ci*s. Notice all of the "s" sounds. The other isomer with one methyl on the top face and the second on the bottom face is called the trans stereoisomer. *Trans* comes from the Latin, meaning "across," as in the word *transatlantic.* In the trans stereoisomer, the methyls are across the ring from each other.

The prefixes *cis-* and *trans-* in italics are included at the beginning of the compound names to denote the arrangement of the substituents in the compound and to give each compound a unique name. In Figure 4.5, the complete name of the isomer on the left is *trans*-1,2-dimethylcyclopentane, and the isomer on the right is *cis*-1,2-dimethylcyclopentane.

Cis–trans isomerism also occurs in alkenes. Consider the carnival trinket shown in **FIGURE 4.6**. If you join the index fingers of each hand with one of these toys (and push your fingers together), your two hands can still rotate independently, despite the fact that they are "bonded" together. The same is true for two carbon atoms joined by a single bond.

Now consider what would happen if you used two of the toys and joined both of your index fingers *and* both of your middle fingers. With two "bonds" now joining your hands, they can no longer rotate independently of each other. The same is true for two carbon atoms joined by a double bond. The double bond between the two atoms will not allow them to rotate independently of each other. Organic chemists say that the carbon–carbon double bond is rigid, or that it has restricted rotation.

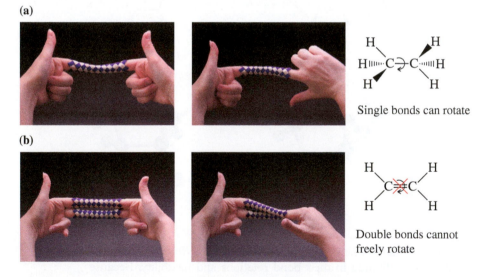

(a)

Single bonds can rotate

(b)

Double bonds cannot freely rotate

FIGURE 4.6 Rotation in single versus double bonds. Like a carnival toy, the rotation about a single bond (one toy) is allowed (a), but the rotation about a double bond (two toys) is restricted (b).

Previously, we saw that the rings of cycloalkanes have both a top face and a bottom face because of their inability to rotate freely. Because of the rigidity of the carbon–carbon double bond, alkenes also have two faces and some can have cis–trans stereoisomers.

Consider the two alkenes in **FIGURE 4.7**. The dashed line separates the two sides of the double bond into a top face (above the line) and a bottom face (below the line). Notice that the compound on the left has both similar groups (often carbon-containing groups) on the same side (face) of the double bond, while the one on the right has them on opposite sides of the double bond. Because there is restricted rotation around a double bond, these two molecules are not identical, nor are they conformers. They cannot be changed into the other by rotating around a bond. These two molecules are cis–trans stereoisomers. The molecules with the similar groups on the same side of the double bond

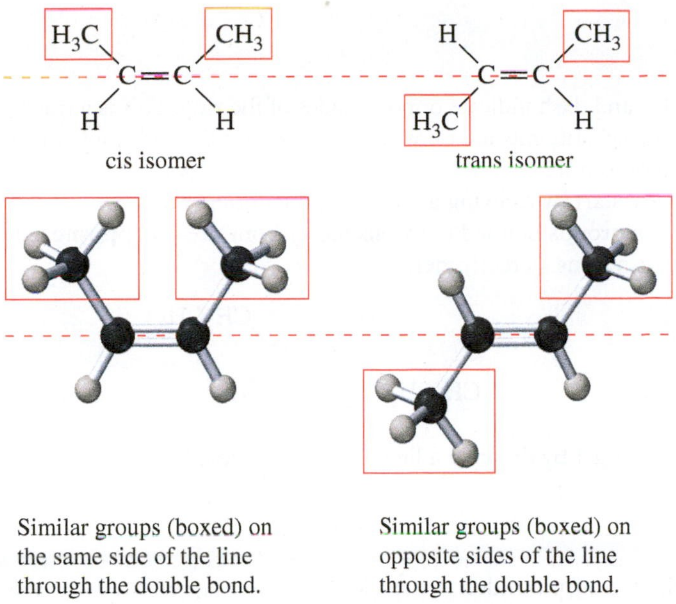

FIGURE 4.7 Cis–trans stereoisomers of alkenes. Alkenes have two sides to them due to restricted rotation around the double bond.

cis isomer

trans isomer

Similar groups (boxed) on the same side of the line through the double bond.

Similar groups (boxed) on opposite sides of the line through the double bond.

is the cis stereoisomer. The molecule with the similar groups on the opposite sides of the double bond is the trans stereoisomer.

FIGURE 4.8 shows some other alkenes and their isomers, as well as the skeletal structures of each. Notice that not all alkenes exhibit cis–trans stereoisomerism. If one of the alkene carbons has two identical groups bonded to it (hydrogen in the case shown), this compound cannot have cis–trans stereoisomers.

(a)

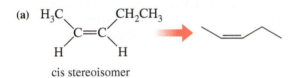

cis stereoisomer

trans stereoisomer

FIGURE 4.8 Structures of cis–trans stereoisomers. (a) Structures of cis–trans stereoisomers and (b) structure showing why some alkenes do not have cis–trans stereoisomers.

(b) Identical groups on the same alkene carbon, so this compound does not have cis-trans stereoisomers.

Sample Problem 4.12 **Identifying Cis–Trans Stereoisomers in Cycloalkanes and Alkenes**

Identify each of the following as either the cis or trans stereoisomer:

a.

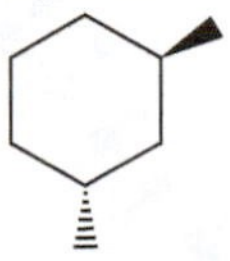

b.

c.

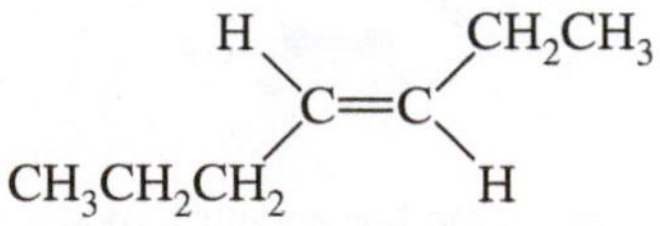

d.

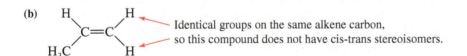

—Continued next page

Continued—

Solution

a. The wedge and dash indicate opposite sides of the ring. This is a trans stereoisomer.
b. Both of the substituents are on wedges and are on the same side of the ring. This is a cis stereoisomer.
c. For alkenes, start by drawing a line along the double bond.
   The carbon groups bonded to the alkene carbons are on opposite sides of the line. This is a trans stereoisomer.

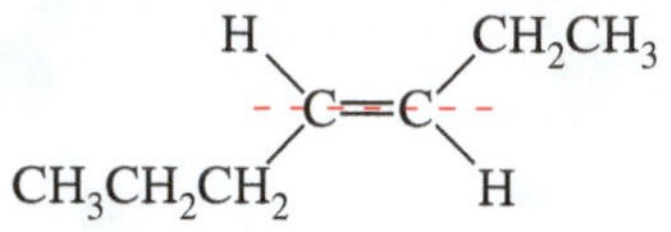

d. For alkenes, start by drawing a line along the double bond.

The carbon groups bonded to the alkene carbons are on the same side of the line. This is a cis stereoisomer.

Unsaturated Fatty Acids Contain Cis Alkenes

Now think back to the fatty acids from Section 4.3. There, we examined saturated fatty acids, a type of lipid. Notice the difference in the structure of oleic acid shown in **FIGURE 4.9** and the structure of the saturated fatty acids found in Table 4.5. Oleic acid and similar fatty acids are **unsaturated fatty acids** because they contain at least one carbon–carbon double bond (alkene), which makes them unsaturated hydrocarbons. Naturally occurring unsaturated fatty acids contain only cis alkenes.

TABLE 4.8 shows the structures and names of six common unsaturated fatty acids. Take particular note of the ω (omega) designations of the unsaturated fatty acids. These designations are commonly used in nutrition literature and in the popular media. In this system, the carbon farthest from the carboxylic acid is numbered 1, the next carbon is 2, and so forth. The **omega number** indicates the carbon positioning of a double bond in the structure of the fatty acid.

The unsaturated fatty acid α-linolenic acid commonly called omega-3 (see Table 4.8) is given a designation of [18:3], ω-3. This notation indicates that the fatty acid has 18 carbons and 3 double bonds, with the first double bond located between carbons 3 and 4. Two more double bonds are understood to occur at intervals of three carbons for a total of three unsaturations. This means that the other two double bonds of α-linolenic acid are located between carbons 6 and 7, and between carbons 9 and 10. Note that there is a $-CH_2-$ between the adjacent double bonds.

The polyunsaturated fatty acids linoleic and α-linolenic are considered **essential fatty acids** because our bodies cannot produce these compounds, so they must be

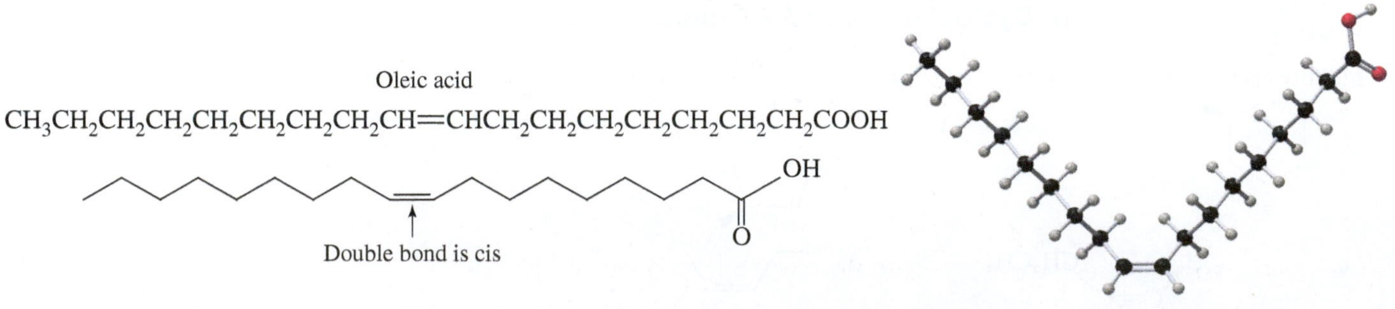

Oleic acid

$CH_3CH_2CH_2CH_2CH_2CH_2CH_2CH_2CH=CHCH_2CH_2CH_2CH_2CH_2CH_2CH_2COOH$

Double bond is cis

FIGURE 4.9 Natural unsaturated fatty acids contain cis alkenes. In the ball-and-stick model, the cis double bond changes the structure by putting a kink in the hydrocarbon chain.

TABLE 4.8 Unsaturated Fatty Acids: Names and Structures

Name	Carbon Designation	Source	Structure
Monounsaturated Fatty Acids			
Palmitoleic acid	[16:1], ω-7	Butter	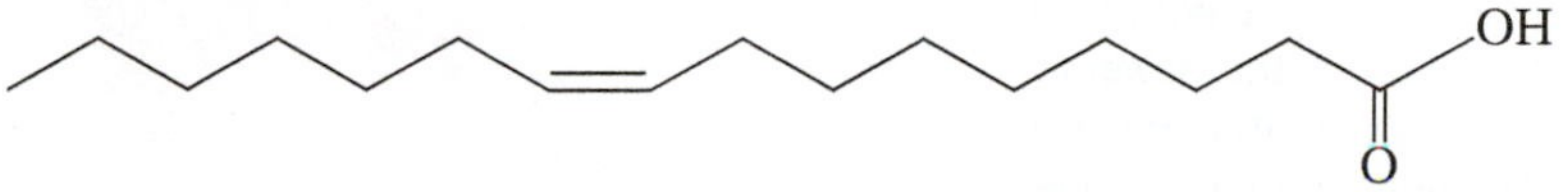
Oleic acid	[18:1], ω-9	Olives, corn	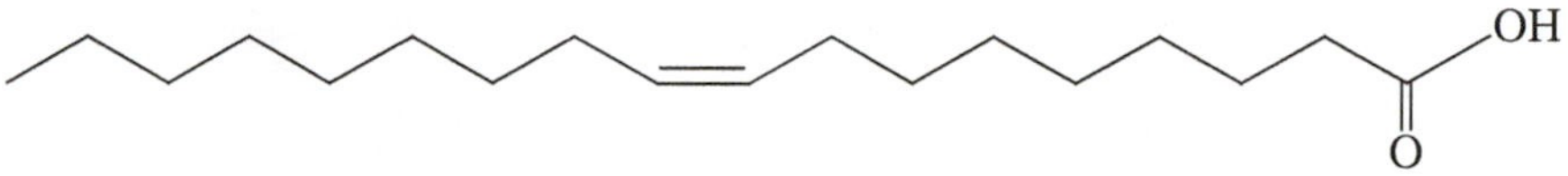
Polyunsaturated Fatty Acids			
Linoleic acid* (commonly called LA or omega-6)	[18:2], ω-6	Soybean, safflower, corn	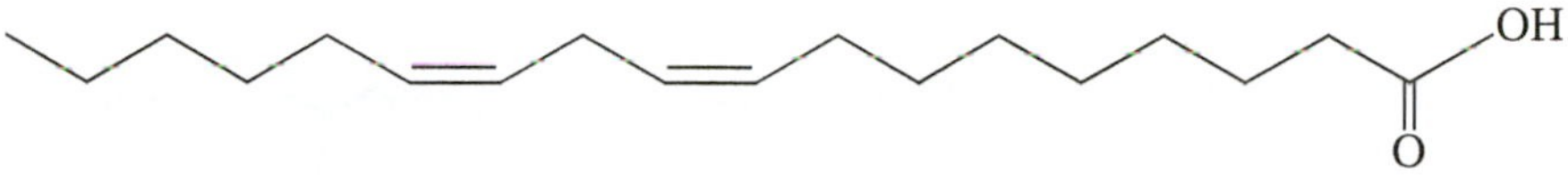
α-Linolenic acid* (commonly called ALA or omega-3)	[18:3], ω-3	Flaxseed, canola	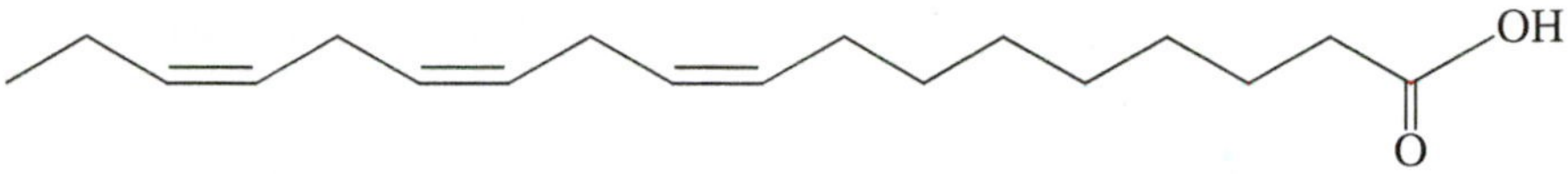
γ-Linolenic acid (commonly called GLA)	[18:3], ω-6	Formed from ALA, evening primrose, spirulina	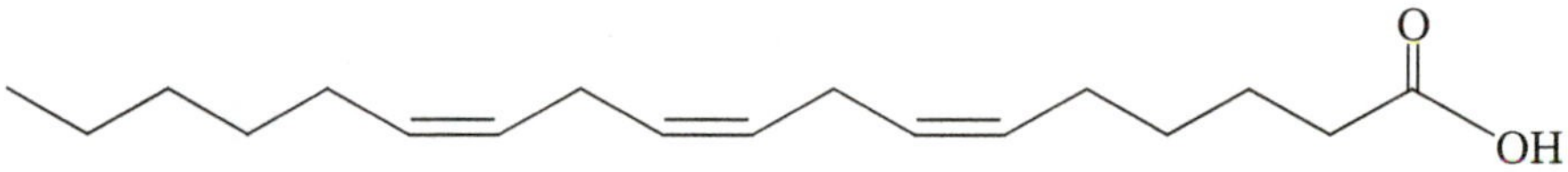
Arachidonic acid (commonly called AA)	[20:4], ω-6	Formed from LA, lard, eggs	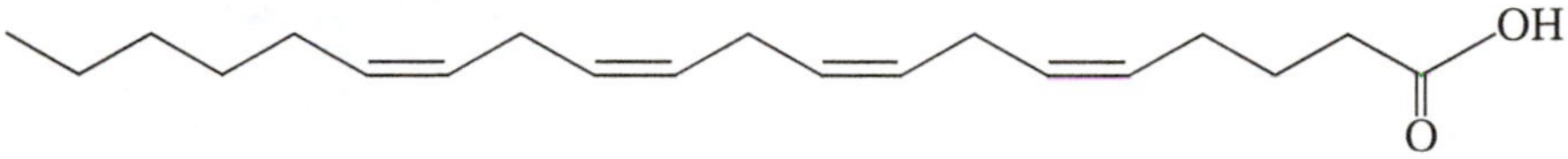

* Indicates essential fatty acid.

obtained in the diet. Fatty acids are mainly taken in as a component of fats and oils in molecules called triglycerides. More on the properties of these molecules can be found in Chapter 7.

Sample Problem 4.13 Drawing Unsaturated Fatty Acids

Draw the skeletal structure for eicosenoic acid [20:1], ω-9.

Solution

Remember that the double bonds in naturally occurring fatty acids are cis double bonds. The omega number tells us that the double bond will be between carbons 9 and 10, counting from the hydrocarbon end. Be sure to include a carboxylic acid group at the other end.

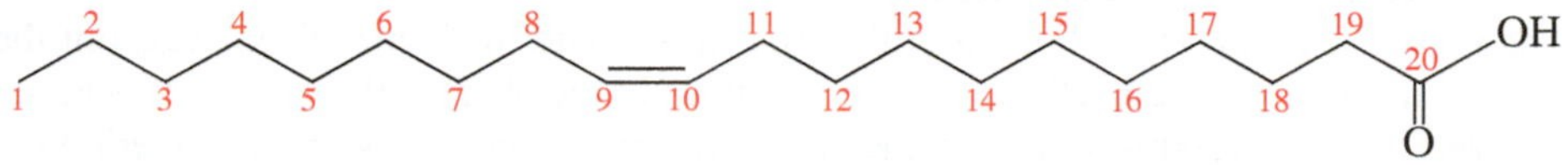

Stereoisomers—Chiral Molecules and Enantiomers

When you experience the pleasant citrus smell of an orange or the harsh piney smell of turpentine, your nose is distinguishing between two stereoisomers of the a compound

FIGURE 4.10 The limonene molecules.
The two stereoisomers found in citrus and turpentine are identical except for their spatial arrangement. Notice the dash and wedge in the skeletal structures for limonene from oranges and turpentine. The group on the bottom of the ring points away from the viewer in limonene from oranges (left) and toward the viewer in limonene from turpentine (right).

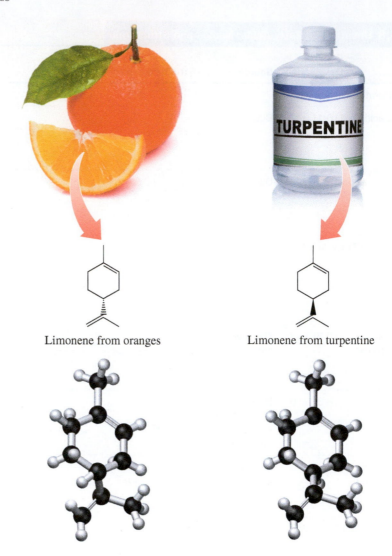

Limonene from oranges   Limonene from turpentine

called limonene. **FIGURE 4.10** shows that the limonene structure in oranges is slightly different from the limonene in turpentine. Even though stereoisomers are identical except for the arrangement of groups in space, because they have different shapes, our bodies can actually distinguish between them.

From Figure 4.10, we can determine that both limonenes have a molecular formula of $C_{10}H_{16}$, so they are isomers. The connectivity of the atoms in the two structures is also the same, meaning that they are not structural isomers. However, the limonenes have only one wedge- or dash-type bond present in each molecule. The two limonenes cannot be cis–trans isomers. As we saw earlier with cycloalkanes, cis–trans isomers must have two such bonds. Are they conformational isomers (the same molecule with the same name)?

In Section 4.4, we said that if two molecules had the same name, they are the same molecule. Besides their names, if molecules are identical, they would be superimposable. In other words, if we placed one directly on top of the other, every atom on each compound would align or stack on top of one another like two baseball caps. Notice that the limonene from oranges has the group at the bottom of the molecule behind the plane of the page, but the limonene from turpentine has the same group in front of the plane of the page. If we stacked the molecules, these groups would not line up. Therefore, the molecules are not superimposable because of the difference of the location in space of this group on each molecule.

Behind→   ≠   ←In front

In fact, if you vertically rotate the ball-and-stick models from Figure 4.10 about 90 degrees counterclockwise (when viewed from above), you can see in the figure below that the molecules are mirror images of each other. Compounds like the two limonenes, which have the same molecular formula and the same connectivity but are nonsuperimposable mirror images of each other, belong to a class of stereoisomers called **enantiomers.**

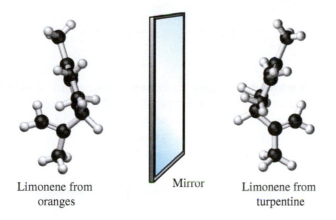

Limonene from
oranges

Mirror

Limonene from
turpentine

In our everyday life, we encounter many common objects that have nonsuperimposable mirror images. Your shoes are one example. Notice that your two shoes look very much the same, but they are not identical. Our feet would be uncomfortable if we put our shoes on the wrong feet. The same holds true for a pair of gloves. The right glove easily fits the right hand and the left glove on the left hand. Although your hands look the same, your two hands are not the same. They are mirror images of each other; that is, holding one up to the mirror shows the image of the other. The same is true of your feet. Objects such as these that have nonsuperimposable mirror images are termed **chiral** (from the Greek *cheir*, meaning "hand").

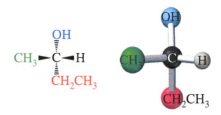

Molecules can also be chiral. The two limonenes, which exist as a pair of enantiomers, can be thought of as a right-handed and left-handed form of the same molecule. Most molecules that exist as enantiomers contain a **chiral center**—a tetrahedral carbon atom bonded to four different atoms or groups of atoms. In **FIGURE 4.11**, the colored spheres around the central carbon highlight the four different groups on the chiral center in the molecule. Notice that the carbon atom (shown in black) is bonded to the four different groups: hydroxyl ($-OH$), hydrogen ($-H$), methyl ($-CH_3$), and ethyl ($-CH_2CH_3$).

There are two forms of this molecule, and they are mirror images that are nonsuperimposable on each other (see **FIGURE 4.12**). The colored spheres do not line up when we try to superimpose them, as shown in the ball-and-stick models to the far right in Figure 4.12. Like the limonenes, this molecule can exist in two forms that are enantiomers of each other. On paper, we represent the attachments to the chiral center with wedges and dashes and indicate the location of a chiral center with an asterisk.

FIGURE 4.11 A chiral center. Four different atoms or groups of atoms are bonded to the carbon (shown in black), forming a chiral center.

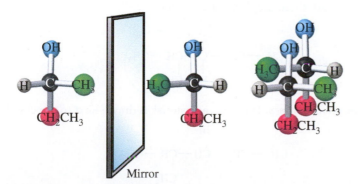

FIGURE 4.12 The mirror-image enantiomers. Enantiomers are nonsuperimposable, so this is a chiral compound.

Identifying Chiral Carbons in a Molecule

Mark the chiral center(s) in the following molecule, if any, with an asterisk (*).

Mastering Video
Mastering the Concepts
Identifying Chiral Centers

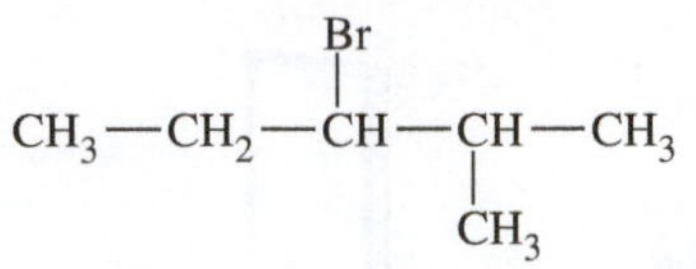

$$Br$$
$$CH_3—CH_2—CH—CH—CH_3$$
$$CH_3$$

STEP 1 **Locate the tetrahedral carbons** (carbons with four atoms bonded to them). Carbons with double or triple bonds are typically not chiral centers because they do not have four groups attached. In this case, all the carbons have four atoms bonded to them (no multiple bonds). They are highlighted in blue.

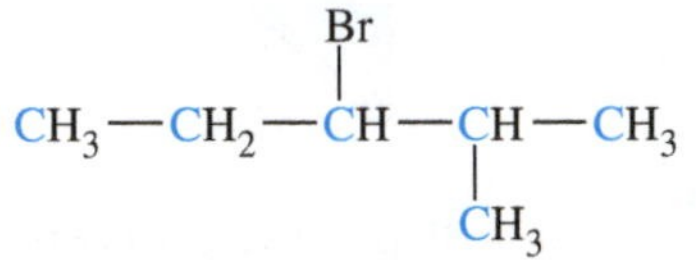

$$Br$$
$$CH_3—CH_2—CH—CH—CH_3$$
$$CH_3$$

STEP 2 **Inspect the tetrahedral carbons.** Determine if the four groups attached to the tetrahedral carbons are different. To determine if the groups are different, you must examine the entire group of atoms bonded to the carbon, not just the atom directly bonded to it.

All carbon atoms bonded to two or more hydrogen atoms cannot be chiral centers because such carbons are not bonded to four different groups. These carbons can be removed from further consideration. In this case, that leaves only two carbon atoms (highlighted in blue) to look at more carefully.

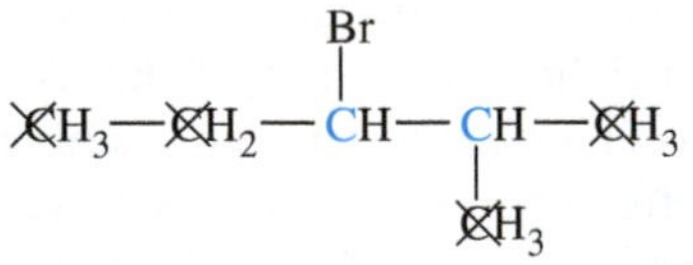

$$Br$$
$$CH_3—CH_2—CH—CH—CH_3$$
$$CH_3$$

The blue-highlighted carbon atom on the right is bonded to two methyl groups. Since these two groups are the same, this carbon cannot be chiral.

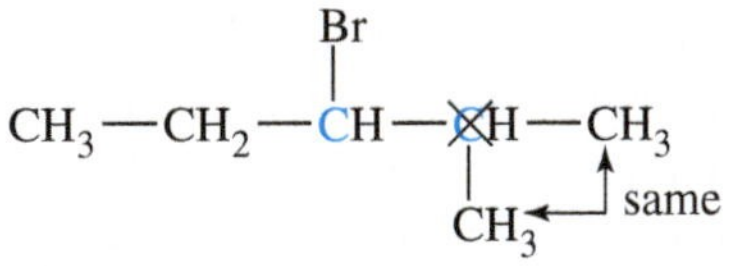

$$Br$$
$$CH_3—CH_2—CH—CH—CH_3$$
$$CH_3 \text{ same}$$

Expanding the structure to show the bond between the carbon and the hydrogen and then comparing the groups, we see that the carbon atom bonded to Br is bonded to four distinct groups and, therefore, is a chiral center.

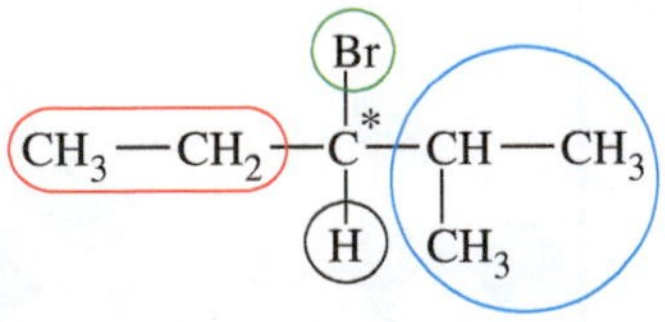

$$Br$$
$$CH_3—CH_2—C—CH—CH_3$$
$$H \quad CH_3$$

STEP 3 **Assign the chiral centers.** Typically, an asterisk is drawn next to the chiral center.

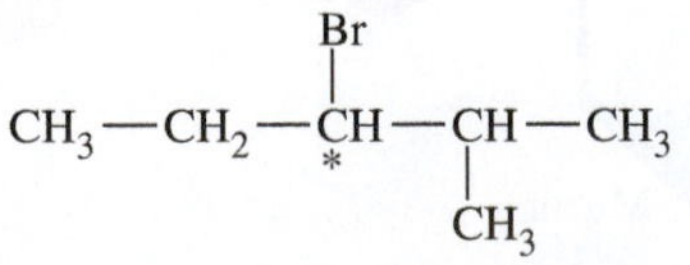

$$Br$$
$$CH_3—CH_2—CH—CH—CH_3$$
$$* \quad CH_3$$

Sample Problem 4.14   Identifying Chiral Centers

Mark the chiral centers in the following molecules, if any, with an asterisk (*):

a. 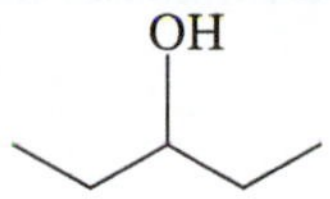   OH   b. 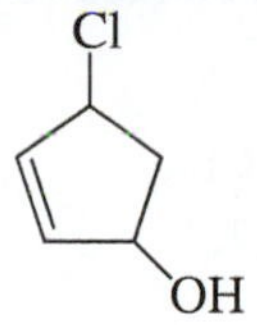   Cl   c.    Br

Solution

a.

STEP 1 Locate the tetrahedral carbons. All the carbons in this molecule are tetrahedral.

STEP 2 Inspect the tetrahedral carbons. Identifying chiral centers on skeletal structures may seem more difficult at first. With practice, looking for these types of structures can make the process quicker. Remember that, in a skeletal structure, a bend or corner is a $-CH_2-$ if no other bonds exist at the corner. You can imme-diately rule out these carbons as well as those at the end of a line ($-CH_3$).

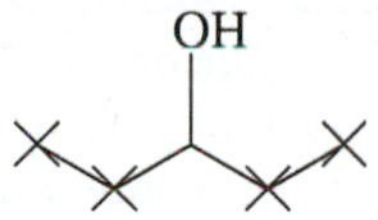

OH

This leaves only one carbon atom to consider further. At this point, you may find it useful to draw in the hydrogen atom that we typically omit in a skeletal structure and perhaps even add the symbol for the carbon itself.

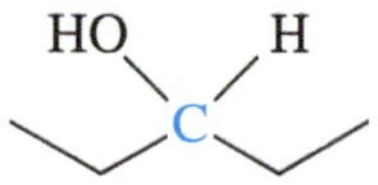

HO   H
C

Now we can see that the OH group and the H are not the same, but notice that the other two groups bonded to this carbon atom are identical: Both are ethyl groups.

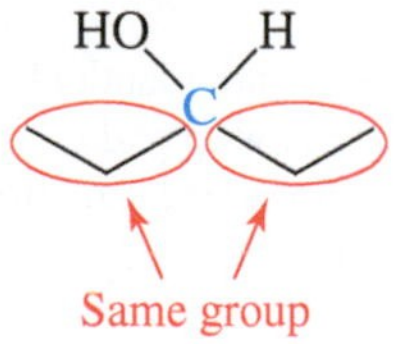

HO   H
C

Same group

STEP 3 Assign the chiral centers. Answer: There are no chiral centers.

b.

STEP 1 Locate the tetrahedral carbons. Any carbons in double or triple bonds can be eliminated.

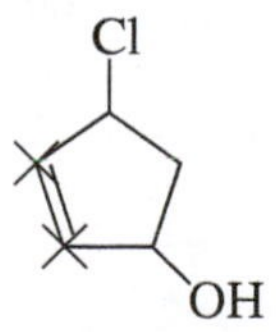

Cl

OH

—Continued next page

Continued—

STEP 2 Inspect the tetrahedral carbons. As before, eliminate those carbon atoms that have two or more hydrogen atoms bonded to them. This also eliminates the carbon on the right that is a —CH$_2$.

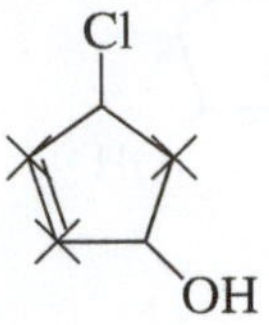

Following the steps from the previous example, consider the remaining carbon atoms one at a time, adding the hydrogen and the C.

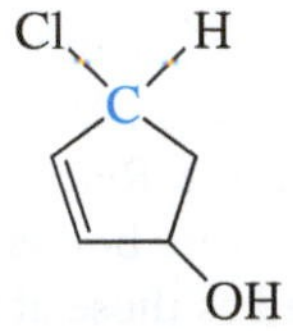

On this carbon atom, the H and Cl are different, but what about the other groups? To determine if these groups are the same or different, list the atoms and what is bonded to them moving clockwise around the ring, starting at the carbon that you are considering. Then do the same in the counterclockwise direction. For this carbon, in the clockwise direction, the atoms are CH$_2$, CH—OH, CH, and CH. In the counterclockwise direction, the atoms are CH, CH, CH—OH and CH$_2$. Notice that the lists are *not* identical. The CH—OH is second in the clockwise direction but third in the counterclockwise direction. The two groups do not have the same connectivity of atoms and, therefore, are not the same. This is a chiral center.

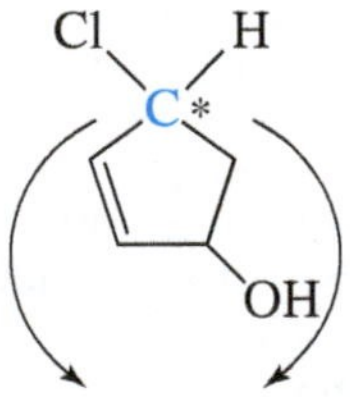

Groups are not identical
in the clockwise and
counterclockwise directions
around the ring.

Following the same procedure for the carbon atom bearing the OH group verifies that it is also a chiral center.

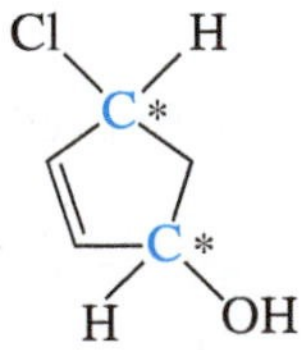

STEP 3 Assign the chiral centers. Answer:

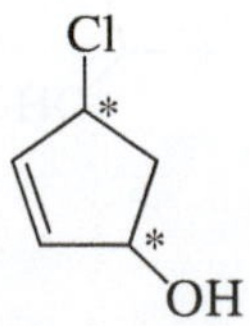

c.

STEP 1 **Locate the tetrahedral carbons.** All the carbons in this molecule are tetrahedral.

STEP 2 **Inspect the tetrahedral carbons.** Crossing out all the "corners" in this molecule, we quickly see that only one carbon remains for further consideration.

Don't let the presence of the wedge-and-dash bonds in this structure lead you to the quick conclusion that this carbon must be chiral. Instead, follow the procedure we used in Part b. Doing this, we discover that in the clockwise direction, the group bonded to the carbon contains five consecutive CH_2's. The same is true for the counterclockwise direction: five consecutive CH_2's. Because the atoms are the same and in the same order in both the clockwise and counterclockwise directions around the ring, the groups are identical. This carbon is not a chiral center.

Groups are identical in the clockwise and counterclockwise directions around the ring.

STEP 3 **Assign the chiral centers.** Answer: There are no chiral centers.

The Consequences of Chirality

Why does one limonene smell like oranges and the other like turpentine? The receptors in your nose also are "handed." Much as your right hand would not fit into a left-handed glove, a chiral molecule can fit only into a complementary receptor—one that accommodates its unique shape. This is also the case with many pharmaceuticals where only a single enantiomer has biological activity. The anti-inflammatory ibuprofen, found in Advil®, has one active enantiomer, whose receptor or target is the enzyme cyclooxygenase. When inflammation occurs, one form of this enzyme, cyclooxygenase-2, is produced in large amounts and activates a pain response. The active enantiomer of ibuprofen fits into a pocket on the enzyme like a hand in a glove. The mirror-image enantiomer won't fit in the pocket and has no effect on the pain response.

In some cases, one enantiomer of a drug can be beneficial and the other harmful. This was the case with the drug thalidomide, which was marketed outside the United States in the late 1950s and early 1960s for use against morning sickness. One enantiomer of the drug was effective in alleviating the symptoms of morning sickness. The mirror image, as was tragically discovered later, was teratogenic—it caused birth defects. Because the drug was initially sold as a 50:50 mixture of the enantiomers, many mothers who took it later gave birth to babies with severe birth defects, including shortened arms or legs and, sometimes, no limbs. Thalidomide was subsequently removed from markets worldwide. Recently, thalidomide has reemerged in the pharmaceutical market as a treatment for HIV, leprosy, and some forms of cancer, but it is prescribed under very tight restrictions.

INTEGRATING Chemistry

◀ Find out how important chirality is to drug interactions.

The active enantiomer of ibuprofen, an anti-inflammatory

Often only one enantiomer of a chiral pharmaceutical is biologically active. Can you draw the inactive enantiomer of ibuprofen?

—*Continued next page*

Continued—

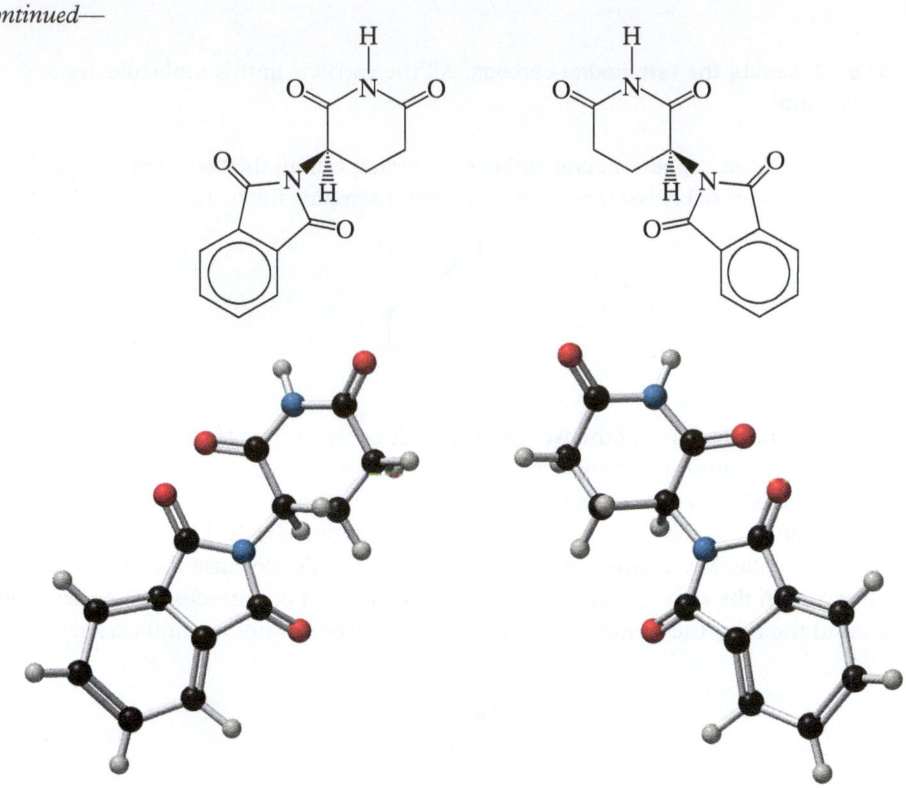

Thalidomide enantiomer used
to treat morning sickness

Thalidomide enantiomer
that causes birth defects

The enantiomers of thalidomide are nonsuperimposable and so have different shapes and different biological effects.

Practice Problems

4.29 What is the difference between a conformational isomer of a compound and a structural isomer of the same compound?

4.30 What do the structural and conformational isomers of a compound have in common?

4.31 Determine the relationship between each of the pairs of the following compounds. Are they structural isomers (different molecules), conformational isomers (the same molecule), or not related?

a. and

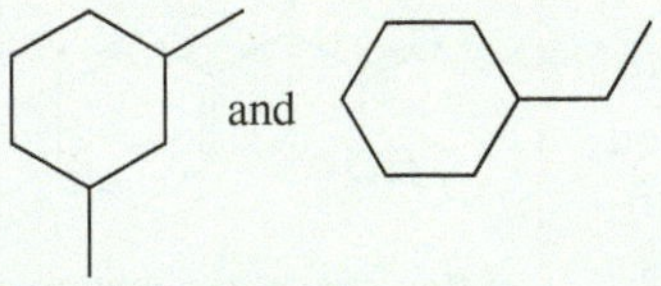

b. $CH_3CH_2CH_2CHCH_3$ and $CH_3CHCHCH_3$
           | | |
           CH_2CH_3 CH_3

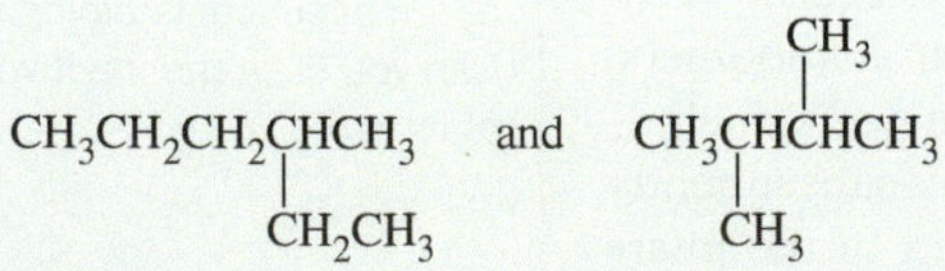

c. and

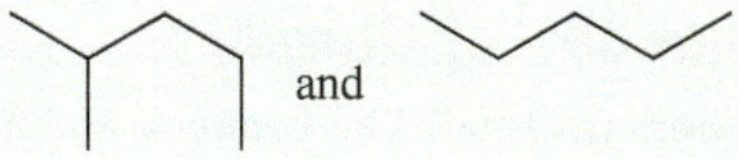

d. $CH_3CH_2CH_2CHCH_3$ and $CH_3CHCH_2CH_2CH_3$
   with CH_3 substituents

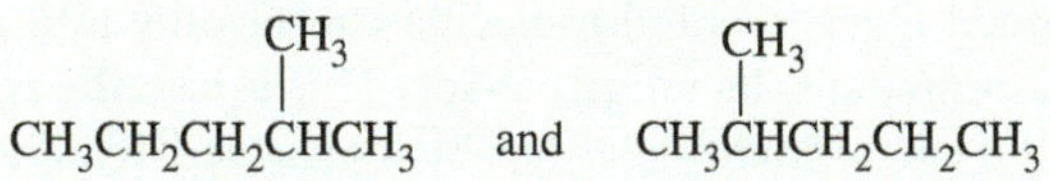

e. and

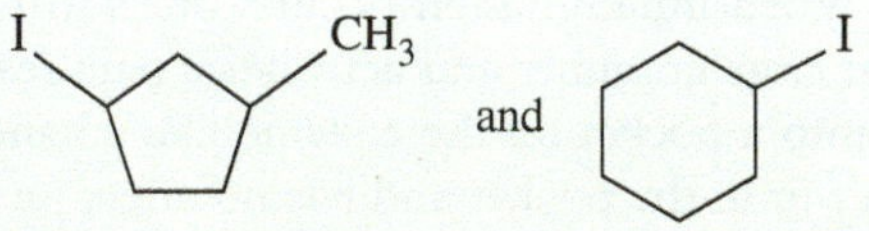

4.32 Determine the relationship between each of the pairs of the following compounds. Are they structural isomers (different molecule), conformational isomers (the same molecule), or not related?

a. 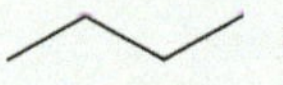 and $CH_3CH_2CH_2CH_3$

b. 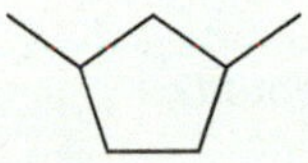 and

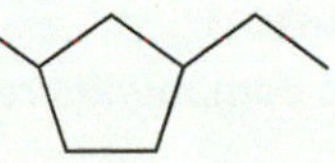

c. 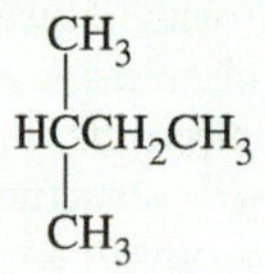
$$\underset{\overset{|}{CH_3}}{\overset{CH_3}{HCCH_2CH_3}}$$
and
$$\underset{\overset{|}{CH_3}}{CH_3CHCH_2CH_3}$$

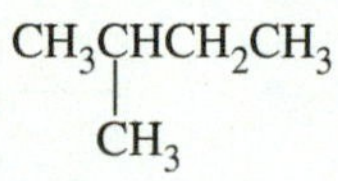

d. 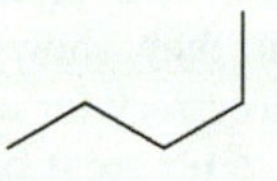 and

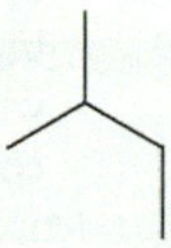

e. 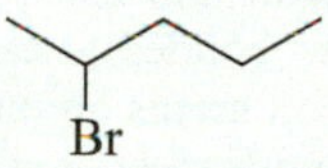 and

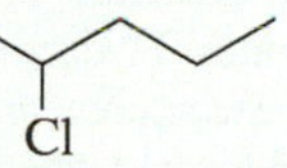

4.33 Determine if each of the following cycloalkanes or alkenes can exist as cis–trans stereoisomers. For those that can, draw the two isomers. Label each of the isomers you drew as the cis stereoisomer or the trans stereoisomer.

a.

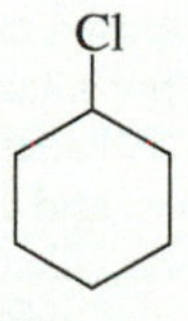

b. $CH_3CH_2CH_2CH_2CH{=}CH_2$

c.

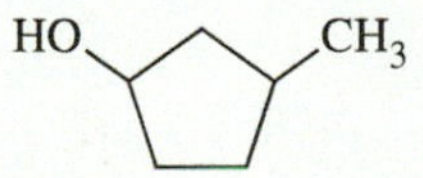

d. $CH_3CH_2CH{=}CHCH_2CH_2CH_3$

4.34 Determine if each of the following cycloalkanes or alkenes can exist as cis–trans stereoisomers. For those that can, draw the two isomers. Label each of the isomers you drew as the cis stereoisomer or the trans stereoisomer.

a.

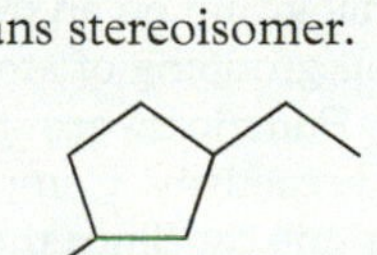

b.

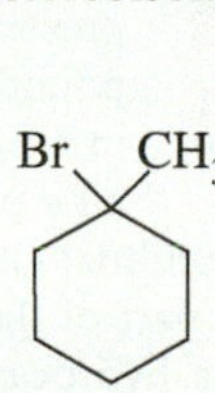

c.

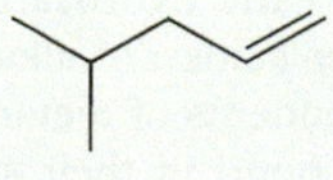

d. $CH_3CH_2CH{=}CHF$

4.35 Mark the chiral centers in the following molecules, if any, with an asterisk (*):

a.

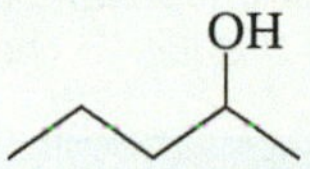

b. 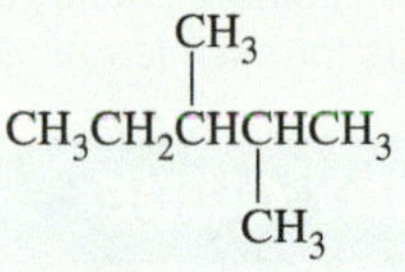
$$\underset{\overset{|}{CH_3}}{CH_3CH_2CHCHCH_3}$$
(with CH_3 above)

c.

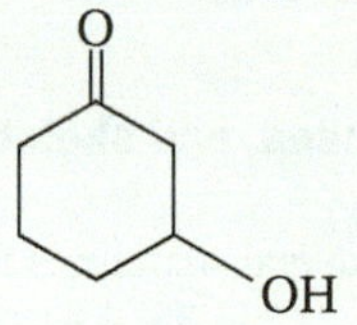

d.

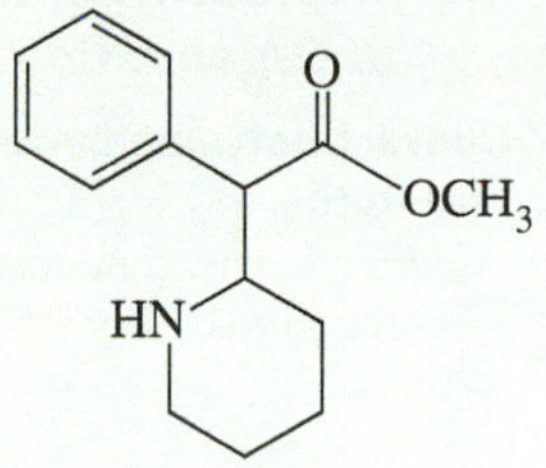

Ritalin®

4.36 Mark the chiral centers in the following molecules, if any, with an asterisk (*):

a.

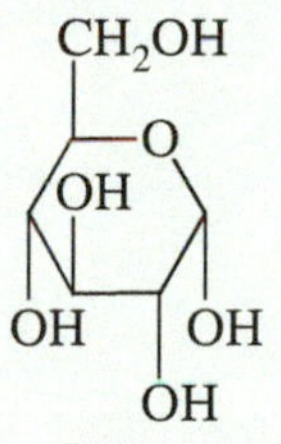

D-Glucose, the primary fuel of brain cells

b.

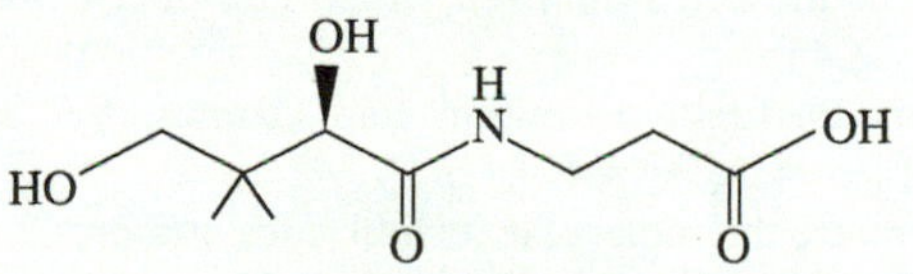

Pantothenic acid, a B vitamin

c.

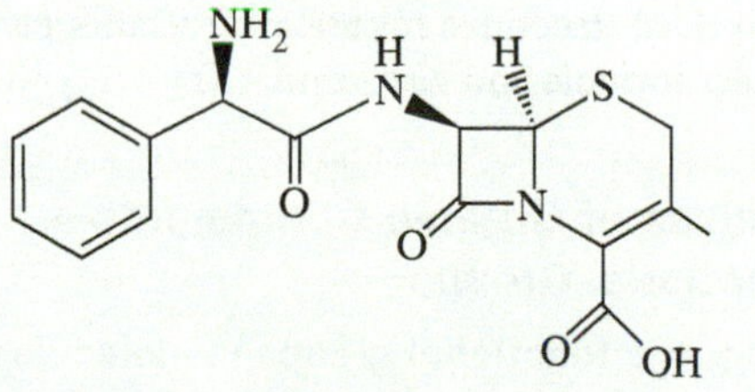

Cephalexin, an antibiotic also called Keflex®

d.

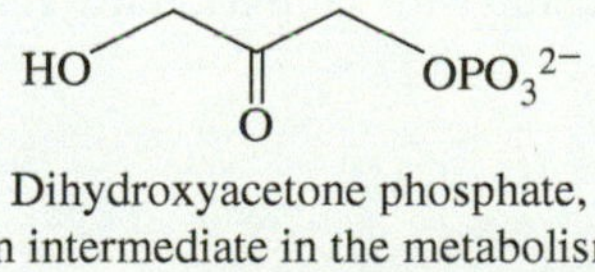

Dihydroxyacetone phosphate, an intermediate in the metabolism of carbohydrates

CHAPTER REVIEW

The study guide will help you check your understanding of the main concepts in Chapter 4. You should be able to answer problems for each learning outcome in this list. To check your mastery, try the problems listed after each.

STUDY GUIDE

4.1 Representing the Structures of Organic Compounds

Convert between Lewis, condensed, and skeletal structures.

- Draw Lewis, condensed, and skeletal structures for organic compounds. (Try 4.5, 4.7, 4.9, 4.37, 4.39, 4.41, 4.43)

4.2 Alkanes: The Simplest Organic Compounds

Characterize simple alkanes.

- Define the terms *saturated* and *unsaturated hydrocarbon*. (Try 4.45)
- Name the first ten straight-chain alkanes. (See Sample Problem 4.5; Try 4.47)
- Compare the molecular formulas for straight-chain alkanes and cycloalkanes. (See Sample Problem 4.4)
- Draw skeletal and condensed structures for the straight-chain alkanes given the molecular formula and vice versa. (Try 4.13, 4.15)
- Draw skeletal structures for the cycloalkanes given the molecular formula and vice versa. (Try 4.11, 4.63)

4.3 Families of Organic Compounds— Functional Groups

Identify organic functional groups in molecules.

- Identify the unsaturated hydrocarbon families—alkenes, alkynes, and aromatics. (Try 4.17)
- Identify common functional groups in organic molecules. (Try 4.19, 4.53, 4.55)
- Draw skeletal structures for saturated fatty acids. (Try 4.21, 4.52)

CHAPTER GUIDE

Inquiry Question

How are organic compounds represented?

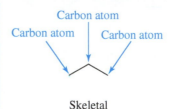

Carbon atom
Carbon atom   Carbon atom

Skeletal structure

Organic compounds can be represented with Lewis structures showing all atoms, bonds, and lone pairs of electrons in a molecule. Condensed structural formulas show all atoms in an organic molecule and their relative positioning, but they show bonds only when necessary for conveying the correct structure. Lone pairs may or may not be a part of a condensed structure. Skeletal structures show the bonding "skeleton" of an organic molecule by showing all carbon–carbon bonds. Hydrogen atoms bonded to carbon are not shown in skeletal structures, and carbon atoms are understood to exist at the corner formed when two bonds meet or at the termination of a bond. Atoms other than carbon and the hydrogens bonded to them are shown in skeletal structures.

Inquiry Question

How are alkanes and cycloalkanes represented?

An organic compound is any compound composed mainly of carbon and hydrogen. The simplest of these are the alkanes, containing *only* single-bonded carbon and hydrogen. Alkanes can exist as straight-chain alkanes, branched-chain alkanes, and cycloalkanes. Because carbon–hydrogen bonds are nonpolar, alkanes are nonpolar molecules and not very reactive molecules, with the exception that they react readily with oxygen in combustion reactions.

Inquiry Question

How are carbon compounds organized?

Organic compounds are grouped into families based on the identity of the functional group(s) present. A functional group on an organic compound is a common grouping of atoms bonded in a particular way. Functional groups have specific properties and reactivity. Compounds with the same functional group behave similarly. Since the functional group is the part of the molecule that is of interest, we typically represent the hydrocarbon portion as *R* (the Rest of the molecule). The alkene, alkyne, and aromatic hydrocarbon families are highlighted in this section. Fatty acids are alkane-like biomolecules that are the primary components of dietary fats. Fatty acids with a carbon–carbon double bond in their structure are referred to as unsaturated, while those without a double bond are referred to as saturated.

4.4 Nomenclature of Simple Alkanes

Name simple alkanes.

• Draw branched-chain alkanes, haloalkanes, and cycloalkanes. Name them using IUPAC naming rules. (Try 4.27, 4.65)

Inquiry Question

How are alkanes named?

Alkanes, cycloalkanes, and haloalkanes can be named by following a simple set of rules developed by the IUPAC. Find the longest continuous chain of carbons, identify the substituents, number the parent chain from the end closest to a substituent, and finally, assign numbers to the substituents and arrange them alphabetically in front of the parent name.

4.5 Isomerism in Organic Compounds

Distinguish the isomerism in various organic compounds.

• Distinguish structural isomers from conformational isomers. (Try 4.29, 4.31, 4.67)
• Identify cis and trans isomers in cycloalkanes and alkenes. (Try 4.33, 4.69, 4.71)
• Draw skeletal structures for unsaturated fatty acids. (4.77, 4.78)
• Locate chiral centers in organic molecules. (Try 4.35, 4.79)

Inquiry Question

How are isomers the same, and how are they different?

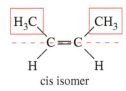
cis isomer

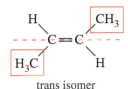
trans isomer

Isomers are compounds with the same molecular formula but with a different arrangement of atoms. Structural isomers are two molecules with the same molecular formula but a different connectivity. Conformational isomers are different representations of the same compound. Stereoisomers are two molecules with the same molecular formula and same connectivity but a different arrangement of the atoms in space. Cis–trans stereoisomers can exist in cycloalkanes with two or more substituents and in alkenes. Naturally occurring unsaturated fatty acids contain cis-alkenes. Organic compounds can also exist as stereoisomers if they contain a chiral center, which occurs when a carbon atom is bonded to four different atoms or groups of atoms. Compounds with a single chiral center can exist as a pair of stereoisomers called enantiomers. Enantiomers are related to each other as nonsuperimposable mirror images. One way to distinguish between isomers and two representations of the same compound (conformer) is to determine the correct name of each compound. Two isomers will have different names.

Mastering Videos

PRACTICING THE CONCEPTS

The following videos can be accessed through the Pearson eText or your Mastering Chemistry course.

• Drawing Skeletal Structures of Organic Molecules
• Identifying Chiral Centers

Additional Problems

4.37 Convert each of the Lewis structures shown into a condensed structural formula:

a.
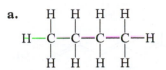

b.
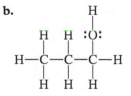

c.
H—C—C—C—C—C—C—C—H (with H's above and below each C)

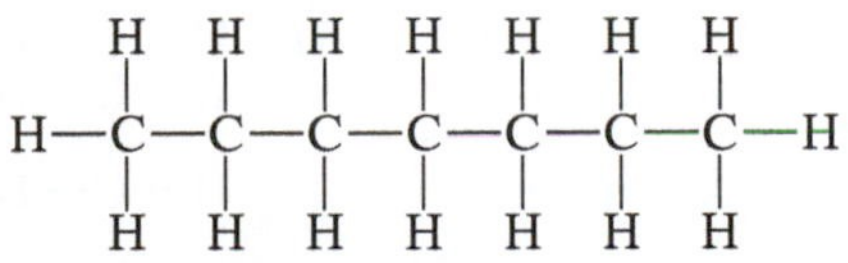

4.38 Convert each of the Lewis structures in Problem 4.37 into a skeletal structure.

4.39 Convert the condensed structures shown to skeletal structures.

a. 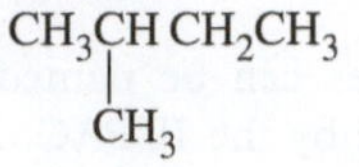 CH$_3$CHCH$_2$CH$_3$
       |
       CH$_3$

b. 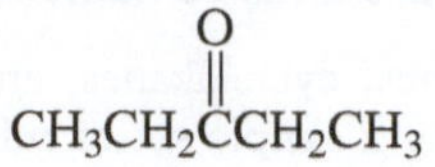
            O
            ||
   CH$_3$CH$_2$CCH$_2$CH$_3$

c. 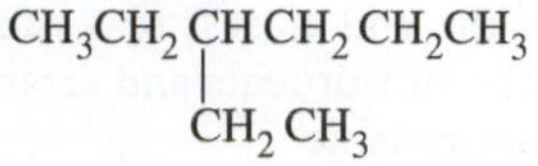 CH$_3$CH$_2$CHCH$_2$CH$_2$CH$_3$
              |
              CH$_2$CH$_3$

4.40 Convert the condensed structures shown to skeletal structures.

a. CH$_3$CH$_2$CH=CH$_2$

b. CH$_3$CH$_2$CH$_2$CH$_2$OH

c. 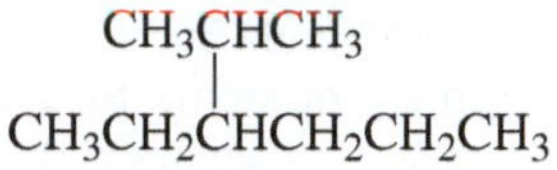   CH$_3$CHCH$_3$
                  |
   CH$_3$CH$_2$CHCH$_2$CH$_2$CH$_3$

4.41 Convert the skeletal structures shown to condensed structures.

a. 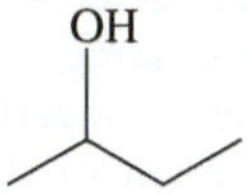 OH

b.  Br

c. 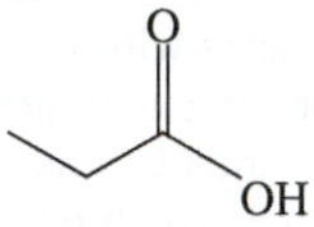 O
       OH

4.42 Convert the skeletal structures shown to condensed structures.

a.

b.

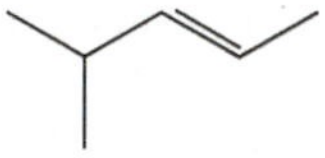

c.

4.43 Lewis structures, condensed structural formulas, and skeletal structures are used to represent the structure of an organic compound. Each of the following compounds is shown in one of these representations. Convert each compound into the other two structural representations not shown.

a. 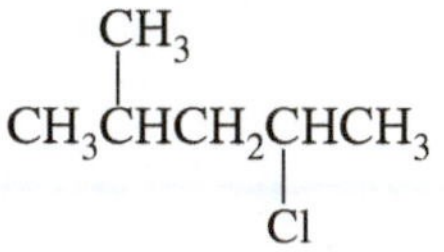   CH$_3$
              |
   CH$_3$CHCH$_2$CHCH$_3$
                  |
                  Cl

b. 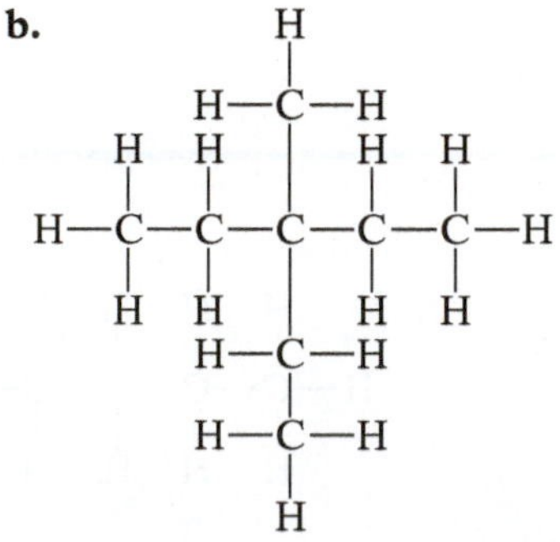
              H
              |
          H—C—H
       H H   H H
       | | | |
   H—C—C—C—C—C—H
       | | | |
       H H   H H
          H—C—H
              |
          H—C—H
              |
              H

c.

4.44 The molecular formula of an organic compound gives the least detail about the compound's structure. Give the molecular formula of each of the compounds in Problem 4.43.

4.45 Alkanes are also referred to as *saturated hydrocarbons*. Explain the meaning of the term *hydrocarbon*. Why are alkanes called saturated hydrocarbons?

4.46 Are alkanes considered polar or nonpolar compounds? Explain.

4.47 Give the skeletal structure and name of the straight-chain alkanes whose molecular formula is shown.

   a. C$_3$H$_8$ **b.** C$_5$H$_{12}$ **c.** C$_8$H$_{18}$

4.48 Give the skeletal structure and name of the straight-chain alkanes whose molecular formula is shown.

   a. C$_6$H$_{14}$ **b.** C$_{10}$H$_{22}$ **c.** C$_4$H$_{10}$

4.49 Give the structure and name of the cycloalkanes described.

   a. A compound whose molecular formula is C$_6$H$_{12}$ and contains a five-membered ring

   b. a compound whose molecular formula is C$_6$H$_{12}$, but has no substituents

4.50 Give the structure and name of the cycloalkanes described.

   a. A compound whose molecular formula is C$_7$H$_{14}$ and contains a six-membered ring

   b. A compound whose molecular formula is C$_5$H$_{10}$, but has no substituents

4.51 Explain the structural difference between a saturated and an unsaturated fatty acid.

4.52 Draw the condensed structural formula and skeletal structure of the saturated fatty acid with 16 carbon atoms. What is the name of this fatty acid?

4.53 Identify all of the functional groups in each of the following molecules:

a.

   H$_2$N

                        O

                              OH

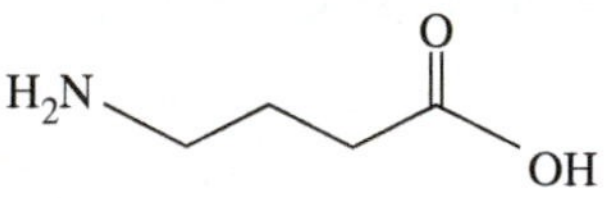

   Gamma-aminobutyric acid

b.

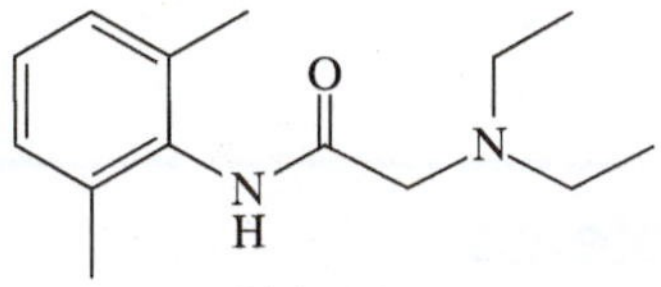

   Lidocaine

4.54 Identify all of the functional groups in each of the following molecules:

a.

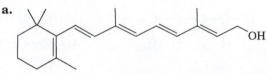

   Vitamin A

b.

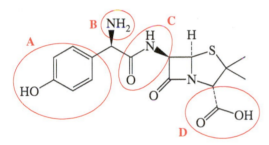

Pseudoephedrine

4.55 Zantac™ is used to treat the overproduction of stomach acid for patients with chronic heartburn or gastric ulcers. Identify the four functional groups circled in the structure of Zantac.

Zantac

4.56 Name the four functional groups circled in the following molecule:

Amoxicillin,
an antibiotic

4.57 Write the condensed formula for each of the following molecules:
  a. 3-ethylhexane
  b. 1,3-dichloro-3-methylheptane
  c. 4-isopropyl-2,3-dimethylnonane

4.58 Draw skeletal structures for each of the following molecules:
  a. ethylcyclopropane
  b. *cis*-1-chloro-3-methylcyclohexane
  c. isopropylcyclohexane

4.59 A widely used general anesthetic is called halothane or Fluothane. Its IUPAC name is 2-bromo-2-chloro-1,1,1-trifluoroethane. Draw the Lewis structure for this compound.

4.60 The refrigerant 1,1,1,2-tetrafluoroethane has been used in air conditioners in cars since the mid-1990s, but is being phased out due to its environmental impact. Draw the skeletal structure for this compound.

4.61 Using condensed structural formulas, draw three conformers of hexane.

4.62 Using skeletal structures, draw two conformers of butane.

4.63 Draw the skeletal structure and give the correct IUPAC name for three cycloalkane structural isomers with the molecular formula C_5H_{10}.

4.64 Draw a condensed structural formula and give the correct IUPAC name for the three alkane structural isomers with the molecular formula C_5H_{12}.

4.65 Draw the two alkane structural isomers with molecular formula C_4H_{10}. Give the IUPAC name of each compound.

4.66 Draw the two cycloalkane structural isomers with the molecular formula C_4H_8. Give the IUPAC name of each compound.

4.67 How many structural isomers are possible for the molecular formula $C_5H_{11}F$? Draw the skeletal structure and give the IUPAC name of each compound.

4.68 Draw the skeletal structure and give the correct IUPAC name for three haloalkane structural isomers with the molecular formula C_4H_9Br.

4.69 Using wedge-and-dash bonds, draw a cis and a trans stereoisomers for each of the following compounds:
  a. 1,3-dimethylcyclohexane
  b. 1-bromo-2-ethylcyclopentane

4.70 Using wedge-and-dash bonds, draw both the cis and trans stereoisomers for each of the following compounds:
  a. 1-chloro-4-fluorocyclohexane
  b. 1,3-diethylcyclobutane

4.71 For each of the following compounds, indicate whether or not it can exist as cis–trans stereoisomers. If it can exist as the two isomers, draw both as a condensed structure.

  a. $H_2C{=}CHCH_2CH_3$ **b.**

  c. $CH_3CH_2CH{=}CHCH_2CH_3$ **d.**

4.72 For each of the following compounds, indicate whether or not it can exist as cis–trans stereoisomers. If it can exist as the two isomers, draw both as a condensed structure.

  a.   **b.**

  c. $CH_3CH{=}CHCHCH_3$ **d.** $H_3C{-}C{=}CHCH_2CH_3$
  $\qquad\qquad\quad\; \underset{CH_3}{|}$ $\qquad\quad \underset{CH_3}{|}$

4.73 The structure of vitamin A is shown in Problem 4.54. Is the double bond nearest the ring on the carbon chain cis or trans?

4.74 The vision process in animals involves the change of one double bond in retinal from its cis form to its trans form. The structures of cis and trans retinal are shown below. Label which is cis and which is trans.

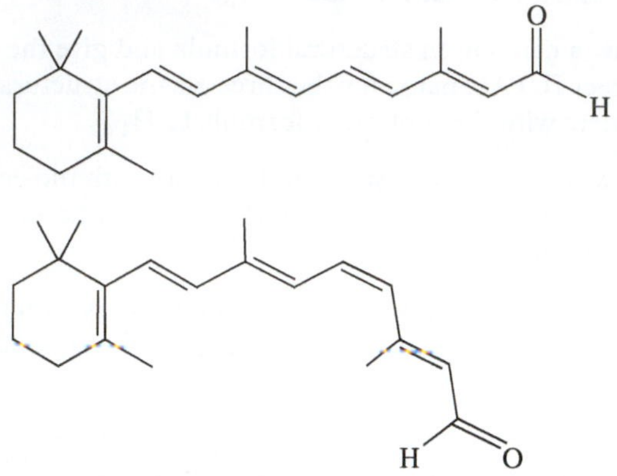

The retinal molecules

4.75 Determine whether each of the following is the cis or the trans stereoisomer:

a.

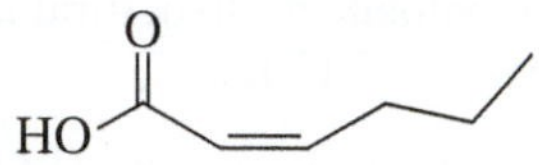

b.

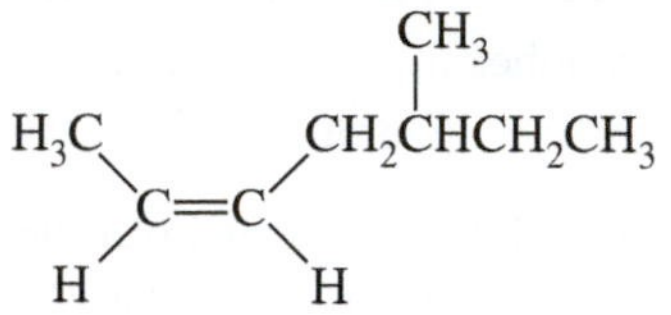

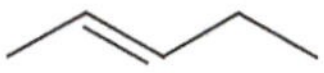

CH₃
H₃C   CH₂CHCH₂CH₃
    C=C
H   H

c.

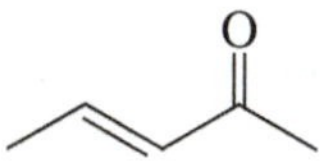

d. CH₃CH₂   H
         C=C
     H   CH₂CH₂CH₂CH₃

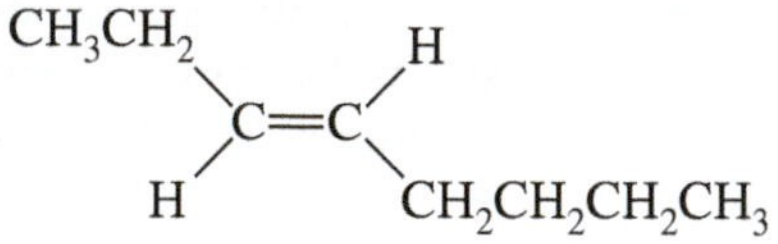

4.76 Determine whether each of the following is the cis or the trans stereoisomer:

a. H₃C   H
         C=C
     H   CH₂CH₂OH

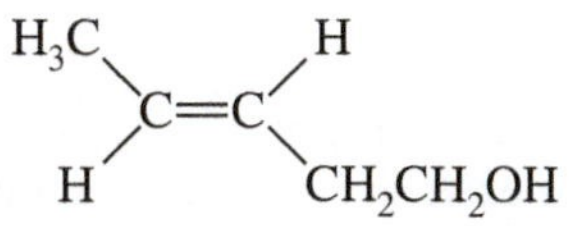

b.

c. CH₃CH₂   CH₂CH₂C—OH
         C=C   ‖
     H   H   O

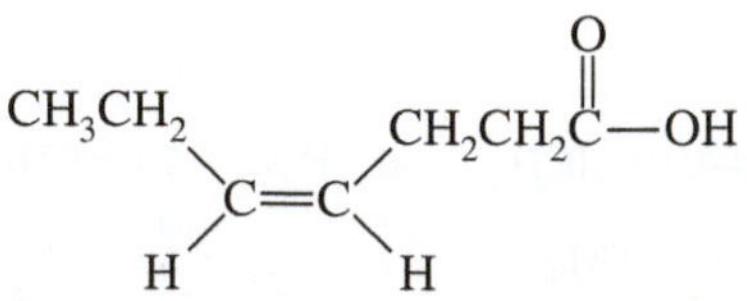

d.

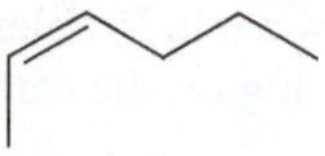

4.77 Draw the skeletal structure and give the name and omega number of the following fatty acid:

$$CH_3CH_2CH_2CH_2CH_2HC=CHCH_2HC=CHCH_2CH_2CH_2CH_2CH_2CH_2CH_2COOH$$

4.78 Draw the structure of γ-linolenic acid [18:3] and give its omega number.

4.79 Mark the chiral centers in the following molecules, if any, with an asterisk (*).

a.   **b.** HO—
         O
H₃N⁺—CH—C—O⁻
    | OH
   CH—OH   OH OH
    | D-ribose,
   CH₃   a carbohydrate
Threonine, an
amino acid

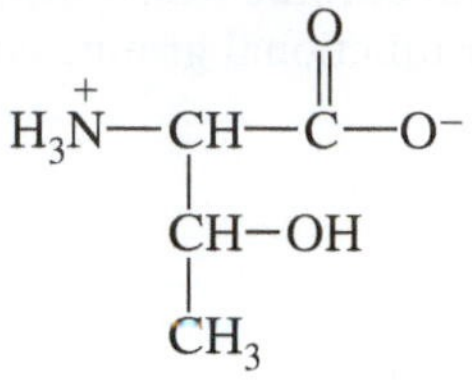

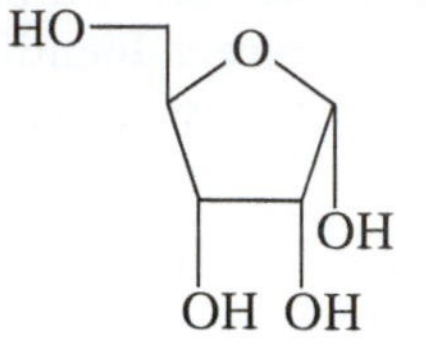

c.   **d.** H₃CO
         O
         ‖
         C—OH
                                 HO⁗
    N   ⁗N
Niacin, vitamin B₃
                                 Tramadol, an analgesic

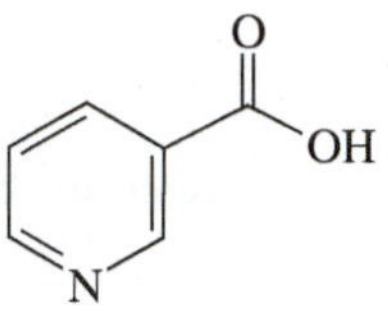

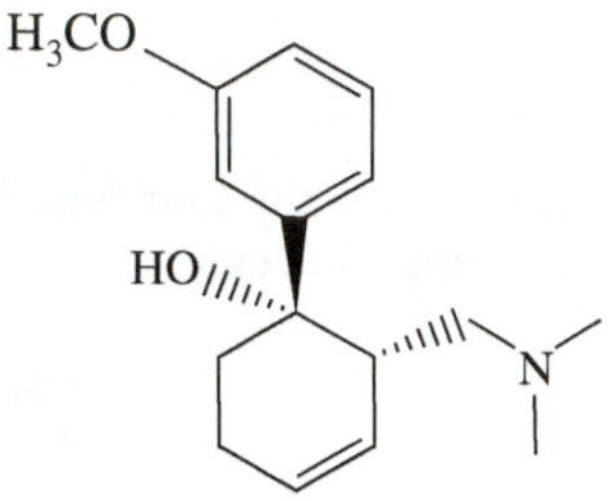

4.80 Mark the chiral centers in the following molecules, if any, with an asterisk (*).

a.   OH
HO   O   O
         HO   OH
         Vitamin C

b.   O
              ‖
H₃N⁺—CH—C—O⁻
    |
   CH—CH₃
    |
   CH₃
Valine, an amino acid

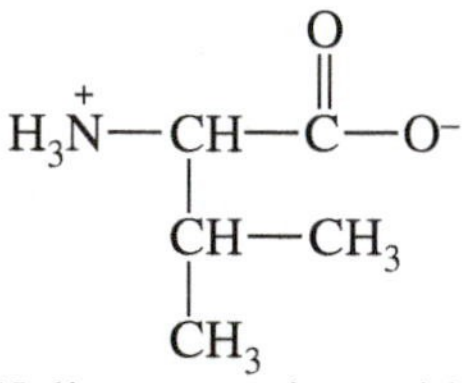

c.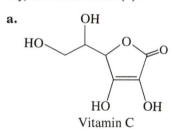
Ibuprofen

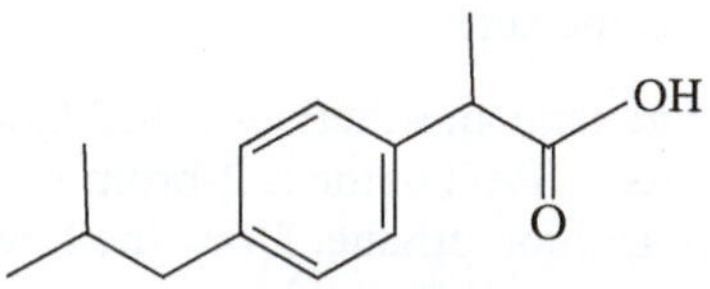

d. 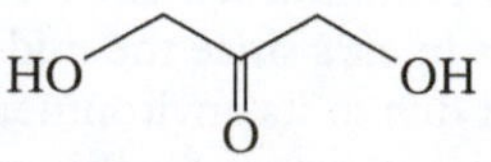
HO   OH
         ‖
         O
Dihydroxyacetone,
used in spray tanning

4.81 Are the following compounds structural isomers, cis–trans isomers, or enantiomers?

a.

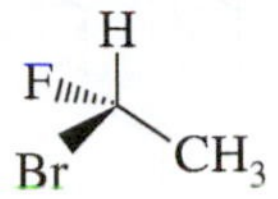

and

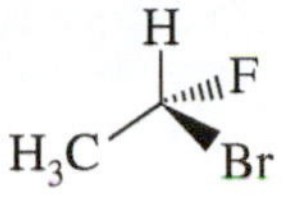

b.

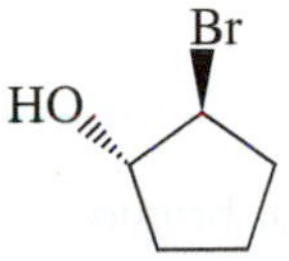

and

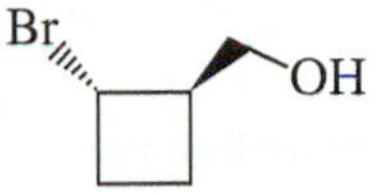

c.

and

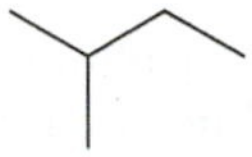

4.82 Draw the enantiomer of each of the following compounds. If the compound is not chiral, state that fact.

a.

b. 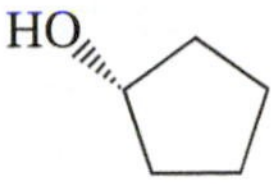 HO

c.

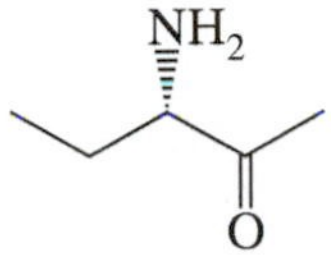

Challenge Problems

4.83 Certain omega-3 fatty acids can be found only in animal sources, such as fatty fish. Two of these are eicosapentaenoic acid (EPA) [20:5] and docosahexaenoic acid (DHA) [22:6], both of which are ω-3 fatty acids. DHA has been shown to be important in healthy brain development, so it has recently been added to infant formulas. Breast milk is rich in DHA as long as the mother maintains a healthy diet that includes fish. Draw skeletal structures of the fatty acids EPA and DHA.

4.84 For each pair of molecules, identify the pair as:

   A. structural isomers.

   B. the same molecule (conformational isomers).

   C. cis–trans stereoisomers.

   D. different molecules.

a.

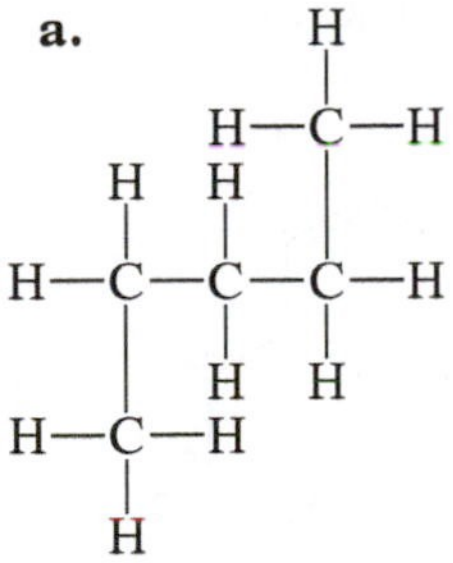

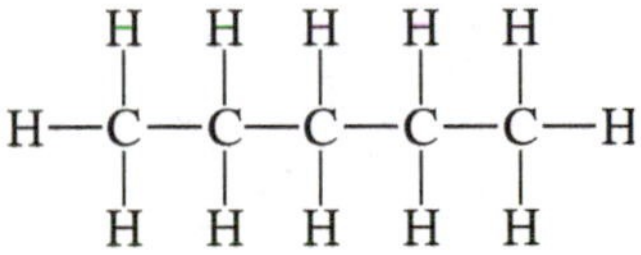

b.

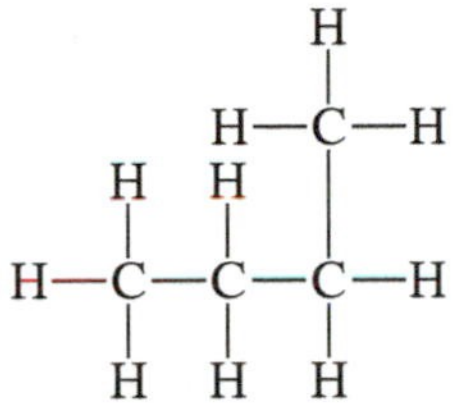

$CH_3CH(CH_3)_2$

c.

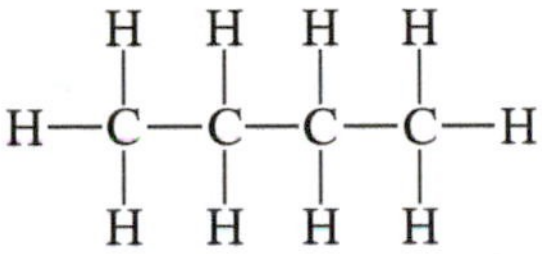

d.

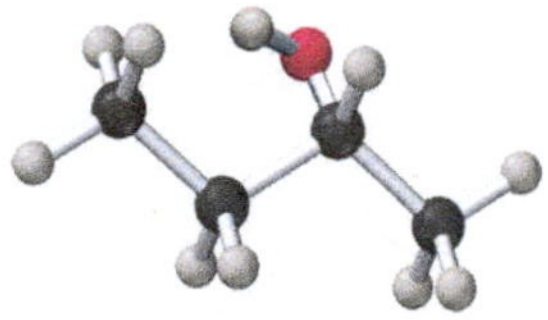

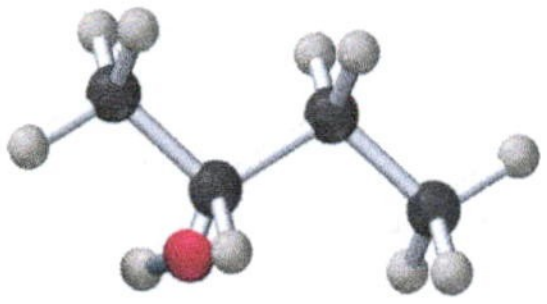

e. $CH_3CH_2CH_2CH_2OH$

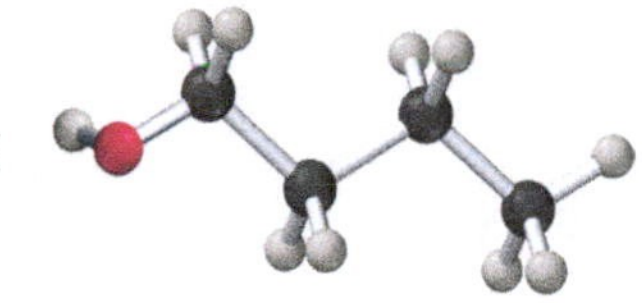

f.

Answers to Odd-Numbered Problems

Practice Problems

4.1 A Lewis structure shows all atoms, bonds, and nonbonding electrons. A condensed structure shows all atoms but as few bonds as possible.

4.3 Skeletal structures show mainly bonds between carbon atoms. Since methane has only one carbon, it is not possible to draw its skeletal structure.

4.5 **a.** 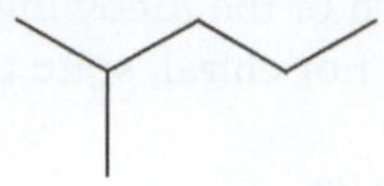 [structure: branched alkane]   **b.** 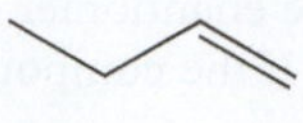 [structure: alkene]

4.7 **a.**

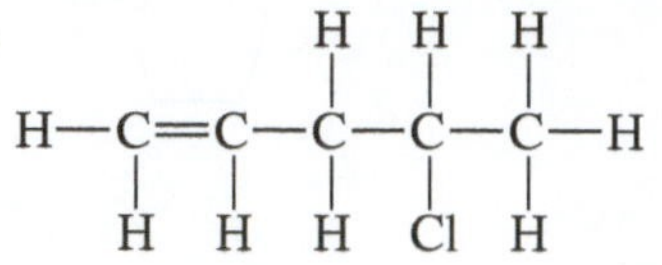

$$H-C=C-\overset{\overset{\displaystyle H}{|}}{C}-\overset{\overset{\displaystyle H}{|}}{\underset{\underset{\displaystyle Cl}{|}}{C}}-\overset{\overset{\displaystyle H}{|}}{\underset{\underset{\displaystyle H}{|}}{C}}-H$$

b.

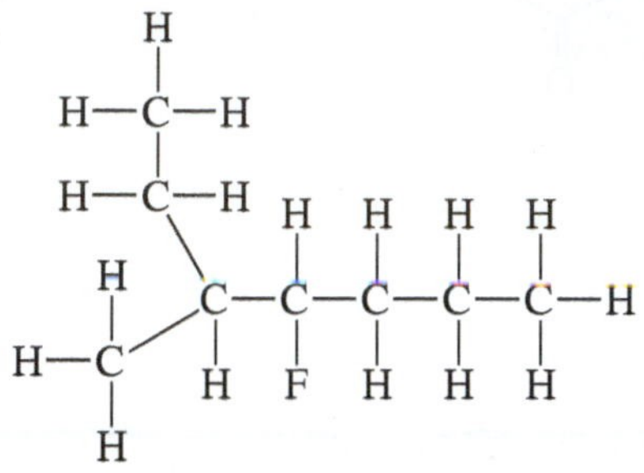

[structure: branched fluorine-containing molecule with central chain C–C–C–C–C–H, substituents include H–C–H, H–C–H, H–C, and F]

4.9 **a.** 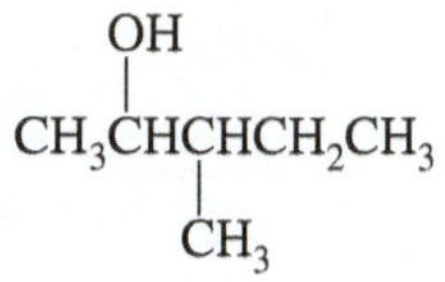   OH

$$CH_3\overset{|}{C}HCHCH_2CH_3$$
$$\qquad\quad\underset{|}{C}H_3$$

b. 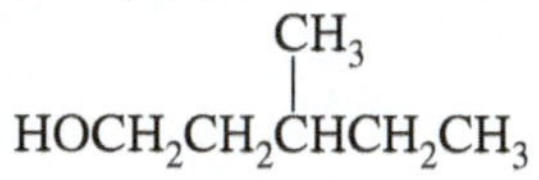 CH_3

$$HOCH_2CH_2\overset{|}{C}HCH_2CH_3$$

4.11 **a.** pentane   **b.** cyclohexane   **c.** heptane

4.13 **a.** $CH_3CH_2CH_2CH_2CH_2CH_2CH_3$ **b.** CH_4
c. $CH_3CH_2CH_2CH_2CH_2CH_2CH_2CH_2CH_2CH_3$

4.15 **a.**  [hexagon structure]   **b.**  [structure]

c.  [structure]

4.17 **a.** alkyne   **b.** alkene
c. alkene (a cycloalkene)

4.19 **a.**

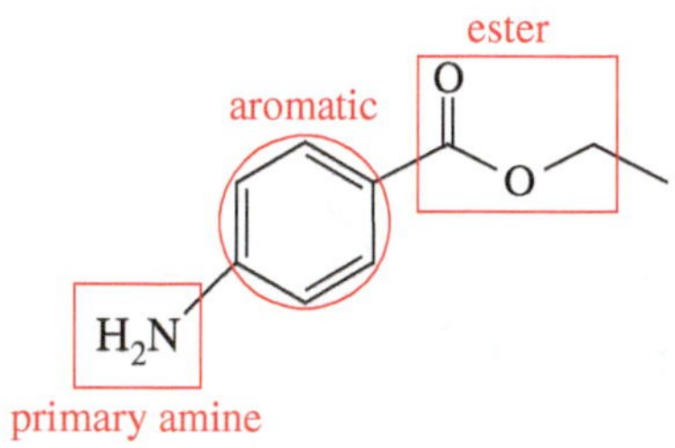

[structure with labels: ester, aromatic, H_2N primary amine]

b.

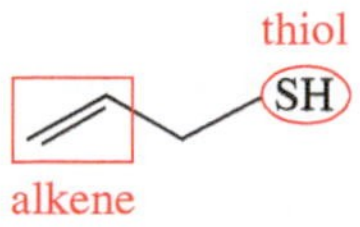

[structure with labels: thiol (SH), alkene]

4.21

[structure: long chain carboxylic acid with O and OH]

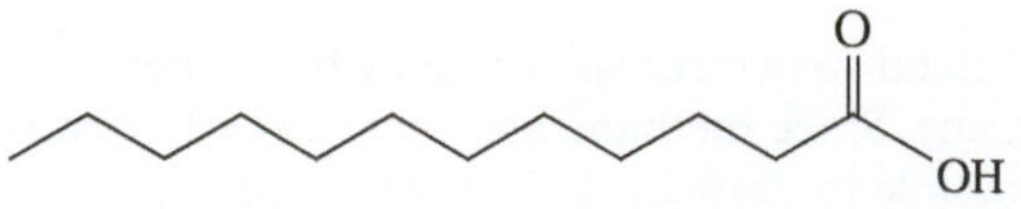

4.23 **a.** $CH_3CH_2CH_2-$ **b.** CH_3-

4.25 **a.** 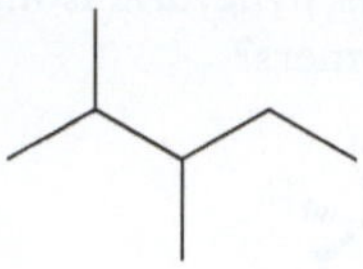 [structure]   **b.** 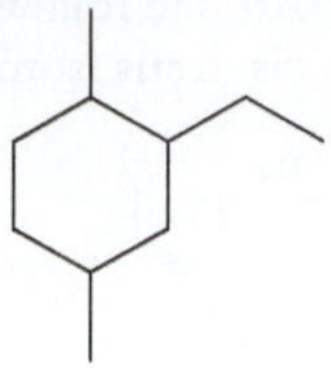 [cyclohexane structure]

c. 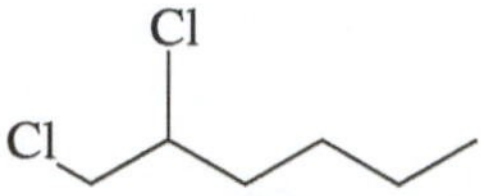   Cl

[structure with two Cl groups]

4.27 **a.** 2,2-dimethylbutane   **b.** bromocyclopentane
c. 2,4-dimethylhexane

4.29 Conformational isomers have the same connectivity of the atoms, but structural isomers do not.

4.31 **a.** structural isomers   **b.** not related
c. not related
d. conformational isomers (the same molecule)
e. structural isomers

4.33 **a.** cis–trans isomers not possible
b. cis–trans isomers not possible
c. 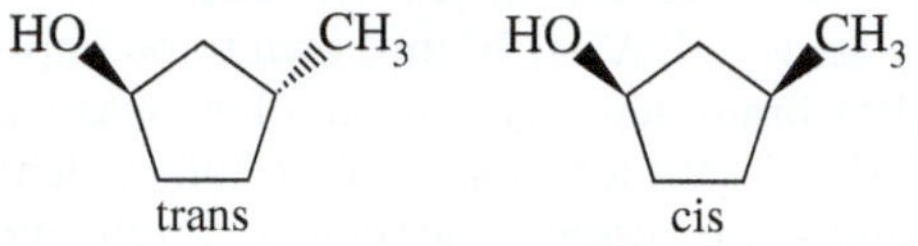 HO [cyclopentane structure] CH_3 HO [cyclopentane structure] CH_3
         trans   cis

d.

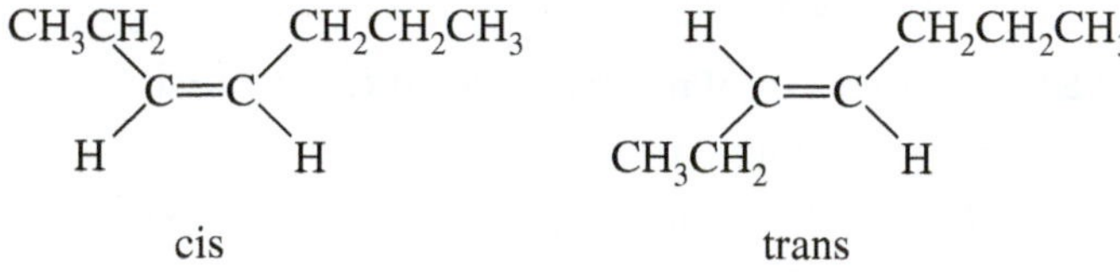

CH_3CH_2 $CH_2CH_2CH_3$ H $CH_2CH_2CH_3$
    C=C   C=C
H   H CH_3CH_2 H
      cis   trans

4.35 **a.** 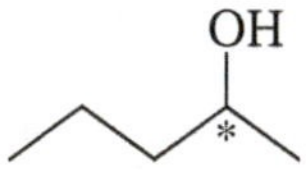   OH   **b.** 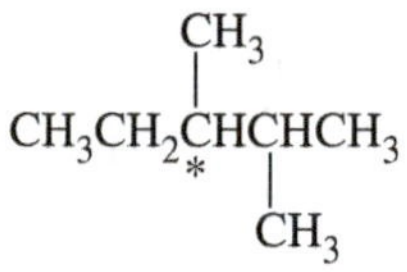 CH_3

[structure with *] $CH_3CH_2\overset{}{C}HCHCH_3$
                                      $\underset{}{C}H_3$

c.   O   **d.**

[cyclohexanone structure with *OH]   [structure with phenyl, *, O, OCH_3, HN, piperidine ring]

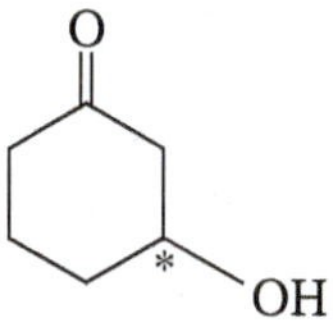

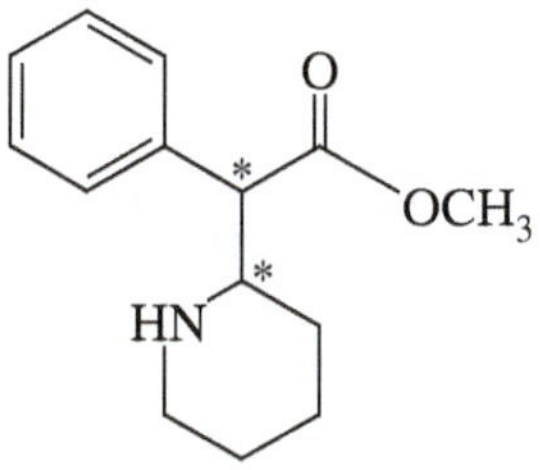

Additional Problems

4.37 **a.** $CH_3CH_2CH_2CH_3$ **b.** $CH_3CH_2CH_2OH$
c. $CH_3CH_2CH_2CH_2CH_2CH_2CH_3$

4.39 **a.** 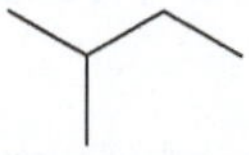 [structure]   **b.** 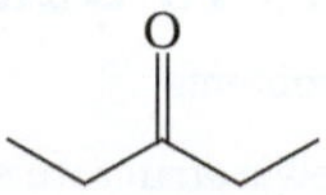 [structure with O]

c. 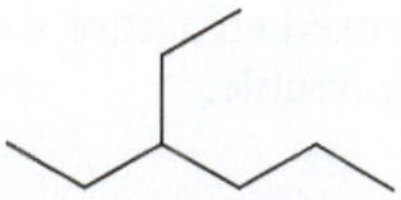 [structure]

4.41 **a.** $CH_3CH(OH)CH_2CH_3$ **b.** $CH_3CH_2CH_2Br$
c. CH_3CH_2COOH

4.43 **a.**

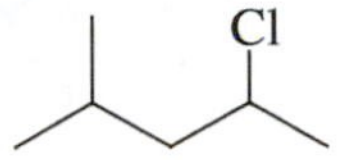
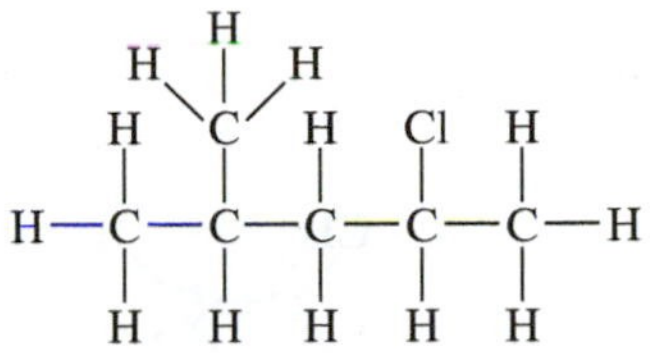

b.

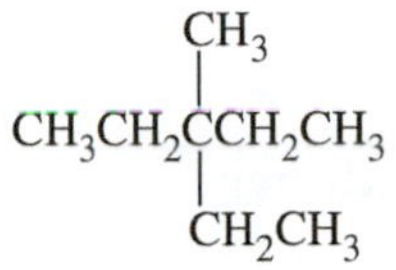

c.

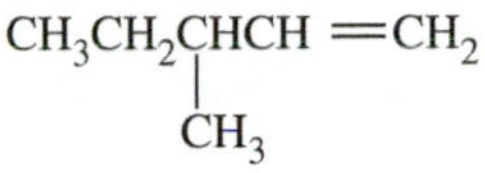
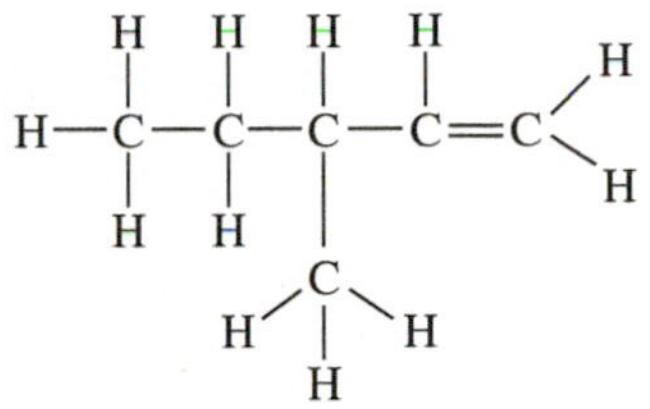

4.45 Hydrocarbons are organic compounds composed only of carbon and hydrogen. Saturated refers to the fact that each carbon is bonded to the maximum number of hydrogens.

4.47 **a.** propane 
b. pentane 
c. octane 

4.49 **a.**

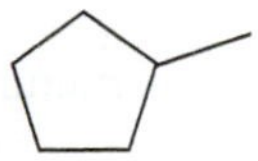

Methylcyclopentane

b.

Cyclohexane

4.51 An unsaturated fatty acid contains one or more carbon–carbon double bonds, but a saturated fatty acid contains no double bonds.

4.53 **a.**

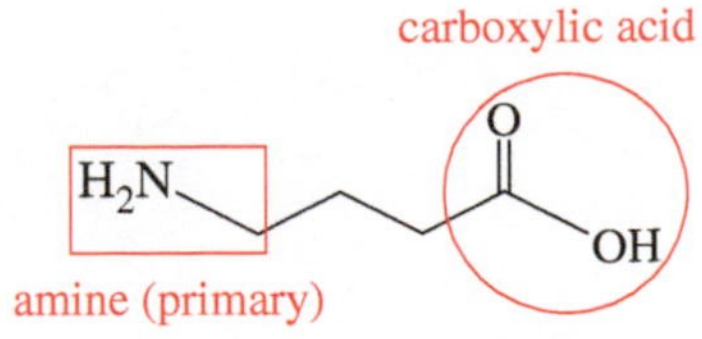
carboxylic acid
amine (primary)

b.

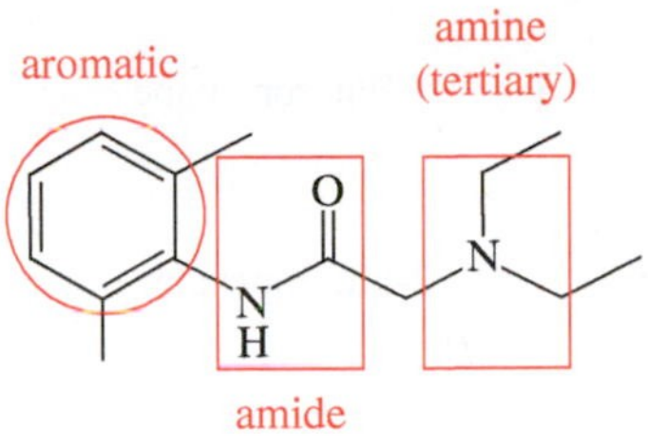
aromatic
amine
(tertiary)
amide

4.55 **A** = protonated amine, **B** = sulfide (thioether), **C** = ether, **D** = amine (tertiary)

4.57 **a.**

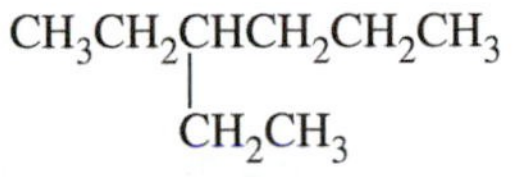

b.

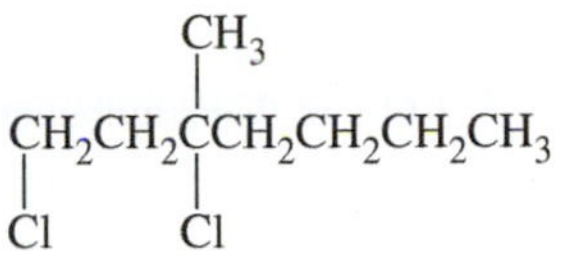

c.

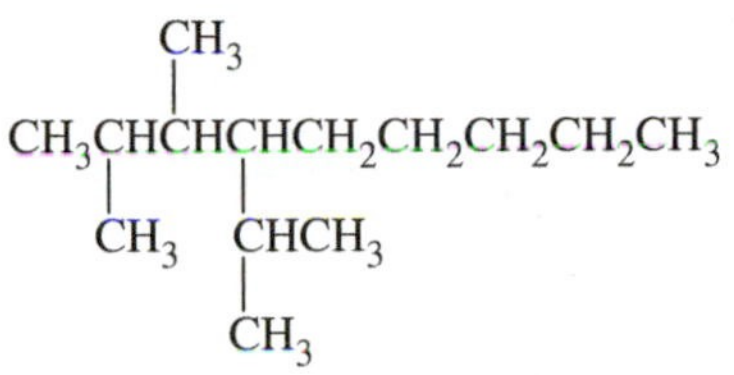

4.59

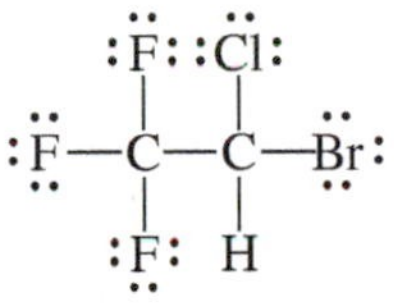

4.61

$CH_3CH_2CH_2CH_2CH_2CH_3$ 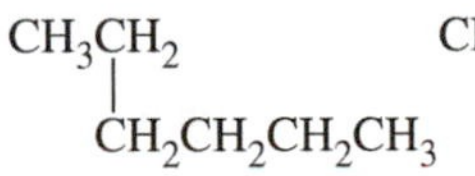 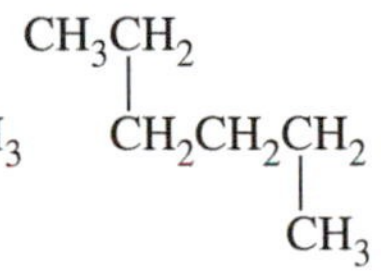

4.63 There are four possible isomers:

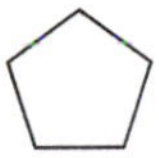

Cyclopentane

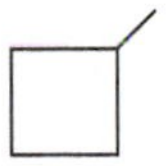

Methylcyclobutane

1,2-Dimethylcylopropane

1,1-Dimethylcylopropane

4.65

Butane

2-Methylpropane

4.67

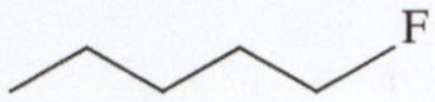

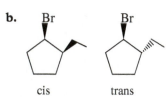

1-Fluoropentane

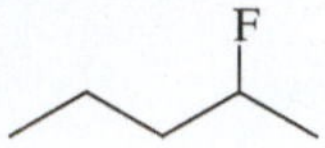

2-Fluoropentane

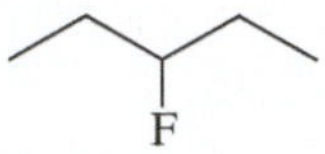

3-Fluoropentane

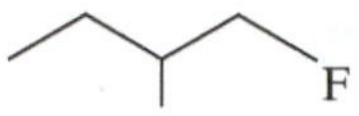

1-Fluoro-2-methylbutane

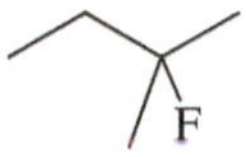

2-Fluoro-2-methylbutane

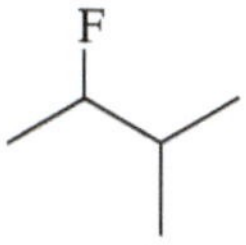

2-Fluoro-3-methylbutane

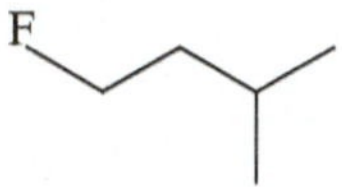

1-Fluoro-3-methylbutane

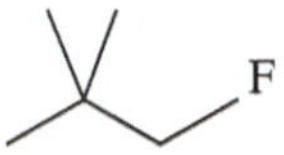

1-Fluoro-2,2-dimethylpropane

4.69 **a.**

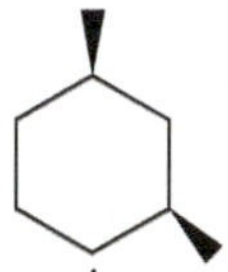

cis 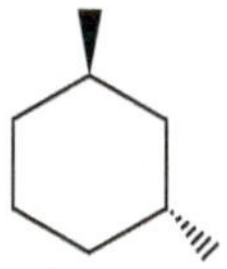   trans

b.   Br   Br

cis   trans

4.71 **a.** no cis–trans isomer

b.

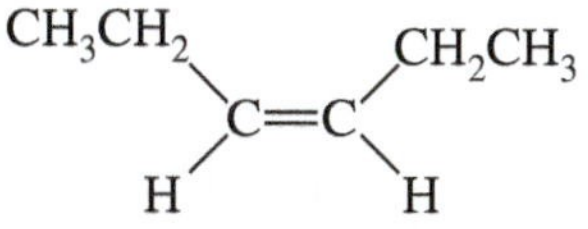

CH_3CH_2 CH_2CH_3 CH_3CH_2 H

C=C   C=C

H   H   H CH_2CH_3

cis 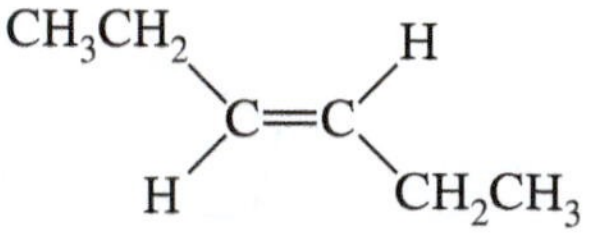   trans

c.

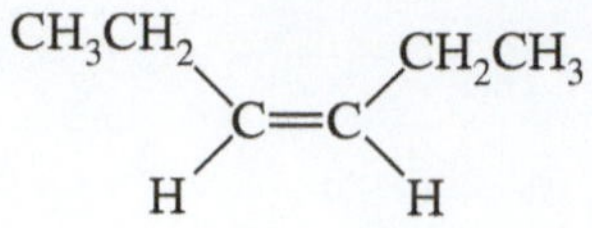

CH_3CH_2 CH_2CH_3 CH_3CH_2 H

C=C   C=C

H   H   H CH_2CH_3

cis 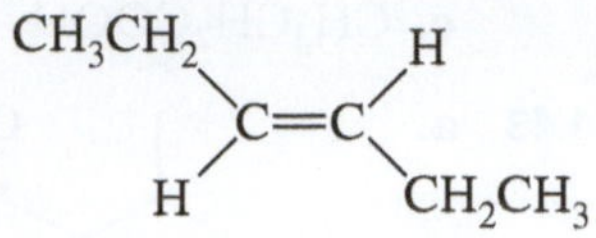   trans

d.

CH_3 CH_2CH_3 CH_3 H

C=C   C=C

H   H   H CH_2CH_3

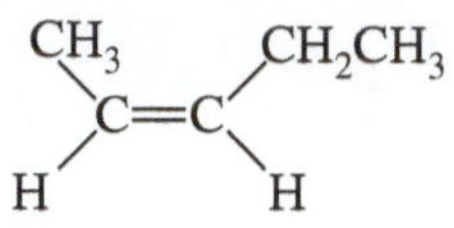

cis 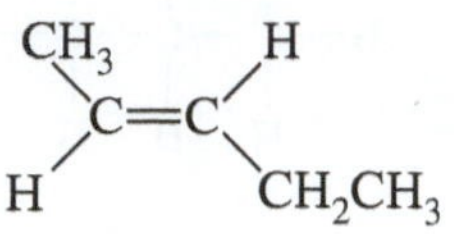   trans

4.73 trans

4.75 **a.** cis   **b.** cis

c. trans   **d.** trans

4.77 linoleic acid, [18:2], ω-6

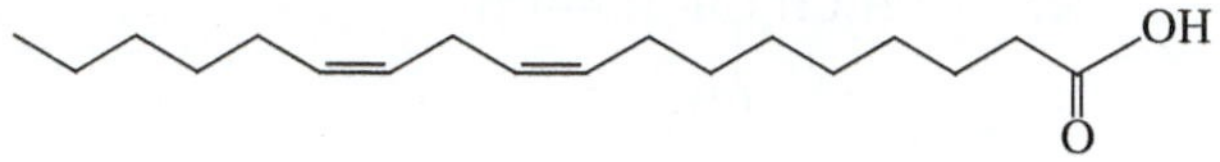

4.79 **a.**

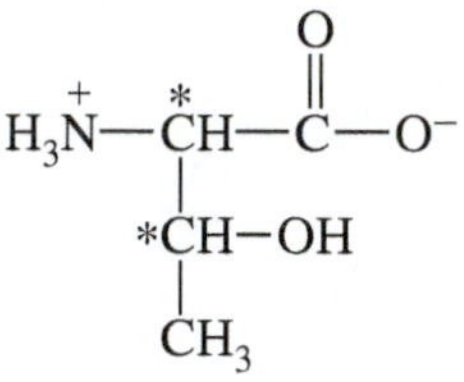

$$H_3\overset{+}{N}-\overset{*}{CH}-\overset{O}{\overset{\|}{C}}-O^-$$
$$\overset{*}{CH}-OH$$
$$CH_3$$

b. HO—

OH
OH OH

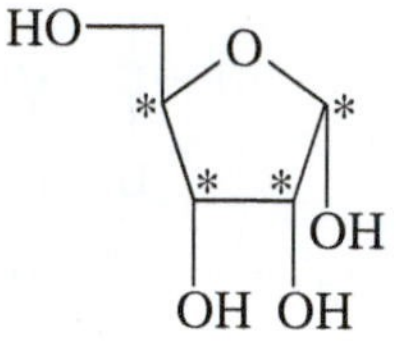

c. no chiral centers

d. H_3CO

HO
*
*
N

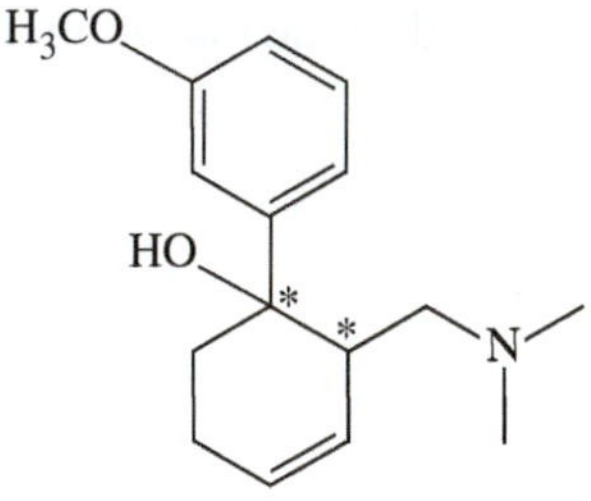

4.81 **a.** enantiomers   **b.** structural isomers

c. structural isomers

4.83

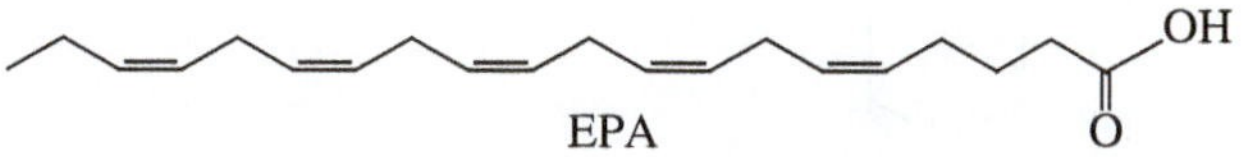

OH
O
EPA

O
OH
DHA

5

Chemical Reactions

A nutrition label provides quite a bit of information—such as the amount of protein, carbohydrate, fat, and other nutrients in a food—but the Calorie content is often the point of interest. The nutritional Calorie is a unit of energy, so the number of Calories is a measure of the energy available in a food. When we eat food, we transform the chemical energy (Calories) from the food into energy that our bodies use. Read Chapter 5 to find out more about energy transformation through reactions and their applications to biochemistry.

CHAPTER OUTLINE

LEARNING TIP   Flash Cards Can Be a Good Study Tool

When you make flash cards, be sure to write the information in your own words instead of copying isolated facts, like the key terms from your book. Keep adding to your stack as the semester moves forward. You will retain more information this way.

NUTRITION LABELS ON packaged foods list, among other things, the number of Calories the food contains, a measure of the energy content of that food. Our bodies capture and convert the energy in the food we eat through many different chemical reactions. Understanding how energy and heat are exchanged in chemical reactions describes an area of chemistry called **thermodynamics**.

When you eat, the food molecules are broken down by acid and enzymes in your digestive system. Enzymes function as catalysts to speed up these digestive reactions. The study of the pace or rate of a chemical reaction is called **reaction kinetics**.

In this chapter, we first explore the thermodynamics and kinetics of chemical reactions. Then we introduce several common organic reactions and consider their biochemical application.

What's an Inquiry Question?

Inquiry Questions are designed to focus your reading on the main concepts by section. As you read each section, see if you can answer each Inquiry Question in your own words.

? 5.1 Inquiry Question

How are energy and heat transferred and measured in a chemical reaction?

5.1 Thermodynamics

Chapter 1 defined energy as the capacity to do work. Chemical reactions involve an exchange of energy between substances. In some chemical reactions, energy is dissipated as heat, such as when we exercise. In other reactions, energy is transformed from potential energy into kinetic energy or vice versa. Energy, heat, and chemical reactions were first investigated in Chapter 1.

Before we begin, remember that when reading chemical reactions, the reactants always appear on the left, the products on the right, and the arrow connecting them means "yields." Let's take a closer look at energy and chemical reactions.

$$\text{Reactants} \xrightarrow[\textit{Yields}]{} \text{Products}$$

Heat of Reaction

Producing heat during chemical reactions is not unique to living things exercising or moving. Many chemical reactions produce or consume energy as heat. A reaction that gives off heat is called an **exothermic reaction** (*exo* means out of, *thermic* means heat; heat flows out). Reactions absorbing heat energy from the surroundings are called **endothermic reactions** (*endo* means inside, and *thermic* means heat; heat flows in).

An infant's heel warmer provides an example of an exothermic reaction. These warmers are used to increase blood flow near the skin surface to make drawing blood easier. The heel warmer contains a highly concentrated liquid solution of sodium acetate, $NaC_2H_3O_2$. When a piece of metal is introduced into the solution by bending a metal disk inside the heel warmer, solid sodium acetate forms, giving off heat. The chemical reaction can be written showing heat as a product for an exothermic reaction.

$$NaC_2H_3O_2(aq) \longrightarrow NaC_2H_3O_2(s) + \text{Heat}$$

An instant cold pack has the opposite effect of a heel warmer. We use cold packs for injuries to slow blood flow to the injured area. These packs typically have two compartments, one containing water and the other a white solid (often ammonium nitrate). When the pack is squeezed, the water compartment breaks and water combines with the solid. The pack gets very cold because the mixing of the two substances absorbs energy in the form of heat from its surroundings.

In an endothermic chemical reaction, the heat is written as a reactant.

$$\text{Heat} + NH_4NO_3(s) \longrightarrow NH_4NO_3(aq)$$

One common endothermic reaction used by the body is the evaporation of sweat from the surface of the skin. The water on the skin absorbs the heat from the body and evaporates from the skin surface. The skin is cooled as a result.

$$H_2O \text{ from sweat}(l) + \text{Heat} \longrightarrow H_2O \text{ from sweat}(g)$$

We can think of the exchange of heat energy in a chemical reaction as the more general thermodynamic term of **change in enthalpy**, represented as $\mathbf{\Delta H}$ (Δ = change, H = enthalpy). In an exothermic reaction, heat is given off as the reaction proceeds, so the products have a lower enthalpy than the reactants. Because ΔH is equal to the enthalpy of the products minus the enthalpy of the reactants, ΔH for an exothermic reaction is a negative number. In an endothermic reaction, energy is absorbed, giving the products a higher energy, which means ΔH is a positive number.

$$\Delta H_{\text{reaction}} = \Delta H_{\text{products}} - \Delta H_{\text{reactants}}$$

Randomness

Why do some reactions give off heat and some reactions require heat? Something else must be happening in the reactions at the molecular level. Consider what happens when you take an ice cube out of the freezer. It spontaneously melts, right? It is absorbing heat from its surroundings (endothermic) as it melts. The water molecules are also becoming less organized, or more randomly arranged, as the ice moves from a solid to a liquid.

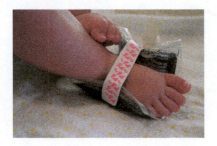

The exothermic reaction occurring in a heel warmer allows more blood to flow to the surface of the infant's skin.

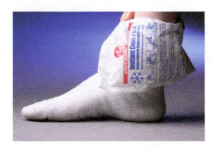

A cold pack produces an endothermic reaction that slows blood flow to an injured area.

As we saw in Chapter 1, the states of matter have differing amounts of randomness. In a solid, the particles are the least random, and in a gas, the particles are the most random (see Table 1.7).

The measure of the change in randomness in a chemical system, such as a reaction, is called the **change in entropy** of the system, and the change in entropy is represented as ΔS. As with enthalpy, the change in entropy is measured by the entropy of the products minus the entropy of the reactants. Thus, in a reaction that has an increase in entropy—for example, a liquid and a solid reacting to produce a gas—ΔS is positive. If the entropy decreases, ΔS is negative.

Free Energy and ΔG

For any chemical reaction, the change in enthalpy and the change in entropy are considered when determining the energy exchanged during a reaction. They are related by the following equation:

$$\Delta G = \Delta H - T(\Delta S)$$

The G in this equation represents the **free energy** or Gibbs free energy, and is defined as the amount energy present in molecules available to do work. The equation demonstrates that the **free energy change (ΔG)** includes contributions from the exchange of heat energy (enthalpy), from randomness (entropy), and from temperature. Because we are considering all of these factors when discussing free energy exchanged, the more general term **exergonic** (gives off energy) is used instead of exothermic. If energy is absorbed from the surroundings in a reaction, the term **endergonic** is used. The terms exergonic reaction and endergonic reaction are more appropriate when considering biochemical reactions because in the body, large amounts of heat are not produced, but large amounts of energy are transferred. We will not use this equation for calculations, but it is important to help us understand how these factors influence the free energy of a reaction.

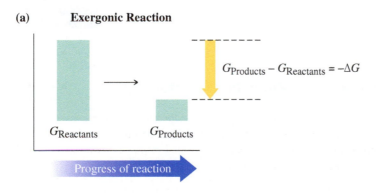

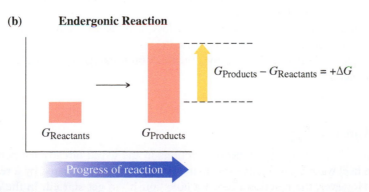

Changes in free energy. (a) Exergonic reactions release energy, meaning that the products are lower in energy than reactants, so ΔG is negative. (b) Endergonic reactions absorb energy, meaning that the products are higher in energy than reactants, so ΔG is positive.

(a)

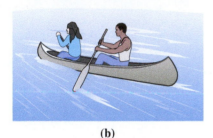

(b)

(a) A canoe traveling downstream with the current is a spontaneous process. (b) A canoe traveling upstream is a nonspontaneous process that requires energy input.

For an exergonic reaction, the free energy of the reactants is greater than the free energy of the products and the value of ΔG is negative. Exergonic reactions do not require energy input and occur spontaneously once initiated. A **spontaneous process** continues to occur once started and does not require energy from the surroundings. For example, a canoe will travel downstream spontaneously with a current with no additional effort from the paddlers. In an endergonic reaction, however, the free energy of the products is greater than that of the reactants, indicating that energy is absorbed and the value of ΔG is positive. Reactions with a positive ΔG value *do* require energy input from their surroundings and are considered nonspontaneous. A **nonspontaneous process** does not occur naturally and requires energy input. In our canoe example, the canoe can only go upstream if the paddler inputs energy by paddling.

A living system's free energy is the energy that can do work under cellular conditions. During a spontaneous change, the free energy decreases ($-\Delta G$) and the stability of the system increases. In a living system, the free energy given off in one chemical reaction is coupled as input to another chemical reaction. In other words, the energy given off by one reaction is transferred to another reaction. The main molecule that transfers energy in the cell in living systems is called **a**denosine **trip**hosphate, abbreviated as **ATP**. For example, our bodies break down food molecules to produce energy, which is used to make ATP. In turn, ATP fuels our muscles for a walk or our brains to ace a chemistry exam! In Chapter 12, we will take an in-depth look at the processes involved in breaking down food molecules and discover the reactions involved in harnessing that energy to produce ATP.

Spontaneous, exergonic reactions

ATP

Nonspontaneous, endergonic reactions

In living systems, free energy is often transferred from exergonic reactions to endergonic reactions through the molecule ATP.

Sample Problem 5.1 **Predicting Spontaneity**

Classify the following as spontaneous or nonspontaneous processes.
  a. a chemical reaction with more energy in the products than reactants
  b. the oxidation of food by the body
  c. a chemical reaction with a negative free energy (ΔG)

Solution

  a. Nonspontaneous. When a reaction requires energy to form products, the value for ΔG is positive and nonspontaneous.
  b. Spontaneous. Overall, the breakdown of food produces energy for the body for other functions such as muscle movement.
  c. Spontaneous. When ΔG is negative, the reaction is exergonic, which is spontaneous.

Activation Energy

So, why haven't all of the spontaneous reactions in the world already happened? Think back to the heel warmer, where we saw that heat energy was generated by a reaction taking place. However, the reaction needed a little "push" to get started. In the case of the heel warmer, the push is provided by bending a metal disk inside the pack to dislodge a small piece of metal so that solid sodium acetate can begin to form. For a reaction to begin, reactants have to collide with enough energy and the correct orientation.

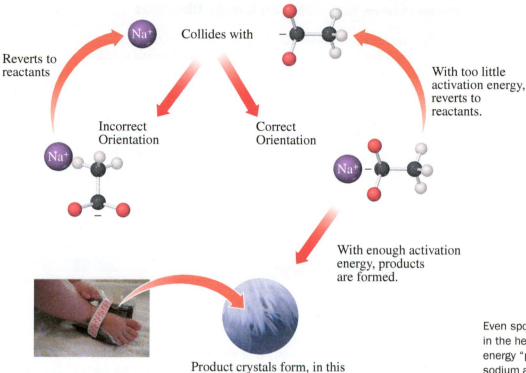

Reverts to reactants

Collides with

With too little activation energy, reverts to reactants.

Incorrect Orientation

Correct Orientation

With enough activation energy, products are formed.

Product crystals form, in this case producing heat

Even spontaneous reactions like that in the heel warmer need an activation energy "push," in this case so that the sodium and acetate ions orient and collide, forming a solid product.

The energy necessary to align the reactant molecules and to cause them to collide with enough energy to form products is known as the **activation energy.** If the energy in the reactant molecules is less than the activation energy, the molecules will bounce off each other without forming any products.

Activation energy is a property of a chemical reaction. We can think of activation energy as the energy hill that molecules must overcome before they change into products. If enough energy is available to allow the molecules to get over the energy hill (also known as the activation barrier), they can react to form products.

A reaction energy diagram is a way to visually examine the energy changes that occur in a chemical reaction. **FIGURE 5.1** shows reaction energy diagrams for an exergonic and an endergonic reaction. Energy appears on the y-axis, and the reaction progress from reactants to products appears on the x-axis. Notice that each reaction has an activation energy hill to climb before proceeding to products.

Each diagram also shows the difference between the energy of the reactants and of the products, giving the overall free energy (ΔG) of the reaction. So, a reaction energy diagram gives information about both the activation energy and free energy.

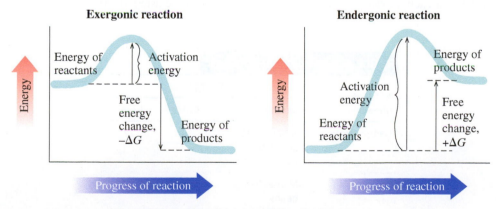

Exergonic reaction

Energy of reactants

Activation energy

Free energy change, $-\Delta G$

Energy of products

Energy

Progress of reaction

Endergonic reaction

Energy of products

Activation energy

Energy of reactants

Free energy change, $+\Delta G$

Energy

Progress of reaction

FIGURE 5.1 Reaction energy diagrams for an exergonic and an endergonic reaction. These diagrams are a compact way of showing how the energy changes in a chemical reaction over time.

Sample Problem 5.2 Reaction Energy Diagrams

Which of the diagrams shown has the following characteristics?

a. exergonic reaction

b. endergonic reaction

c. (ΔG) is negative

d. (ΔG) is positive

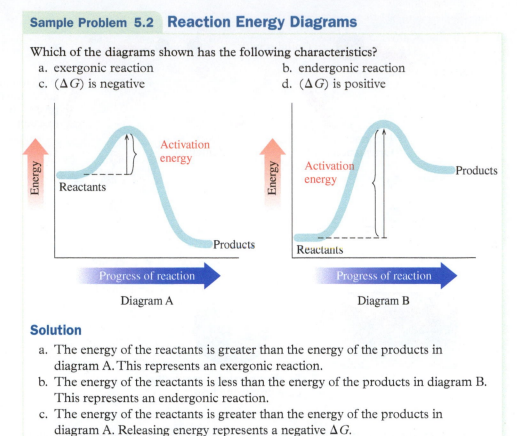

Diagram A

Diagram B

Solution

a. The energy of the reactants is greater than the energy of the products in diagram A. This represents an exergonic reaction.

b. The energy of the reactants is less than the energy of the products in diagram B. This represents an endergonic reaction.

c. The energy of the reactants is greater than the energy of the products in diagram A. Releasing energy represents a negative ΔG.

d. The energy of the reactants is less than the energy of the products in diagram B. Absorbing energy represents a positive ΔG.

Energy Content in Food

When trying to lose weight, we talk about *burning* Calories, although we are not lighting a match to ourselves. Whether burning outside the body (where we can see flames with oxygen in the air) or inside the body (where food combines with oxygen), the combustion reaction is the same: The molecules in food react with oxygen, producing carbon dioxide, water, and energy. The food we eat contains different amounts of energy. In our bodies, we extract energy from the nutrients carbohydrate, protein, and fat. Energy (caloric) values for the main nutrient molecules we extract from food are given in **TABLE 5.1**. You can review the energy units, nutritional Calorie (Cal) and kilojoule (kJ), in Section 1.5.

The Nutrition Facts labels on packaged food list the total grams of each nutrient present per serving. The caloric content can be calculated by summing the Calories present for each nutrient. Values for some common foods are listed in **TABLE 5.2**.

TABLE 5.1 Caloric Values in Food

Nutrient Molecule	Example	Energy	
		Cal/g	kJ/g
Carbohydrate	Table sugar, potatoes, flour	4	17
Protein	Meats, fish, beans	4	17
Fat	Oil, butter	9	38

TABLE 5.2 Composition and Energy Content of Some Foods

Food	Carbohydrate (g)	Protein (g)	Fat (g)	Calories (kJ)
Apple	21	0.27	0.49	89 (370)
Bread, white, 1 slice	14.1	2.4	0.9	74 (310)
Butter, 1 Tbsp	0.008	0.12	11.5	104 (435)
Corn, ½ cup	20.6	2.72	1.05	103 (430)
Egg (chicken), 1 large	0.6	6.07	5.58	77 (320)
Orange juice, 1 cup	27	2	0	116 (485)
Peanut butter, 1 Tbsp	3.55	3.98	7.80	100 (420)
Rice, brown, cooked, ½ cup	24.8	2.45	0.78	116 (485)
Steak, 3 oz	0	19	27	320 (1300)

$$12g\ Fat \times \frac{9\ Cal}{1g\ Fat} = 108\ Cal$$

$$15g\ Carb \times \frac{4\ Cal}{1g\ Carb} = 60\ Cal$$

$$7g\ Protein \times \frac{4\ Cal}{1g\ Protein} = 28\ Cal$$

Total = 196 Cal

How to calculate the number of Calories from a nutrition label for peanut butter. Manufacturers round the Calories to remain in compliance with FDA labeling regulations.

Sample Problem 5.3 Calculating the Energy Content in Food

Calculate the number of Calories present in a slice of pepperoni pizza that contains 34 g carbohydrate, 13 g protein, and 12 g fat.

Solution

Use Table 5.1 to determine the number of Calories for each molecule type and sum them.

$$\text{Carbohydrate: } 34\ g \times \frac{4\ Calories}{1\ g} = 136\ Calories$$

$$\text{Protein: } 13\ g \times \frac{4\ Calories}{1\ g} = 52\ Calories$$

$$\text{Fat: } 12\ g \times \frac{9\ Calories}{1\ g} = 108\ Calories$$

Total: 296, which rounds to 300 Calories (1 significant figure)

The measured Calories per gram for each molecule type have only one significant figure present. The values must be rounded to one significant figure at the conclusion of the problem. If you spend time reading nutrition labels, you may have observed that most caloric values are rounded to one or two significant figures.

Low-Calorie Foods

Why does a regular soft drink have more Calories than a diet soft drink? What is in low-fat foods?

Most of the Calories in a regular soft drink come from the sugar in it. Sugar is a carbohydrate, which can be converted to energy in the body. Diet soft drinks contain substances that are not sugar but taste sweet. These sweet-tasting molecules, or sweeteners, undergo very few chemical reactions in the body and so do not produce much energy (Calories). Molecules that are unreactive in the body are not converted to energy.

Fats produce the highest amount of energy per gram when they undergo chemical reactions in the body (9 Cal/g). Manufacturers of baked goods have introduced fat substitutes into their products to produce low-fat alternatives. Substances like dextrins, celluloses, and gums are carbohydrate-based molecules that produce less energy (4 Cal/g) when eaten. The consumer gets a product that tastes similar but contains fewer Calories because less fat is present.

Diet soft drinks and low-fat foods produce less energy (Calories) when consumed.

Practice Problems

Find the answers to the odd-numbered Practice Problems at the end of the chapter.

5.1 When vinegar (CH_3COOH) and baking soda ($NaHCO_3$) are combined, the mixture spontaneously undergoes an endothermic reaction while releasing CO_2 gas. Would the test tube feel hot or cold? Do you think this reaction would have a + or − value for ΔG? Explain your reasoning.

5.2 In your own words, define free energy change, ΔG.

   a. How does the ΔG differ in exergonic and endergonic reactions?

   b. Which is spontaneous, a reaction with a + or − value for ΔG?

5.3 Classify the following as exothermic or endothermic reactions:

   a. When two solids are combined in a test tube, the test tube gets hot.

   b. A reaction must be heated for it to continue.

5.4 Classify the following as exergonic or endergonic reactions:

   a. On a reaction energy diagram, the reactants show lower energy than the products.

   b. On a reaction energy diagram, the products are lower in energy than the reactants.

5.5 Give the sign of ΔH for each of the reactions in Problem 5.3.

5.6 Give the sign of ΔG for each of the reactions in Problem 5.4.

5.7 Would the sign of ΔS be + or − for the following changes?

   a. water freezing to form ice

   b. a protein being formed from amino acids

   c. digestion of a meal

5.8 Would the sign of ΔS be + or − for the following changes?

    a. liquid gasoline burning to form carbon dioxide and steam

    b. cleaning up your bedroom

    c. a stick of dynamite exploding

5.9 Classify the following as spontaneous or nonspontaneous processes:

    a. a hot bowl of oatmeal cooling on the table

    b. a chemical reaction that gives off free energy

5.10 Classify the following as spontaneous or nonspontaneous processes:

    a. walking on a treadmill at the gym

    b. a chemical reaction that requires free energy

5.11 Calculate the number of Calories present in a chocolate chip cookie that contains 17 g carbohydrate, 1 g protein, and 7 g fat.

5.12 Use Table 5.2 to determine how many Calories would be found in a breakfast consisting of two eggs, two pieces of toast buttered with 1 Tbsp of butter, and one 6 fl. oz. glass of orange juice.

DISCOVERING THE CONCEPTS

⁇ INQUIRY ACTIVITY—Reaction Energy Diagrams

Information

The diagrams shown are called reaction energy diagrams and graphically show the progress of two different chemical reactions on the x-axis and the amount of energy required as the reaction moves forward on the y-axis. The *activation energy* is the amount of energy required to get the reactants into position for collision with enough energy so that they react. Reactions with larger activation energies occur more slowly than reactions with smaller activation energies.

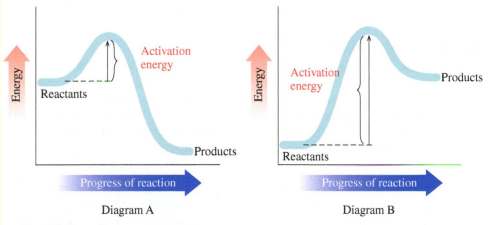

Diagram A   Diagram B

Questions

1. Which reaction has the larger activation energy?
2. Based on the diagrams, which reaction can form products more quickly?
3. A *catalyst* speeds up a chemical reaction by lowering the activation energy. Sketch diagram A and draw a second line on the same diagram for the reaction when a catalyst is present. Label the two traces as "uncatalyzed" and "catalyzed."
4. As we explored in Section 5.1, chemical reactions that release heat energy upon forming product are called *exergonic* and reactions that require heat energy are called *endergonic*.
   a. Which reaction (A or B) has more energy in the reactants than in the products?
   b. Which reaction is exergonic?
5. Examine your sketch from question 3. Does a catalyst change the amount of energy produced or required in a chemical reaction? How did you decide?
6. Draw a reaction energy diagram for a slow, exergonic reaction and contrast it to a diagram for a fast, endergonic reaction.

The conversion of carbon from diamond to graphite is a spontaneous, albeit slow, process.

5.2 Chemical Reactions: Kinetics

When we think of spontaneous reactions, we may be inclined to think that such reactions, once given a little nudge to get them started, will proceed rapidly to completing. But consider this—both graphite and diamond are solid forms of carbon. One can convert into the other.

$$C(s, diamond) \longrightarrow C(s, graphite)$$

This reaction has a negative ΔG value, meaning that diamonds spontaneously convert to graphite. Yet we do not see diamonds turning into graphite before our eyes or even over our lifetime. While the reaction is spontaneous, it takes place very slowly. In this section, we will explore chemical kinetics, which is the study of how quickly or slowly a chemical reaction takes place.

Factors Affecting Reaction Rates

We can measure the **rate of reaction** by determining the amount of product formed (or reactant used up) in a certain period of time. For a chemical reaction to occur, the molecules of the reactant(s) must collide with each other with enough energy and the proper orientation. Any factor that affects the rate of collision or the energy available to the reaction affects the rate of the reaction. The rate of a reaction is affected by several factors, three of which are discussed here: temperature, amount of reactants, and the presence of a catalyst.

For a visual summary of the factors affecting chemical reaction rates, see **FIGURE 5.2**.

Temperature

The faster that cars drive on the highway, the more likely an accident is to occur. If you think of chemical reactants as the cars on the highway and a collision as the chemical reaction, this example implies that the rate of chemical reactions increases with temperature. Increasing temperature increases the kinetic energy (motion) of the reacting molecules, so they collide with each other more often, causing more reactions to occur (see **FIGURE 5.2b**).

Amount of Reactant

The more cars on the highway, the more likely an accident will occur. If the cars are the reactants and the accident is the chemical reaction, this example implies that the rate of chemical reactions increases with the amount of reactants. Having more reactant available allows the molecules to collide with each other more often, causing more reactions to occur (see **FIGURE 5.2c**). As reactants are consumed and form products, chemical reactions slow down due to a decrease in the amount of reactants.

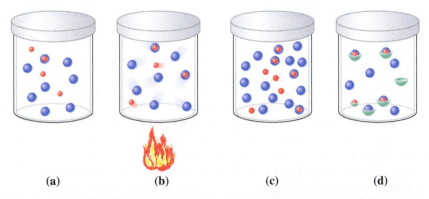

(a)   (b)   (c)   (d)

FIGURE 5.2 Factors affecting reaction rates. At the molecular level, the rate of a chemical reaction increases because the probability of a collision between the reactants increases. (a) Reactants in a chemical reaction (red fits in blue to form product). (b) Increasing the temperature increases the movement of the reactants and their collision frequency. (c) Doubling the amount of reactants increases the collision frequency. (d) A catalyst (green cup) can increase the likelihood of collision by providing a surface with optimal orientation for the reaction to occur.

Presence of a Catalyst

Next, consider driving either over a mountain road or through a tunnel carved in the mountain to get to the other side. Either route will get you there, but you arrive sooner if you drive through the tunnel. Similarly, a **catalyst** speeds up a reaction by *lowering* the activation energy. With a lower energy hill to climb, the reaction progresses to form products more quickly. A catalyst participates in a chemical reaction but remains unchanged at the end of the reaction; in other words, it is regenerated and can be used again. Whether a reaction is exergonic or endergonic, a catalyst does not affect the energy of products or reactants—a catalyst only speeds up formation of the products. A catalyst is neither a product nor a reactant (see **FIGURE 5.2d**). Acids, bases, and metal ions often act as catalysts. Many catalysts immobilize reactants that are near to each other, increasing the likelihood of collision and therefore the rate.

As an example, consider hydrogen peroxide, H_2O_2, which is often used to clean cuts. When hydrogen peroxide is put on a cut, the area bubbles because oxygen gas is produced by the reaction of hydrogen peroxide with a catalyst called *catalase* found in the blood. Catalase is a protein that acts as a biological catalyst called an **enzyme**.

The balanced chemical reaction is shown below:

$$2H_2O_2(l) \xrightarrow{\text{catalase}} 2H_2O(l) + O_2(g)$$

Because catalysts are not changed during a chemical reaction, they are written above or below the arrow in the reaction equation. In the absence of the catalyst, the hydrogen peroxide reaction occurs very slowly. In fact, if you pour hydrogen peroxide on your uncut skin, nothing will happen. Hydrogen peroxide cleans a wounded area by reacting with catalase to produce oxygen quickly at the wound site.

Sample Problem 5.4	**Factors Affecting Reaction Rate**

Determine if the following changes in condition would increase or decrease the rate of the chemical reactions in which they are involved:
   a. baking cookies at 400 °F instead of 325 °F
   b. taking medication every other day instead of every day
   c. spraying an enzyme treatment on stained carpet fibers to clean them

Solution

   a. Increase. Raising the temperature of baking will decrease the cooking time.
   b. Decrease. Taking half the recommended medication means that the amount of medication in the body is lowered and any chemical reactions will be less affected.
   c. Increase. Adding an enzyme to a stain catalyzes breaking up the stain from the carpet fibers.

Enzymes Are Biological Catalysts

Because the body maintains a constant temperature and the amount of reactant present in cells does not fluctuate greatly, most **biochemical reactions**—chemical reactions that occur in living systems—use enzymes to increase reaction rates. Most enzymes are proteins. Enzymes can enhance the rate of a chemical reaction by a factor of more than ten million (1×10^7). This rate is much faster than most chemical catalysts. Enzymes increase the speed of reactions by immobilizing the reactants at a site on the enzyme called the **active site** and orienting them correctly for the reaction to occur.

Enzymes are highly specific to the reactions that they catalyze in the body. They are so important that when they are defective, the result is often an altered or diseased state in the organism. For example, albinism, the lack of pigment in the skin, hair, and eyes, is caused by a defect in tyrosinase, an enzyme in the pathway that produces the pigment melanin. **TABLE 5.3** shows several diseases caused by enzyme deficiency. More information on proteins, enzymes, and their catalysis is presented in Chapter 10.

Enzymes increase the rate of biological chemical reactions by providing a site called the active site where the position of the reactants is optimized, lowering the activation energy.

TABLE 5.3 Some Enzyme-Based Diseases

Disease	Enzyme	Clinical Symptoms
Albinism	Tyrosinase	Absence of pigment in skin, hair, and eyes
Gout	Phosphoribosyl pyrophos-phate synthetase	Renal problems, joint problems, high levels of uric acid in the bloodstream
Lactose intolerance	Lactase	Diarrhea, bloating, flatulence, nausea, abdominal cramping
Glycogen storage deficiency	Glycogen synthase	Enlarged fatty liver, low sugar levels during fasting
Phenylketonuria (PKU)	Phenylalanine hydroxylase	Neurologic symptoms, mental retardation
Tay–Sachs disease	Hexosaminidase A	Mental retardation, blindness, muscular weakness, death

Practice Problems

5.13 a. How does increasing the temperature increase the rate of a chemical reaction?

b. How does increasing the amount of reactants increase the rate of a chemical reaction?

5.14 a. Describe activation energy for a chemical reaction.

b. How does adding a catalyst increase the rate of a chemical reaction?

5.15 Why does the rate of a chemical reaction decrease as the reaction progresses?

5.16 Slow chemical reactions have _____ activation energies.

5.17 Enzymes increase the rate of a biological chemical reaction by lowering the _____.

5.18 In an enzyme, the location where catalysis occurs is called the _____.

5.19 Determine if the following changes in condition would increase or decrease the rate of the chemical reactions in which they are involved:

a. storing leftovers in the refrigerator

b. increasing the kinetic energy of the reactant molecules

c. increasing the frequency of collision

5.20 Determine if the following changes in condition would increase or decrease the rate of the chemical reactions in which they are involved:

a. adding a metal surface for two reactants to locate each other

b. decreasing the amount of one of the reactants

c. caramelizing onions on a medium-high stove setting versus a low setting

DISCOVERING THE CONCEPTS

 INQUIRY ACTIVITY—Types of Chemical Reactions

Data Set

Reaction Type	Sample Chemical Reactions
Synthesis	$HCl(aq) + NaOH(aq) \longrightarrow NaCl(aq) + H_2O(l)$
	$H_2(g) + Br_2(g) \longrightarrow 2HBr(g)$
Decomposition	$Br_2(g) + BaI_2(s) \longrightarrow BaBr_2(s) + I_2(g)$
Exchange	$2H_2O(l) \longrightarrow 2H_2(g) + O_2(g)$

Questions

1. Using the data set on the previous page, match the sample chemical reactions with the reaction type. The types can be used more than once. Be prepared to support your choices.
2. How are a synthesis and a decomposition reaction the same, and how are they different?
3. If the general reaction for a synthesis can be written as A + B $\longrightarrow$ AB, how would you write a general reaction for a decomposition?
4. One of the reactions in the data set is a single exchange, and another is a double exchange. How are these the same, and how are they different?
5. Complete the following general reactions for a single and a double exchange:
   Single exchange: AB + C $\longrightarrow$
   Double exchange: AB + CD $\longrightarrow$
6. Categorize the following reactions as a synthesis, decomposition, or exchange reaction:

   $N_2(g) + O_2(g) \longrightarrow 2NO(g)$

   $C_4H_9Br(g) \longrightarrow C_4H_8(g) + HBr(g)$

   $Mg(s) + 2HCl(aq) \longrightarrow MgCl_2(aq) + H_2(g)$

5.3 Overview of Chemical Reactions

Now that we have explored the thermodynamics and kinetics of chemical reactions, we look at some types of chemical reactions. Most chemical reactions fit into one of three general types of reactions—decomposition, synthesis, or exchange. These reactions can be either reversible or irreversible. Conditions that affect the reaction, like temperature or a catalyst, will appear above or below the reaction arrow.

Types of Chemical Reactions

FIGURE 5.3 shows a schematic of the types of chemical reactions: synthesis, decomposition, and exchange. **Synthesis reactions** are like adding letters together to make words, except that in chemistry, smaller molecules or atoms combine to create larger molecules. An example is the combination of amino acids to form a protein. Synthesis reactions often require energy because products often have higher free energy (less random) than reactants (more random) when smaller molecules join to build larger ones. These reactions typically have only one product.

Decomposition reactions are the opposite. These reactions are just what they sound like. In decomposition reactions, larger molecules break apart into smaller molecules. As an example, animals break down starch (found in plants) into the sugar glucose. Decomposition reactions often give off energy because product compounds form stronger bonds than the reactants. These reactions typically have only one reactant.

Exchange reactions, sometimes called *displacement* reactions, involve both synthesis and decomposition. Bonds are broken, and new bonds are formed as substances decompose and then swap partners, synthesizing different product compounds. When you use an antacid like Milk of Magnesia to combat heartburn, you break down the excess acid (HCl) in the stomach with magnesium hydroxide ($Mg(OH)_2$). The two substances exchange ions with each other, forming $MgCl_2$ and HOH (H_2O).

5.3 Inquiry Question

What are some general categories of chemical reactions?

Amino acids → Protein

(a) Synthesis—The formation of a protein from component amino acids is a synthesis reaction.

Starch → Glucose

(b) Decomposition—The breakdown of starch is a decomposition reaction.

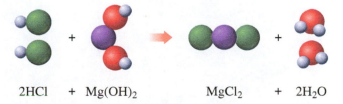

$2HCl$ + $Mg(OH)_2$ → $MgCl_2$ + $2H_2O$

(c) Exchange—Using an antacid to combat heartburn is an exchange reaction.

FIGURE 5.3 Types of chemical reactions. (a) In a synthesis reaction, larger compounds are synthesized from smaller ones. (b) In a decomposition reaction, larger compounds are broken down into smaller ones. (c) In an exchange reaction, the reactants trade partners.

These three basic reaction types, summarized in **TABLE 5.4**, allow us to begin to detect patterns as we look at chemical reactions more closely. It should be noted that not all reactions clearly fit into one of these general categories. These reaction types represent a first step in understanding the more specific chemical reactions presented later in the chapter.

TABLE 5.4 Types of Chemical Reactions	
Reaction Type	**General Reaction Scheme**
Synthesis	$A + B \longrightarrow AB$
Decomposition	$AB \longrightarrow A + B$
Exchange	$AB + C \longrightarrow AC + B$ (single)
	$AB + CD \longrightarrow AD + CB$ (double)

Sample Problem 5.5   Distinguishing Reaction Types

Categorize the following reactions as synthesis, decomposition, or exchange reactions:

   a. $2Fe_2O_3(s) + 3C(s) \longrightarrow 3CO_2(g) + 4Fe(s)$
   b. $2Al_2O_3(s) \longrightarrow 4Al(s) + 3O_2(g)$
   c. $H_2(g) + I_2(g) \longrightarrow 2HI(g)$

Solution

To answer this problem, apply the patterns in Table 5.4 to the chemical equations.

   a. Exchange. Two reactants (Fe_2O_3 and C) form two products (CO_2 and Fe). The oxygen is exchanged between the Fe and the C.
   b. Decomposition. A single reactant (Al_2O_3) breaks into two smaller parts (Al and O_2).
   c. Synthesis. Two reactants (H_2 and I_2) combine to form one product (HI).

Reversible and Irreversible Reactions

During a basketball game, there is a balance between the number of players playing on the court and players on the teams' benches. When a player is tired or in foul trouble, he goes to the bench and a player from the bench moves onto the court. The number of players on the court does not change nor does the number of players on the bench.

Just as players can move from the court to the bench and back, some chemical reactions are reversible. In this type of reaction, called a **reversible reaction**, reactants react to form products and products react to form reactants all at the same time. Reversible chemical reactions are indicated with a pair of arrows between the reactants and products pointing in opposite directions.

$$A + B \rightleftharpoons AB$$

In this example, the forward reaction occurs when A and B form AB. When a sufficient amount of AB builds up, the reverse reaction occurs and A and B re-form. At the point when both the forward and the reverse reaction are occurring at the same rate, and neither reaction dominates, the reaction is said to be in a state of **chemical equilibrium**. This does not mean that there are equal amounts of reactants and products, just as there are

Reversible chemical reactions are like the players in a basketball game. Some players are on the court and some are on the bench. The players can change places, but the number in each place does not change.

not equal numbers of players on the benches and on the court. A reaction is at equilibrium when it reaches a point where the rate of formation of products and the rate of re-formation of reactants are the same. In terms of free energy, when a reaction reaches chemical equilibrium, there is no net exchange of free energy and ΔG equals zero. Chemical equilibrium is discussed in more depth in Chapter 9.

If you think about chemical reactions such as a candle burning, you know that it is not easy to reverse this reaction and bring the candle back to its original state. Reactions that are highly exothermic are, for practical purposes, considered **irreversible reactions**. Rejoining the combusted molecules from the candle wax to re-form the candle once they are dissipated into the atmosphere requires too much energy. When we break down food, these chemical reactions are never directly reversed to re-form the actual food. Instead, we use the energy produced to form other molecules and maintain bodily functions.

The combustion of butane (C_4H_{10}) as seen in this Zippo® lighter app is an example of an exothermic, irreversible reaction.

Sample Problem 5.6 **Reversible and Irreversible Reactions**

Determine whether the following reactions are most likely reversible or irreversible:
a. $2N_2O(g) \rightleftharpoons N_2O_4(g)$
b. $HBr(aq) + NaOH(aq) \longrightarrow NaBr(aq) + H_2O(l)$
c. burning gasoline in an automobile engine

Solution

By inspecting these reactions, we can determine if they are reversible or irreversible.

a. Reversible. In this reaction, the formation of two gases is indicated by the equilibrium arrow, N_2O_4 in the forward direction and NO_2 in the reverse direction.
b. Irreversible. In this reaction, water is formed as a product. Notice that the reaction arrow only goes in the forward direction.
c. Irreversible. In the burning of gasoline, it is impossible to capture the exhaust from the reaction and re-form gasoline from it.

Combustion

Next, we take a more in-depth look at one of the reactions we encountered earlier, combustion. In a **combustion** reaction, organic compounds react with oxygen to produce carbon dioxide and water. Earlier, we noted that combustion reactions tend to be highly exothermic. Because of the large amount of energy produced as heat, they are considered irreversible reactions. Combustion reactions are a type of decomposition reaction because larger molecules are broken down into smaller ones. They are also considered a special type of exchange reaction called an oxidation–reduction reaction because oxygen is reacted, something that we explore further in Section 5.4.

Combustion reactions use oxygen and produce energy whether it dissipates as heat or whether it performs work. They are an important part of understanding fuels, including fossil fuels used to heat homes and drive cars and foods that fuel our bodies with energy. Because the products of complete combustion are always the same (carbon dioxide and water), these reactions are a useful place to start an exploration of organic chemical reactions.

Alkanes

Although alkanes are generally considered unreactive, they do undergo combustion. Alkanes are excellent fuel sources because they give off a lot of energy as heat when they are burned. Unfortunately, internal combustion engines like those found in gasoline-powered cars do not completely combust their hydrocarbon fuels. Such engines emit partially reacted hydrocarbons back to the environment in their exhaust. Other hydrocarbons are not completely oxidized, and forms of carbon soot result. The term *clean burning* refers to the efficiency of the combustion. The cleaner a fuel burns, the more complete the combustion.

Sample Problem 5.7 **Combustion of Alkanes**

Provide the products and balance the following reaction for the complete combustion of propane, C_3H_8.

$$C_3H_8(g) + O_2(g) \longrightarrow ?$$

Solution

In combustion, the products are CO_2 and H_2O, so we add those as products first, and then we use the steps we learned in Section 1.3 to balance the equation.

$$C_3H_8(g) + O_2(g) \longrightarrow CO_2(g) + H_2O(g)$$

STEP 1 Examine.

Element	Number in Reactants	Number in Products
C	3	1
H	8	2
O	2	3

STEP 2 Balance. Because they are found in both products, it is best to balance the oxygen atoms last. Starting with carbon, there are three carbons on the reactant side, so placing a 3 in front of the CO_2 will balance the carbon atoms. The eight hydrogen atoms in propane can be balanced in the products by placing a 4 in front of the H_2O.

$$C_3H_8(g) + O_2(g) \longrightarrow 3CO_2(g) + 4H_2O(g)$$

At this point, we examine the number of oxygen atoms present, and we notice that there are two on the reactant side and ten on the product side. Ten atoms of oxygen are necessary on the reactant side, so a coefficient of 5 is added in front of the O_2.

$$C_3H_8(g) + 5O_2(g) \longrightarrow 3CO_2(g) + 4H_2O(g)$$

STEP 3 Check.

Element	Number in Reactants	Number in Products
C	3	3
H	8	8
O	10	10

The equation is now balanced. In this equation, the propane is oxidized (oxygen added) to form CO_2, and the oxygen is reduced (hydrogen added) to form water.

Distinguishing Chemical Reactions

Up to this point, our understanding of chemical reactions has been limited to examining the formulas and the arrows, and categorizing and balancing the reaction. Recognizing the type of chemical reaction and balancing a chemical equation are important actions in determining how much product might be obtained and how much reactant is required in a chemical reaction.

In contrast, for many organic reactions, the interesting part of the reaction is *how* the functional groups change as reactants form products. When writing an organic chemical equation, the structures of organic molecules are often written out because, as we saw earlier, there are many structural isomers for the same molecular formula. Often, the organic reactant is the only reactant shown. Small molecules that participate in the reaction are shown above the reaction arrow, and reaction conditions like a catalyst are often shown below the arrow. Here is an example of an organic reaction called **hydrogenation**, which is discussed a little later in this chapter.

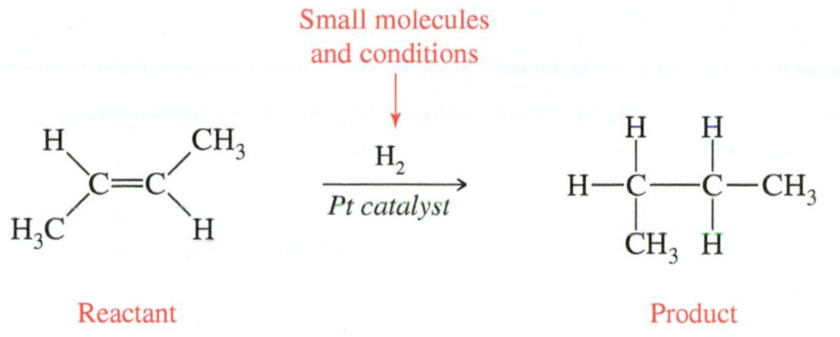

Organic reactions focus on changes in the functional groups. Here, an alkene is changed to an alkane through a hydrogenation.

Small molecules and conditions

Reactant

Product

Biochemical reactions are written like organic reactions, and the reactant, product, and reaction conditions are shown. Sometimes biochemical reactions are *coupled* to each other through energy transfer; that is, two reactions occur at the same time, and energy is transferred from one reaction to fuel the other. The two reactions are often written together, with the energy transfer reaction written at the arrow. As noted earlier, ATP is commonly used to transfer energy in biochemical reactions. Typically, the structure of this common energy-providing molecule is not shown.

For example, in the first reaction of **glycolysis,** the breakdown of the sugar molecule glucose in the body, glucose is converted to the molecule glucose-6-phosphate. The phosphate comes from the energy molecule ATP, which transfers a phosphate group and becomes **a**denosine **dip**hosphate, **ADP**, in the process. The energy transfer reaction is drawn at the arrow. The enzyme that catalyzes the reaction is shown below the arrow.

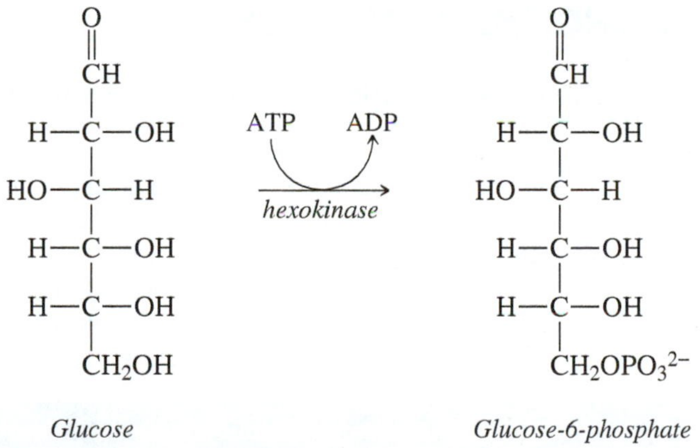

Biochemical reactions show changes in functional groups coupled with a transfer of energy. Here, the energy is transferred through ATP.

ATP   ADP

hexokinase

Glucose

Glucose-6-phosphate

In summary, the way a chemical equation is written depends on the purpose for writing the chemical reaction. These differences are summarized in **TABLE 5.5** for a general reaction, an organic reaction, and a biochemical reaction.

In this section, we began by categorizing some general types of reactions and then balanced some combustion reactions. In the next sections, we examine several organic reactions and provide some biochemical examples. More examples of biochemical reactions occur throughout the book as we examine biological molecules.

TABLE 5.5 Differences in Notation in Chemical Equations

Type of Chemical Reaction	Notation of Reactants and Products	Use of "Yields" Arrow	Key Information Provided
General reaction	Chemical formula only	Indicates reversibility and conditions	Moles of reactants and products
Organic reaction	Drawn structure of organic molecule of interest	Indicates reversibility, small reactants, and conditions	Changes in functional groups
Biochemical reaction	Drawn structure of organic molecule of interest	Indicates reversibility, enzyme, and any common reactions coupled to reaction of interest; for example, ATP $\longrightarrow$ ADP	Changes in functional groups and possibly energy-coupled reactions

Practice Problems

5.21 Categorize the following reactions as synthesis, decomposition, or exchange reactions in the forward direction:

a. $CuO(s) + 2HCl(aq) \longrightarrow CuCl_2(aq) + H_2O(l)$

b. $C_6H_{12}O_6(aq) \longrightarrow 2C_2H_6O(aq) + 2CO_2(g)$

c. $2H_2(g) + O_2(g) \rightleftharpoons 2H_2O(g)$

5.22 Categorize the following reactions as synthesis, decomposition, or exchange reactions in the forward direction:

a. $N_2(g) + 3H_2(g) \rightleftharpoons 2NH_3(g)$

b. $CH_4(g) + 2O_2(g) \longrightarrow CO_2(g) + 2H_2O(g)$

c. $Al_2(SO_4)_3(aq) + 6KOH(aq)$
$\longrightarrow 2Al(OH)_3(s) + 3K_2SO_4(aq)$

5.23 Determine if the reactions in Problem 5.21 are reversible or irreversible.

5.24 Determine if the reactions in Problem 5.22 are reversible or irreversible.

5.25 Write the products and balance the following reaction for the complete combustion of ethane, C_2H_6.

$$C_2H_6(g) + O_2(g) \longrightarrow ?$$

5.26 Write the products and balance the following reaction for the complete combustion of butane, C_4H_{10}.

$$C_4H_{10}(g) + O_2(g) \longrightarrow ?$$

5.27 List the similarities between chemical equations used in organic chemistry and chemical equations used in biochemistry.

5.28 List the differences between general chemical equations and organic chemical equations.

DISCOVERING THE CONCEPTS

 INQUIRY ACTIVITY—Oxidation–Reduction Reactions

Information

Oxidation–reduction reactions, also called *redox* reactions, can be identified in one of two ways: (1) they contain metal ions and involve the exchange of electrons (inorganic), or (2) they can involve the addition or removal of oxygen or hydrogen (organic). Oxidation and reduction reactions always occur together. One substance is oxidized at the same time that the other substance is reduced. Sometimes, the reactants can be mixed (a metal and an organic molecule).

TABLE 1 Characteristics of Oxidation and Reduction (Redox) Reactions

Oxidation		Reduction	
Always Involves	**May Involve**	**Always Involves**	**May Involve**
Loss of electrons	Addition of oxygen	Gain of electrons	Loss of oxygen
	Loss of hydrogen		Gain of hydrogen

Questions

Sample Reaction 1: Inorganic Redox Reaction Containing Metals

$$4Fe(s) + 3O_2(g) \longrightarrow 2Fe_2O_3(s)$$
$$\text{Rust}$$

1. Based on your previous knowledge of ionic compounds, assign charges to Fe(s), $O_2(g)$, and the cation and anion in $Fe_2O_3(s)$.

2. Complete the following table based on **TABLE 1** and your answers to question 1.

Reaction	Did reactant gain or lose electrons to form product?	Oxidation or reduction?
Fe(s) forms iron(III) ion		
$O_2(g)$ forms oxide		

Sample Reaction 2: Biological Redox Reaction

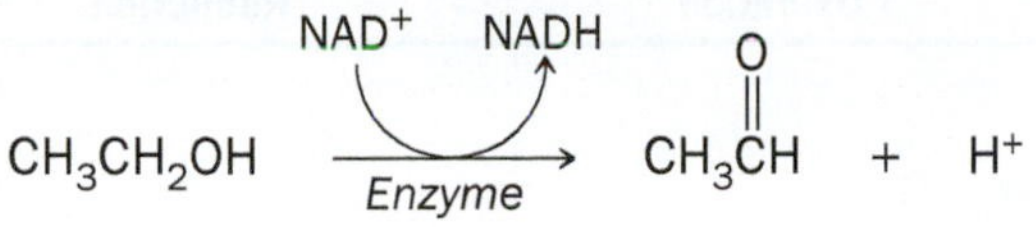

$$CH_3CH_2OH \xrightarrow[\text{Enzyme}]{\text{NAD}^+ \quad \text{NADH}} CH_3\overset{\overset{\displaystyle O}{\|}}{C}H \ + \ H^+$$

3. Complete the following table based on Table 1 and sample reaction 2.

Reaction	Did the reactant gain O (or lose H) or gain H (or lose O) to form product?	Oxidation or Reduction?
CH_3CH_2OH forms CH_3CHO		
NAD^+ forms NADH		

Sample Reaction 3: Organic and Inorganic Mixed

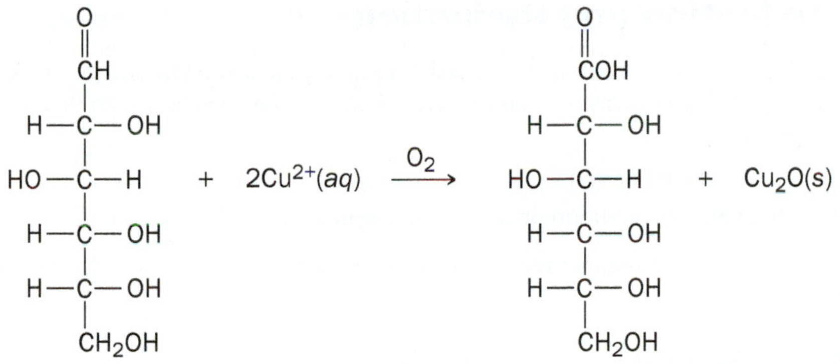

D-Glucose   D-Gluconic acid

4. Complete the following table based on Table 1 and sample reaction 3. *Note that the "always involves" column in Table 1 takes precedence over the "may involve" column.*

Reaction	Oxidation or Reduction?
D-Glucose forming D-gluconic acid	
Cu^{2+} forming Cu_2O	

5. In each of the following reactions, determine which reactant is undergoing oxidation and which is undergoing reduction.

Reaction	Oxidation	Reduction
Formation of salt: $2Na(s) + Cl_2(g) \longrightarrow 2NaCl(s)$		
Burning of coal: $C(s) + O_2(g) \longrightarrow CO_2(g)$		
Complete combustion of glucose: $C_6H_{12}O_6 + 6O_2 \longrightarrow 6CO_2 + 6H_2O$ D-Glucose		

—Continued next page

Continued—

Reaction	Oxidation	Reduction

Reaction 8 of the citric acid cycle:

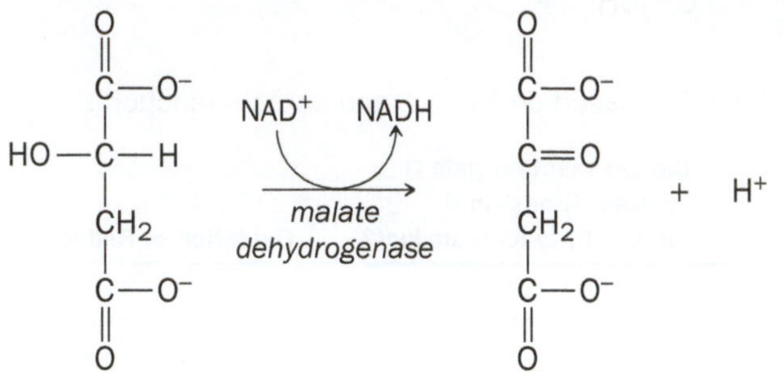

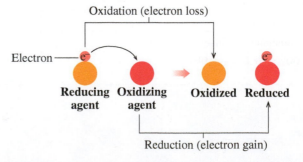

Malate   *Oxaloacetate*

5.4 Oxidation and Reduction

 5.4 Inquiry Question

How are oxidation and reduction reactions identified?

Two chemical reactions called *oxidation* and *reduction* are essential to energy production and transfer in living systems. Oxidation–reduction, or *redox*, reactions can be identified in one of two ways:

1. they involve metal ions and the movement of electrons (inorganic) or
2. they can involve the addition or removal of oxygen and hydrogen (organic).

Both inorganic and organic redox reactions are present in biological systems, so we examine both here.

Inorganic Oxidation and Reduction

 Mastering Video
Practicing the Concepts
Oxidation–Reduction (Redox) Reactions

Rusting is caused by the oxidation of iron atoms to iron(III) ions in the presence of oxygen. The reaction is shown.

$$4Fe(s) + 3O_2(g) \longrightarrow 2Fe_2O_3(s)$$
$$\text{Rust}$$

Let's examine the reactants and product more closely to see how the reactants (iron and oxygen) changed into products during this reaction.

The iron atoms (no charge) in the reactant formed the Fe^{3+} ions of the product. (To determine the charges on the ions of a compound, review Checking a Chemical Formula in Section 3.3.) The iron atoms *lost electrons*. Atoms that lose electrons during a chemical reaction are said to undergo **oxidation**. Where did the electrons go? In this case, the O_2 (no charge) in the reactant formed the O^{2-} anions in the product Fe_2O_3. The reactant oxygen atoms *gained electrons*. Atoms that gain electrons during a chemical reaction are said to undergo **reduction**.

In other words, while the iron was being oxidized, it was reducing the oxygen. In this sense, iron is the **reducing agent** (undergoing oxidation itself), and oxygen is the **oxidizing agent** (undergoing reduction).

Oxidation (electron loss)

Electron —

Reducing Oxidizing   Oxidized Reduced
agent   agent

Reduction (electron gain)

In oxidation–reduction reactions, metal atoms lose electrons, forming cations (get oxidized), and nonmetals gain electrons, forming anions (get reduced). Oxidation and reduction reactions are always coupled to each other because if a reactant is oxidized, losing one or more electrons, another reactant is simultaneously reduced, gaining one or more of those electrons. A mnemonic device that is helpful for remembering what is happening to the electrons in a redox reaction is to remember the letters in the words "OIL RIG," which stand for Oxidation Is Loss (of electrons), Reduction Is Gain (of electrons).

In biological systems, metals are also oxidized and reduced. For example, a protein called cytochrome *c* plays an important role in the electron transport chain of the cell during ATP production. This protein contains an Fe^{2+} that undergoes oxidation to Fe^{3+} followed by reduction back to Fe^{2+} as it transports single electrons through the mitochondrial membrane. The iron ion is held in place on the protein by nitrogen atoms on a small organic molecule called a *heme* that is associated with the protein, as shown in **FIGURE 5.4**.

O I L	R I G
x s o	e s a
i o s	d s i
d s s	u n
a s	c
t	t
i	i
o	o
n	n

Mnemonic (mind-jogging) device for remembering oxidation–reduction

FIGURE 5.4 Biochemical redox of a metal ion. The iron (shown in orange) of cytochrome *c* is surrounded by an organic heme group (rest of structure). The iron is oxidized to Fe^{3+} when an electron is lost from Fe^{2+}.

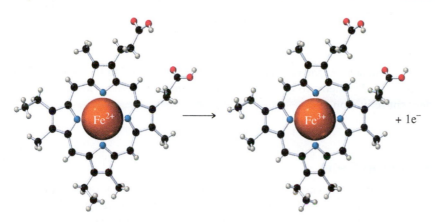

$+ 1e^{-}$

Sample Problem 5.8 Inorganic Redox

In the following reaction, determine which of the reactants is undergoing oxidation and which is undergoing reduction.

$$Ca(s) + S(s) \longrightarrow CaS(s)$$

Solution

In this reaction, the ionic compound calcium sulfide is formed. The calcium atom (no charge) formed the cation Ca^{2+}. The calcium atom lost electrons and underwent oxidation. The sulfur atom gained electrons to form the sulfide anion, S^{2-}. The sulfur underwent reduction.

Organic Oxidation and Reduction

In organic redox reactions, it is more straightforward to see how functional groups change by looking at hydrogen and oxygen rather than seeing where electrons are being gained or lost. An organic molecule is oxidized if it gains oxygen *or* loses hydrogen and is reduced if it gains hydrogen *or* loses oxygen (see **TABLE 5.6**). One organic group that readily undergoes oxidation and reduction is the carbonyl, C=O. Carbonyls in functional groups such as aldehydes can be reduced to alcohols and can be oxidized to carboxylic acids.

TABLE 5.6 Characteristics of Oxidation and Reduction Reactions

Oxidation	
Always Involves	**May Involve**
Loss of e^{-}	Addition of oxygen Loss of hydrogen

Reduction	
Always Involves	**May Involve**
Gain of e^{-}	Loss of oxygen Gain of hydrogen

R—C(=O)—OH ⟵ Oxidation R—C(=O)—H Reduction ⟶ R—CH(OH)—H

[Add oxygen]   [Add hydrogen]

Carboxylic acid   Aldehyde   Alcohol

Aldehydes can be reduced to alcohols and oxidized to carboxylic acids.

As we move through the textbook, we will see examples of oxidation and reduction in reducing sugars (Chapter 6) and metabolic reactions (Chapter 12).

Sample Problem 5.9   Organic Redox

In the following reaction, determine if the organic reactant (benzoic acid) is undergoing oxidation or reduction.

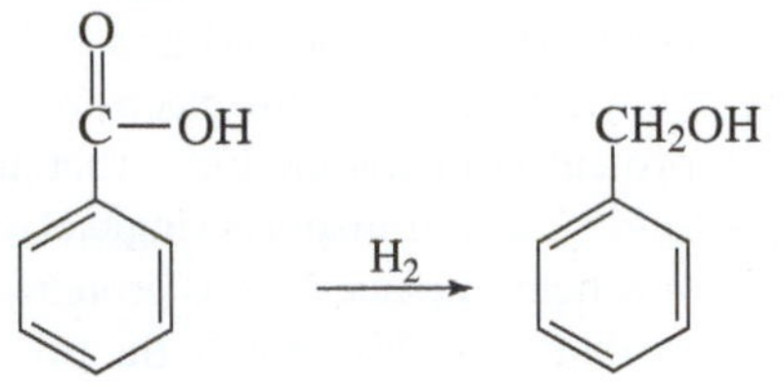

Benzoic acid

Solution

The carboxylic acid in the reactant is changed to an alcohol in the product. It lost one oxygen and gained two hydrogen atoms, so the carboxylic acid in benzoic acid underwent reduction.

Oxidation in Cells

We fuel our bodies with nutrients that are broken down through oxidation. In the body, the process of combustion does not occur in one chemical reaction but through many steps in a chemical pathway, transferring energy as it goes. For example, one molecule of glucose ($C_6H_{12}O_6$) undergoes complete oxidation through combustion in the body, but this process takes several steps and metabolic pathways to accomplish (see Chapter 12). Collectively, the series of reactions through which glucose is combusted, yielding carbon dioxide, water, and energy, is a type of **cellular respiration**. We breathe in oxygen and exhale carbon dioxide as a product. We can write the overall reaction as

$$C_6H_{12}O_6(s) + 6O_2(g) \longrightarrow 6CO_2(g) + 6H_2O(l)$$

In addition to breaking down nutrients and producing energy, the body performs many other organic oxidation–reduction reactions. For example, when we metabolize alcohol (ethanol) from alcoholic beverages, an oxidation–reduction reaction occurs. An enzyme called liver alcohol dehydrogenase (LAD) catalyzes the reaction. Ethanol is oxidized to the aldehyde, ethanal, also called acetaldehyde. What is reduced? We noted earlier that oxidation–reduction reactions are always coupled, so if ethanol is oxidized, another substance is reduced. Remember that we also said that most biochemical reactions are coupled so that the energy is transferred efficiently. In the case of ethanol oxidation, the molecule undergoing reduction is nicotinamide adenine dinucleotide, abbreviated NAD^+. The coupled redox reaction for the biological oxidation of ethanol is shown at left.

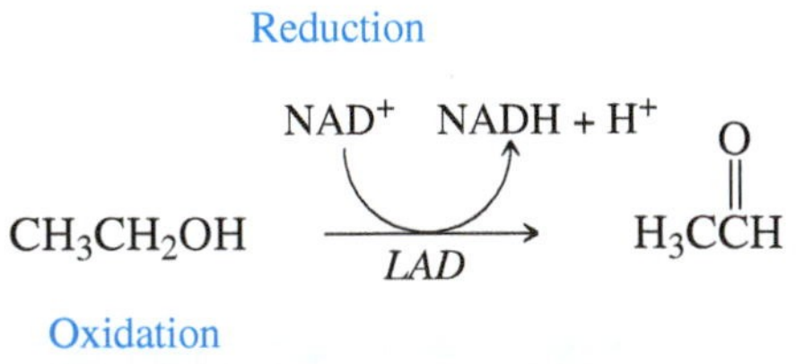

NAD^+ and flavin adenine dinucleotide (FAD) are two important molecules used in metabolic oxidation–reduction reactions. Each of these has an oxidized and reduced form and participates in many biological redox reactions (see **TABLE 5.7**) as energy transfer molecules. Their structures are shown in Chapter 12. Because most of the molecule stays the same, it is useful to use its acronym in a reaction.

Reduction

Oxidation

NAD^+ undergoes reduction to NADH, while ethanol is oxidized to ethanal.

TABLE 5.7 Common Biological Redox Molecules and Their Abbreviations		
Name	**Oxidized form**	**Reduced form**
Nicotinamide adenine dinucleotide	NAD^+	NADH
Flavin adenine dinucleotide	FAD	$FADH_2$

Practice Problems

5.29 Are the substances shown in italics undergoing oxidation or reduction?

a. *Copper(II) ions* in solution form solid copper metal on a surface.

b. *Aluminum* metal forms aluminum oxide (Al_2O_3).

c. *Wine* (containing ethanol, CH_3CH_2OH) sours to vinegar (CH_3COOH).

5.30 Are the substances shown in italics undergoing oxidation or reduction?

a. *Silver ions* (Ag^+) are electroplated as silver atoms on flatware.

b. *D-Glucose* reacts with hydrogen to form the sugar alcohol D-sorbitol.

c. The biomolecule *FADH$_2$* loses hydrogen becoming FAD.

5.31 The reaction shown occurs in fuel cells and produces great quantities of energy. Identify the reactant that is oxidized and the reactant that is reduced.

$$2H_2(g) + O_2(g) \longrightarrow 2H_2O(g)$$

5.32 Although we mine most sodium chloride, it can be synthesized by the reaction shown. Identify the reactant oxidized and the reactant reduced.

$$2Na(s) + Cl_2(g) \longrightarrow 2NaCl(s)$$

5.33 When we exercise vigorously, pyruvate is converted to lactate in our cells through the redox reaction shown. Identify the reactant oxidized and the reactant reduced.

5.34 Ketone bodies are molecules formed in the liver during periods when carbohydrate intake is restricted. Acetoacetate and beta-hydroxybutyrate can be converted into energy in the absence of glucose. The two molecules are interconverted through the redox reaction shown. Identify the reactant oxidized and the reactant reduced.

5.5 Organic Reactions: Condensation and Hydrolysis

Water forming on the outside of a glass of ice water displays the physical process called condensation. Similarly, in a **condensation** reaction, water is produced as two organic molecules are joined. Because water is produced, this type of reaction can also be called a **dehydration** reaction; however, we will use the more general term *condensation* in this textbook. Condensation reactions can occur among a number of functional groups that contain an —H in a polar bond (like O—H or N—H) and an —OH group that can be removed to form water.

A condensation reaction operating in reverse is a **hydrolysis** reaction (*hydro* means "water" and *lysis* means "break"), where water is consumed as a reactant and the reactant molecule is split into two smaller molecules.

? 5.5 Inquiry Question

What are the characteristics of the organic reactions called condensation and hydrolysis?

Condensation and hydrolysis reactions are common in biochemistry. The energy molecule ATP is hydrolyzed to ADP through a hydrolysis reaction.

Adenosine triphosphate (ATP)

Adenosine diphosphate (ADP) + Energy

Because ATP is so large and is often shown in biochemical reactions only to indicate energy transfer, the reaction is often written symbolically as

$$ATP \underset{Condensation}{\overset{Hydrolysis}{\rightleftharpoons}} ADP + P_i$$

where P_i symbolizes the hydrogen phosphate anion, HPO_4^{2-}.

Many functional groups are added and removed from molecules through condensation and hydrolysis reactions, respectively. Two common examples are the carboxyl (or carboxylic acid) group and the phosphate group.

Carboxylation reactions are important condensation reactions. They involve, just as it sounds, the addition of a carboxyl, or carboxylic acid, group. As carbon dioxide moves through the cell it is added to and removed from small molecules by two enzymes called carboxylase and decarboxylase, respectively. As an example, the small molecule pyruvate gets carboxylated to oxaloacetate in one of the steps of the pathway called *gluconeogenesis* ("the creation of new glucose") in the cell. The energy for this reaction is provided by the hydrolysis of ATP.

Bicarbonate   Pyruvate   Oxaloacetate

The functional group phosphate, symbolized as P_i, also moves around the cell through ATP. The addition and removal of phosphate from a molecule is one way cells regulate chemical pathways. These reactions are called **phosphorylation** and **dephosphorylation** and are condensation and hydrolysis reactions, respectively. The enzymes that catalyze the condensation are called phosphorylases, and the enzymes that hydrolyze phosphate are called phosphatases. One important phosphatase, alkaline phosphatase (ALP), is routinely screened in patients susceptible to liver or bone disease (**FIGURE 5.5**). When these cells are damaged, ALP levels will rise in the bloodstream. This enzyme catalyzes phosphate hydrolysis.

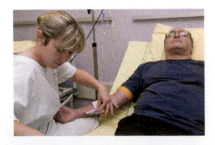

FIGURE 5.5 Enzymes in diagnosing disease. The enzyme alkaline phosphatase (ALP) can be screened in a routine blood test. Patients suspected of bone or liver disease have higher than normal values for this enzyme.

Hydrolyzable and Nonhydrolyzable Lipids

In Chapter 4, we introduced lipids, a diverse class of nonpolar biomolecules. Despite their large structural diversity, lipids can be divided into two broad classifications—those that can undergo a hydrolysis reaction and those that cannot. **Hydrolyzable lipids** can react in a hydrolysis reaction to form simpler compounds, whereas **nonhydrolyzable lipids** cannot undergo hydrolysis. The hydrolyzable lipids—such as waxes, fats, and oils—contain an ester functional group that can react with water to form a carboxylic acid and an alcohol. The reaction below shows the hydrolysis of the ester that is the major component of beeswax into its component carboxylic acid and alcohol.

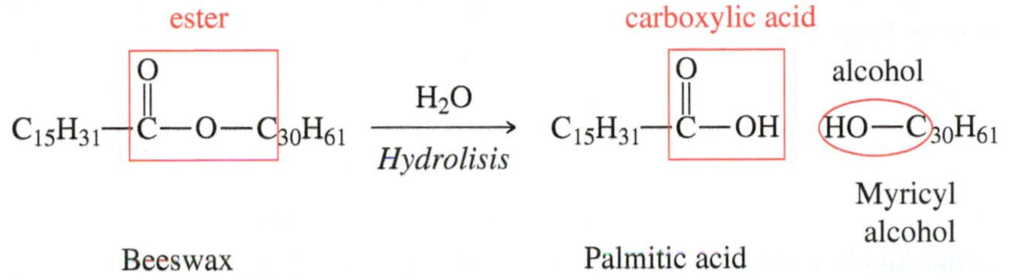

Similarly, fats and oils can be hydrolyzed to their component fatty acids and glycerol, which has three alcohol functional groups.

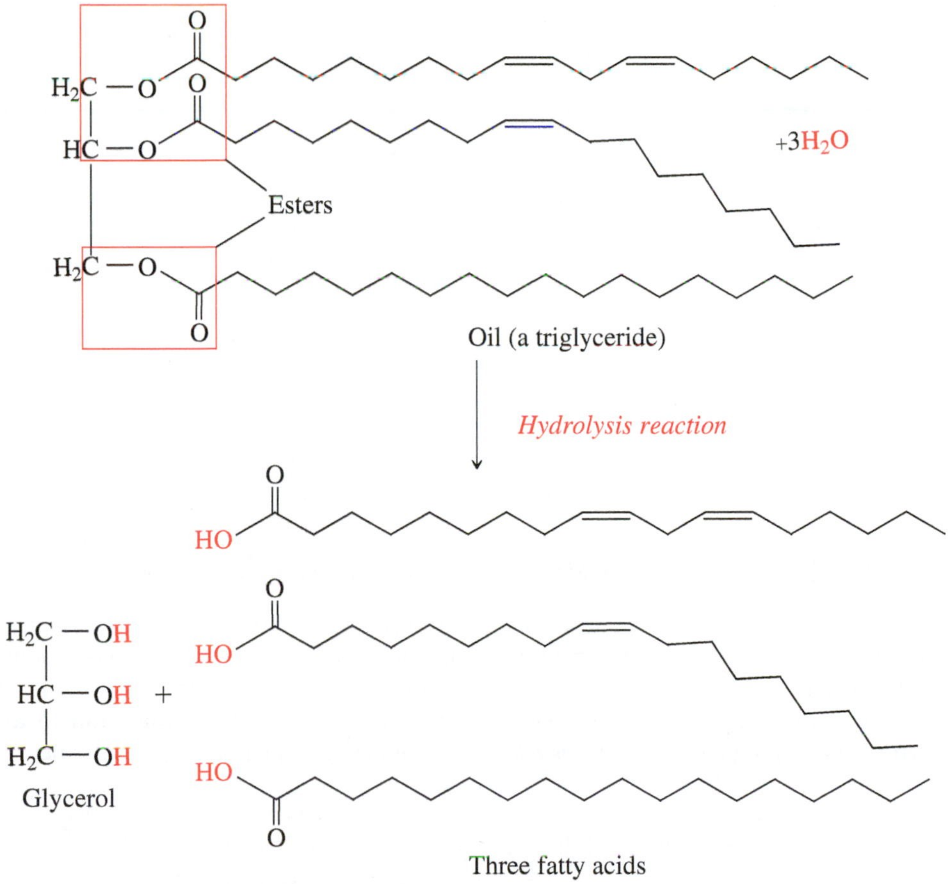

Nonhydrolyzable lipids like vitamin E do not contain an ester functional group and therefore cannot be hydrolyzed.

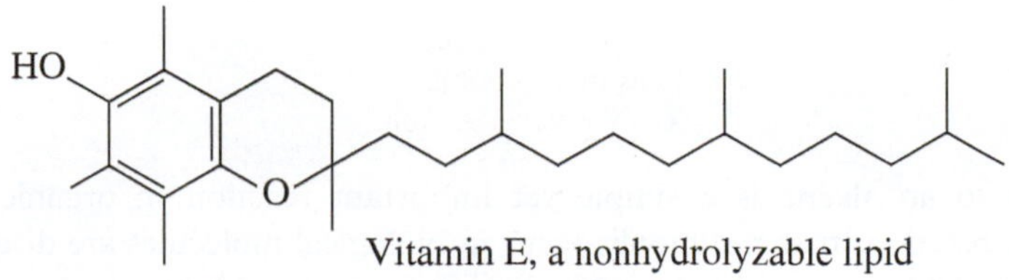

Vitamin E, a nonhydrolyzable lipid

Sample Problem 5.10 **Distinguishing Condensation and Hydrolysis**

Identify the following reactions as representations of a condensation or a hydrolysis:

a.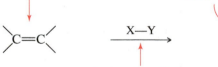

b.

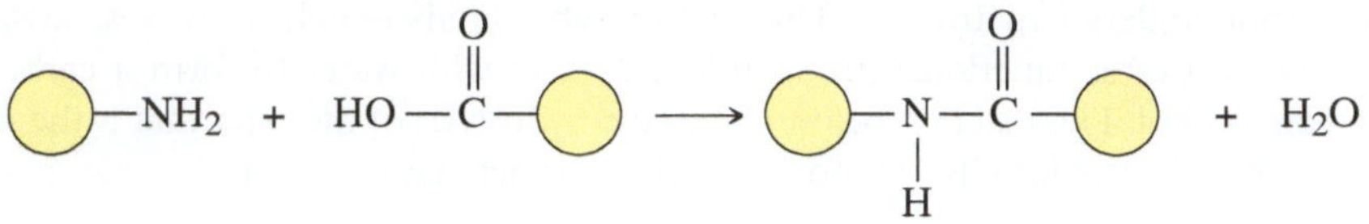

$$\bigcirc \!\!-\! NH_2 \;+\; HO\!-\!\overset{\overset{\displaystyle O}{\|}}{C}\!-\!\bigcirc \;\longrightarrow\; \bigcirc\!\!-\!\underset{\underset{\displaystyle H}{|}}{N}\!-\!\overset{\overset{\displaystyle O}{\|}}{C}\!-\!\bigcirc \;+\; H_2O$$

c. $ADP + P_i \rightarrow ATP$

Solution

a. This representation indicates water as a reactant, breaking a larger molecule. This is a *hydrolysis*.
b. This representation produces water in the reaction when the functional groups amine and carboxylic acid are combined to form an amide. This is a *condensation*.
c. This reaction connects ADP and a phosphate, P_i, the reverse reaction of ATP hydrolysis. This is a *condensation*.

Practice Problems

5.35 Which type of chemical reaction is hydrolysis: synthesis, decomposition, or exchange?

5.36 Which type of chemical reaction is condensation: synthesis, decomposition, or exchange?

5.37 Draw the products for the ester formed through the condensation reaction shown.

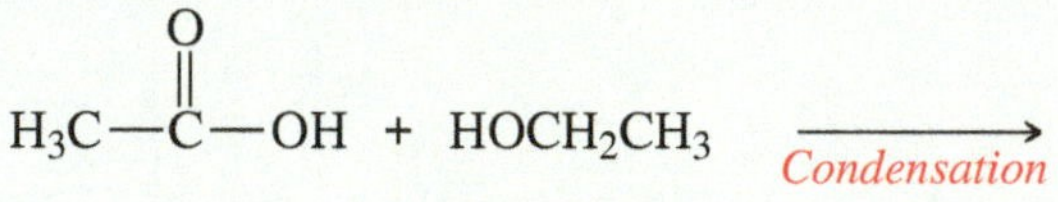

$$H_3C\!-\!\overset{\overset{\displaystyle O}{\|}}{C}\!-\!OH \;+\; HOCH_2CH_3 \;\xrightarrow{\textit{Condensation}}$$

5.38 Draw the products formed from the ester hydrolysis reaction shown.

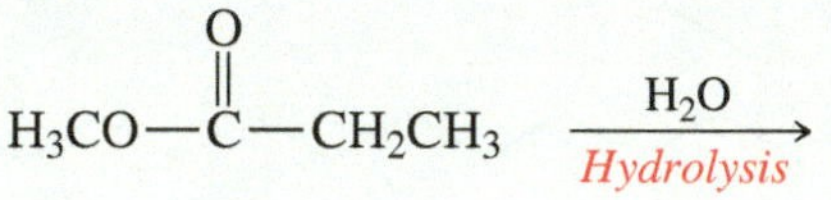

$$H_3CO\!-\!\overset{\overset{\displaystyle O}{\|}}{C}\!-\!CH_2CH_3 \;\xrightarrow[\textit{Hydrolysis}]{H_2O}$$

5.6 Organic Addition Reactions to Alkenes

So far we have seen exchange reactions that transfer electrons (oxidation–reduction), synthesis reactions that combine smaller organic molecules (condensation), and decomposition reactions that break larger molecules apart (hydrolysis). Often, the name of the reaction can tell us something about what the reaction does. In an **addition to an alkene,** an atom or group of atoms is *added* to a double bond in an organic molecule.

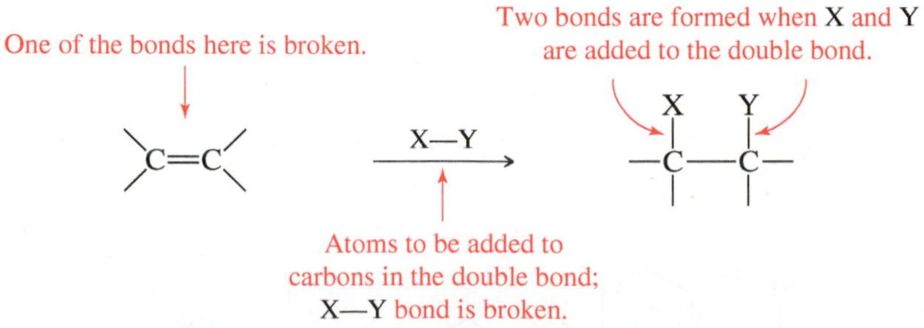

In an addition to an alkene, the second bond in a double bond and the bond between the atoms to be added are broken and two single bonds are formed. Atoms are added to the carbons that were in the double bond. Are additions synthesis or decomposition?

One of the bonds here is broken.

Two bonds are formed when **X** and **Y** are added to the double bond.

$$\overset{}{>}\!C\!=\!C\overset{}{<} \quad\xrightarrow{X\!-\!Y}\quad -\!\overset{\overset{\displaystyle X}{|}}{C}\!-\!\overset{\overset{\displaystyle Y}{|}}{C}\!-$$

Atoms to be added to carbons in the double bond; X—Y bond is broken.

Addition to an alkene is a simple yet important reaction in organic chemistry. Two addition reactions that commonly occur in biological molecules are discussed here: hydrogenation (hydrogen is added) and hydration (water is added).

Hydrogenation and Trans Fats

If you read labels on processed foods, you may see partially hydrogenated soybean oil as an ingredient. This is soybean oil that has undergone hydrogenation, an organic addition reaction where two hydrogen atoms are added, converting an alkene double bond to an alkane single bond. A catalyst such as platinum (Pt), nickel (Ni), or palladium (Pd) is used. Oils derived from plants have long been considered a healthier alternative for the human diet because they have a higher percent of unsaturated fatty—fats containing carbon–carbon double bonds (alkene functional groups).

$$\text{C=C} \xrightarrow[\text{Pt, Ni, or Pd}]{\text{H—H}} \underset{|}{\overset{\overset{\displaystyle H}{|}}{-C}} \underset{|}{\overset{\overset{\displaystyle H}{|}}{-C-}}$$

Complete hydrogenation of an oil converts all of the double bonds into single bonds and changes the unsaturated oil into a fully saturated compound. This changes the physical properties of an oil, causing it to have properties more like those of a solid fat. Recall from Chapter 4 that the fatty acid components of fats and oils also contain a carboxylic acid functional group. The carbonyl double bond is not affected in a hydrogenation.

Because the reaction is difficult to control, hydrogenation of oils often does not affect every alkene double bond. Some of the double bonds that are not completely hydrogenated will re-form, but when they do, they switch from the natural cis form to the more stable trans form, resulting in compounds known as *trans fats*. The reaction is called partial hydrogenation. Such partially hydrogenated oils are found in many processed foods (see **FIGURE 5.6**). Manufacturers use hydrogenated oils because they give processed foods a longer shelf life than foods prepared using unsaturated oils.

Currently, all nutritional labeling must give the amounts of trans fat present in food. There is enough recent evidence that trans fats boost the amount of low-density lipoprotein ("bad" cholesterol) in the bloodstream and lower the amount of high-density lipoprotein ("good" cholesterol) that the FDA recently ruled that food companies that want to use trans fats in processed foods must get their approval. Otherwise, food manufacturers can no longer use trans fats in their products. Many manufacturers have considered natural saturated fat alternatives like palm or coconut oil.

INTEGRATING Chemistry

◀ Find out how unsaturated fats are saturated and can form trans fats.

FIGURE 5.6 Partially hydrogenated soybean oil is found in many processed foods. The fatty acids in soybean oil undergo partial hydrogenation, forming trans fats. Because of significant figures, manufacturers can claim 0 g trans fat on a label if the serving size contains less than 0.5 g.

Naturally occuring cis double bonds.

H₂ | *Pt catalyst*

The double bonds that were here were fully hydrogenated.

This double bond was partially hydrogenated, changing to its trans form.

Sample Problem 5.11 Predicting Products of Hydrogenation

Provide the products of the complete hydrogenation for each of the reactions shown.

a. $CH_3CH_2CH = CHCH_3 \xrightarrow[\text{catalyst}]{H_2 \ \text{Pt}}$

b. 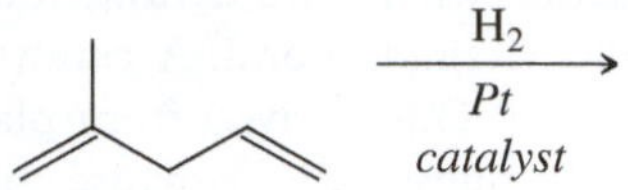 $\xrightarrow[\text{catalyst}]{H_2 \ \text{Pt}}$

c. 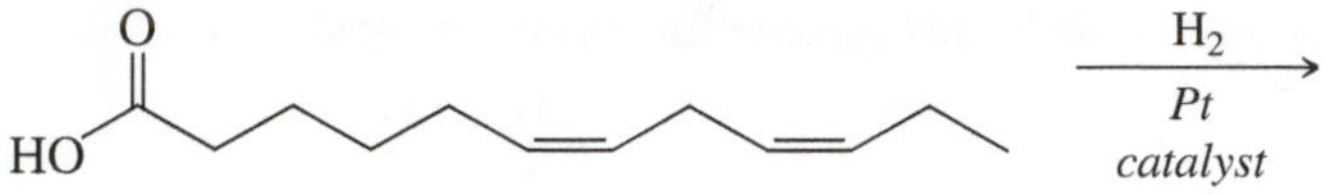 $\xrightarrow[\text{catalyst}]{H_2 \ \text{Pt}}$

Solution

In each case, if the alkenes are completely hydrogenated, they will form saturated hydrocarbons, or alkanes. The products are as follows.

a. $CH_3CH_2CH_2CH_2CH_3$

b.

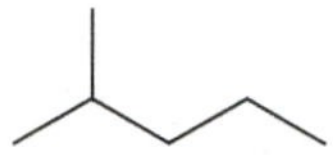

c.

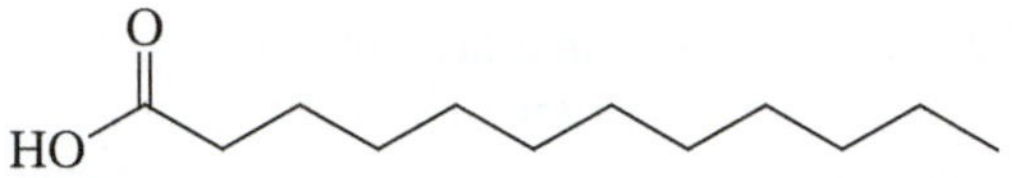

Hydration

Hydration is the addition of water to the double bond in an alkene. The water gets added as —H and —OH. This reaction often requires a catalyst such as an acid or an enzyme.

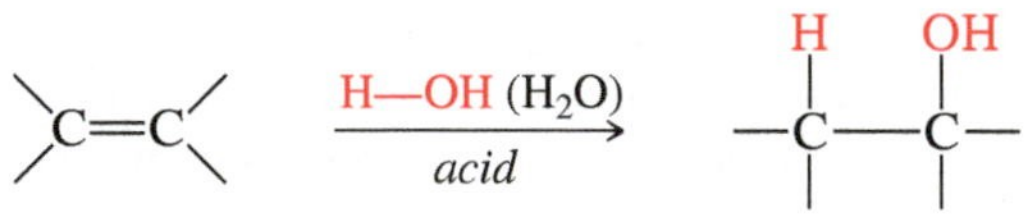

Hydration reactions occur in biochemistry, too. One example occurs at Reaction 7 in the citric acid cycle (also called the Krebs cycle; see Chapter 12). The reaction of fumarate to malate is a hydration reaction. This reaction is catalyzed by the enzyme fumarase.

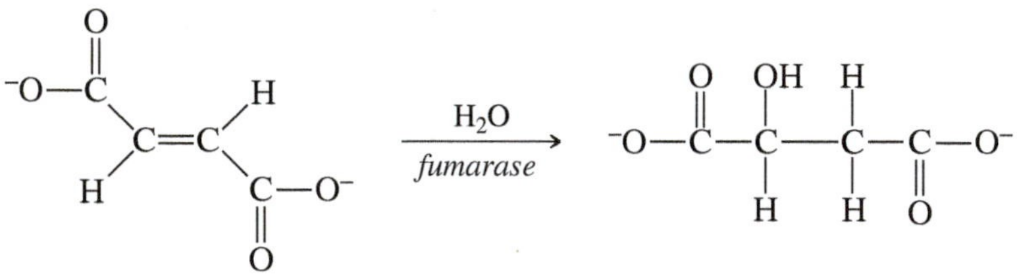

Fumarate   Malate

In contrast to hydrogenation, this reaction has two *different* groups adding to the carbons that were in the double bond—an OH on one and an H on the other. What products might be produced when the alkene does not have the same groups attached to the double bond?

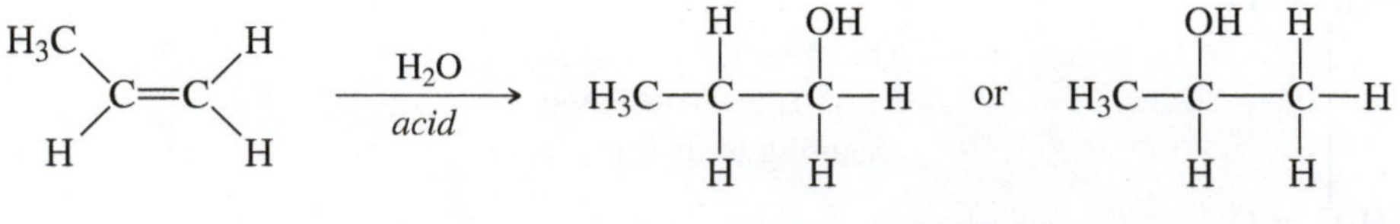

or both?

If the alkene double bond is asymmetric—that is, it does not have the same attachments (also called substituents) on the carbons at the double bond—the H will usually bond to the carbon with more hydrogen atoms, forming the major product. That is, the OH will bond to the carbon with more carbon groups attached. A Russian chemist named Vladimir Markovnikov was the first chemist to notice this in addition reactions of alkenes. This observation is called **Markovnikov's rule.**

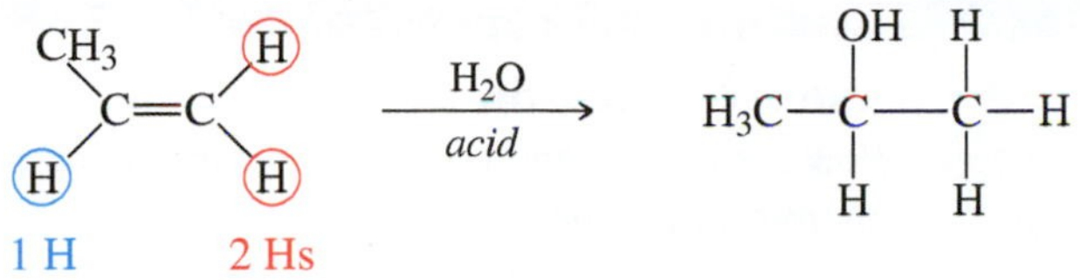

1 H   2 Hs

Sample Problem 5.12 · Predicting Products of Hydration

Provide the main product for the hydration of each of the alkenes shown.

a. $CH_3CH_2CH{=}CH_2$ $\xrightarrow[acid]{H_2O}$

b. 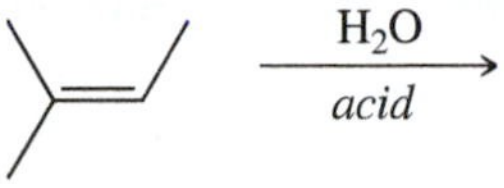 $\xrightarrow[acid]{H_2O}$

c. 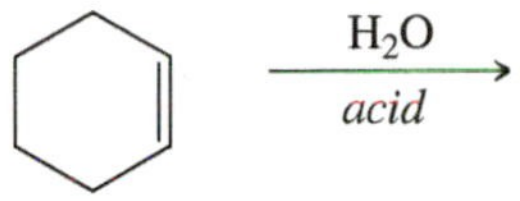 $\xrightarrow[acid]{H_2O}$

Solution

Remember that if the carbons in the alkene are unequally substituted, Markovnikov's rule applies; that is, most of the product will have the H placed on the carbon with more Hs. The main products are as follows.

a. 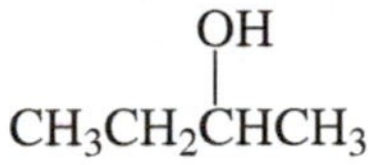 $CH_3CH_2\overset{OH}{\underset{}{C}}HCH_3$

b.

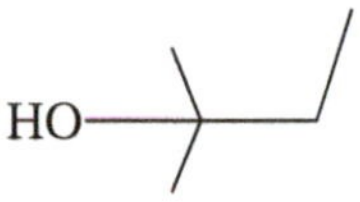

c.

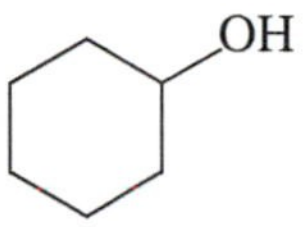

Practice Problems

5.39 Write the products for the following hydrogenation reactions:

a. 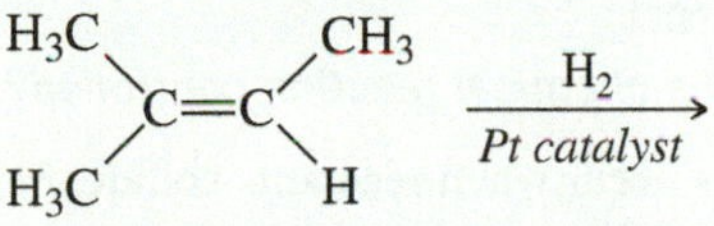 $\xrightarrow{H_2 \text{ Pt catalyst}}$

b. 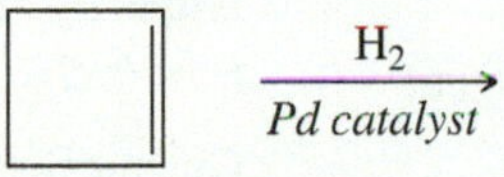 $\xrightarrow{H_2 \text{ Pd catalyst}}$

c. 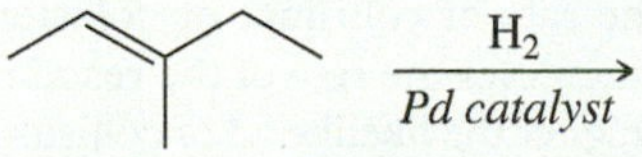 $\xrightarrow{H_2 \text{ Pd catalyst}}$

5.40 Write the products for the following hydrogenation reactions:

a. 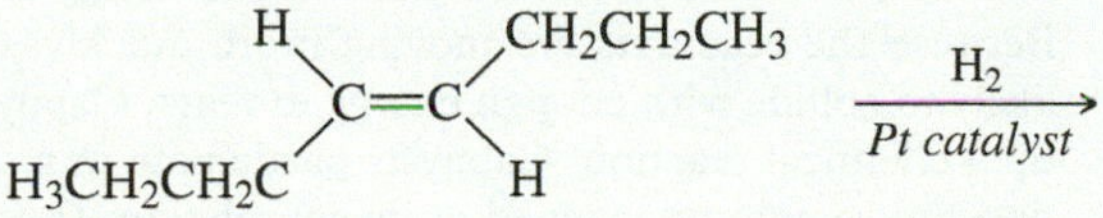 $\xrightarrow{H_2 \text{ Pt catalyst}}$

b. 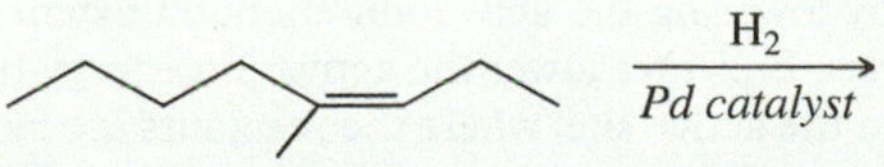 $\xrightarrow{H_2 \text{ Pd catalyst}}$

c. 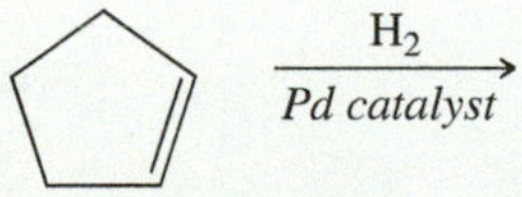 $\xrightarrow{H_2 \text{ Pd catalyst}}$

5.41 Write the main product of hydration for the alkenes shown in Problem 5.39.

5.42 Write the main product of hydration for the alkenes shown in Problem 5.40.

CHAPTER REVIEW

The study guide will help you check your understanding of the main concepts in Chapter 5. You should be able to answer problems for each learning outcome in this list. To check your mastery, try the problems listed after each.

STUDY GUIDE	CHAPTER GUIDE

5.1 Thermodynamics

Investigate the properties of heat and energy.

- Predict the spontaneity of a reaction based on the ΔG value. (Try 5.9, 5.45c)
- Draw reaction energy diagrams for exergonic and endergonic reactions. (See Sample Problem 5.2. Try 5.51, 5.45d)
- Calculate the energy content in foods from its nutrient molecules. (Try 5.11, 5.53)

Inquiry Question

How are energy and heat transferred and measured in a chemical reaction?

Chemical reactions involve energy exchange. Reactions that give off heat are exothermic, and those that require heat are endothermic. For a chemical reaction to occur, the reactants must collide with each other with enough energy and the correct orientation to react. This initial energy required for a reaction to occur is called the activation energy. The thermodynamics of a chemical reaction can be represented on a reaction energy diagram showing the energy of the reactants, and the products, the activation energy, and the free energy change, ΔG.

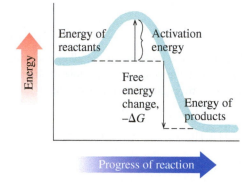

If ΔG is a negative value, a given chemical reaction is spontaneous, and if ΔG is positive, the reaction is nonspontaneous.

5.2 Chemical Reactions: Kinetics

Examine the factors that affect the rate of a chemical reaction.

- Predict relative activation energies and speed of reactions using a reaction energy diagram. (Try 5.49)
- Determine the effect that temperature, amount of reactants, and a catalyst have on the rate of a reaction. (Try 5.13, 5.15)
- Recognize that enzymes are biological catalysts. (Try 5.17)

Inquiry Question

How is the rate of a chemical reaction controlled?

Chemical reactions occur when reactants collide. Several factors control the rate (how fast reactants form products) of a chemical reaction. Increasing the rate of collision, or collision frequency, of chemical reactants increases the rate of the reaction. The more reactants present, the higher the likelihood of collision, so increasing the amount of reactants increases the rate. Temperature increases the rate by increasing the kinetic energy of the reactants. Because the reactants are moving more quickly, they are more likely to collide with enough energy to react. Catalysts also speed up a chemical reaction. Catalysts participate in a chemical reaction but remain unchanged at its completion. Catalysts increase the rate by lowering the activation energy. Enzymes are biological catalysts. Enzymes lower the activation energy by providing a site, called the active site, where the reactants are close and in the correct orientation to react instead of being positioned randomly in a reaction mixture.

5.3 Overview of Chemical Reactions

Categorize chemical reactions by type.

- Classify reactions as synthesis, decomposition, or exchange reactions. (Try 5.21)
- Predict the products of a synthesis, decomposition, or exchange reaction. (Try 5.55)
- Distinguish reversible and irreversible reactions. (Try 5.23)
- Predict the products and balance the chemical equation for a hydrocarbon undergoing combustion. (Try 5.25, 5.59)
- Contrast a general chemical equation and an organic chemical equation. (Try 5.27)

Inquiry Question

What are some general categories of chemical reactions?

There are three basic types of chemical reactions: synthesis, decomposition, and exchange. Depending on the amount of energy released during a chemical reaction, these reactions can be further classified as either reversible or irreversible. Reversible reactions can reach a point called *chemical equilibrium* where the rates of the forward and reverse reactions are constant and no net products are formed in either direction. Organic hydrocarbons like alkanes can undergo combustion (reaction with oxygen, O_2) to form the products carbon dioxide (CO_2) and water (H_2O). Chemical equations are written for various purposes. General reactions are written to determine amounts of reactants consumed or products made, so balancing them is important. Organic reactions are usually written to show how functional groups change during the reaction, so their structures are written out. Biochemical reactions show organic structures and can also show energy coupling and enzymes used to catalyze the reactions.

5.4 Oxidation and Reduction

Distinguish oxidation–reduction reactions in both inorganic and organic chemical reactions.

- Identify the substance oxidized and the substance reduced in an inorganic oxidation–reduction reaction. (Try 5.29a and b, 5.32, 5.61)
- Identify the substance oxidized and the substance reduced in an organic oxidation–reduction reaction. (Try 5.30b and c, 5.33)
- Predict the products of an organic oxidation or reduction. (Try 5.65b and c)

Inquiry Question

How are oxidation and reduction reactions identified?

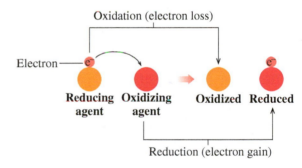

Oxidation and reduction (redox) reactions always occur simultaneously. Oxidation always involves a loss of electrons and may involve the addition of oxygen or removal of hydrogen. Reduction always involves a gain of electrons and may involve the addition of hydrogen or removal of oxygen. If one substance in a reaction is oxidized, another substance in the reaction is reduced. If metals are involved, it is easier to determine which substance is oxidized or reduced by determining the charge on the metal. In organic reactions, it is often easier to look for the movement of oxygen and hydrogen. Combustion (reaction with oxygen) is an example of a redox reaction.

5.5 Organic Reactions: Condensation and Hydrolysis

Characterize the organic reactions of condensation and hydrolysis.

- Predict the products of an organic condensation reaction. (Try 5.37, 5.65b)
- Predict the products of an organic hydrolysis reaction. (Try 5.38, 5.66b, 5.67)
- Identify organic reactions as oxidation, reduction, condensation, or hydrolysis. (Try 5.57, 5.75)

Inquiry Question

What are the characteristics of the organic reactions called condensation and hydrolysis?

Organic condensation and hydrolysis reactions occur in opposite directions. A condensation reaction joins molecules and often produces water. Hydrolysis reactions break molecules, and water is a reactant. Carboxylations and phosphorylations are examples of biological condensation reactions. Dephosphorylation is an example of a biological hydrolysis reaction.

—Continued next page

Continued—

5.6 Organic Addition Reactions to Alkenes

Demonstrate how small molecules are added to alkenes.

- Predict the products of a hydrogenation reaction, an addition reaction of an alkene. (Try 5.39, 5.69)
- Predict the products of a hydration reaction, an addition reaction of an alkene. (Try 5.41, 5.73)

Inquiry Question

How are small molecules added to alkenes?

The addition reactions known as hydrogenation and hydration in alkenes are also very common reactions in biological molecules. In addition reactions, two atoms or groups of atoms are added to the alkene double bond, forming a carbon–carbon single bond.

Two bonds are formed when X and Y are added to the double bond.

$$-\underset{|}{\overset{|}{C}}-\underset{|}{\overset{|}{C}}-$$

In a complete hydrogenation, one hydrogen atom is added to each carbon of the double bond. In hydration, an H and an OH are added to the carbons. In a hydration of an alkene with different groups attached to the alkene, the hydrogen will bond to the carbon in the double bond with more hydrogens bonded directly to the alkene. This demonstrates Markovnikov's rule.

 Mastering Videos

PRACTICING THE CONCEPTS

The following videos can be accessed through the Pearson eText or your Mastering Chemistry course.

- Oxidation-Reduction (Redox) Reactions
- Addition Reactions of Alkenes

Summary of Reactions

Types of Chemical Reactions

Reaction Type	General Reaction Scheme
Synthesis	A + B ⟶ AB
Decomposition	AB ⟶ A + B
Exchange	AB + C ⟶ AC + B (single)
	AB + CD ⟶ AD + BC (double)

Combustion

$$C_xH_y + O_2 \longrightarrow CO_2 + H_2O$$

Condensation/Hydrolysis

Reactions are the reverse of each other.

Condensation

Two smaller molecules form a larger one.

⬤—OH + HO—⬤   ⬤—O—⬤ + H_2O

Hydrolysis

One large molecule forms two smaller ones.

Oxidation–Reduction

Reactions occur simultaneously.

Oxidation	
Always Involves	**May Involve**
Loss of electrons	Addition of oxygen; Loss of hydrogen

Reduction	
Always Involves	**May Involve**
Gain of electrons	Loss of oxygen; Gain of hydrogen

Addition Reactions to Alkenes

Hydrogenation

$$\overset{\diagup}{\underset{\diagdown}{}}C=C\overset{\diagdown}{\underset{\diagup}{}} \xrightarrow[Pt,\ Ni,\ or\ Pd]{H\!-\!H} -\underset{|}{\overset{|}{C}}-\underset{|}{\overset{|}{C}}-$$

Hydration

$$\overset{\diagup}{\underset{\diagdown}{}}C=C\overset{\diagdown}{\underset{\diagup}{}} \xrightarrow[acid]{H\!-\!OH\ (H_2O)} -\underset{|}{\overset{|}{C}}-\underset{|}{\overset{|}{C}}-$$

Additional Problems

5.43 Consider the Gibbs free energy equation, $\Delta G = \Delta H - T\Delta S$. Explain how an increase in temperature can affect the spontaneity of a reaction.

5.44 Give the correct symbol (ΔG, ΔH, or ΔS) and sign for each of the following:

a. represents the free energy of a spontaneous reaction

b. represents the change in enthalpy for an endothermic reaction

c. represents the increase in entropy in a reaction

5.45 Methane (natural gas) can react with oxygen in a combustion reaction as shown:

$$CH_4(g) + 2O_2(g) \longrightarrow 2H_2O(g) + CO_2(g) + Heat$$

a. Is the reaction exothermic or endothermic?

b. Considering that heat is a form of energy, predict whether the reactants or products have more free energy.

c. Is the reaction spontaneous?

d. Sketch a reaction energy diagram for this reaction. Label the axes, the energy of reactants, the energy of products, and ΔG.

5.46 In the following reaction energy diagram, label 1, 2, 3, and 4 as one of the following: (a) energy of reactants, (b) energy of products, (c) activation energy, or (d) free energy change.

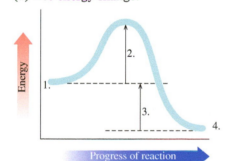

5.47 Which reaction occurs at a faster rate, an exergonic reaction with a low activation energy or an endergonic reaction with a high activation energy? Explain.

5.48 What is measured by the free energy change?

5.49 Use the following reaction energy diagram to answer the questions:

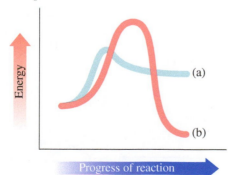

a. Which curve represents the faster reaction, and which represents the slower?

b. Which curve represents an endergonic reaction, and which curve represents an exergonic reaction?

c. Which curve represents a spontaneous reaction?

d. Which reaction has a positive ΔG value?

5.50 Two curves for the same reaction are shown in the following reaction energy diagram. Which curve represents the uncatalyzed reaction, and which curve represents the reaction with catalyst present?

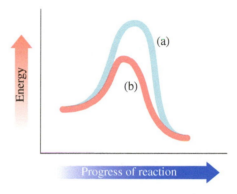

5.51 Draw and label a reaction energy diagram for an endergonic reaction in which the activation energy is two times greater than its free energy change (ΔG).

5.52 Draw and label a reaction energy diagram for an exergonic reaction in which the free energy change (ΔG) is three times greater than its activation energy.

5.53 Calculate the number of Calories present in a Burger King Whopper® with Cheese that contains 53 g carbohydrate, 35 g of protein, and 48 g of fat.

5.54 One gram of alcohol provides 7 Calories. Calculate the number of Calories in a 5 fluid oz glass of red wine that contains 2.5 g carbohydrate and 13.7 g alcohol.

5.55 Write the products that would result from the following reactions, and balance each equation:

a. synthesis : $Mg(s) + Cl_2(g) \longrightarrow$

b. decomposition: $HI(g) \longrightarrow$

c. exchange (single): $Ca(s) + Zn(NO_3)_2(aq) \longrightarrow$

d. exchange (double): $K_2S(aq) + Pb(NO_3)_2(aq) \longrightarrow$

5.56 Write the products that would result from the following reactions, and balance each equation:

a. synthesis: $Mg(s) + O_2(g) \longrightarrow$

b. decomposition: $PbO_2(s) \longrightarrow$

c. exchange (single): $KI(s) + Br_2(g) \longrightarrow$

d. exchange (double): $CuCl_2(aq) + Na_2S(aq) \longrightarrow$

5.57 Identify the main organic reaction shown as condensation, hydrolysis, oxidation, or reduction:

a.

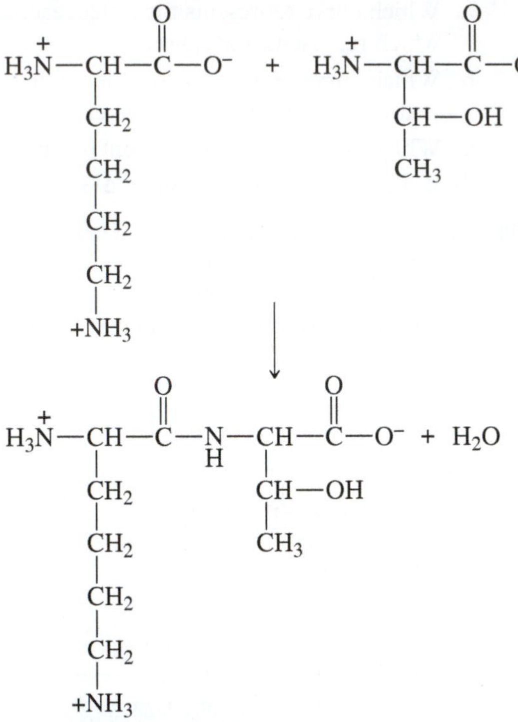

$$H_3\overset{+}{N}-CH-\overset{O}{\overset{\|}{C}}-O^- \quad + \quad H_3\overset{+}{N}-CH-\overset{O}{\overset{\|}{C}}-O^-$$

with side chains:

$$\begin{array}{c} CH_2 \\ | \\ CH_2 \\ | \\ CH_2 \\ | \\ CH_2 \\ | \\ +NH_3 \end{array} \qquad \begin{array}{c} CH-OH \\ | \\ CH_3 \end{array}$$

↓

$$H_3\overset{+}{N}-CH-\overset{O}{\overset{\|}{C}}-\overset{}{\underset{H}{N}}-CH-\overset{O}{\overset{\|}{C}}-O^- + H_2O$$

$$\begin{array}{c} CH_2 \\ | \\ CH_2 \\ | \\ CH_2 \\ | \\ CH_2 \\ | \\ +NH_3 \end{array} \qquad \begin{array}{c} CH-OH \\ | \\ CH_3 \end{array}$$

b.

$$^-O\overset{O}{\overset{\|}{C}}-\overset{H}{\underset{H}{C}}-\overset{H}{\underset{CO^-}{C}}-H \xrightarrow[\text{fumarase}]{FAD \quad FADH_2} $$

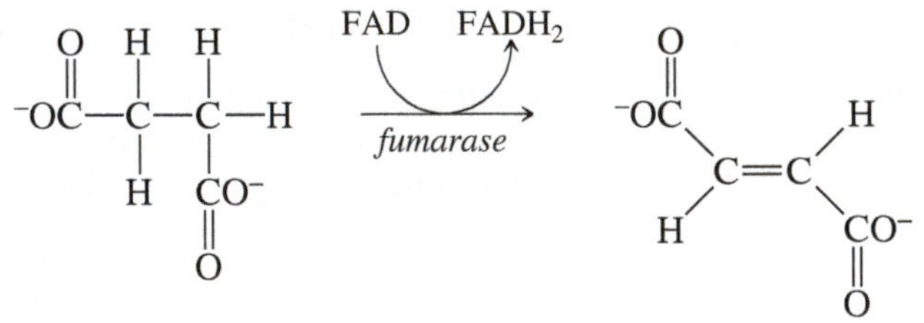

Succinate → Fumarate

5.58 Identify the main organic reaction shown as condensation, hydrolysis, oxidation, or reduction:

a.

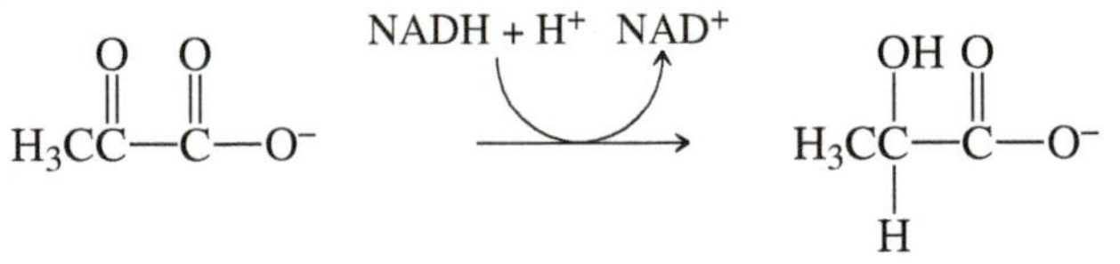

$$H_3C\overset{O}{\overset{\|}{C}}\overset{O}{\overset{\|}{C}}-O^- \xrightarrow[]{NADH + H^+ \quad NAD^+} H_3C\overset{OH}{\underset{H}{C}}\overset{O}{\overset{\|}{C}}-O^-$$

Pyruvate → Lactate

b. (disaccharide hydrolysis reaction with CH$_2$OH sugar rings + H$_2$O)

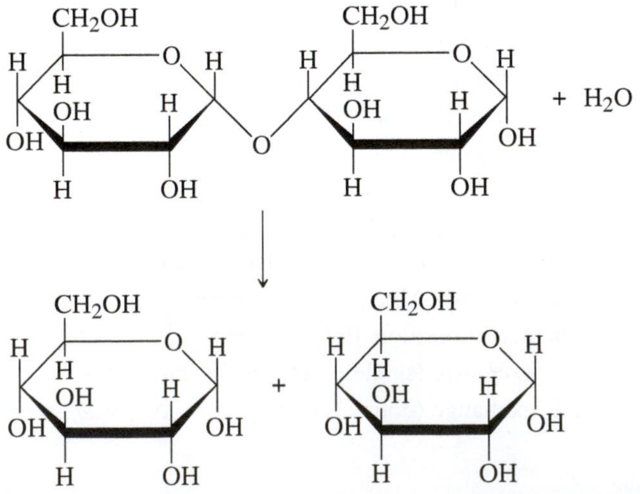

5.59 Provide the products and balance the following reaction for the complete combustion of pentane, C_5H_{12}. Identify the reactant that is oxidized and the reactant that is reduced.

$$C_5H_{12}(g) + O_2(g) \longrightarrow ?$$

5.60 Provide the products and balance the following reaction for the complete combustion of the fatty acid octanoic acid, $C_8H_{16}O_2$. Identify the reactant that is oxidized and the reactant that is reduced.

$$C_8H_{16}O_2(s) + O_2(g) \longrightarrow ?$$

5.61 Identify the reactant that is oxidized and the reactant that is reduced in the following reactions:
   a. $Cu(s) + 2AgNO_3(aq) \longrightarrow Cu(NO_3)_2(aq) + 2Ag(s)$
   b. $4Al(s) + 3O_2(g) \longrightarrow 2Al_2O_3(s)$
   c. $2AgBr(s) \longrightarrow 2Ag(s) + Br_2(g)$

5.62 Identify the reactant that is oxidized and the reactant that is reduced in the following reactions:
   a. $Mg(s) + FeCl_2(aq) \longrightarrow MgCl_2(aq) + Fe(s)$
   b. $2PbO(s) \longrightarrow 2Pb(s) + O_2(g)$
   c. $2Li(s) + F_2(g) \longrightarrow 2LiF(s)$

5.63 Determine whether each of the following organic reactions is an oxidation or a reduction reaction. (Only the organic compounds are shown.)
   a.

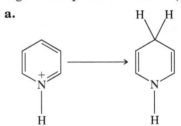

   b. (cyclohexanol → cyclohexanone)

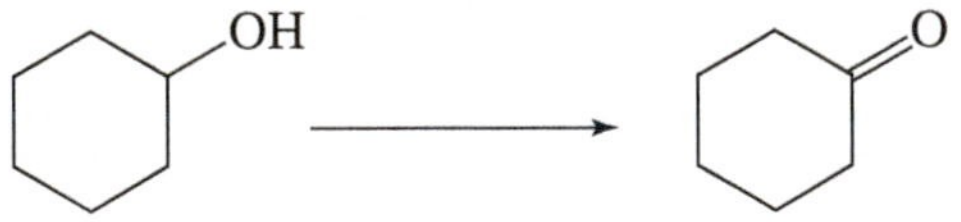

5.64 Determine whether each of the following organic reactions is an oxidation or a reduction reaction. (Only the organic products are shown.)
   a.

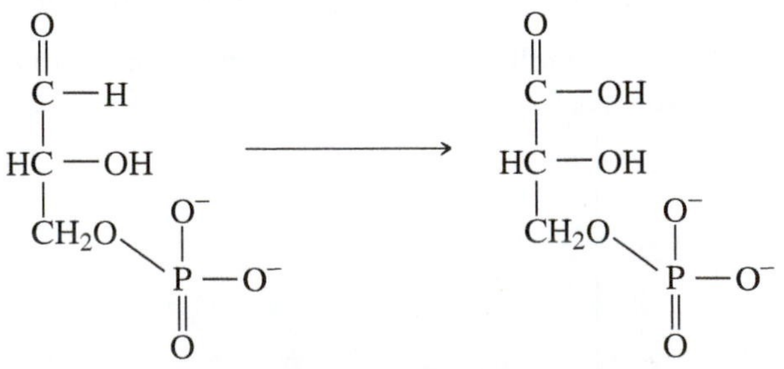

$$\begin{array}{c} \overset{O}{\overset{\|}{C}}-H \\ | \\ HC-OH \\ | \\ CH_2O \end{array}\overset{O^-}{\underset{O}{\overset{}{P}}}-O^- \longrightarrow \begin{array}{c} \overset{O}{\overset{\|}{C}}-OH \\ | \\ HC-OH \\ | \\ CH_2O \end{array}\overset{O^-}{\underset{O}{\overset{}{P}}}-O^-$$

   b. (2-methyl-2-butanol type structure with OH → alkene)

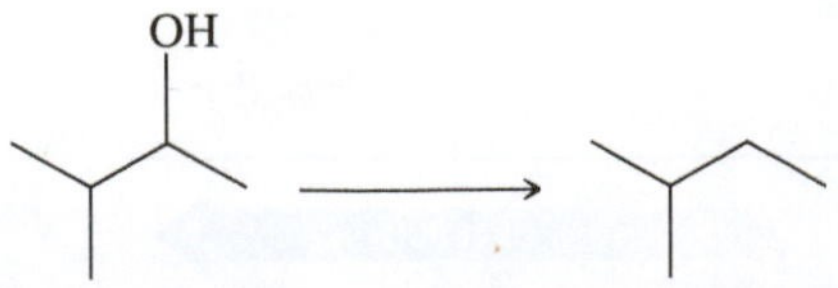

5.65 Write the products of the following reactions:

a.

$$CH_3OH + HO\overset{\overset{\displaystyle O}{\|}}{C}-CH_3 \longrightarrow H_2O + ?$$

b.

$$H-\overset{\overset{\displaystyle O}{\|}}{C}-CH_2CH_3 \xrightarrow{\text{Reduction}}$$

c.

$$\text{(structure)} \xrightarrow[\text{OH}]{\text{Oxidation}}$$

5.66 Write the products of the following reactions:

a.

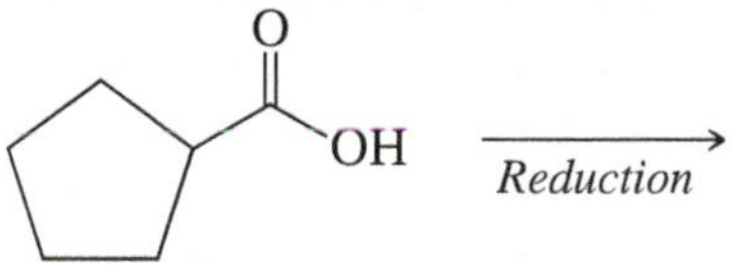

(cyclopentane with carboxylic acid) $\xrightarrow{\text{Reduction}}$

b.

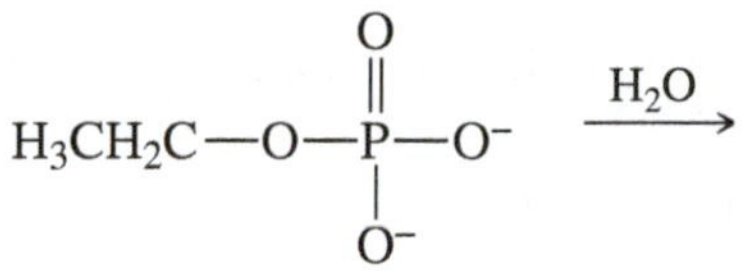

$$H_3CH_2C-O-\overset{\overset{\displaystyle O}{\|}}{P}-O^- \xrightarrow{H_2O}$$
$$\underset{\displaystyle O^-}{|}$$

c.

(cyclopentane with aldehyde) $\xrightarrow{\text{Oxidation}}$

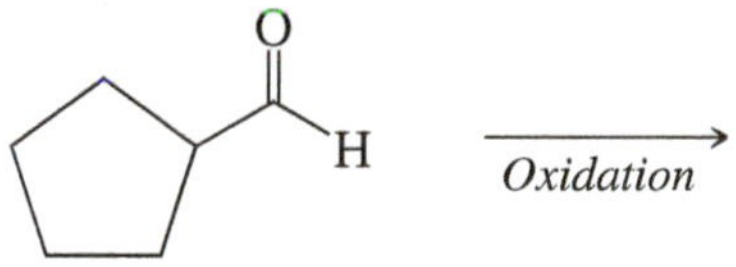

5.67 Several digestive enzymes break down proteins through hydrolysis. If the molecule below represents two amino acids linked through an amide bond, draw the products of the hydrolysis reaction of pepsin.

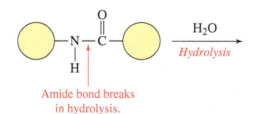

$$\xrightarrow[\text{Hydrolysis}]{H_2O}$$

Amide bond breaks in hydrolysis.

5.68 Acetylsalicylic acid (aspirin) can be synthesized by combining salicylic acid and acetic acid through a condensation reaction. The —OH group from phenol on the salicylic acid condenses with the carboxylic acid group of acetic acid, forming acetylsalicylic acid. Draw the structure of acetylsalicylic acid.

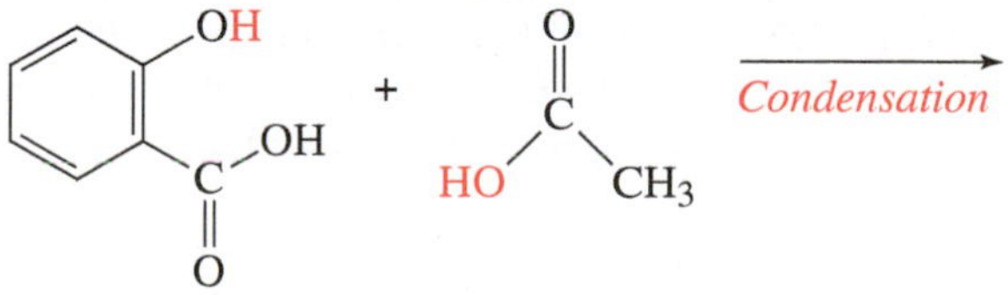

(salicylic acid structure) + (acetic acid structure) $\xrightarrow{\text{Condensation}}$

5.69 Fill in the missing organic products or reactants for the following hydrogenation reactions:

a.

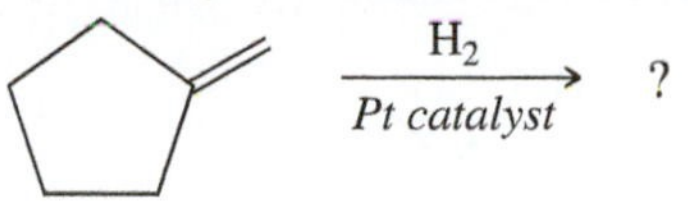

(methylenecyclopentane) $\xrightarrow[\text{Pt catalyst}]{H_2}$?

b.

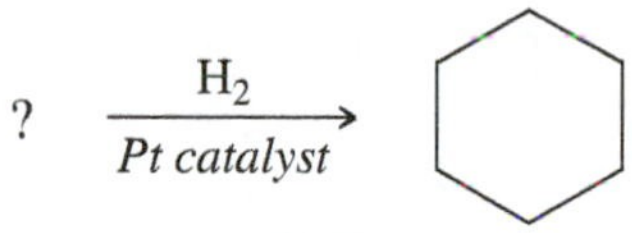

? $\xrightarrow[\text{Pt catalyst}]{H_2}$ (cyclohexane)

c.

(3-methyl-1-butene) $\xrightarrow[\text{Pd catalyst}]{H_2}$?

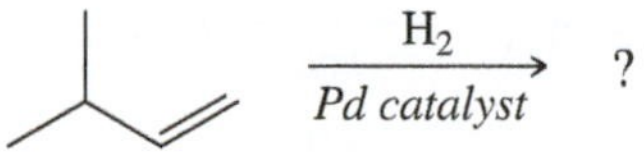

5.70 Epinephrine is the active ingredient in the EpiPen® used to treat severe allergic reactions. EpiPens expire due to the oxidation of the epinephrine. One of these reactions is shown below. Circle the groups in the product that were oxidized.

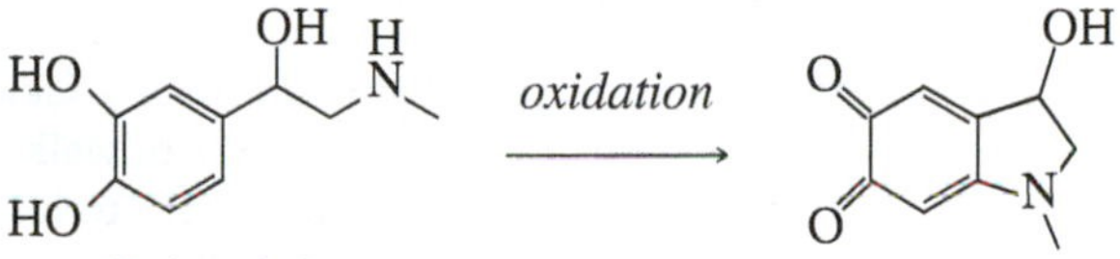

(Epinephrine structure) $\xrightarrow{\text{oxidation}}$ (product structure)

Epinephrine

5.71 When it is exposed to the oxygen in air for prolonged periods of time, wine will sour due to the oxidation of the ethanol to form acetic acid, which has a sour taste. Look up the structures for the reactants and products and write the equation for the reaction that occurs when wine sours.

5.72 Fill in the missing organic products for the complete hydrogenation of the following:

a.

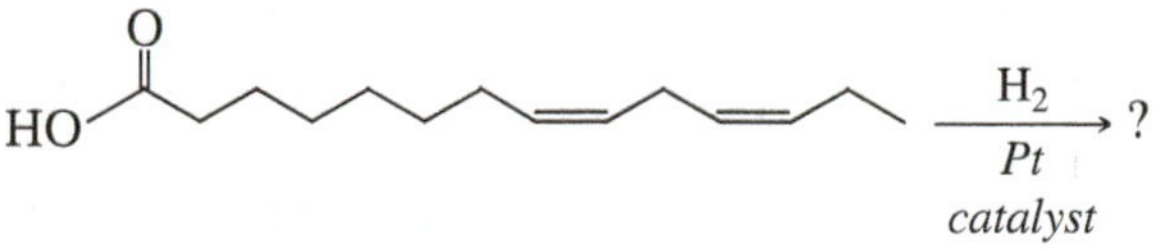

(unsaturated carboxylic acid structure) $\xrightarrow[\substack{\text{Pt} \\ \text{catalyst}}]{H_2}$?

b.

(diene structure) $\xrightarrow[\text{Pd catalyst}]{H_2}$?

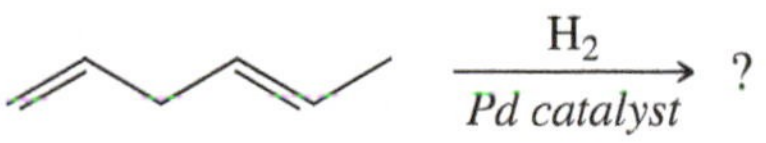

c.

$$CH_3CH_2CH{=}CHCH{=}CHCH_3 \xrightarrow[\text{Pt catalyst}]{H_2} ?$$

5.73 Fill in the missing organic product or reactant for the following hydration reactions:

a.

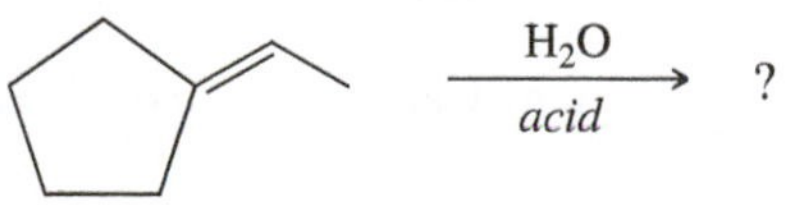

(ethylidenecyclopentane) $\xrightarrow[\text{acid}]{H_2O}$?

b.

? $\xrightarrow[\text{acid}]{H_2O}$ (cyclohexanol)

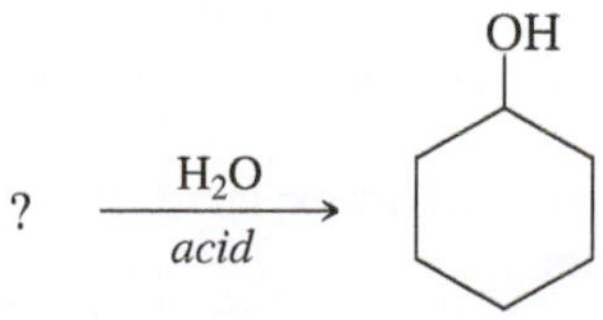

c.

? $\xrightarrow[\text{acid}]{H_2O}$ $CH_3\overset{\overset{\displaystyle OH}{|}}{C}HCH_3$

5.74 Fill in the missing organic product or reactant for the following hydration reactions:

a.

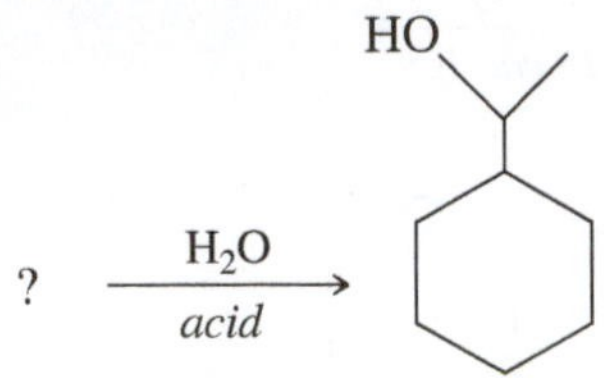

$$? \xrightarrow[acid]{H_2O}$$ [cyclohexanol structure with HO]

b.

$$CH_3CH_2CH{=}CH_2 \xrightarrow[acid]{H_2O} ?$$

c.

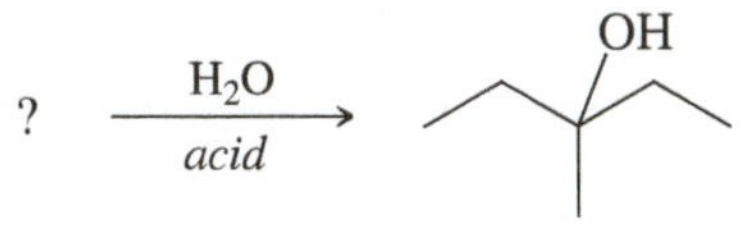

$$? \xrightarrow[acid]{H_2O}$$ [2-methyl-2-pentanol structure with OH]

5.75 How do low-carb diets work? We store glucose molecules in our muscles and liver as glycogen, which consists of thousands of glucose molecules linked together. During periods of fasting, we can activate glycogen to provide glucose.

    a. Determine which of the following reactions below would be a condensation and which would be a hydrolysis.

$$\text{Thousands of glucose} \underset{?}{\overset{?}{\rightleftharpoons}} \text{Glycogen} + H_2O$$

    b. Individuals who do not eat carbohydrates do not store the same levels of glycogen as people who do. Explain the weight loss associated with storing less glycogen.

Challenge Problems

5.76 If you begin an exercise program and burn 5 Calories per minute by walking, how many hours of walking per day would you have to do in two weeks to lose 5 pounds? You must burn an extra 3500 Calories to lose 1 pound of body weight.

5.77 Your daily latte contains 290 Calories. You want to lose weight by giving up your morning latte and drinking water instead. Assuming the rest of your diet remains constant, how many days will it take you to lose 5 pounds? You must burn an extra 3500 Calories to lose 1 pound of body weight.

5.78 Your basal metabolic rate (BMR) is the amount of energy your body uses to function at rest in a day. This includes the energy required to keep the heart beating, the lungs breathing, the kidneys functioning, and the body temperature stabilized. Approximately 70% of an individual's energy expenditure goes into basal metabolism. The other 30% goes into digestion and absorption of nutrients (10%) and physical movement (about 20%).

    a. Calculate your basal metabolic rate in Calories using the appropriate equation here:

        Females: BMR = 655 + (9.6 × weight (kg))
                + (1.8 × height (cm)) − (4.7 × age (yrs))
        Males: BMR = 66 + (13.7 × weight (kg))
                + (5 × height (cm)) − (6.8 × age (yrs))

    b. Calculate how many Calories you burn overall in a day.

5.79 Several neurotransmitters are made through decarboxylation of amino acids. Serotonin, a neurotransmitter involved in mood response, is made in the brain when 5-hydroxytryptophan is decarboxylated. Provide the structure of serotonin.

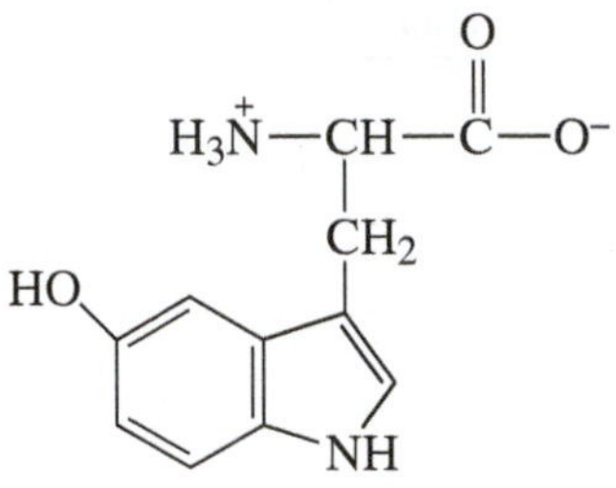

[structure of 5-Hydroxytryptophan]

$$+ \ H_2O \xrightarrow{decarboxylase}$$

5-Hydroxytryptophan

5.80 The reverse reaction of hydration is dehydration. The dehydration of an alcohol involves removing an OH from one carbon and an H from the carbon next to it to form an alkene. In glycolysis, the enzyme enolase catalyzes the dehydration of 2-phosphoglycerate to form phosphoenolpyruvate (PEP), which contains a carbon–carbon double bond. Complete the reaction below by drawing the structure of PEP.

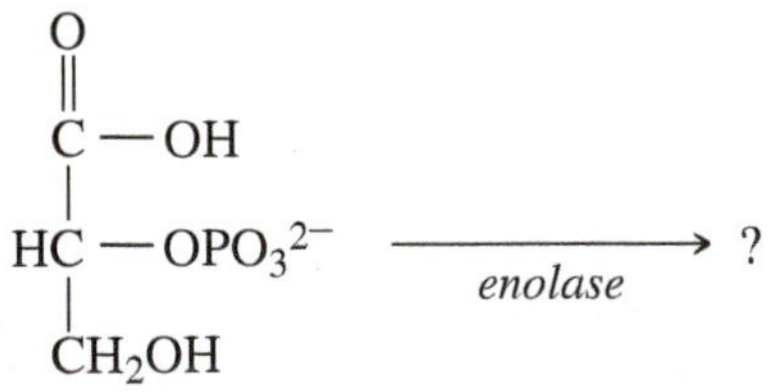

[structure of 2-phosphoglycerate]

$$\xrightarrow{enolase} \ ?$$

5.81 Pharmaceuticals are labeled with an expiration date to ensure maximum potency. A bottle of aspirin that has expired is said to smell like vinegar due to a hydrolysis reaction that occurs with moisture in the air. Predict the products of the hydrolysis reaction of aspirin and explain the vinegar odor of the expired drug.

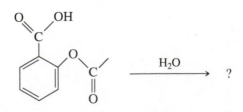

$$\xrightarrow{H_2O} \ ?$$

Answers to Odd-Numbered Problems

Practice Problems

5.1 Cold. If the reaction is endothermic, it absorbs heat from its surroundings, which cools the reaction. This reaction occurs spontaneously, so it would have a $-\Delta G$.

5.3 **a.** exothermic   **b.** endothermic

5.5 **a.** negative   **b.** positive

5.7 **a.** negative   **b.** negative   **c.** positive

5.9 **a.** spontaneous   **b.** spontaneous

5.11 140 Calories

5.13 **a.** Increasing the temperature increases the rate by increasing the movement of the reactants. The faster they move, the more likely they are to collide and react.

b. Increasing the amount of reactant introduces more reactant molecules, which increases the likelihood of collision and reacting.

5.15 The amount of reactants decreases as the reaction progresses, so the rate slows.

5.17 activation energy

5.19 **a.** decrease   **b.** increase   **c.** increase

5.21 **a.** exchange   **b.** decomposition   **c.** synthesis

5.23 **a.** irreversible   **b.** irreversible   **c.** reversible

5.25 $2C_2H_6(g) + 7O_2(g) \rightarrow 4CO_2(g) + 6H_2O(g)$

5.27 In both organic and biochemistry reactions, the reactants appear on the left, the products appear on the right, and an arrow separates them. The arrows indicate the reversibility of the reactions. In both cases, the structure of the organic reactant may be shown.

5.29 **a.** reduction   **b.** oxidation   **c.** oxidation

5.31 Hydrogen is oxidized, and oxygen is reduced.

5.33 Pyruvate is reduced to lacate; NADH is oxidized to NAD^+.

5.35 decomposition

5.37

$$H_3C-\overset{\overset{\displaystyle O}{\|}}{C}-OCH_2CH_3 + H_2O$$

5.39 **a.**

$$H_3C-\underset{\underset{\displaystyle H}{|}}{\overset{\overset{\displaystyle H_3C}{|}}{C}}-\underset{\underset{\displaystyle H}{|}}{\overset{\overset{\displaystyle CH_3}{|}}{C}}-H$$

b. (square)   **c.** (branched chain structure)

5.41 **a.**

$$H_3C-\underset{\underset{\displaystyle OH}{|}}{\overset{\overset{\displaystyle H_3C}{|}}{C}}-\underset{\underset{\displaystyle H}{|}}{\overset{\overset{\displaystyle CH_3}{|}}{C}}-H$$

b. (square with OH)

c. (structure with OH)

Additional Problems

5.43 An increase in temperature will make the $T\Delta S$ term more negative, which means the spontaneity will increase.

5.45 **a.** exothermic   **b.** reactants

c. Yes, the reaction is spontaneous.

d.

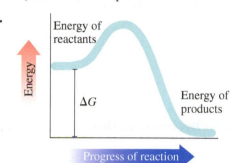

5.47 An exergonic reaction with a low activation energy occurs faster. The low activation energy allows the reactants to react more quickly.

5.49 **a.** Curve (a) is faster, and curve (b) is slower.

b. Curve (a) is endergonic, and curve (b) is exergonic.

c. Curve (b) is spontaneous.

d. Curve (a) has a positive ΔG value.

5.51

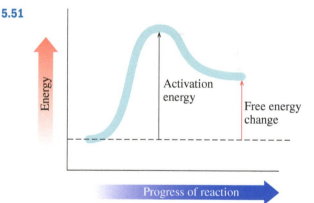

5.53 780 Calories

5.55 **a.** $Mg + Cl_2 \longrightarrow MgCl_2$

b. $2HI \longrightarrow H_2 + I_2$

c. $Ca + Zn(NO_3)_2 \longrightarrow Ca(NO_3)_2 + Zn$

d. $K_2S + Pb(NO_3)_2 \longrightarrow 2KNO_3 + PbS$

5.57 **a.** condensation   **b.** oxidation

5.59 $C_5H_{12}(g) + 8O_2(g) \longrightarrow 5CO_2(g) + 6H_2O(g)$; pentane is oxidized, and oxygen is reduced.

5.61 **a.** Copper is oxidized, and silver ion is reduced.

b. Aluminum is oxidized, and oxygen is reduced.

c. Bromide is oxidized, and silver ion is reduced.

5.63 **a.** reduction   **b.** oxidation

5.65 **a.**

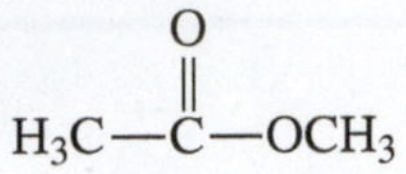

H₃C—C(=O)—OCH₃

b. HOCH₂CH₂CH₃

c.

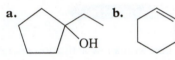

a butyric acid structure with OH

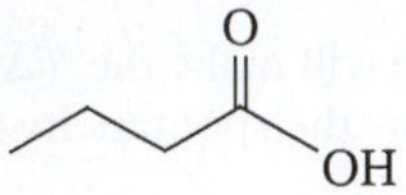

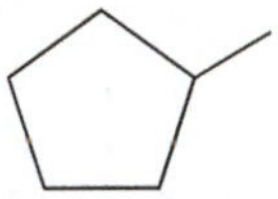

5.67

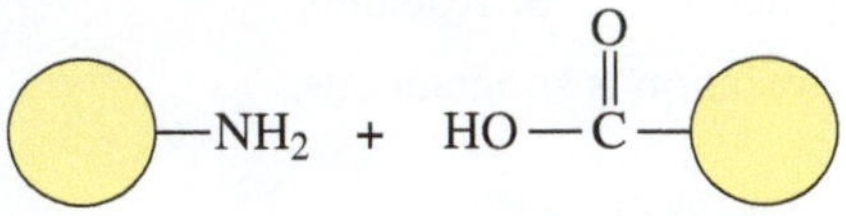

○—NH₂ + HO—C(=O)—○

5.69 **a.** methylcyclopentane **b.** cyclohexene **c.** 2-methylbutane branched structure

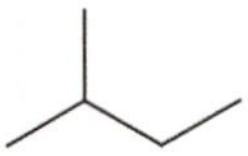

5.71 CH₃CH₂OH ⟶ CH₃COOH

5.73 **a.** 1-ethylcyclopentanol with OH **b.** cyclohexene **c.** CH₂=CHCH₃

5.75 **a.** The forward reaction (glucose to glycogen) is condensation; the reverse reaction (glycogen to glucose) is hydrolysis.

b. A person who does not store as much glycogen does not produce water when forming glycogen from glucose. The initial weight lost is water weight.

Challenge Problems

5.77 60 days

5.79

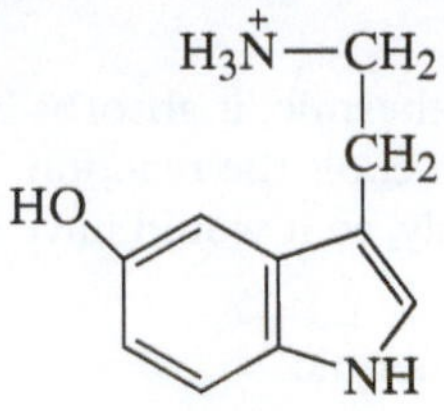

H₃N⁺—CH₂
       |
       CH₂
HO—(indole ring system)—NH

Serotonin

5.81

O=C(—OH)—(benzene ring with OH)

+

HO—C(=O)—CH₃

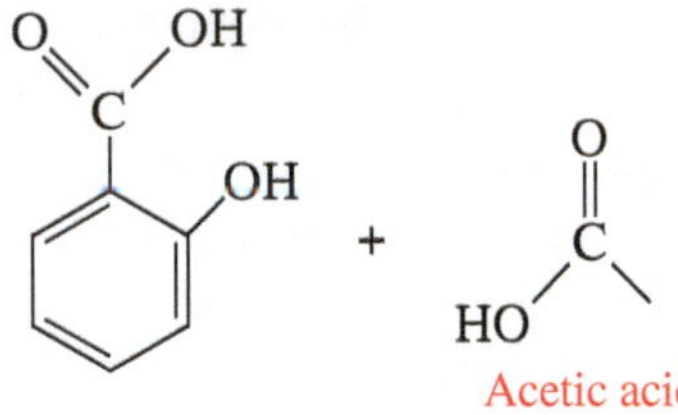

Acetic acid
(the primary
component of vinegar)

Carbohydrates—Life's Sweet Molecules

People with diabetes have trouble getting glucose, a carbohydrate, into their cells. Because of this difficulty, it builds up in their bloodstream. Diabetics regularly monitor their blood-sugar (glucose) levels to keep them in a normal range. Glucose is found in all types of edible carbohydrates, from sugar to starch to fiber, but it is not the only carbohydrate. What do carbohydrates look like, and how are their structures different? Chapter 6 explores the unique structures and functions of carbohydrates.

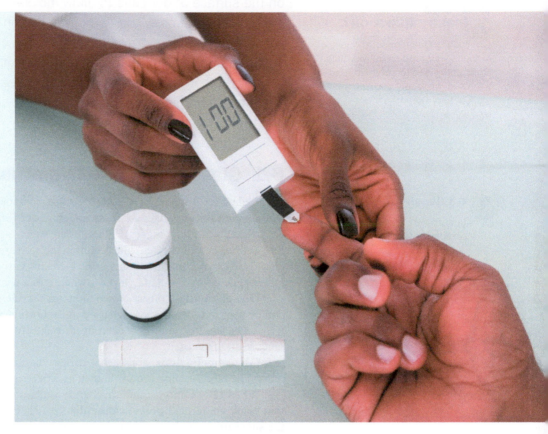

CHAPTER OUTLINE

LEARNING TIP   Give Your Brain a Break

Did you know that your brain functions on glucose, and glucose needs oxygen to metabolize? If you get tired while studying, getting up and moving around, lightly stretching, and eating a piece of fruit can help to re-energize your studying.

WE OFTEN HEAR that we should curb sugar consumption and that "carbs" are the reason that so many people are overweight. What purposes do carbohydrates serve in our bodies? How should we include them in our daily diets?

Carbohydrates are commonly known as sugars, and they do provide energy. The carbohydrate glucose is the main nutrient that our body uses to produce energy, especially in our brain and red blood cells. Our bodies break down glucose and other sweet-tasting *simple sugars* through oxidation to produce the energy that we require. When we consume more carbohydrates than we need for energy, our bodies convert them to fat to store the energy. So, consumption of carbohydrates in excess *can* result in weight gain.

Complex carbohydrates include starches and insoluble fibers found in plants, the most common of which is cellulose. As we will see, starch and cellulose are larger molecules composed of glucose. We can digest and absorb starch, but we cannot digest and absorb cellulose. This may seem strange because both contain glucose. Starch and cellulose combine glucose differently into their structures, accounting for this difference (see Section 6.5).

Besides their essential role in nutrition, carbohydrates have other important biological functions. Simple sugars serve important structural roles in our genetic material, DNA and RNA. Other sugars serve as markers on the surface of our cells to allow molecules to distinguish different types of cells. For example, the carbohydrates on the surface of your red blood cells determine your blood type. Complex carbohydrates such as heparin serve an important role in the prevention of blood clots. Finally, some carbohydrates can actually play a role in diseases like diabetes and lactose intolerance. In this chapter, we explore the different types of carbohydrates and their roles.

What's an Inquiry Question?

Inquiry Questions are designed to focus your reading on the main concepts by section. As you read each section, see if you can answer each Inquiry Question in your own words.

6.1 Inquiry Question

How do we classify carbohydrates?

6.1 Classes of Carbohydrates

A carbohydrate is a simple or complex sugar composed of carbon, hydrogen, and oxygen. The simplest carbohydrates are **monosaccharides** (*mono* is Greek for "one," *sakkharis* is the early Greek word for "sugar"). These often sweet-tasting sugars cannot be broken down into smaller carbohydrates. The common carbohydrate glucose, $C_6H_{12}O_6$, is a monosaccharide. Monosaccharides contain the elements carbon, hydrogen, and oxygen, and they have the general formula $C_n(H_2O)_n$, where n is a whole number 3 or higher.

Disaccharides consist of two monosaccharide units joined together. A disaccharide can be split into two monosaccharide units. Ordinary table sugar, sucrose, $C_{12}H_{22}O_{11}$, is a disaccharide that can be broken up through hydrolysis into the two monosaccharides glucose and fructose.

Oligosaccharides are carbohydrates containing three to nine monosaccharide units. The blood-typing groups known as ABO are oligosaccharides.

When ten or more monosaccharide units are joined together, the large molecules that result are termed **polysaccharides** (*poly* is Greek for "many"). In polysaccharides, the sugar units can be connected in one continuous chain, or the chain can be branched. Starch, a polysaccharide in plants, contains large chains of glucose that can be broken down to produce energy.

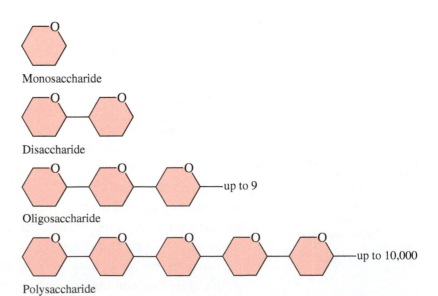

The various classes of carbohydrates contain different numbers of monosaccharides.

Fiber in Your Diet

Dietary fiber consists of carbohydrates that we cannot digest with our own enzymes. Dietary fiber is divided into two classes called soluble and insoluble fiber. **Soluble fiber** can mix with water, forming a gel-like substance that swells in the stomach and digestive tract. This gives a sense of fullness, and slows sugar absorption into the bloodstream. Soluble fiber has also been shown to help lower blood cholesterol by interfering with cholesterol absorption. Foods high in soluble fiber include oatmeal, legumes (peas, beans, and lentils), apples, psyllium husk, and carrots. Fruit pectins used in making jellies (giving a gel consistency) contain soluble fiber.

In contrast, **insoluble fiber** does not mix with water, although it plays a critical role in the digestive tract. Insoluble fiber has a laxative effect and adds bulk to the diet, thus preventing constipation. The polysaccharide cellulose (Section 6.5) is an insoluble fiber. Sources include whole grains, seeds, brown rice, cabbage, and vegetable stalks and skins.

Many processed foods contain processed fiber. Although it adds more fiber to the diet, processed fiber doesn't provide the feeling of fullness. Selecting fiber from natural sources instead of processed foods with added fiber is a better choice.

INTEGRATING Chemistry

◀ Find out how your body uses fiber.

Dietary fiber consists of carbohydrates that we cannot digest and that can be found in whole grains, oats, fruits, and vegetables. Fibers from natural sources are important for a healthy digestive tract.

Practice Problems

Find the answers to the odd-numbered Practice Problems at the end of the chapter.

6.1   Classify the following carbohydrates as a monosaccharide, disaccharide, oligosaccharide, or polysaccharide:

   a. carageenan, a seaweed extract containing up to 25,000 carbohydrate units

   b. levans, soluble fiber containing three to six carbohydrate units

   c. maltose, containing two glucose units

6.2   Classify the following carbohydrates as a monosaccharide, disaccharide, oligosaccharide, or polysaccharide:

   a. raffinose, a soluble fiber containing three carbohydrate units

   b. starch, a storage carbohydrate of plants that contains thousands of glucose units

   c. fructose, a simple sugar found in fruit with the formula $C_6H_{12}O_6$

6.3   Identify the following as characteristics of soluble or insoluble fiber:

   a. can mix with water

   b. can lower blood cholesterol

   c. has a laxative effect

6.4   Identify the following as containing soluble or insoluble fiber:

   a. oatmeal   b. brown rice   c. celery

6.2 Functional Groups in Monosaccharides

Monosaccharides are organic molecules that contain several functional groups. The functional groups of the monosaccharide glucose are shown in **FIGURE 6.1**. Note that glucose contains several alcohol (or hydroxyl) groups represented as —OH. Glucose also includes the carbonyl-containing aldehyde functional group. Some monosaccharides, like fructose, contain a ketone functional group instead of an aldehyde. Carbohydrates are considered polyhydroxyaldehydes or polyhydroxyketones because they contain several hydroxyl (alcohol) groups and either an aldehyde or a ketone group. Before discussing monosaccharides further, let's look at these functional groups and some common alcohols, aldehydes, and ketones more closely.

Alcohol

Ethanol, a commercially important compound produced from the fermentation of the simple sugars in grains and fruits, is one of the simplest members of the family of organic compounds known as **alcohols**. Ethanol is the alcohol present in liquor, beer, and wine. It is also the main additive in alternative fuel blends such as flex fuels, which have formulations of 15–90% ethanol.

As we discussed in Chapter 4, alcohols are classified as primary (1°), secondary (2°), or tertiary (3°) by the number of alkyl groups attached to the carbon atom that is bonded to the hydroxyl group (see Table 4.4). The number of alkyl groups attached to the alcoholic carbon directly impacts the reactivity of the alcohol. A **primary (1°) alcohol** has one alkyl group attached to the alcoholic carbon, a **secondary (2°) alcohol** has two such alkyl groups, and a **tertiary (3°) alcohol** has three. Monosaccharides contain both primary and secondary alcohols. Organic compound names ending in *ol* contain an alcohol functional group.

Aldehyde

Alcohol (hydroxyl)

FIGURE 6.1 Functional groups in the monosaccharide glucose. Glucose is a polyhydroxyaldehyde because it includes the carbonyl-containing functional group aldehyde and several hydroxyl (alcohol) groups.

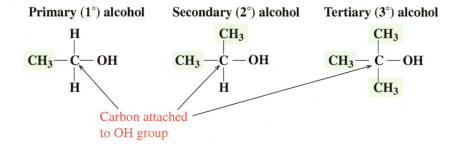

Primary (1°) alcohol   Secondary (2°) alcohol   Tertiary (3°) alcohol

Carbon attached to OH group

Sample Problem 6.1   Classifying Alcohols

Classify each of the following alcohols as a primary (1°), secondary (2°), or tertiary (3°) alcohol:

a.   OH

Isopropanol, rubbing alcohol

b.   OH

1-Butanol, used in the extraction of essential oils

Solution

To determine if an alcohol is primary, secondary, or tertiary, answer the following question: How many carbons are directly bonded to the C—OH carbon?

a. Two carbons are bonded to the C—OH; therefore, this is a secondary (2°) alcohol.

b. One carbon is bonded to the C—OH; therefore, this is a primary (1°) alcohol.

Aldehyde

Benzaldehyde, a compound that smells like almonds and is used in foods as a cherry flavoring, is a member of the simplest family of carbonyl-containing organic compounds, known as the **aldehydes**. The presence of the benzene ring in its structure further classifies benzaldehyde as an aromatic aldehyde. Organic compounds whose IUPAC names end in *al* contain an aldehyde functional group.

Members of the aldehyde family always have a carbonyl group with a hydrogen atom bonded to one side of the carbonyl and an alkyl or aromatic group bonded to the other. The lone exception to this is formaldehyde, which has a hydrogen bonded to each side of the carbonyl. Formaldehyde was once used as a preservative for biological specimens, but it is no longer used because it was found to cause cancer in some animals. Many monosaccharides contain an aldehyde functional group at one end of the molecule (in addition to multiple hydroxyl groups).

The aldehyde functional group is found in many scented compounds. The simplest aldehyde is formaldehyde, in which the R group is a hydrogen.

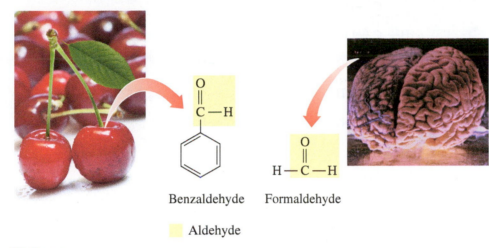

Benzaldehyde   Formaldehyde

☐ Aldehyde

Ketone

The **ketone** family of organic compounds is structurally similar to the aldehydes. The difference is that ketones have an alkyl or aromatic group on *both* sides of the carbonyl. The simplest ketone is acetone, which was once the main component of fingernail polish remover. Because of its tendency to cause dry skin, acetone has now been largely replaced in many formulations of nail polish remover. Organic compound names ending in *one* contain a ketone functional group.

$$H_3C - \overset{\overset{\textstyle O}{\|}}{C} - CH_3$$

☐ Ketone   Acetone

In addition to monosaccharides, ketones are found in a wide variety of biologically relevant compounds. For example, pyruvate is a ketone-containing compound formed during the metabolic breakdown of carbohydrates. Butanedione, the chemical responsible for the flavor of butter, contains two ketone groups. Note that in the structures of pyruvate and butanedione, the carbonyl has a carbon as the *first* atom on each side. This arrangement distinguishes the functional group as a ketone regardless of the other atoms that may be present in the structure.

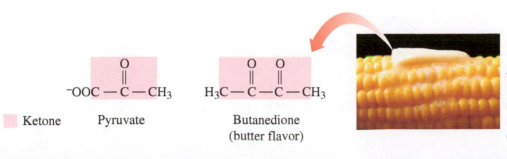

☐ Ketone   Pyruvate   Butanedione
                                     (butter flavor)

The ketone functional group is found in several natural flavors like butter, raspberries, spearmint, and caraway. It consists of a carbonyl bonded to two carbons.

A monosaccharide that contains an aldehyde functional group is referred to as an **aldose**, and one that contains a ketone functional group is called a **ketose**. The most common monosaccharides contain three to six carbons. A monosaccharide with three carbons is a *triose,* one with four carbons is a *tetrose,* one with five carbons is a *pentose,* and one with six carbons is a *hexose.*

For instance, **glucose**, the most abundant monosaccharide found in nature, is an aldose containing six carbons and is therefore classified as an aldohexose, while **fructose**, a common monosaccharide found in fruits, is a ketose containing six carbons and is classified as a ketohexose. Look at the monosaccharides shown here and convince yourself that these are examples of an aldohexose and ketopentose by counting the carbons and identifying the functional groups.

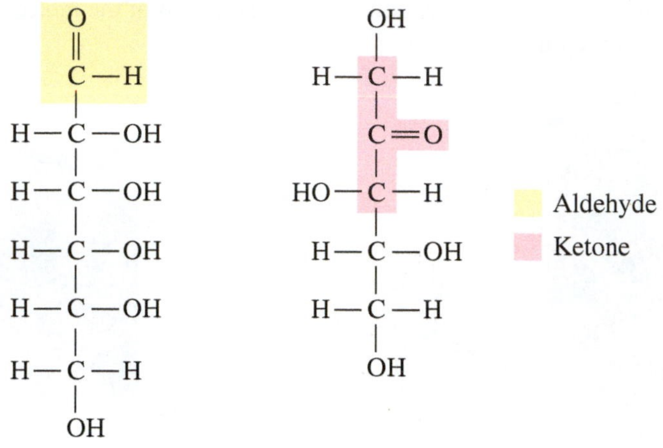

An aldohexose   A ketopentose

Sample Problem 6.2   Identifying Aldehydes and Ketones

Circle and label the aldehyde or ketone functional group in each of the following compounds:

a.

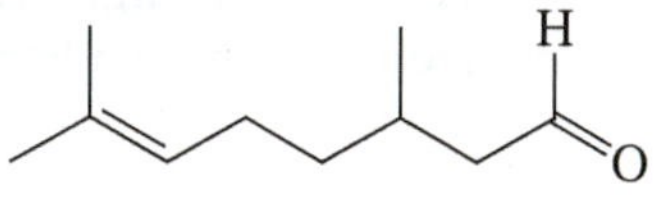

Citronellal, the main component of citronella oil

b.

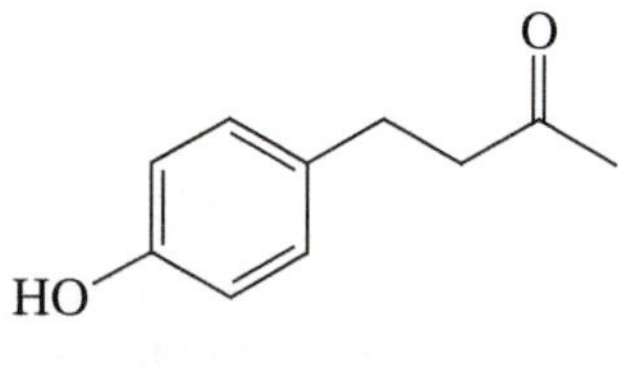

Raspberry ketone, the primary aroma compound in red raspberries

c.

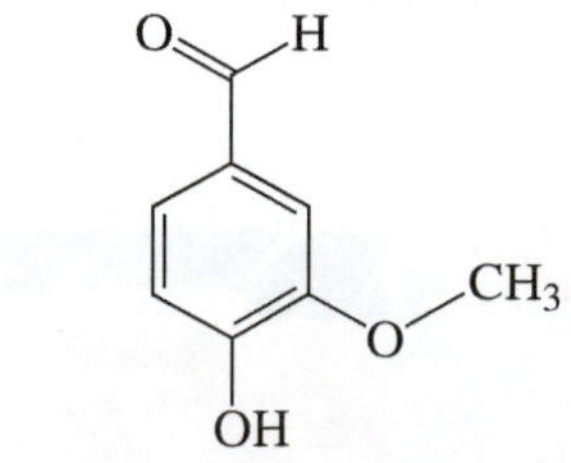

Vanillin, the primary component in vanilla extract

Solution

a.

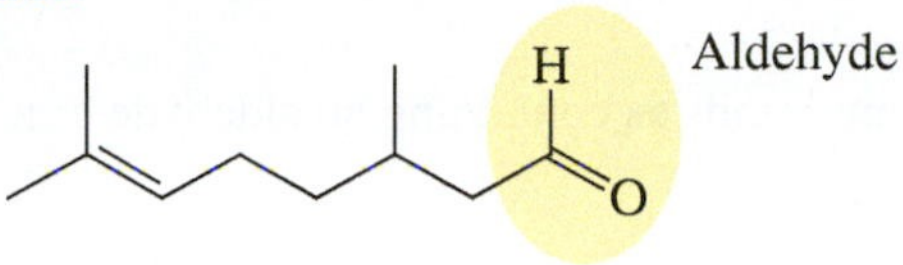

Aldehyde

Citronellal, the main component of citronella oil

b.

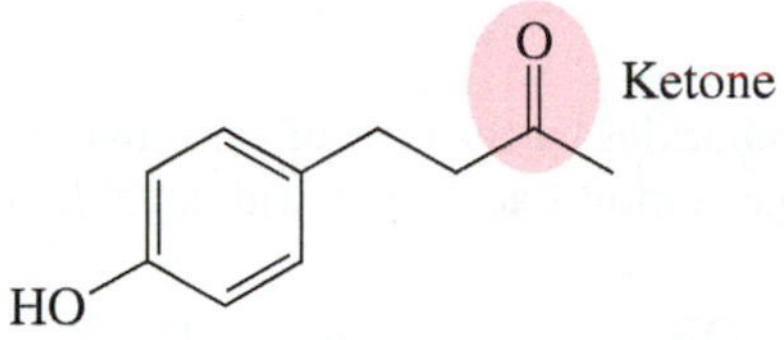

Ketone

Raspberry ketone, the primary aroma compound in red raspberries

c.

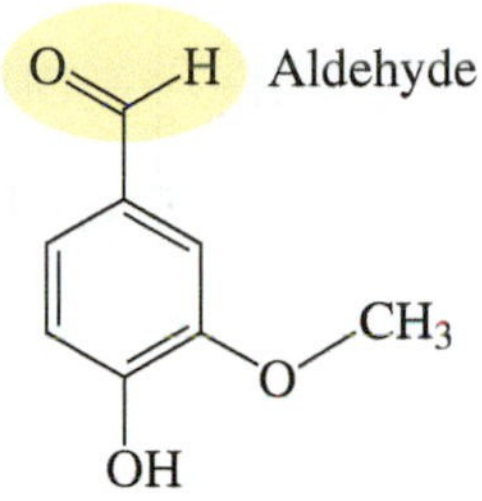

Aldehyde

Vanillin, the primary component in vanilla extract

Practice Problems

6.5 Classify each of the following alcohols as a primary (1°), secondary (2°), or tertiary (3°) alcohol:

a.

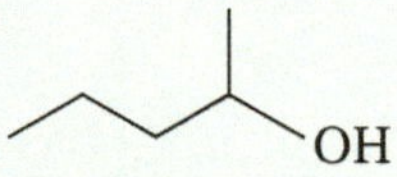

b. $CH_3CH_2CH_2CH_2OH$

c.
$$CH_3\overset{\overset{\displaystyle OH}{|}}{\underset{\underset{\displaystyle CH_3}{|}}{C}}CH_2CH_3$$

d.

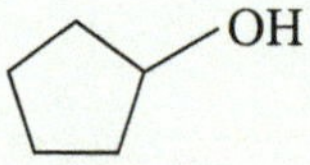

6.6 Classify each of the following alcohols as a primary (1°), secondary (2°), or tertiary (3°) alcohol:

a.

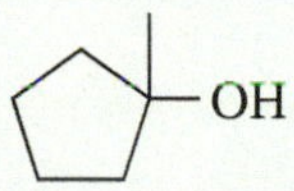

b.

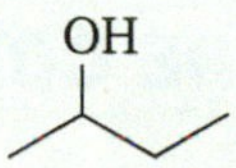

c.

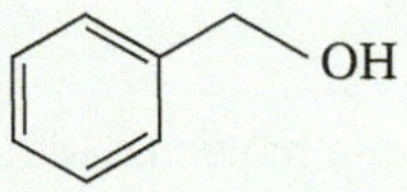

d.

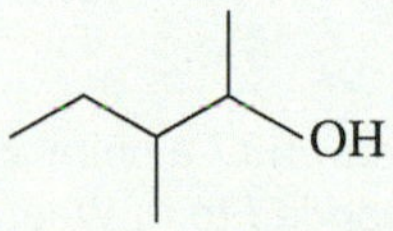

6.7 Identify each of the following compounds as containing an aldehyde or a ketone:

a.

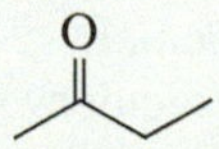

b.

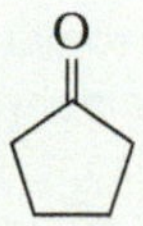

c.

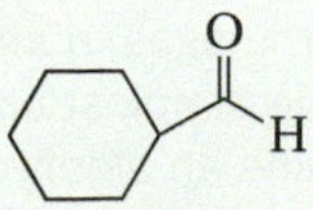

—Continued next page

Continued—

6.8   Identify each of the following compounds as containing an aldehyde or a ketone:

a.

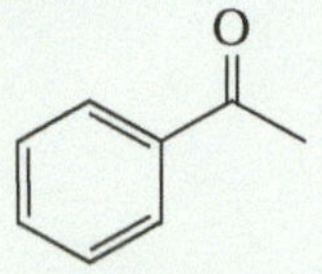

b.

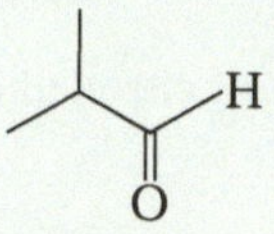

c.

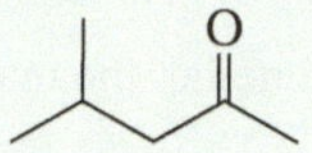

6.9   Classify each of the following monosaccharides by the type of carbonyl group and the number of carbons (for example, a monosaccharide with an aldehyde and three carbons is an aldotriose).

a.

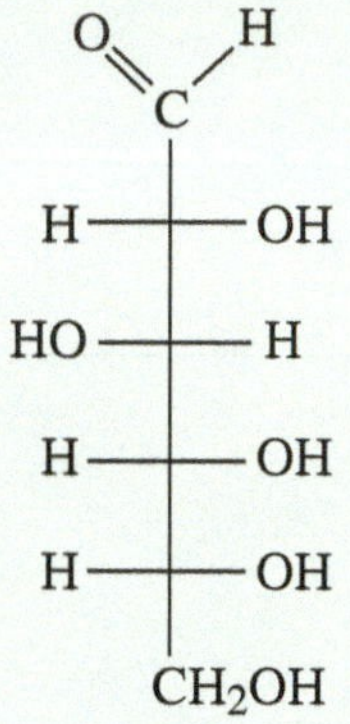

b.

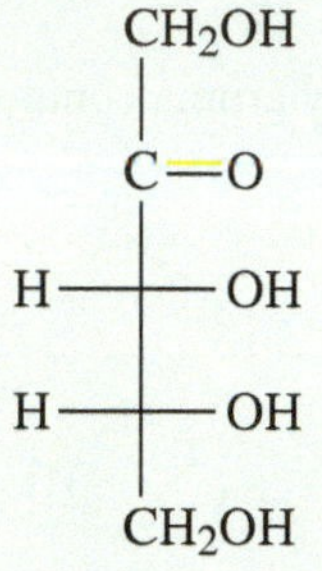

c.

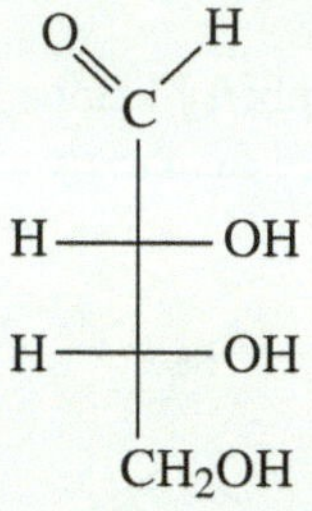

6.10 Classify each of the following monosaccharides by the type of carbonyl group and the number of carbons (for example, a monosaccharide with an aldehyde and three carbons is an aldotriose).

a.

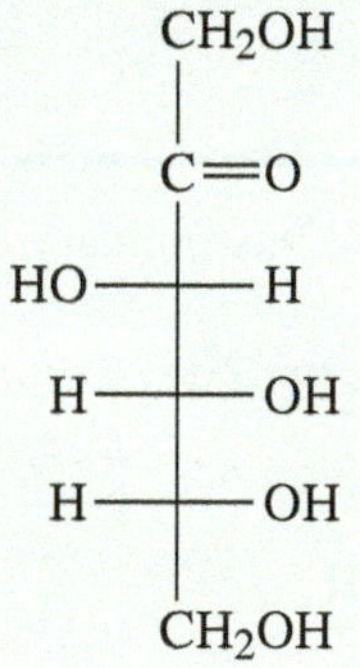

b.

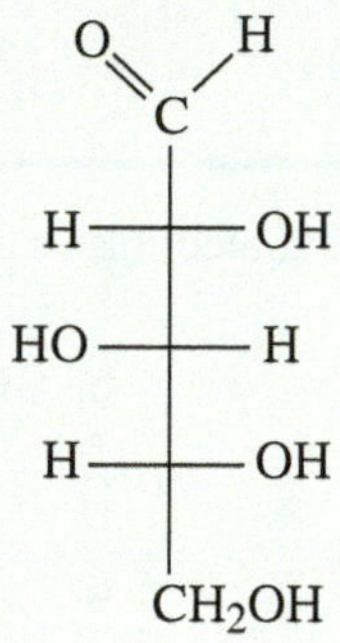

c.

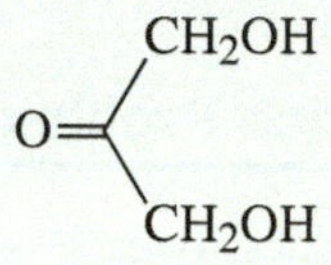

DISCOVERING THE CONCEPTS

❓ INQUIRY ACTIVITY—Fischer Projections

Information

In Chapter 4, we saw that a chiral center is a carbon in a molecule with four different groups bonded to it. When a chiral center appears in a molecule, the four groups can arrange themselves around the carbon in two different ways. The two resulting molecules are related to each other as nonsuperimposable mirror images, a special type of stereoisomer called an *enantiomer*.

The simplest carbohydrate is glyceraldehyde, $C_3H_6O_3$. Glyceraldehyde contains an aldehyde functional group, so it is referred to as an *aldose*. In carbohydrates, a pair of enantiomers is designated by writing either D- or L- in front of the name. The D-isomer has the chiral center farthest from the carbonyl arranged like D-glyceraldehyde, and the L-isomer has the chiral center farthest from the carbonyl arranged like L-glyceraldehyde.

Model	L-Glyceraldehyde	D-Glyceraldehyde
Wedge-and-dash	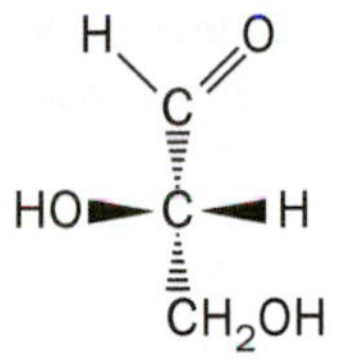 H C=O, HO►C◄H, CH₂OH	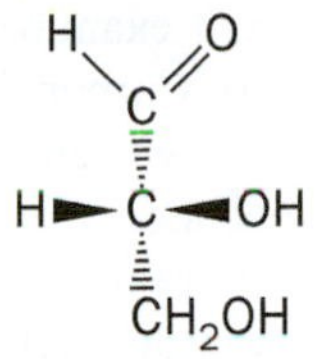 H C=O, H►C◄OH, CH₂OH
Fischer projection	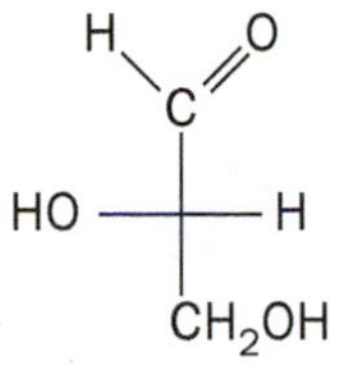 H C=O, HO—H, CH₂OH	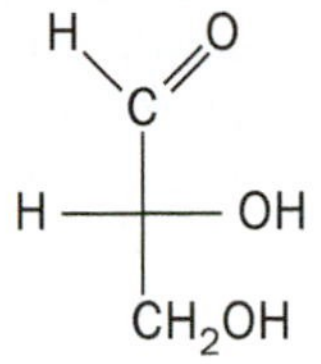 H C=O, H—OH, CH₂OH
Ball-and-stick	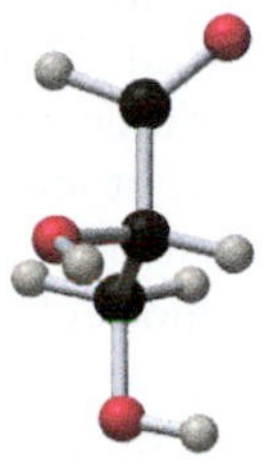	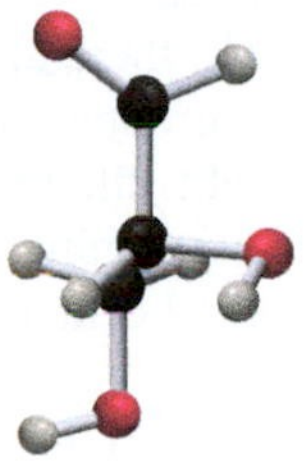

FIGURE 1 The enantiomers of glyceraldehyde (L- on left, D- on right) shown in three different representations: wedge-and-dash, Fischer projection, and ball-and-stick.

Questions

1. In D-glyceraldehyde (**FIGURE 1**), list the groups attached to the chiral center that are in front of the plane of the paper. Which groups attached to the chiral center are behind the plane of the paper?
2. Where is the chiral center in the Fischer projection? Identify it on the glyceraldehyde molecules (Figure 1) with an asterisk.
3. What does the horizontal line in the center of the Fischer projections shown represent?
4. What does the vertical line in the center of the Fischer projections shown represent?
5. How are D- and L-glyceraldehyde different?
6. Identify the chiral centers on D-glucose (**FIGURE 2**) with an asterisk.
7. Draw the Fischer projection of D-glucose.
8. Draw the Fischer projection of L-glucose (the mirror image of D-glucose) next to the D-glucose that you drew in question 7.
9. D-Galactose is an *epimer* of D-glucose: All the chiral centers are identical, except for carbon 4, which is reversed. Draw the Fischer projection of D-galactose.

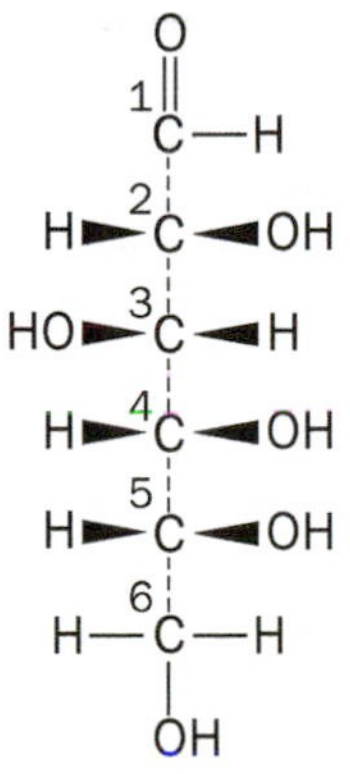

O
‖
1 C—H
H►2C◄OH
HO►3C◄H
H►4C◄OH
H►5C◄OH
6
H—C—H
|
OH
D-Glucose

FIGURE 2 D-Glucose shown in wedge-and-dash representation.

6.3 Stereochemistry in Monosaccharides

Let's look at the structure of glucose a little more closely and see if we recognize any other properties of carbon-containing molecules seen earlier in Chapter 4. The carbons bonded to the alcohol (—OH) groups all have a tetrahedral geometry. Remember that a carbon atom with tetrahedral geometry that has four different atoms or groups of atoms

6.3 Inquiry Question

What structural properties do monosaccharides display?

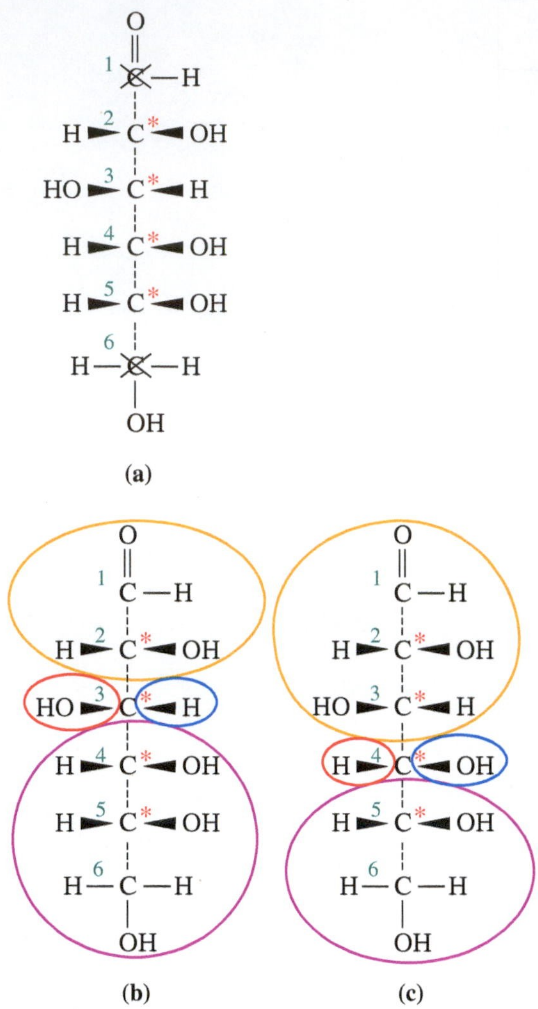

(a)

(b)   **(c)**

FIGURE 6.2 Monosaccharides have chiral centers.
(a) The chiral centers for glucose are designated with an asterisk; notice that carbons 1 and 6 are not chiral centers. The four different groups for carbon 3 (b) and carbon 4 (c) are called out with colored circles. Can you identify the four different groups for carbons 2 and 5?

attached to it is a chiral center. As we also saw in Chapter 4, a compound with a single chiral center can exist in two different forms called *enantiomers*.

How many chiral centers does a glucose molecule contain (see **FIGURE 6.2**)? Let's examine the carbons in glucose starting with carbon 1, which is at the top. Carbon 1, containing an aldehyde group, is not tetrahedral, so it cannot be a chiral center. Carbons 2 to 5 are tetrahedral and have four different atoms or groups of atoms attached—notice the four colored circles in Figure 6.2b and c, indicating the groups and atom attached to carbons 3 and 4, respectively. Recall from Chapter 4 that a carbon atom bonded to four different groups is a chiral center. Carbon 6 is tetrahedral but does not have four *different* groups of atoms attached to it because two of the atoms attached are hydrogen, so it is not chiral. Therefore, glucose has a total of four chiral centers. The groups bonded to each chiral center of a glucose molecule could have two different arrangements (a pair of mirror images), so how many possible arrangements around the chiral centers, or stereoisomers, are possible?

The number of stereoisomers possible increases with the number of chiral centers present in a molecule. For molecules with one chiral center, there are only two different ways the attached atoms or groups can be arranged spatially (and these are mirror images). For molecules with two chiral centers, there are a total of four different ways to arrange the attached atoms or groups differently, and for molecules with three chiral centers, there are eight different ways to attach atoms or groups. The general formula for determining the number of stereoisomers is 2^n, where n is the number of chiral centers present in the molecule. Because glucose has four chiral centers ($2^4 = 2 \times 2 \times 2 \times 2 = 16$), 16 stereoisomers are possible. It is amazing then that only one of these stereoisomers is our preferred energy source.

Representing Stereoisomers— The Fischer Projection

Considering all the stereoisomers possible for a molecule like glucose and the need for designating the position of the attachments on chiral centers, it would be convenient to have a way of representing molecules without having to draw all those wedges and dashes on all the chiral centers. The Fischer projection provides a simpler way of indicating chiral molecules by showing their three-dimensional structure in two dimensions.

Consider the simplest aldose, glyceraldehyde ($C_3H_6O_3$), shown in the following figure. Because it has one chiral center, it can exist as one of two enantiomers. These are designated as a D-enantiomer and an L-enantiomer.

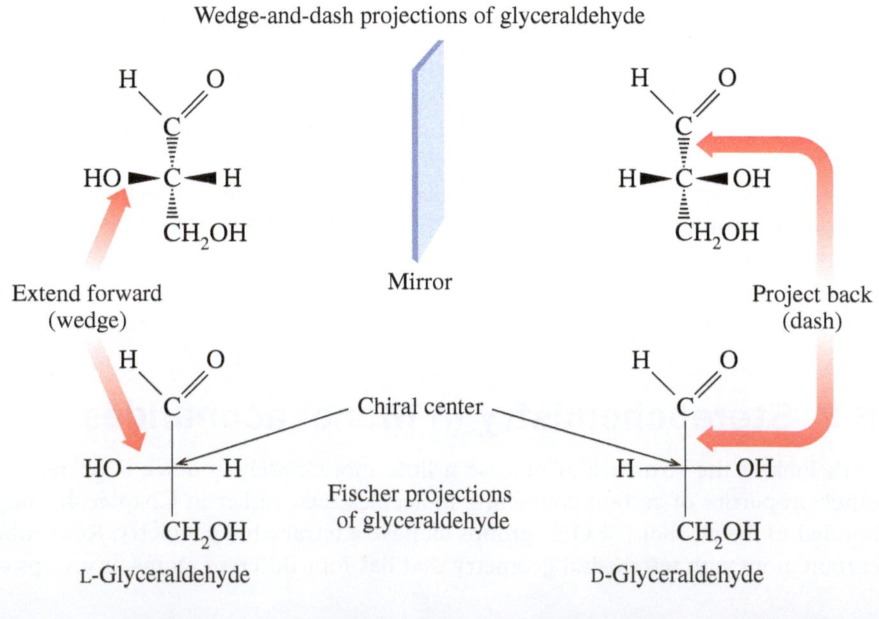

Wedge-and-dash projections of glyceraldehyde

Extend forward
(wedge)

Mirror

Project back
(dash)

Chiral center

Fischer projections
of glyceraldehyde

L-Glyceraldehyde   D-Glyceraldehyde

If we were to represent the enantiomers of glyceraldehyde with wedges and dashes to show their shape, we could represent the molecules as shown in the top part of the figure. In this representation, the only atom that is on the plane of the paper is the carbon at the chiral center, with all other atoms either in front of (wedges) or behind (dashes) the plane of the page.

In the **Fischer projection**, *horizontal lines on a chiral center represent wedges, and vertical lines on a chiral center represent dashes.* A chiral center is not shown as a "C" on a Fischer projection but is implied at the intersection of the lines. This gives the viewer a quick and easy way of identifying the number of chiral centers. The designation of D and L for glyceraldehyde and all other carbohydrates is based on the Fischer projection positioning in glyceraldehyde, used as a reference molecule for this designation.

All **D-sugars** have the —OH on the chiral center farthest from the carbonyl (C=O), on the *right side* of the molecule in the Fischer projection. The enantiomer of this is the **L-sugar**, which has the —OH group on the chiral center farthest from the C=O, on the *left side* of the Fischer projection. Most of the carbohydrates commonly found in nature and the ones we use for energy are D-sugars. D- and L-glyceraldehyde are represented by the Fischer projections shown in the bottom part of the figure on the previous page.

Similarly, **FIGURE 6.3** shows the molecule D-glucose transformed into a Fischer projection.

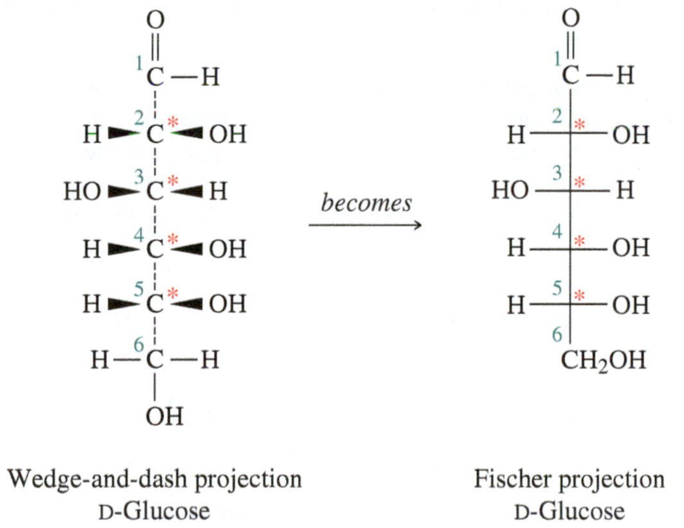

Wedge-and-dash projection
D-Glucose

Fischer projection
D-Glucose

FIGURE 6.3 D-Glucose represented in a Fischer projection. Attachments to chiral centers on horizontal bonds point toward the viewer, and attachments to chiral centers on vertical bonds point away from the viewer. Note: The vertical dashes in the wedge-and-dash projection are drawn as vertical lines for clarity.

When we draw enantiomers in a Fischer projection, they are written as if there is a mirror placed between the two molecules. The reflection of one molecule is identical to the second molecule, its enantiomer. When viewed in a Fischer projection, all chiral centers in the enantiomer have horizontal groups that are switched. Attached atoms or groups on the right in one enantiomer appear on the left side of the other. The enantiomers D- and L-glucose are shown in the following figure:

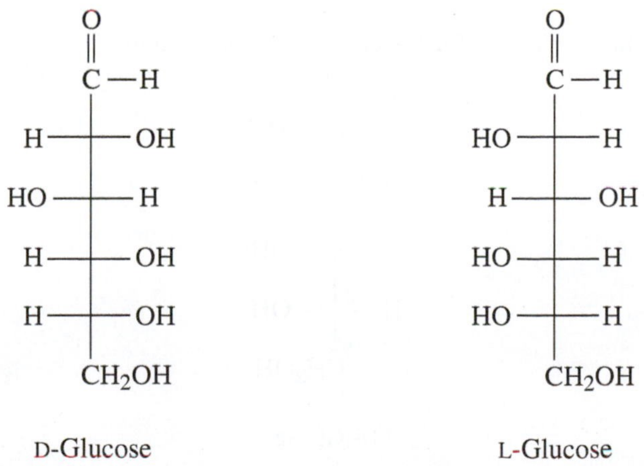

D-Glucose

L-Glucose

Drawing an Enantiomer in a Fischer Projection

Draw the Fischer projection for the enantiomer of D-galactose.

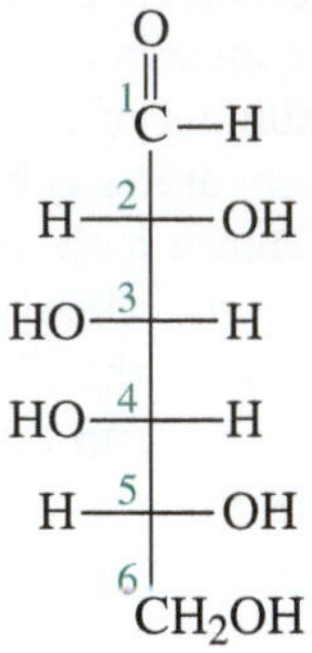

$$
\begin{array}{c}
\overset{O}{\underset{\parallel}{}} \\
\overset{1}{C}\!-\!H \\
H\!-\!\overset{2}{|}\!-\!OH \\
HO\!-\!\overset{3}{|}\!-\!H \\
HO\!-\!\overset{4}{|}\!-\!H \\
H\!-\!\overset{5}{|}\!-\!OH \\
\overset{6}{C}H_2OH
\end{array}
$$

D-Galactose

STEP 1 Locate the chiral centers. In a Fischer projection, the chiral centers are located at the intersections of the vertical and horizontal lines. D-Galactose has four chiral centers at positions 2, 3, 4, and 5.

STEP 2 Switch horizontal groups on the chiral centers. If you imagine that a mirror exists between the pair of molecules so that one enantiomer's reflection will be the other enantiomer, what appears on the right of one enantiomer will appear on the left of the other. The positioning of the atoms attached to carbons that are not chiral centers is not considered when drawing stereoisomers. The L-enantiomer is shown in blue to the right of the original.

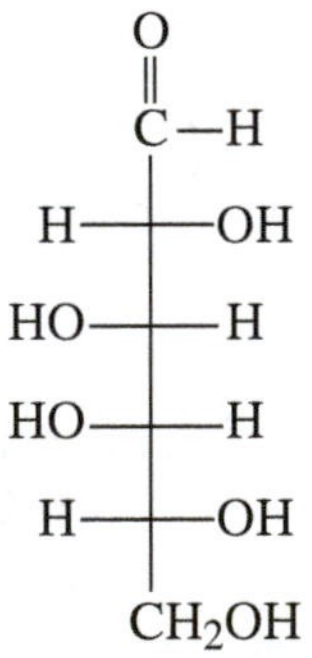

$$
\begin{array}{cc}
\begin{array}{c}
\overset{O}{\underset{\parallel}{}} \\
C\!-\!H \\
H\!-\!OH \\
HO\!-\!H \\
HO\!-\!H \\
H\!-\!OH \\
CH_2OH
\end{array}
&
\begin{array}{c}
\overset{O}{\underset{\parallel}{}} \\
C\!-\!H \\
HO\!-\!H \\
H\!-\!OH \\
H\!-\!OH \\
HO\!-\!H \\
CH_2OH
\end{array}
\\
\text{D-Galactose} & \text{L-Galactose}
\end{array}
$$

Sample Problem 6.3   **Drawing an Enantiomer in a Fischer Projection**

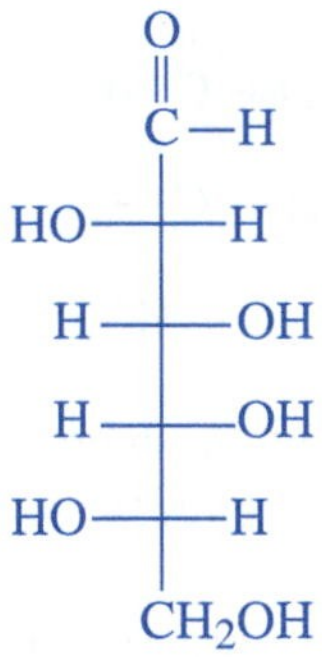

Draw the Fischer projection for the enantiomer of D-ribose.

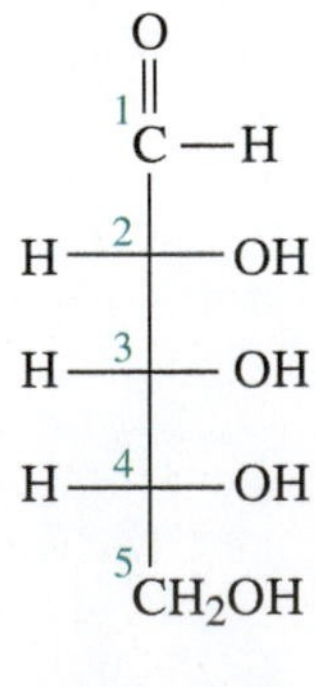

$$
\begin{array}{c}
\overset{O}{\underset{\parallel}{}} \\
\overset{1}{C}\!-\!H \\
H\!-\!\overset{2}{|}\!-\!OH \\
H\!-\!\overset{3}{|}\!-\!OH \\
H\!-\!\overset{4}{|}\!-\!OH \\
\overset{5}{C}H_2OH
\end{array}
$$

D-Ribose

Solution

STEP 1 Locate the chiral centers. D-Ribose has three chiral centers at carbons 2, 3, and 4.

STEP 2 Switch horizontal groups on the chiral centers. The L-enantiomer is shown in blue to the right of the original.

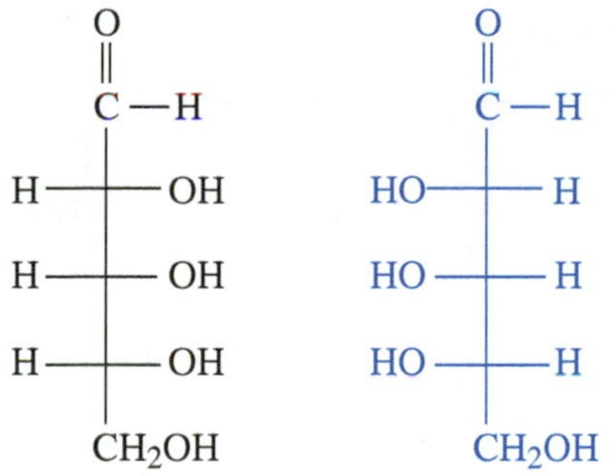

D-Ribose   L-Ribose

Stereoisomers That Are Not Enantiomers

So far, we have seen one stereoisomer of D-glucose: its enantiomer, L-glucose. Because there can be only one mirror image for any stereoisomer, how are the other 14 of 16 possible stereoisomers related to D-glucose? Stereoisomers that are not enantiomers are called **diastereomers**. Diastereomers are stereoisomers that are not exact mirror images. **FIGURE 6.4** shows the monosaccharides D-galactose and D-talose in a Fischer projection. Both of them are diastereomers of D-glucose.

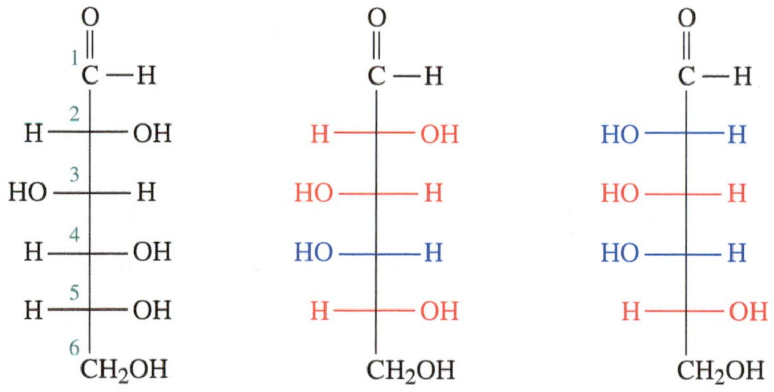

D-Glucose   D-Galactose   D-Talose

FIGURE 6.4 Diastereomers of D-glucose. The monosaccharides D-galactose and D-talose are classified as diastereomers of D-glucose because some chiral centers are oriented the same as in D-glucose (shown in red) and some chiral centers are mirror images (shown in blue). Notice that all three monosaccharides are D-sugars.

Important Monosaccharides

Several of the more common monosaccharides are hexoses (containing six carbons) produced and used in nature only as the D-isomers. The D-form is discussed here.

The most abundant monosaccharide found in nature is glucose. We commonly refer to D-glucose as dextrose on food labels, or as blood sugar in medicine. It is found in fruits, vegetables, and corn syrup. As we noted in Chapter 5, glucose can be broken down inside cells to produce energy through **glycolysis**. People with diabetes have difficulty getting glucose from the bloodstream into their cells so that glycolysis can occur. This is why they must regularly monitor their blood glucose levels. Glucose is also a sugar unit in the disaccharides sucrose (table sugar) and lactose (milk sugar) as well as the polysaccharides amylose, amylopectin, glycogen, and cellulose. The stereoisomers of the D-sugars with three to six carbons are shown in **TABLE 6.1**. Note the highest numbered chiral center is outlined in yellow in each structure.

TABLE 6.1 Fischer Projections and Names of D-Monosaccharides Containing 3 to 6 Carbons

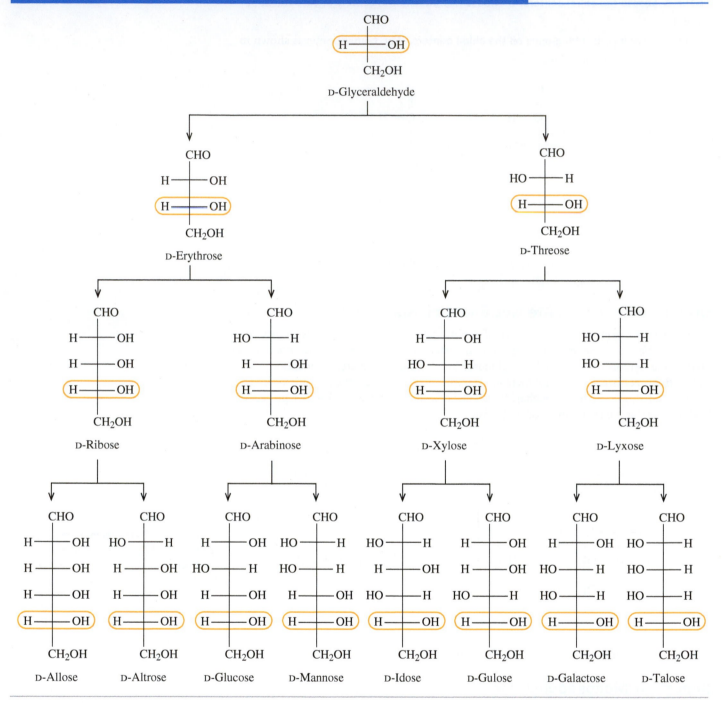

Galactose (Figure 6.4) is found in nature combined with glucose in the disaccharide lactose, which is present in milk and other dairy products. Galactose has one of its chiral centers (carbon 4) opposite that of glucose. Diastereomers that differ in just one chiral center (as compared with more than one chiral center) are called **epimers**. The body can chemically convert galactose into glucose for use in glycolysis through a series of steps that include an enzyme called an *epimerase*.

Mannose is a monosaccharide found in some fruits and vegetables. It is not easily absorbed by the body. The most notable fruit that contains high amounts of mannose is the cranberry. For decades, cranberry juice has been promoted for treatment of urinary tract infections (UTIs) due to excess mannose being flushed through the urinary tract. Recent studies now question this practice.

CH₂OH
|
C=O
|
HO——H
|
H——OH
|
H——OH
|
CH₂OH

D-Fructose

Apples and honey contain high amounts of the monosaccharide D-fructose.

The ketose fructose is also commonly referred to as fruit sugar or levulose and is found in fruits, vegetables, and honey. In combination with glucose, it gives us the disaccharide sucrose (table sugar). Fructose is the sweetest monosaccharide, one-and-a-half times sweeter than table sugar, making it popular with dieters who can get the same sweet taste with fewer calories. Even though it is not an epimer of glucose, fructose can be broken down for energy production in the body by the chemical reactions of glycolysis, as we will see in Chapter 12.

The pentoses (five-carbon sugars) ribose and 2-deoxyribose are a part of larger biomolecules called nucleic acids that make up our genetic material, a topic we will address in Chapter 11. The nucleic acids are distinguished in their name by the monosaccharide they contain. *Ribo*nucleic acid (RNA) contains the sugar ribose, and *deoxyribo*nucleic acid (DNA) contains the sugar deoxyribose. Structurally, the only difference between the two pentoses is the absence of an oxygen atom on carbon 2 of deoxyribose. Ribose is also found in the vitamin *ribo*flavin and other biologically important molecules, as we will see in Chapter 12. Take note that carbon 2 of deoxyribose is shown as a "C" instead of a cross, indicating that it is not a chiral center.

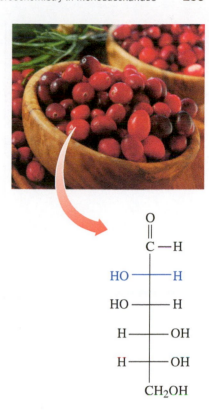

O
||
C—H
|
HO——H
|
HO——H
|
H——OH
|
H——OH
|
CH₂OH

D-Mannose

Cranberries are a rich source of D-mannose, the 2-epimer of D-glucose. See Table 6.1 to compare the structures of these two monosaccharides.

O
||
¹C—H
|
H——²OH
|
H——³OH
|
H——⁴OH
|
⁵CH₂OH

D-Ribose

O
||
¹C—H
|
H——²C—H
|
H——³OH
|
H——⁴OH
|
⁵CH₂OH

D-2-Deoxyribose

D-2-Deoxyribose has the—OH group removed from carbon 2.

Glucose and Diabetes

Glucose levels in our blood and cells are strictly regulated by signaling hormones that control the movement of glucose inside and outside of the cell. After a meal is eaten and blood glucose levels rise, the hormone insulin signals cells to take up glucose, where it is either broken down through oxidation or stored. During periods of fasting (when we are asleep, for example), the hormone glucagon is released, signaling the liver to make glucose and send it out of the cell to the bloodstream when glucose levels are low.

Most people know someone with diabetes. According to the Centers for Disease Control and Prevention, almost 10% of the U.S. population has the disease. Diabetes

—*Continued next page*

INTEGRATING Chemistry

◀ Find out what happens to glucose in diabetics.

Continued—

is characterized by prolonged periods of high blood sugar (glucose). In diabetics, the insulin signal to cells is compromised and the cells do not take up glucose. In type 1 diabetes, also called juvenile onset or insulin-dependent diabetes, insulin production is low. In type 2 diabetes, also called insulin-resistant diabetes, the cells do not receive the signal from the insulin, and transport of glucose into the cells does not occur. In either type, the results are the same. Blood glucose levels are elevated. Pets can also be diabetic for the same reasons. In the short term, a person or animal with diabetes will feel weak, tired, and thirsty and will urinate more than usual. Over the long term, however, high blood sugar levels can lead to blindness, kidney disease, heart disease, or nerve damage due to weak circulation that can further lead to amputations.

People with type 1 diabetes must have regular insulin delivered by injection or pump device. Up to 95% of the adult cases of diabetes are type 2. Often, people with type 2 diabetes can control blood-glucose levels through healthy eating, regular physical activity, weight loss, and medications that lower blood-sugar levels.

Individuals with diabetes must regularly test their blood glucose levels using a glucose monitor.

Sample Problem 6.4   Distinguishing Stereoisomers

Classify structures A, B, and C in the figure as being either an enantiomer or a diastereomer of D-mannose.

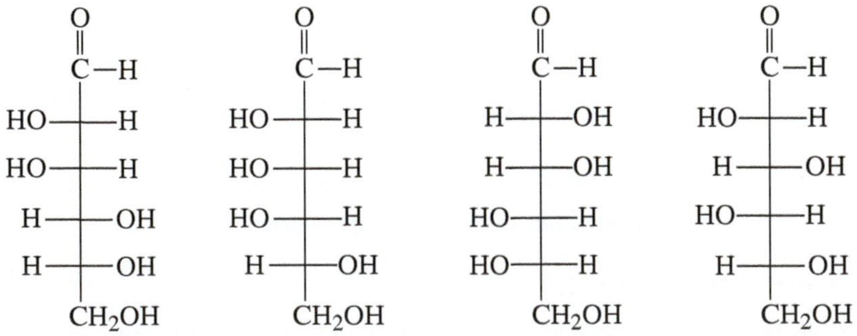

D-Mannose	A	B	C
$C{=}O$, C—H	$C{=}O$, C—H	$C{=}O$, C—H	$C{=}O$, C—H
HO—H	HO—H	H—OH	HO—H
HO—H	HO—H	H—OH	H—OH
H—OH	HO—H	HO—H	HO—H
H—OH	H—OH	HO—H	H—OH
CH_2OH	CH_2OH	CH_2OH	CH_2OH

Solution

Determine the differences between D-mannose and structures A, B, and C. Structure A has one chiral center oriented differently from D-mannose, so it is a diastereomer (epimer). Structure B is the mirror image enantiomer of D-mannose, L-mannose. Structure C has two chiral centers oriented differently from D-mannose, so it is a diastereomer.

Practice Problems

6.11 Identify the following monosaccharides as the D- or the L-isomer:

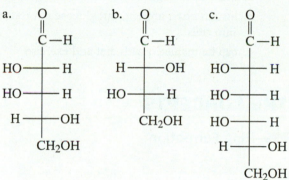

6.12 Identify the following monosaccharides as the D- or the L-isomer:

6.13 Draw the Fischer projection for the enantiomer (mirror image) of each of the following:

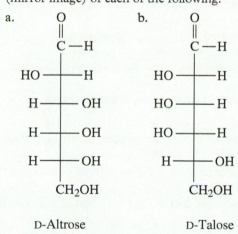

D-Altrose   D-Talose

6.14 Draw the Fischer projection for the enantiomer (mirror image) of each of the following:

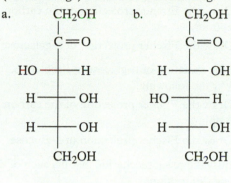

D-Fructose   D-Sorbose

6.15 Classify structures A, B, and C in the figure as being either an enantiomer or a diastereomer of D-galactose.

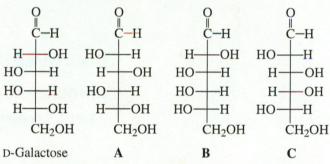

D-Galactose   A   B   C

6.16 Classify structures A, B, and C in the figure as being either an enantiomer or a diastereomer of D-glucose.

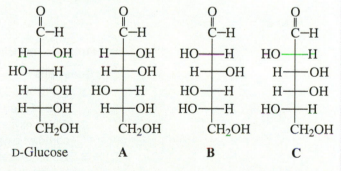

D-Glucose   A   B   C

—*Continued next page*

Continued—

6.17 Use the structure of D-galactose in Problem 6.15 to answer the following:

   a. Draw the Fischer projection of the carbon 3 epimer.

   b. Draw the Fischer projection of L-galactose.

6.18 Use the structure of D-glucose in Problem 6.16 to answer the following:

   a. Draw the Fischer projection of the carbon 3 epimer.

   b. Draw the Fischer projection of L-glucose.

6.19 Identify the monosaccharide that fits each of the following descriptions:

   a. also referred to as dextrose

   b. also called fruit sugar

   c. used to treat urinary tract infections

6.20 Identify the monosaccharide that fits each of the following descriptions:

   a. in combination with glucose produces the disaccharide lactose

   b. also called blood sugar

   c. also called levulose

6.21 Indicate whether the statements below apply to the glucose-regulating hormone insulin or to glucagon:

   a. signals cells to take up glucose

   b. signals liver cells to produce glucose

   c. not produced in people with type 1 diabetes

6.22 Indicate whether the following statements apply to type 1 or type 2 diabetes:

   a. most cases begin in youth

   b. insulin often present, but glucose not transported into cells

   c. can be managed with diet and exercise

DISCOVERING THE CONCEPTS

❓ INQUIRY ACTIVITY—Ring Formation

Part 1. Information

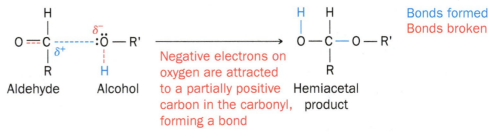

FIGURE 1 The formation of a hemiacetal from an aldehyde and an alcohol.

Questions

1. In **FIGURE 1**, name the functional groups of the compounds on the reactant side of the chemical equation.
2. Which oxygen in the hemiacetal product in Figure 1 (right or left) was the carbonyl oxygen from the aldehyde?
3. Which oxygen in the hemiacetal product in Figure 1 (right or left) was from the alcohol, R'OH?

Part 2. Information

The carbonyl carbon (C1) in a monosaccharide is referred to as the *anomeric* carbon.

> ### *Conventions for Drawing the Ring Form of a Six-Carbon D-Aldose*
> * Carbon 6 will always be on the top side of the ring.
> * Hydroxyls that do not participate in ring formation and are on the right of a Fischer projection will be on the bottom side of the ring.
> * Hydroxyls that do not participate in ring formation and are on the left of a Fischer projection will be on the top side of the ring.

Questions

4. Monosaccharides chemically react with themselves (intramolecularly) due to polar opposites strongly attracting each other within the molecule.
   a. Considering the structure of D-glucose, if the O on carbon 5 of D-glucose (O5) is attracted to the carbonyl carbon 1 (C1), reacting to form a bond between them (making a hemiacetal functional group), how many atoms would be enclosed in a ring?
   b. How many of those atoms are carbon?
   c. How many atoms are oxygen? Recall from Section 4.5 that a carbon ring can be represented with a top and bottom face.
   d. Considering that rings have two sides, a top and a bottom, draw the ring form of D-glucose.
5. Where did you place the OH for C1 (top or bottom)?
6. If you placed it on the bottom, you drew α-D-glucose. If you placed the OH on C1 on the top, you drew β-D-glucose. Which one did you draw?
7. Can you devise a rule for identifying these two ring forms (α and β anomers) of D-glucose relative to the position of C6?

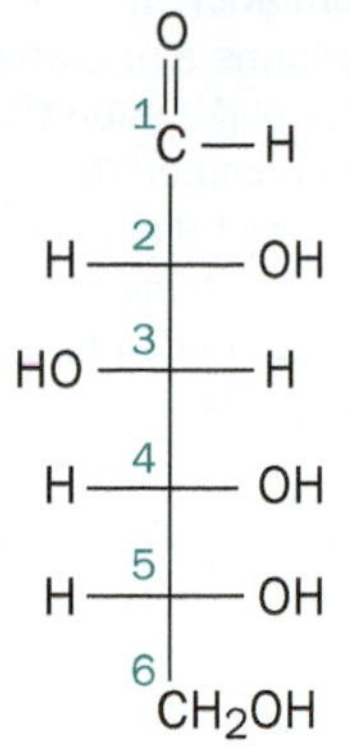

D-Glucose

6.4 Reactions of Monosaccharides

Now that we have established the structural characteristics of monosaccharides, we look at some of their common reactions: ring formation and oxidation–reduction. Both of these reactions are in part due to the highly polar nature of the carbonyl found in the aldehyde and ketone functional groups.

Ring Formation—The Truth About Monosaccharide Structure

The linear structures that helped us understand functional groups and chirality of monosaccharides do not show how most monosaccharides actually are structured. Pentoses and hexoses (five- and six-carbon monosaccharides) bend back on themselves to form rings. When opposite charges on different functional groups within the monosaccharide attract strongly enough, new bonds are formed and other bonds are broken.

This reaction is illustrated for the functional groups in an aldose in **FIGURE 6.5**. The carbonyl group present in aldehydes and ketones is very polar and highly reactive. The partially positive (δ^+) carbon in the carbonyl attracts the lone pairs of electrons on the oxygen of a hydroxyl that is partially negative (δ^-). If a bond forms between this carbon and oxygen, the new functional group is called a **hemiacetal** (pronounced hem-ee-ass′-i-tal).

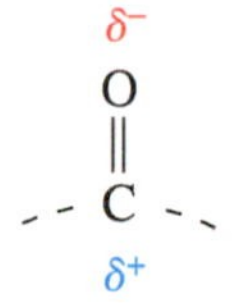

? **6.4 Inquiry Question**

How are the products of monosaccharide reactions drawn?

The carbonyl group in aldoses and ketoses is highly polar, making the group reactive.

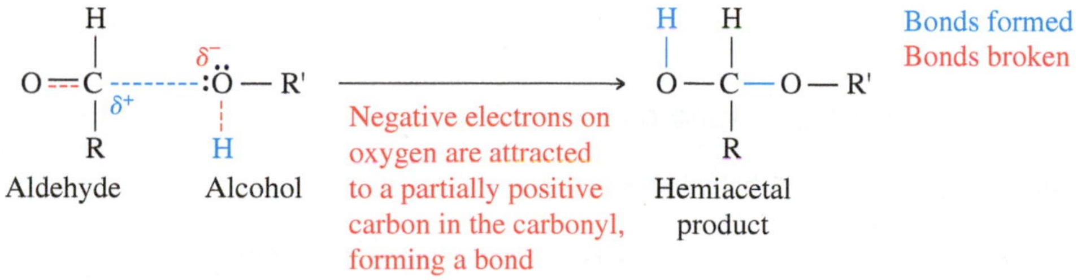

FIGURE 6.5 Formation of the hemiacetal group. The electrons on the oxygen of the alcohol group (δ^-) are attracted to the partial positive charge (δ^+) of the carbon in the carbonyl (blue dashed line). To maintain an octet, the alcohol's hydrogen moves to the oxygen of the carbonyl. The joining of these two functional groups results in a new functional group called a hemiacetal.

Negative electrons on oxygen are attracted to a partially positive carbon in the carbonyl, forming a bond

Bonds formed
Bonds broken

Aldehyde   Alcohol   Hemiacetal product

Monosaccharides contain both a carbonyl and several hydroxyl (alcohol) functional groups, and these two functional groups can react within the same molecule or intramolecularly. Using D-glucose as an example, we find that the hydroxyl on carbon 5 (C5) can bend around and react, forming a new bond with the carbonyl at carbon 1 (C1) and

FIGURE 6.6 Ring formation in
D-glucose. D-Glucose forms a ring when
the oxygen of the alcohol group on C5
reacts with the carbonyl carbon (C1),
forming a new bond. α and β anomers
are formed. C6 remains outside the
ring. Dashed lines show bonds breaking
and forming to create a ring

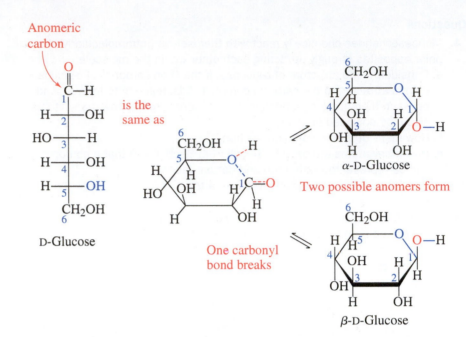

D-Glucose

One carbonyl
bond breaks

α-D-Glucose

Two possible anomers form

β-D-Glucose

Mastering Videos
Practicing the Concepts
Monosaccharide Ring Formation

producing two possible ring structures as shown in **FIGURE 6.6**. When this occurs, five
carbons and an oxygen form a ring, and one of the carbons (carbon 6) remains outside
the ring. Monosaccharides exist in a ring form most of the time because the carbonyl
group reacts readily with a hydroxyl to form this hemiacetal ring.

Two structures are possible during ring formation. Recall that a carbonyl group
is trigonal planar (flat), so the oxygen in the hydroxyl group on carbon 5 can form its
bond on either the top or the bottom side of the carbonyl. Two possible ring forms can
be produced from a single linear monosaccharide chain. The two forms are intercon-
vertible and are termed **anomers**. In any monosaccharide, the carbonyl carbon that
reacts to form the hemiacetal in the reaction is referred to as the **anomeric carbon**.
Note that in the ring form, the anomeric carbon is the only carbon bonded directly to
two oxygen atoms.

The two hemiacetal anomers of D-glucose are referred to as the alpha (α) and the
beta (β) anomers. Anomers are distinguished by the positioning of the —OH group
on the anomeric carbon relative to the position of the carbon outside the ring (for
D-glucose carbon 6). In the six-membered ring form of D-isomers called a **pyranose**,
carbon 6 (C6) is always drawn on the top side of the ring. The following describes how to
determine whether a monosaccharide is the α or the β anomer:

- In the α anomer, the —OH on the anomeric carbon is trans to carbon 6 (the carbon outside the ring). They are on opposite sides of the ring.
- In the β anomer, the —OH on the anomeric carbon is cis to carbon 6 (the carbon
  outside the ring). They are on the same side of the ring.

Sample Problem 6.5   Alpha and Beta Forms

Identify the following carbohydrates as the α or β anomer:

a.

b.

c.

Solution

a. The —OH on the anomeric carbon (carbon 1) is on the opposite side of the ring (trans) from carbon 6, so this is the α anomer.
b. The —OH on the anomeric carbon (carbon 1) is on the same side of the ring (cis) as carbon 6, so this is the β anomer.
c. The —OH on the anomeric carbon (carbon 2) is on the same side of the ring (cis) as carbon 6, so this is the β anomer.

Let's practice drawing monosaccharides in the pyranose ring form (six-membered ring) from the linear Fischer projection.

Drawing Pyranose Rings from Linear Monosaccharides

Solving a Problem

Draw the β anomer of D-galactose.

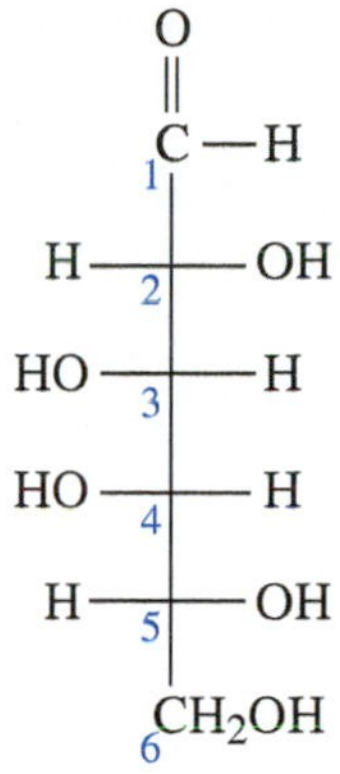

O
C—H
1
H—2—OH
HO—3—H
HO—4—H
H—5—OH
CH_2OH
6

D-Galactose

STEP 1 **Draw the ring.**
Create a scaffolding of a ring with C6 on the top side.

6
CH_2OH
O
5
4
1
3
2

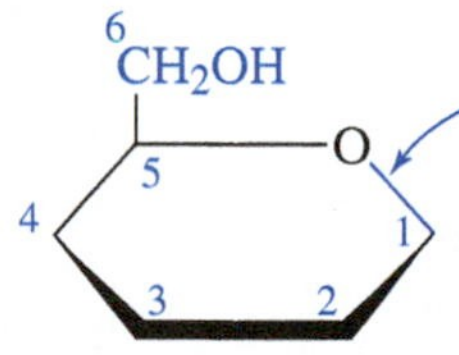

This is the new bond formed between O of C5 and C1.

STEP 2 **Assign α or β.**
Put the —OH of C1 either α or β as directed in the problem.

6
CH_2OH
O
OH
5
4
1
3
2
H

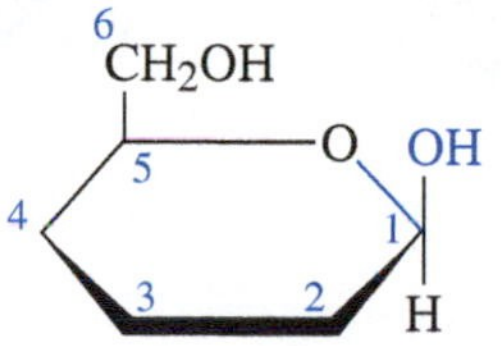

This problem requests the β anomer (—OH cis to C6).

STEP 3 **Fill in the remaining atoms.**
For C2, C3, and C4, atoms or groups on the right side of the Fischer projection are placed on the bottom of the ring, and those on the left of the Fischer projection are placed on the top of the ring.

6
CH_2OH
OH
O
OH
5
4
H
OH
H
1
3
2
H
H
H
OH

β-D-Galactose

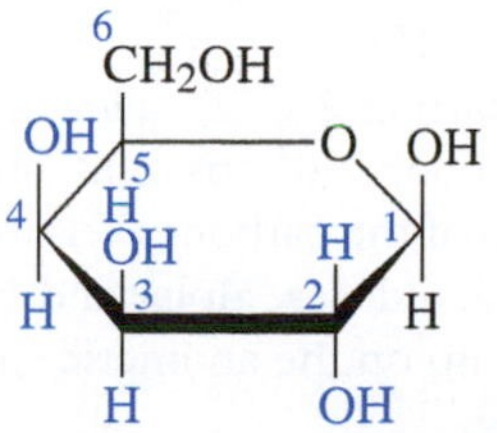

Sample Problem 6.6   **Drawing Pyranose Rings from Linear Monosaccharides**

Draw the α anomer of D-mannose in pyranose ring form.

Solution

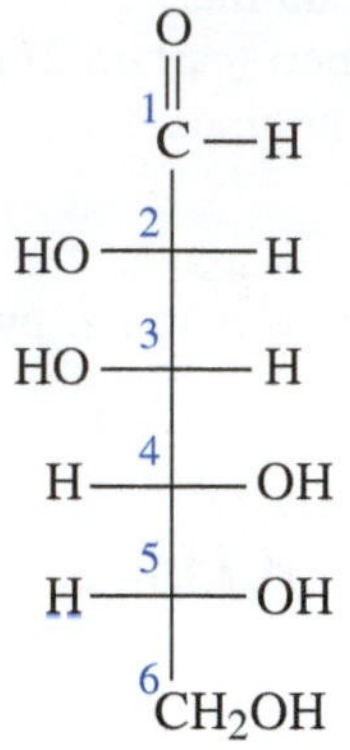

D-Mannose

STEP 1 **Draw the ring.**
Create a scaffolding of a
ring with C6 on the top side.

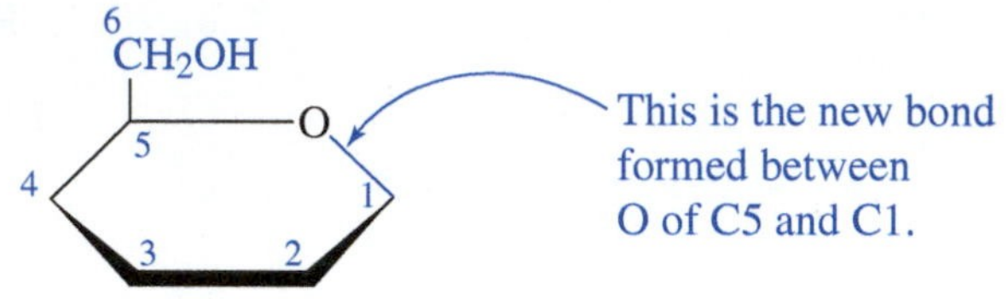

This is the new bond
formed between
O of C5 and C1.

STEP 2 **Assign α or β.**
Put the —OH of C1 either
α or β as directed in the problem.

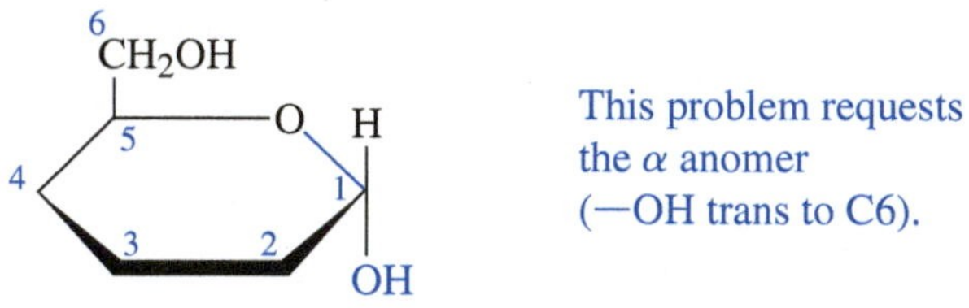

This problem requests
the α anomer
(—OH trans to C6).

STEP 3 **Fill in the remaining atoms.**
For C2, C3, and C4, atoms or
groups on the right side of the
Fischer projection are placed
on the bottom of the ring, and
those on the left of the Fischer
projection are placed on the
top of the ring.

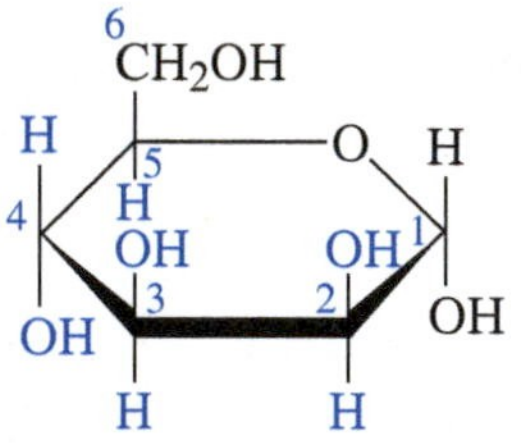

α-D-Mannose

D-Fructose contains a ketone group, instead of an aldehyde, and several hydroxyl groups. The —OH on carbon 5 (C5) can curl around and react with the carbonyl positioned at carbon 2 (C2), allowing two possible ring structures, as in D-glucose (see **FIGURE 6.7**). Four carbons and an oxygen form the five-membered ring called a **furanose**. Two of the carbons (C1 and C6) remain outside the ring.

As with the aldoses, alpha and beta anomers are determined by the positioning of the —OH group on the anomeric carbon (in D-fructose, this is C2) relative to carbon 6 outside the ring.

- In the α anomer, the —OH on the anomeric carbon is trans C6 (on opposite sides).
- In the β anomer, the —OH on the anomeric carbon is cis to C6 (on the same side).

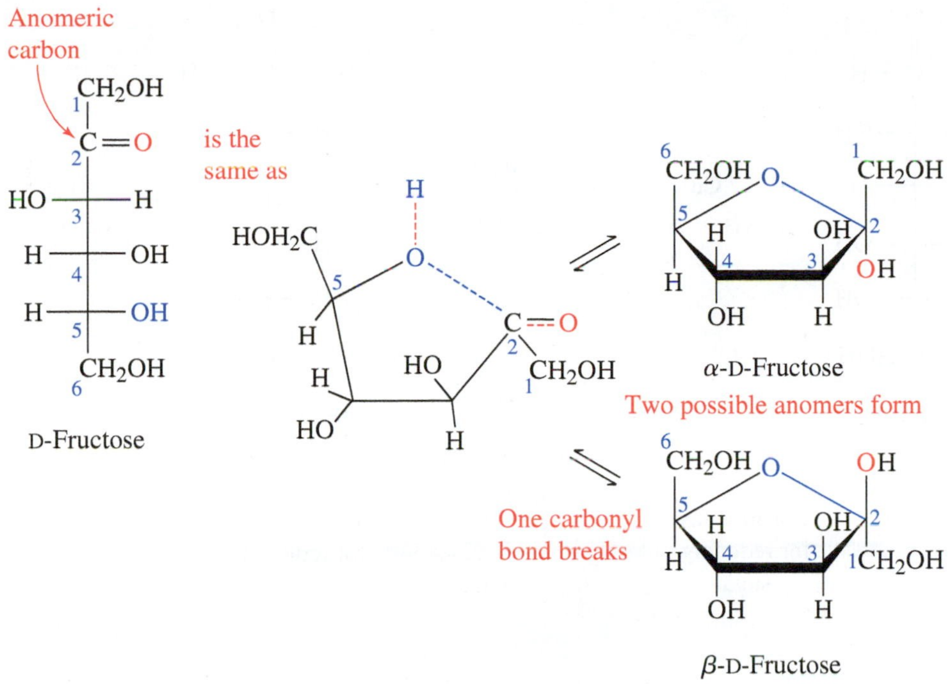

FIGURE 6.7 Ring formation in D-fructose. D-Fructose forms a ring when the alcohol group on C5 reacts with the carbonyl C2. α and β anomers are formed. C6 and C1 remain outside the ring. Dashed lines show bonds breaking and forming to create the ring.

Anomeric carbon

D-Fructose

is the same as

One carbonyl bond breaks

Two possible anomers form

α-D-Fructose

β-D-Fructose

Oxidation–Reduction and Reducing Sugars

Besides forming rings, the carbonyl group (the anomeric carbon) in aldoses can also undergo organic oxidation and reduction, as discussed in Chapter 5. This occurs by adding oxygen (or losing hydrogen) during oxidation and adding hydrogen (or losing oxygen) during reduction. Aldoses can undergo both oxidation *and* reduction. The aldehyde functional group can oxidize to a carboxylic acid or can reduce to an alcohol. In monosaccharides, an oxidation produces a sugar acid and a reduction produces a sugar alcohol.

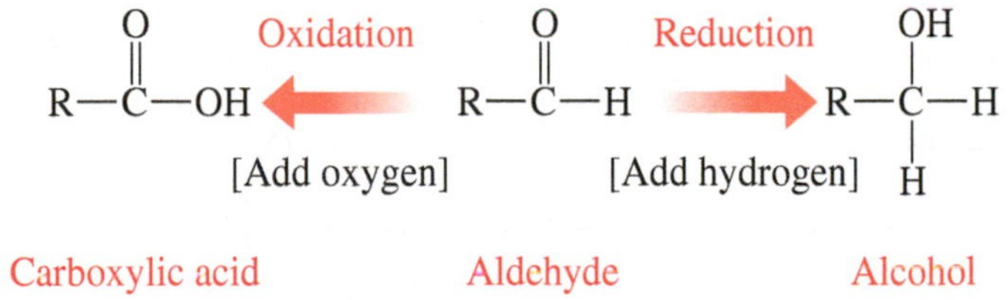

Carboxylic acid   Aldehyde   Alcohol

When molecules are oxidized, they act as a reducing agent and reduce a second reactant. One useful oxidation reaction for sugars occurs in **Benedict's test**, which tests for the presence of an aldose in a solution. Using Benedict's test, an aldehyde group undergoes oxidation while reducing copper ions from Cu^{2+} to Cu^{+} (copper gained one electron). The Cu^{2+} ions are soluble, coloring the initial reaction solution blue. The aldehyde group in turn is oxidized by Cu^{2+}, forming a sugar acid, while Cu^{2+} undergoes reduction by the aldehyde, forming Cu^{+} ions. The resulting copper(I) oxide (Cu_2O) is not soluble and forms a brick-red precipitate in solution (see **FIGURE 6.8**).

Because aldoses are easily oxidized and can therefore serve as reducing agents, sugars capable of reducing substances such as Cu^{2+} are referred to as **reducing sugars**. Aldoses are easily oxidized because they contain an aldehyde functional group. The anomeric ring forms of the aldoses are in equilibrium with a small amount of the open-chain form, which has an aldehyde. Fructose and other ketoses are also reducing sugars, even though they do not contain an aldehyde group, because in the presence of oxidizing agents, they can rearrange to aldoses.

People with diabetes must monitor their glucose levels regularly. Excess glucose in urine suggests high levels of glucose in the bloodstream. Benedict's test can be used to

FIGURE 6.8 The oxidation of D-glucose. In the presence of Benedict's reagent (containing Cu^{2+} ions), D-glucose reacts to produce the sugar acid D-gluconic acid and a brick-red precipitate, Cu_2O.

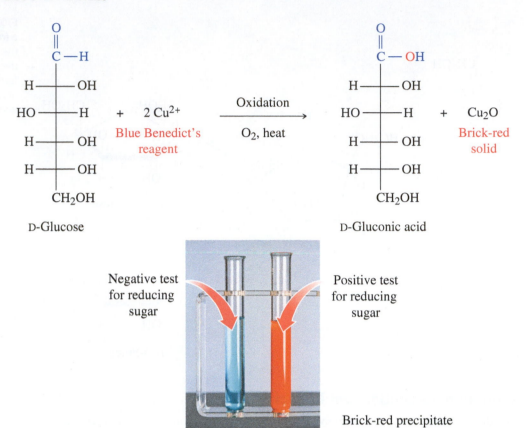

D-Glucose   D-Gluconic acid

Negative test for reducing sugar

Positive test for reducing sugar

Brick-red precipitate

Cu^{2+} $Cu_2O(s)$

Even though fructose is a ketose, it can rearrange upon heating to a mixture of glucose and mannose; therefore, fructose is a reducing sugar.

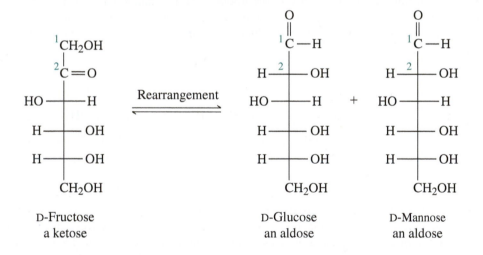

D-Fructose   D-Glucose   D-Mannose
a ketose   an aldose   an aldose

The color change in a glucose test strip measures the amount of glucose in the urine through an oxidation–reduction reaction.

monitor glucose levels in urine. Glucose test strips, like the one shown at left, produce a range of color changes if glucose is present.

An aldose or ketose can also be reduced to an alcohol when the carbonyl reacts with hydrogen in a hydrogenation reaction (see **FIGURE 6.9**). Sugar alcohols still have a sweet taste, but they do not have the calories found in monosaccharides. When eaten, a sugar alcohol provides a cooling sensation and clean finish in the mouth. Sugar alcohols like sorbitol (six carbons), xylitol (five carbons), and erythritol (four carbons) are often added to low-calorie drinks and candies.

Sugar alcohols can also be produced in the body when glucose levels remain high in the bloodstream. An enzyme called aldose reductase acts to reduce excessive glucose to the sugar alcohol sorbitol, which at high concentration can contribute to cataracts (clouding of the lens of the eye). These so-called sugar cataracts are commonly seen in those with diabetes.

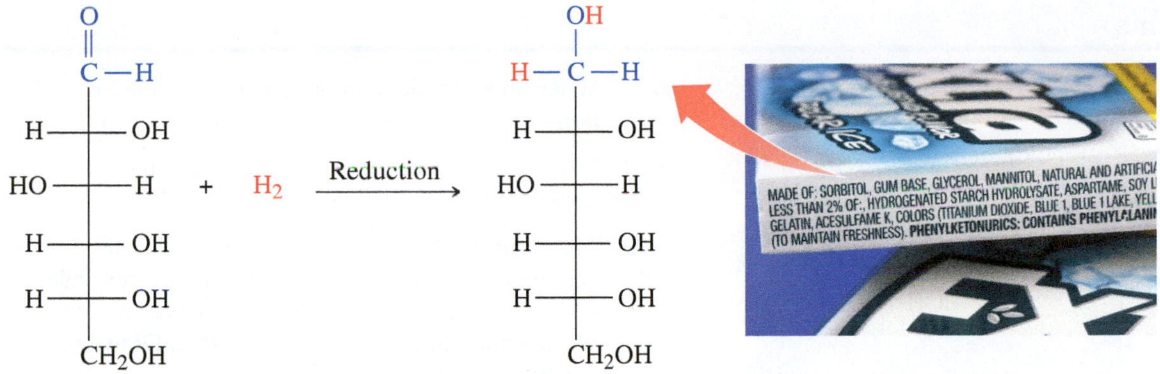

D-Glucose

D-Glucitol, also called D-sorbitol

FIGURE 6.9 The reduction of D-glucose. In the presence of a hydrogen source, D-glucose reacts to produce the sugar alcohol D-glucitol, which is better known as D-sorbitol.

Sample Problem 6.7 Redox Reactions in Monosaccharides

Draw the product when D-galactose undergoes oxidation at carbon 1.

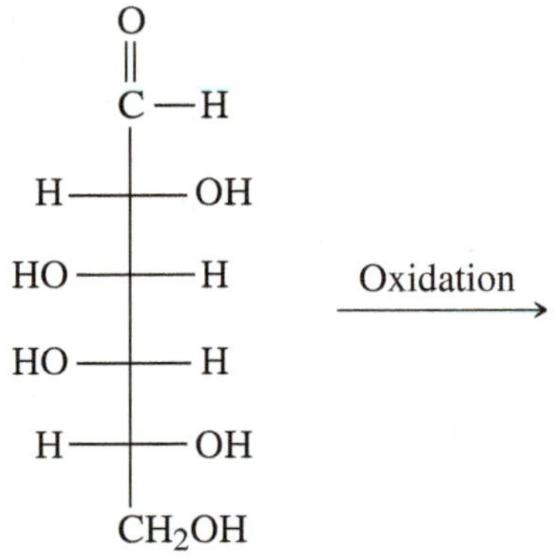

D-Galactose

Solution

The monosaccharides can be oxidized or reduced at the carbonyl group. An aldehyde functional group is oxidized to a carboxylic acid. When undergoing oxidation, D-galactose is oxidized to the following:

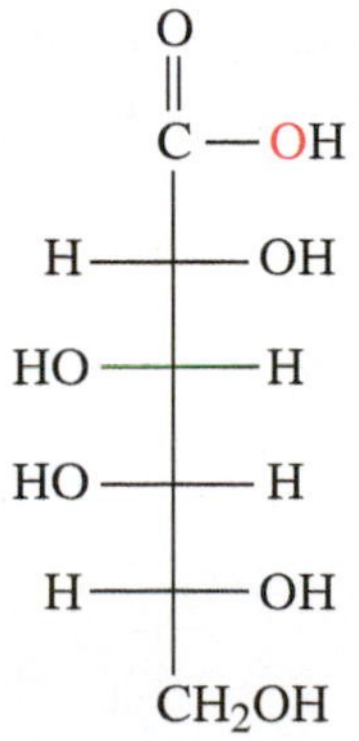

Practice Problems

6.23 Identify the following carbohydrates as the α or β anomer:

a.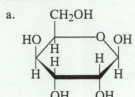
b.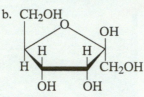

6.24 Identify the following carbohydrates as the α or β anomer:

a.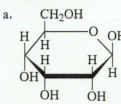
b.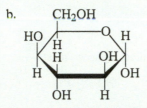

6.25 Draw the α and β anomer of D-talose in pyranose ring form:

$$
\begin{array}{c}
\text{O} \\
\parallel \\
\text{C-H} \\
\text{HO---H} \\
\text{HO---H} \\
\text{HO---H} \\
\text{H---OH} \\
\text{CH}_2\text{OH}
\end{array}
$$

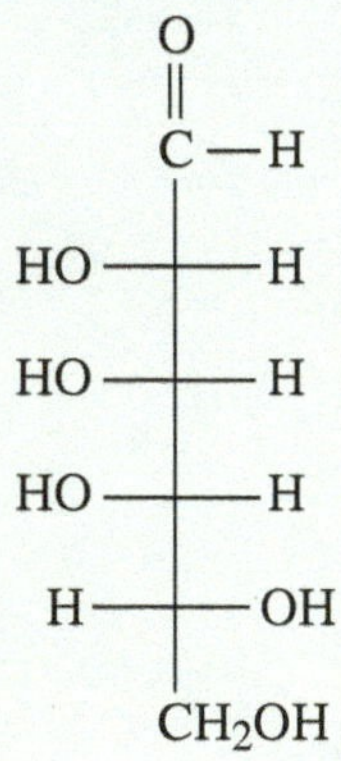

D-Talose

6.26 Draw the α and β anomer of D-altrose in pyranose ring form:

$$
\begin{array}{c}
\text{O} \\
\parallel \\
\text{C-H} \\
\text{HO---H} \\
\text{H---OH} \\
\text{H---OH} \\
\text{H---OH} \\
\text{CH}_2\text{OH}
\end{array}
$$

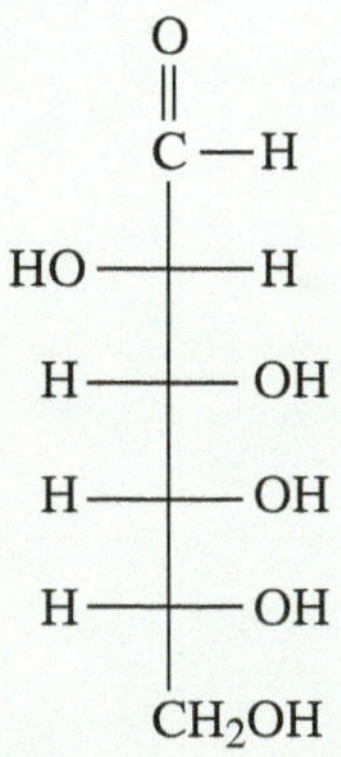

D-Altrose

6.27 When an aldehyde undergoes oxidation, the functional group _____ results.

6.28 When an aldehyde undergoes reduction, the functional group _____ results.

6.29 The sugar alcohol ribitol is a component of the vitamin riboflavin and the energy transfer molecule FAD. Ribitol is formed when the monosaccharide ribose undergoes reduction at carbon 1. Draw the structure of ribitol.

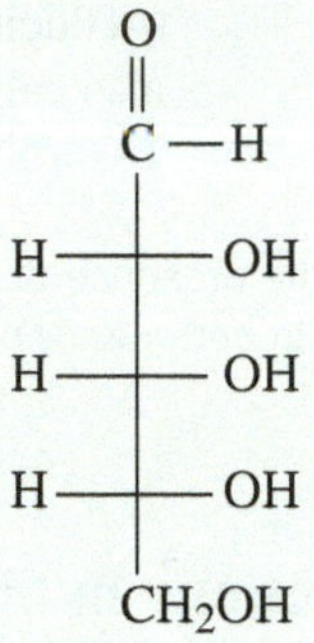

$$
\begin{array}{c}
\text{O} \\
\parallel \\
\text{C-H} \\
\text{H---OH} \\
\text{H---OH} \\
\text{H---OH} \\
\text{CH}_2\text{OH}
\end{array}
$$

D-Ribose

6.30 The sugar alcohol erythritol is often included in low-calorie sweeteners. It is 70% as sweet as table sugar. Erythritol is the reduced form of the aldotetrose erythrose. Draw erythritol.

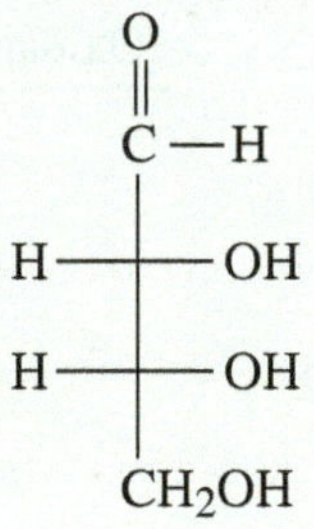

$$
\begin{array}{c}
\text{O} \\
\parallel \\
\text{C-H} \\
\text{H---OH} \\
\text{H---OH} \\
\text{CH}_2\text{OH}
\end{array}
$$

D-Erythrose

6.31 Pentoses also exist in a ring form, but they most commonly occur as furanose rings. D-Ribose exists in its furanose ring form in the nucleic acid RNA. Using the structure of D-ribose from Table 6.1, draw the furanose form of β-D-ribose.

6.5 Disaccharides

So far, we have seen that the carbonyl at the anomeric carbon reacts to form a ring structure in monosaccharides and reacts in oxidation and reduction reactions. Even when enclosed in a ring, the anomeric carbon is still reactive. Disaccharides are formed when two monosaccharides are joined together through a condensation reaction occurring at an anomeric carbon.

6.5 Inquiry Question

How are disaccharides formed and identified?

Condensation and Hydrolysis—Forming and Breaking Glycosidic Bonds

In a monosaccharide in ring form, the anomeric carbon has the most reactive —OH in the molecule (C1 in an aldose). When the hydroxyl on the anomeric carbon reacts with a hydroxyl on another monosaccharide in a condensation reaction, a **glycoside** is formed. The bond that connects the two is called a **glycosidic bond**. Glycosidic bonds join monosaccharides to each other and, in a more general sense, connect monosaccharides to any alcohol. In **FIGURE 6.10**, two glucose molecules are joined through a condensation reaction to form the disaccharide maltose.

Glycosidic bond formed
(α position)

CH$_2$OH   CH$_2$OH   CH$_2$OH   CH$_2$OH

Monosaccharide
α-D-Glucose

Monosaccharide
α-D-Glucose

A disaccharide containing
two D-glucose units
Maltose

+ H$_2$O

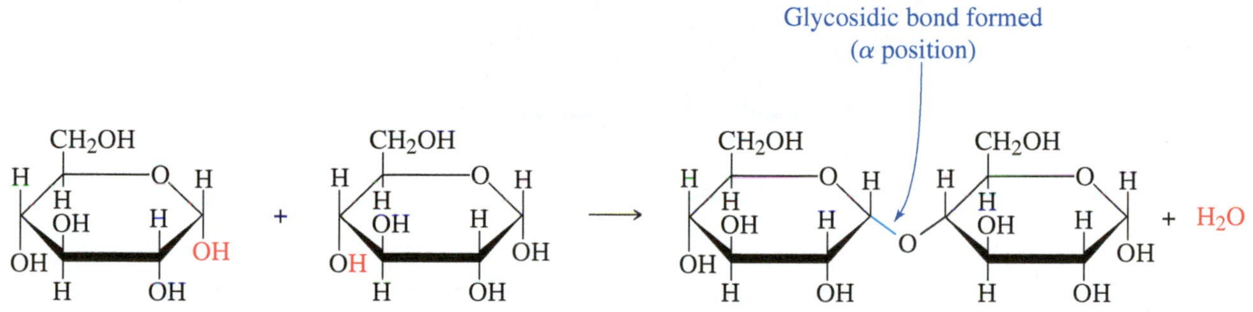

FIGURE 6.10 Formation of the disaccharide maltose. Two molecules of glucose are joined, forming a glycoside. The loss of H and OH from the glucose molecules produces a molecule of water in this condensation reaction.

Sample Problem 6.8   Locating a Glycosidic Bond

Find the glycosidic bond in the following carbohydrates:

a.   CH$_2$OH

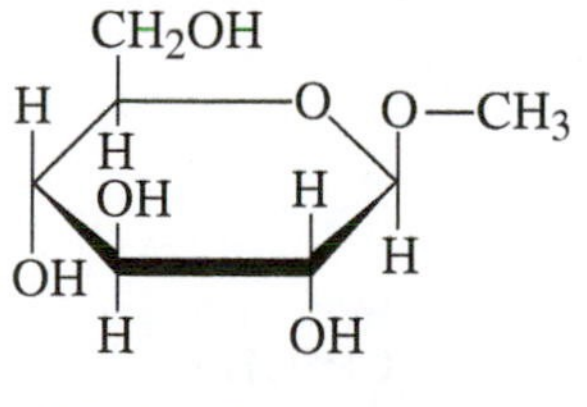

Methyl β-D-glucoside

b.   CH$_2$OH

CH$_2$OH

β-D-Lactose

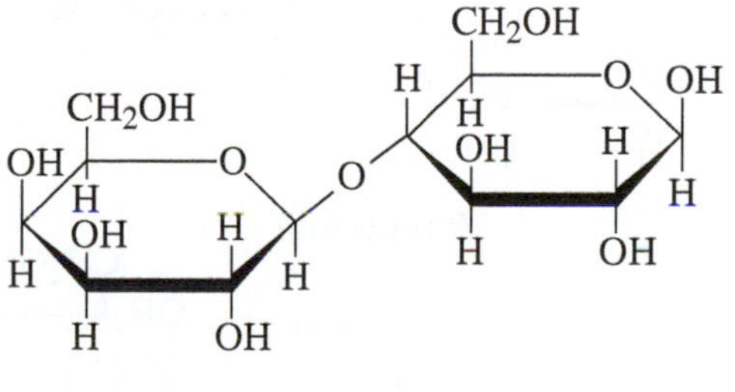

c.   CH$_2$OH   CH$_2$OH

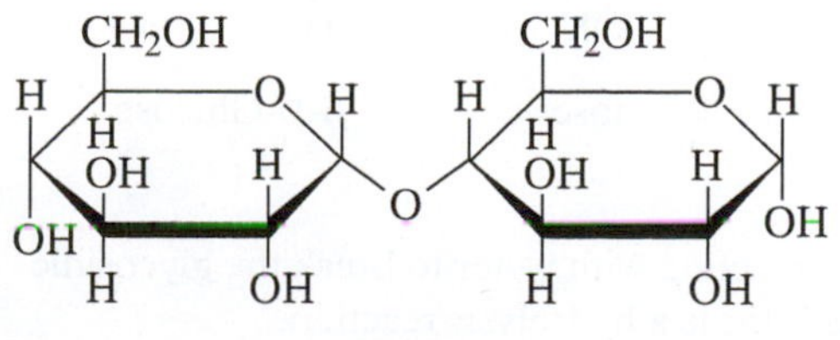

α-D-Maltose

—*Continued next page*

Continued—

Solution

The glycosidic bond joins a monosaccharide through a condensation reaction to another alcohol group. One of the carbons bonded to the oxygen must be the anomeric carbon, and the glycosidic bond forms there. The glycosidic bonds are blue in the following figures.

a.

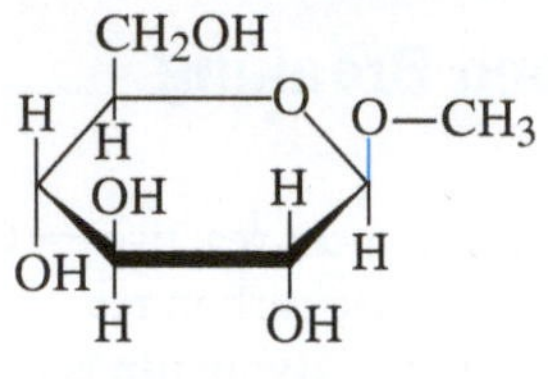

Methyl β-D-glucoside

b.

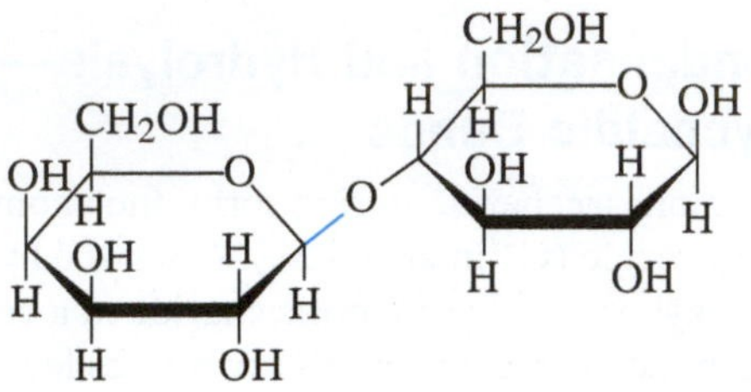

β-D-Lactose

c.

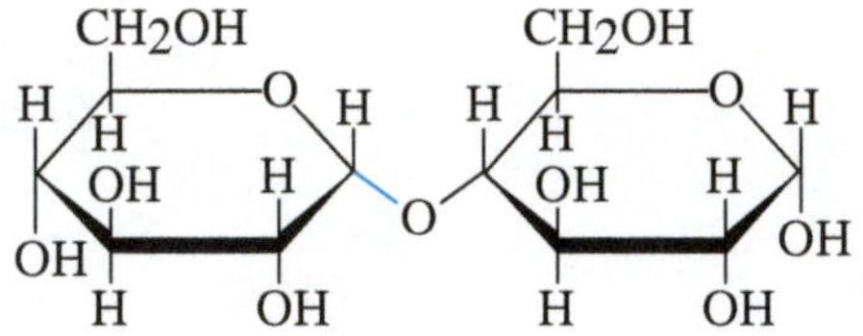

α-D-Maltose

The formation of a glycoside is an example of a condensation reaction. Recall from Section 5.5 that in a condensation reaction, a molecule of water is eliminated as two molecules are joined. During digestion, we break up disaccharides at the glycosidic bond through the opposite reaction, hydrolysis, where water is consumed as a reactant.

Sample Problem 6.9   Identifying Condensation and Hydrolysis

Identify the formation of a disaccharide as a condensation or a hydrolysis.

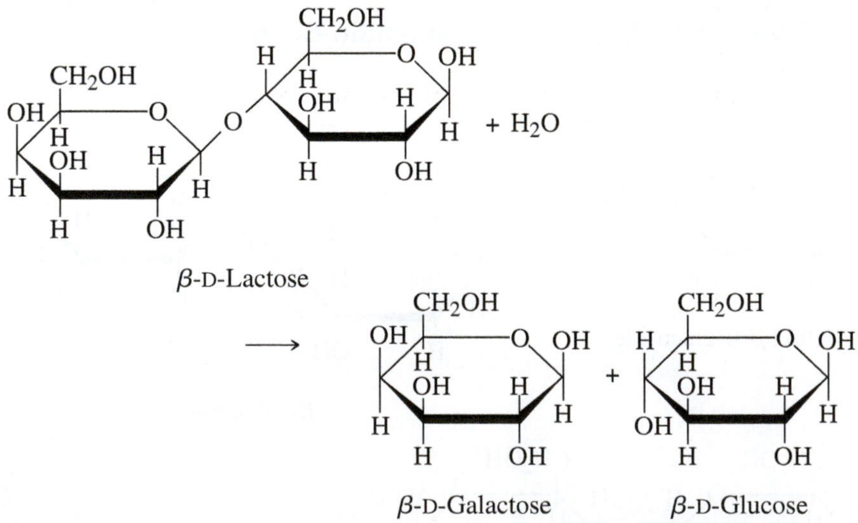

β-D-Lactose

β-D-Galactose   β-D-Glucose

Solution

This reaction involves a disaccharide reacting with water to break the glycosidic bond, forming two monosaccharides. This is a hydrolysis reaction.

Naming Glycosidic Bonds

If we look back at the disaccharide maltose that was formed from two glucose units, we see that the glycosidic bond was formed in the α position. If instead the second mono-saccharide had bonded to the β side (in this case the top) of the first glucose, the glyco-sidic bond would have created a different disaccharide molecule with a different shape (see **FIGURE 6.11**). Because of the two possible anomers, it becomes necessary to specify how the monosaccharides are bonded, that is, α or β. Besides the specific configuration bonded (α or β), the carbons that were joined must also be specified. Any —OH group on the second monosaccharide can, in principle, react with the first monosaccharide's anomeric carbon.

Because a monosaccharide has many hydroxyl groups that can undergo conden-sation, it is necessary to specify the carbon atoms joined by the glycosidic bond. The convention for naming glycosidic bonds is to specify the anomeric form and the carbons bonded. For maltose, the glycosidic bond is specified as $\alpha(1 \rightarrow 4)$ (stated "alpha-one-four"); whereas for celliobiose, the glycosidic bond is specified as $\beta(1 \rightarrow 4)$.

Mastering Video
Practicing the Concepts
Naming Glycosidic Bonds

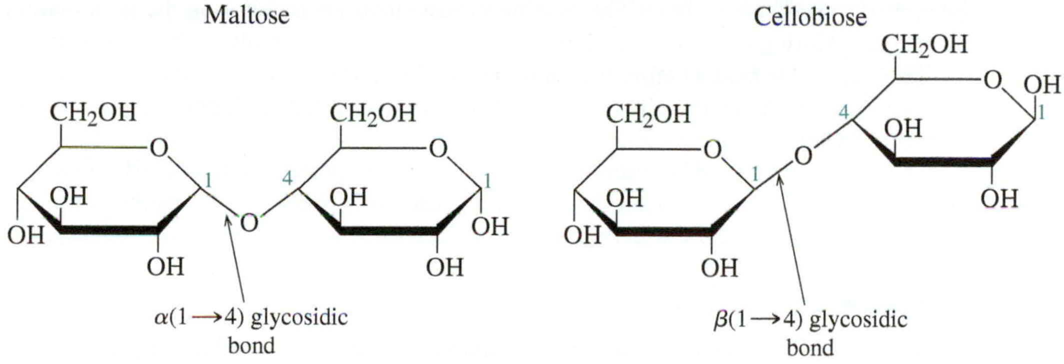

FIGURE 6.11 Comparison of maltose and cellobiose. Maltose and cellobiose are disaccha-rides of D-glucose bonded $\alpha(1 \rightarrow 4)$ and $\beta(1 \rightarrow 4)$, respectively. The glycosidic bond positioning (alpha versus beta) makes the shape of these two molecules different. Hydrogen atoms are removed from the structures for clarity.

Sample Problem 6.10 Naming Glycosidic Bonds

Name the glycosidic bond in the following disaccharides:

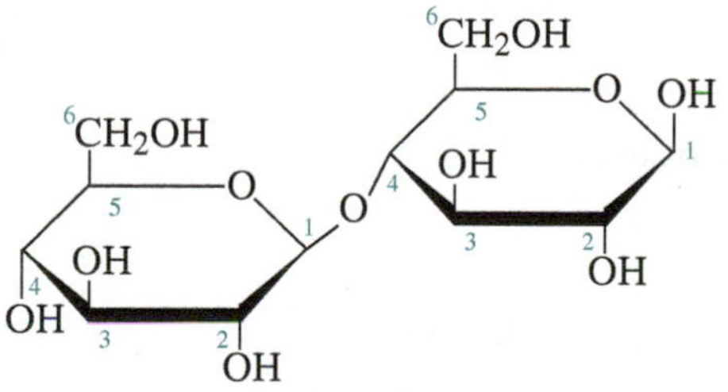

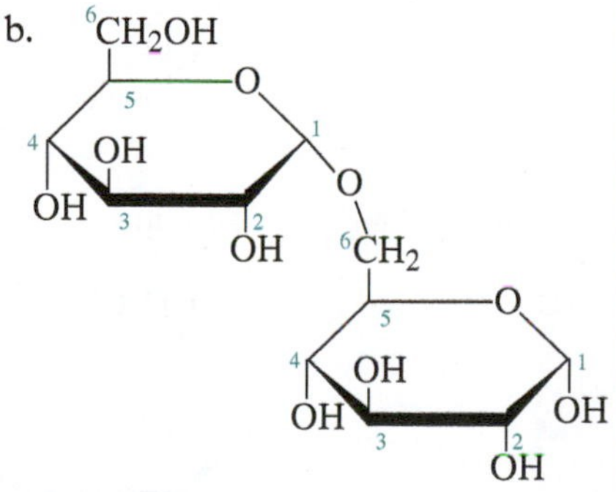

Solution

a. The anomeric carbon (C1) in this glycosidic bond is in the β anomer form. The carbons that are connected are C1 on the left and C4 on the right. The glycosidic bond is designated $\beta(1 \rightarrow 4)$.

b. The anomeric carbon (C1) in this glycosidic bond is in the α anomer form. The carbons that are connected are C1 on the left and C6 on the right. The glycosidic bond is designated $\alpha(1 \rightarrow 6)$.

Three Important Disaccharides—Maltose, Lactose, and Sucrose

Three common disaccharides formed through the condensation of two monosaccharides are maltose, lactose, and sucrose. Their formation is as follows:

D-glucose + D-glucose → maltose + H_2O [glycosidic bond $\alpha(1 \rightarrow 4)$]
D-galactose + D-glucose → lactose + H_2O [glycosidic bond $\beta(1 \rightarrow 4)$]
D-glucose + D-fructose → sucrose + H_2O [glycosidic bond $\alpha, \beta(1 \rightarrow 2)$]

In discussing saccharides larger than the monosaccharides, the "D" designation is often dropped from the name and assumed unless noted. For clarity, it is also common to see the ring form with the hydrogens not shown but implied. These shortcuts will be used from this point forward.

Maltose

Maltose, also known as malt sugar, is a disaccharide formed in the breakdown of starch. Malted barley, a key ingredient in beer, contains high levels of maltose. Many grains can be malted to produce maltose. The malting process involves monitoring the germination of the grain, during which the starch in the grain is converted to maltose through hydrolysis. This process is halted before the grains sprout by drying and roasting the grains. The glucose in the maltose of malted barley can then be converted to alcohol by yeast in the fermentation process.

The glycosidic bond between the glucose units in maltose is $\alpha(1 \rightarrow 4)$. Because one of the anomeric carbons (C1 on the second glucose unit) is not in a glycosidic bond and is considered a free anomeric carbon, maltose is a reducing sugar (see **FIGURE 6.12**).

Lactose

Lactose, or milk sugar, is found in mammalian milk. Intolerance to lactose can occur in people who do not inherit or who lose the ability to produce the enzyme that hydrolyzes this disaccharide into its two monosaccharides. The hydrolytic enzyme that digests lactose, commonly known as lactase, is sometimes added to commercial products to assist in the digestion of milk products for lactose-intolerant people. When lactose remains undigested, intestinal bacteria break down undigested lactose and, in doing so, produce abdominal gas and cramping.

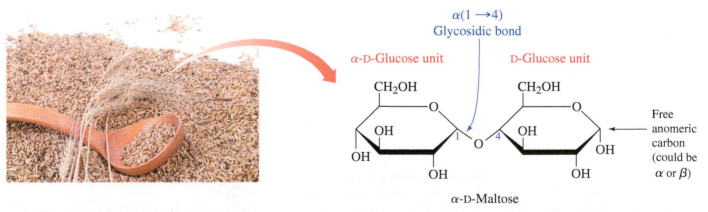

FIGURE 6.12 The disaccharide maltose. One of the main sugars in malted barley (grains in photo) is maltose formed from hydrolyzed starch. The bonded anomeric carbon is in the α position, so this glycosidic bond is $\alpha(1 \rightarrow 4)$.

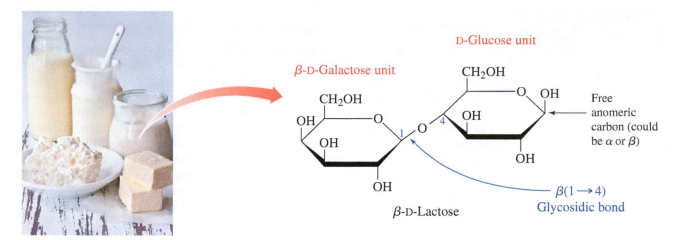

FIGURE 6.13 The disaccharide lactose. Lactose, found in the milk of mammals, consists of galactose and glucose. The bonded anomeric carbon is in the β position, so this glycosidic bond is $\beta(1 \rightarrow 4)$.

The glycosidic bond in lactose is $\beta(1 \rightarrow 4)$ because it occurs between C1 of a β-galactose and C4 of a glucose unit. Because the anomeric carbon on the glucose unit is free (not in a glycosidic bond), lactose is a reducing sugar (see **FIGURE 6.13**).

Sucrose

Sucrose, ordinary table sugar, is the most abundant disaccharide in nature. High quantities are found in sugar cane and sugar beets, which are the main commercial sources of sucrose.

When glucose and fructose join in an $\alpha, \beta(1 \rightarrow 2)$ glycosidic bond, sucrose is formed. In sucrose, both anomeric carbons are bonded (carbon 1 of glucose and carbon 2 of fructose), so both are noted in the name of the glycosidic bond. Because there is no free anomeric carbon, sucrose does not react with Benedict's reagent and is *not* a reducing sugar (see **FIGURE 6.14**).

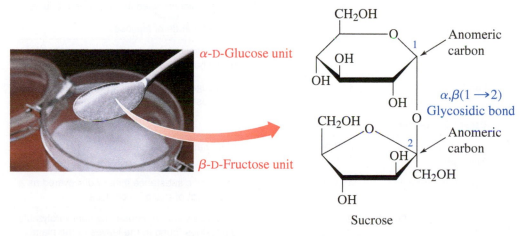

FIGURE 6.14 The disaccharide sucrose. Sucrose, obtained from sugar beets and sugar cane, contains glucose and fructose. Both anomeric carbons are bonded in the glycosidic bond. Glucose is in the α position and fructose is in the β position, so this glycosidic bond is $\alpha, \beta(1 \rightarrow 2)$.

INTEGRATING
Chemistry

Find out why ▶
high-fructose corn
syrup is used in
beverages.

Relative Sweetness of Sugars and Artificial Sweeteners

Dieters looking for a sweet taste and fewer calories can use artificial sweeteners like NutraSweet®, Splenda®, or Truvia™. Sweetness is perceived by our taste buds, where sugar molecules and other artificial sweeteners bind and register this taste in the brain.

Sucrose (table sugar) is recognized as the standard for sweetness and is assigned a numerical value of 100 (see **TABLE 6.2**). Fructose has a sweetness value at least one-and-a-half times that of sucrose. This means that it takes less fructose to achieve the same sweet taste as sucrose. High-fructose corn syrup, used in many beverage formulations, contains a mixture of 55% fructose and 45% glucose, making it sweeter tasting than an equal amount of sucrose or corn syrup, which is composed entirely of glucose. This allows manufacturers to use less sweetener to achieve the same level of sweetness.

Artificial sweeteners are hundreds, even thousands, of times sweeter than sucrose, so a much smaller amount can be used to achieve a similar sweet taste. Moreover, many artificial sweeteners lack calories because they cannot produce energy, as many monosaccharides do through glycolysis. The sweetness of some sugars and other sweeteners compared to sucrose is shown in Table 6.2.

TABLE 6.2 Relative Sweetness of Sugars and Artificial Sweeteners

Sweetener	Sweetness Relative to Sucrose (= 100)	Description
Simple Sugars		
Fructose	140–175	Fruit sugar, a monosaccharide that is a component of sucrose
Invert sugar (hydrolyzed sucrose)	120	Found in honey
Sucrose	100	Table sugar, a disaccharide composed of glucose and fructose
Xylitol	100	A sugar alcohol, used in sugar-free products
Glucose	75	Dextrose, the most common monosaccharide that is a component of sucrose, lactose, and maltose
Erythritol	70	A sugar alcohol used in sugar-free products
Sorbitol	36–55	A sugar alcohol used in sugar-free products
Maltose	32	A disaccharide of glucose
Galactose	30	A monosaccharide that is a component of the disaccharide lactose
Lactose	15	Milk sugar, a disaccharide containing galactose and glucose
Other Sweeteners		
Sucralose (found in Splenda)	60,000	Chlorinated disaccharide
Saccharin (found in Sweet'N Low®)	45,000	An organic substance initially discovered as a by-product of research on dyes
Stevia (found in Truvia)	25,000	Natural sweetener containing nonhydrolyzable glucosides found in the leaves of the plant *Stevia rebaudiana*
Acesulfame K	200	Used in combination with other sweeteners due to bitter aftertaste
Aspartame (found in Equal®)	18,000	Contains two amino acids, aspartic acid and phenylalanine

Practice Problems

6.32 Identify the following reactions as condensation or hydrolysis:

a. two monosaccharides reacting to form a disaccharide

b. the formation of a glycosidic bond

c. a reaction in which one molecule breaks into two and an —H and —OH are added

6.33 Identify the following reactions as condensation or hydrolysis:

a. a disaccharide breaking into two monosaccharides

b. a reaction in which two molecules combine, forming one, and a molecule of water is produced

c. the breaking of a glycosidic bond

6.34 Name the glycosidic bond present in mannobiose, shown in the following figure:

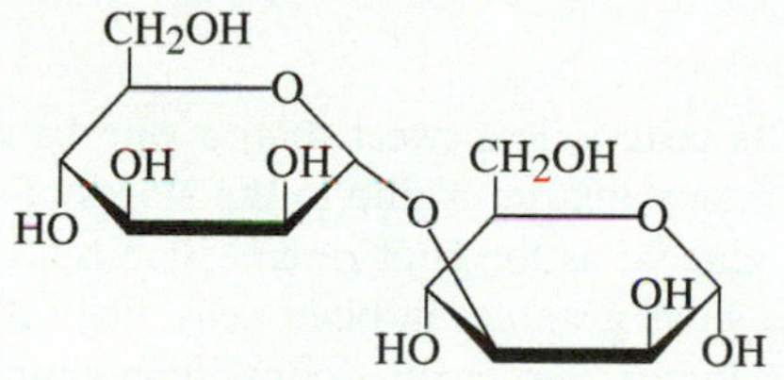

Mannobiose

6.35 Name the glycosidic bond present in melibiose, a disaccharide that has the sweetness of about 30 compared with sucrose.

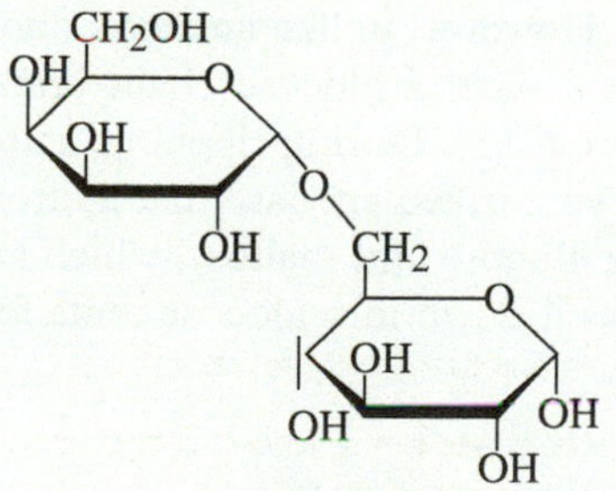

Melibiose

6.36 For each of the following disaccharides, name the glycosidic bond and draw the monosaccharide units produced by hydrolysis:

a.

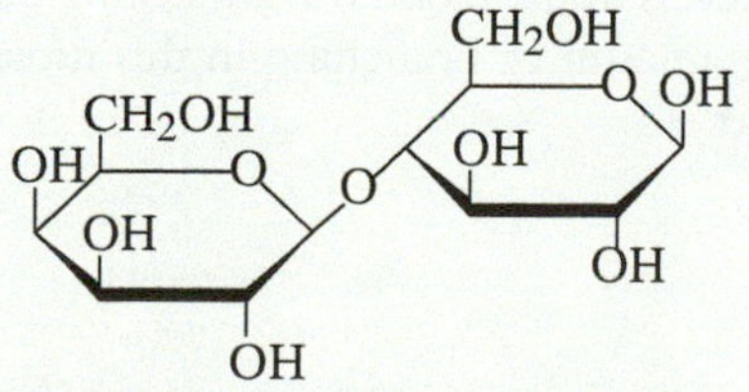

b.

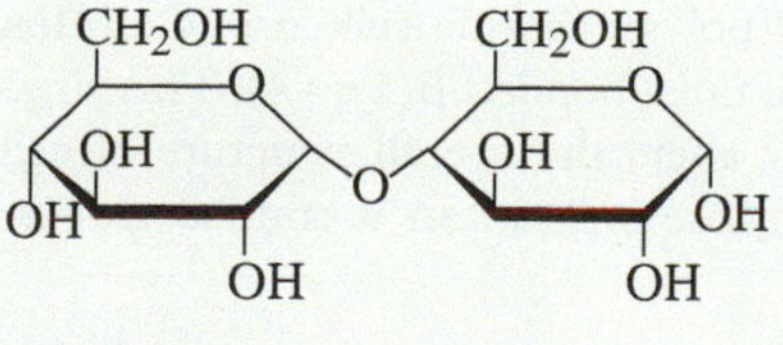

6.37 For each of the following disaccharides, name the glycosidic bond and draw the monosaccharide units produced by hydrolysis:

a.

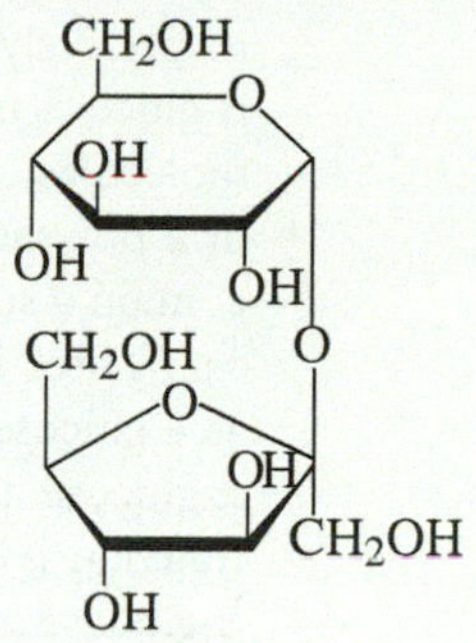

b.

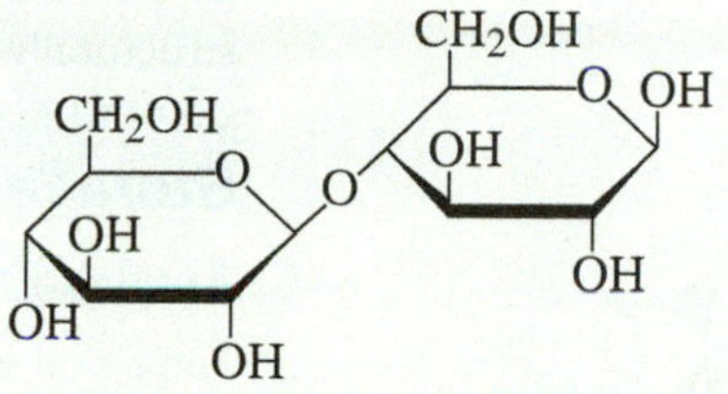

6.38 Lactulose is a disaccharide used in the treatment of chronic constipation. Its formal name is galactose $\beta(1 \rightarrow 4)$ fructose.

a. Draw the structure of lactulose.

b. Is lactulose a reducing or nonreducing sugar?

6.39 Identify a disaccharide that fits each of the following descriptions:

a. ordinary table sugar

b. found in milk and milk products

c. also called malt sugar

d. hydrolysis gives galactose and glucose

6.40 Identify a disaccharide that fits each of the following descriptions:

a. not a reducing sugar

b. composed of two glucose units

c. also called milk sugar

d. hydrolysis gives glucose and fructose

6.41 Based on the sweetness index in Table 6.2, if you tasted a drop of each of the syrups below, which would taste the sweetest?

a. light corn syrup (100% glucose)

b. agave syrup (10:90 glucose:fructose)

c. honey (90:10 invert sugar:maltose)

d. high-fructose corn syrup (45:55 glucose:fructose)

6.42 If one sweetener packet of Splenda, Sweet'N Low, or Equal has the same sweetness as two tablespoons of sugar, according to Table 6.2, which of the packets contains the smallest amount of the sweetener?

6.6 Polysaccharides

6.6 Inquiry Question

How are polysaccharides characterized?

Plants use the sun's energy to make glucose, their food. Some days get more sunlight than other days—can plants store glucose for a rainy day? What happens if we eat more simple sugars (monosaccharides and disaccharides) than we need for energy—can we store that energy? Glucose can be stored in both plants and animals as starch and glycogen, respectively, by connecting α-glucose units through glycosidic bonds. Connecting many β-glucose units produces the molecule cellulose, a structural material in plants. These molecules, called polysaccharides, are very large. Because most of the monosaccharides in a polysaccharide are bonded through their anomeric carbon, polysaccharides do not contain a sufficient number of reducing ends to give a positive Benedict's test.

Three important **storage polysaccharides**—polysaccharides that are stored in cells as a glucose energy reserve—are amylose, amylopectin, and glycogen; each contains only α-glucose units. Two important **structural polysaccharides**—polysaccharides whose function is to provide structure for an organism—are cellulose and chitin; each contains β-glucose or glucose-derived units. Chitin contains a modified β-glucose unit. Even though all of these polysaccharides have a similar chemical makeup (glucose units), they differ structurally and functionally because of their glycosidic bonds and differences in branching.

Storage Polysaccharides

Amylose and Amylopectin—Starch

Have you ever eaten an unripe banana? Its taste is less sweet than a ripe banana, and the texture is rather mealy. The mealy or grainy texture is due to the stores of starch in the fruit that have yet to be hydrolyzed to glucose as the fruit ripens. Starch is a glucose storage polysaccharide that accumulates in small granules in plant cells. Plant foods that we think of as starchy—including potatoes, grains, and beans—contain an abundance of these polysaccharides. Starch is a mixture of two polysaccharides, amylose and amylopectin (see **FIGURE 6.15**). **Amylose**, which makes up about 20% of starch, is made up of anywhere from 250 to 4000 D-glucose units bonded $\alpha(1 \rightarrow 4)$ in a continuous chain. Long chains of D-glucose bonded $\alpha(1 \rightarrow 4)$ will tend to coil like a spring.

Amylopectin makes up about 80% of plant starch. It also contains D-glucose units connected by $\alpha(1 \rightarrow 4)$ glycosidic bonds. However, unlike amylose, about every 25 glucose units along a linear glucose chain, a second glucose chain branches off through an $\alpha(1 \rightarrow 6)$ glycosidic bond (see Figure 6.15). During ripening, starch granules undergo hydrolysis by the action of an enzyme called amylase that hydrolyzes the $\alpha(1 \rightarrow 4)$ glycosidic bonds in starch, producing glucose and maltose, which are sweet. When we eat starch, our digestive system breaks it down into glucose units for use by our bodies.

Glycogen

Animals store glucose as a polysaccharide, too. **Glycogen** is the storage polysaccharide found in animals. Most glycogen stores are located in the liver and in muscles. Glycogen is identical in structure to amylopectin except that $\alpha(1 \rightarrow 6)$ branching occurs about every 12 glucose units. Glycogen is hydrolyzed to glucose in the liver and sent into the bloodstream to maintain constant glucose levels in the blood during fasting periods when sugars are not being consumed. The large amount of branching in this molecule allows for quick hydrolysis when glucose is needed.

Structural Polysaccharides

Cellulose

Trees stand tall because of the structural polysaccharide cellulose. **Cellulose** also contains glucose units, but they are D-glucose units bonded $\beta(1 \rightarrow 4)$. This single change in glycosidic bond configuration completely alters the overall structure of cellulose compared with that of amylose. Whereas amylose coils in an α-bonded glucose chain, the

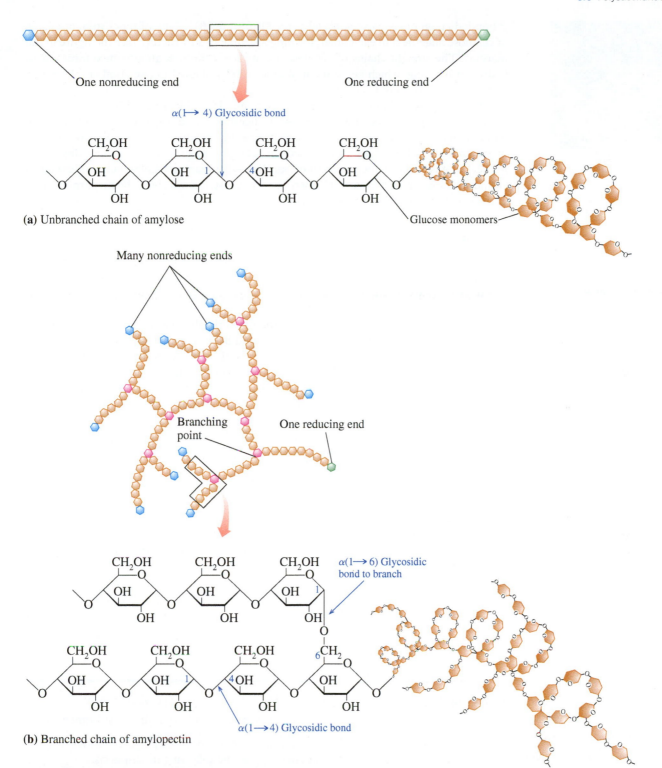

(a) Unbranched chain of amylose

(b) Branched chain of amylopectin

FIGURE 6.15 The storage polysaccharide—starch. Both components of starch are polymers of glucose. (a) Amylose is an unbranched chain that forms a linear spiral. (b) Amylopectin is branched and has a shape that is more globular. Like most polysaccharides, each starch molecule contains a single reducing end. Amylose has one nonreducing end while amylopectin has many. Both are considered nonreducing sugars.

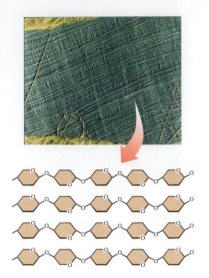

Cellulose molecules align to form rigid, tough fibers.

β-bonded chain of cellulose is straight (see **FIGURE 6.16**). Many of these straight chains of cellulose align next to each other, forming a strong, rigid structure (see the figure in the margin). The straight chains of cellulose are attracted through an attractive force called hydrogen bonding, which keeps the molecules rigidly aligned to one another. Hydrogen bonding is discussed in Chapter 7.

Nutritionally, cellulose is the most abundant insoluble fiber. Some animals and insects—like sheep, cows, and termites—can digest cellulose because their digestive system contains bacteria and other microorganisms that produce cellulase, an enzyme that hydrolyzes the $\beta(1 \rightarrow 4)$ glycosidic bonds in cellulose. As humans, we cannot digest cellulose, but it is still an important part of our diet—abundant in whole-grain foods such as oatmeal, whole-wheat bread, and brown rice—because it assists with digestive movement in the small and large intestines.

Chitin

Chitin is a polysaccharide that makes up the exoskeleton of insects and crustaceans and the cell walls of most fungi. This polysaccharide is made up of a modified β-D-glucose called *N*-acetylglucosamine containing $\beta(1 \rightarrow 4)$ glycosidic bonds. An amide group replaces the hydroxyl group at carbon 2.

Like cellulose, chitin is a structurally strong material that has many uses, one of which is as a surgical thread that biodegrades as a wound heals. Chitin is present in the exoskeletons of arthropods such as insects and spiders and serves to protect them from water. Because of this property, chitin can be used to waterproof paper. When ground up, chitin becomes a powder that holds in moisture, and it can be added to cosmetics and lotions. Figure 6.16 shows a comparison of the structure and shape of four of the polymers of glucose. Note that the $\alpha(1 \rightarrow 4)$ linked polymers have quite different overall shapes than the $\beta(1 \rightarrow 4)$ linked polymers. The elongated, linear structures of the β linked polymers are well-suited for the structural roles they play in plants and animal exoskeletons.

Sample Problem 6.11 Structures of Polysaccharides

Identify the polysaccharide described by each of the following:
  a. a polysaccharide that is stored in the liver and muscle tissues
  b. a component of starch containing only $\alpha(1 \rightarrow 4)$ glycosidic bonds
  c. a polysaccharide found in the exoskeleton of insects

Solution

  a. glycogen   b. amylose   c. chitin

Practice Problems

6.43 Describe the similarities and differences of the following polysaccharides:

a. amylose and amylopectin

b. amylopectin and glycogen

6.44 Describe the similarities and differences of the following polysaccharides:

a. amylose and cellulose

b. cellulose and chitin

6.45 Give the name of one or more polysaccharides that matches each of the following descriptions:

a. not digestible by humans

b. the storage form of carbohydrates in plants

c. contains only $\alpha(1 \rightarrow 4)$ glycosidic bonds

d. glucose polysaccharide with the most branching

6.46 Give the name of one or more polysaccharides that matches each of the following descriptions:

a. the storage form of carbohydrates in animals

b. contains only $\beta(1 \rightarrow 4)$ glycosidic bonds

c. contains both $\alpha(1 \rightarrow 4)$ and $\alpha(1 \rightarrow 6)$ glycosidic bonds

d. produces maltose during digestion

Amylose (unbranched starch)

Amylopectin (branched starch)

Cellulose

Chitin

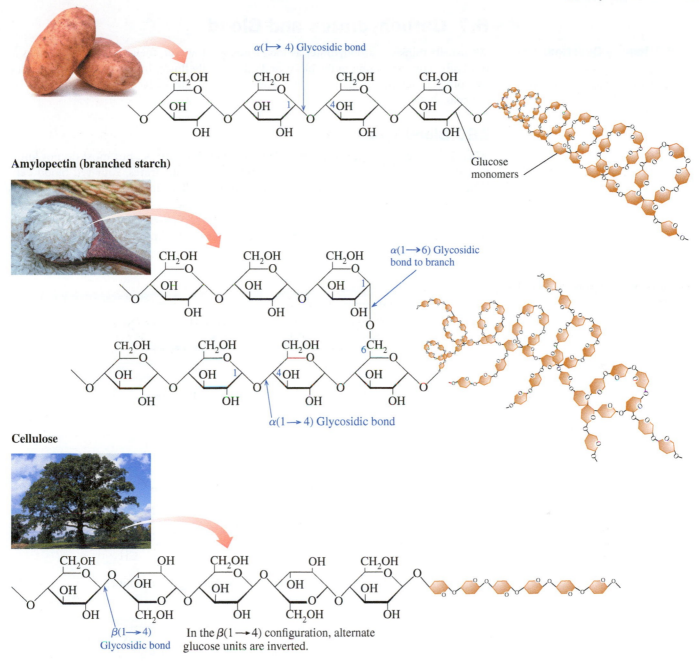

$\alpha(1 \longrightarrow 4)$ Glycosidic bond

Glucose monomers

$\alpha(1 \longrightarrow 6)$ Glycosidic bond to branch

$\alpha(1 \longrightarrow 4)$ Glycosidic bond

$\beta(1 \longrightarrow 4)$ Glycosidic bond

In the $\beta(1 \longrightarrow 4)$ configuration, alternate glucose units are inverted.

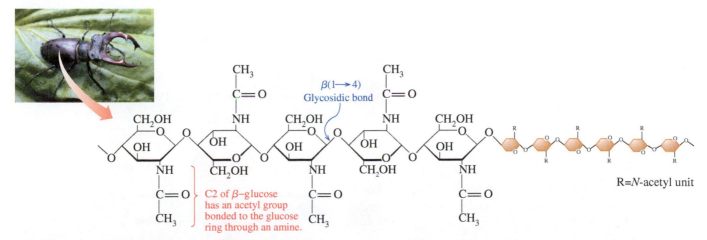

$\beta(1 \longrightarrow 4)$ Glycosidic bond

C2 of β–glucose has an acetyl group bonded to the glucose ring through an amine.

R=*N*-acetyl unit

FIGURE 6.16 Comparing the polysaccharides of glucose. Notice the structural differences in these polymers and the resulting changes in the shapes of the molecules. For the $\alpha(1 \rightarrow 4)$ linked polymers, (a) amylose is linear but has a spiral shape, while (b) amylopectin has a more globular shape due to its $\alpha(1 \rightarrow 6)$ branches. Both of the $\beta(1 \rightarrow 4)$ linked polymers, (c) cellulose and (d) chitin, are linear straight chains.

6.7 Carbohydrates and Blood

6.7 Inquiry Question

What other roles can carbohydrates play in the body?

We usually think of carbohydrates as food sources. However, they also play other roles in our body. They are found in the blood and other bodily fluids. In this section, we describe two other functions of carbohydrates—as signals on the surface of blood cells and as anticoagulants.

ABO Blood Types

What is your blood type? Many people know their blood type, but most do not know that the blood types of A, B, AB, and O refer to carbohydrates. Your red blood cells have a number of chemical markers bonded to the cell surface identifying the cells as yours as they circulate through the bloodstream. One set of those chemical markers is made up of oligosaccharides known as the ABO blood markers, which contain either three or four monosaccharides. Some of these are modified monosaccharides and contain other functional groups.

Structures of the monosaccharides that are in the ABO blood groups.

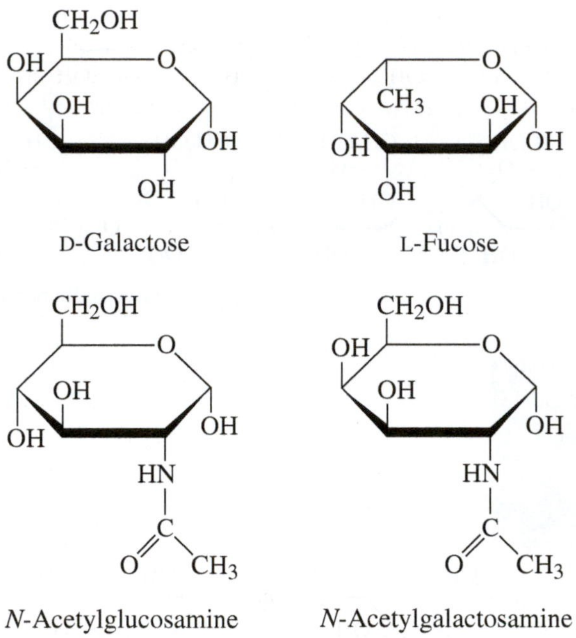

D-Galactose
L-Fucose
N-Acetylglucosamine
N-Acetylgalactosamine

Comparison of the sugar linking in the O, A, and B blood groups.

Type O
Red blood cell
N-Acetyl-glucosamine
Galactose
Fucose
Type A
N-Acetyl-galactosamine
Type B

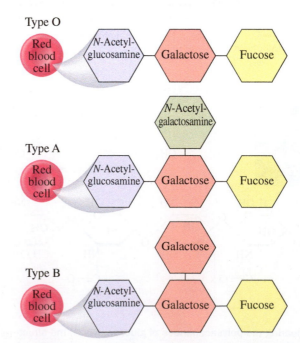

All the blood types include the carbohydrates *N*-acetylglucosamine, galactose, and **fucose**, a carbohydrate that often appears in nature as its L-enantiomer. Type O blood contains only these three carbohydrates attached to the surface of the red blood cell through a glycosidic bond. Type A and type B blood contain these three plus a fourth carbohydrate. In type A, the fourth carbohydrate is an *N*-acetylgalactosamine bonded to the galactose unit, while in type B blood, the fourth carbohydrate is a second galactose bonded to the galactose. People with type AB blood have both the type A carbohydrate set and the type B carbohydrate set on their red blood cells.

Universal Donors and Acceptors

Everyone has a unique blood type. One of most important markers is the ABO blood marker group. The O blood type is considered the universal donor blood type, but why? Each person's immune system can recognize only its own carbohydrate set (A, B, or O) and will try to destroy what it considers a foreign blood type by producing protein molecules called antibodies that act to destroy the foreign blood type. (Antibodies are described in more detail in Chapter 10.) Because the trisaccharide on the cells of O-type blood (*N*-acetylglucosamine—galactose—L-fucose) is present on cells of *all* blood types (A, B, and AB), O-type blood can be donated to any blood type and is considered the universal donor. No blood type recognizes the O carbohydrate set as foreign.

The AB blood type is considered the universal acceptor blood type for a similar reason. AB blood contains all possible ABO combination types, so any blood type transfused will be accepted by the body. The compatibility of blood types is summarized in **TABLE 6.3**.

> **INTEGRATING**
> Chemistry
>
> ◀ Find out which ABO blood types are compatible.

TABLE 6.3 Compatibility of ABO Blood Groups		
Blood Group	**Can Receive Blood Types**	**Can Donate to Blood Types**
A	A, O	A, AB
B	B, O	B, AB
AB[a]	A, B, AB, O	Cannot donate to other blood types
O[b]	O	A, B, AB, O (all types)

[a] Type AB is the universal acceptor.
[b] Type O is the universal donor.

Sample Problem 6.12 ABO Blood Types

Explain whether the following blood types could be donated to a person with type A blood:

a. A   b. B   c. AB   d. O

Solution

a. Two matched blood types can always be donated to each other; the transfused blood will not be recognized as foreign. Type A can be donated to type A.
b. Type B cannot be donated to type A because the person with type A blood would recognize type B blood as foreign and try to eliminate it.
c. Even though type AB contains the A carbohydrates, it also contains the B carbohydrates, which a person with type A blood would recognize as foreign. Type AB cannot be donated to type A.
d. Because the O carbohydrate set is present within the type A carbohydrates, type O blood will not be recognized as foreign and will be accepted. Type O can be donated to type A.

INTEGRATING Chemistry

Find out why the ▶ carbohydrate heparin is critical in health care.

Heparin

Heparin is a medically important polysaccharide that prevents clotting, acting as an anticoagulant in the bloodstream. Test tubes, tubing, and needles used for drawing blood are routinely coated with heparin to prevent the blood from clotting. Heparin is a highly ionic polysaccharide of many repeating disaccharide units. The disaccharide includes an oxidized monosaccharide and a D-glucosamine. This polysaccharide belongs to a group of polysaccharides called **glycosaminoglycans**, all of which have highly charged repeating disaccharide units. Negative charges are mainly due to the presence of sulfate groups.

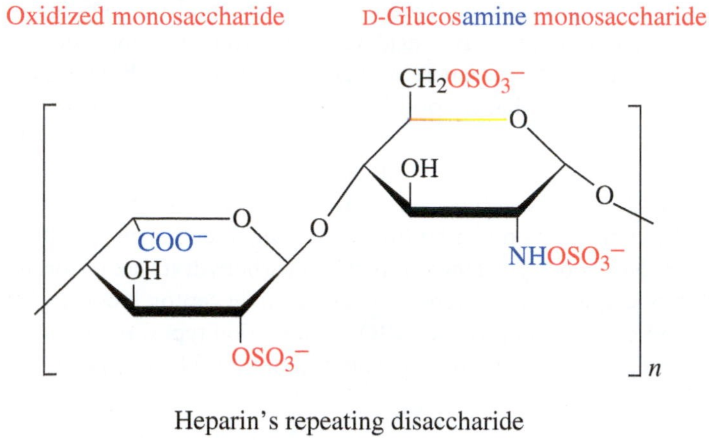

Oxidized monosaccharide   D-Glucosamine monosaccharide

Heparin's repeating disaccharide

Heparin is a modified polysaccharide that serves as an anticoagulant in the bloodstream.

Practice Problems

6.47 Explain whether the following blood types could be donated to a person with type B blood:

a. A   b. AB

6.48 Explain whether the following blood types could be donated to a person with type O blood:

a. B   b. AB

6.49 How is the polysaccharide heparin different from the glucose polysaccharides?

6.50 What is the role of heparin in the bloodstream?

6

CHAPTER REVIEW

The study guide will help you check your understanding of the main concepts in Chapter 6. You should be able to answer problems for each learning outcome in this list. To check your mastery, try the problems listed after each.

STUDY GUIDE	CHAPTER GUIDE

6.1 Classes of Carbohydrates

Classify carbohydrates.

- Classify carbohydrates as mono-, di-, oligo-, or polysaccharides. (Try 6.1, 6.53)
- Distinguish soluble and insoluble fiber. (Try 6.3, 6.55)

Inquiry Question

How do we classify carbohydrates?

Carbohydrates are classified as monosaccharides (simple sugars), disaccharides (two monosaccharide units), oligosaccharides (three to nine monosaccharide units), and polysaccharides (many monosaccharide units). The simplest carbohydrates are the monosaccharides with a molecular formula of $C_n(H_2O)_n$, where $n = 3-6$. Edible carbohydrates that cannot be broken down by the body's enzymes are classified as either soluble or insoluble fiber based on their ability to mix with water.

6.2 Functional Groups in Monosaccharides

Locate organic functional groups in monosaccharides.

- Distinguish primary, secondary, and tertiary alcohols. (Try 6.5, 6.59)
- Recognize and draw the functional groups alcohol, aldehyde, and ketone. (Try 6.7, 6.57)

Inquiry Question

What functional groups are present in monosaccharides?

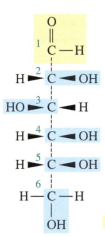

Monosaccharides contain several alcohol (hydroxyl) groups and either an aldehyde or a ketone functional group. Alcohols can be primary, secondary, or tertiary depending on the number of carbons attached to the alcohol carbon. Aldehydes and ketones are carbonyl-containing functional groups. Because of the functional groups present, monosaccharides are called aldoses or ketoses and are referred to as polyhydroxyaldehydes and polyhydroxyketones.

☐ Aldehyde

☐ Alcohol (hydroxyl)

6.3 Stereochemistry in Monosaccharides

Characterize the structural properties of monosaccharides.

- Distinguish D- and L-stereoisomers of monosaccharides. (Try 6.11, 6.63)
- Define enantiomer, epimer, and diastereomer. (Try 6.15, 6.63)
- Draw enantiomers and diastereomers of linear monosaccharides. (Try 6.13, 6.17b)
- Draw epimers of monosaccharides. (Try 6.17a, 6.65)
- Characterize common monosaccharides. (Try 6.19)

Inquiry Question

What structural properties do monosaccharides display?

Monosaccharides can be drawn linearly in a representation called a Fischer projection, which highlights their chiral centers. Most carbohydrates found in nature are the D-isomer. D-Isomers have the —OH on the chiral carbon farthest from the carbonyl on the right side. Stereoisomers that have multiple chiral centers can be related to each other as enantiomers (mirror images of each other) or as diastereomers (not enantiomers). Some important monosaccharides are glucose, galactose, mannose, fructose, deoxyribose, and ribose.

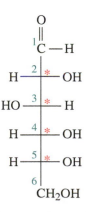

Fischer projection
D-Glucose

—Continued next page

Continued—

6.4 Reactions of Monosaccharides

Draw products of reactions at the anomeric carbon.

- Draw the cyclic α and β anomers from linear monosaccharide structures. (Try 6.25, 6.69)
- Draw oxidation and reduction products of aldoses. (See Sample Problem 6.7. Try 6.25, 6.29, 6.71, 6.73)

Inquiry Question

How are the products of monosaccharide reactions drawn?

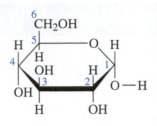

α-D-Glucose
in ring form

A hydroxyl group and the carbonyl functional group of a linear monosaccharide can react to enclose the hydroxyl's oxygen in a ring. Because the carbonyl group is a planar structure, this reaction produces two possible ring arrangements about the anomeric carbon termed the α and the β anomers when a ring is formed. The anomeric carbon of carbohydrates (C1 of an aldose) is highly reactive and can undergo oxidation to a carboxylic acid or reduction to an alcohol. Monosaccharides are considered reducing sugars because their anomeric carbon can react to reduce another compound in a redox reaction.

6.5 Disaccharides

Characterize glycosidic bonds in disaccharides.

- Distinguish condensation and hydrolysis reactions of simple sugars. (Try 6.33, 6.77)
- Locate and name glycosidic bonds in disaccharides. (Try 6.34, 6.35)
- Characterize common dissacharides. (Try 6.39)
- Apply the sweetness scale to various sweet products. (Try 6.41)

Inquiry Question

How are disaccharides formed and identified?

Carbohydrates form glycosides when an anomeric carbon reacts with a hydroxyl on a second organic molecule. This condensation results in the formation of a glycosidic bond. Glycosidic bonds are named by designating the anomer of the reacting monosaccharide and the carbons that are bonded, for example, $\alpha(1 \rightarrow 4)$. Some important disaccharides formed through condensation reactions are maltose, lactose, and sucrose. Monosaccharides and disaccharides make up simple sugars, many of which are sweet tasting. The sweetness of carbohydrates and other carbohydrate substitutes is indexed relative to the sweetness of sucrose (see Table 6.2).

6.6 Polysaccharides

Distinguish common polysaccharides.

- Identify polysaccharides by glycosidic bond and sugar subunit. (Try 6.43, 6.45)

Inquiry Question

How are polysaccharides characterized?

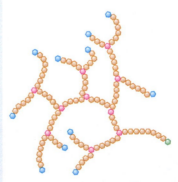

Polysaccharides consist of many monosaccharide units bonded together through glycosidic bonds. Glucose can be stored as a polysaccharide called starch in plants and glycogen in animals. Starch consists of two polysaccharides: amylose, a linear chain of glucose, and amylopectin, a branched chain of glucose. Glycogen is also a branched polysaccharide of glucose, but it contains more branching than amylopectin. Two polysaccharides that are structurally important in nature are cellulose in plants (wood) and chitin in arthropods (exoskeleton) and fungi (cell wall). Cellulose is a linear chain of glucose, but in cellulose the glycosidic bonds are β, whereas in starch they are bonded α. Chitin is also a linear chain of a modified glucose, *N*-acetylglucosamine bonded $\beta(1 \rightarrow 4)$. Both of these structural polysaccharides form strong, water-resistant materials when the linear chains are aligned with each other.

6.7 Carbohydrates and Blood

Predict ABO blood group compatibility.

- Predict ABO compatibility. (Try 6.47, 6.85)
- Describe the structure and role of heparin. (Try 6.49)

Inquiry Question

What other roles can carbohydrates play in the body?

Carbohydrates are also used as recognition markers on the surfaces of cells and in other bodily fluids. The ABO blood groups are oligosaccharides of which one of the carbohydrate units is L-fucose, one of the few L-sugars found in nature. These ABO oligosaccharides are found on the surface of red blood cells. The A and B blood groups look like the O blood group except that they contain an additional monosaccharide. For this reason, the O blood type is considered the universal donor. Heparin is a polysaccharide consisting of a repeating disaccharide containing an oxidized monosaccharide and a glucosamine. Heparin functions in the blood as an anticoagulant and is commonly found as a coating on medical tubing and syringes used during blood transfusions.

Mastering Videos

PRACTICING THE CONCEPTS

The following videos can be accessed through the Pearson eText or your Mastering Chemistry course.

- Monosaccharide Ring Formation
- Naming Glycosidic Bonds

Summary of Reactions

Ring Formation (D-Glucose)

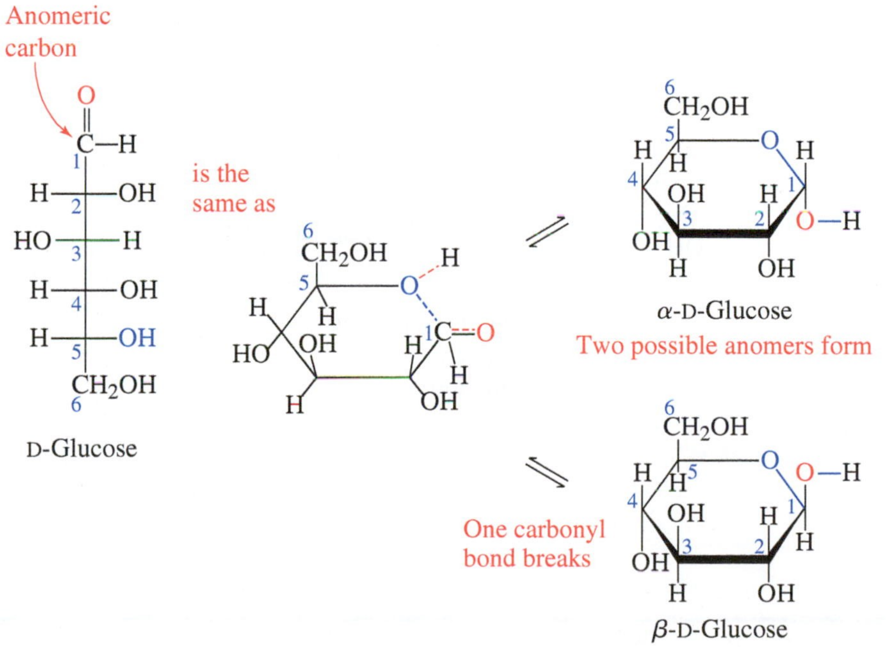

Anomeric carbon

D-Glucose   is the same as

Two possible anomers form

One carbonyl bond breaks

α-D-Glucose

β-D-Glucose

Oxidation of Monosaccharides (D-Glucose)

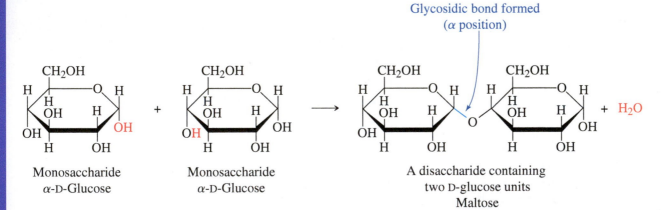

D-Glucose + 2 Cu^{2+} Blue Benedict's reagent $\xrightarrow{\text{Oxidation}}_{\text{O}_2, \text{heat}}$ D-Gluconic acid + Cu$_2$O Brick-red solid

Reduction of Monosaccharides (D-Glucose)

D-Glucose + H$_2$ $\xrightarrow{\text{Reduction}}$ D-Glucitol also called D-sorbitol

Glycosidic Bond Formation (D-Glucose)

Glycosidic bond formed (α position)

Monosaccharide α-D-Glucose + Monosaccharide α-D-Glucose $\longrightarrow$ A disaccharide containing two D-glucose units Maltose + H$_2$O

Disaccharide Formation Through Condensation

Glucose + glucose $\longrightarrow$ maltose + H$_2$O
Glucose + galactose $\longrightarrow$ lactose + H$_2$O
Glucose + fructose $\longrightarrow$ sucrose + H$_2$O

Additional Problems

6.51 Write the molecular formula for a carbohydrate containing three carbons.

6.52 What would be the molecular formula of a monosaccharide characterized as an aldopentose?

6.53 Explain the difference between an oligosaccharide and a polysaccharide.

6.54 Explain the difference between an aldose and a ketose.

6.55 Describe the properties of soluble fiber.

6.56 Describe the properties of insoluble fiber.

6.57 Name the functional groups present in aldoses.

6.58 Name the functional groups present in ketoses.

6.59 Classify each of the following as primary, secondary, or tertiary alcohols:

a. 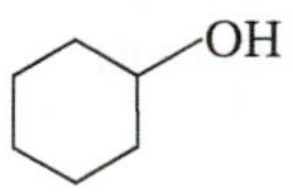 [structure: cyclohexane with OH]   b. 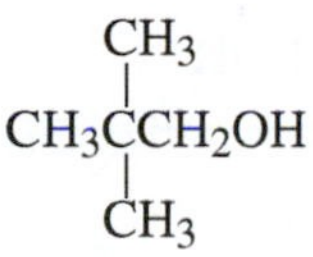 CH₃
CH₃CCH₂OH
CH₃

c. 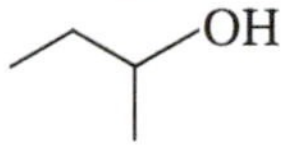 [structure with OH]   d. 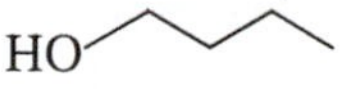 HO[structure]

e. 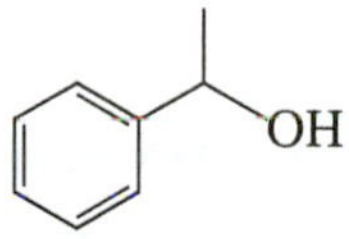 [structure: benzene ring with CH(CH₃)OH]

6.60 Classify each of the following as primary, secondary, or tertiary alcohols:

a. 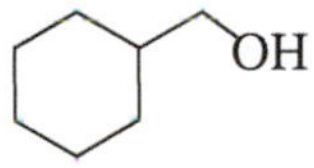 [structure: cyclohexane with CH₂OH]   b.  [structure with OH]

c. 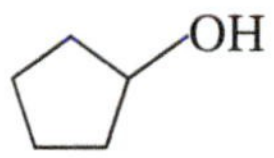 [structure: cyclopentane with OH]   d. 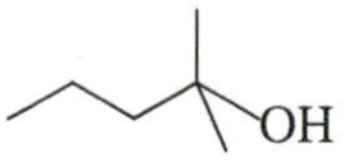 [structure with OH]

e. 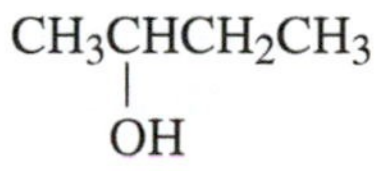 CH₃CHCH₂CH₃
OH

6.61 Classify the —OH groups in D-glucose as primary, secondary, or tertiary.

[Fischer projection of D-Glucose:
1 C—H (=O top)
2 H—OH
3 HO—H
4 H—OH
5 H—OH
6 CH₂OH]

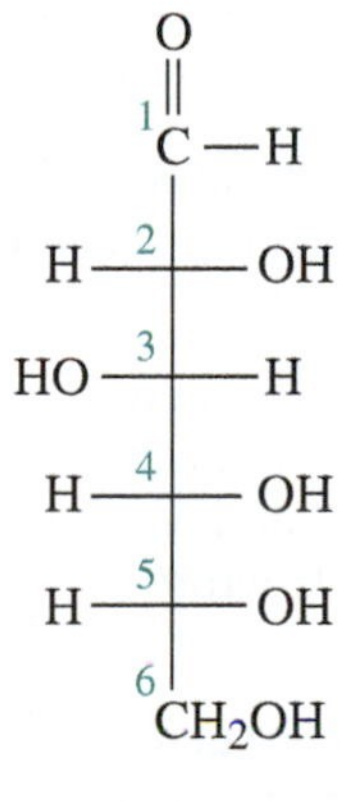

D-Glucose

6.62 Classify the —OH groups in the pyranose form of D-glucose as primary, secondary, or tertiary.

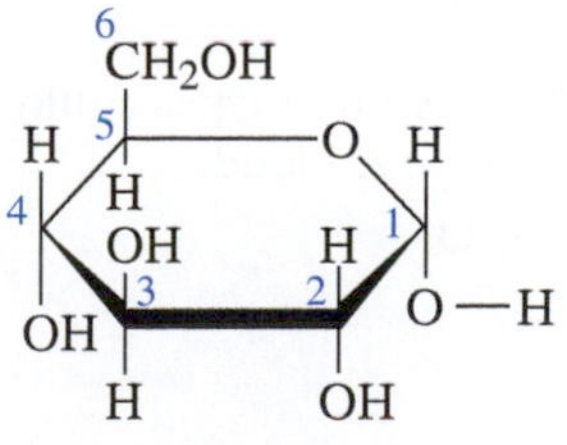

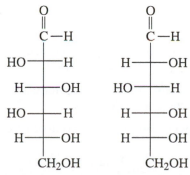

α-D-Glucose

6.63 How are the following pairs of carbohydrates, shown in a Fischer projection, related to each other? Are they structural isomers, enantiomers, diastereomers, or epimers? Identify each as the D- or L-isomer.

a. 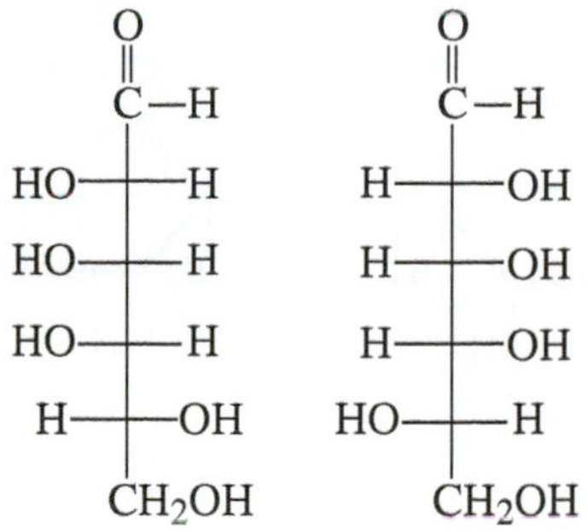
```
     O                  O
     ||                 ||
     C—H                C—H
HO——H            H——OH
HO——H            H——OH
HO——H            H——OH
 H——OH          HO——H
   CH₂OH            CH₂OH
```

b. 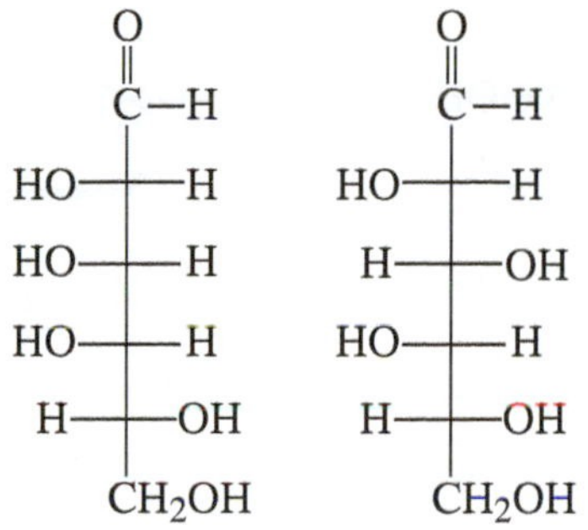
```
     O                  O
     ||                 ||
     C—H                C—H
HO——H            HO——H
HO——H            H——OH
HO——H            HO——H
 H——OH           H——OH
   CH₂OH            CH₂OH
```

6.64 How are the following pairs of carbohydrates, shown in a Fischer projection, related to each other? Are they structural isomers, enantiomers, diastereomers, or epimers?

a. 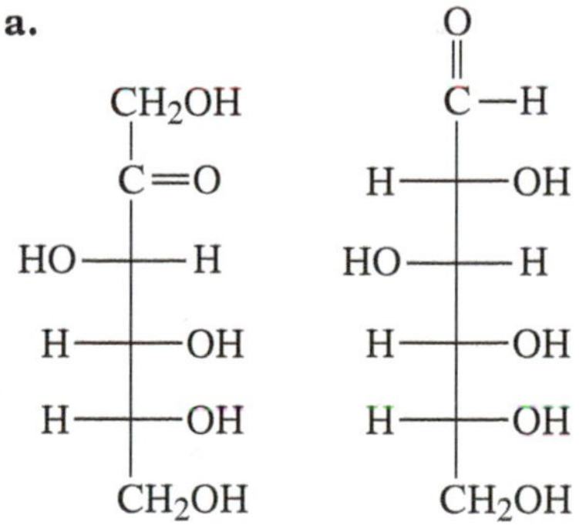
```
   CH₂OH                O
                        ||
    C=O                 C—H
HO——H            H——OH
 H——OH          HO——H
 H——OH           H——OH
   CH₂OH            CH₂OH
```

b.
```
     O                  O
     ||                 ||
     C—H                C—H
HO——H            H——OH
 H——OH          HO——H
HO——H            H——OH
 H——OH           H——OH
   CH₂OH            CH₂OH
```

6.65 Draw the Fischer projection of the C3 epimer of D-glucose. Compare your structure with those in Table 6.1 and give the name of this compound.

6.66 Draw the Fischer projection of the C3 epimer of D-ribose. Compare your structure with those in Table 6.1 and give the name of this compound.

6.67 Identify the following carbohydrates as the α or β anomer:

a.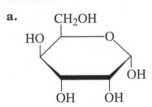

b.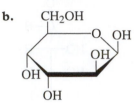

6.68 Identify the following carbohydrates as the α or β anomer:

a.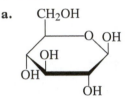

b.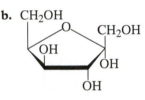

6.69 Draw the α and β anomer of D-mannose.

6.70 Draw the α and β anomer of D-fructose.

6.71 Draw the Fischer projection of the product of the oxidation of D-galactose at C1.

6.72 Draw the Fischer projection of the product of the oxidation of the monosaccharide D-talose at C1.

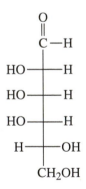

D-Talose

6.73 Draw the Fischer projection of the product of reduction reaction of D-galactose at C1.

6.74 Draw the Fischer projection of the product of the reduction reaction of D-talose at C1.

6.75 Will the following carbohydrates produce a positive Benedict's test?
a. D-glucose
b. lactose
c. sucrose
d. starch

6.76 Will the following carbohydrates produce a positive Benedict's test?
a. D-mannose
b. L-fucose
c. maltose
d. glycogen

6.77 Draw the product of the following $1 \rightarrow 4$ condensation and name the glycosidic bond:

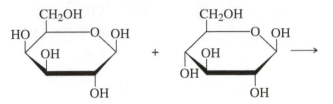

6.78 Draw the product of the following $1 \rightarrow 4$ condensation:

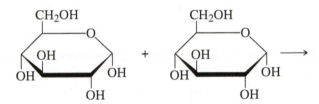

6.79 Isomaltose, a disaccharide formed during caramelization in cooking, contains two glucose units bonded $\alpha(1 \rightarrow 6)$. Draw the structure of isomaltose.

6.80 The glycosidic bond in a disaccharide was determined to be $\alpha(1 \rightarrow 6)$. Hydrolysis of the disaccharide produced one galactose and one fructose. Draw the structure of the disaccharide.

6.81 The glycosidic bond in a disaccharide was determined to be $\beta(1 \rightarrow 4)$. Hydrolysis of the disaccharide produced two mannoses. Draw the structure of the disaccharide.

6.82 Our bodies cannot digest cellulose because we lack the enzyme cellulase. Why is cellulose an important part of a healthy diet if we cannot digest it?

6.83 The shell of a shrimp is composed of chitin. If you eat a boiled shrimp without removing the shell, will your body break the shell down into its component sugars? Explain. (*Hint:* Compare chitin's structure to that of amylose and cellulose.)

6.84 Glycogen and amylopectin are both branched polymers of glucose. Read the descriptions of each in Section 6.6. Which molecule has a more compact structure? Explain.

6.85 Explain whether the following blood types could be accepted by a person with type O blood:
a. B
b. A

6.86 Explain whether the following blood types could be accepted by a person with type A blood:
a. AB
b. O

Challenge Problems

6.87 On an exam, a student was asked to draw the Fischer projection of L-glucose, but he had only memorized the structure of D-glucose. He wrote the structure of D-glucose and switched the hydroxyl group on C5 from the right to the left. Was his answer correct? If not, what is the name of the aldose that he drew?

6.88 Carbohydrates are abbreviated using a three-letter abbreviation followed by their glycosidic bond type. For example, maltose and sucrose can be written respectively as

Glcα(1 → 4)Glc   Glcα(1 → 2)βFru
Maltose   Sucrose

Provide the structure for the O-type blood carbohydrate set given the following abbreviation:

L-Fucα(1 → 2)Galβ(1 → 4)GlcNAc

6.89 The structure of sucralose, found in the artificial sweetener Splenda, is shown in the figure. It consists 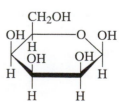 of a chlorinated disaccharide made up of galactose and fructose. In its structure shown,

a. identify the galactose unit and the fructose unit.

b. identify the type of glycosidic bond present.

c. determine if Splenda is a reducing sugar.

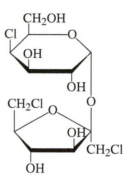

6.90 Which of the components in starch is more likely to be broken down more quickly in plants, amylose or amylopectin? Why?

6.91 How much energy is produced if a person eats 50 g of digestible carbohydrate (not fiber) in a day? In this case, what percent of a 2200 Calorie diet would be digestible carbohydrate? Recall that carbohydrates provide four Calories of energy per gram consumed.

6.92 D-Fructose can also form a six-membered ring. Draw the β anomer of D-fructose in the six-membered ring form.

6.93 Trehalose is a naturally occurring disaccharide used in cosmetics because of its ability to retain moisture. The formal name of trehalose is glucose α, α(1 → 1) glucose. Draw the structure of trehalose. Is it a reducing or nonreducing sugar?

Answers to Odd-Numbered Problems

Practice Problems

6.1 **a.** polysaccharide   **b.** oligosaccharide
c. disaccharide

6.3 **a.** soluble   **b.** soluble   **c.** insoluble

6.5 **a.** secondary   **b.** primary
c. tertiary   **d.** secondary

6.7 **a.** ketone   **b.** ketone   **c.** aldehyde

6.9 **a.** aldohexose   **b.** ketopentose   **c.** aldotetrose

6.11 **a.** D-isomer   **b.** L-isomer   **c.** D-isomer

6.13 **a.**   **b.**

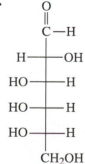

6.15 A—diastereomer; B—diastereomer; C—enantiomer

6.17 **a.**   **b.**

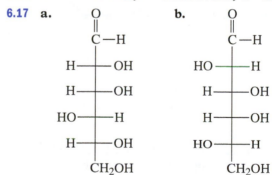

6.19 **a.** D-glucose   **b.** D-fructose   **c.** D-mannose

6.21 **a.** insulin   **b.** glucagon   **c.** insulin

6.23 **a.** beta   **b.** beta

6.25

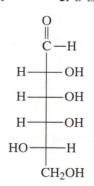

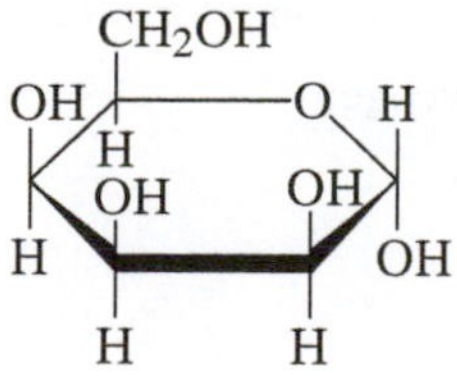

α Anomer   β Anomer

6.27 carboxylic acid

6.29

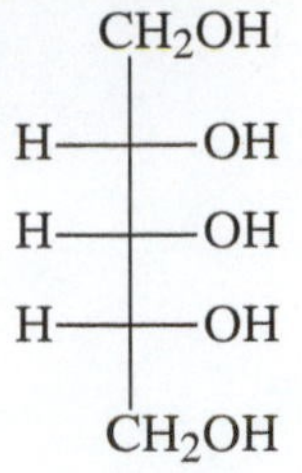

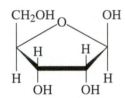

D-Ribitol

6.31

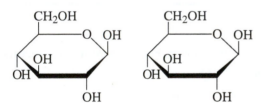

6.33 **a.** hydrolysis   **b.** condensation   **c.** hydrolysis

6.35 $\alpha(1 \rightarrow 6)$

6.37 **a.** bond is $\alpha(1 \rightarrow 2)$

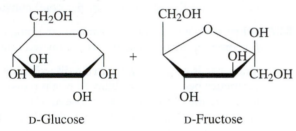

D-Glucose   D-Fructose

b. bond is $\beta(1 \rightarrow 4)$

Both monosaccharides are D-glucose

6.39 **a.** sucrose   **b.** lactose

c. maltose   **d.** lactose

6.41 **a.** agave syrup

6.43 **a.** Both contain $\alpha(1 \rightarrow 4)$ glycosidic bonds and only D-glucose. Amylopectin also contains branching $\alpha(1 \rightarrow 6)$.

b. Both contain $\alpha(1 \rightarrow 4)$ glycosidic bonds and branching $\alpha(1 \rightarrow 6)$ and only D-glucose. Branching occurs more often in glycogen than amylopectin.

6.45 **a.** cellulose, chitin   **b.** amylose, amylopectin

c. amylose   **d.** glycogen

6.47 **a.** no   **b.** no

6.49 In contrast to the glucose polysaccharides that contain only glucose as the repeating unit, heparin contains two monosaccharides as the repeating unit. The monosaccharides in heparin are also charged, whereas glucose is not.

Additional Problems

6.51 $C_3H_6O_3$

6.53 An oligosaccharide is smaller, containing between 3 and 9 monosaccharide units, while a polysaccharide contains 10 or more monosaccharide units.

6.55 Soluble fiber mixes with water and forms a gel-like substance that gives a feeling of fullness when eaten.

6.57 aldehyde, hydroxyls (alcohol)

6.59 **a.** secondary   **b.** primary   **c.** secondary

d. primary   **e.** secondary

6.61 Carbons 2–5 in glucose are secondary, C6 is primary.

6.63 **a.** enantiomers; left molecule is D-isomer, right is L-isomer

b. epimers (differ only at C3); both isomers are D

6.65

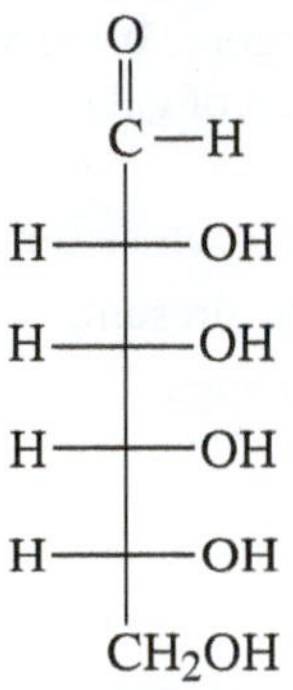

```
        O
        ‖
        C—H
H———OH
H———OH
H———OH
H———OH
      CH₂OH
```

D-Allose

6.67 **a.** alpha   **b.** beta

6.69

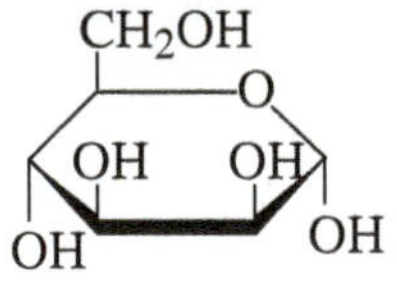

α-D-Mannose β-D-Mannose

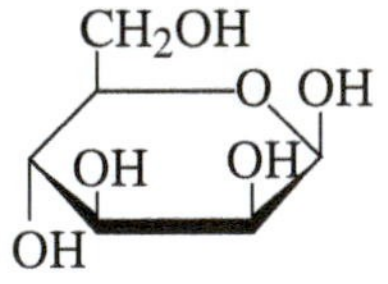

6.71

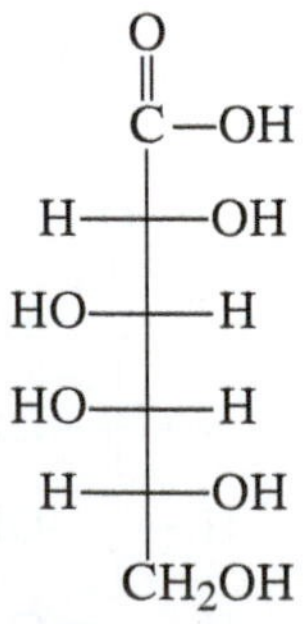

```
        O
        ‖
        C—OH
H———OH
HO———H
HO———H
H———OH
      CH₂OH
```

Galactose oxidized at C1

6.73

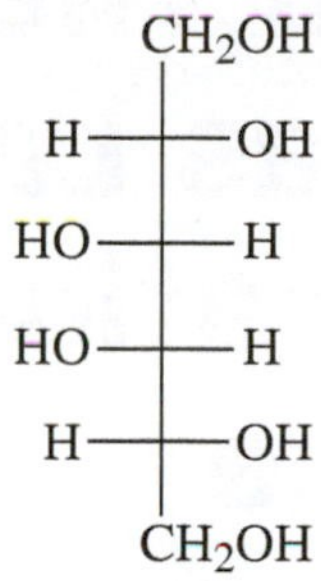
CH2OH
H
OH
HO
H
HO
H
H
OH
CH2OH

Galactose reduced at C1

6.75 **a.** yes **b.** yes
c. no **d.** no

6.77

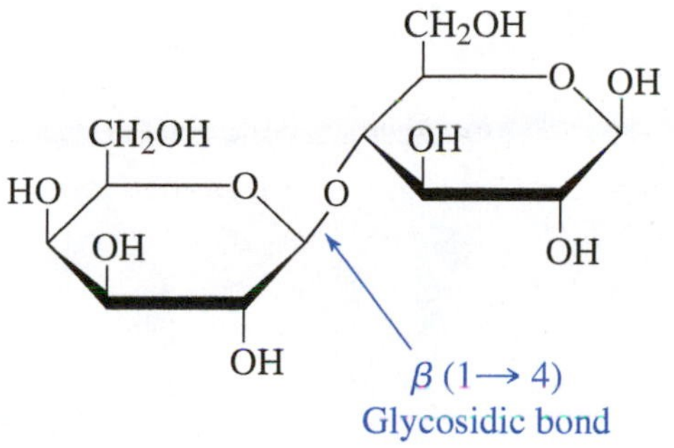
CH2OH
O
OH
CH2OH
OH
HO
O
O
OH
OH
OH
β (1→ 4)
Glycosidic bond

6.79

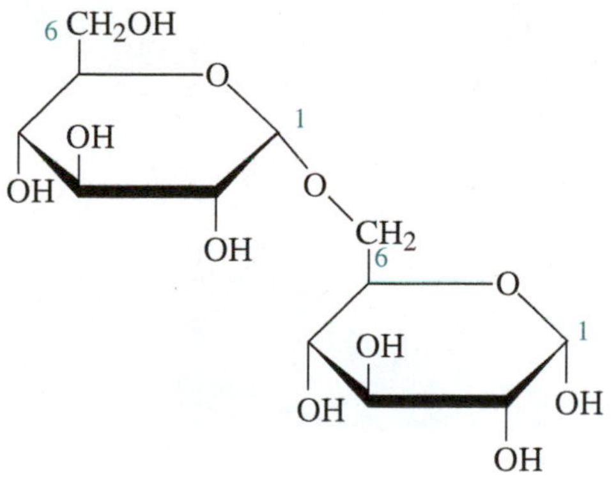
6 CH2OH
O
OH
1
OH
O
OH
CH2
6
O
OH
1
OH
OH
OH

Isomaltose

6.81

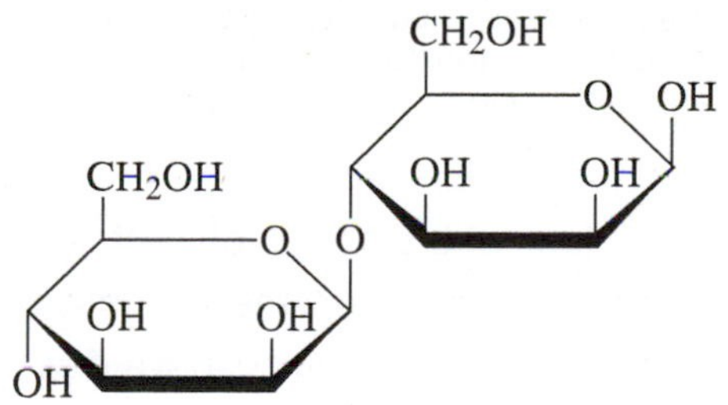
CH2OH
O
OH
CH2OH
OH
OH
O
O
OH
OH
OH

6.83 Like cellulose, chitin has $\beta(1 \rightarrow 4)$ glycosidic linkages. Given this similarity and the fact that our bodies can't break down cellulose into its component sugars, a good hypothesis would be that our bodies will most likely not break down chitin.

6.85 **a.** no **b.** no

6.87 His structure was not correct. L-glucose is the mirror image of D-glucose, not simply the C5 epimer. The student drew L-idose.

6.89 **a.** and **b.**

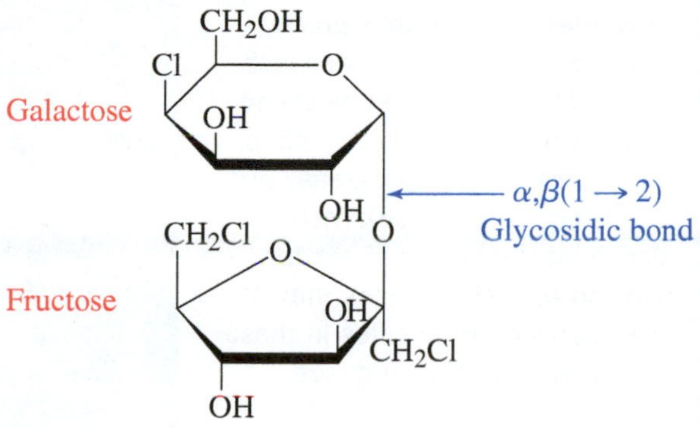
CH2OH
Cl
O
Galactose
OH
OH
O
α,β(1 → 2)
Glycosidic bond
CH2Cl
O
Fructose
OH
CH2Cl
OH

c. Splenda is not a reducing sugar.

6.91 200 Calories, 9%

6.93 Trehalose is difficult to draw in the pyranose ring forms we are accustomed to because both monosaccharides are in the alpha configuration, so we have to "extend" the glycoside bond to beable to represent the structure. Because the two anomeric carbons are joined, meaning there is no free anomeric hydroxyl, trehalose is a nonreducing sugar.

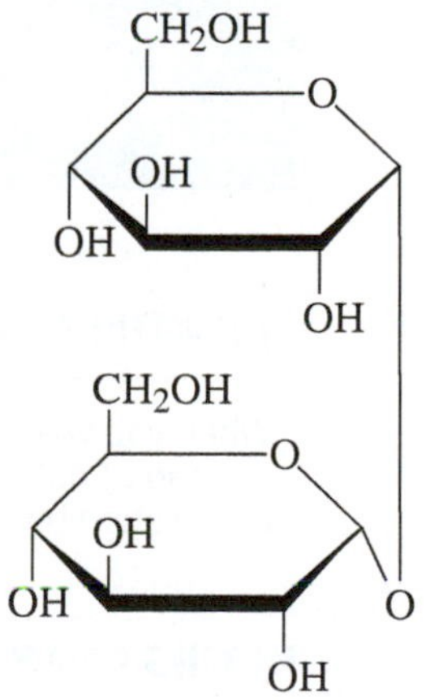
CH2OH
O
OH
OH
OH
CH2OH
O
OH
OH
O
OH

How does soap remove greasy dirt from clothes when we wash them? The answer can be found in the attractions between soap molecules, water, and greasy dirt. In this chapter, we look more closely at the states of matter and the attractive forces that exist between molecules in these states. We begin with gases.

CHAPTER OUTLINE

LEARNING TIP   Mix Up Your Practice Problems

When you work problems on the different gas laws while studying, mix up the problems instead of doing a whole set on one gas law at a time. Your ability to discriminate the type of question being asked on an exam will get better if you mix problems when studying.

IN THIS CHAPTER, we take a more in-depth look at the states of matter. Whether matter is in a solid, liquid, or gaseous state depends on the attractive forces between compounds. The chapter starts with gases, which have few attractive forces. However, gases have some unique properties and are key to our survival—gases move in and out of our bodies through respiration and dissolve in liquids like blood—so they deserve a look. From gases, we move to the liquid and solid states and examine attractive forces between molecules. We apply attractive forces to the physical properties of boiling, melting, vapor pressure, and solubility in water. From there, we apply attractive forces to further understand several lipids, including soaps, dietary fats and oils, phospholipids, and cholesterol, and finish the chapter with the structure of a cell membrane.

7.1 Gases and Gas Laws

Let's begin our study of the states of matter with gases, which have very few attractive forces. Recall that in gases, particles of matter are extremely far apart (see Section 1.5) so they don't typically interact with each other. However, gases have other properties, such as pressure and solubility, that play roles in respiration and the dissolving of gases in the bloodstream, respectively.

Gases and Pressure

One of the most important quantities to consider with gases is pressure. To illustrate the concept of pressure, imagine an oral medicine syringe with no needle attached.

If the plunger of the syringe is drawn all the way out, the syringe fills with air. If you place your finger firmly over the open tip of the syringe and depress the plunger, what happens? The plunger moves in slightly, and the sample of air is "squeezed." Remember that the particles of a gas are usually far apart and not closely associated with each other. In other words, a sample of gas is mostly empty space. When the air in the syringe is squeezed, the space between the particles is decreased; the particles of the air are forced closer together and have less room to move about. By depressing the plunger, you are compressing the gas by applying pressure to it. **Pressure** is a force exerted against a given area—in this case, the sides of the syringe cylinder.

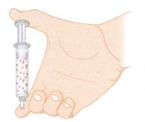

The thumb pressing on the plunger creates pressure on the air particles inside the syringe.

Pressure measurements are expressed in a variety of units. Air pressure is often measured in the unit **atmosphere (atm).** The SI unit of pressure is the **pascal (Pa).** For our discussion, we will use two of the more common units, pounds per square inch (psi) and millimeters of mercury (mmHg). The latter unit is often used when measuring blood pressure.

Suppose you placed a 5-pound bag of sugar on top of your thumb. You would feel pressure on your thumb. Your thumbnail area is about 1 square inch, so the pressure would be about 5 pounds per square inch, or 5 psi. The unit of **pounds per square inch (psi)** measures pressure as the force (measured in pounds) applied to an area of 1 square inch. The pressure of the atmosphere at sea level is about 14.7 pounds per square inch, or 14.7 psi. The units atmosphere, pascal, pounds per square inch, and mmHg can be expressed as equivalent units as

$$1.00 \text{ atm} = 101{,}325 \text{ Pa} = 14.7 \text{ psi} = 760 \text{ mmHg}$$

The unit **millimeters of mercury (mmHg)** comes from the mercury barometer which measures the pressure of the Earth's atmosphere. A simple barometer consists of a long, sealed glass tube filled with liquid mercury inverted into a dish of mercury without letting air into the tube (see **FIGURE 7.1**). The force of the atmosphere pushing down on the mercury in the dish prevents the mercury in the tube from draining out. In fact, if the tube is long enough, a column of mercury 760 mm high (29.92 in.) will remain inside the tube at sea level. Because the pressure exerted by the atmosphere determines the height of the column of mercury in the tube, measuring the height of the mercury column in millimeters gives us the pressure in mmHg.

As mentioned, the typical pressure exerted by the atmosphere at sea level supports a column of mercury 760 mm high, equivalent to 1.00 atmosphere, 14.7 pounds per square inch, or 101,325 pascals. Atmospheric pressure varies with altitude and changes in the weather.

What's an Inquiry Question?

Inquiry Questions are designed to focus your reading on the main concepts by section. As you read each section, see if you can answer each Inquiry Question in your own words.

7.1 Inquiry Question

How do the gas laws demonstrate the properties of gases?

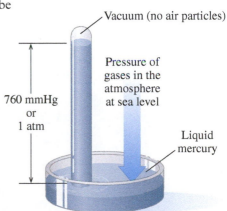

FIGURE 7.1 The mercury barometer. The pressure of the gases in Earth's atmosphere is able to support a column of mercury 760 mm high, equivalent to a pressure of 1 atm.

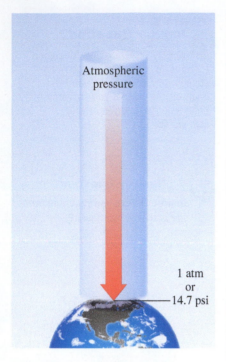

At sea level, the atmosphere exerts an average pressure of 1 atmosphere, or 14.7 pounds per square inch, on the Earth's surface.

Sample Problem 7.1 **Converting Between Pressure Units**

Convert the following barometric readings to the units requested.
  a. 1.08 atm to mmHg   b. 750. mmHg to psi   c. 0.94 atm to psi

Solution

Use conversion factors to calculate equivalent pressure units, being sure the desired unit is in the numerator and the other units cancel.

$$1.00 \text{ atm} = 14.7 \text{ psi} = 760 \text{ mmHg}$$

a. $\dfrac{760 \text{ mmHg}}{1.00 \text{ atm}} \times 1.08 \text{ atm} = 821 \text{ mmHg}$

The answer has three significant figures. The conversion factor 760 mmHg is an exact number.

b. $\dfrac{14.7 \text{ psi}}{760 \text{ mmHg}} \times 750. \text{ mmHg} = 14.5 \text{ psi}$

The answer has three significant figures. The conversion factor 760 mmHg is an exact number.

c. $\dfrac{14.7 \text{ psi}}{1.00 \text{ atm}} \times 0.94 \text{ atm} = 14 \text{ psi}$

The answer has two significant figures

Gas Behavior

Whether we are considering oxygen gas (O_2), nitrogen gas (N_2), carbon dioxide gas (CO_2), or any other gas, they all behave similarly. Scientists studying gases in the seventeenth and eighteenth centuries noticed some of their unique behaviors and developed the kinetic molecular theory of gases to explain them. A gas that perfectly adheres to the kinetic molecular theory of gases is said to be an **ideal gas.** For our initial encounter with gases, we make the following assumptions as outlined in the **kinetic molecular theory of gases:**

- Gas particles are far apart from each other. Because of this, most of the volume of a gas is empty space.
- Gas particles are in constant, random motion. The particles have a range of speeds.
- Gas particles have no attractive forces between them. When gas particles collide, energy is conserved.
- Gas particles are moving and therefore have kinetic energy. This energy is directly proportional to the absolute temperature. Gas particles move more quickly at higher temperatures.

These assumptions about gas behavior allow us to quantify relationships as gases change conditions like pressure, volume, temperature, and amount of gas (number of moles, n). Next we take a brief look at several gas laws that hold for an ideal gas.

Pressure and Volume—Boyle's Law

In Chapter 1, we saw that when a gas is enclosed in a container, it will occupy the entire volume of the container. The gas particles exert a certain force, or pressure, against the walls of the container. If we think back to the plugged syringe, as the plunger is depressed, the applied pressure increases and the volume of the air decreases (see **FIGURE 7.2**).

In the mid-1600s, the Irish chemist Robert Boyle first noted this relationship in gases and began to experiment with the effect of pressure on the volume of a gas. In his experiments, Boyle discovered that the volume of a gas and the pressure applied to it are related if the temperature and amount of the gas are not allowed to change. He found that when the pressure on a gas was doubled, the volume of the gas was reduced to one-half of its

FIGURE 7.2 Pressure versus volume of a gas. When the pressure on a gas is doubled, the volume is halved. When the pressure is tripled, the volume decreases to one-third of the initial volume.

Initial pressure   Twice the pressure   Triple the pressure

initial volume. Similarly, if he tripled the pressure on the gas, the volume was reduced to one-third of its initial volume. Figure 7.2 shows how these changes in pressure affect the volume of a gas.

Boyle found, and his law states, *that the volume of a fixed amount of gas at constant temperature is inversely proportional to the pressure.* Simply stated, Boyle's law says that an increase in the pressure on a gas will result in a decrease of its volume and vice versa.

Mathematically, Boyle's law allows us to determine what will happen to a sample of gas of known volume and pressure if we make changes to either the volume or the pressure while the temperature stays the same. The following formula gives the relationship for such a determination:

$$P_i \times V_i = P_f \times V_f$$

where P_i = the initial (or starting) pressure
  V_i = the initial (or starting) volume
  P_f = the final (or ending) pressure
  V_f = the final (or ending) volume

Boyle's Law

Solving a Problem

The lungs of a normal adult female hold 4.2 L of air under typical atmospheric pressure (1.00 atm). If a diver holds her breath and dives to a depth of 5.0 m and the pressure increases to 1.5 atm, what is the volume of the lungs at this pressure?

STEP 1 Determine the given information. In this problem, we are given P_i = 1.00 atm, V_i = 4.2 L, and P_f = 1.5 atm. The depth of 5.0 m is given, but this information is not necessary to solve the problem.

STEP 2 Solve for the missing variable using the Boyle's law relationship. In this case, we are looking for V_f.

$$P_i \times V_i = P_f \times V_f$$

Divide both sides by P_f.

$$\frac{P_i \times V_i}{P_f} = \frac{P_f \times V_f}{P_f}$$

This gives us

$$\frac{P_i \times V_i}{P_f} = V_f$$

STEP 3 Substitute the given information into the equation and solve.

$$\frac{1.00 \text{ atm} \times 4.2 \text{ L}}{1.5 \text{ atm}} = V_f$$

$$V_f = 2.8 \text{ L}$$

STEP 4 Check your answer. When you complete the calculation (taking care to cancel the units on your numbers), check your answer by seeing if Boyle's law holds. Did the volume go down when the pressure went up? Yes, just as you may have predicted.

Boyle's Law and Breathing

Breathing is an everyday application of Boyle's law. When you breathe in, you are not forcibly drawing air into your lungs. Instead, the muscles of your rib cage and your diaphragm contract to cause the volume of your chest cavity to increase (see **FIGURE 7.3**). When this happens, the air pressure inside your lungs decreases; this is called negative pressure (volume increase results in pressure decrease). The higher pressure of the atmosphere causes air to rush into your lungs to equalize the internal and external pressures. When the

INTEGRATING
Chemistry

◀ Find out how you use Boyle's law every day.

muscles relax, the volume of your chest cavity decreases, increasing the pressure in your lungs above that of the outside, and air flows out to the lower pressure (atmosphere).

Boyle's law also helps explain how we can get more oxygen into the lungs using hyperbaric oxygen therapy. In a hyperbaric chamber, the pressure is higher than normal atmospheric pressure, so more pressure presses against the alveoli in the lungs. When the external pressure is higher, the volume of the alveoli decreases and more oxygen is present to be transferred to the blood through diffusion.

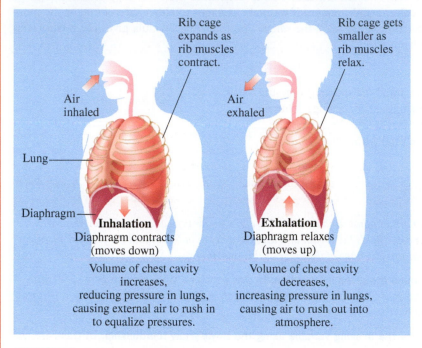

FIGURE 7.3 Boyle's law and breathing. Breathing is controlled by contraction and relaxation of the diaphragm muscles.

Sample Problem 7.2   Applying Boyle's Law

A pilot undergoes altitude training. He starts his training at an atmospheric pressure of 762 mmHg, where his lungs hold 6.00 L of air. He changes altitude, and his lungs expand to hold 6.25 L of air. What is the new atmospheric pressure in mmHg?

Solution

Changes in the volume and pressure of a gas are related using Boyle's law. First, check the problem to determine what is given and what information is requested.

STEP 1 Determine the given information. We are given $P_i = 762$ mmHg, $V_i = 6.00$ L, and $V_f = 6.25$ L.

STEP 2 Solve for the missing variable using the Boyle's law relationship. In this case, we are looking for P_f.

$$\frac{P_i \times V_i}{V_f} = P_f$$

STEP 3 Substitute the given information into the equation and solve.

$$\frac{762 \text{ mmHg} \times 6.00 \text{ L}}{6.25 \text{ L}} = P_f$$

$$P_f = 732 \text{ mmHg}$$

STEP 4 Check your answer. If the volume increases, Boyle's law tells us that the pressure should decrease. Is the final pressure lower than the initial pressure given? Yes.

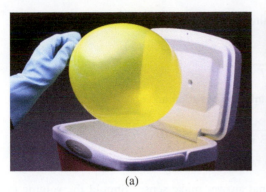 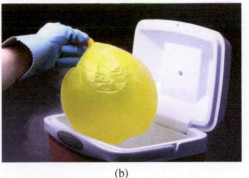

(a)   (b)

FIGURE 7.4 Investigating a balloon's volume versus its temperature. The balloon (a) is considerably smaller after being stored in a cooler (b). This demonstrates that as the temperature of a gas decreases, so does its volume.

Temperature and Volume—Charles's Law

To see how temperature affects the volume of a gas, consider the photos in **FIGURE 7.4**. In the first photo, you see a fully inflated balloon just before it was placed in a cooler at very low temperature. In the second photo, you see the same balloon just after it was taken out of the cooler. Notice the difference? After being in the cooler, the balloon is smaller—its volume has decreased. This relationship between the temperature of a gas and its volume was discovered in the late 1700s by French scientist Jacques Charles.

Why does this occur? Changing the temperature of a gas directly affects the motion of the particles. When the temperature of a gas decreases, the motion of the particles (speed) decreases and they take up less space. Similarly, adding heat energy increases the temperature of the gas and causes the gas particles to move more rapidly, requiring more space. In each case, the change in the temperature results in a change in the motion of the particles of the gas.

In his experiments with gases, Charles found that if the pressure and amount of a gas are not allowed to change, the volume of the gas is directly proportional to its absolute temperature. We'll discover why he used absolute temperature a little later. Charles discovered that when the absolute temperature of a gas was doubled, the volume of the gas also doubled.

Charles's law states that *the volume of a fixed amount of gas at constant pressure is directly proportional to its absolute temperature*. In other words, an increase in the temperature of the gas will result in an increase in its volume, whereas a decrease in temperature will result in a decrease in its volume.

We can use Charles's law to determine what will happen to a sample of gas of known volume and temperature if we make changes to either its volume or its temperature when the pressure remains the same. Charles's law can be applied as shown for making such determinations:

$$\frac{V_i}{T_i} = \frac{V_f}{T_f}$$

where T is the absolute temperature in units of Kelvin, V is volume, and the subscripts i and f stand for "initial" and "final," respectively (as they did in the Boyle's law discussion).

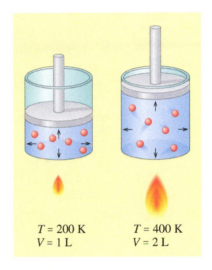

$T = 200 \text{ K}$ $T = 400 \text{ K}$
$V = 1 \text{ L}$ $V = 2 \text{ L}$

The effect of temperature on the volume of a gas can be shown in an expandable container. When the temperature doubles, its volume doubles. The increased motion of the gas particles causes the volume to increase.

Charles's Law

Solving a Problem

If you remove a balloon with a volume of 105 mL of air from a freezer at 0 °C and allow the balloon to warm to a room temperature of 25 °C, what would its final volume be?

STEP 1 Determine the given information. We are given $V_i = 105$ mL, $T_i = 0$ °C, and $T_f = 25$ °C. Temperatures of 0 °C cannot be put into Charles's law because, mathematically, division by zero is not allowed. In gas law problems, temperatures must be converted to the absolute temperature scale, Kelvin. (Refer to Chapter 1, Section 1.5, to review how to convert temperature units.) So $T_i = 273$ K and $T_f = 298$ K.

STEP 2 Solve for the missing variable using the Charles's law relationship. In this case, we are looking for V_f.

$$\frac{V_i \times T_f}{T_i} = V_f$$

STEP 3 Substitute the given information into the equation and solve.

$$\frac{105 \text{ mL} \times 298 \text{ K}}{273 \text{ K}} = V_f$$

$$115 \text{ mL} = V_f$$

STEP 4 Check your answer. Does the final answer make sense? Should the volume of the balloon increase as the temperature increases? Yes. That is exactly what Charles's law predicts.

Sample Problem 7.3 Applying Charles's Law

Determine the volume of the balloon we just discussed (105 mL in the freezer) if it is removed from the freezer at 0 °C and warmed to 145 °F.

Solution

STEP 1 Determine the given information. We cannot complete our calculation with the temperature in degrees Fahrenheit. The temperature must be converted to the Kelvin scale. We convert the temperature first to °C and then to K as follows.

$$°C = (145 \text{ °F} - 32) \times \frac{1 \text{ °C}}{1.8 \text{ °F}}$$

$$°C = 63$$

$$K = 63 + 273 = 336 \text{ K}$$

The final temperature T_f is 336 K. We are also given $V_i = 105$ mL and $T_i = 273$ K (after converting from Celsius).

STEP 2 Solve for the missing variable using the Charles's law relationship. In this case, we are looking for V_f.

$$\frac{V_i \times T_f}{T_i} = V_f$$

STEP 3 Substitute the given information into the equation and solve.

$$\frac{105 \text{ mL} \times 336 \text{ K}}{273 \text{ K}} = V_f$$

$$V_f = 129 \text{ mL}$$

STEP 4 Check your answer. The volume increased because the temperature went up, so our answer follows Charles's law.

Temperature and Pressure—Gay-Lussac's Law

What happens to the air pressure inside a gas tank (fixed volume) when the temperature increases? From our earlier discussion, we can reason that the gas particles move faster and hit the walls of the tank more frequently, increasing the pressure. This phenomenon was first observed in various gases by Joseph Louis Gay-Lussac.

We can use Gay-Lussac's law to determine what happens to a sample of gas of known pressure and temperature if we make changes to either its pressure or its temperature when the volume remains the same. Gay-Lussac's law can be applied as shown:

$$\frac{P_i}{T_i} = \frac{P_f}{T_f}$$

where P is pressure, T is absolute temperature in units of Kelvin, and subscripts i and f stand for "initial" and "final," respectively (as they did in Boyle's and Charles's laws).

Temperature, Pressure, and Volume—The Combined Gas Law

Charles's, Boyle's, and Gay-Lussac's laws can be combined since pressure, volume, and temperature hold the same relationship, even if all three change for the same amount of gas so that:

$$\frac{P_i V_i}{T_i} = \frac{P_f V_f}{T_f}$$

Sample Problem 7.4   Applying the Combined Gas Law

Mastering Video
Practicing the Concepts
Behavior of Gases

A 1.20-L sample of a gas at 30. °C and 1.00 atm is heated to 225 °C and 1.50 atm final pressure. What is the final volume of the gas?

Solution

STEP 1 Determine the given information. We cannot complete our calculation with the temperature in degrees Celsius. The temperature must be converted to the Kelvin scale.

$$T_i = 30 + 273 = 303 \text{ K}; \ T_f = 225 + 273 = 498 \text{ K}$$

From the information in the problem, we are also given the initial and final pressures, $P_i = 1.00$ atm and $P_f = 1.50$ atm, and the initial volume $V_i = 1.20$ L. Because the temperature, pressure, and volume all change, we use the combined gas law to solve this problem.

STEP 2 Solve for the missing variable using the combined gas law relationship. In this case, we are looking for V_f.

$$\frac{P_i V_i \times T_f}{T_i \times P_f} = V_f$$

STEP 3 Substitute the given information into the equation and solve.

$$\frac{1.00 \ \cancel{\text{atm}} \times 1.20 \text{ L} \times 498 \ \cancel{\text{K}}}{303 \ \cancel{\text{K}} \times 1.50 \ \cancel{\text{atm}}} = V_f$$

$$V_f = 1.31 \text{ L}$$

STEP 4 Check your answer. Combined gas law problems can be tricky. Depending on which variable changes the most, the temperature or the pressure, the volume may go either up or down. Because the temperature increased more than the pressure (which has the effect of decreasing the volume), the volume increased. In these problems, it is best to double-check how you arranged the equation and your math.

Using the Ideal Gas Law to Recall Other Gas Laws

Many students can recite the ideal gas law, $PV = nRT$. Knowing this simple formula can assist in remembering all the previously noted gas laws. All the relationships from Boyle's, Charles's, and Gay-Lussac's laws can be found in this one simple equation. If the number of moles of gas present, *n*, is held constant and R is the ideal gas constant, the ideal gas law becomes

$$\frac{PV}{T} = \text{constant}$$

All the previous gas laws are identifiable in this relationship. It is easier to recall the relationships between the gas variables in this way than to memorize several gas law equations.

Practice Problems

Find the answers to the odd-numbered Practice Problems at the end of the chapter.

7.1 Convert the following barometric pressure readings to psi.

a. 0.988 atm   b. 748 mmHg   c. 101,278 Pa

7.2 Convert the following barometric pressure readings to atm.

a. 101,331 Pa   b. 762 mmHg   c. 14.6 psi

7.3 Suppose you had the same amount of gas in two identical containers, A and B. Container A is at 300 K and container B is at 350 K. Which gas is at higher pressure? Explain.

7.4 Suppose you had the same amount of gas in two containers, C and D. Both containers are at the same temperature. Container C is half the size of container D. Which gas is at higher pressure? Explain.

7.5 A child's balloon containing 8.5 L of helium gas at sea level is released and floats away. What will the volume of the helium be when the balloon rises to the point where the atmospheric pressure is 380 mmHg?

7.6 When you swim underwater, the pressure of the water pushes against your chest cavity and subsequently reduces the size of your lungs. For every 10 m below the water surface, the pressure exerted by the water adds 14.7 psi to atmospheric pressure, so the total pressure 10 m below the surface would be 29.4 psi. If you have a lung volume of 6 L at sea level (atmospheric pressure), what would the volume of your lungs be when you are at the bottom of a pool that is 5 m deep?

7.7 A balloon is blown up to a volume of 0.50 L at 72 °F. The balloon sits in a hot car, which reaches 103 °F. Assuming constant pressure, what is the new volume of the balloon?

7.8 A gas has a volume of 10 L at 0 °C. What is the final temperature of the gas (in °C) if its volume increases to 25 L?

7.9 Nitrous oxide (N_2O), or laughing gas, is a common anesthetic agent. As a cylinder containing N_2O gas is emptied, what happens to the pressure within the tank?

7.10 Assuming no air can move in or out, what happens to the volume of a hot-air balloon as the air is heated?

7.11 A balloon is filled with helium gas, which is lighter than air. Which of the diagrams (A, B, or C) looks most like the new volume of the balloon when the following changes are made?

Initial volume   A   B   C

a. The balloon floats to a higher altitude where the outside pressure is lower, but the temperature is the same.

b. The balloon is placed in a hyperbaric chamber where the pressure is increased, but the temperature stays the same.

c. The temperature of the balloon changes from 5 °C to 40 °C at constant pressure.

d. The balloon is warmed and then cooled to its initial temperature and the pressure remains the same.

7.12 An Ambu bag is used in emergency resuscitation. Squeezing an Ambu bag decreases its volume. What happens to the pressure inside the bag? How does this help a patient who is not breathing on his own?

A manual breathing unit, the Ambu bag

7.13 A 25.0-mL bubble is released from a diver's air tank at a pressure of 4.00 atm and a temperature of 11 °C. What is the volume (mL) of the bubble when it reaches the ocean surface where the pressure is 1.00 atm and the temperature is 18 °C?

7.14 A 100.0-mL bubble of hot gases at 225 °C and 1.80 atm escapes from an active volcano. What is the new volume of the bubble outside the volcano where the temperature is −25 °C and the pressure is 0.80 atm?

7.15 Suppose you have two identical containers at the same temperature. One contains 4 moles of H_2, and the other contains 4 moles of O_2. According to the ideal gas law, which one has the higher pressure?

7.16 As a cylinder of compressed gas empties, the pressure inside the cylinder decreases. Explain why this happens, using the ideal gas law.

DISCOVERING THE CONCEPTS

? INQUIRY ACTIVITY—The Attractive Forces

Information

Molecules "stick" together in the liquid and solid states due to attractions between the molecules. This accounts for some of their physical properties such as boiling point, melting point, viscosity, and surface tension. In solids, each molecule is fixed (moves very little) relative to the neighboring molecules, and the attractive forces are optimized.

Attractive forces are often referred to as "intermolecular forces" because they occur between (*inter-* is a prefix meaning "between") molecules, not within molecules. Technically, ions are not molecules, so the term *attractive* force is used here to encompass the five forces shown in **TABLE 1**.

Questions

(Examine Table 1, on the next page, to answer the following questions.)

1. What is common about the attractions between the forces shown circled in Table 1?
2. Which is stronger, a formal $+/-$ attraction (caused by a gain or loss of electrons) or a partial δ^+/δ^- attraction (caused by a shift in electron position)?
3. Which of the attractive forces occurs between nonpolar organic molecules?
4. Hydrogen bonds are a unique attractive force (not an actual covalent bond) present between some molecules. The attraction occurs between a δ^+ (also known as *polarized*) hydrogen in a polar bond with O, N, or F and a lone pair of electrons (:) on a neighboring O, N, or F. The hydrogen is called the *donor*, and the pair of electrons is referred to as the *acceptor*. This attraction is indicated with a dashed line between the polarized H and O, N, or F.
   a. How many polarized Hs are in water?
   b. How many lone pairs of electrons are in a molecule of H_2O?
   c. Based on your answers to a and b, how many H bonds can one water molecule make to neighboring water molecules?
   d. Carefully re-read the information given at the beginning of question 4. Draw all possible hydrogen bonds that can be made between one water molecule and neighboring water molecules.
5. How many hydrogen bonds can one methanol molecule (CH_3OH) make to neighboring methanol molecules? Draw them.
6. The boiling point of pure water is 100 °C, and the boiling point of pure methanol is 65 °C. Provide an explanation for this difference based on the strength or number of attractive forces present in each.
7. Identify the strongest attractive force operating between molecules of the given compound.
   a. methane, CH_4
   b. water, H_2O
   c. chloroform, $CHCl_3$
   d. ethanol, CH_3CH_2OH

—Continued next page

Continued—

TABLE 1 Examples of the Attractive Forces Between Molecules (from weakest to strongest)

Name of Attractive Force	Example
London forces	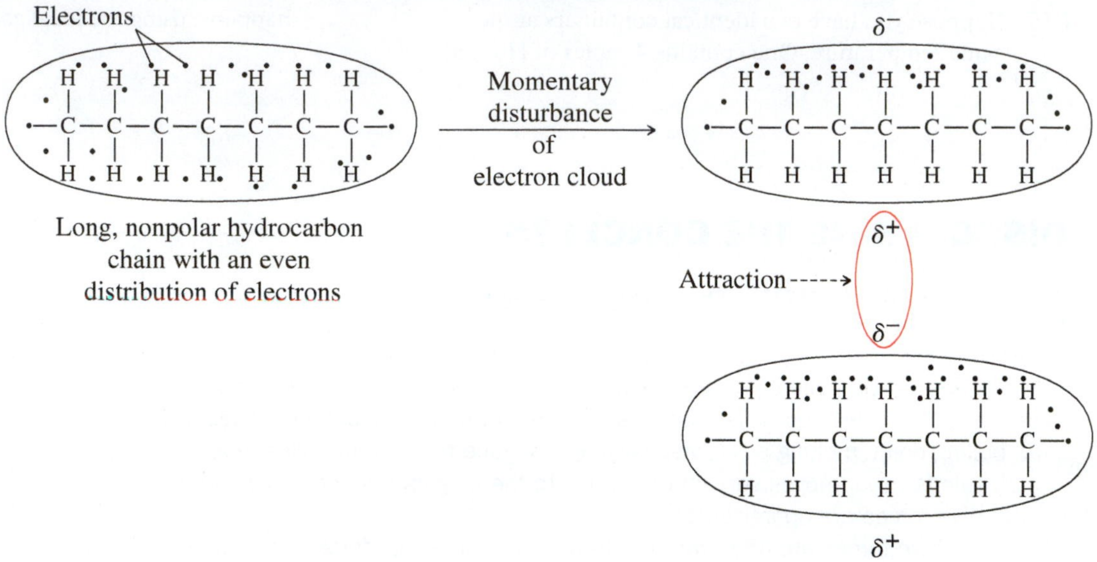 Long, nonpolar hydrocarbon chain with an even distribution of electrons
Dipole–dipole	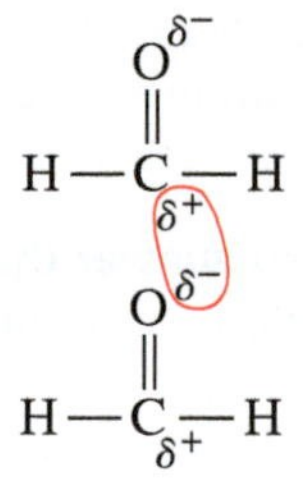
Hydrogen bonding	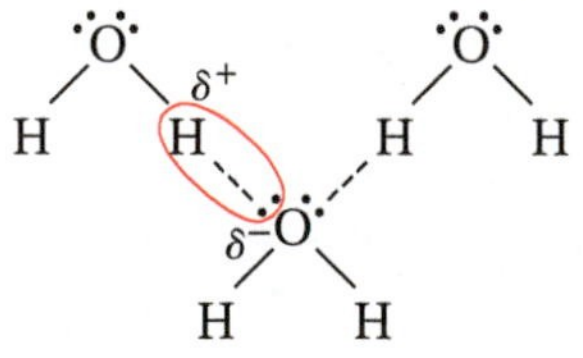
Ion–dipole	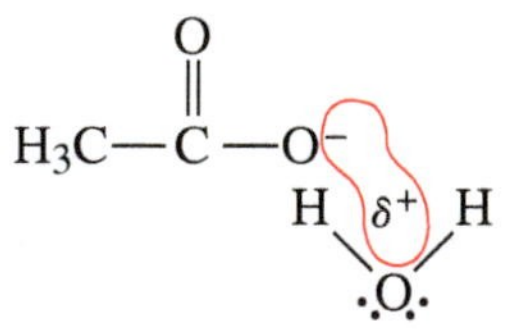
Ionic (also known as a salt bridge)	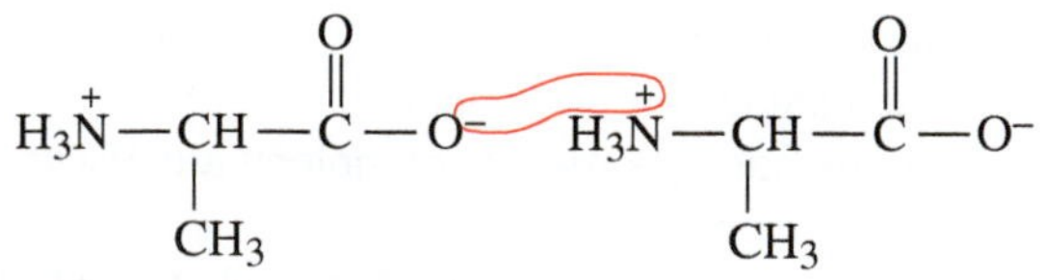

7.2 Liquids and Solids: Predicting Properties Through Attractive Forces

In the classic children's Christmas movie *Frosty the Snowman,* Frosty takes his little friend Karen into a greenhouse to help her warm up. Once they get inside, Professor Hinkle, an evil magician, locks them in, and Frosty changes from a snowman into a puddle of water. In scientific terms, he went from being a solid (snow or ice) to a liquid (water).

How did this happen? How did the greenhouse cause this major change in Frosty's life? Of course, you know that it was the heat in the greenhouse that changed Frosty from a solid to a liquid, but how did the heat cause that change?

Did the heat in the greenhouse cause Frosty's water molecules to move faster? Yes, and by doing so, it disrupted attractive forces between the water molecules. The state in which a given substance exists depends largely on the motion of its molecules—molecules of a liquid move faster and with more types of motion than the molecules of a solid. The attractive forces present in compounds dictate many of the physical properties of substances, like boiling point, vapor pressure, and solubility. We can identify the attractive forces present by looking at a compound's structure. Next, we explore attractive forces by examining some of these physical properties.

7.2 Inquiry Question

What are the attractive forces and how can they be used to predict the relative vapor pressures and boiling points of compounds?

Changes of State and Attractive Forces

Just as holding hands with your sweetie is much easier when you are sitting together on a couch than when you are dancing, attractive forces are more likely between two particles (molecules or ions) moving slowly than between two bouncing wildly about. When two particles with lower energy come near each other, there is a much greater possibility that an attraction will develop than when two particles with higher energy move quickly past each other. In other words, as a substance is heated, the particles begin to move faster and faster. In turn, the attractions between them start to become less important—they are no longer held together as tightly. Remember that in an ideal gas, there are no attractions between molecules.

When a gas is cooled, it can condense, becoming a liquid, and when a liquid is cooled, it can freeze and form a solid. These transitions are called **changes of state.** The changes of state between a solid, liquid, and gas are summarized in **FIGURE 7.5.** These include **freezing** and **melting** (between liquids and solids), **vaporization** and **condensation** (between liquids and gases), and **sublimation** and **deposition** (between solids and gases).

Removing heat slows the movement of the particles in a substance down, and the attractive forces are more likely to hold particles together as matter. Conversely, adding energy as heat to a substance causes its particles to move faster, creating less order among them, which disrupts the attractive forces holding the matter together.

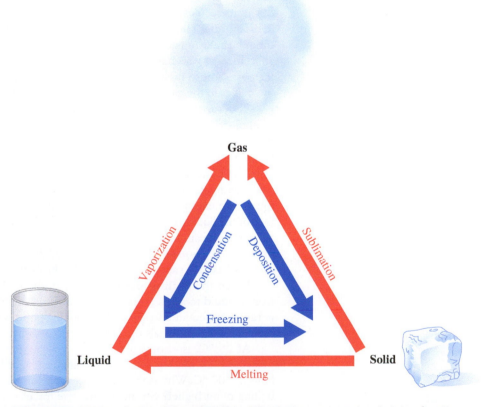

FIGURE 7.5 Changes in the states of matter. Changes to a more ordered state of matter (blue) result in more attractive forces between particles than changes to a less ordered state of matter (red).

Sample Problem 7.5 **Changes in States of Matter**

Name the change of state occurring for each.
  a. ice forming on a puddle of water during the winter
  b. dew forming on grass
  c. water boiling

Solution

Use Figure 7.5 as a guide for this problem. Establish the initial state and final state of matter—solid, liquid, or gas. The change of state will be on the arrow connecting the initial state to the final state.
  a. Water in the liquid state (initial state) forms ice, a solid (final state): freezing.
  b. Water as a gas (initial state) deposits on blades of grass, forming a liquid (final state): condensation.
  c. Water in the liquid state (initial state) boils, forming water in the gaseous state (final state): vaporization.

Vapor Pressure and Boiling Points

FIGURE 7.6 Vapor pressure and physical equilibrium. When water is enclosed in a container, some of the molecules vaporize, forming a characteristic pressure above the liquid. Eventually, the rates of movement into the vapor phase and movement back into the liquid phase are the same (orange arrows). When this occurs, physical equilibrium is reached.

Which dries faster when applied to the skin, water or rubbing alcohol (isopropyl alcohol)? If you tried this experiment, you would notice that the rubbing alcohol dries faster than the water, but why? Understanding this difference requires a deeper look at vaporization and vapor pressure of a liquid.

Consider a liquid like water sitting in a closed container (**FIGURE 7.6**). In this situation, some of the water molecules contain enough energy to push back molecules in the air space above the liquid and change phase to water vapor. Eventually, the movement of the molecules between the vapor and the liquid phase, and vice versa, happens at a constant rate. The state called **physical equilibrium** is reached. The pressure caused by the molecules pushing into the space above the liquid is called **vapor pressure**. Every substance has a characteristic vapor pressure at a given temperature. As temperature increases, more molecules from the liquid gain enough energy to become a gas. More molecules in the gas phase increases the vapor pressure.

To return to the question posed earlier about water and rubbing alcohol, the rubbing alcohol dries on the skin (evaporates) faster from the skin than does the water. This is because rubbing alcohol has a higher vapor pressure at a given temperature than water. It vaporizes more readily.

The same attractive forces that are defining a compound's vapor pressure are also at work defining a compound's boiling point. During **boiling**, *all* the molecules in a liquid must have enough energy to push back the gas molecules of the atmosphere at the surface of the liquid, allowing gas molecules of the liquid to escape. Each liquid molecule must overcome the attractive forces to neighboring molecules in the liquid and move into the gas phase as a *single molecule*. The heat supplied during boiling provides the energy necessary for each molecule to vaporize, moving individually from the liquid into the gas phase.

When the **boiling point** is reached, the molecules have enough energy to change from a liquid to a gas (vaporize). Because all the molecules can turn to a gas at this temperature, we could also say that the boiling point is the temperature where the vapor pressure of the liquid equals the atmospheric pressure.

At 25 °C, isopropyl alcohol has a vapor pressure of 44 mmHg and water's vapor pressure is 24 mmHg, while the boiling point for isopropyl alcohol is 82.6 °C and water boils at 100 °C. Why is the vapor pressure for water lower than isopropyl alcohol and its boiling point higher? Stronger attractive forces predict lower vapor pressures and higher boiling points at a given temperature. Let's take a deeper look at the types of attractive forces that can be present in molecules and see how this applies to water and isopropyl alcohol to predict their boiling points and vapor pressure.

Types of Attractive Forces

In Chapter 3, we saw that atoms form compounds either by giving or taking electrons (ionic compounds) or by sharing electrons (covalent compounds) to form bonds. In both cases, electrons are distributed unevenly and charge builds up with the compound. Because opposite charges attract, compounds with either an ionic charge $(+/-)$ or a partial charge (δ^+/δ^-) can attract each other.

Attractive forces are caused by the attraction of an electron-rich area of one compound (where electrons spend more time) to an electron-poor area of another compound (where electrons spend less time). Attractive forces can occur between two molecules, between two ions, or between an ion and a molecule. If the attraction is between two molecules, it is called an **intermolecular force**.

We will consider five attractive forces, shown in order of strength in **FIGURE 7.7**. We will start with the weakest, London forces, and work up to the strongest, ionic attraction.

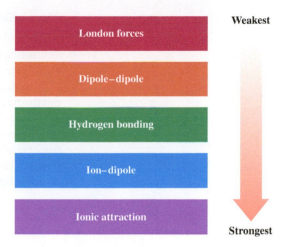

Weakest

| London forces |
| Dipole–dipole |
| Hydrogen bonding |
| Ion–dipole |
| Ionic attraction |

Strongest

FIGURE 7.7 Types of attractive forces. Attractive forces can occur between molecules, ionic compounds, or a combination of the two. Intermolecular forces occur between molecules. Ionic attractions are the strongest attractive force, and London forces are the weakest.

London Forces

If you have ever wrestled with a piece of plastic wrap that clings to itself rather than to the food you are trying to wrap, you have experienced **London forces**. These attractive forces occur momentarily between all molecules when electrons become unevenly distributed over a molecule's surface. When this happens, the partially positive side of this temporary dipole attracts the electrons of the second molecule, creating an attraction between these two molecules and inducing a temporary dipole in the second molecule. The result is the attractive force that you feel as you pull apart plastic wrap (see **FIGURE 7.8**). While *all compounds exhibit London forces*, these forces are significant only in the case of nonpolar molecules because these are the only attractive force present between nonpolar molecules.

Many plastic wraps contain molecules that have long, nonpolar hydrocarbon chains that interact with each other, creating temporary dipoles throughout the material. The attraction between the temporary dipoles causes the wrap to stick to itself. Although this is the weakest attractive force, many of these weak forces acting together can become quite strong. The terms **induced dipole** and *dispersion force* describe the same attractive force as London forces.

FIGURE 7.8 An example of London forces. When plastic wrap's even distribution of electrons is disturbed, the temporary dipoles (red represents partially negative, blue partially positive) induce temporary dipoles in neighboring hydrocarbon chains, resulting in attractions between molecules called London forces.

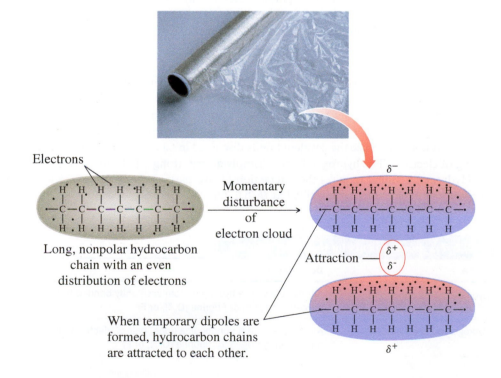

Electrons

Long, nonpolar hydrocarbon chain with an even distribution of electrons

Momentary disturbance of electron cloud

Attraction

When temporary dipoles are formed, hydrocarbon chains are attracted to each other.

Dipole–Dipole Attraction

Polar molecules have a permanently uneven distribution of electrons caused by electronegativity differences in the atoms that make up the molecules. See Section 3.7 for a review of polarity. One area of the molecule is partially positive and another partially negative. This charge separation is permanent in polar molecules, so it is called a **permanent dipole**.

Because the dipole in these molecules does not come and go as it does in the case of London forces, the attraction of the partially positive end of one molecule for the partially negative end of another molecule is stronger and more pronounced than in London forces. This type of attraction is caused by the interaction of two dipoles and is called a **dipole–dipole attraction** (see **FIGURE 7.9**).

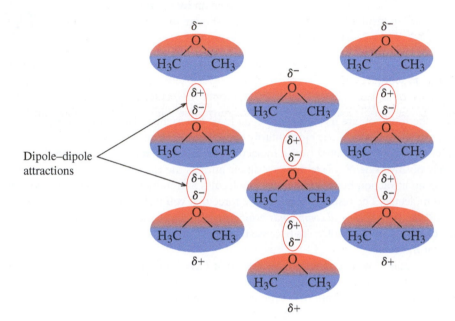

Dipole–dipole attractions

FIGURE 7.9 An example of a dipole–dipole attraction. Molecules of dimethyl ether are polar. The partially negative end of the dipole of one molecule of dimethyl ether (shaded red) is attracted to the partially positive end of the dipole of the second molecule (shaded blue).

Dipole–dipole attractions do not exist between nonpolar molecules because these molecules do not have permanent dipoles. Molecules with permanent dipoles will also have London forces, but the attraction of the dipoles is much stronger, making the contribution from London forces negligible.

Hydrogen Bonding

The attractive force called hydrogen bonding is so prevalent in nature that it has been given its own name, even though it is just a very strong dipole–dipole attraction. **Hydrogen bonding** involves a polarized hydrogen (hydrogen in a polar bond) and is much stronger than other dipole–dipole forces. It is important to understand that this attraction is not a bond like the covalent bonds discussed in Chapter 3 because there is no sharing of electrons. The hydrogen bond is simply a very strong dipole–dipole attraction.

Hydrogen bonding requires the interaction of two parts, a donor hydrogen and an acceptor pair of electrons, as described in **TABLE 7.1**.

TABLE 7.1 Requirements for Hydrogen Bonding	
Name	**Description**
Hydrogen-bond donor (δ^+)	A molecule with a hydrogen atom covalently bonded to an oxygen, nitrogen, or fluorine (O, N, or F)
Hydrogen-bond acceptor (δ^-)	A molecule with a nonbonding (lone) pair of electrons on an oxygen, nitrogen, or fluorine (O, N, or F)

The high electronegativity of O, N, and F polarizes the hydrogen atom of the donor, giving the hydrogen a high partial positive charge (δ^+). The partial positive strongly attracts the high partial negative charge (δ^-) centered on the nonbonding electron pair of the O, N, or F of the acceptor.

Let's look at a glass of pure water as our first example of hydrogen bonding (see **FIGURE 7.10**). A molecule of water has two hydrogen atoms that are bonded to the oxygen. Both of these hydrogen atoms are attached to O by polar covalent bonds (because O is more electronegative than H), polarizing the H so that it is partially positive (δ^+). Each of the two hydrogen atoms in water can act as a hydrogen-bond donor in a hydrogen-bonding interaction. A water molecule also contains two nonbonding (lone) pairs of electrons on the oxygen. Because of its nonbonding pairs, each water molecule can also accept two hydrogen bonds. So, one water molecule can make up to four hydrogen bonds to other water molecules at any one time, as shown in Figure 7.10. Hydrogen bonds are always illustrated using a dashed line.

Now let's return to the boiling point and vapor pressure of water and isopropyl alcohol, $CH_3CH(OH)CH_3$, to see why they are different. Which attractive forces are present in these two molecules? Both water and isopropyl alcohol contain an —OH and are polar molecules. In Figure 7.10, we saw that each water molecule can form up to four hydrogen bonds in the liquid phase. What about isopropyl alcohol? Isopropyl alcohol can form up to three hydrogen bonds, as shown in **FIGURE 7.11**, and also contains nonpolar hydrocarbons not found in water. Because water can make more hydrogen bonds than isopropyl alcohol (four versus three hydrogen bonds), the water molecules are attracted to each other more than isopropyl alcohol molecules, which raises the boiling point and lowers the vapor pressure of water when compared with isopropyl alcohol.

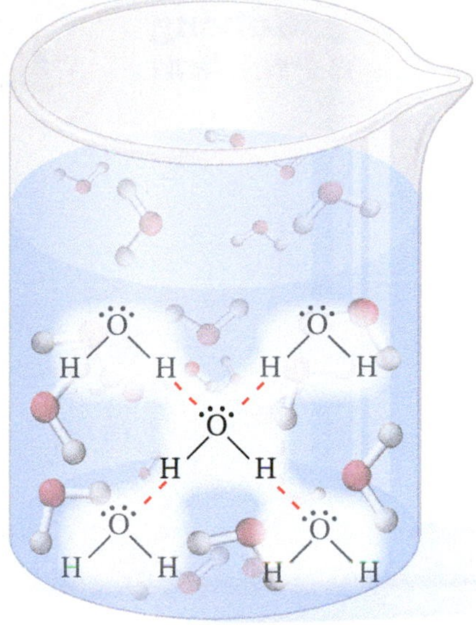

FIGURE 7.10 Hydrogen bonding in water. A single water molecule (center) can form up to four hydrogen-bonding interactions with neighboring water molecules. Hydrogen bonds are indicated with red dashed lines.

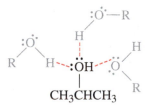

FIGURE 7.11 Isopropyl alcohol can make up to three hydrogen bonds to other isopropyl alcohol molecules.
Molecules shown in gray have hydrocarbon portions represented by "R."

Hydrogen bonds can occur between the same molecules (as seen in the water and isopropyl alcohol example), between two different polar molecules (as seen in the interactions of ethanol and acetone with water), or even between different parts of the same molecule (intramolecularly, as seen in the molecule ethylene glycol monomethyl ether).

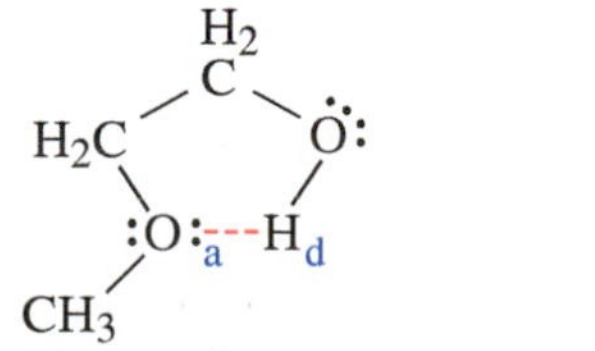

Intramolecular hydrogen bonding in a molecule containing an ether and alcohol group

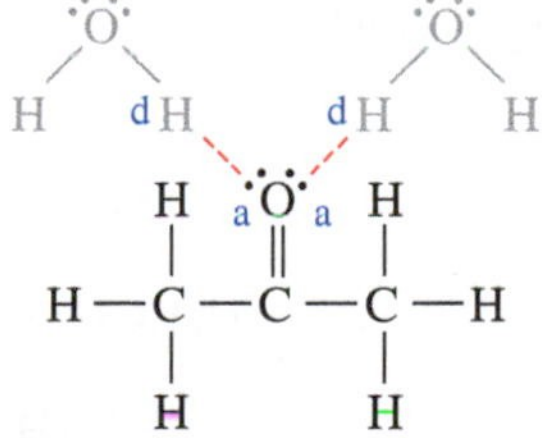

Intermolecular hydrogen bonding between acetone and water

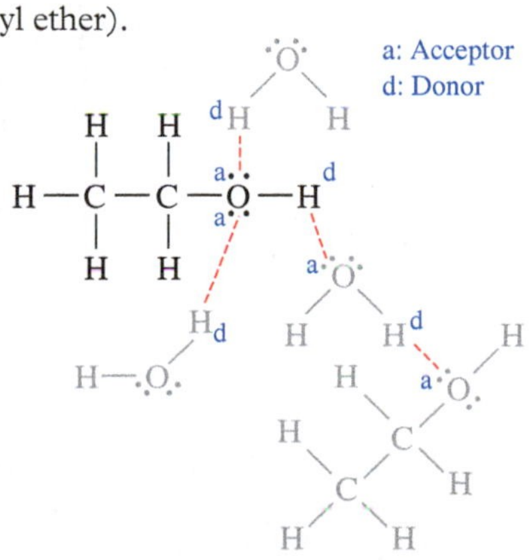

Intermolecular hydrogen bonding between ethanol and water

a: Acceptor
d: Donor

Hydrogen bonding can occur between any acceptor lone pair of electrons (a) and donor hydrogen atom (d).

Solving a Problem

Drawing Hydrogen Bonds

How many hydrogen bonds can the molecule shown make to water molecules? Draw them.

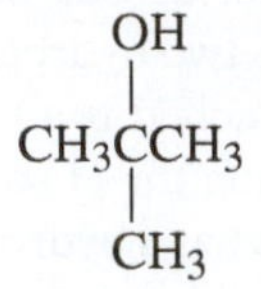

$$\begin{array}{c} \text{OH} \\ | \\ \text{CH}_3\text{CCH}_3 \\ | \\ \text{CH}_3 \end{array}$$

STEP 1 Determine the number of hydrogen-bond donors and acceptors present in the molecule. Only hydrogen atoms bonded to an O, N, or F are polarized enough to form hydrogen bonds. Remember that C—H bonds are nonpolar, so these hydrogens do not participate in hydrogen bonding. This molecule is both a hydrogen-bond donor (O—H hydrogen) and hydrogen-bond acceptor (two electron pairs on O). This molecule can form three hydrogen bonds with water.

STEP 2 Draw hydrogen bonds. Connect donors and acceptors from the molecule to acceptors and donors on water, respectively, with a dashed line.

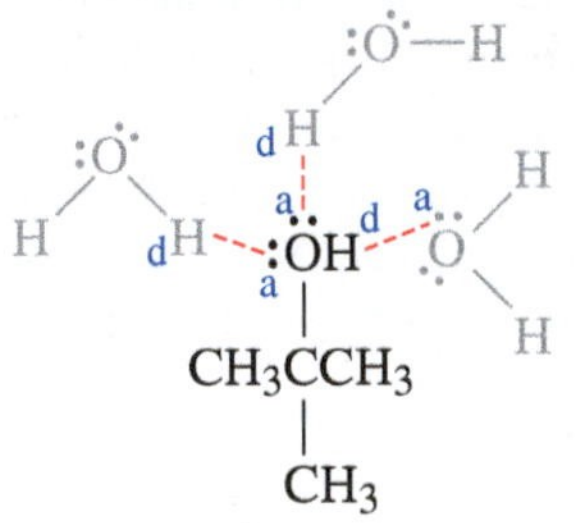

Sample Problem 7.6 Drawing Hydrogen Bonds

How many hydrogen bonds can the molecules shown make with water molecules? Draw them.

a. $\text{CH}_3\text{CH}_2\text{NHCH}_2\text{CH}_3$ b. $\text{CH}_3\text{CH}_2\text{Cl}$ c. CH_3CN

Solution

To hydrogen bond with water, the molecule must have either a hydrogen that is bonded to an N, O, or F (donor) or a nonbonding pair of electrons on an O, N, or F (acceptor). Remember that C—H bonds are nonpolar, so these hydrogens do not participate in hydrogen bonding.

a.

STEP 1 Determine the number of hydrogen-bond donors and acceptors present in the molecule. This molecule has one N—H bond, so there is one donor hydrogen and one lone pair of electrons that can act as an acceptor. Two hydrogen bonds can be formed to water.

STEP 2 Draw hydrogen bonds.

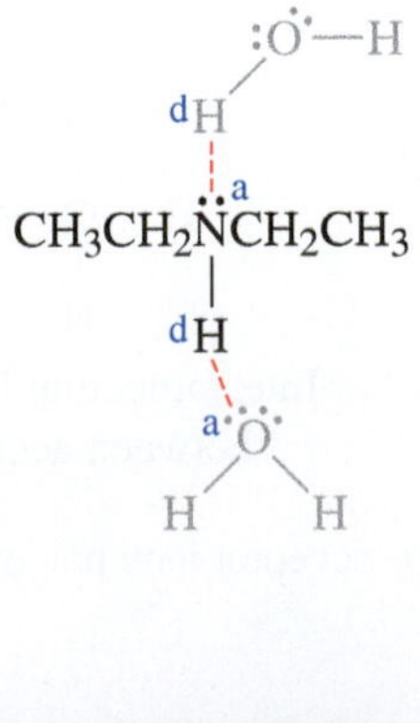

b.

STEP 1 Determine the number of hydrogen-bond donors and acceptors present in the molecule. This molecule has no O, N, or F atoms, so there are no hydrogen-bond acceptors present. The H atoms in this molecule are bonded to carbon, making them nonpolar. Therefore, there are no hydrogen-bonding donors present in this molecule. This molecule cannot form hydrogen bonds to water.

c.

STEP 1 Determine the number of hydrogen-bond donors and acceptors present in the molecule. This molecule has one nitrogen atom with one lone pair of electrons present that can act as a hydrogen-bond acceptor. The H atoms in this molecule are bonded to carbon, making them nonpolar. Therefore, no hydrogen-bonding donors are present in this molecule. This molecule can make one hydrogen bond to water.

STEP 2 Draw hydrogen bonds.

$$CH_3CN\colon\!\text{---}\,H\overset{\ddot{\ddot{O}}}{\diagup}\;H$$
$$\quad\quad\;\; a \quad d$$

Ion–Dipole Attraction

When you add a teaspoon of salt to a glass of water and stir, you observe that the salt dissolves. What you are witnessing is another very strong attractive force in action. The **ion–dipole attraction** occurs between ionic charges like those found in salt and polar molecules such as water. Ion–dipole attractions are an important attractive force often seen in biological systems because water is present. This attractive force is stronger than hydrogen bonding.

The interaction between sodium or chloride ions and water molecules results from the attraction of an ion to the opposite partial charge on a polar molecule. As shown in **FIGURE 7.12**, a cation like sodium ion is attracted to the partially negative end of the dipole of a polar molecule like water. Similarly, an anion like chloride is attracted to the partially positive end of the dipole of water. As we will see later in the chapter, this interaction plays an important role in the solubility of ionic compounds.

Ionic Attraction

As we saw in Chapter 3, there are two types of ions in ionic compounds—anions and cations. Anions have more electrons than their neutral atom and possess a − charge. Cations have fewer electrons than their neutral atom and possess a + charge. When these opposite charges attract each other, an **ionic attraction** exists. Ionic attractions are also called ionic bonds because they occur between two ions with opposite charges. An ionic attraction is the strongest attractive force because it involves more than just an uneven distribution of electrons as seen in a dipole–dipole (δ^+/δ^-) attraction.

FIGURE 7.12 An ion–dipole attraction. The cation Na^+ is attracted to the negative dipole of water, and the anion Cl^- is attracted to the positive dipole of water, allowing NaCl to dissolve in water.

*Ionic bond,
or salt bridge*

When two functional groups like carboxylate and protonated amine contain opposite charges, an ionic attraction can result.

An ionic attraction is sometimes called a salt bridge because **salts** are ionic compounds. More information about how salts form can be found in Section 9.2. Ionic charges are also found in the organic functional groups carboxylate and protonated amine, containing − and + charges, respectively. These groups are found in the amino acids that form proteins and can form salt bridges (ionic bonds) when they come into contact with each other. For example, the oxygen transport protein hemoglobin forms salt bridges between different areas of the protein as it releases oxygen at the tissues. For more information on hemoglobin, go to Chapter 10.

Recognizing the attractive forces present in a compound from its structure will allow you to predict the physical properties of compounds. The flow chart in **FIGURE 7.13** will help you to predict the strongest attractive force present.

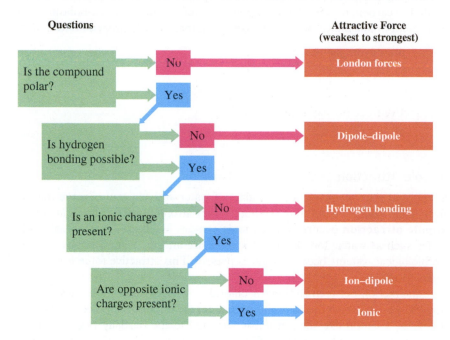

FIGURE 7.13 Flow chart for determining the strongest attractive force present in a compound.

Solving a Problem

Identifying Attractive Forces in Compounds

Name the strongest attractive force present in the pure substances shown.

   a. ammonia, NH_3

   b. hexane, C_6H_{14}

   c.

$$H_3\overset{+}{N}-\overset{\overset{\textstyle H}{|}}{\underset{\underset{\textstyle H}{|}}{C}}-\overset{\overset{\textstyle O}{\parallel}}{C}-O^-$$

Glycine, an amino acid

Use the flow chart in Figure 7.13 to determine each.

   a. Ammonia has N—H bonds, which are polar. N—H bonds can participate in hydrogen bonding. An ionic charge is not present. The strongest attractive force between NH_3 molecules is hydrogen bonding.

   b. Hexane contains only carbon and hydrogen, so it is a nonpolar molecule. The strongest attractive force between hexane molecules is London forces.

   c. Glycine contains opposite charges, so glycine can form ionic bonds. The strongest attractive force between glycines is ionic.

Sample Problem 7.7	**Identifying Attractive Forces**

Identify the strongest attractive force present in the pure substances shown.
   a. one of the first anesthetics, diethyl ether, $CH_3CH_2OCH_2CH_3$
   b. rubbing alcohol, isopropanol, $CH_3CH(OH)CH_3$
   c. the ozone-depleting substance trichloroethane, CCl_3CH_3

Solution

Use the flow chart in Figure 7.13 to determine each.
   a. Diethyl ether has $C—O$ bonds, which are polar. It cannot form hydrogen bonds because no hydrogens are bonded to O, N, or F. The strongest attractive force in diethyl ether is dipole–dipole.
   b. Rubbing alcohol has $C—O$ and $O—H$ bonds, which are polar. The $O—H$ bonds allow isopropanol to hydrogen bond. It does not contain an ionic charge. The strongest attractive force is hydrogen bonding.
   c. Trichloroethane has $C—Cl$ bonds, which are polar. It cannot form hydrogen bonds because none of its hydrogens are bonded to O, N, or F. The strongest attractive force present is dipole–dipole.

Predicting Vapor Pressure and Boiling Points for Alkanes

Now that we have seen an example of each type of attractive force, let's apply this to predicting vapor pressures and boiling points of organic compounds.

Keep in mind that the stronger the attractive forces, the lower the vapor pressure and the higher the boiling point. Because alkanes are nonpolar, the only attractive force present is London forces.

Let's compare pentane and octane, shown in **FIGURE 7.14**. When interacting in the liquid phase, a single octane molecule has more carbon atoms and, therefore, a greater surface area, allowing for more interactions with neighboring molecules than for molecules of pentane.

Recall that London forces between molecules are a result of a disturbance in the distribution of electrons over the surface of a molecule. A molecule with a larger surface area has more surface contact with other molecules as well as more electrons to disturb. In other words, straight-chain alkanes with more carbons have stronger attractions between molecules. Therefore, because these attractions must be overcome for the compound to vaporize and boil, more heat is necessary to disrupt these attractions, which means the boiling point is higher.

TABLE 7.2 shows the boiling points for the first ten straight-chain alkanes. Because vapor pressure is temperature dependent, vapor pressures are not shown on the table but follow similar trends. Higher boiling points imply lower vapor pressures because it takes more energy to vaporize. Notice that as the number of carbon atoms in the chain (and, therefore, the surface area of the molecule) increases, the boiling point also increases. This trend to higher boiling points with increasing carbon chain length is also true for **melting points**, the temperature at which molecules move from the solid to the liquid phase.

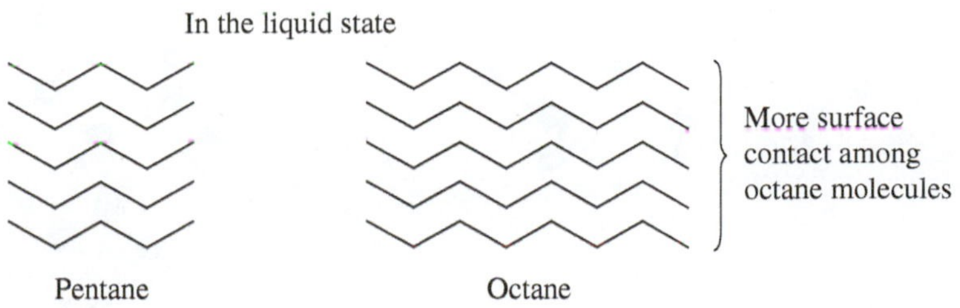

In the liquid state

More surface contact among octane molecules

Pentane   Octane

FIGURE 7.14 More surface contact between molecules raises the boiling point. Octane has a larger surface area and more potential surface for contact than pentane, so its boiling point is higher.

TABLE 7.2 Boiling Points of Common Straight-Chain Alkanes

Name	Structure	Boiling Point (°C)
Methane	CH_4	−161
Ethane	CH_3CH_3	−89
Propane	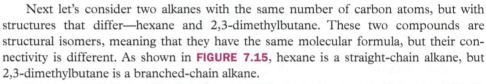	−42
Butane		−0.5
Pentane		36
Hexane		69
Heptane		98
Octane		125
Nonane		151
Decane		174

The strands of spaghetti have more surface contact between them than the meatballs do, just as straight-chain alkanes have more points of contact between them than branched ones.

Next let's consider two alkanes with the same number of carbon atoms, but with structures that differ—hexane and 2,3-dimethylbutane. These two compounds are structural isomers, meaning that they have the same molecular formula, but their connectivity is different. As shown in **FIGURE 7.15**, hexane is a straight-chain alkane, but 2,3-dimethylbutane is a branched-chain alkane.

Hexane has a boiling point of 69 °C and 2,3-dimethylbutane has a boiling point of 58 °C. If two molecules have the same number of carbon atoms, why do their boiling points have an 11 °C difference? The answer lies in attractive forces and in the surface area of the molecules.

Consider a plate of spaghetti and an accompanying bowl of meatballs. The long noodles have more points of contact with each other and can line up with each other if you stretch them out on your plate. Similarly, the molecules of hexane, a straight-chain alkane, can "stack" close together like the spaghetti. In contrast, molecules of 2,3-dimethylbutane are, like meatballs, more spherical in their overall shape. When next to each other on your plate, two meatballs touch at only one small point. Likewise, the molecules of 2,3-dimethylbutane have less surface contact than do the hexane molecules. As mentioned earlier, the more contact between two molecules, the greater the attraction caused by the London forces between them. In turn, the greater London forces attraction means that the vapor pressure of hexane at a given temperature is lower and hexane's boiling point is higher than that of 2,3-dimethylbutane. In fact, as a general rule, *for alkanes with the same number of carbon atoms, straight-chain alkanes have higher boiling points and lower vapor pressures than do branched alkanes.*

Hexane   2,3-Dimethylbutane

FIGURE 7.15 Two structural isomers of C_6H_{14}. Hexane and 2,3-dimethylbutane are two alkanes with the same molecular formula but different connectivity, which affects their boiling points.

Boiling Points   69 °C   58 °C

Melting Points

Melting points follow the same trends as boiling points in that the stronger and more numerous the forces between molecules, the higher the melting point. For example, H_2S has a similar structure to H_2O but cannot form hydrogen bonds. The melting points of the two compounds are extremely different, with H_2S having a melting point of $-80\ °C$ compared to $0\ °C$ for H_2O.

TABLE 7.3 summarizes the trends in vapor pressure, boiling points, and melting points previously discussed.

TABLE 7.3 Vapor Pressure, Boiling Point, and Melting Point Trends		
Nonpolar Molecules	**Similar-Sized Molecules, Different Attractive Forces**	**Molecules with Same Attractive Forces**
The greater the surface area, the lower the vapor pressure and the higher the boiling or melting point.	The stronger the attractive forces, the lower the vapor pressure and the higher the boiling or melting point.	The stronger and more numerous the attractive forces, the lower the vapor pressure and the higher the boiling or melting point.

Predicting Vapor Pressures and Boiling Points

Solving a Problem

Predict which substance in the pair would have the lower vapor pressure and the higher boiling point. Justify your prediction using attractive forces.

a. 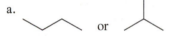 or

b. H_2O or CH_3OCH_3

Solution

a.

STEP 1 Determine the strongest attractive force present in each. Notice that both molecules, butane and 2-methylpropane, are alkanes and contain only hydrogen and carbon and are nonpolar. These molecules contain the same attractive force, London forces.

STEP 2 Predict the lower vapor pressure and higher boiling point based on the strength and number of forces present. The trends in Table 7.3 indicate that if two molecules have the same forces present, the one with more of that force will have a lower vapor pressure and a higher boiling point. Because both have only London forces, we must consider the number of forces present. The straight-chain alkane, butane, makes more contacts with its neighboring molecules in the liquid state, so more London forces would be present. The straight-chain alkane, butane, is predicted to have the lower vapor pressure and higher boiling point.

b.

STEP 1 Determine the strongest attractive force present in each. The strongest attractive force present in H_2O is hydrogen bonding and the strongest attractive force present in CH_3OCH_3 is the dipole–dipole attraction.

STEP 2 Predict the lower vapor pressure and the higher boiling point based on the strength and number of forces present. Because hydrogen bonding is a stronger attractive force than dipole–dipole, H_2O is predicted to have the lower vapor pressure and the higher boiling point.

Sample Problem 7.8 **Predicting Vapor Pressures and Boiling Points**

Predict which substance in the pair would have the lower vapor pressure and the higher boiling point. Justify your prediction using attractive forces.

a. CH_3OH or CH_3CH_3

b. 〜〜〜〜   or   〜〜〜〜

c. CH_3NHCH_3 or CH_3CH_2OH

Solution

a.

STEP 1 **Determine the strongest attractive force present in each.** The strongest attractive force present in CH_3OH is hydrogen bonding, and the strongest attractive force present in CH_3CH_3 is London forces.

STEP 2 **Predict the higher boiling point based on the strength and number of forces present.** Because hydrogen bonding is a stronger attractive force than the London force, CH_3OH is predicted to have the higher boiling point.

b.

STEP 1 **Determine the strongest attractive force present in each.** Notice that both molecules are hydrocarbons containing only hydrogen and carbon so are therefore nonpolar. These molecules contain the same attractive force, London forces.

STEP 2 **Predict the higher boiling point based on the strength and number of forces present.** If two molecules have the same attractive force present, the one with more of that force will have a lower vapor pressure and a higher boiling point. Because both of these molecules have only London forces, we must consider the number of forces present. The straight-chain alkane, octane (left molecule), makes more contacts with its neighboring molecules than the cis-alkene (right molecule). The straight-chain alkane is predicted to have the lower vapor pressure and the higher boiling point.

c.

STEP 1 **Determine the strongest attractive force present in each.** The strongest attractive force present in CH_3NHCH_3 is hydrogen bonding. The strongest attractive force in CH_3CH_2OH is hydrogen bonding.

STEP 2 **Predict the higher boiling point based on the strength and number of forces present.** The trends in Table 7.3 indicate that if two molecules have the same forces present, the one with more of that force will have a higher boiling point. CH_3NHCH_3 has one hydrogen-bond donor and one hydrogen-bond acceptor present, forming up to two hydrogen bonds. CH_3CH_2OH has one donor and two acceptors present, forming up to three hydrogen bonds. Because CH_3CH_2OH can form more hydrogen bonds per molecule than CH_3NHCH_3, CH_3CH_2OH is predicted to have the higher boiling point.

INTEGRATING
Chemistry

Find out how big ▶
molecules maintain
their shape.

Attractive Forces Keep Biomolecules in Shape

The attractive forces discussed in this section are used extensively in nature to hold biological molecules together in their characteristic shapes. In Chapter 6, we noted the formation of cellulose fibers from the alignment of zigzagging $\beta(1 \rightarrow 4)$ bonded glucose units. The linear cellulose molecules are held tightly together through hydrogen bonding between the glucose units in neighboring molecules.

London forces hold cell membranes together (see Section 7.5). Hydrogen bonding holds a DNA double helix in its twist (see **FIGURE 7.16**). Protein structures are held together by combinations of all the attractive forces discussed. We explore attractive forces in DNA and in protein structures in more detail as they are discussed in later chapters.

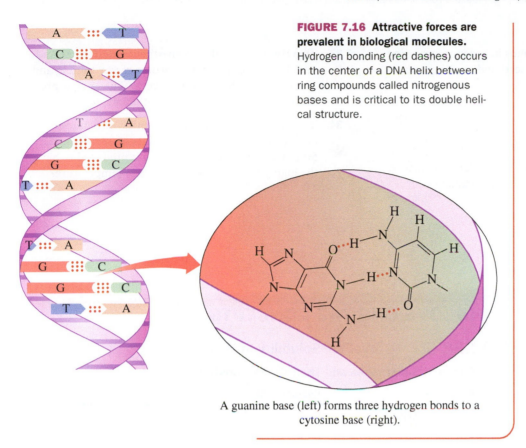

FIGURE 7.16 Attractive forces are prevalent in biological molecules. Hydrogen bonding (red dashes) occurs in the center of a DNA helix between ring compounds called nitrogenous bases and is critical to its double helical structure.

A guanine base (left) forms three hydrogen bonds to a cytosine base (right).

Practice Problems

7.17 Explain the difference between a covalent bond and an attractive force.

7.18 What type(s) of attractive forces exist between all molecules?

7.19 Explain the difference between the dipole in London forces and the dipole in dipole–dipole attractions.

7.20 Given that only polar molecules can participate as donors in hydrogen bonding, is it true that all polar molecules can be hydrogen-bond donors? Explain.

7.21 Explain the requirements for a molecule to be a hydrogen-bond acceptor.

7.22 Considering your answers to Problems 7.20 and 7.21, can water and carbon dioxide form a hydrogen bond? Explain.

7.23 An ion–dipole attraction often occurs when ionic compounds mix with water. Why do you think this is so?

7.24 Name two functional groups that contain ionic charges.

7.25 Identify the strongest attractive force present in each of the pure substances shown: London forces, dipole–dipole, hydrogen bonding, ion–dipole, or ionic attractions. Refer to the flow chart in Figure 7.13.

a. NaF
b. $CH_3CH_2CH_3$
c. CH_3NH_2
d. CH_3F

7.26 Identify the strongest attractive force present in the pure substances shown: London forces, dipole–dipole, hydrogen bonding, ion–dipole, or ionic attractions.

a. $CH_3CH_2CH_2CH_2OH$
b. KCl
c. CH_2Cl_2
d.

$$\begin{array}{c} O \\ \parallel \\ C \\ H \quad \quad H \end{array}$$

7.27 How many hydrogen bonds can CH_3NH_2 make to water? Draw them.

7.28 How many hydrogen bonds can CH_2O make to water? Draw them.

7.29 Predict which pure substance will have the lower vapor pressure, water (H_2O) or ethanol (CH_3CH_2OH). Explain your answer.

7.30 Predict which pure substance will have the lower melting point, baking soda ($NaHCO_3$) or phenol (C_6H_5OH). Explain your answer.

—*Continued next page*

Continued—

7.31 Rank the following compounds in order of increasing boiling point (lowest to highest) and explain your ranking:

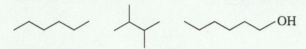

7.32 Rank the following compounds in order of increasing boiling point (lowest to highest) and explain your ranking:

7.33 Two common inhalation anesthetics, desflurane and sevoflurane, have vapor pressures of 669 mmHg and 170 mmHg at 20 °C, respectively. Which one is predicted to have the higher boiling point?

7.34 Two common inhalation anesthetics, desflurane and halothane, have boiling points of 23.5 °C and 50.2 °C, respectively. Which one is predicted to have the higher vapor pressure at a given temperature?

DISCOVERING THE CONCEPTS

? **INQUIRY ACTIVITY**—Solubility in Water

Common Substances and Their Characteristics

Substance	Characteristics
NaCl (table salt)	Ionic compound
Sucrose (table sugar)	Polar covalent compound
Vegetable oil	Nonpolar covalent compound
Soap	Contains a highly polar part (often ionic) and a nonpolar part

Questions

1. Based on your experience, which of the compounds in the table will dissolve in water? To dissolve is to separate a substance into single particles in solution.
2. Which of the characteristics listed applies to water?
3. The *golden rule of solubility* states that "like dissolves like." What is *alike* about the compounds you listed in question 1 and water?
4. Based on their characteristics, predict whether each of the following compounds might be soluble in water.

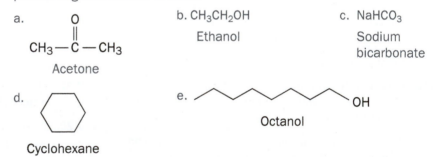

a.

   $CH_3-\overset{\overset{\displaystyle O}{\|}}{C}-CH_3$

   Acetone

b. CH_3CH_2OH

   Ethanol

c. $NaHCO_3$

   Sodium bicarbonate

d.

   Cyclohexane

e.

   Octanol

5. Draw and describe how a polar covalent compound like formaldehyde might interact with water through dipole–dipole attractions.

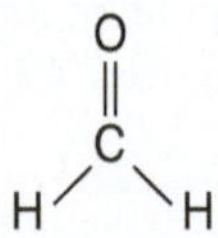

6. Ionic compounds containing sodium will separate into cations and anions in solution. Draw and describe how an ionic compound like NaCl might interact with water through ion–dipole attractions.

7. Many pharmaceuticals contain aromatic rings or large hydrocarbon areas (nonpolar). Quite often, pharmaceuticals are synthesized in an ionic form to increase their solubility in aqueous solution. Ionic charges on organic molecules dramatically increase the compound's solubility. Predict whether the following pharmaceuticals are likely to be soluble in water based on their polarity and the distribution of polarity throughout the compound.

Pentolinium—an antihypertensive

Methamphetamine—mainly recreational usage yet sometimes prescribed for ADHD

Remeron™—an antidepressant

8. Devise a rule to predict the solubility of an organic compound in water.

7.3 Attractive Forces and Solubility

Before adding oil-and-vinegar dressing to a salad, you must vigorously shake the mixture and then quickly pour it. Why? Because oil and water (vinegar is mostly water) don't mix. Have you ever wondered why oil and water don't mix? The answer lies in the attractive forces between the molecules.

 7.3 Inquiry Question

How can we predict the solubility of a compound in water?

The Golden Rule of Solubility

Oil and water don't mix, but sugar and water do. In both cases, we are witnessing the solubility (or lack thereof) of one substance in another. The maximum amount of a substance that can dissolve in a specified amount of water at a given temperature defines a substance's **solubility** in water. As we discussed in Section 7.2, the polarity of molecules (polar or ionic versus nonpolar) dictates the types of attractive forces between the molecules.

The **golden rule of solubility**—*like dissolves like*—means that molecules that are similar will dissolve in each other. The "similarity" mentioned here relates to the character of the molecules—are they polar or nonpolar? In short, the golden rule of solubility says that *molecules that have similar polarity and participate in the same types of attractive forces will dissolve each other.*

The terms hydrophilic and hydrophobic may be more familiar to you than polar and nonpolar when discussing molecular polarity and solubility. Technically, the terms hydrophobic and hydrophilic are only used when discussing solubility in water (the prefix *hydro*- refers to water). **Hydrophilic** literally means *water-loving* and **hydrophobic** means *water-fearing*, referring to substances that are or are not soluble in water, respectively.

Oil does not mix with water because their polarities are too dissimilar.

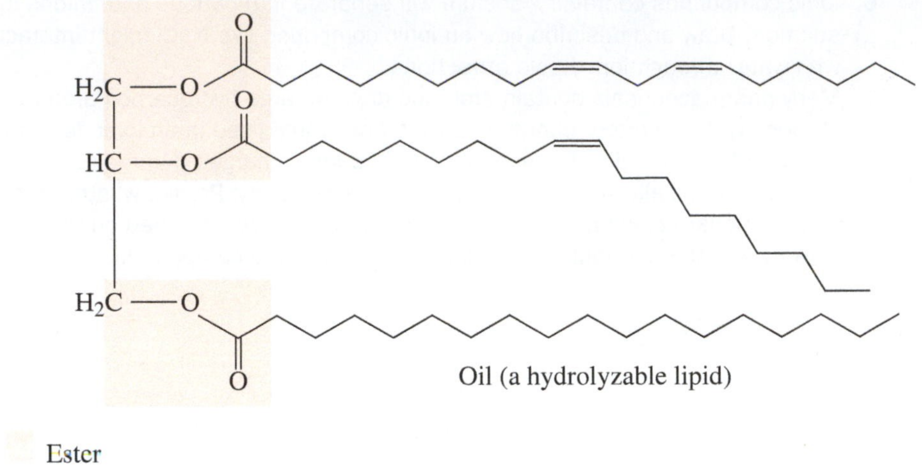

**FIGURE 7.17 Oils contain triglyc-
erides, nonpolar molecules formed
through esterification.** Although the
triglyceride shown contains some polar
oxygen bonds in the three ester functional groups (shaded), the nonpolar
areas of the molecule dominate the
polarity, making the molecule overall
nonpolar.

Oil (a hydrolyzable lipid)

Ester

Predicting Solubility

Nonpolar Compounds

Let's look more closely at oil and water to see why they don't mix. In Chapter 5 we were introduced to the dietary oils, **triglycerides** that, as hydrolyzable lipids, can undergo hydrolysis to break down into the component parts—a glycerol molecule and three fatty acids. Conversely, triglycerides form through a condensation reaction specifically called an **esterification**, since three ester bonds are formed during the condensation (see the ester bonds highlighted in **FIGURE 7.17** above). If we further examine this structure for its polarity, we see three ester functional groups and a lot of nonpolar hydrocarbon. The large amount of hydrocarbon present outweighs the few polar bonds present, making this molecule nonpolar.

As nonpolar compounds, oils are attracted to neighboring molecules through London forces. Water, in contrast, is a polar molecule and interacts with other substances through dipole–dipole, hydrogen bonding, and ion–dipole attractions. This means that, in terms of attractive forces, oil and water are very *unlike* each other. Yet for one molecule to dissolve another, the molecules must interact with each other. The attractions among the water molecules are much greater than the attraction between a water molecule and an oil molecule. Because oil and water do not share attractive forces with each other, they are not soluble in each other. Even after a bottle of oil and vinegar salad dressing is shaken vigorously, the contents of the bottle will separate back into two layers—a watery layer (vinegar) and an oil layer.

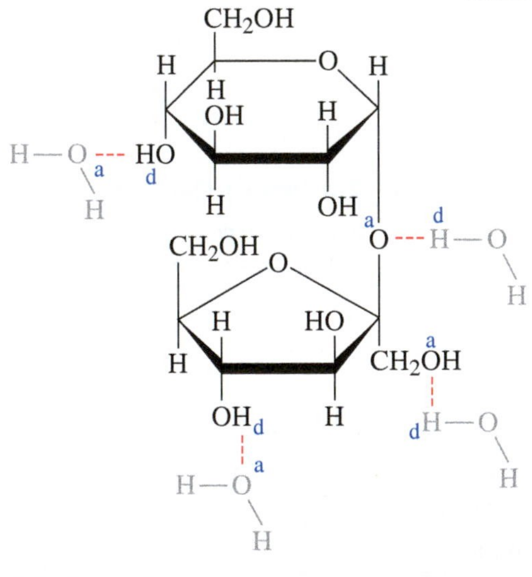

Polar Compounds

Why does table sugar, also an organic compound, dissolve in water, but oil does not? In Chapter 6, we learned that the substance that we call table sugar is the disaccharide sucrose. Like all other carbohydrates, sucrose has multiple hydroxyl (—OH) groups. The key to the interaction of sucrose and water lies in these functional groups.

The hydroxyl groups of sucrose make it a polar compound and give it the ability to interact with water not only through dipole–dipole attractions, but also through the stronger hydrogen-bonding interactions, as shown in **FIGURE 7.18**. Because table sugar and water are both polar and share these attractive forces, table sugar is an organic compound that is soluble in water.

**FIGURE 7.18 Sucrose hydrogen bonding with
water.** A few of the many hydrogen bonds (shown
in red) possible between water and sucrose
(acceptors and donors labeled). The attractions
make sugar soluble in water. The lone pair electrons on O are implied, not shown, for the sake
of clarity.

Ionic Compounds

What about the solubility of ionic compounds in water? In Section 7.2, we saw that ion–dipole attractions can exist between water and ions. These attractions are much stronger than the partial dipole attractions that exist between water and polar covalent compounds. Because ion-dipole attractions are so strong, ionic compounds are often more soluble in water than are covalent compounds.

For most ionic compounds, the ion–dipole attractions between the ions of an ionic compound and water are so strong that when enough water molecules interact with each ion, the ionic bond between the two ions is disrupted.

Individual ion–dipole attractions are not stronger than an ionic bond, but when multiple water molecules interact with an ion, the sum of these attractive forces is greater than the strength of the ionic bonds. The result, as shown in **FIGURE 7.19** for sodium chloride, is that in water many ionic compounds break apart into their component ions, and each ion is surrounded by water molecules. This process is called **hydration**.

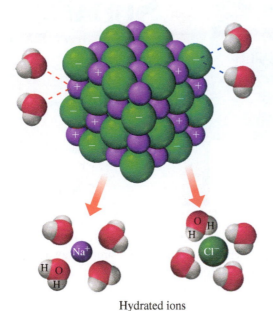

Hydrated ions

FIGURE 7.19 Hydration of NaCl. When NaCl dissolves in water, the multiple ion–dipole attractions between water and the Na^+ and Cl^- ions cause the compound to break apart into its component ions.

Ion–Dipole Attractions Increase Drug Solubility

About 85% of all drugs are administered orally to be absorbed in the digestive tract, so these drugs must be soluble in water. In contrast, more than 40% of the new chemical entities discovered by the pharmaceutical industry as potential therapeutic agents are insoluble in water. As an example, consider the amphetamine pseudoephedrine, used as a nasal decongestant. Based on its overall polarity, do you think it is likely soluble in water? Pseudoephedrine has a polar O and a polar N, but the amount of hydrocarbon present makes this molecule mainly nonpolar and only slightly soluble in water. Creating stronger attractive forces between water and pseudoephedrine increases its solubility. Pharmaceutical companies often try to create molecules that have an amine or carboxylic acid group in them so that at the pH of the body, a charged form is present, increasing the molecule's solubility. (We discuss more about charges and pH in Chapter 9.) Creating the charged form of a molecule is readily done in the lab by reacting molecules like pseudoephedrine with HCl. This adds a proton (H^+) to the amine group.

INTEGRATING
Chemistry

◄ Find out how drugs are made more soluble in water.

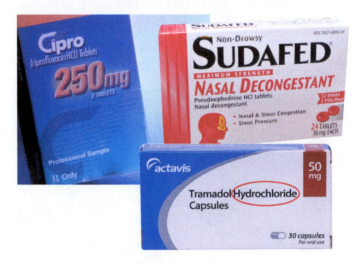

Pseudoephedrine → Pseudoephedrine hydrochloride

Adding a charge to the amine yields pseudoephedrine HCl, which is easily hydrated in water.

Many over-the-counter drugs that are taken orally contain the hydrochloride (salt) version as the active ingredient. Salt formation is a simple way to increase the solubility of organic molecules in water.

Amphipathic Compounds

In Chapter 4, we saw that fatty acids are mostly hydrocarbons but contain a carboxylic acid group, $R-\overset{\overset{\displaystyle O}{\parallel}}{C}-OH$. Despite the presence of a carboxylic acid group containing two electronegative oxygen atoms, these compounds behave similarly to nonpolar alkanes. The carboxylic acid part of a fatty acid is polar, but relative to the large nonpolar hydrocarbon chain, it does not contribute as much to the overall polarity of the molecule. The nonpolar part of the fatty acid is so much larger than the polar carboxylic acid group that it dominates the character of the compound, making fatty acids mainly nonpolar. Molecules like fatty acids that have both polar and nonpolar parts are called **amphipathic** (from the Greek *amphi* meaning "both" and *pathic* meaning "condition") compounds.

Similar in structure to fatty acids, soaps are also composed of fatty acids, although soaps are ionic, increasing the ion–dipole attractive force with water. Soap is composed of fatty acid salts. Fatty acid salts contain the carboxylate form, $R-\overset{\overset{\displaystyle O}{\parallel}}{C}-O^-$, of the carboxylic acid functional group (hydrogen removed) at one end as shown in **FIGURE 7.20**. The ionic, polar end is commonly called the polar *head*. The rest of the molecule behaves like a nonpolar compound. Fatty acid salts have long nonpolar hydrocarbon *tails* and extremely polar (ionic) *heads*, so they are amphipathic. Recall from Chapter 4 that the number of carbons and double bonds in a fatty acid tail can be designated by numbers in brackets separated by a colon; for example, the 16-carbon saturated fatty acid palmitic acid, found in palm oil, is given a designation of [16:0].

Amphipathic compounds like soaps are not soluble in water because of the large nonpolar tails present. The nonpolar tails are hydrophobic and will be excluded from the water, associating with each other and interacting through London forces. The ionic heads, which are hydrophilic, interact with the water mainly through ion–dipole attractions.

Because the water interacts only with the ionic heads, the tails associate with each other, creating the core of a spherical structure in the water called a **micelle** (see **FIGURE 7.21**). The polar heads are attracted to the water and form the shell of the micelle. Micelles form because the ion–dipole and hydrogen-bonding attractions between water molecules and the ionic heads are stronger than (and preferred to) water interactions with the hydrocarbon tails.

Most stains, including grease and dirt, are nonpolar. Based on the golden rule, greasy dirt is not soluble in water. When skin or clothing with a greasy dirt stain is washed with soapy water, the stain is attracted to the nonpolar hydrocarbon tails of the soap and is dissolved in the interior of the micelle formed by the soap molecules. The ionic head groups surround the exterior surface of the micelle and form ion–dipole interactions with the water. Because the surface of the micelle is covered with the polar head groups, the entire micelle, with the greasy stain molecules now dissolved and "hidden" in its interior, is soluble in water and is washed down the drain. Amphipathic compounds like soaps are called **emulsifiers** because they allow nonpolar and polar compounds to be suspended in the same mixture.

Many organic substances such as pharmaceuticals are soluble in water to some extent, making them fall into a gray area of solubility based on how polar (or like water) they are. Let's try to predict the solubility of some molecules in water.

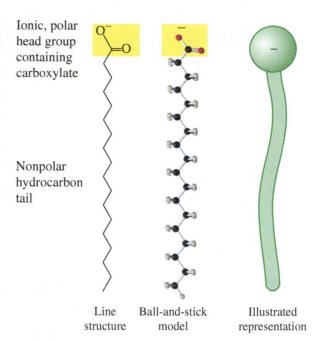

Ionic, polar head group containing carboxylate

Nonpolar hydrocarbon tail

Line structure   Ball-and-stick model   Illustrated representation

FIGURE 7.20 Fatty acids and soaps are amphipathic. Amphipathic compounds like a fatty acid salt can be illustrated by representing the polar head as a sphere and nonpolar tails with a wavy line.

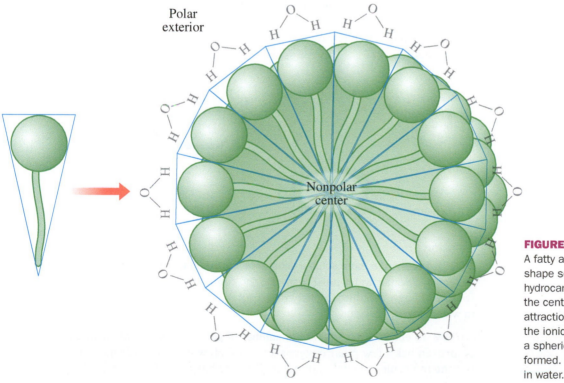

Polar exterior

Nonpolar center

FIGURE 7.21 A spherical micelle. A fatty acid salt has a conical shape so that when the nonpolar hydrocarbon tails are pushed to the center by the strong ion–dipole attraction between the water and the ionic charge on the head group, a spherical-shaped micelle is formed. Soaps form micelles in water.

Solving a Problem

Predicting Solubility in Water

Predict which of the following compounds is likely to be more soluble in water. Justify your answer.

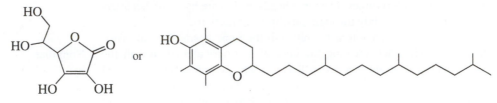

Vitamin C, ascorbic acid   Vitamin E, alpha-tocopherol

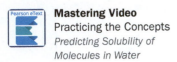

Mastering Video
Practicing the Concepts
Predicting Solubility of Molecules in Water

Solution

STEP 1 Predict the more soluble substance based on the overall polarity of the entire molecule. Because water must surround the compound, molecules that are more soluble will have polar functional groups spread throughout. Vitamin C has polar bonds distributed throughout its structure, whereas the polar areas in vitamin E are limited. Vitamin C is more soluble in water.

STEP 2 To confirm, determine the type and number of polar attractive forces in each. Because we are considering solubility in water, a polar compound, we examine the number of polar interactions present in the compounds. Vitamin C contains several oxygens and hydrogens available for hydrogen bonding and is, therefore, considered a highly polar molecule. Vitamin E has a few polar areas but overall is mostly hydrocarbon and mainly nonpolar. This justifies the prediction made in step 1.

Sample Problem 7.9 Predicting Solubility in Water

Predict which of the following compounds is likely to be more soluble in water. Justify your answer.

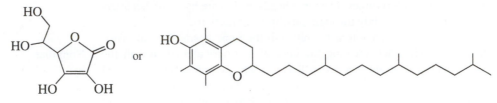

Naproxen, the active ingredient in Aleve®   β-D-Glucose

Solution

STEP 1 Predict the more soluble substance based on the overall polarity of the entire molecule. Because water must surround the compound, molecules that are more soluble will have polar functional groups spread throughout. Glucose is more soluble in water than naproxen.

STEP 2 To confirm, determine the number of polar attractive forces in each. Naproxen has a few polar oxygens and an OH, which can hydrogen bond, but the molecule is mainly nonpolar. Glucose has many OH groups that can hydrogen bond.

Practice Problems

7.35 Based on their polarity and attractive forces, predict whether the following compounds are soluble, are insoluble, or form a micelle in water.

a. heptane, $CH_3CH_2CH_2CH_2CH_2CH_3$

b. acetate anion, CH_3COO^-

c. Lauryl alcohol, found in shampoos and skin cleansers,

 OH

7.36 Based on their polarity and attractive forces, predict whether the following compounds are soluble, are insoluble, or form a micelle in water.

a. NaCl

b. sodium stearate [18:0], a fatty acid salt

c. isopropyl alcohol, $CH_3CH(OH)CH_3$

7.37 Predict which of the following compounds is likely to be more soluble in water. Justify your answer.

a. fatty acid or fatty acid salt

b. KCl or $(CH_3)_3N$

7.38 Predict which of the following compounds is likely to be more soluble in water. Justify your answer.

a. CH_3CH_2OH or CH_3CH_3

b. CH_3NH_2 or H_2S

7.4 Dietary Lipids

Up to this point, we have examined a compound's state of matter and attractive forces present. Establishing this information allows us to correctly predict some of the compound's physical properties. Next, we turn our attention to applying these principles to some biologically relevant liquids and solids, the lipids. Recall that in Section 7.3 we identified the lipid called a triglyceride as a nonpolar (hydrophobic) molecule that is not soluble in water. Triglycerides exist as solids or liquids. In this section, we examine structural changes that affect their melting points.

Fats Are Solids; Oils Are Liquids

A juicy steak, a tub of lard, and a stick of butter all contain animal fats. In contrast, dietary oils are derived from plants like corn, soybeans, and peanuts. Both fats and oils belong to a class of hydrolyzable lipids called triglycerides. Determining a fat versus an oil is as simple as examining the physical state of the triglycerides at room temperature—**oils** are liquids and **fats** are solids. How can two molecules that are so similar in structure exist in different states?

In **FIGURE 7.22**, a fat molecule and an oil molecule are shown both in skeletal structure and as a ball-and-stick model. Compare the hydrocarbon tails. Do you see differences? Let's look at the fat first (Figure 7.22a). The three hydrocarbon chains (tails) are mainly saturated single carbon–carbon bonds. In the ball-and-stick model, notice that these tails are physically close to each other. This proximity creates disturbances in the electron distribution around each of the tails as they interact with other tails. These disturbances result in London forces—the tails of the fat are attracted to one another. These mostly straight-chain tails allow for many surface contacts, which increases the attractions between the tails and slows down (restricts) the molecular motions, allowing the molecules to form a solid.

In contrast, look next at the oil molecule in Figure 7.22b. It has many unsaturated cis carbon–carbon double bonds in the hydrocarbon tails. Many oils are referred to as **polyunsaturated** since they contain many unsaturated hydrocarbons. Looking at the ball-and-stick model, notice how much room the cis double bonds take up when compared to the single bonds in the fat. How does this simple difference translate to a difference in the physical state between a fat and an oil (solid versus liquid)? Notice how the cis double bonds in the tails of the oil creates kinks in the otherwise straight hydrocarbon chain. The tails of the unsaturated oil cannot stack together as closely as those in the fat can. The kinked tails interact less with each other through London forces because they are not as physically close to one another. Less interaction means less London force attractions, which in turn means that the hydrocarbon chains in the oil move more freely. The greater molecular motion among the hydrocarbon tails in an oil does not allow adequate stacking of the tails for a solid to form at room temperature.

7.4 Inquiry Question
How are fats and oils different?

Tails are closer together in fat

(a) Fat

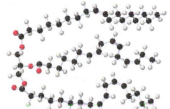

Tails are farther apart in oil

(b) Oil

FIGURE 7.22 **(a) A typical fat molecule and (b) an oil molecule.**

So fats are solids at room temperature because the motion of their hydrocarbon tails is restricted by the London forces holding them together. Fats are typically solid substances but with low melting points. Even the heat of your hand (37 °C) is enough to melt the cocoa butter in chocolate. Why is the melting point of a fat low enough to melt at body temperature? When a substance melts, it changes from a solid to a liquid. As we saw with boiling, when heat energy is added to a substance, attractive forces are disrupted and molecular motion increases. For a fat to melt, the forces that have to be disrupted are weak London forces. Because the melting points are low, fats are often said to be semisolid.

Now that we have distinguished triglycerides as fats and oils, let's apply what we learned about drawing saturated and unsaturated fatty acids in Section 4.5 to drawing saturated triglycerides.

Sample Problem 7.10   Drawing Triglycerides

Draw the skeletal structure for a saturated triglyceride with 18 carbons on each tail.

Solution

Every triglyceride contains a three-carbon backbone, glycerol, linked to the fatty acid tails through ester bonds.

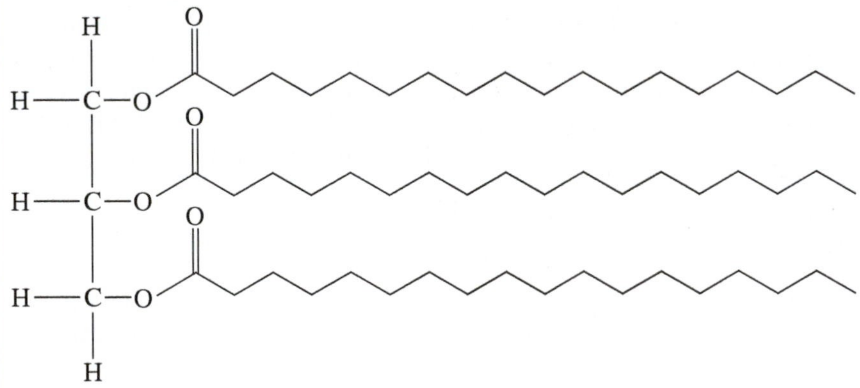

Practice Problems

7.39 In terms of their attractive forces, explain why fats are solids but oils are liquids under normal conditions, despite the fact that both are triglycerides.

7.40 What component molecules make up a triglyceride?

7.41 Considering their chemical structure, why are the melting points of oils lower than those of fats?

7.42 Olive oil is a monounsaturated oil with a melting point of −6 °C. Soybean oil is a polyunsaturated oil with a melting point of −16 °C. Explain their difference in melting points.

7.43 One of the triglycerides found in coconut oil contains three fatty acids with carbon designations of [10:0], [8:0], and [12:0]. (Refer to Section 4.5 for a review of fatty acid carbon designations.) The fatty acid with carbon designation [8:0] appears on the middle carbon of the three-carbon glycerol. Draw the triglyceride.

7.44 One of the main triglycerides in palm oil is tripalmitin. It contains three fatty acids with carbon designations of, [16:0]. Draw the structure of tripalmitin. Would you expect this molecule to be a solid or liquid at room temperature?

7.5 Attractive Forces and the Cell Membrane

The cells of our bodies are encased in an amazing membrane. The structure of the cell membrane, which is similar in cells throughout the body, has the ability to be flexible yet firm. As we'll see in Chapter 8, the cell membrane serves to allow some molecules to move into and out of the cell while inhibiting passage of other molecules. What is it about the structure and composition of the cell membrane that enables it to be selective?

? 7.5 Inquiry Question

How do attractive forces apply to the structure of a cell membrane?

A Look at Phospholipids

The main structural components of cell membranes are lipid molecules called **phospholipids**. Structurally similar to triglycerides, phospholipids have a glycerol backbone with fatty acids linked to it through an ester bond. Like triglycerides, phospholipids are hydrolyzable lipids. Unlike triglycerides, phospholipids have only two fatty acids on their glycerol backbone, arranged as shown in **FIGURE 7.23a**. The third OH group of the glycerol is bonded to a phosphate-containing group. The phosphate-containing group is ionic (polar), which is in contrast to the fatty acid tails (nonpolar).

There are many different phospholipids, but they all share this similar structure: a glycerol backbone with two nonpolar fatty acid tails and a polar phosphate-containing head. Because these molecules have both polar and nonpolar parts, they are amphipathic.

As shown in **FIGURE 7.23b**, phospholipids are often illustrated to highlight the dual character of these molecules (strongly polar head and two nonpolar tails). Earlier, we saw that the fatty acid salts in soap have a similar structure, with an ionic polar head but just one nonpolar tail. Having two nonpolar tails and a larger ionic head enhances both the nonpolar and polar characteristics of phospholipids as compared to soap. The two tails of the phospholipid also affect the overall shape of the molecule. While soap molecules tend to be more conical in their overall shape and form micelles, phospholipids generally have a more cylindrical shape with a much larger polar head group. This cylindrical shape of the phospholipids hinders their ability to arrange themselves into micelles. So how do phospholipids behave in water?

The Cell Membrane Is a Bilayer

Consider the cell and its environment: Its surroundings are watery, *and* its contents function in an aqueous environment. With water environments both inside and outside the cell, how can phospholipids, molecules that have nonpolar hydrophobic tails, serve to surround and protect cells? The cylindrical shape of a phospholipid constrains them to

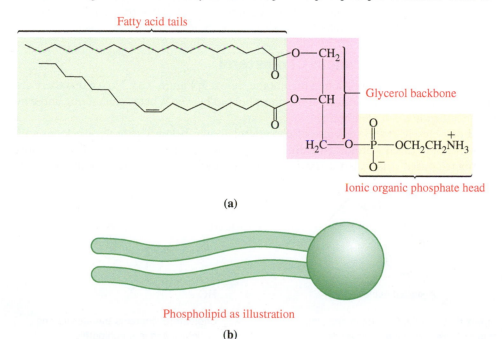

Fatty acid tails

Glycerol backbone

Ionic organic phosphate head

(a)

Phospholipid as illustration

(b)

FIGURE 7.23 Phospholipid (a) in line structure and (b) as an illustration. Phospholipids have a polar phosphate-containing head group and two nonpolar hydrocarbon tails.

FIGURE 7.24 The cell membrane's phospholipid bilayer. The hydrophobic hydrocarbon tails are excluded from the aqueous environment in the center of the bilayer, and the hydrophilic head groups interact with the aqueous environment both inside and outside the cell.

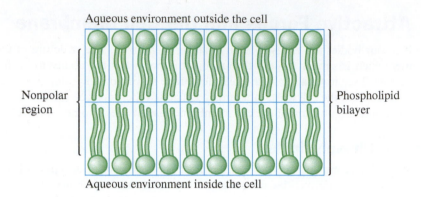

Aqueous environment outside the cell

Nonpolar region

Phospholipid bilayer

Aqueous environment inside the cell

lining up in a layer instead of a sphere like soaps do. But no matter how the tails orient themselves—toward the inside or outside of the cell—the hydrocarbon portion would be forced into an aqueous environment.

A cell membrane composed of phospholipids cannot exist as a single layer. Instead, the phospholipids form a double layer called a bilayer, as shown in **FIGURE 7.24**. The phospholipids of the "outer" layer of the membrane are oriented such that the polar heads are directed out into the surrounding aqueous environment. In turn, the phospholipids of the "inner" layer of the membrane are oriented with the polar heads directed into the aqueous environment of the interior of the cell. This arrangement leaves the hydrophobic tails of both layers directed toward each other, creating a nonpolar interior region. The phospholipid bilayer, as it is known, is the structural foundation for a cell's membrane. Next let's examine the other molecules found in the membrane.

Among the most important components in the phospholipid bilayer of the cell membrane are large biomolecules called proteins. Protein molecules can span the lipid bilayer, extending outward on either side (integral membrane proteins), or they can associate with one hydrophilic surface of the bilayer (peripheral membrane proteins). While the phospholipids provide the basis for the cell membrane's structure, the proteins are the membrane's functional components, allowing selected molecules to move into and out of the cell. (Read more about this in Chapter 8.) The exterior surface of the cell membrane also contains carbohydrates (like the ABO blood groups mentioned in Chapter 6) that act as cell signals.

The current model for the cell membrane, called the **fluid mosaic model,** shown in **FIGURE 7.25**, puts all these pieces together. This model creates what can be thought of as "icebergs" of protein floating in a "sea" of lipids, implying that the membrane itself is fluidlike because the phospholipids are able to move freely within their bilayer.

Steroids in Membranes: Cholesterol

Take a closer look at the portion of membrane shown in Figure 7.25. Notice that there are cholesterol molecules positioned in the phospholipid bilayer. What is cholesterol, and why is it included in a cell membrane?

Look closely at the structure of cholesterol shown below. Cholesterol belongs to a class of lipids called steroids. **Steroids** are lipids with a structure that contains a four-membered fused ring called a *steroid nucleus*. Steroids are considered nonhydrolyzable

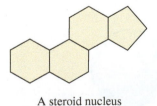

A steroid nucleus

Lipids that contain four fused rings like this one are classified as steroids.

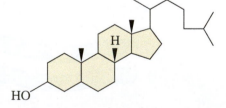

HO

Cholesterol contains the steroid ring structure and is amphipathic.

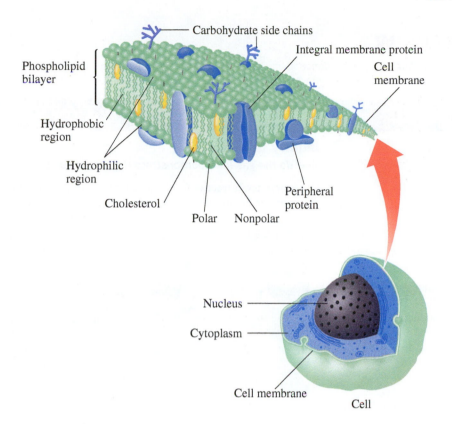

FIGURE 7.25 The fluid mosaic model of the cell membrane. Proteins and cholesterol are embedded in a lipid bilayer of phospholipids. The bilayer forms a hydrophobic barrier around the cell contents. Polar heads line the membrane surface, and the nonpolar tails face the center away from water.

lipids because they cannot be broken into smaller components through hydrolysis. Even though the steroid structure differs greatly from fatty acids and triglycerides, steroids are classified as lipids based on their nonpolar nature.

Steroids have a variety of functions in the body, including regulating sexual development (testosterone and estrogens) and emulsifying dietary fats (bile acids). One of the vitamins, vitamin D, also has a steroid form, ergosterol.

Cholesterol contains the steroid nucleus. Cholesterol, though, has a polar end (the OH group) and a nonpolar portion (the rest of the molecule), so like a phospholipid, it is amphipathic. Cholesterol situates itself in the phospholipid bilayer so that the OH group protrudes out into the surrounding aqueous environment, while the rest of the rigid hydrocarbon rings of the molecule are nestled in the nonpolar interior of the membrane.

Cholesterol is an important component of the cell membrane. It modulates the fluidity or flexibility of a cell membrane depending on the amount present. Cholesterol can slip in between the hydrocarbon tails of the phospholipids, disrupting the London forces and increasing the membrane's fluidity. Cholesterol can also interact with the unsaturated hydrocarbon tails via London forces, decreasing their motion and increasing the rigidity of the membrane.

Practice Problems

7.45   How will phospholipids (mainly nonpolar) arrange themselves when they are put in water?

7.46   Draw a possible hydrogen bond between a molecule of cholesterol and a molecule of water.

The study guide will help you check your understanding of the main concepts in Chapter 7. You should be able to answer problems for each learning outcome in this list. To check your mastery, try the problems listed after each.

STUDY GUIDE

7.1 Gases and Gas Laws

Apply gas laws.

- Convert pressure units. (Try 7.1, 7.47)
- Apply the Kinetic Molecular Theory of Gases. (Try 7.3, 7.49)
- Apply and solve problems using Boyle's law. (Try 7.5, 7.51, 7.53)
- Apply and solve problems using Charles's law. (Try 7.7, 7.55)
- Apply and solve problems using Gay-Lussac's law. (Try 7.59)
- Apply and solve problems using the combined gas law. (Try 7.13, 7.61)
- Apply the relationships found in the ideal gas law. (Try 7.15, 7.50, 7.93)

7.2 Liquids and Solids: Predicting Properties Through Attractive Forces

Identify the attractive forces and predict boiling points and vapor pressures for liquids based on attractive forces.

- Describe the five types of attractive forces present in compounds. (Try 7.19, 7.21, 7.65)
- Determine the attractive forces present in a compound from its chemical structure. (Try 7.25, 7.69)
- Draw hydrogen bonds between compounds that can hydrogen bond. (Try 7.27, 7.67)
- Describe changes in the states of matter. (Try 7.63)
- Predict relative boiling points for liquids based on the attractive forces present. (Try 7.31, 7.33, 7.71)
- Predict relative vapor pressures for liquids based on the attractive forces present. (Try 7.29, 7.34, 7.75)

CHAPTER GUIDE

Inquiry Question

How do the gas laws demonstrate the properties of gases?

Gases have few attractive forces because gas particles are often too far apart for an attraction to occur. Pressure is force exerted on

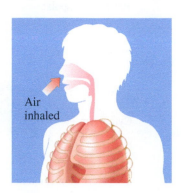

Air inhaled

a given area and is typically measured in units of psi or mmHg. The ideal behavior of a gas is outlined in the kinetic molecular theory of gases. Changing the pressure or temperature of a gas causes a change in the volume of the gas. Boyle's law compares volume and pressure when temperature is unchanged. As pressure increases, volume decreases. Charles's law compares volume and temperature when pressure is constant. As temperature increases, volume increases. Gay-Lussac's law compares temperature and pressure when volume is constant. As temperature increases, pressure increases. These three gas laws form the combined gas law, comparing two sets of conditions where temperature, pressure, and volume vary. When using the gas law equations, temperatures must be in Kelvin. All the gas laws can be derived from the relationships in the ideal gas law, $PV = nRT$.

Inquiry Question

What are the attractive forces and how can they be used to predict the relative vapor pressures and boiling points of compounds?

The attractive forces between compounds predict the physical properties of that compound. These forces arise from the polarity of the compound. The types of attractive forces present between compounds (from weakest to strongest) are London forces, dipole–dipole, hydrogen bonding, ion–dipole, and ionic attractions. Each of these forces involves the attraction of an area of negative charge (either partial or full) on one molecule or ion to an area of positive charge on a second.

In a closed container, evaporation into the space above a liquid builds up a pressure, called vapor pressure. Vapor pressures are different at different temperatures; higher temperatures cause higher vapor pressure as more molecules escape from the liquid into the gas. The boiling point is the temperature at which the change of state from liquid to gas occurs. For a compound to boil, the compound must overcome the attractive forces between molecules. At the boiling point, the vapor pressure equals the atmospheric pressure. The vapor pressures, boiling points, and melting points of compounds can be predicted by examining the attractive forces present. Compounds with stronger attractive forces require more heat energy to disrupt the attractions, resulting in lower vapor pressures and higher boiling points. Water has a comparatively high boiling point and low vapor pressure. For alkanes that have only London forces between molecules, the vapor pressure and boiling point depend largely on the surface area of the molecule—the greater the surface area, the lower the vapor pressure and higher the boiling point.

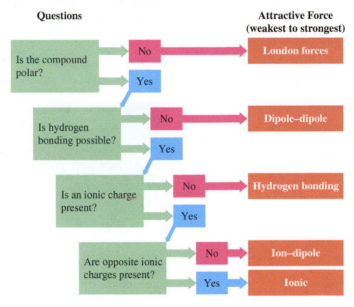

7.3 Attractive Forces and Solubility

Predict solubility in water based on attractive forces.

- State the golden rule of solubility. (Try 7.78)
- Predict the solubility of a molecule in water. (Try 7.35, 7.37, 7.75)
- Recognize an amphipathic molecule. (Try 7.35, 7.83)
- Define the role of an emulsifier. (Try 7.77, 7.97)
- Draw a fatty acid micelle using an illustration. (Try 7.92, 7.99)

Inquiry Question

How can we predict the solubility of a compound in water?

The solubility of one substance in another is described by the golden rule of solubility, *like dissolves like*, which means polar compounds dissolve other polar compounds and nonpolar compounds dissolve other nonpolar compounds. In other words, the more attractive forces that two compounds have in common, the more soluble they will be in each other. Soap is an amphipathic compound. In water, soap molecules form spherical structures called micelles with the hydrophobic tails gathered together in the nonpolar interior of the micelle while the hydrophobic heads cover the surface and interact with the aqueous external environment. Most organic compounds containing functional groups are soluble to various extents in water based on the number and distribution of polar or ionic groups.

—Continued next page

Continued—

7.4 Dietary Lipids

Apply attractive forces to distinguish a fat from an oil.
- Distinguish a fat from an oil. (Try 7.39)
- Describe the differences in melting points of fats and oils based on their attractive forces. (Try 7.41, 7.100)
- Draw a triglyceride. (Try 7.43, 7.87)

Inquiry Question

How are fats and oils different?

Melting is the change of state from solid to liquid. This change occurs when the motion of the molecules in the substance increases, causing the attractive forces between the molecules to lessen. This is often due to temperature changes. Fats are solid triglyceride compounds that have low melting points. Oils are also triglycerides, but they are liquids at room temperature. The fatty acid tails of oils are more unsaturated, and they have fewer London forces between their fatty acid tails than fats.

7.5 Attractive Forces and the Cell Membrane

Apply attractive forces to the structure of a cell membrane.
- Draw a phospholipid bilayer. (Try 7.45, 7.88)
- Locate the polar and nonpolar regions of a phospholipid and cholesterol. (Try 7.46, 7.90)
- Describe the structure of a cell membrane. (Try 7.91)

Inquiry Question

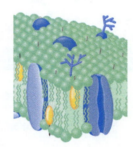

How do attractive forces apply to the structure of a cell membrane?

Cell membranes are mainly composed of phospholipids, which are amphipathic. Phospholipids have a strongly polar head and two nonpolar tails. Because of their shape, phospholipids of the cell membrane arrange themselves in a bilayer around the cellular contents with the polar heads of one layer oriented outward into the surrounding aqueous environment and the polar heads of the second layer directed inward toward the aqueous interior of the cell. This results in a hydrophobic interior layer formed by the nonpolar tails. The fluid mosaic model describes the complete structure of the cell membrane. Cell membranes also contain cholesterol, a steroid lipid, which helps modulate the fluidity of the membrane.

 Mastering Videos

PRACTICING THE CONCEPTS

The following videos can be accessed through the Pearson eText or your Mastering Chemistry course.
- Gas Laws: Behavior of Gases
- Predicting Solubility of Molecules in Water

Important Equations

Boyle's Law

$$P_i \times V_i = P_f \times V_f$$

Charles's Law

$$\frac{V_i}{T_i} = \frac{V_f}{T_f}$$

Gay-Lussac's Law

$$\frac{P_i}{T_i} = \frac{P_f}{T_f}$$

The Combined Gas Law

$$\frac{P_i V_i}{T_i} = \frac{P_f V_f}{T_f}$$

The Ideal Gas Law

$$PV = nRT$$

Additional Problems

7.47 A hyperbaric oxygen chamber can be used in the treatment of bone damage caused by radiation therapy. Pressures in the chamber can be raised to 2280 mmHg. What is this pressure in atmospheres?

7.48 Tire pressure for a popular auto tire is 40 psi. What is this pressure in atm?

7.49 Is the pressure of a gas caused by gas molecules hitting a surface or gas molecules hitting each other? Explain.

7.50 If the flasks below are the same size, are at the same temperature, and contain the same number of molecules, which flask has the highest pressure? Explain.

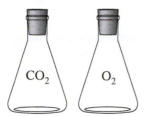

7.51 A sealed syringe contains 5 cc of oxygen gas under 14.7 psi of pressure. If the pressure on the syringe is increased to 36.8 psi, but the temperature remains constant, what is the volume of the oxygen gas in the syringe?

7.52 Underwater pressure increases by 14.7 psi for every 10 m of depth. Standing on a boat in scuba gear, you fill a balloon with helium.

    a. If you take the balloon with you as you dive to a depth of 40 m, what, if anything, will happen to the size of the balloon (assume temperature stays the same)?

    b. If the balloon slips out of your hand and floats up into the sky before you dive, what, if anything, will happen to the size of the balloon (assume temperature stays the same)?

    c. If the balloon had a volume of 2 L at a depth of 40 m, what was the original volume of the balloon if we assume the pressure at the surface of the water is 14.7 psi?

7.53 The lowest recorded atmospheric pressure occurred during a typhoon in the Pacific Ocean in 1979 and measured 652 mmHg. The highest atmospheric pressure on record occurred in Mongolia in 2001 and measured 814 mmHg. If a balloon were inflated with air to a volume of 10. L under the conditions in Mongolia and then rapidly transported to the conditions described for the typhoon in the Pacific, what would the volume of the balloon be under these new conditions, assuming no change in temperature?

7.54 Scuba divers wear tanks of compressed air to help them breathe underwater. Based on your knowledge of Boyle's law, explain why the air in a scuba tank must be pressurized for a deep-sea diver to breathe while underwater.

7.55 A beach ball is filled with 10.0 L of air in a hotel room where the temperature is 65 °F. What is the volume of the air in the beach ball out on the beach where the temperature is 38 °C?

7.56 The night before a big January "polar bear" bike ride, you take your bike inside to clean it up and pump up the tires. The next morning, after spending the night in your cold garage, the tires on your bike feel "squishy." Assuming they did not leak overnight, why do the tires feel less firm this morning than they did last night?

7.57 If a hot-air balloon has an initial volume of 1000 L at 50 °C, what is the temperature (in °C) of the air inside the balloon if the volume expands to 2500 L?

7.58 Liquid nitrogen is an extremely cold liquid (−196 °C) used in chemistry "magic" shows to "shrink" balloons (and do other fun things). If a balloon is inflated to 2 L at 26 °C and then cooled in liquid nitrogen, what is the final volume of the balloon?

7.59 A full cylinder of compressed gas is moved to a loading dock (85 °F) from an air-conditioned truck (72 °F) where the pressure was 28.2 psi. What is the new pressure on the loading dock?

7.60 A steel cylinder has a volume of 12.0 L. If it is filled with a gas at a pressure of 175 atm when the temperature is 20.0 °C and the cylinder sits in the sun until it reaches a temperature of 97.0 °C, what is the final pressure of the gas in the cylinder?

7.61 Your friend breathes in 1.5 L of helium from a balloon at 22 °C and 1.00 atm and talks funny. You want to give it a try, but you are breathing from a small tank that is at 17 °C and 1.50 atm. At this temperature and pressure, how much He will you have to breathe to talk funny?

7.62 An oxygen tank contains 2.5 L at 29.4 psi at a temperature of 27 °C. A diver takes the tank to 58.8 psi and the temperature drops to 12 °C. If the tank has a maximum volume of 4.0 L, is the tank in danger of exploding at these depths? Explain.

7.63 Name the change of state occurring for each.

    a. alcohol drying after being rubbed on the skin

    b. dry ice forming a spooky mist on a Halloween display

    c. fat solidifying on the top of leftover food after being placed in the refrigerator

7.64 Name the change of state occurring for each.

    a. water vapor from the air forming a pattern of ice on a window pane in sub-freezing weather

    b. rocks in a volcano forming molten lava when heated to extreme temperatures

    c. water droplets forming on the outside of a glass containing a cold drink

7.65 Compare and contrast a dipole–dipole attraction and an ion–dipole attraction.

7.66 Compare and contrast an ion–dipole attraction and an ionic attraction.

7.67 Show a possible hydrogen bond between acetone (CH_3COCH_3) and water. Label the donor and the acceptor. Is there more than one possible hydrogen bond between these two molecules? Explain.

7.68 Show a possible hydrogen bond between methanol (CH_3OH) and dimethyl ether (CH_3OCH_3). Label the donor and the acceptor. Is there more than one possible hydrogen bond between these two molecules? Explain.

7.69 List all attractive forces found between molecules in each of the pure substances:

a. Na_2CO_3 b.

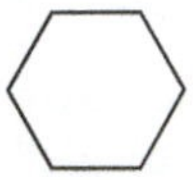

c.

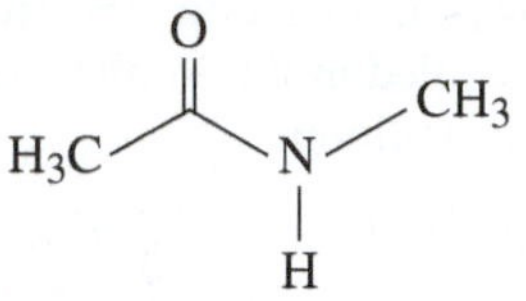

d.

e. CH_2Cl_2 f.

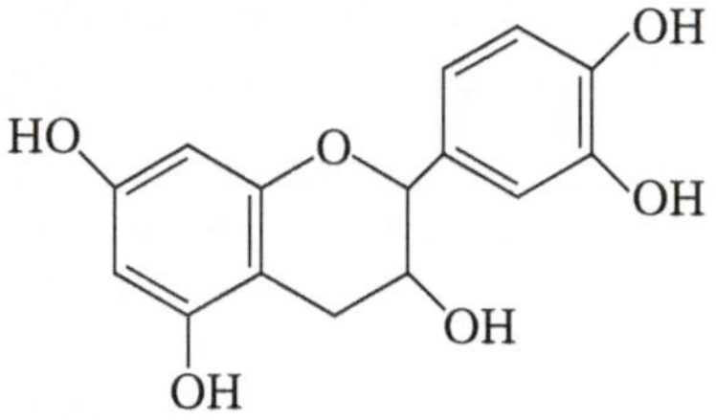

7.70 List all attractive forces found between each of the pure substances:

a. CF_4 b.

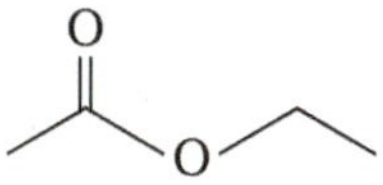

c.

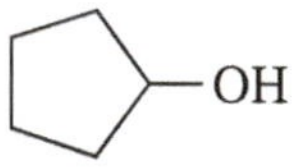

d.

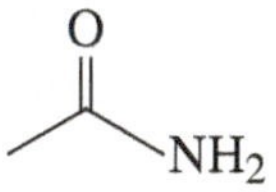

e.

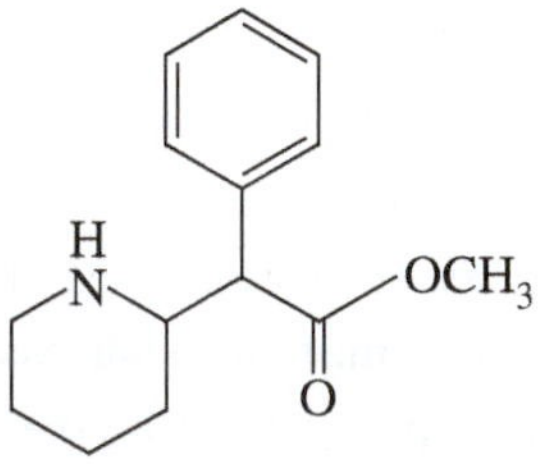

f. KBr

7.71 Rank the four compounds in the figure in decreasing order of boiling point. Explain the reasoning behind your ranking.

a.

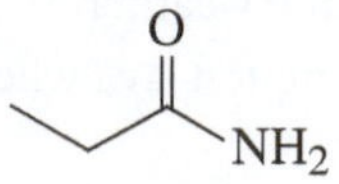

b.

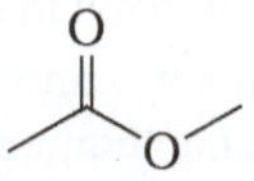

c.

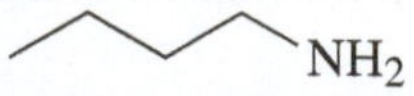

d.

7.72 Rank the four compounds in the figure in decreasing order of boiling point. Explain the reasoning behind your ranking.

a.

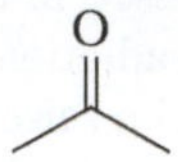

b.

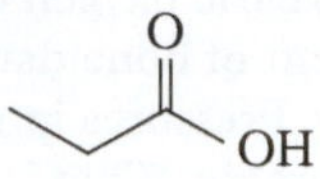

c.

d.

7.73 The boiling point of octane is much greater than that of hexane (see Table 7.2). Explain the difference in their boiling points.

7.74 The boiling point of heptane is over 60 °C higher than the boiling point of pentane. Draw the structure of both compounds and explain the difference in their boiling points.

7.75 Explain why a straight-chain alkane has a lower vapor pressure than a branched-chain alkane with the same molecular formula.

7.76 Based on your knowledge of vapor pressure trends in alkanes, predict which compound has the higher boiling point, butane or 2-methylpropane (isobutane).

7.77 Explain how an emulsifier can combine a nonpolar substance with a polar substance.

7.78 In terms of attractive forces, explain why oil and water do not mix.

7.79 Predict which member of each of the following pairs of compounds will be more soluble in vegetable oil:
a. fatty acid or fatty acid salt
b. fat or water
c. CH_3NH_2 or CS_2
d. KCl or $CH_3CH_2CH_2CH_2CH_2CH_3$

7.80 Predict which member of each of the following pairs of compounds will be more soluble in vegetable oil:
a. NH_3 or $CH_3CH_2CH_2 CH_2CH_3$
b. NaCl or $CH_3CH_2OCH_2CH_3$
c. cyclopentane or $CH_3CH_2NH_2$
d. CH_3OH or CH_3OCH_3

7.81 A stain on your shirt will not come out when you wipe it with a wet cloth. What clue does this give you to the nature of the stain?

7.82 To remove stains from delicate fabrics without using soap and water, dry cleaners once used perchloroethylene, which was banned in 1998 for environmental reasons. Its structure is shown in the figure. Explain why perchloroethylene makes a good stain remover.

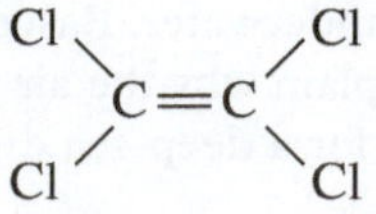

7.83 Predict which of the following compounds is likely to be more soluble in water.

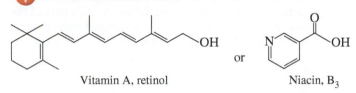

Vitamin A, retinol   or   Niacin, B₃

7.84 Predict which of the following compounds is likely to be more soluble in water.

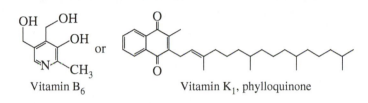

Vitamin B₆   or   Vitamin K₁, phylloquinone

7.85 Predict whether each of the following pharmaceuticals is likely to be soluble in water based on its polarity and the distribution of polarity throughout the compound.

   a. the ADHD medication, amphetamine (Adderall®)

   b. the antinausea medication promethazine HCl (Phenergan™)

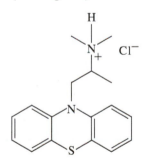

7.86 Predict whether each of the following pharmaceuticals is likely to be soluble in water based on their polarity and the distribution of polarity throughout the compound.

a. the blood pressure medication atenolol (Tenormin®)

b. the laxative bisacodyl (Dulcolax®)

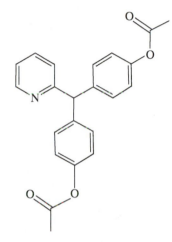

7.87 One of the main triglycerides in hydrogenated soybean oil is tristearin. It contains three fatty acids [18:0]. Draw the structure of tristearin. Would you expect this molecule to be a solid or liquid at room temperature?

7.88 Sketch an illustrated model of a phospholipid bilayer of a cell membrane and label the polar and nonpolar parts.

7.89 Compare the structure of a soap molecule to a phospholipid and explain why a soap's polar head is smaller than that of a phospholipid.

7.90 Soap, phospholipids, and cholesterol are all amphipathic molecules. Provide a drawing of each and label the polar and nonpolar areas.

7.91. Describe other components present in a cell membrane and their relative location.

7.92. Sketch an illustrated model of a fatty acid micelle and label the polar and nonpolar parts.

Challenge Problems

7.93 Nitrous oxide, N₂O, is a common anesthetic also known as laughing gas, or simply "nitrous." If a cylinder of nitrous contains 3.4 kg of N₂O, what volume of N₂O can be obtained from this cylinder at a standard temperature of 20 °C and standard pressure of 1.00 atm? ($R = 0.0821$ L·atm/mole·K)

7.94 Ideal gases obey Avogadro's law, which states that 1.00 mole (*n*) of an ideal gas occupies 22.4 L at 1.00 atm and 0 °C. Show mathematically that this is true. ($R = 0.0821$ L·atm/mole·K)

7.95 Consider the situation shown. Two identical balloons are at the same temperature, pressure, and volume. One balloon contains helium; the other contains carbon dioxide. Which statement is true?

    a. The helium balloon weighs more than the carbon dioxide balloon.

    b. The carbon dioxide balloon weighs more than the helium balloon.

    c. The two balloons weigh the same.

    d. There is not enough information given to determine the weight.

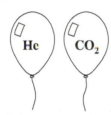

7.96 Desflurane and sevoflurane are common inhalation anesthetics that are liquids at room temperature. Their vapor pressures are 669 mmHg and 170 mmHg, respectively, at 20 °C. If desflurane were accidentally put into a sevoflurane vaporizer, would the patient receive more or less anesthetic than intended? Explain your answer.

7.97 Mayonnaise is a thick mixture containing oil, vinegar (water-based), and eggs; the eggs contain a phospholipid molecule called lecithin. Mayonnaise cannot be made without lecithin. Explain why lecithin is a critical ingredient.

7.98 Fats and oils are both triglycerides. Consider the shortening Crisco®. Would scientists consider this a fat or an oil? What about margarine?

7.99 If very small amounts of soap are dissolved in water, the soap molecules can form a structure known as a monolayer on the surface of the water, with part of the soap molecule dissolved in the water and part sticking into the air. Using illustrated representations for soap molecules, draw a small part of a monolayer showing the water, air, and soap molecules as they would be arranged.

7.100 The structures and boiling points of cyclopentane and tetrahydrofuran are shown in the following figure. Explain why the boiling point of tetrahydrofuran is higher than that of cyclopentane.

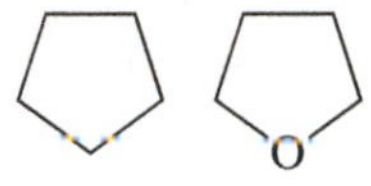

    Boiling point   49 °C   65 °C

7.101 Some additional physical properties of liquids, other than those discussed in this chapter, include viscosity, the resistance to flow, which can be thought of as the "thickness of a liquid," and surface tension, which measures the forces at a liquid surface. Consider the liquids honey (mainly glucose and fructose) and water. (You may need to think back to Chapter 6 to consider the structures of glucose and fructose.)

    a. In your experience, which liquid has a higher viscosity?

    b. What attractive forces are present? Which one can form more hydrogen bonds?

    c. Explain viscosity in terms of attractive forces.

    d. It is possible to float a small paper clip on the surface of a cup of water. Try it. What happens if you add soap to the water? Can you float the same paper clip?

Answers to Odd-Numbered Problems

Practice Problems

7.1   **a.** 14.5 psi   **b.** 14.5 psi   **c.** 14.7 psi

7.3   The gas in container B is at higher pressure because the gas particles have a higher kinetic energy at higher temperatures. The particles hit the sides of the container more often, creating more pressure.

7.5   17 L

7.7   0.53 L

7.9   The pressure decreases.

7.11 **a.** balloon C   **b.** balloon A

       c. balloon C   **d.** balloon B

7.13 102 mL

7.15 Because the moles of gas, volume, and temperature are the same in each container, the pressure will be the same in each container according to the ideal gas law.

7.17 Covalent bonds involve the sharing of electrons within a molecule, creating a chemical bond. Attractive forces involve attractions between positive and negative areas of different molecules. In attractive forces, electrons are not shared between atoms as in a chemical bond.

7.19 The dipole in London forces is temporary; the dipole in dipole–dipole attractions is permanent.

7.21 A hydrogen bond acceptor is a lone pair of electrons on an O, N, or F.

7.23 The cation and anion of the ionic compound become hydrated as water molecules surround the ions, breaking them apart. The partial negative end of water (the O) is attracted to the cation, while the partial positive end of water (the H) is attracted to the anion.

7.25 **a.** ionic   **b.** London forces

       c. hydrogen bonding   **d.** dipole–dipole

7.27 Three hydrogen bonds can be formed.

7.29 Water. Because water can make four hydrogen bonds with other water molecules and ethanol molecules can only make three hydrogen bonds with each other, there are stronger attractions between the water molecules. The stronger the attractive forces, the lower the vapor pressure (less molecules in the vapor phase).

7.31 Of the two alkanes, the one with the more branching has the lower boiling point because of its decreased surface area. The alcohol molecule on the right has the highest boiling point because it has the strongest attractive forces present. It is an alcohol and has both hydrogen bonding and dipole–dipole attractions.

7.33 sevoflurane

7.35 a. insoluble   **b.** soluble   **c.** forms a micelle

7.37 a. The fatty acid salt is more soluble in water. The salt can form more attractions to water through ion–dipole interactions not available to the fatty acid.

   b. KCl. The ions in KCl readily dissolve in water and form ion–dipole interactions, which are the strongest attraction that a molecule can have with water.

7.39 Even though fats and oils are both triglycerides, the large number of cis double bonds in oils lessens the number of attractive forces between the unsaturated hydrocarbon tails compared to the saturated tails of the fats. More attractions mean that the tails pack more tightly, more like a solid, forming a fat.

7.41 The melting points are lower for oils because it takes less energy (as heat) to move the loosely packed tails of an oil than it does to move the more tightly packed tails of a fat.

7.43 A triglyceride is a dietary lipid containing three fatty acids bonded to a glycerol molecule.

7.45 Phospholipids will form two layers called a bilayer (see Figure 7.24) with the nonpolar tails of the two layers facing each other and their polar heads facing the water.

Additional Problems

7.47 3.00 atm

7.49 Gas molecules hitting a surface cause pressure. Pressure is defined as force over an area.

7.51 2 cc

7.53 12 L

7.55 10.7 L

7.57 535 °C

7.59 28.9 psi

7.61 No, you would only breathe in 0.98 L of He at this temperature and pressure.

7.63 a. evaporation   **b.** sublimation   **c.** freezing

7.65 Both are attractive forces that occur between positive and negative areas. A dipole–dipole attraction occurs between two opposite partial charges. An ion–dipole attraction occurs between opposite ionic charges and partial charges.

7.67

The molecules can form more than one hydrogen bond. Acetone can form two hydrogen bonds, acting as an acceptor twice (lone pairs on oxygen).

7.69 a. ionic attraction

   b. London forces

   c. London forces, dipole–dipole, hydrogen bonding

   d. London forces

   e. London forces, dipole–dipole

   f. London forces, dipole–dipole, hydrogen bonding

7.71 Highest to lowest boiling point:

   a. five possible H bonds per molecule (strongest)

   c. three possible H bonds per molecule

   b. polar molecule; has dipole–dipole, but no H bonding

   d. nonpolar molecule (weakest)

7.73 Because the hydrocarbon chain is longer in octane (meaning the molecule has a greater surface area), there are more London forces between molecules, which makes the boiling point higher for octane.

7.75 A branched alkane has less surface contact with neighboring molecules than does a straight-chain alkane. There will be more London forces between the straight-chain alkane molecules so at a given temperature, there are less molecules of straight-chain alkane in the vapor phase and the vapor pressure is lower.

7.77 An emulsifier can attract both a nonpolar and a polar substance, allowing both substances to be suspended in a mixture.

7.79 a. fatty acid **b.** fat

 c. CS_2 **d.** $CH_3CH_2CH_2CH_2CH_2CH_3$

7.81 The stain must be hydrophobic (nonpolar) because it is not soluble in water.

7.83 Niacin. The niacin overall contains more polar groups versus nonpolar areas than the vitamin A.

7.85 a. likely not soluble due to the mostly nonpolar hydrocarbon and no charge

 b. likely soluble due to the charge of the HCl salt

7.87 Likely a solid at room temperature due to the saturated fatty acid tails having strong interactions with each other.

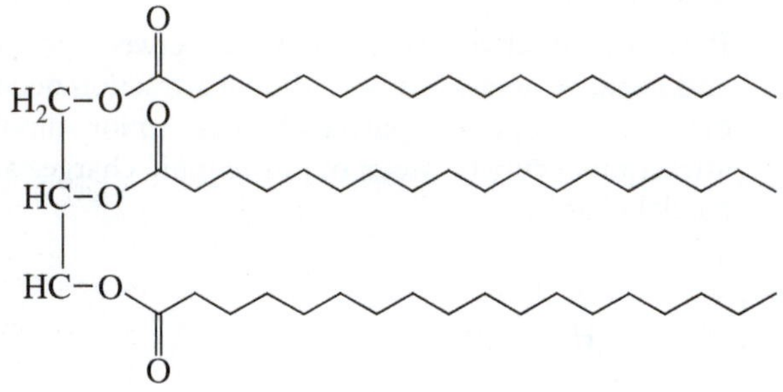

7.89 The phospholipid head group contains many more atoms than the three atoms in the carboxylate head group of a soap and is much larger.

7.91 Cell membranes contain cholesterol inserted into the phospholipid layer, carbohydrates on the outer surface, and proteins that can either span the membrane or sit on either surface of the membrane.

Challenge Problems

7.93 1900 L

7.95 b.

7.97 Phospholipids are amphipathic molecules. They act to emulsify the nonpolar oil and the polar water in the mayonnaise, forming a thick mixture.

7.99

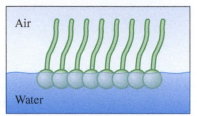

Hydrocarbons would interact with nonpolar air, polar heads with water at surface.

7.101 a. honey

 b. Hydrogen bonding, dipole–dipole attractions, and London forces. A glucose or fructose in ring form can make many more hydrogen bonds to itself than water molecules can make to themselves.

 c. Because there are so many more hydrogen bonds possible between sugar molecules in honey, the honey is more viscous than water.

 d. By adding soap to the water, the hydrogen bonding network between the water molecules is disrupted, and the paper clip can no longer float on the surface.

Solution Chemistry—
Sugar and Water Do Mix

Sports drinks help athletes replenish water and electrolytes more quickly than just drinking water or juice. How does a sports drink solution transport glucose and electrolytes inside a cell so quickly? Read Chapter 8 to learn more about solutions and movement of molecules into the cell.

LEARNING TIP   Write Your Notes, Don't Type

Did you know that writing out notes by hand during class can better prepare you for conceptual questions than typing your notes? When you write out notes in your own words, your brain more deeply processes the information. Typing information verbatim during class doesn't provide that same level of processing.

SPORT DRINKS like Gatorade® contain sucrose, glucose, citric acid, and potassium phosphate, which all dissolve in water, making it a mixture. Because such drinks are uniform throughout, they are homogeneous mixtures. Gatorade also contains an emulsifier that helps to suspend nonpolar flavor molecules in the solution, making the drink slightly cloudy. This is a different kind of mixture than apple juice, which you can see through clearly. We begin Chapter 8 by exploring these differences.

If your juice is too sweet for you, you can change the taste by diluting it with water, which decreases the sugar concentration. Understanding solutions and concentration helps us to better understand concentrated medications, which need to be precisely diluted before dispensing. Also, body fluids are complex aqueous solutions containing dissolved substances like glucose and ions like K$^+$, Na$^+$, Cl$^-$, and HCO$_3^-$. Our bodies have sophisticated ways of keeping the concentration of dissolved molecules and ions constant both inside and outside of cells. We examine solutions and how concentrations are regulated in the body in Chapter 8.

8.1 Solutions Are Mixtures

A cup of coffee with sugar represents a type of homogeneous mixture called a **solution.** A solution consists of at least one substance—the **solute**—evenly dispersed throughout a second substance—the **solvent.** The components in a solution do not react with each other, so the sugar mixed into the coffee is still sugar and still tastes sweet. The solute is the substance in a solution present in the smaller amount, and the solvent is the substance present in the larger amount. In the case of sweetened coffee, the sugar is one of the solutes, and the water is the solvent. A cup of coffee, though dark brown in color, is transparent; if held up to a light, you can see through the liquid. Once the sugar is dissolved in the water, it will not undissolve over time. These properties of solutions provide a quick way for you to determine if a substance is a solution. See **TABLE 8.1** for some examples of other properties of solutions.

States of Solutes and Solvents

We tend to think of solutions as being composed of solids, like sugar, dissolved in liquids like water. However, this does not have to be the case. Solutions can also be homogeneous mixtures containing gases or solids. The air we breathe, for example, is a homogeneous mixture of gases, so it is also a solution where nitrogen is the solvent, and oxygen and other gases are the solutes. Brass is a solution of solids where the solute is metal zinc and the solvent is metal copper. The solute and solvent can be a solid, liquid, or gas—the three states of matter—in a solution (see **FIGURE 8.1**).

Our study of solutions will focus on dilute solutions where water is the solvent. These solutions are referred to as **aqueous solutions.**

The Unique Behavior of Water

Let's look more closely at water, H$_2$O. Water has some properties that distinguish it from many other solvents. Remember that each water molecule can form up to four hydrogen bonds with neighboring water molecules. These strong attractive forces give water some unique properties not found in other solvents. For example, we saw in Chapter 1 that water has a high specific heat, allowing it to trap energy for longer periods of time than many other substances. In Chapter 7 we saw that it has a high boiling point and a low vapor pressure.

We have all seen that solid ice floats on liquid water. This does not seem unusual to us, but compared to most organic molecules, this property is actually unique. As water freezes, its molecules move farther apart to make the optimal hydrogen-bonding contacts. Ice has a lower density than liquid water and floats on the surface. Compare this to a substance like an oil. When an oil freezes, it packs the molecules more tightly together than in the liquid phase, making the solid more dense. Solid olive oil, therefore, sinks in liquid olive oil.

Solute: The substance present in lesser amount

Solvent: The substance present in greater amount

Solutions are homogeneous mixtures.

TABLE 8.1 Some Properties of Solutions

Particles are evenly distributed.
Components do not chemically react with each other.
Aqueous solutions are transparent.
Components do not separate upon standing.
Concentration can be changed.

FIGURE 8.1 Solutions can be solid, liquid, or gas. (a) Brass is a solution of copper and zinc solids, (b) air is a solution of gases, and (c) carbonated beverages represent a solution of a gas in a liquid.

(a)   (b)   (c)

Water is often referred to as the universal solvent. While that may be true for life because living things are mostly water, most nonpolar and organic substances do not dissolve in water. When considering solvents and the solubility of solutes, remember to be consistent with the golden rule of solubility—*like dissolves like.*

Water   **Olive oil**

One of the unique properties of water is that the solid form is less dense than the liquid. The solid form (red arrow) of most substances, like olive oil, are more dense than the liquid and so sink.

Sample Problem 8.1   **Identifying Solute and Solvent**

Identify the solute and the solvent of the two components given in the mixtures below.
  a. A teaspoon of honey is in a cup of hot tea.
  b. A packet of sweetener is added to a cup of coffee.
  c. A small amount of the anesthetic nitrous oxide mixes with air (80% N_2) during a patient's surgery.

Solution

Remember that the solute is the substance in the smaller amount and the solvent is the substance in the larger amount.
  a. Water is the solvent, and honey is the solute.
  b. Water is the solvent, and the sweetener is the solute.
  c. Nitrogen in the air is the solvent (present in the largest amount) and nitrous acid is the solute.

Colloids and Suspensions

Is coffee with cream a solution? If we look back to the properties of solutions in Table 8.1, we would have to say that it is not a solution because coffee with creamer is not a transparent liquid. The milk or cream that you put in your coffee is not a transparent liquid, so it is not a solution. If it is not a solution, what is it?

Homogenized milk or cream is uniform throughout and is therefore a homogeneous mixture, yet both contain protein and fat molecules that do not dissolve in water, the solvent. Instead, these molecules aggregate to form larger particles in solution much as the soap molecules, discussed in Chapter 7, formed micelles. The nonpolar areas collect in the center, and the polar areas of the milk proteins face the exterior and interact with the solvent.

Homogenized milk and cream are **colloids** (or colloidal mixtures) because of the proteins and fats that do not dissolve. By definition, the particles in a colloid must be between 1 nanometer (1×10^{-9} meter, abbreviated nm) and 1000 nm in diameter. Particles of this size remain suspended in solution, so a colloid does not separate over time.

What kind of mixture is muddy water? The dirt in the muddy-water mixture will separate from the water upon standing. If the diameter of the particles in a mixture is greater than 1000 nm (1 micrometer, or μm), the mixture is considered a **suspension.**

Blood is another example of a suspension. Blood contains blood cells that are larger than 1 μm, and these cells will settle to the bottom of a test tube upon standing. Blood cells are routinely separated from plasma through centrifugation, a spinning process that accelerates settling (see **FIGURE 8.2**).

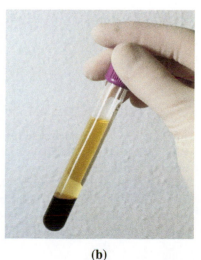

colloid

Cream is a colloid because the fat and protein molecules do not dissolve in the liquid or separate upon standing.

Blood after centrifugation

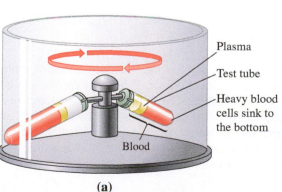

Plasma

Test tube

Heavy blood cells sink to the bottom

Blood

(a)   **(b)**

FIGURE 8.2 Blood is a suspension. The cells in blood can be isolated using centrifugation (a) into the soluble components (plasma) and cells (b).

In medicine, intravenous fluids tend to be classified as crystalline and colloid solutions. Crystalline solutions are solutions that contain small solutes that completely dissolve in aqueous solution. Some examples are normal saline (NS) and 5% dextrose in water (D5W). Colloid solutions are colloids and contain large molecules like proteins that do not dissolve in water, but remain suspended in in the solution. Some examples are albumin and dextran fluids.

Sample Problem 8.2   Type of Mixture

Are the following solutions or colloids?
a. Gatorade   b. cranberry juice   c. a cloud

Solution

a. Gatorade is opaque and uniform throughout. It contains molecules that aggregate with flavor molecules that are suspended throughout the solution. Therefore, it is a colloid.
b. Cranberry juice is transparent and uniform throughout. Therefore, it is a solution.
c. A cloud is uniform throughout and contains small water droplets, but it is not transparent. Therefore, it is a colloid, not a solution.

Practice Problems

Find the answers to the odd-numbered Practice Problems at the end of the chapter.

8.1 Identify the solute(s) and solvent in solutions composed of the following:

a. a cleaning solution containing a tablespoon of vinegar and a cup of water
b. a 3-gallon gas tank filled with the natural gas methane and a small amount of methyl mercaptan, which gives an odor should any gas leak
c. a kilo bar of 18-karat rose gold that contains 750 g of gold, 220 g of copper, and 30 g of silver

8.2 Identify the solute and solvent in solutions composed of the following:

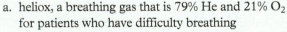

a. heliox, a breathing gas that is 79% He and 21% O_2 for patients who have difficulty breathing
b. Children's Tylenol® oral suspension, containing 160 mg acetaminophen per 5 g of water
c. a tincture containing a small amount of an herbal remedy dissolved in alcohol

8.3 Identify the following as solutions or colloids:

a. seltzer water   b. yogurt   c. saltwater

8.4 Identify the following as solutions or colloids:

a. 5% albumin, a protein IV replacement fluid
b. the electrolyte replacement IV fluid lactated Ringer's solution
c. D5W, a 5% dextrose IV fluid

8.2 Formation of Solutions

8.2 Inquiry Question

How do temperature and pressure affect the formation of a solution?

In Chapter 7, the golden rule of solubility, *like dissolves like,* was introduced to explain why some compounds dissolve in water and others do not. Polar covalent compounds, like glucose, and ionic compounds, like NaCl, will dissolve in water because strong attractive forces exist between them. Nonpolar covalent compounds, like hydrocarbons and oils, are not soluble in the polar molecules of water. Amphipathic molecules, like fatty acids that contain polar and nonpolar parts, can interact with water through their polar part, but these molecules typically do not form solutions.

The dissolving process requires that individual solvent particles surround the solute molecules and interact through attractive forces. This process is referred to as **solvation** for all solutions and, more specifically, as **hydration** in the case of an aqueous solution. Because the particles are just redistributing themselves, dissolving is a physical change, not a chemical change. For a review of this topic, see Section 7.3.

Factors Affecting Solubility and Saturated Solutions

How much sugar can you dissolve in a glass of iced tea before the sugar stops dissolving and starts to settle to the bottom of your glass? You may have observed that it is easier to dissolve sugar in hot tea than in cold tea. Solubility depends on a number of factors, including the polarity of the solute and the solvent (the attractive forces present), as well as the temperature of the solvent.

If a solution does *not* contain the maximum amount of the solute that the solvent can hold, it is referred to as an **unsaturated solution.** If a solution contains all the solute that can possibly dissolve, the solution is referred to as a **saturated solution.** If more solute is added to a saturated solution, the additional solute will remain undissolved. A solution that is saturated reaches a physical equilibrium between the dissolved solute and undissolved solute. At equilibrium, the speed or rate of dissolving solute and the rate of dissolved solute re-forming crystals (called recrystallization) are the same. This can be represented in an equation in which a double arrow, or *equilibrium arrow*, is used between the products and reactants in the chemical equation.

$$\text{Undissolved solute} \underset{\text{crystallization}}{\overset{\text{solvation}}{\rightleftharpoons}} \text{Dissolved solute}$$

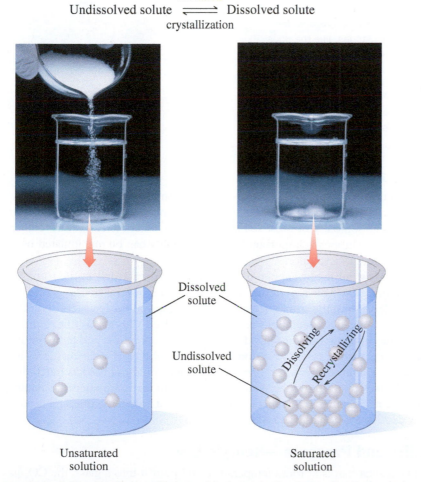

In a saturated solution, physical equilibrium exists between the states of solvation and crystallization of the solute.

Gout, Kidney Stones, and Solubility

The conditions of gout and kidney stones occur when compounds build up in the body and exceed their solubility limits. In the case of gout, the solid compound is uric acid. In some individuals, the release of uric acid by the kidneys into the urine is reduced, causing a buildup in the bodily fluids. Insoluble needlelike crystals form in cartilage and tendons at the joints, often in the ankles and feet. Kidney stones contain uric acid or the ionic compounds calcium phosphate or calcium oxalate. They can form in the urinary tract, kidneys, ureter, or bladder when the compounds do not remain dissolved in the urine. Both gout and kidney stones can be treated through changes in diet and drug therapy (see **FIGURE 8.3**).

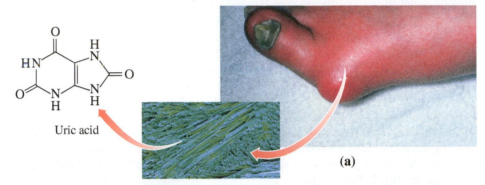

Uric acid

(a)

FIGURE 8.3 Exceeding solubility limits. (a) When uric acid levels in the blood exceed the solubility limit, the excess forms the needlelike crystals of uric acid responsible for gout. (b) Kidney stones, like the ones shown here (next to a match for size comparison), are typically composed of the ionic compounds calcium phosphate $(Ca_3(PO_4)_2)$ or calcium oxalate (CaC_2O_4). Even small crystals are painful when passing through the urethra.

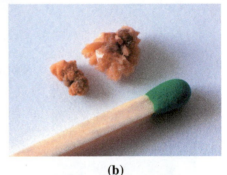

(b)

Solubility and Temperature

It is easier to dissolve sugar in a cup of hot tea than in a glass of iced tea. In fact, more sugar will dissolve in hot tea than in cold tea. This demonstrates the effect of temperature on solubility. The solubility of most solids dissolved in water *increases* with temperature. This property of solutions is important because solubility can be manipulated by changing the temperature of a solution.

Think about what might happen to a bottle of fizzy water left in the trunk of a car on a hot summer day. You may return to your car after several hours to find that the container exploded and the trunk of your car is soaking wet! Fizzy waters and other carbonated drinks are aqueous solutions containing dissolved CO_2 gas as one of the solutes. The reason the container burst is related to gas solubility, pressure, and temperature. (For a review of gas pressure, refer to Section 7.1.) The solubility of a gas dissolved in water *decreases* with a rise in temperature. The warmer drink could not hold as much dissolved gas. As the CO_2 escaped from the solution, it moved into the space above the liquid in the container. Because there are now more molecules in that space, the pressure is increased, ultimately causing the container to rupture.

T = lower T = higher

At higher temperature, the solubility of a gas in a liquid decreases.

Solubility and Pressure—Henry's Law

If you open a soda that is at room temperature and pour it into a glass, the CO_2 gas will still fizz and escape from the liquid even though the temperature is not changing. Why is the soda fizzing? When you first open the can of soda or untwist the cap on a bottle, you hear a hissing sound. The hiss is because the little space above the beverage in the container is filled with CO_2 at a pressure higher than atmospheric pressure, and the CO_2

escapes quickly once the seal is broken. The CO_2 that was dissolved in the drink at the higher pressure (before opening) will not stay dissolved at the lower pressure of the atmosphere, and the CO_2 begins to bubble out of the solution.

This relationship between gas solubility and pressure was summarized by the English chemist William Henry (1775–1836) in **Henry's law,** which states that the solubility of a gas in a liquid is directly related to the pressure of that gas over the liquid. That is, the amount of a gas that can dissolve in a liquid increases as the pressure of the gas in the space above the liquid increases.

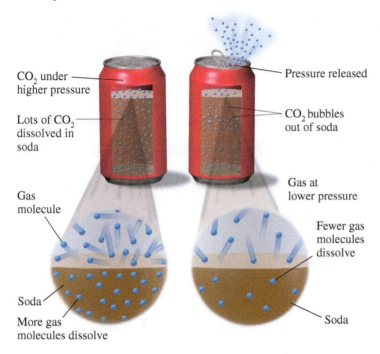

If the pressure above a liquid is higher (unopened can), more gas can be dissolved in the liquid. Once the can is opened, the pressure above the can is lowered and the gas bubbles out of the liquid.

In Section 7.1 we discussed the relationship between pressure and volume (Boyle's law) and its application to breathing. Here the relationship between gas solubility and pressure in Henry's law explains gas exchange while breathing. The pressure exerted by carbon dioxide (P_{CO_2}) produced in the tissues or the pressure of oxygen (P_{O_2}) inhaled at the lungs results in an exchange of gases. Just as dissolved carbon dioxide bubbles out of an open can of soda, if the pressure of CO_2 is higher in the blood delivered back to the lungs (coming from the tissues) than the pressure of CO_2 found at the lungs, the gaseous CO_2 will pass out of the bloodstream and into the lungs (see **FIGURE 8.4**). Similarly, oxygen dissolves into the blood at the lungs because, like the unopened soda can, the pressure of oxygen in the air is higher, allowing it to dissolve in the bloodstream. This oxygenated blood then circulates throughout the body.

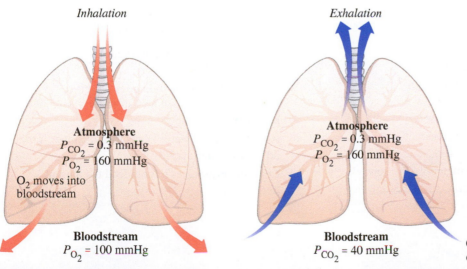

FIGURE 8.4 Henry's law in the lungs. During inhalation, the O_2 levels in the atmosphere are higher than in the returning blood, allowing more O_2 to be dissolved in the blood. The CO_2 levels in blood returning to the lungs from the tissues are higher than in the atmosphere, which allows it to be released in the expired air during exhalation.

Sample Problem 8.3   **Effect of Temperature on Solubility**

Will the solubility of the solute increase or decrease in each of the following situations?
  a. Sucrose is dissolved in water at 95 °C instead of 20 °C.
  b. A can of soda containing dissolved CO_2 is stored at room temperature instead of in the refrigerator.

Solution

  a. The solubility of sucrose, a solid, will increase if dissolved at 95 °C.
  b. The solubility of the CO_2, a gas, will decrease at room temperature.

Sample Problem 8.4   **Effect of Pressure on Gas Solubility**

Will the pressure of oxygen (P_{O_2}) in the bloodstream increase or decrease in each of the following situations?
  a. A person climbs a mountain 10,000 ft high, where the pressure of oxygen in the air is lower.
  b. A person in shock receives mouth-to-mouth resuscitation, and the pressure of air in the lungs increases.

Solution

  a. The pressure of oxygen (P_{O_2}) in the lungs will be lower than normal; therefore, the solubility of oxygen in the bloodstream will decrease.
  b. The pressure of oxygen (P_{O_2}) increases because it is being forced into the lungs; therefore, the amount of oxygen in the bloodstream will increase.

Practice Problems

8.5   Do the following represent a saturated solution?
  a. a pinch of salt thrown into a pot of boiling water
  b. a layer of sugar forming at the bottom of a glass of tea as ice is added

8.6   Do the following represent a saturated solution?
  a. a piece of rock salt that does not change in size when added to a salt solution
  b. a teaspoon of honey that dissolves in hot tea

8.7   Fill in the blank in the statements with the words "increase," "decrease," or "stay the same."
  a. If the temperature of a solution increases, the solubility of most solid solutes will

  _____.

  b. If the temperature of a solution increases, the solubility of a gaseous solute will _____.
  c. If the pressure above a solution increases, the solubility of a gaseous solute will _____.

8.8   Fill in the blank in the statements with the words "increase," "decrease," or "stay the same."
  a. If the temperature of a solution decreases, the solubility of a gaseous solute will

  _____.

  b. If the temperature of a solution decreases, the solubility of a solid solute will _____.
  c. If the pressure above a solution decreases, the solubility of a gaseous solute will _____.

8.9   Explain what is happening in the following situations in terms of solubility:
  a. An opened bottle of soda goes flat more slowly when recapped and stored in the refrigerator.
  b. A bottle of fizzy water keeps its fizz longer if the lid is screwed on tightly.
  c. Less sugar dissolves in cold iced tea than in hot tea.

8.10 Explain what is happening in the following situations in terms of solubility:

a. Fish die in a lake that is heated with water from a power plant cooling tower.

b. Crystals form in a jar of honey that has been sitting for several months.

c. A person receives oxygen through a breathing device.

8.11 Where would you expect a freshly poured glass of champagne to hold its bubbles longer—in a mountain resort at an elevation of 5000 ft or on a beach at sea level? Assume that the champagne is at the same temperature in both locations.

8.12 Hyperbaric oxygen chambers contain 100% oxygen instead of the 20% oxygen found in air. These chambers can be used to reverse the effects of carbon monoxide poisoning in the bloodstream faster than would occur in air. Based on Henry's law, how is this possible?

DISCOVERING THE CONCEPTS

❓ INQUIRY ACTIVITY—Electrolytes

Part 1. Information

nonelectrolyte—A substance that does not conduct electricity because it does not ionize (dissociate forming ions) in aqueous solution. A polar covalent compound like glucose is a nonelectrolyte.

strong electrolyte—A substance that conducts electricity because it completely dissociates into ions in aqueous solution. Ionic compounds that dissolve in water are strong electrolytes.

weak electrolyte—A substance that weakly conducts electricity because it partially dissociates in aqueous solution. A weak acid like a carboxylic acid is a weak electrolyte.

The warning on the label in the figure above appears on some electrical appliances.

Questions

1. Electrical current is carried by a flow of charged particles, which often occurs through electrons in metal or ions in solution. This flow can only happen in an aqueous solution if whole (not partial) charges are present. Based on this information,

   a. does tap water conduct electricity? Consider whether there are enough charges for ions to flow in a solution of tap water.

   b. does pure water conduct electricity? Consider whether there are enough charges for ions to flow in a solution of pure water.

2. Are the following compounds likely to dissociate, partially dissociate, or not dissociate when dissolved in solution? Classify each as a nonelectrolyte, strong electrolyte, or weak electrolyte.

   a. $NaNO_3(s)$

   b. Sucrose, $C_{12}H_{22}O_{11}(s)$

   c. Formic acid, $HCOOH(l)$

8.3 Chemical Equations for Solution Formation

A warning symbol on hair dryers and other small electrical appliances used in the bathroom cautions us that we should not use them near the bathtub. Why? Tap water conducts electricity, so if an electrical appliance falls into a tub while plugged in, a person in the tub could be electrocuted. Pure water does not conduct electricity, but tap water does. What makes the difference? Tap water contains dissolved ions, and those dissolved ions can conduct electricity. Solutes that produce ions in solution are called **electrolytes.**

We can place all substances that dissolve in water (hydrate) into one of three categories based on their ability to conduct electricity. Any aqueous solution can be tested to see whether or not it will conduct electricity with a simple circuit consisting of a battery, wires to two electrodes, and a light bulb (see **FIGURE 8.5**). If a solution conducts electricity, the light bulb glows when the electrodes are immersed in the solution. A soluble ionic compound completely dissociates into its charged ions when dissolved in water. The large number of charged ions present makes the light bulb in the simple circuit burn brightly.

Ionic compounds that dissolve in water are considered **strong electrolytes.** Covalent compounds like sugar do not **ionize** (dissociate to form ions) when they dissolve in solution. They exist as molecules, and no ions are formed upon dissolving. Soluble covalent compounds do not conduct electricity and are, therefore, termed **nonelectrolytes.** Some covalent compounds *can* partially ionize in water. These compounds contain a highly polar bond that can dissociate, forming some ions in water. These covalent compounds weakly conduct electricity and are termed **weak electrolytes.** We can write balanced chemical equations to describe the hydration of each of these substances.

In Chapters 1 and 5, we balanced chemical equations. Here we balance chemical equations for hydration reactions as solutes dissolve into solution. Remember that reactants (represented by their chemical formula) always appear on the left side, and the products (also represented by their chemical formula) always appear on the right side. These two are separated by an arrow meaning *react to form* or *yield*.

$$\text{Reactants} \xrightarrow[\text{(react to form)}]{} \text{products}$$

Strong Electrolytes

Let's look at the parts of a balanced chemical equation for the hydration of the ionic compound magnesium chloride ($MgCl_2$), a strong electrolyte. The following equation can be stated in words as "solid magnesium chloride dissolves in water to yield magnesium ions and chloride ions."

$$MgCl_2(s) \xrightarrow[H_2O]{} Mg^{2+}(aq) + 2Cl^-(aq)$$

FIGURE 8.5 Simple experiments for determining if a solution contains strong electrolytes, nonelectrolytes, and weak electrolytes. A pair of electrodes connected to a light bulb and inserted in the solution will conduct electricity strongly, not at all, or weakly based on dissociation into ions by the solute. Notice that for the strong electrolyte, each $MgCl_2$ dissociates into three particles, while the weak electrolyte, acetic acid, only partially dissociates into H^+ and acetate (CH_3COO^-).

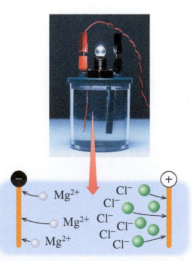

Strong electrolyte

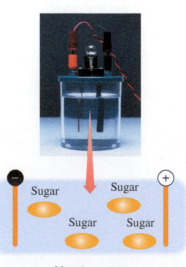

Nonelectrolyte

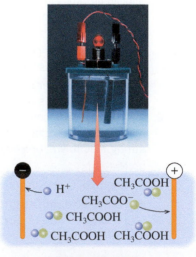

Weak electrolyte

First, notice that the number of magnesium and chloride ions formed as products is the same as the number of each found in the reactant, regardless of whether the ions are together in solid form or dissolved in solution. This was first observed and stated by the French chemist Antoine Lavoisier (1743–1794) as the **law of conservation of matter**: Matter can neither be created nor destroyed. Matter merely changes forms. In other words, atoms do not just appear or disappear.

As outlined in earlier chapters, it is important to check an equation after balancing to make sure the same number of atoms of each element appears on each side of the equation. In the preceding sample equation, the red coefficient 2 in front of the chloride indicates that two chlorides are produced for every one $MgCl_2$ that dissociates. Besides balancing the number of atoms of each element, the charges in a hydration equation must be the same on both sides of the chemical equation. In this example, $MgCl_2$ has no net charge (reactant) and one Mg^{2+} and two Cl^- sum to a total charge of zero (products), so the charges also are balanced.

$$MgCl_2(s) \xrightarrow[H_2O]{} Mg^{2+}(aq) + 2Cl^-(aq)$$

Charge in reactants = 0

Charge in products $(2^+) + 2(1^-) = 0$

Second, note that the reaction arrow points in one direction, implying that the process occurs in only one direction; that is, the reaction is irreversible.

Third, note the phases of matter. For ionic compounds, the reactant will usually be a solid that dissolves. In the products, the phases will always be aqueous because the substance is hydrated; that is, it is dissolved in water.

Finally, notice the water (H_2O) appearing below the reaction arrow. Substances that are not involved in the balanced equation, like the solvent in this case, are often placed at the arrow to give the reader more information regarding the conditions of the reaction. For the hydration equations in this chapter, water will appear here.

Nonelectrolytes

Nonelectrolytes are polar compounds that dissolve in water but do not ionize in water. For example, examine the chemical equation for glucose $(C_6H_{12}O_6)$ dissolving in water:

$$C_6H_{12}O_6(s) \xrightarrow[H_2O]{} C_6H_{12}O_6(aq)$$

Covalent compounds such as glucose do not dissociate. The only difference between the reactant and the product is the phase. Because glucose dissolves in water, we indicate this on the product side by changing the phase of the glucose molecules to aqueous.

Weak Electrolytes

Which covalent compounds act as weak electrolytes? If we reexamine the functional group list on the inside back cover, we see that two functional groups contain a form with a charge, the carboxylic acid/carboxylate group and the amine/protonated amine.

Neutral form   Charged form

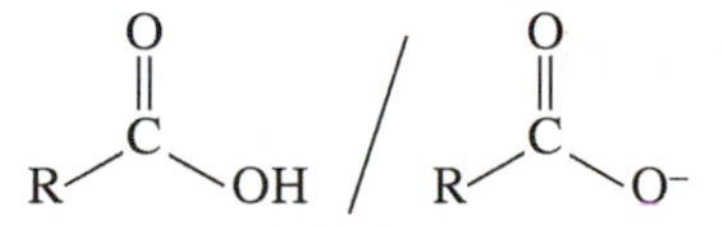

Carboxylic acid   Carboxylate

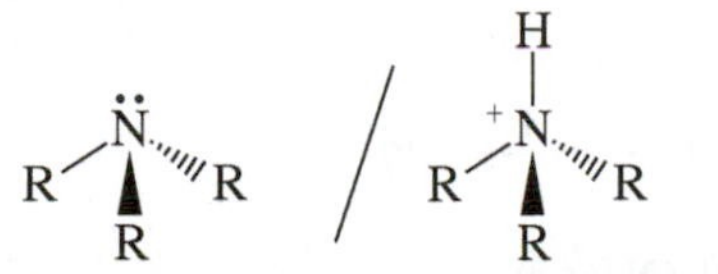

Amine   Protonated amine

The functional groups carboxylic acid and amine are weak electrolytes that have charged forms differing from the neutral form by a hydrogen. Recall that amines can be primary, secondary, or tertiary and have protonated forms. The tertiary amine and its protonated (charged) form are shown.

A carboxylic acid group contains a very polar O—H bond that can dissociate to form carboxylate ions and H^+ ions when dissolved. Carboxylic acids and amines are used as examples of weak electrolytes in this chapter. Consider the partial ionization of acetic acid, the compound in vinegar:

$$CH_3-\overset{\overset{\displaystyle O}{\|}}{C}-OH(l) \underset{H_2O}{\rightleftharpoons} CH_3-\overset{\overset{\displaystyle O}{\|}}{C}-O^-(aq) + H^+(aq)$$

As the number of H^+ and CH_3COO^- ions builds up in the solution, some of them will recombine, re-forming CH_3COOH. Eventually, the rates of the forward and reverse reactions equalize, and the ratio of products to reactants no longer changes. Thus, a chemical equilibrium exists. An equilibrium arrow, $\rightleftharpoons$, is used in this chemical equation to indicate this. When inspecting the equation, we see that the number of atoms of each element and the total charge on each side are balanced, so the equation is balanced. The phase of the weak electrolyte before hydration may be solid, liquid, or gas.

Sample Problem 8.5 **Characterizing Strong, Weak, and Nonelectrolyte Solutions**

Predict if the following will fully dissociate, partially dissociate, or not dissociate when dissolved in solution:
- a. $MgSO_4(s)$, used to treat seizures
- b. lactose, $C_{12}H_{22}O_{12}(s)$, milk sugar
- c. lactic acid, $CH_3CH(OH)COOH(s)$, found in Ringer's lactate, used to restore fluid balance

Solution

- a. The ionic compound $MgSO_4$ will fully dissociate, forming Mg^{2+} and SO_4^{2-} ions when dissolved in solution.
- b. The covalent compound $C_{12}H_{22}O_{12}$ will not dissociate when dissolved in solution.
- c. The lactic acid will *partially* dissociate, forming the ions H^+ and $CH_3CH(OH)COO^-$. $CH_3CH(OH)COOH$ molecules will also be present.

Solving a Problem

Balancing Hydration Equations

Write a balanced equation for the hydration of the following:
- a. $CaCl_2(s)$, a strong electrolyte
- b. $CH_3OH(l)$, a nonelectrolyte
- c. $NH_4OH(aq)$, a weak electrolyte

Solution

STEP 1 **Write the reactants and products with their state of matter.** The arrow is irreversible for strong electrolytes and nonelectrolytes and reversible for weak electrolytes. Because they dissolve, the products will all be aqueous. For our examples,

a. $CaCl_2(s) \xrightarrow{H_2O} Ca^{2+}(aq) + Cl^-(aq)$

b. $CH_3OH(l) \xrightarrow{H_2O} CH_3OH(aq)$

c. $NH_4OH(aq) \underset{H_2O}{\rightleftharpoons} NH_4^+(aq) + OH^-(aq)$

STEP 2 Examine and balance the equations. Be sure to balance the number of atoms of each element on both sides. Upon examination, only part a is unbalanced. If there are two chlorides in the solid, $CaCl_2$, there must be two Cl^- present on the product side. By adding the coefficient 2 in front of the Cl^-, we have balanced both the elements and the charge.

a. $CaCl_2(s) \xrightarrow{H_2O} Ca^{2+}(aq) + 2Cl^-(aq)$

b. $CH_3OH(l) \xrightarrow{H_2O} CH_3OH(aq)$

c. $NH_4OH(aq) \underset{H_2O}{\rightleftharpoons} NH_4^+(aq) + OH^-(aq)$

STEP 3 Check the equations by confirming equal charge on each side. Checking the charges of the reactants and products in each of the equations confirms that the charges are equal on either side.

a. $CaCl_2(s) \xrightarrow{H_2O} Ca^{2+}(aq) + 2Cl^-(aq)$

   Charge in reactants = 0   Charge in products $(2^+) + 2(1^-) = 0$

b. $CH_3OH(l) \xrightarrow{H_2O} CH_3OH(aq)$

   Charge in reactants = 0   Charge in products = 0

c. $NH_4OH(aq) \underset{H_2O}{\rightleftharpoons} NH_4^+(aq) + OH^-(aq)$

   Charge in reactants = 0   Charge in products $(1^+) + (1^-) = 0$

Sample Problem 8.6 Balancing Hydration Equations

Write a balanced equation for the hydration of the following:
   a. $KCl(s)$, a strong electrolyte
   b. $NH_2CONH_2(l)$, a nonelectrolyte
   c. $HF(l)$, a weak electrolyte

Solution

a.

STEP 1 Write the reactants and products with their state of matter.

$$KCl(s) \xrightarrow{H_2O} K^+(aq) + Cl^-(aq)$$

STEP 2 Examine and balance the equations. Examination indicates that the equation is balanced.

STEP 3 Check the equations by confirming equal charge on each side. The charges are balanced to zero on both sides.

b.

STEP 1 Write the reactants and products with their state of matter. A nonelectrolyte will dissolve but not ionize, merely changing states. Because the formulas did not change, the equation will be balanced after STEP 1.

$$NH_2CONH_2(l) \xrightarrow{H_2O} NH_2CONH_2(aq)$$

—*Continued next page*

Continued—

c.

STEP 1 **Write the reactants and products with their state of matter.** HF is a common weak acid. (Other properties of weak acids are discussed in detail in Chapter 9.)

$$HF(l) \underset{H_2O}{\rightleftharpoons} H^+(aq) + F^-(aq)$$

STEP 2 **Examine and balance the equations.** Examination indicates that the equation is balanced.

STEP 3 **Check the equations by confirming equal charge on each side.** The charges are balanced to zero on both sides.

Ionic Solutions and Equivalents

Blood and other bodily fluids contain many electrolytes as dissolved ions, like Na^+, K^+, Cl^-, and HCO_3^-. Because the body transfers energy through charges, it is sometimes useful to know the number of charges in a solution. The amount of a dissolved ion found in fluids can be expressed by the unit **equivalent (Eq).** An equivalent relates the charge in a solution to the number of ions or the moles of ions present (see **TABLE 8.2**). Recall from Chapter 3 that moles can be related to the measurable quantity grams through the atomic mass.

For example, 1 mole of Na^+ has one equivalent of charge because the charge on a sodium ion is plus 1. One mole of Ca^{2+} has two equivalents of charge because one mole of calcium contains two charges (or equivalents) per mole. *The number of equivalents present per mole of an ion equals the charge on that ion per mole.*

TABLE 8.2 The Relationship Between Equivalents and Moles Expressed as Conversion Factors

Ion	Conversion Factor
Na^+	$\dfrac{1 \text{ Eq } Na^+}{1 \text{ mole } Na^+}$
Ca^{2+}	$\dfrac{2 \text{ Eq } Ca^{2+}}{1 \text{ mole } Ca^{2+}}$
Cl^-	$\dfrac{1 \text{ Eq } Cl^-}{1 \text{ mole } Cl^-}$
SO_4^{2-}	$\dfrac{2 \text{ Eq } SO_4^{2-}}{1 \text{ mole } SO_4^{2-}}$

Sample Problem 8.7 Electrolytes in Solution

How many equivalents of Cl^- are present in a solution that contains 2.50 moles of Cl^-?

Solution

Use the conversion factor that relates the number of moles of Cl^- to the number of equivalents.

$$2.50 \text{ moles } Cl^- \times \frac{1 \text{ Eq } Cl^-}{1 \text{ mole } Cl^-} = 2.50 \text{ Eq } Cl^-$$

Sample Problem 8.8 Electrolytes in Solution

How many equivalents of Ca^{2+} are present in a solution that contains 2.50 moles of Ca^{2+}?

Solution

Use the conversion factor that relates the number of moles of Ca^{2+} to the number of equivalents.

$$2.50 \text{ moles } Ca^{2+} \times \frac{2 \text{ Eq } Ca^{2+}}{1 \text{ mole } Ca^{2+}} = 5.00 \text{ Eq } Ca^{2+}$$

Electrolytes in Blood Plasma

The amounts of electrolytes present in bodily fluids and typical intravenous fluid replacements are often represented as the number of milliequivalents per liter of solution (mEq/L).

Ionic solutions have a balance in the number of positive and negative charges present because they are formed by dissolving ionic compounds that have no net charge. Typical blood plasma has a total electrolyte concentration of 150 mEq/L, meaning that the total concentration of both positive and negative ions is 150 mEq/L (see **TABLE 8.3**). Examples of the concentrations of other ionic solutions used as intravenous fluids are shown in **TABLE 8.4**. These also show a balance of total charge.

INTEGRATING
Chemistry

◀ Find out the electrolyte concentrations in IV fluids.

TABLE 8.3 Typical Concentrations of Electrolytes in Blood Plasma

Electrolyte (mEq/L)	Concentration
Cations	
Na^+	138
K^+	5
Mg^{2+}	3
Ca^{2+}	4
Total	150
Anions	
Cl^-	110
HCO_3^-	30
HPO_4^{2-}	4
Proteins	6
Total	150

TABLE 8.4 Electrolyte Concentrations in Intravenous Replacement Solutions

Solution	Electrolytes (mEq/L)	Use
Normal saline (0.90%)	Na^+: 154, Cl^-: 154	Replacement of fluid loss
KCl (0.15%) in 5% dextrose	K^+: 20, Cl^-: 20	Treatment of malnutrition and dehydration
Lactated Ringer's solution	Na^+: 130, K^+: 4, Ca^{2+}: 3, Cl^-: 109, lactate: 28	Replacement of fluids and electrolytes; buffers pH
Normosol® -R	Na^+: 140, K^+: 5, Mg^{2+}: 3, Cl^-: 98, acetate 27, gluconate 23	Fluid replacement in acute fluid loss from trauma and during surgery
5% albumin in normal saline	Na^+: 154, Cl^-: 154	Blood plasma expansion

Sample Problem 8.9 Using Equivalents

A lactated Ringer's solution for intravenous fluid replacement has a concentration of 109 mEq Cl^- per liter of solution. If a patient receives 1750 mL of Ringer's solution, how many equivalents of Cl^- were given?

Solution

This problem asks for the number of equivalents of Cl^- present, which can be determined from the number of mEq given through the conversion factor $\dfrac{1\ \text{Eq}}{1000\ \text{mEq}}$.

Using conversion factors and the information given (concentration and volume), and canceling appropriate units, we see that the problem is solved as

$$\frac{109\ \cancel{\text{mEq}}\ Cl^-}{1\ \cancel{\text{L solution}}} \times \frac{1\ \text{Eq}}{1000\ \cancel{\text{mEq}}} \times \frac{1\ \cancel{\text{L}}}{1000\ \cancel{\text{mL}}} \times 1750\ \cancel{\text{mL solution}} = 0.191\ \text{Eq}\ Cl^-$$

Practice Problems

8.13 Predict if the following IV electrolytes will fully dissociate, partially dissociate, or not dissociate when placed in solution:

　　a. potassium chloride, KCl(s), a strong electrolyte

　　b. sodium lactate, $NaC_3H_5O_3$ a weak electrolyte

　　c. dextrose (D-glucose), $C_6H_{12}O_6(s)$, a nonelectrolyte

8.14 Predict if the following will fully dissociate, partially dissociate, or not dissociate when dissolved in solution:

　　a. sodium sulfate, $Na_2SO_4(s)$, a strong electrolyte

　　b. ethyl alcohol, $CH_3CH_2OH(l)$, a nonelectrolyte

　　c. hypochlorous acid, $HClO(aq)$, a weak electrolyte

—*Continued next page*

Continued—

8.15 Provide a balanced equation for the hydration of each compound in Problem 8.13.

8.16 Provide a balanced equation for the hydration of each compound in Problem 8.14.

8.17 For the following ionic compounds, write a balanced equation for their hydration in water:

a. $CaCl_2$

b. $NaOH$

c. KBr

d. $Fe(NO_3)_3$

8.18 For the following ionic compounds, write a balanced equation for their hydration in water:

a. $FeCl_3$

b. $NaHCO_3$

c. K_2SO_4

d. $Ca(NO_3)_2$

8.19 How many equivalents of K^+ are present in a solution that contains 4.25 moles of K^+?

8.20 How many equivalents of Mg^{2+} are present in a solution that contains 2.50 moles of Mg^{2+}?

8.21 A sodium chloride intravenous (IV) fluid contains 154 mEq of Na^+ per liter of solution. If a patient receives exactly 500 mL of the IV solution, how many moles of Na^+ were given?

8.22 A 5% dextrose IV fluid contains 35 mEq of K^+ per liter of solution. If a patient receives 235 mL of the IV solution, how many moles of K^+ were delivered?

8.4 Concentration

8.4 Inquiry Question

How is the concentration of a solution expressed?

Whether you drink a can of soda or a glass of the same soda from a 2-liter bottle, the soda tastes the same. Beverage manufacturers go to great lengths to maintain a consistent product. Each beverage must have exactly the same amount of ingredients (see **FIGURE 8.6**).

The amount of each ingredient dissolved in the soda determines its **concentration.** The concentration of a solution can be expressed in many different units, but the units we will encounter in this chapter relate the amount of each solute dissolved to the total amount of solution present.

$$\text{Concentration} = \frac{\text{amount of solute}}{\text{amount of solution}}$$

Note that the denominator is the *total* amount of solution (includes the solute), not just the amount of solvent.

TABLE 8.5 shows the normal concentration ranges for some substances typically tested in a blood profile. Notice that the units vary depending on the type of solute. Why all the different units? Different units are used so that all the numbers fall in convenient ranges readable without using numbers after a decimal. For example, if the units for cholesterol were g/dL instead of mg/dL, the numbers in the normal range would be 0.120–0.240 g/dL, which is not as easy to read and discuss with a patient as a number ranging in the hundreds.

In this section, the common concentration units of molarity, percent, parts per million, and parts per billion are introduced.

FIGURE 8.6 Concentration and volume. The concentration of the sugar and the flavor of the beverage are the same in all containers, even though the volume differs.

TABLE 8.5 Normal Concentration Ranges for Five Common Substances Measured in a Blood Test	
Substance	**Normal Range**
Na^+	136–146 mEq/L
Cl^-	97–110 mEq/L
Glucose	65–110 mg/dL
Cholesterol	120–240 mg/dL
Hemoglobin (a protein)	12–16 g/dL

Millimoles per Liter (mmol/L) and Molarity (M)

Sometimes, the units for electrolytes are given in mmoles/L instead of the previously mentioned mEq/L. The charge on an ion is the number of equivalents present in 1 mole (Section 8.3). So, for an ion with a charge like sodium, the number of equivalents is equal to the number of moles while the value for an ion like calcium is two times the number of equivalents per mole. For sodium, the value expressed as mEq/L and mmole/L is the same while the value for calcium ion is twice the number of mmoles/L as mEq/L. We can show this mathematically by converting mEq/L to mmole/L using one of the conversion factors shown:

$$\frac{1\ \text{mmole Na}^+}{1\ \text{mEq Na}^+}, \quad \frac{2\ \text{mmole Ca}^{2+}}{1\ \text{mEq Ca}^{2+}}$$

Using the Na^+ conversion factor, we can determine the concentration of a 135 mEq/L Na^+ solution in mmole/L:

$$\frac{135\ \cancel{\text{mEq Na}^+}}{1\ L} \times \frac{1\ \text{mmole Na}^+}{1\ \cancel{\text{mEq Na}^+}} = \frac{135\ \text{mmole Na}^+}{1\ L}$$

Chemists use a unit related to mmole/L to describe the concentrations of solutions prepared in the laboratory. This unit is called **molarity, M**, and it is defined as

$$M = \frac{\text{mole solute}}{\text{L solution}}$$

Remember that the *mole* is directly related to the number of particles present (ions or molecules) through Avogadro's number. A 5.0 M (read "five molar") solution of ethanol (CH_3CH_2OH) has the same number of molecules present as a 5.0 M glucose ($C_6H_{12}O_6$) solution, even though the *mass* of ethanol and *mass* of glucose used to make the two solutions are different. The unit molarity is useful for chemists because they want to know how many particles of a solute are available to react in a chemical reaction.

Calculating Molarity

Solving a Problem

What is the molarity of an electrolyte solution prepared by dissolving 5.6 g of NaCl (about 1 teaspoon) in water, giving a total volume of 250 mL (about 1 cup)?

Solution

STEP 1 Examine the problem. Decide what information is given and what information is being sought. In this problem, you are solving for molarity (M), so your final units should be moles/L. To arrive at moles/L, you must use some conversion factors because you are given grams of NaCl and the volume in mL.

STEP 2 Find appropriate conversion factors. To get NaCl in moles, you must find the molar mass of NaCl, which is the sum of the atomic masses from the periodic table:

Na:	22.99
Cl:	35.45
NaCl:	58.44 g/mole

The conversion factor for mL to L is

$$\frac{1000\ \text{mL}}{1\ L}$$

STEP 3 Solve the problem. Be sure that the units you don't want cancel and you are left with the units of moles/L, which is molarity.

$$\frac{5.6\ \cancel{g}\ \text{NaCl}}{250\ \cancel{\text{mL}}\ \text{solution}} \times \frac{1000\ \cancel{\text{mL}}}{1\ L} \times \frac{1\ \text{mole}}{58.44\ \cancel{g}\ \text{NaCl}} = \frac{0.38\ \text{mole}}{1\ L} = 0.38\ M,\ \text{or}\ 380\ mM$$

Sample Problem 8.10   Molarity

What is the molarity of a solution prepared by dissolving 5.00 g of $MgCl_2$ in water to give a total volume of 500. mL?

Solution

STEP 1 Examine the problem. Decide what information is given and what information is being sought. In this problem, you are solving for molarity (M), so your final units should be moles/L. To arrive at moles/L, you must use some conversion factors because you are given grams of $MgCl_2$ and the volume in mL.

STEP 2 Find appropriate conversion factors. To get $MgCl_2$ in moles, you must find the molar mass of $MgCl_2$, which is the sum of the atomic masses from the periodic table:

$$
\begin{array}{lr}
\text{Mg:} & 24.31 \\
\text{2 Cl:} & 35.45 \times 2 = 70.90 \\
\hline
\text{MgCl}_2\text{:} & 95.21 \text{ g/mole}
\end{array}
$$

The conversion factor for mL to L is

$$\frac{1000 \text{ mL}}{1 \text{ L}}$$

STEP 3 Solve the problem. Be sure that the units that you don't want cancel and that you are left with the units of moles/L, which is molarity.

$$\frac{5.00 \text{ g MgCl}_2}{500. \text{ mL solution}} \times \frac{1000 \text{ mL}}{1 \text{ L}} \times \frac{1 \text{ mole}}{95.21 \text{ g MgCl}_2} = \frac{0.105 \text{ mole}}{1 \text{ L}} = 0.105 \text{ M}$$

Sample Problem 8.11   Molarity

How many moles of Mg^{2+} are found in 750 mL of blood plasma with a concentration of 0.001 M Mg^{2+}?

Solution

In this problem, the concentration of the electrolyte Mg^{2+} in the plasma is given, so the number of moles can be found. Often, the concentration of an electrolyte is given for just a particular ion when several different ones are dissolved in the same solvent (see Table 8.4). Whether the concentration is for a compound or an ion, the concentration is still the number of particles (moles) per volume (liter). The unit on the final answer is moles. By definition, a 0.001 M Mg^{2+} concentration is 0.001 moles/L. Set up an equation to solve for moles with the given information:

$$\frac{0.001 \text{ moles Mg}^{2+}}{L} \times \frac{1 \text{ L}}{1000 \text{ mL}} \times 750 \text{ mL} = 0.00075 \text{ moles, or } 0.75 \text{ mmoles Mg}^{2+}$$

Percent (%) Concentration

The concentration unit percent is no different in definition from the everyday definition of percent discussed in Chapter 1. Percent concentration is the number of *parts out of the whole times 100*. As an example, a dime (10 pennies) is 10% of a dollar (100 pennies), or (10/100) × 100. There are three common concentration units that use percent: mass/mass percent, volume/volume percent, and mass/volume percent.

$$\% \text{ Concentration} = \frac{\text{parts solute}}{\text{parts solution}} \times 100$$

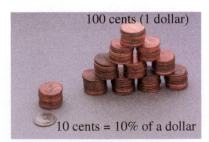

100 cents (1 dollar)

10 cents = 10% of a dollar

A dime is 10% of a dollar.

Percent Mass/Mass, % (m/m)

A solution can easily be prepared in a hospital pharmacy or laboratory with the unit percent mass/mass by measuring both the solute and the solvent on a balance and mixing, realizing that mass of solute + mass of solvent = mass of solution. The concentration unit percent mass/mass, or % (m/m), can be determined by

$$\% \, (m/m) = \frac{\text{g solute}}{\text{g solution}} \times 100\%$$

This unit is also known as percent weight/weight [% (wt/wt)]. The percent of fat in milk is determined using this unit. For example, one type of milk is 2%, which means that there are 2 g of milkfat in 100 g of the milk solution.

Percent milkfat is calculated as g of milkfat per 100 g of milk solution.

Two percent milk is prepared by skimming off the milkfat and then adding the appropriate amount back to 2%. If we wanted to prepare 2% milk, we would add 2 g of milkfat to 98 g of skim milk (0 g milkfat). The total mass of the solution is the mass of the solute and the solvent.

$$\frac{2 \text{ g milkfat}}{100 \text{ g milk solution}} \times 100 = 2\%$$

2 g milkfat + 98 g skimmed milk
(Solute)   (Solvent)

Percent Volume/Volume, % (v/v)

The unit percent volume/volume is typically used when liquids or gases are the solute and solvent, for example, ethanol and water. A bottle of wine that is 14% (v/v) alcohol means that 14 mL of alcohol is present in 100 mL of the wine. The concentration unit percent volume/volume, or % (v/v), can be determined by

$$\% \, (v/v) = \frac{\text{mL solute}}{\text{mL solution}} \times 100\%$$

Percent Mass/Volume, % (m/v)

The unit percent mass/volume is often used in the preparation of intravenous fluids. Normal saline (NS) solution used to replace body fluids is 0.90% (m/v) NaCl, which means the solution contains 0.90 g of NaCl in 100 mL of solution. The concentration unit percent mass/volume, or % (m/v), can be determined by

$$\% \, (m/v) = \frac{\text{g solute}}{\text{mL solution}} \times 100\%$$

The percent of alcohol in wine is a % (v/v) concentration.

This unit is also known as percent weight/volume [% (wt/v)]. In product labeling, the type of percent concentration (m/m, v/v, or m/v) is often not indicated. The state of matter of the solute indicates the percent concentration unit.

In a hospital pharmacy, a solution can be prepared by weighing the solute on a balance and mixing it with a solvent to form a final total volume of solution. For this reason, it is important to determine the amount of solute you might add to a solution as well as calculating the concentration of a solution.

Mastering Video
Practicing the Concepts
Calculating Percent Concentration

Sample Problem 8.12   Calculating Percent Concentration

What is the % (m/v) concentration of a NaCl solution prepared when 35 g of NaCl is dissolved to a total volume of 500. mL of solution?

Solution

The mass of the solute is 35 g of NaCl. The volume of the solution is 500. mL.

$$\% \ (m/v) = \frac{35 \text{ g NaCl}}{500. \text{ mL solution}} \times 100\% = 7.0\%$$

Sample Problem 8.13   Solute in Percent Concentration

How many grams of fat are present in 250 g of 2% milk?

Solution

Remember the definition of this concentration unit. The percent for milkfat is grams/100 g milk. A simple ratio can be set up, and the number of grams in 250 g of milk determined:

$$\frac{2 \text{ g milkfat}}{100 \text{ g milk}} = \frac{? \text{ g milkfat}}{250 \text{ g milk}}$$

$$\frac{2 \text{ g milkfat}}{100 \text{ g milk}} \times 250 \text{ g milk} = ? \text{ g milkfat}$$

$$? = 5 \text{ g milkfat}$$

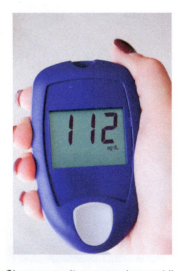

Glucose monitor measuring mg/dL.

Relationship to Other Common Units

The unit typically used for measuring the oxygen-carrying protein hemoglobin in the blood is g/dL, which is the same unit as % (m/v). A deciliter is equal to 100 mL, so the unit g/dL is the same as g/100 mL. Hemoglobin levels of 13–18 g/dL (13–18 g per 100 mL of blood) for men and 12–16 g/dL for women are considered in the normal range.

A common unit used when measuring molecules such as glucose and cholesterol levels in the blood is milligrams per deciliter (mg/dL). This unit is also referred to as mg% (milligram percent). The mg in front of the % symbol indicates that the definition is mg per 100 mL instead of the usual g per 100 mL definition of % (m/v). Glucose levels over 110 mg% (110 mg per 100 mL blood) after fasting are considered higher than normal. Total cholesterol levels below 200 mg/dL (200 mg per 100 mL blood) are considered desirable.

Parts per Million (ppm) and Parts per Billion (ppb)

Parts per million (ppm) and parts per billion (ppb) are convenient concentration units used to describe very dilute solutions. To get an idea of how small a part per million is, note that there are 1 million pennies in $10,000, so a penny could be considered 1 ppm of $10,000. In terms of volume, about 5 drops of food coloring in a bathtub of water represents a part per million, and about 1 drop of food coloring in an Olympic-sized swimming pool of water is about a part per billion—a very dilute solution.

Ppm and ppb are often used to discuss water treatment. For example, fluoride is often added to tap water at a level of less than 4 ppm to promote strong teeth. In contrast, the maximum contaminant level of lead in drinking water is 15 ppb. In the case of the fluoride, 4 ppm means 4 g of fluoride in every million mL of tap water. In the case of the lead, 15 ppb means 15 g of lead in every billion mL of tap water. It is easier to think in terms of liters of water rather than millions or billions of mL of water, so the unit ppm is sometimes referred to as 1 mg/L and ppb as 1 μg/L.

$$1 \text{ ppm} = \frac{1 \text{ g solute}}{1,000,000 \text{ mL solution}} \times \frac{1000 \text{ mg}}{1 \text{ g}} \times \frac{1000 \text{ mL}}{1 \text{ L}} = \frac{1 \text{ mg solute}}{1 \text{ L solution}}$$

Alternatively, recall that a percent mass/volume [% (m/v)] is *parts per hundred*, so similarly, parts per million and parts per billion can be determined by multiplying by a million or a billion, respectively:

$$\text{ppm} = \frac{\text{g solute}}{\text{mL solution}} \times 1,000,000 \longleftarrow 1 \times 10^6$$

$$\text{ppb} = \frac{\text{g solute}}{\text{mL solution}} \times 1,000,000,000 \longleftarrow 1 \times 10^9$$

Sample Problem 8.14 **Calculating ppm and ppb**

What is the concentration in (a) parts per million (ppm) and (b) parts per billion (ppb) of a solution that contains 15 mg of lead in 1250 mL of solution?

Solution

a. A part per million is equivalent to a mg per L, so convert the units given to mg/L:

$$\frac{15 \text{ mg lead}}{1250 \text{ mL solution}} \times \frac{1000 \text{ mL}}{1 \text{ L}} = 12 \text{ mg/L} = 12 \text{ ppm}$$

b. A part per billion is equivalent to a μg per L (it is 1000 times smaller than a ppm), so convert the units given to μg/L. There are 1000 μg per 1 mg.

$$\frac{15 \text{ mg lead}}{1250 \text{ mL solution}} \times \frac{1000 \text{ mL}}{1 \text{ L}} \times \frac{1000 \text{ }\mu\text{g}}{1 \text{ mg}} = 12 \text{ }\mu\text{g/L} = 12,000 \text{ ppb}$$

Practice Problems

8.23 What is the concentration in mmole/L of a Ca^{2+} solution that is 2.50 mEq/L?

8.24 What is the concentration in mEq/L of a Mg^{2+} solution that is 100 mmole/L?

8.25 What is the molarity of 250 mL of solution containing 4.20 moles of $ZnCl_2$?

8.26 What is the molarity of 750 mL of solution containing 2.10 moles of Na_2SO_4?

8.27 What is the chloride molarity of 400. mL of blood plasma containing 44 mmoles of Cl^-?

8.28 What is the potassium ion molarity of 850 mL of blood plasma containing 10. mmoles of K^+?

8.29 How many grams of KCl must be added to water to create 5.0 L of an 0.020 M KCl electrolyte solution?

8.30 How many grams of NaCl must be added to water to create 2.0 L of a 0.154 M normal saline solution?

8.31 How many grams of D-glucose (dextrose) must be added to water to prepare 500. mL of a 5.0% (m/v) dextrose solution (D5W)?

8.32 Calculate the percent mass/volume [% (m/v)] for the solute in each of the following solutions:

a. 75 g of Na_2SO_4 in 250 mL of Na_2SO_4 solution

b. 39 g of sucrose in 355 mL of a carbonated drink

8.33 Calculate the percent mass/volume [% (m/v)] for the solute in each of the following solutions:

a. 2.50 g of KCl in 50.0 mL of solution

b. 7.5 g of casein in 120 mL of low-fat milk

8.34 What is the concentration in % (m/m) of a solution prepared by mixing 10.0 g of KCl with 100.0 g of distilled water?

8.35 How many grams of insulin are present in 26.0 g of a 0.450% (m/m) solution?

8.36 How many grams of NaCl must be combined with water to prepare 500 mL of a 0.90% (m/v) physiological saline solution?

8.37 What is the concentration in ppm of a solution containing 0.30 mg of fluoride and 60 mL of tap water? What is the concentration in ppb?

8.38 What is the concentration in ppm of a solution containing 1.50 mg of copper and 100.0 mL of tap water?

8.5 Dilution

? 8.5 Inquiry Question

How is the concentration of a solution determined if a dilution is made?

If you have ever made flavored water using drink drops, you have made a dilute solution from a more concentrated one. One way to prepare solutions of lower concentration is to dilute a solution of higher concentration by adding more solvent.

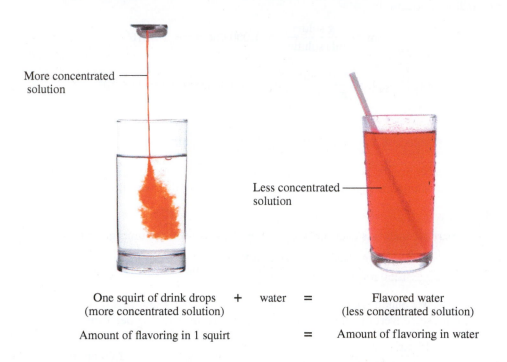

More concentrated solution

Less concentrated solution

| One squirt of drink drops (more concentrated solution) | + | water | = | Flavored water (less concentrated solution) |

| Amount of flavoring in 1 squirt | | | = | Amount of flavoring in water |

When you add the concentrated flavoring to the water, the amount of flavoring present does not change even though you have a lot more solution present—the amount of flavoring in one squirt is equal to the amount of flavoring in the water. Thus, the *amount of solute* stays the same, but the *volume* of solution increases; as a consequence, the *concentration* of the solution decreases.

The following dilution equation represents this mathematically, where $C_{initial}$ represents the initial concentration, C_{final} represents the final concentration, $V_{initial}$ represents the initial volume, and V_{final} represents the final volume. If three of the four variabls are known, the fourth one can be determined:

$$C_{initial} \times V_{initial} = C_{final} \times V_{final}$$

For example, if you diluted 150 mL of a normal saline solution [0.90% (m/v)] to a final volume of 450 mL with water, what would be the concentration of the final, diluted solution?

$C_{initial} = 0.90\%$ m/v $V_{final} = 450$ mL dilute solution

$V_{initial} = 150$ mL $C_{final} = ?$

$$C_{initial} \times V_{initial} = C_{final} \times V_{final}$$

$$\frac{C_{initial} \times V_{initial}}{V_{final}} = C_{final}$$

$$\frac{(0.90\%) \times (150 \text{ mL})}{(450 \text{ mL})} = C_{final}$$

$$0.30\% \text{ m/v} = C_{final}$$

The dilution equation works with any concentration unit where the amount of solution (the denominator) is expressed in volume units. Therefore, the equation does not work with % (m/m). The dilution equation is useful in the health fields because many pharmaceuticals are often prepared as concentrated aqueous solutions and must be diluted before being administered to the patient.

Using the Dilution Equation

*Hydrocortisone is used as an anti-inflammatory for localized pain. You must prepare a 50 mg/mL
solution of hydrocortisone for an injection. You have 5 mL of a 200 mg/mL stock solution of hydrocortisone available. How many milliliters of the 50 mg/mL solution can you prepare?*

Solution

STEP 1 Establish the given information. In this problem, two concentrations and one
volume are given. The volume of the initial stock solution is given. You will solve for the
final volume.

$C_{initial} = 200$ mg/mL $C_{final} = 50$ mg/mL

$V_{initial} = 5$ mL $V_{final} = ?$

STEP 2 Arrange the dilution equation to solve for the unknown quantity.

$$C_{initial} \times V_{initial} = C_{final} \times V_{final}$$

$$\frac{C_{initial} \times V_{initial}}{C_{final}} = V_{final}$$

STEP 3 Solve for the unknown quantity.

$$\frac{(200 \text{ mg/mL}) \times (5 \text{ mL})}{(50 \text{ mg/mL})} = V_{final}$$

$$20 \text{ mL} = V_{final}$$

Does your answer make sense? Because you are diluting, the final volume determined
should be more than the initial volume.

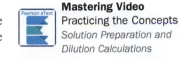

Sample Problem 8.15   Preparing a Solution

How would you prepare 500 mL of a 5% (m/v) glucose solution from a 25% (m/v)
stock solution?

Solution

STEP 1 Establish the given information. In this problem, two concentrations and one
volume—the volume of the final solution—are given. You will solve for the initial volume.

$C_{initial} = 25\%$ $C_{final} = 5\%$

$V_{initial} = ?$ $V_{final} = 500$ mL

STEP 2 Arrange the dilution equation to solve for the unknown quantity.

$$C_{initial} \times V_{initial} = C_{final} \times V_{final}$$

$$V_{initial} = \frac{C_{final} \times V_{final}}{C_{initial}}$$

STEP 3 Solve for the unknown quantity.

$$V_{initial} = \frac{(5\%) \times (500 \text{ mL})}{(25\%)}$$

$$V_{initial} = 100 \text{ mL}$$

The answer to the equation tells us that 100 mL of the stock solution should be used.
The question asks how we would prepare the solution, so the final answer requires
that we interpret our answer to the dilution equation by stating that to prepare the 5%
solution we should combine 100 mL of the 25% solution and enough water for a total
solution volume of 500 mL.

Dilution Factors and Concentrated Stock Solutions

The use of dilution factors is a modification of the dilution equation often used in microbiology and medicine. This is useful if the concentrations are not known or not important in the calculation. The dilution equation can be modified as

$$\text{Dilution factor} = \frac{V_f}{V_i}$$

As another example, the directions on a can of condensed soup indicate that 1 can of water or milk should be added to the contents of the can to prepare the soup. In this case, the dilution factor would be 1 part soup + 1 part water for a 1:2 dilution, or a dilution factor of 2.

$$\text{Dilution factor} = \frac{2 \text{ cans prepared soup}}{1 \text{ can soup concentrate}} = 2$$

When large volumes of a solution are needed, laboratories often store solutions as concentrated stock solutions. For our soup example, the concentrated soup in the can could be considered a 2× stock solution (two times more concentrated) since to dilute this to usable form, we would be putting one part of the stock solution (the condensed soup) with one part of the solute (water) for a total of 2 parts.

The use of concentrated stock solutions is common practice for storage of medications in the pharmacy. The easiest way to understand these manipulations of the dilution equation is to solve some problems.

1 can soup + 1 can water is a 2× dilution

Sample Problem 8.16   Using a Dilution Factor

Suppose a pharmacist has a 200× stock solution in the pharmacy and needs to prepare 500 mL of a 1:200 solution for a patient. How much of the stock solution should be used?

Solution

In this case, the dilution factor is 200 and the final volume is 500 mL. We can rearrange the dilution factor equation to solve for our initial volume of stock solution.

$$V_i = \frac{V_f}{\text{Dilution factor}}$$

$$V_i = \frac{500 \text{ mL}}{200}$$

$$V_i = 2.5 \text{ mL}$$

To prepare the correct dilution, the pharmacist would add 2.5 mL of the stock solution to enough solvent to produce a final volume of 500 mL.

Practice Problems

8.39 How many liters of a 0.15% (m/v) KCl IV solution can be prepared from 2.0 L of a 1.0% (m/v) stock solution?

8.40 How many liters of half-normal saline IV solution with a concentration of 0.45% (m/v) can be prepared from 1.0 L of a 9.0% (m/v) stock solution?

8.41 What is the concentration of the resulting solution if 450 mL of a 15% (m/v) glucose solution is diluted to a final volume of 2.0 L with distilled water?

8.42 A patient order requires 1.0 L of a 0.5% benzalkonium chloride antiseptic. You have a 12% stock solution. How much stock solution will you need to prepare the order?

8.43 How would you prepare 250 mL of a 0.225% (m/v) NaCl solution from a 0.90% (m/v) NaCl stock solution?

8.44 How would you prepare 2 L of 1 M $MgCl_2$ from a 5 M $MgCl_2$ stock solution?

8.45 You must dispense a povidone iodine solution for an abrasion. You have a 12× stock solution. How much of the stock solution will you need to prepare 250 mL of a 1× solution?

8.46 You must prepare 400 mL of a disinfectant that requires a 1:8 dilution of a stock solution. How much of the stock solution will be required?

DISCOVERING THE CONCEPTS

? INQUIRY ACTIVITY—Osmosis

Part 1. Information

Cells and the solutions surrounding them are separated by the cell's membrane. The cell membrane is semipermeable. This means that some molecules are able to pass through while others are not. Water can spontaneously flow through the cell membrane in a direction that will equalize the concentrations inside and outside the cell. Three possibilities exist for a cell and its surrounding solution: The concentrations of both are the same, the exterior solution is of higher concentration than the solution inside the cell, or the interior solution is a higher concentration than the solution outside the cell. The solution types are outlined in the following table and figure:

Solution Type	Amount of Solute (amount of water) Outside the Cell	Amount of Solute (amount of water) Inside the Cell	Movement of Water	Effect on the Cells
Isotonic	The same as inside	The same as outside	No net movement	No effect
Hypotonic	Lower than inside (higher than inside)	Higher than outside (lower than outside)	Water moves into the cells	Cells swell and can burst (lyse)
Hypertonic	Higher than inside (lower than inside)	Lower than outside (higher than outside)	Water moves out of the cells	Cells shrivel (crenate)

Isotonic solution

Hypotonic solution

Hypertonic solution

Red blood cells are isotonic with a NaCl solution that is 0.90% (m/v), also known as normal saline, and 5% glucose, also known as D5W (dextrose 5% in water). Intravenous fluids must be isotonic to avoid cell lysis (by bursting) or the opposite effect, shriveling, also called *crenation*.

Demonstration—Common Osmosis

You will need two carrots and two lemon slices. Dissolve as much table salt as possible into two glasses of tap water to make a saturated salt solution. Place one carrot and one lemon slice into separate glasses of salt solution and one carrot and one lemon into separate glasses of tap water. Allow them to sit overnight. Then observe each and answer the following questions.

Questions

1. Describe the appearance of the foods in (a) the tap water and (b) the saltwater.
2. How did the foods change when they were placed in (a) the tap water and (b) the saltwater?
3. In which glass did the water in the solution move into the food cells?
4. In which glass did the solution remove water from the food cells?

—*Continued next page*

Continued—

5. Which of the solutions (tap water or saltwater) is (a) hypotonic or (b) hypertonic to the vegetables?
6. A cell is placed in a 10% (m/v) glucose solution.
   a. In which direction will water flow to equalize the concentrations: into the cell or out of the cell?
   b. Will the cell swell or crenate?
7. If a person pours a concentrated saltwater solution on an earthworm, the worm will wriggle and die. What type of solution (isotonic, hypertonic, or hypotonic) is the salt solution? What are the earthworm's cells doing?
8. If a person drinks too much water too quickly, a condition known as hyponatremia will occur. In this condition, there is not enough sodium in the body fluids outside the cell. What type of solution (isotonic, hypertonic, or hypotonic) would a person's cells be in? What are the cells likely to do?

8.6 Osmosis and Diffusion

Osmosis

8.6 Inquiry Question

How is the direction of osmosis predicted?

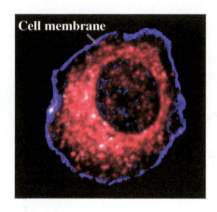

The cell membrane is a semipermeable membrane.

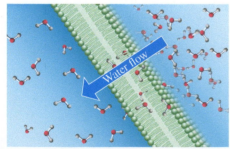

Water will pass through a cell membrane to equalize the concentration on either side.

Because our bodies are mostly water, we could consider ourselves to be composed of a set of specialized aqueous solutions. These solutions are found both inside and outside of cells. The concentration of these solutions is highly controlled by the cells. The solutions are separated by a semipermeable barrier called the *cell membrane*, whose structure was described in Chapter 7. A **semipermeable membrane** allows some molecules to pass through the barrier but not others. Under normal physiological conditions, these solutions are considered to be **isotonic solutions,** meaning that the concentration of dissolved solutes is the same on both sides of the membrane.

Consider this: Why is it that there is a limit to the amount of water that we can drink? And why is it that we cannot survive by drinking sea water? The short answer is that tap water is not as concentrated as the body's solutions, and sea water is more concentrated. Let's look more closely at these two questions.

A person can drink too much fresh water. It is recommended that we drink between 2 and 3 liters of water a day. A normal adult kidney can process up to 15 L of water a day, so the average person would really have to overdo it to drink too much water.

Yet people engaging in vigorous exercise and infants whose formula is overly diluted can drink too much water. What is happening in the body? In this instance, the concentration of dissolved solutes in the body's internal solutions is higher than the concentration of dissolved solutes in tap water or dilute formula. When a person drinks large quantities of tap water, it dilutes the blood. The concentration of solutes in the blood goes down, resulting in an imbalance between the concentration of the solution outside the cells (lower concentrated solution, less dissolved solutes) and the concentration of solution inside the cells (higher concentrated solution, more dissolved solutes). In this case, the solution outside of the cells is said to be a **hypotonic solution** (*hypo*- is a prefix meaning "lower than").

When the solution concentrations inside and outside the cells are different, water will travel across the cell membrane to equalize the concentrations. This passage of *water* through a semipermeable membrane such as a cell membrane is called **osmosis.** So, if we drink too much water, a condition known as hyponatremia (low sodium concentration caused by excess water) can result from the bloodstream becoming diluted by all the ingested water. Through osmosis, too much water enters the cells in an attempt to equalize the concentrations, and as a result, the cells swell up and could even burst (a phenomenon called *lysing*).

As water flows through a semipermeable membrane during osmosis, the water molecules in the more concentrated solution exert a certain amount of pressure (recall that pressure is force over an area) on the membrane as they attract the water molecules from the less concentrated solution to the more concentrated solution. This pressure is termed **osmotic pressure.** The greater the relative number of solute molecules present in a solution (that is, the more concentrated the solution), the greater the number of water molecules that will pass into the more concentrated solution to equalize the concentrations, and, as a result, the higher the osmotic pressure. Pure water has an osmotic pressure of zero. Applying pressure in opposition to the osmotic pressure will stop osmosis.

If we were marooned on a desert island without fresh water, why could we not drink sea water to quench our thirst? The concentration of dissolved ions in saltwater is about three times that in the blood, so when sea water is consumed, it draws water out of the cells through osmosis, to equalize the concentrations. If a person were to drink sea water, the concentration of solutes in the bloodstream would go up, resulting in an imbalance between the concentration outside the cells (higher concentrated solution, more dissolved solutes) and inside the cells (lower concentrated solution, less dissolved solutes). This drawing of water outside of the cells dehydrates them, and as a consequence, the person suffers from dehydration. The solution outside the cells is said to be a **hypertonic solution** (*hyper-* is a prefix meaning "higher than"). During dehydration, the cells shrivel in a process known as **crenation.** A summary of how cells behave in different types of solutions is shown in **FIGURE 8.7**.

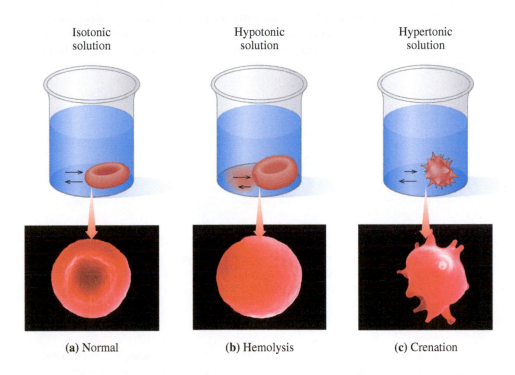

Isotonic solution Hypotonic solution Hypertonic solution

(a) Normal **(b)** Hemolysis **(c)** Crenation

FIGURE 8.7 How cells behave in different types of solutions. (a) Cells in isotonic solution. (b) Cells in hypotonic solution will swell. (c) Cells in hypertonic solution will shrink (crenate).

Let's look more closely at how osmosis occurs across a cell membrane. Given that cell membranes separate solutions of different concentrations, consider the situation in the illustration shown in the margin. In which direction would osmosis occur (water flow) to equalize the concentration?

During osmosis, water flows through the cell membrane from an area where the solution is less concentrated to where it is more concentrated, to even out the concentrations.

The net flow of water will be from the solution with the lower solute concentration into the solution of higher solute concentration, diluting the solution that is more concentrated. Water moves from a solution containing more water molecules (dilute solution) into a solution containing fewer water molecules (more concentrated solution), equalizing the concentrations.

Because cell membranes are semipermeable, osmosis is an ongoing process and is used by the cells to maintain the concentrations inside and outside the cells at about the same level. Body fluids like blood, plasma, and lymph all exert osmotic pressure on the cell membrane. Intravenous (IV) solutions that are delivered into patients' bloodstreams are isotonic—they have solute concentrations equal to the solute concentrations inside of

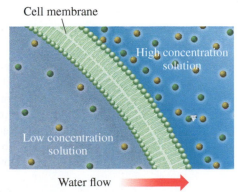

Cell membrane

High concentration solution

Low concentration solution

Water flow

cells. Isotonic solutions minimize osmosis, which is desirable when introducing IV solutions into the blood and eventually to cells. Common isotonic IV solutions used in hospitals include 0.90% (m/v) NaCl (normal saline, NS) and a 5% (m/v) D-glucose (dextrose) solution commonly referred to as D5W ("dextrose 5% in water"). These solutions are called **physiological solutions** because they exert the same osmotic pressure as the cells and are isotonic with cells.

Instead of using a concentration of molarity, physiological solutions are often expressed in terms of **osmolarity**, the molarity (moles/L) times the number of particles in solution. Osmolarity takes into account the number of particles exerting osmotic pressure against a membrane. Blood plasma, for example, has an osmolarity of 0.3 osmoles/L.

Diffusion and Dialysis

Suppose we put a drop of green food coloring into a large beaker of water. Over time, the green dye molecules (solute) will mix with the water (solvent), and the resulting solution will have a uniform light green tinge to it. The two solutions (one with a high concentration of green molecules and one with no green molecules) spontaneously mix, and the green solute molecules *diffuse* into the water to form one dilute solution with a final green color intermediate between green food coloring from the dropper bottle and water.

Time

Over time, a drop of concentrated dye will diffuse through water, making the whole container a uniform color.

INTEGRATING
Chemistry

Find out why ▶
dialysis is a
diffusion process,
not osmosis.

Kidney Dialysis

Small molecules and ions can also pass through the cell membrane. If a membrane is permeable to other molecules in addition to water, those molecules will also move across the membrane by a process similar to osmosis to equalize the concentrations of the two solutions.

The movement of *solute* molecules across a semipermeable membrane is a diffusion process, not osmosis. **Diffusion** is the movement of molecules in a direction that equalizes the concentration. One notable place where diffusion occurs in the body is in the kidneys. The kidneys act to remove small waste molecules from the blood through diffusion across membranes in the kidneys. Cells and larger molecules, such as proteins found in the blood, are too large to pass through the membrane. These larger molecules are reabsorbed into the bloodstream after passing through the kidney. At the same time, small molecules like urea diffuse out of the blood (higher concentration) and move into urine (lower concentration) in a process called **dialysis.**

If the kidneys cannot dialyze waste products out of the bloodstream, increased levels of urea and other wastes in the bloodstream can become life-threatening. A person whose kidneys are failing can undergo artificial dialysis—called *hemodialysis*—to cleanse the blood (see **FIGURE 8.8**). In this process, blood is removed from the patient and passes through one side of a semipermeable membrane in contact on the opposite side with a dialyzing solution that is isotonic with normal blood solute concentrations. Urea and small waste molecules are present in greater concentration in the patient's blood and diffuse out of the passing blood and into the dialyzing solution, and the dialyzed blood returns to the patient.

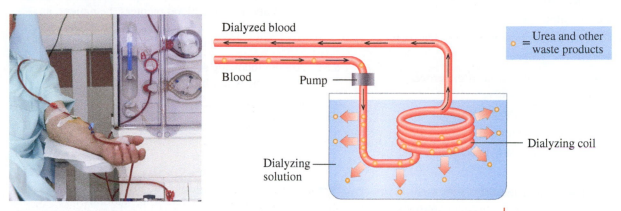

FIGURE 8.8 Hemodialysis. Hemodialysis removes waste products from the blood while outside the body, much as a healthy kidney removes waste products from the blood inside the body.

Sample Problem 8.17   **Osmosis and Osmotic Pressure**

A cell with concentration equal to 5% (m/v) glucose is placed in a 10% (m/v) glucose solution.
   a. In which direction will water flow to equalize the concentrations (into the cell or out of the cell)?
   b. Which solution is exerting the greater osmotic pressure?
   c. Will the cell swell or crenate?

Solution

   a. Water will flow from the area where there is more water—the 5% (m/v) solution— to the area where there is less water—the 10% (m/v) solution—so water will flow out of the cell.
   b. The 10% (m/v) glucose solution has the higher solute concentration, so it has the greater osmotic pressure.
   c. Because water is moving out of the cell, the cells will crenate.

Sample Problem 8.18   **Isotonic, Hypotonic, and Hypertonic Solutions**

Are each of the following solutions considered isotonic, hypotonic, or hypertonic with respect to body fluids?
   a. 3% (m/v) NaCl   b. 0.90% (m/v) NaCl   c. 0.09% (m/v) NaCl

Solution

Compare each solution to physiological saline, which is 0.90% (m/v) NaCl.
   a. hypertonic   b. isotonic   c. hypotonic

Practice Problems

8.47 A cucumber placed in a briny (saltwater) solution makes a pickle.

   a. Does water leave or enter the cucumber's cells?

   b. Does the cucumber swell or crenate?

   c. Is the saltwater solution hypertonic or hypotonic to the cucumber?

8.48 A lettuce leaf in the produce aisle at the grocery is accidentally submerged in water.

   a. Will water exit or enter the lettuce cells?

   b. Will the cells swell or crenate?

   c. Is tap water hypertonic or hypotonic to the lettuce leaf cells?

8.49 A patient is undergoing hemodialysis. As the blood leaves the patient, its solute concentrations are _____ (hypertonic, hypotonic, isotonic) to the dialyzing solution. As the blood reenters the patient, its solute concentrations are _____ (hypertonic, hypotonic, isotonic) to the dialyzing solution.

8.50 A patient is undergoing hemodialysis. As the blood leaves the patient, it has _____ (higher, lower, the same) osmotic pressure than the dialyzing solution. As the blood reenters the patient, its osmotic pressure is _____ (higher, lower, the same as) the dialyzing solution.

8.51 Are the following solutions isotonic, hypotonic, or hypertonic with physiological solutions?

   a. 10% (m/v) NaCl

   b. 0.90% (m/v) glucose

   c. 5% (m/v) NaCl

   d. 5% (m/v) glucose

8.52 Will a red blood cell undergo crenation, lysis, or no change in each of the following solutions?

   a. 10% (m/v) NaCl

   b. distilled water

   c. 5% (m/v) NaCl

   d. 5% (m/v) glucose

8.7 Transport Across Cell Membranes

? 8.7 Inquiry Question

How does the cell membrane control diffusion in and out of the cell?

We saw in Section 8.6 that water can cross the cell membrane through a process called osmosis, but how do polar molecules like glucose or ions like Na^+ move across? As we discussed in Section 7.5, the main structural component of the cell membrane is the phospholipid, which has a polar (hydrophilic) "head" containing a phosphate group and nonpolar (hydrophobic) hydrocarbon "tails." Recall that because the phospholipid head is hydrophilic and the tails are hydrophobic, phospholipids will spontaneously organize into a phospholipid bilayer, with the polar heads oriented toward the aqueous environments on the inside and outside of the cell, and with the tails oriented toward the interior of the cell membrane (**FIGURE 8.9**). This organization of the hydrophobic tails toward the interior of the membrane thus acts as a barrier to ions and polar molecules, which are hydrophilic. How, then, are such atoms and molecules able to cross the membrane?

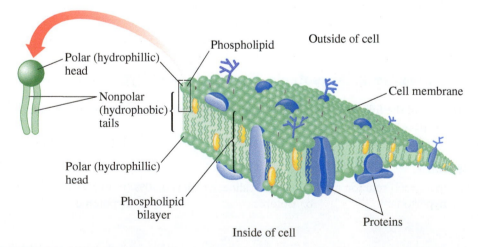

FIGURE 8.9 The cell membrane. Phospholipids create a nonpolar barrier to the passage of polar molecules.

Ions, nonpolar molecules, and polar molecules move across cell membranes in different ways. We will focus on three main forms of transport across cell membranes: passive diffusion, facilitated transport, and active transport.

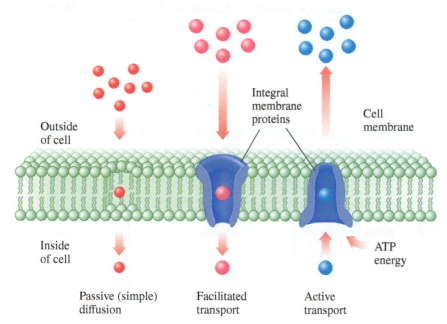

Transport through a cell membrane can occur with diffusion (no energy required) or against diffusion (energy required). Integral membrane proteins assist in both facilitated transport (transport with the concentration gradient) and active transport (transport against the concentration gradient).

Small molecules that are in high concentration—like water and the nonpolar molecules O_2, N_2, and CO_2—can diffuse directly through the cell membrane. Diffusion moves solutes in a direction that equalizes the concentrations on either side of a membrane. This process does not require any additional energy, so this simple diffusion process is also called **passive diffusion.** Other nonpolar molecules like steroids can also passively diffuse through cell membranes.

Because of their polar character, ions and small polar molecules diffuse very slowly across the nonpolar barrier (refer to Figure 8.9) found in the center of the cell membrane, water being one exception. To enable small molecules and ions to pass through the cell membrane, some proteins found in the cell membrane have polar channels that open and close, allowing small polar molecules and ions to be transported across the cell membrane, to equalize concentrations. These proteins are often integral membrane proteins, spanning the phospholipid bilayer (Chapter 7). This type of transport is called **facilitated transport,** and such transport does not require energy. Glucose transporter proteins are found in virtually all cell membranes and facilitate the transport of glucose into the cell during times when glucose concentrations in the bloodstream are high (for example, after a meal).

Sometimes ions or small molecules must be moved across the cell membrane in the opposite direction of diffusion (making the concentrations more different, less equal), but this cannot occur without the use of energy. Transporting ions or small polar molecules across the cell membrane in a direction that is opposite to equalizing concentrations requires the assistance of a protein channel or pump. This pumping in the opposite direction is called **active transport,** and it requires energy, usually in the form of the energy molecule adenosine triphosphate (ATP). ATP and active transport are discussed further in Chapter 12.

One of these active transport pumps, known as H^+/K^+-ATPase, controls the concentration of potassium and hydrogen ions in the stomach and is responsible for keeping the acid concentration in your stomach constant. Medications such as Tagamet®, Prilosec®, and Pepcid® prevent the effects of heartburn (high levels of stomach acid) by blocking the normal production of stomach acid through these pumps. A summary of the types of membrane transport is found in **TABLE 8.6**.

TABLE 8.6 Types of Membrane Transport

Transport Type	Energy Required?	Channel Required?	Movement of Molecules Across Membrane
Passive diffusion	No	No	With diffusion
Facilitated transport	No	Yes	With diffusion
Active transport	Yes	Yes	Against diffusion

Sample Problem 8.19 **Transport Across Membranes**

In the figure, in which direction would passive diffusion occur across the cell membrane (to the right or to the left)?

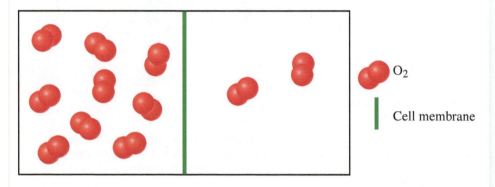

Solution

Because the O_2 will diffuse to equalize the concentrations, the O_2 molecules will move to the right side to increase the concentration on that side.

Practice Problems

8.53 Identify the type of transport (passive diffusion, facilitated transport, or active transport) that will occur for the following molecules:

a. oxygen

b. glucose, no energy required

c. Na^+, energy required

d. K^+, no energy required

8.54 Identify the type of transport (passive diffusion, facilitated transport, or active transport) that will occur for the following molecules:

a. cholesterol, no energy required

b. nitrogen

c. Cl^-, no energy required

d. H^+, energy required

8

CHAPTER REVIEW

The study guide will help you check your understanding of the main concepts in Chapter 8. You should be able to answer problems for each learning outcome in this list. To check your mastery, try the problems listed after each.

STUDY GUIDE	CHAPTER GUIDE

8.1 Solutions Are Mixtures

Identify a solution based on its physical properties.
- Distinguish solute and solvent. (Try 8.1, 8.55)
- Identify solutions, colloids, and suspensions. (Try 8.3, 8.57)

Inquiry Question

How is a solution identified?

A solution forms when a solute dissolves in a solvent. In a solution, the particles of a solute are evenly distributed in the solvent. The solute and solvent may be solid, liquid, or gas. Aqueous solutions contain water as the solvent. Solutions are transparent. Homogeneous mixtures with suspended particles are colloids and are usually not transparent. Mixtures that contain particles that settle upon standing are suspensions.

8.2 Formation of Solutions

Predict the effects of temperature and pressure on solution formation.
- Define saturated and dilute solutions. (Try 8.5)
- Predict the effect of temperature on the solubility of a solute. (Try 8.7, 8.59)
- Predict the effect of pressure on the solubility of a gas in a liquid. (Try 8.9, 8.11, 8.61)

Inquiry Question

How do temperature and pressure affect the formation of a solution?

An increase in temperature increases the solubility of most solids in water but decreases the solubility of gases in water. Henry's law describes the relationship between pressure and gas solubility. Increasing the pressure above a solution with a dissolved gas in it increases the solubility of the gas. A solution that contains the maximum amount of dissolved solute is a saturated solution. A solution that is saturated reaches an equilibrium state between the dissolved solute and undissolved solid solute where the rate of dissolving and re-forming solid is the same.

8.3 Chemical Equations for Solution Formation

Write chemical equations for hydration of electrolytes, nonelectrolytes, and weak electrolytes.
- Predict if different types of electrolytes will fully, partially, or not dissociate when dissolved in solution. (Try 8.13, 8.63)
- Write chemical equations for hydration of electrolytes, nonelectrolytes, and weak electrolytes. (Try 8.17, 8.65)
- Calculate the number of milliequivalents present for an ionic compound that fully dissociates in solution. (Try 8.69)
- Convert from Eq to moles. (Try 8.19, 8.67)

Inquiry Question

How are hydration equations written for electrolytes, nonelectrolytes, and weak electrolytes?

Hydration equations can be written for solutes dissolving in solvents. The form of this equation depends on the ability of the solute to dissociate in solution. Substances that release ions when they dissolve in water are called electrolytes because the solution will conduct an electrical current. Strong electrolytes are ionic compounds that completely dissociate in water. Nonelectrolytes are substances (usually polar covalent compounds) that dissolve in water but do not dissociate. Weak electrolytes only partially dissociate into ions. The unit known as an equivalent expresses the amount of dissolved ions in fluids. The number of equivalents per mole of an ion equals the charge on that ion.

—Continued next page

Continued—

8.4 Concentration

Determine the concentration of solutions using common units.

- Express concentration in molarity units. (Try 8.23, 8.73, 8.75)
- Express concentration in percent units. (Try 8.33, 8.79)
- Express concentration in parts per million and parts per billion. (Try 8.37, 8.83)

Inquiry Question

How is the concentration of a solution expressed?

The concentration of a solution is the amount of solute dissolved in a certain amount of solution. Fluid replacement solutions are often expressed in units of mEq/L or in some cases mmole/L. Molarity is the moles of solute per liter of solution. Percent mass/volume expresses the ratio of the mass of solute (in g) to the volume of solution (in mL) multiplied by 100. Percent mass/volume is equivalent to the unit g/dL. Percent concentration is also expressed as mass/mass and volume/volume ratios. Parts per million and parts per billion describe very dilute solutions.

8.5 Dilution

Apply the dilution equation to concentration units.

- Calculate concentrations or determine volumes using the dilution equation. (Try 8.41, 8.43, 8.85)
- Calculate initial volumes of concentrated stock solution from a dilution factor. (Try 8.45, 8.93)

Inquiry Question

How is the concentration of a solution determined if a dilution is made?

When a solution is diluted, the amount of solute stays the same while the volume of solution increases. The concentration of the solution decreases. Solutions can be diluted by adding more solvent to a stock solution. A dilution factor can be determined as the ratio of the final volume to the initial volume.

8.6 Osmosis and Diffusion

Predict the direction of osmosis.

- Predict the direction of osmosis or diffusion given the concentration on both sides of a semipermeable membrane. (Try 8.47, 8.95)

Inquiry Question

How is the direction of osmosis predicted?

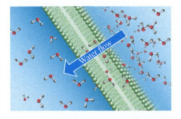

In osmosis, solvent (water) passes through a semipermeable membrane from a solution of lower solute concentration to a solution of higher solute concentration. The osmotic pressure exerted on the membrane is directly related to the number of water molecules pushing against that membrane. Isotonic solutions have osmotic pressures equal to those of bodily fluids. Cells maintain their volume in an isotonic solution, but they swell and may burst in a hypotonic solution and shrivel in a hypertonic solution. In dialysis, water and small solute particles pass through a dialyzing membrane in a related process called *diffusion* while large particles like proteins are retained.

8.7 Transport Across Cell Membranes

Distinguish types of transport across a cell membrane.

- Characterize three forms of transport across a cell membrane. (Try 8.53, 8.101)

Inquiry Question

How does the cell membrane control diffusion in and out of the cell?

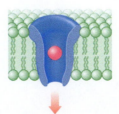

The semipermeable membrane surrounding cells separates the cellular contents from the external fluids. Molecules can be transported across the cell membrane by passive diffusion, facilitated transport, or active transport, depending on their concentration inside and outside the cell and their polarity.

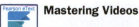

Mastering Videos

PRACTICING THE CONCEPTS

The following videos can be accessed through the Pearson eText or your Mastering Chemistry course.

- Calculating Percent Concentration
- Solution Preparation and Dilution Calculations

Important Equations

$$\text{Concentration} = \frac{\text{amount of solute}}{\text{amount of solution}}$$

$$\text{Molarity, M} = \frac{\text{mole solute}}{\text{L solution}}$$

$$\%\text{Concentration} = \frac{\text{parts solute}}{100 \text{ parts solution}}$$

$$\text{Dilution factor} = \frac{V_f}{V_i}$$

$$\%(\text{m/m}) = \frac{\text{g solute}}{\text{g solution}} \times 100$$

$$\%(\text{v/v}) = \frac{\text{mL solute}}{\text{mL solution}} \times 100$$

$$\%(\text{m/v}) = \frac{\text{g solute}}{\text{mL solution}} \times 100$$

$$C_{\text{initial}} \times V_{\text{initial}} = C_{\text{final}} \times V_{\text{final}}$$

$$\text{ppm} = \frac{\text{g solute}}{\text{mL solution}} \times 1{,}000{,}000$$

$$\text{ppb} = \frac{\text{g solute}}{\text{mL solution}} \times 1{,}000{,}000{,}000$$

Additional Problems

8.55 To protect the skin from allergic reactions before applying casts, a Compound Tincture of Benzoin (CTB) is applied to the skin. This solution consists of 10% benzoin and 90% ethanol. Identify the solvent and the solute.

8.56 Drysol is an antiperspirant solution that contains 7.5 g of aluminum chloride hexahydrate in 37.5 g of alcohol. Identify the solvent and the solute.

8.57 Identify the following as solutions or colloids.
a. normal saline, an IV fluid replacement containing 0.9% NaCl
b. a dextran IV fluid replacement containing large carbohydrate molecules
c. a hazelnut latte

8.58 Identify the following as solutions or colloids.
a. whipped cream
b. hair conditioner
c. iced tea

8.59 Does the solubility of the solute increase or decrease in each of the following situations?
a. Sugar is dissolved in iced tea instead of hot tea.
b. A bottle of soda (solute is CO_2 gas) is placed in the refrigerator instead of in a pantry at room temperature.
c. An opened bottle of champagne (solute is CO_2 gas) is allowed to sit open to the air for hours.

8.60 Does the solubility of the solute increase or decrease in each of the following situations?
a. Ice is placed in a glass of cola (solute is CO_2 gas).
b. Honey is dissolved in hot water instead of cold water.
c. The pressure of O_2 over a solution is increased (solute is O_2 gas).

8.61 When an asthmatic has an attack, a smaller amount of air enters the lungs, so a lower oxygen pressure (concentration) exists in the lungs. How is the concentration of oxygen in the blood affected by an asthma attack?

8.62 Would you expect the concentration of oxygen in your blood to be higher or lower if you were at an altitude of 10,000 ft above sea level?

8.63 Predict whether the following will fully dissociate, partially dissociate, or not dissociate when dissolved in solution:

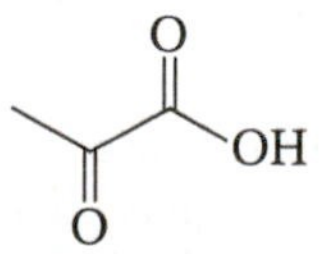

a. sodium iodide, $NaI(s)$, a strong electrolyte
b. pyruvic acid (as shown above), a weak electrolyte
c. glucose, $C_6H_{12}O_6(s)$, a nonelectrolyte

8.64 Predict whether the following will fully dissociate, partially dissociate, or not dissociate when dissolved in solution:

    a. potassium nitrate, $KNO_3(s)$, a strong electrolyte

    b. isopropyl alcohol, $CH_3CH(OH)CH_3(l)$, a nonelectrolyte

    c. boric acid, $H_3BO_3(s)$, a weak electrolyte

8.65 Provide a balanced equation for the hydration of each compound in Problem 8.63.

8.66 Provide a balanced equation for the hydration of each compound in Problem 8.64.

8.67 How many mmoles of Mg^{2+} are present in 1 L of a Normosol-R solution that contains 3.0 mEq of Mg^{2+}?

8.68 How many mmoles of lactate $(C_3H_6O_3^-)$ are present in 1 L of lactated Ringer's solution containing 28 mEq of lactate?

8.69 A Ringer's solution for intravenous fluid replacement contains 4 mEq Ca^{2+} per liter of solution. If a patient receives 1750 mL of Ringer's solution, how many milliequivalents of Ca^{2+} were given?

8.70 An extracellular replacement fluid contains 3 mEq/L citrate. The conversion factor for citrate is 3 Eq citrate/ 1 mole of citrate. If a patient receives 550 mL of this solution, how many millimoles of citrate were delivered?

8.71 What is the concentration in mmole/L of a SO_4^{2-} solution that is 25.0 mEq/L?

8.72 What is the concentration in mEq/L of a Na^+ solution that is 115 mmole/L?

8.73 What is the calcium ion molarity of 400 mL of lactated Ringer's containing 0.8 mmoles of Ca^{2+}?

8.74 What is the sodium ion molarity in 250 mL of a Normosol-R fluid replacement containing 35 mmoles of Na^+?

8.75 What is the molarity of 0.50 L of solution containing 40.0 g of $CaCO_3$?

8.76 What is the molarity of 450 mL of solution containing 15 g NaCl?

8.77 A 750 mL bottle of wine contains 12% (v/v) ethanol. How many milliliters of ethanol are in the bottle of wine?

8.78 What is the concentration in % (v/v) of 650 mL of an aqueous hydrogen peroxide solution containing 20 mL of hydrogen peroxide?

8.79 Gentian violet solution is used as an antifungal. What is the concentration in % (m/v) if 1.50 g of gentian violet is combined to a total solution volume of 300 mL?

8.80 How many grams of phenylephrine are needed to prepare 20 mL of a 10% (m/v) solution for a glaucoma patient?

8.81 How many grams of dextrose are in 800 mL of a 5% (m/v) D5W solution?

8.82 The normal range for blood hemoglobin in females is 12–16 g/dL. What is this value in % (m/v)?

8.83 The normal level of urea nitrogen in adults is 7–18 mg/dL. What is this value in ppm?

8.84 The normal creatinine levels in adults are 0.6–1.2 mg/dL. What is this value in ppm?

8.85 Calculate the final concentration of each of the following diluted solutions:

    a. 2.0 L of a 3.0 M HNO_3 solution is added to water so that the final volume is 6.0 L.

    b. Water is added to 0.50 L of a 6.0 M KOH solution to make 4.0 L of a diluted KOH solution.

8.86 Calculate the final concentration of each of the following diluted solutions:

    a. A 50.0-mL sample of 4.0% (m/v) NaOH is diluted with water so that the final volume is 100.0 mL.

    b. A 15.0-mL sample of 35% (m/v) acetic acid (CH_3COOH) solution is added to water to give a final volume of 25 mL.

8.87 What is the final volume in liters that can be prepared from each of the following concentrated solutions?

    a. a 1.00 M HCl solution prepared from 150.0 mL of a 6.0 M HCl solution

    b. a 4.00% (m/v) $NaHCO_3$ solution prepared from 250. mL of a 20.0% (m/v) $NaHCO_3$

8.88 What is the initial volume in liters necessary to prepare each of the following diluted solutions?

    a. 2.0 L of a 0.90% NaCl using a 18.0% (m/v) NaCl stock solution

    b. 1.5 L of a 5.0% glucose solution using a 15.0% glucose solution

8.89 How would you prepare 1.0 L of a normal saline solution [(0.9% (m/v)] from a stock solution that is 18% (m/v)?

8.90 How would you prepare 500 mL of a 5% D5W (dextrose in water) solution from a 25% stock solution?

8.91 Aluminum acetate can be used to treat contact dermatitis such as poison ivy and rashes from skin sensitivity. If 100 mL of a 50% solution is diluted to 1 L, what is the dilution factor? What is the concentration of the final solution?

8.92 You must dilute a 4× stock solution of hydrogen peroxide to 800 mL to dispense a 3% solution. How much of the 4× stock do you need? What is the concentration of the 4× stock solution?

8.93 A physician orders 0.5 L of an 8% (m/v) solution of a drug in aqueous solution. The pharmacy has 300 mL of a 10% (m/v) solution of this drug on hand. Can the pharmacist prepare this prescription? How would it be prepared?

8.94 Suppose the pharmacist has a 1.5 L of a 20× concentrate of an antibiotic and must dispense 250 mL of a 1× solution to customers. What volume of the 1× antibiotic can the pharmacist prepare? How many customers can receive the 1× solution?

8.95 Consider a cell placed in solution as shown in the following figure:

If the inside of the cell has a concentration equivalent to 0.90% (m/v) NaCl, is the cell in a hypertonic, hypotonic, or isotonic solution if the solution outside of the cell is the following:

a. 5% NaCl   **b.** 0.090% NaCl

c. 0.90% NaCl

8.96 Under the conditions in Problem 8.95a–c, in which direction will water move by osmosis: into the cell, out of the cell, or will no net movement occur?

8.97 Edema, commonly referred to as water retention, is characterized by a swelling of the tissues. Individuals with kidney disease will not excrete normal amounts of Na^+, leading to higher levels of Na^+ in the tissues. In terms of osmosis, explain how high levels of Na^+ in the tissues can lead to edema.

8.98 Many people gain relief from swollen feet at the end of the day by soaking their feet in Epsom salts, which creates a hypertonic solution of $MgSO_4$. In terms of osmosis, explain how soaking in Epsom salts can reduce swelling in the feet.

8.99 A process called reverse osmosis purifies tap water by pushing water through a set of semipermeable membrane filters. Explain why this process is called *reverse* osmosis.

8.100 Consider the cell in the following figure.

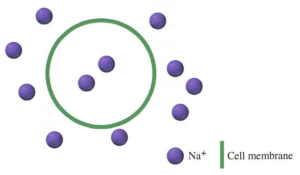

Na^+ | Cell membrane

In which direction (into the cell or out of the cell) would the Na^+ be transported if the transport protein in the cell membrane was

a. an ATPase pump?

b. an ion transport protein operating under facilitated transport?

8.101 Do the following processes require energy when transporting molecules across the cell membrane?

a. passive diffusion   **b.** facilitated transport

c. active transport

8.102 Describe the concentration of the solution outside the cell as hypertonic or hypotonic if that solute is being transported across the cell membrane by

a. passive diffusion.   **b.** facilitated transport.

c. active transport.

Challenge Problems

8.103 A scuba diver diving 100 m down in the ocean experiences ten times more pressure on her body than a person swimming at sea level experiences from the atmosphere. The high pressure of the air in a scuba diver's air tank affects the solubility of gases in the bloodstream. A condition called "the bends," which is related to nitrogen gas solubility, can occur in a scuba diver who ascends too quickly to the surface from a deep dive. Can you give an explanation for this condition?

8.104 How would you prepare 500 mL of a 6.0 M NaOH solution using solid NaOH pellets? What is the % (m/v) concentration of a 6.0 M NaOH solution?

8.105 Two containers of equal volume are separated by a membrane that allows free passage of water but completely restricts passage of solute molecules. Solution A has 3 molecules of the protein albumin (molecular weight 66,000) and solution B contains 16 molecules of glucose (molecular weight 180). Into which compartment would water flow, or will there be no net movement of water? Which solution is at the higher concentration? Which chamber is exerting the higher osmotic pressure?

Solution A | Solution B

Semipermeable membrane

8.106 Proteinuria is a condition in which excessive protein is found in the urine. What must be happening to the kidney's membrane filters if such large molecules are found in the urine?

8.107 You are asked to prepare 240 mL of a 3% (m/v) morphine sulfate solution. You have four 50-mL vials of morphine sulfate labeled as 50 mg/mL. Do you have enough stock solution? How will you prepare the aqueous solution?

8.108 A 5% dextrose in ½ normal saline (D5 ½ NS) solution is commonly administered to patients needing post-operative IV fluids. How could you prepare 500 mL of this solution? Dextrose is 5% (m/v) and the NaCl is 0.45% (m/v).

Answers to Odd-Numbered Problems

Practice Problems

8.1 **a.** solute, vinegar; solvent, water

    b. solute, methyl mercaptan; solvent, methane

    c. solutes, copper and silver; solvent, gold

8.3 **a.** solution   **b.** colloid   **c.** solution

8.5 **a.** no, not saturated   **b.** yes, saturated

8.7 **a.** increase   **b.** decrease   **c.** increase

8.9 **a.** Decreasing the temperature increases the solubility of the gas in the soda, so more of the gas stays in the solution. Capping the bottle increases the CO_2 pressure over the solution and increases the solubility of CO_2.

    b. If the lid is on tight, no gas can escape from the container and gas that escapes from the solution will build up pressure above the solution. At some point, an equilibrium is reached in which no more gas will escape from the solution.

    c. The solubilities of most solid solutes decrease with lower temperature, so less sugar will dissolve in the iced tea.

8.11 at sea level because the atmospheric pressure is higher at sea level

8.13 **a.** fully dissociate

    b. partially dissociate

    c. not dissociate

8.15 **a.** $KCl(s) \xrightarrow[H_2O]{} K^+(aq) + Cl^-(aq)$

    b. $NaC_3H_5O_3(s) \underset{H_2O}{\rightleftharpoons} Na^+(aq) + C_3H_5O_3{}^-(aq)$

    c. $C_6H_{12}O_6(s) \xrightarrow[H_2O]{} C_6H_{12}O_6(aq)$

8.17 **a.** $CaCl_2(s) \xrightarrow[H_2O]{} Ca^{2+}(aq) + 2Cl^-(aq)$

    b. $NaOH(s) \xrightarrow[H_2O]{} Na^+(aq) + OH^-(aq)$

    c. $KBr(s) \xrightarrow[H_2O]{} K^+(aq) + Br^-(aq)$

    d. $Fe(NO_3)_3(s) \xrightarrow[H_2O]{} Fe^{3+}(aq) + 3NO_3^-(aq)$

8.19 4.25 Eq

8.21 0.0770 mole Na^+

8.23 1.25 mmoles Ca^{2+}/L

8.25 17 M

8.27 110 mM or 0.11 M

8.29 7.5 g KCl

8.31 25 g dextrose

8.33 **a.** 5.00%   **b.** 6.3% (m/v)

8.35 0.117 g insulin

8.37 5 ppm, 5000 ppb

8.39 13 L

8.41 3.4% (m/v)

8.43 Add 62.5 mL of the 0.90% NaCl stock solution to enough distilled water for a total volume of 250 mL solution.

8.45 21 mL

8.47 **a.** leave   **b.** crenate   **c.** hypertonic

8.49 hypertonic, isotonic

8.51 **a.** hypertonic   **b.** hypotonic

    c. hypertonic   **d.** isotonic

8.53 **a.** passive diffusion   **b.** facilitated transport

    c. active transport   **d.** facilitated transport

Additional Problems

8.55 Benzoin is the solute, and ethanol is the solvent.

8.57 **a.** solution   **b.** colloid   **c.** colloid

8.59 **a.** decrease   **b.** increase   **c.** decrease

8.61 The person has less oxygen in the lungs, so less can move into the bloodstream and the concentration of oxygen in the blood is therefore less.

8.63 **a.** fully dissociate   **b.** partially dissociate

    c. not dissociate

8.65 **a.** $NaI(s) \xrightarrow[H_2O]{} Na^+(aq) + I^-(aq)$

    b.

    c. $C_6H_{12}O_6(s) \xrightarrow[H_2O]{} C_6H_{12}O_6(aq)$

8.67 1.5 mmoles Mg^{2+}

8.69 7 mEq Ca^{2+}

8.71 12.5 mmole/L

8.73 0.002 M or 2 mM Ca^{2+}

8.75 0.80 M

8.77 90. mL ethanol

8.79 0.5% (m/v)

8.81 40 g dextrose

8.83 70–180 ppm urea nitrogen

8.85 **a.** 1.0 M HNO_3 **b.** 0.75 M KOH

8.87 **a.** 0.90 L   **b.** 1.25 L

8.89 Combine 50 mL of the stock solution [(18% (m/v)] with enough water to make 1 L of solution.

8.91 10×; 5%

8.93 No. This dilution requires 400 mL of the 10% (m/v) solution brought to volume with water.

8.95 **a.** hypertonic

    b. hypotonic

    c. isotonic

8.97 The cells will attempt to dilute the higher than normal concentration of Na^+ in the tissues by moving more water into the tissues, causing fluid retention.

8.99 If osmosis is water moving from a lower concentrated solution to a higher concentrated solution, balancing out the concentrations, the reverse would require water moving in the opposite direction. Less pure water (higher concentration of solutes) is forced (requires energy) through a filter (semipermeable membrane), removing more impurities, making drinkable water.

8.101 a. no   **b.** no   **c.** yes

8.103 According to Henry's law, more nitrogen would dissolve in the bloodstream at lower depths (higher pressure) than at the surface (lower pressure). When a diver ascends to the surface quickly, the pressure of the air in the tank (and therefore the air in the lungs) lessens more rapidly than a person can expel the nitrogen. Unable to stay dissolved in the bloodstream, nitrogen gas bubbles begin forming in the bloodstream, causing the condition.

8.105 Water will move from solution A to solution B because solution B is a higher concentrated solution (more solute/solution). It does not matter that the albumin particles are huge; osmotic flow depends on the concentration, not the size of the particles. Since solution B has a higher concentration, it exerts a higher osmotic pressure.

8.107 Yes. Add 144 mL from the stock vials (200mL is available) to water for a total volume of 240 mL.

9

Acids, Bases, and Buffers in the Body

Shampoos claim to be "pH balanced" to keep your hair looking healthy and shiny. The naturally high pH of soaps can strip oils from your scalp, which has a lower pH, so a buffer is added to the shampoo to prevent this. What is pH? What is a buffer? What is acidity and how is it calculated? Read Chapter 9 for an introduction to acid–base chemistry, pH, and buffers.

LEARNING TIP   Cramming Only Works in the Short Term

Studying only right before an exam might get you good results on that exam, but you won't retain as much in the long term. Research shows that the more times you come in contact with the information you're studying, the more strengthened the neural connections become and the easier it is to remember the information.

WHEN YOU EAT a slice of lemon or lime, how does it taste? Sour, right? Vinegar, found in pickles and vinaigrette salad dressings, also has a sour taste. Citrus fruits, pickles, and even some candies taste sour because they all contain acid. Our stomachs produce acid to help digest the food we eat, and our muscles produce lactic acid when we exercise. Acids can be neutralized by substances called bases. Soaps and toothpastes are mild bases and, like other bases, feel smooth or slippery to the touch.

pH refers to the level of acidity of a solution. Life operates under very strict pH conditions. For example, amino acids, the building blocks of proteins, will change form if the acidity of their environment changes. Proteins change their shape and their ability to function if the pH of their surroundings changes. Our body fluids, including blood and urine, contain compounds called buffers that maintain pH as we exercise and our breathing rates change. In this chapter, we discuss the bicarbonate buffer system in the blood and the physiological conditions of acidosis and alkalosis, but first we consider acids, bases, and equilibrium.

9.1 Acids and Bases—Definitions

Acids

The first definition of an acid was provided by the Swedish chemist Svante Arrhenius (pronounced *Ar-RAY-nee-us*) in the late 1800s. He described **acids** as substances that dissociate, producing hydrogen ions (H^+) when dissolved in water. The presence of hydrogen ions (H^+) gives acids their sour taste and allows acids to corrode some metals. In the early twentieth century, Johannes Brønsted and Thomas Lowry, working independently, expanded the definition of an acid to include the concept that an acid is a compound that *donates* a proton.

$$HCl(g) \xrightarrow[H_2O]{} H^+(aq) + Cl^-(aq)$$

Dissociates into ions

An acid is a substance that dissociates in water, producing a hydrogen ion, H^+.

How is the hydrogen ion (H^+) described by Arrhenius related to the proton used in the Brønsted–Lowry definition of an acid? We know that a proton is a subatomic particle and that H^+ is an ion. Because most hydrogen atoms contain one proton and one electron, a hydrogen ion (H^+), which is a hydrogen atom that has lost its electron, and a proton are one and the same. In an aqueous solution, free protons (H^+) seldom exist. The partial negative charge on the oxygen atom in water is strongly attracted to the positive charge of a proton. This attraction is so strong that the proton and the oxygen atom in water form a covalent bond, creating a **hydronium ion, H_3O^+**.

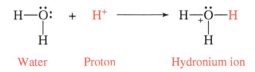

| Water | Proton | Hydronium ion |

Bases

According to Arrhenius, **bases** are ionic compounds that, when dissolved in water, dissociate to form a metal ion and a hydroxide ion (OH^-). Most Arrhenius bases are formed from Group 1A and 2A metals, such as NaOH, KOH, LiOH, and $Ca(OH)_2$. These hydroxide bases are characterized by a bitter taste and a slippery feel. The Brønsted–Lowry definition of a base mirrors the acid definition in that a Brønsted–Lowry base is a compound that *accepts* a proton.

Acids and Bases Are Both Present in Aqueous Solution

The Brønsted–Lowry definition states that acids *donate* protons and that bases *accept* protons, implying that a proton is usually transferred in an **acidic** or **basic** solution. Often, water can act as an acid or a base by donating or accepting a proton. We can write the formation of a hydrochloric acid solution as a transfer of a proton from hydrogen chloride to water. As shown in the figure below, by accepting a proton in the reaction, water is acting as a base, according to the Brønsted–Lowry theory.

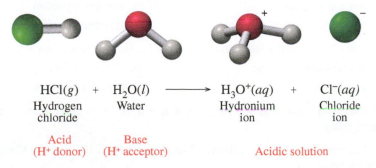

HCl(g)	+	$H_2O(l)$	$\longrightarrow$	$H_3O^+(aq)$	+	$Cl^-(aq)$
Hydrogen chloride		Water		Hydronium ion		Chloride ion
Acid (H^+ donor)		Base (H^+ acceptor)		Acidic solution		

In this reaction, HCl is a Brønsted–Lowry acid because it donates H^+ to water, forming hydronium ion.

What's an Inquiry Question?

? Inquiry Questions are designed to focus your reading on the main concepts by section. As you read each section, see if you can answer its Inquiry Question in your own words.

? 9.1 Inquiry Question

How do acids and bases differ?

NaOH(s)

○ OH^-
⊕ Na^+

— Water

$$NaOH(s) \xrightarrow{H_2O} Na^+(aq) + OH^-(aq)$$
Ionic compound ... Hydroxide ion

Dissociation

An Arrhenius base is an ionic compound that dissociates, producing hydroxide, OH^-.

In contrast, let's look at another reaction in which ammonia (NH_3) reacts with water. Because the nitrogen atom of NH_3 has a stronger attraction for a proton than the oxygen of water, water acts as an acid in this case by donating a proton.

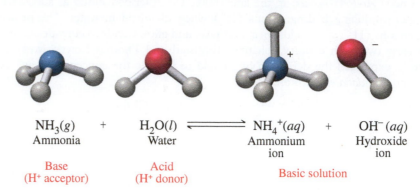

$$NH_3(g) \quad + \quad H_2O(l) \rightleftharpoons NH_4^+(aq) \quad + \quad OH^-(aq)$$

Ammonia	Water	Ammonium ion	Hydroxide ion

Base
(H⁺ acceptor)   Acid
(H⁺ donor)   Basic solution

In this reaction, NH_3 is a Brønsted–Lowry base because it is accepting H⁺ from water, producing hydroxide.

Sample Problem 9.1 Identifying Acids and Bases

In each of the equations, identify the reactant that is an acid (H⁺ donor) and the reactant that is a base (H⁺ acceptor):
  a. $HNO_3(aq) + H_2O(l) \longrightarrow NO_3^-(aq) + H_3O^+(l)$
  b. $ClO^-(aq) + H_2O(l) \rightleftharpoons HClO(aq) + OH^-(l)$

Solution

  a. Examine what is happening to each of the reactants as it changes to product. When HNO_3 forms NO_3^-, a proton is donated (H⁺), so HNO_3 is acting as the acid in this equation. When H_2O forms H_3O^+, it is accepting an H⁺, which makes water a base in this equation.
  b. Examine what is happening to each of the reactants as it changes to product. When H_2O forms OH^-, a proton is donated (H⁺), forming OH^-, so H_2O is acting as the acid in this equation. When ClO^- forms $HClO$, it is accepting an H⁺ (+ and − charges make $HClO$ a neutral compound), making ClO^- a base in this equation.

Practice Problems

Find the answers to the odd-numbered Practice Problems at the end of the chapter.

9.1   Indicate if each of the following statements is characteristic of an acid or a base:

  a. has a sour taste   b. accepts a proton

  c. produces H⁺ ions in water   d. is named potassium hydroxide

9.2   Indicate if each of the following statements is characteristic of an acid or a base:

  a. neutralized by a base   b. produces OH⁻ in water

  c. has a slippery feel   d. donates a proton

9.3   In your own words, explain how a proton and the hydrogen ion represent the same thing.

9.4   In your own words, explain how, in aqueous solution, H⁺ and H_3O^+ represent similar things.

9.5 In each of the following equations, identify the reactant that is an acid (H$^+$ donor) and the reactant that is a base (H$^+$ acceptor):

a. $HBr(aq) + H_2O(l) \longrightarrow H_3O^+(aq) + Br^-(aq)$

b. $HCO_3^-(aq) + H_2O(l) \rightleftharpoons H_2CO_3(aq) + OH^-(aq)$

9.6 In each of the following equations, identify the reactant that is an acid (H$^+$ donor) and the reactant that is a base (H$^+$ acceptor):

a. $H_3PO_4(aq) + H_2O(l) \rightleftharpoons H_2PO_4^-(aq) + H_3O^+(aq)$

b. $PO_4^{3-}(aq) + H_2O(l) \rightleftharpoons HPO_4^{2-}(aq) + OH^-(aq)$

9.2 Strong Acids and Bases

In Section 9.1, we saw that acids and bases are classified by their ability to donate or accept protons, respectively. **TABLE 9.1** shows the six **strong acids.** Strong acids completely (~100%) dissociate, which means they break up completely into ions when placed in water, forming hydronium ions and anions.

$$HCl(g) + H_2O(l) \longrightarrow H_3O^+(aq) + Cl^-(aq)$$

Acids that only partially dissociate (~5%) are considered **weak acids** and will be examined more closely in Section 9.4. One such example is acetic acid (CH_3COOH), which is the main component of vinegar (see **FIGURE 9.1**).

$$CH_3COOH(l) + H_2O(l) \rightleftharpoons CH_3COO^-(aq) + H_3O^+(aq)$$

Strong bases, like NaOH (also known as lye), are used in household products such as oven cleaners and drain openers. The Arrhenius bases LiOH, KOH, NaOH, and Ca(OH)$_2$ are **strong bases** that dissociate completely (100%) in water. Because they are

9.2 Inquiry Question

What happens when an acid reacts with a base?

TABLE 9.1 Six Common Strong Acids

Acid Name	Formula
Perchloric acid	HClO$_4$
Sulfuric acid[a]	H$_2$SO$_4$
Hydroiodic acid	HI
Hydrobromic acid	HBr
Hydrochloric acid	HCl
Nitric acid	HNO$_3$

[a]Only the first proton is 100% dissociated. The product, HSO$_4^-$, is a weak acid.

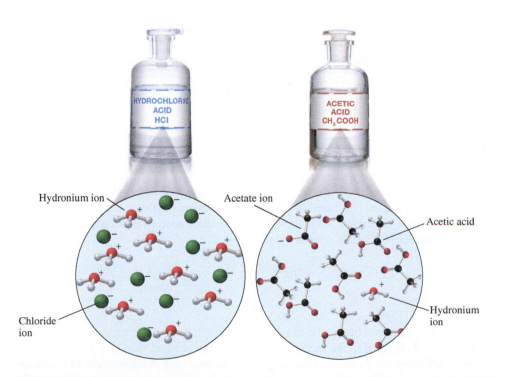

FIGURE 9.1 Strong versus weak acid. A strong acid such as HCl (left) is completely dissociated (100%), whereas a weak acid such as CH_3COOH (right) contains mostly molecules and a few dissociated ions in solution.

TABLE 9.2 Corresponding Anion and Acid Suffixes

Anion name ends in	Acid name is
-ide	Hydro[anion name minus ide]-ic acid
-ate	[anion name minus ate]-ic acid
-ite	[anion name minus ite]-ous acid

ionic compounds, they dissociate in water to give an aqueous solution containing a metal ion and a hydroxide ion.

Bases that only partially dissociate (~5%) are considered **weak bases.** Many common weak bases, such as those found in household cleaners, contain ammonia (NH_3).

Naming Acids

Why is HCl called hydrogen chloride as a gas and hydrochloric acid when dissolved in water? Naming acids follows a systematic set of rules. The name of any acid formed in water is related to the anion's name when the proton is donated. An anion ending in *ide*, will be called *hydro*(name)*ic acid* where (name) is the anion name minus *ide*. In the case of hydrochloric acid, the anion is chloride, so when it accepts a proton, the acid would be called hydrochloric acid. See **TABLE 9.2** for a summary.

Sample Problem 9.2   Naming Acids

Using Table 9.2, name the acids formed from the following anions.
   a. cyanide, a poison, CN^-
   b. hypochlorite, found in bleach, ClO^-
   c. nitrate, used as a preservative in foods, NO_3^-

Solution

   a. The name of the acid formed when CN^- accepts a proton is hydrocyanic acid, HCN.
   b. The name of the acid formed when ClO^- accepts a proton is hypochlorous acid, HClO.
   c. The name of the acid formed when NO_3^- accepts a proton is nitric acid, HNO_3.

Neutralization

What happens when a strong acid and a strong base are mixed? Consider the HCl and NaOH discussed earlier. Because both completely dissociate to form ions in water, the water contains sodium ions and chloride ions as well as hydronium ions and hydroxide ions. The protons in the hydronium ions are strongly attracted to the hydroxide ions and combine to form water molecules. This chemical reaction produces a lot of energy as heat and is therefore an exothermic reaction.

$$H_3O^+(aq) + OH^-(aq) \longrightarrow 2H_2O(l) + \text{Heat}$$

The sodium ions and the chloride ions remain in solution. Suppose the water were removed by boiling. The ionic compound, sodium chloride (NaCl), would remain. The reaction of a strong acid and strong base therefore always produces water and an ionic compound called a **salt.** This reaction is called a **neutralization** reaction because the acid and base neutralize each other when they react to produce water. We can write the neutralization reaction for HCl and NaOH as follows:

$$\underset{\substack{\text{Strong} \\ \text{acid}}}{HCl(aq)} + \underset{\substack{\text{Strong} \\ \text{base}}}{NaOH(aq)} \longrightarrow \underset{\text{Salt}}{NaCl(aq)} + \underset{\text{Water}}{H_2O(l)}$$

Note that the chemical equation is balanced as written; that is, the number of atoms on the reactant side is equal to the number of atoms on the product side.

Neutralization reactions can proceed even if the base hydroxide is not in the aqueous phase. A strong acid like HCl can react with iron(III) hydroxide, a component in rust, in a neutralization to remove the solid rust:

$$3HCl(aq) + Fe(OH)_3(s) \longrightarrow 3H_2O(l) + FeCl_3(aq)$$

Completing a Neutralization Reaction

Solving a Problem

Complete and balance the following neutralization reaction.

$$HNO_3(aq) + KOH(s) \longrightarrow$$

Solution

STEP 1 Form the products. The products will always be (a) a salt and (b) H_2O. The salt produced must be a neutral ionic compound (refer to Section 3.3 for a refresher on forming ionic compounds like salts). The potassium ion has a 1+ charge (potassium is a Group 1A element), and the nitrate anion has a 1− charge (see **TABLE 3.2**), so the salt formed has a 1:1 ratio of cations to anions with a formula of KNO_3.

$$\underset{\substack{\text{Strong} \\ \text{acid}}}{HNO_3(aq)} + \underset{\substack{\text{Strong} \\ \text{base}}}{KOH(s)} \longrightarrow \underset{\text{Salt}}{KNO_3(aq)} + \underset{\text{Water}}{H_2O(l)}$$

STEP 2 Balance the chemical equation. Balancing is done by adding coefficients in front of the product or reactant compounds where appropriate. Remember that the same number of atoms must appear in both the reactants and products. If we inspect this reaction after forming the salt, we see that the same number of each atom is present (two H, four O, one N, and one K), so no further balancing is needed. ■

Sample Problem 9.3   Completing a Neutralization Reaction

Complete and balance the following neutralization reaction.

$$HBr(aq) + Ca(OH)_2(s) \longrightarrow$$

STEP 1 Form the products. The calcium ion has a 2+ charge (a Group 2A element), and the bromide anion has a 1− charge (a Group 7A element), so the salt formed combines in a 1:2 ratio of cations to anions with a formula of $CaBr_2$.

$$\underset{\substack{\text{Strong} \\ \text{acid}}}{HBr(aq)} + \underset{\substack{\text{Strong} \\ \text{base}}}{Ca(OH)_2(s)} \longrightarrow \underset{\text{Salt}}{CaBr_2(aq)} + \underset{\text{Water}}{H_2O(l)}$$

STEP 2 Balance the chemical equation. Two atoms of Br appear on the product side; therefore, a coefficient of 2 is needed on the reactant side in front of the HBr. Adding the coefficient produces a total of 4 H atoms and 2 O atoms on the reactant side, so a coefficient of 2 is also needed in front of the H_2O on the product side. Now the equation is balanced.

$$2HBr(aq) + Ca(OH)_2(s) \longrightarrow CaBr_2(aq) + 2H_2O(l)$$

Antacids

Antacids are substances that are used to neutralize excess stomach acid (HCl). Some antacids are mixtures of aluminum hydroxide and magnesium hydroxide. These hydroxides are not very soluble in water and therefore are only weak bases, and the levels of available OH^- are not damaging to the intestinal tract. However, aluminum hydroxide has the side effects of producing constipation and binding phosphate in the intestinal tract, which may cause weakness and loss of appetite. Magnesium hydroxide has a laxative effect. These side effects are less likely when a combination is used.

$$Mg(OH)_2(s) + 2HCl(aq) \rightarrow MgCl_2(aq) + 2H_2O(l)$$

INTEGRATING Chemistry

◄ Find out how antacids neutralize stomach acid and their side effects.

—*Continued next page*

Continued—

Some antacids use carbonates to neutralize excess stomach acid. When carbonates are used to neutralize acid, the reaction produces a salt, water, and carbon dioxide gas.

$$CaCO_3(s) + 2HCl(aq) \rightarrow CaCl_2(aq) + H_2O(l) + CO_2(g)$$

When calcium carbonate is used, about 10% of the calcium is absorbed into the bloodstream, where it elevates the levels of serum calcium. Calcium carbonate is not recommended for people who have peptic ulcers or a tendency to form kidney stones.

Sodium bicarbonate can affect the acidity level of the blood and elevate sodium levels in body fluids. It is also not recommended in the treatment of peptic ulcers.

The neutralizing substances in some antacid preparations are shown in **TABLE 9.3**.

TABLE 9.3 Basic Compounds in Some Antacids

Antacid	Base(s)
Amphojel®	$Al(OH)_3$
Milk of magnesia	$Mg(OH)_2$
Mylanta®, Maalox®, Di-Gel™, Gelusil®, Riopan®	$Mg(OH)_2$, $Al(OH)_3$
Bisodol®	$CaCO_3$, $Mg(OH)_2$
Titralac™, Tums®, Pepto-Bismol®	$CaCO_3$
Alka-Seltzer®	$NaHCO_3$, $KHCO_3$

Practice Problems

9.7 Which of the following are strong acids?
a. H_2SO_4 b. HCl   c. HF

9.8 Which of the following are strong acids?
a. HNO_2 b. HNO_3 c. H_3PO_4

9.9 Which of the following are strong bases?
a. NH_3 b. KOH   c. LiOH

9.10 Which of the following are strong bases?
a. $Mg(OH)_2$ b. NaOH   c. $Fe(OH)_3$

9.11 Which of the following compounds completely ionize in water?
a. HNO_3 b. NH_4OH c. HF

9.12 Which of the following compounds completely ionize in water?
a. LiOH
b. H_2O
c. benzoic acid, C_6H_5COOH

9.13 Name and write the formula for the acids formed from the following anions.
a. bromide, Br^-
b. chlorate, ClO_3^-
c. nitrite, NO_2^-

9.14 Name and write the formula for the acids formed from the following anions.
a. iodide, I^-
b. perchlorate, ClO_4^-
c. acetate, CH_3COO^-

9.15 Complete and balance the following neutralization reactions:
a. $HNO_3(aq) + LiOH(s) \longrightarrow$
b. $H_2SO_4(aq) + Ca(OH)_2(s) \longrightarrow$

9.16 Complete and balance the following neutralization reactions:
a. $HNO_3(aq) + Mg(OH)_2(s) \longrightarrow$
b. $HCl(aq) + NaHCO_3(s) \longrightarrow$

9.17 Complete and balance the following neutralization reactions:
a. $HBr(aq) + Al(OH)_3(s) \longrightarrow$
b. $HI(aq) + CaCO_3(s) \longrightarrow$

9.18 Complete and balance the following neutralization reactions:
a. $H_2SO_4(aq) + NaOH(s) \longrightarrow$
b. $HBr(aq) + LiOH(s) \longrightarrow$

9.3 Chemical Equilibrium

 9.3 Inquiry Question

How do the properties of equilibrium apply to Le Châtelier's principle?

Before we continue our discussion of weak acids and bases, let's consider the general principles of chemical equilibrium. Recall that we have touched on physical equilibrium when discussing vapor pressure in Section 7.2 and dissolution in Section 8.2. While chemical equilibrium is similar, it involves a chemical reaction instead of a physical change. Let's look at a simple example to better understand the state of equilibrium.

Imagine a concert with general-admission seating. Everyone tries to arrive in plenty of time to get a seat. When the doors open, people rush in and fill up all the seats in the concert hall. When the hall is full, event security stops people from entering the hall even though some people who waited in line did not get a seat. During the concert, the security personnel allow people to leave and others to enter to keep the hall full but not overfilled. This steady flow of people in and out of the concert results in a state of dynamic equilibrium where the rate at which people entering the hall and the rate at which people leaving are the same. Note that the number of people in the concert hall stays the same once the hall fills, even though the individuals in the concert hall differ.

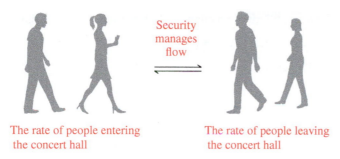

The rate of people entering the concert hall

The rate of people leaving the concert hall

When the rate of people entering (like reactants forming products) and leaving (like products forming reactants) a concert hall is the same, a state of equilibrium exists, even if the individuals inside and outside of the concert hall (individual molecules reacting) at any point in time are different.

Like a concert, some chemical reactions will, after forming product (people entering the concert), reverse and re-form reactants (people leaving the concert). These are reversible reactions. To explore these reversible reactions more, let's start with a chemical reaction for the generation of ammonia:

$$N_2(g) + 3H_2(g) \rightleftharpoons 2NH_3(g)$$

Reactants   Equilibrium   Products
                      arrow

The generation of ammonia is a reversible reaction. Once ammonia is formed, the reaction will reverse, re-forming nitrogen and hydrogen. Eventually, the rate of the formation of ammonia (the forward reaction) and the rate of re-formation of nitrogen and hydrogen gases (the reverse reaction) become equal. This balance of the rates of the reactions is called **chemical equilibrium.** A special type of reaction arrow, called an equilibrium arrow (shown in the equation), is used in this chemical equation to indicate that both the forward and reverse reactions can take place simultaneously. This does not mean that the *amounts* of ammonia and of nitrogen and hydrogen are equal, but because the rates of the forward and reverse reactions are equal, there is no net change in amounts. In other words, the amounts of products and reactants stay the same once equilibrium is reached.

The Equilibrium Constant, *K*

Going back to the concert example, suppose that the concert hall remains completely full for the duration of the concert and the number of people waiting to get in remains the same. The concert attendance has reached equilibrium, and while there may be may be more people in the concert hall than those who are waiting to get in, we could observe that the ratio of people inside the concert hall to people outside the concert hall would be constant.

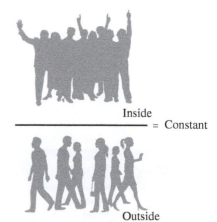

Inside
──────────── = Constant
Outside

Like the value for the equilibrium constant, *K*, the ratio of people inside the concert hall (products) to those outside (reactants) is constant for a particular concert (reaction).

Similarly, the ammonia reaction shown reaches equilibrium. If we measure the concentrations of ammonia, nitrogen, and hydrogen present, the ratio of products to reactants would be constant. This value is called the **equilibrium constant, K**, and it is a characteristic of equilibrium reactions at a given temperature. K is defined as

$$K = \frac{[\text{Products}]}{[\text{Reactants}]}$$

The brackets, [ ], mean "molar concentration of," or molarity (see Section 8.4). So, the equation states that the equilibrium constant, K, is equal to the molar concentration of the products divided by that of the reactants. If there is more than one reactant or product, the concentrations are multiplied. The expression for K is shown for the generation of ammonia. The superscripts in the expression come from the coefficients (number of moles of each) found in the balanced chemical equation.

$$K = \frac{[\text{NH}_3]^2}{[\text{N}_2][\text{H}_2]^3}$$

In general, for an equilibrium reaction of the form

$$a\text{A} + b\text{B} \rightleftharpoons c\text{C} + d\text{D}$$

the general equilibrium expression is given as

$$K = \frac{[\text{C}]^c[\text{D}]^d}{[\text{A}]^a[\text{B}]^b}$$

Only substances whose concentrations change appear in an equilibrium expression. Solids (s) and pure liquids (l) have constant concentrations and do not appear in the equilibrium expression.

Sample Problem 9.4 Equilibrium Constant Expressions

Write an equilibrium constant expression for the following reactions:
a. $\text{CH}_4(g) + \text{H}_2\text{O}(g) \rightleftharpoons \text{CO}(g) + 3\text{H}_2(g)$
b. $\text{CaCO}_3(s) \rightleftharpoons \text{CO}_2(g) + \text{CaO}(s)$

Solution

K is defined as the ratio of product concentrations to reactant concentrations. Any coefficients in front of molecules appear as superscripts in the expression.

a. $K = \dfrac{[\text{CO}][\text{H}_2]^3}{[\text{CH}_4][\text{H}_2\text{O}]}$

H$_2$O appears in this expression because it appears as a gas in the equilibrium equation.

b. $K = [\text{CO}_2]$

Because CaCO$_3$ and CaO are solids, they have constant concentrations and do not appear in the equilibrium expression.

TABLE 9.4 Interpreting Values of K

Value of K	Predominant at Equilibrium
$K = 1$	Equal amounts of products and reactants
$K > 1$	Products predominate
$K < 1$	Reactants predominate

The value of K for the formation of ammonia is 9.6 at 300 °C. A value of 9.6 indicates a ratio of 9.6:1 for products:reactants. Values for K vary greatly depending on temperature and reaction. If K has a value equal to 1, the ratio of products:reactants is 1:1, or the $[\text{products}] = [\text{reactants}]$. A value of K greater than ($>$) 1 indicates that the amount of products (numerator) is larger than the amount of reactants (denominator), or $[\text{products}] > [\text{reactants}]$. A value of K less than ($<$) 1 indicates that the amount of reactants (denominator) is larger than the amount of products (numerator), or $[\text{products}] < [\text{reactants}]$ (see **TABLE 9.4**).

Sample Problem 9.5 **Interpreting Values of _K_**

K for the following reaction at 25 °C is 1×10^2.

$$CO(g) + H_2O(g) \rightleftharpoons H_2(g) + CO_2(g)$$

Based on the value for K, at equilibrium are the products or reactants greater?

Solution

Because the value of $K > 1$, the products are present in greater amounts.

Effect of Concentration on Equilibrium— Le Châtelier's Principle

Suppose that we had our reaction for the generation of ammonia sitting in a reaction vessel and that it was at equilibrium. What would happen if we decided to inject more nitrogen into the vessel?

Mastering Video
Practicing the Concepts
Le Châtelier's Principle

$$N_2(g) + 3H_2(g) \rightleftharpoons 2NH_3(g)$$

According to **Le Châtelier's principle** (pronounced *leh-shat-lee-AYS*), if we disturb an equilibrium—chemists refer to this as applying stress to the equilibrium—the rate of the forward or reverse reaction will change to offset the stress and regain equilibrium. Applying this principle, if we add more N_2 to our system—on the reactant side of the equation below—the rate of the forward reaction will increase, shifting the equilibrium to produce more products (see the orange arrows). This rate increase occurs because adding more N_2 molecules increases the chance that N_2 will collide with H_2 molecules, forming NH_3 more rapidly.

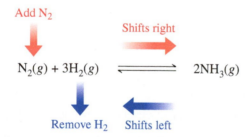

As a contrast, suppose that we remove one of the reactants, H_2. To regain equilibrium, the reverse reaction must be faster than the forward reaction, allowing the H_2 to be replenished. The equilibrium shifts to the left, forming more of the reactants and reestablishing the equilibrium between the two rates (see the blue arrows in the equation above).

In general, we can think of equilibrium as a balancing act between forward and reverse reactions. If one side of the reaction gains a substance, the reaction shifts to the other side to reestablish equilibrium. If one side of the reaction loses a substance, the reaction shifts toward that side to regain its equilibrium.

Effect of Temperature on Equilibrium

What will happen to the equilibrium of our ammonia reaction if we change the temperature of the reaction? First, we have to know whether the reaction itself is one that produces heat, an exothermic reaction, or one that absorbs heat from its surroundings, an endothermic reaction. The generation of ammonia is known to be an exothermic reaction. The chemical equation can be written as follows to signify that heat is produced:

$$N_2(g) + 3H_2(g) \rightleftharpoons 2NH_3(g) + Heat$$

Because energy in the form of heat is a product of the reaction, if the temperature of the reaction is raised (i.e., if heat is added), the rate of the reverse reaction increases to offset the stress of adding heat. This causes the equilibrium to shift to the left. If the reaction is cooled down (i.e., if heat is removed), the rate of the forward reaction increases to replenish the heat energy produced, shifting the equilibrium to the right.

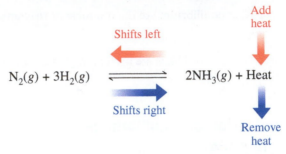

In an endothermic reaction (one that absorbs heat energy from its surroundings), like the reaction for production of NO gas shown, heat appears as a reactant, so the opposite shifts occur. **TABLE 9.5** summarizes the effects of stress on a chemical equilibrium.

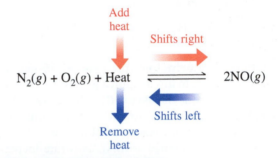

TABLE 9.5 Effects of Changes on Equilibrium

Factor	Change (stress)	Reaction Shifts Toward
Concentration	Add reactant	Right
	Remove reactant	Left
	Add product	Left
	Remove product	Right
Temperature	Raise temperature of endothermic reaction	Right
	Lower temperature of endothermic reaction	Left
	Raise temperature of exothermic reaction	Left
	Lower temperature of exothermic reaction	Right

Sample Problem 9.6 **Factors Affecting Equilibrium**

How does each of the following actions affect the equilibrium of the reaction shown?

$$2SO_2(g) + O_2(g) \rightleftharpoons 2SO_3(g) + Heat$$

a. add O_2 b. lower the temperature   c. add SO_3

Solution

a. The addition of O_2 increases the forward reaction; the equilibrium shifts to the right.
b. Lowering the temperature removes heat, which favors the exothermic direction, so the equilibrium shifts to the right.
c. The addition of SO_3 favors the reverse reaction toward reactants; the equilibrium shifts to the left.

Practice Problems

9.19 What is meant by the term *reversible reaction*?

9.20 When does a reversible reaction reach equilibrium?

9.21 Write an equilibrium constant expression for the following reactions:

a. $CO(g) + H_2O(g) \rightleftharpoons H_2(g) + CO_2(g)$

b. $CH_3COOH(aq) + H_2O(l) \rightleftharpoons H_3O^+(aq) + CH_3COO^-(aq)$

9.22 Write an equilibrium constant expression for the following reactions:

a. $N_2(g) + 3Br_2(g) \rightleftharpoons 2NBr_3(g)$

b. $C(s) + O_2(g) \rightleftharpoons CO_2(g)$

9.23 For the following values of K, indicate whether the products or reactants are present in larger amounts:

a. 1×10^{-5} b. 156   c. 1

9.24 For the following values of K, indicate whether the products or reactants are present in larger amounts:

a. 1×10^7 b. 0.0045   c. 0.00000079

9.25 Oxygen and hydrogen gas can be generated if electrical energy is added through the electrolysis of water.

$$2H_2O(l) + Energy \rightleftharpoons 2H_2(g) + O_2(g)$$

In which direction will the equilibrium shift (right or left) if the following changes occur to the reaction at equilibrium?

a. add O_2 b. add more electrical energy

c. remove H_2 d. add H_2O

9.26 Lead(II) iodide can be produced upon cooling lead(II) and iodide ions from a solution.

$$2I^-(aq) + Pb^{2+}(aq) \rightleftharpoons PbI_2(s) + Heat$$

In which direction will the equilibrium shift (right or left) if the following changes occur to the reaction at equilibrium?

a. add iodide (I^-)   b. heat the reaction

c. remove lead(II) iodide   d. cool the reaction

9.27 In the lower atmosphere, oxygen is converted to ozone (O_3) by the energy provided from lightning.

$$3O_2(g) + Heat \rightleftharpoons 2O_3(g)$$

In which direction will the equilibrium shift (right or left) if the following changes occur to the reaction at equilibrium?

a. add O_3

b. add O_2

c. raise temperature

d. lower temperature

9.28 When heated, carbon reacts with water to produce carbon monoxide and hydrogen.

$$C(s) + H_2O(g) + Heat \rightleftharpoons CO(g) + H_2(g)$$

In which direction will the equilibrium shift (right or left) if the following changes occur to the reaction at equilibrium?

a. add heat

b. lower temperature

c. remove CO

d. add H_2O

9.29 After you open a bottle of soda, the drink eventually goes flat. How can Le Châtelier's principle explain this using the following reaction?

$$H^+(aq) + HCO_3^-(aq) \rightleftharpoons CO_2(g) + H_2O(l)$$

9.30 When you exercise, energy is produced by increasing the rate of the following reaction involving glucose. Why do you breathe faster when you exercise?

$$C_6H_{12}O_6(aq) + 6O_2(g) \rightleftharpoons$$
$$6CO_2(g) + 6H_2O(l) + Energy$$

DISCOVERING THE CONCEPTS

INQUIRY ACTIVITY—Weak Acids

Information

Weak acids and bases establish an equilibrium in solution. In contrast to strong acids, only a few molecules are dissociated (<5%). Some weak acids can donate more than one proton. In this case, more than one equilibrium is established. These acids are called *polyprotic* (literally "many proton") acids. The Brønsted–Lowry definition is useful to describe the behavior of weak acids and bases.

—*Continued next page*

Continued—

Acid—Any substance that can donate a proton (H^+) to another substance. When an acid donates a proton, a conjugate base is produced.
Base—Any substance that can accept a proton (H^+) from another substance. When a base accepts a proton, a conjugate acid is produced.

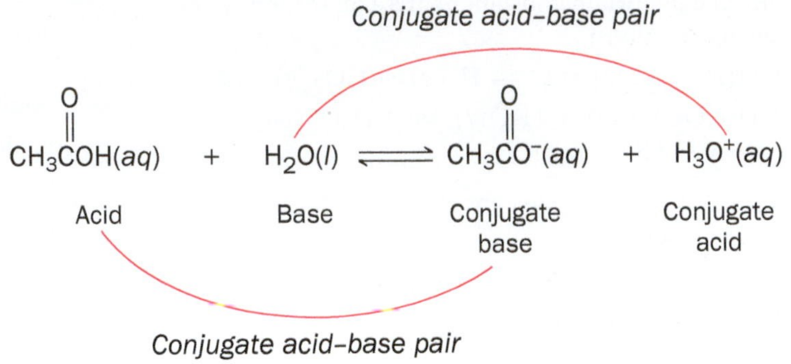

Conjugate acid–base pair

$$CH_3COH(aq) \ + \ H_2O(l) \ \rightleftharpoons \ CH_3CO^-(aq) \ + \ H_3O^+(aq)$$

| Acid | Base | Conjugate base | Conjugate acid |

Conjugate acid–base pair

Questions

1. Does the acid in the example donate or accept a proton when forming the conjugate base?
2. Does the conjugate base in the example donate or accept a proton when re-forming the acid?
3. Label the acid, base, conjugate acid, and conjugate base in the reaction of ammonia with water shown.

$$NH_3(aq) + H_2O(l) \rightleftharpoons NH_4^+(aq) + OH^-(aq)$$

4. Provide a definition for conjugate *acid–base pair*.
5. Provide the acid–base reaction for lactic acid ($CH_3CH(OH)COOH$), a weak acid, when it reacts with water. Label the acid, base, conjugate acid, and conjugate base.

$$\text{Lactic acid} + H_2O(l) \rightleftharpoons$$

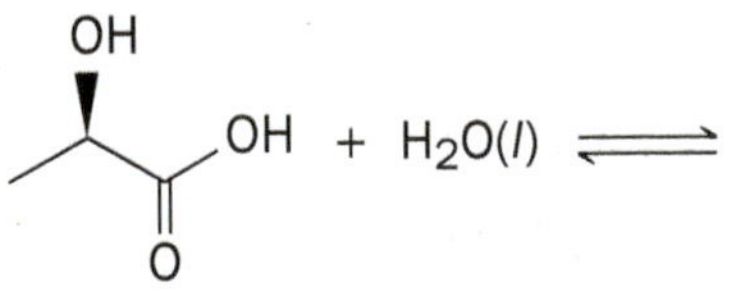

Lactic acid (*aq*)

6. Provide the
   a. conjugate base of H_2S. _____
   b. conjugate acid of OH^-. _____
   c. conjugate acid of HPO_4^{2-}. _____
   d. conjugate base of HCO_3^-. _____

9.4 Weak Acids and Bases

9.4 Inquiry Question

How are weak acid–base equations written?

The principles of equilibrium from Section 9.3 apply to weak acids and bases because weak acids and bases only partially dissociate into ions, establishing an equilibrium in aqueous solution. For example, the dissociation of the weak acid acetic acid (CH_3COOH) into acetate anions (CH_3COO^-) and hydronium ions is as follows:

$$CH_3COH(aq) + H_2O(l) \rightleftharpoons CH_3CO^-(aq) + H_3O^+(aq)$$

The equilibrium constant expression representing this reaction is

$$K = \frac{[CH_3COO^-][H_3O^+]}{[CH_3COOH]}$$

Remember that pure liquids like $H_2O(l)$ are present in large amounts that do not change significantly as a reaction approaches equilibrium. Therefore, they are considered constant and do not appear in the equilibrium constant expression.

The Equilibrium Constant K_a

All weak acids dissociate similarly in water by donating a proton, forming hydronium ion. However, because dissociation is much less than 100%, a weak acid has a special equilibrium constant for the dissociation of H^+ called the **acid dissociation constant**, K_a. The K_a value for acetic acid is 1.75×10^{-5}. Notice that this number is less than 1, which indicates that more acetic acid molecules are present at equilibrium than acetate anions. Because all weak acids can attain equilibrium in solution, they all have a K_a value at a given temperature. Each weak acid dissociates to a different extent, so the K_a values are different from acid to acid (see **TABLE 9.6**).

We can predict the strength of weak acids from K_a values. The larger the K_a, the stronger the acid (the more protons dissociated). Recall the two weak acids mentioned in Section 8.3 that are weak electrolytes: carboxylic acids and protonated amines. Their dissociation reactions with water are shown in **FIGURE 9.2**.

TABLE 9.6 K_a Values for Substances Acting as Weak Acids (25 °C)

Name	Formula	K_a
Hydrogen sulfate ion	HSO_4^-	1.0×10^{-2}
Phosphoric acid	H_3PO_4	7.5×10^{-3}
Hydrofluoric acid	HF	6.5×10^{-4}
Nitrous acid	HNO_2	4.5×10^{-4}
Formic acid	$HCOOH$	1.8×10^{-4}
Lactic acid	$CH_3CH(OH)COOH$	1.4×10^{-4}
Acetic acid	CH_3COOH	1.75×10^{-5}
Carbonic acid	H_2CO_3	4.5×10^{-7}
Dihydrogen phosphate ion	$H_2PO_4^-$	6.6×10^{-8}
Ammonium ion	NH_4^+	6.3×10^{-10}
Hydrocyanic acid	HCN	6.2×10^{-10}
Bicarbonate ion	HCO_3^-	4.8×10^{-11}
Hydrogen phosphate ion	HPO_4^{2-}	1.0×10^{-12}
Water	H_2O	1.0×10^{-14}

Increasing acid strength

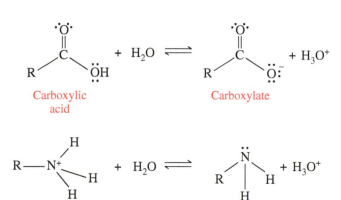

FIGURE 9.2 Functional groups as acids and bases. Two common organic functional groups act as weak acids in aqueous solution: carboxylic acids and protonated amines.

Sample Problem 9.7 Strength of Weak Acids

Of the following acids—H_3PO_4, HCO_3^- and CH_3COOH—which is (a) the strongest? (b) the weakest?

Solution

Use the K_a values in Table 9.6. The larger values (closer to 1) are stronger acids.
  a. The strongest acid of the group is H_3PO_4, with a K_a value of 7.5×10^{-3}.
  b. The weakest acid of the group is HCO_3^-, with a K_a value of 4.8×10^{-11}.

Conjugate Acids and Bases

According to the Brønsted–Lowry theory, the reaction between an acid and a base involves a proton transfer, so if a weak acid mixes with water, water will act as a base and accept a proton. Consider the dissociation of acetic acid, CH_3COOH:

When two weak acids such as acetic acid and water are mixed, the stronger weak acid (acetic acid) donates a H^+ to the stronger weak base (H_2O). The acetate ion formed is the weaker of the two bases.

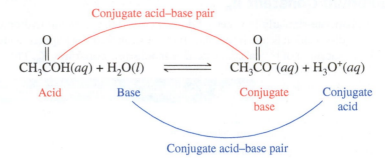

The acid CH_3COOH donates a proton to a molecule of water that accepts the proton, forming a hydronium ion, H_3O^+. The anion that remains after the donation, CH_3COO^- —a carboxylate called acetate anion—is the **conjugate base** of CH_3COOH. The term *conjugate base* comes from the fact that in the reverse reaction, the CH_3COO^- acts as a base and accepts the proton from the hydronium ion to form CH_3COOH. Likewise, when the water acts as a base, accepting a proton from acetic acid, a hydronium ion is formed. Hydronium ion is called the **conjugate acid** in this reaction because, during the reverse reaction, hydronium ion acts as an acid, donating its proton to the acetate anion. Molecules or ions related by the loss or gain of one H^+ are referred to as **conjugate acid–base pairs.** The functional groups carboxylic acid and carboxylate are conjugate acid–base pairs. Weak acids are generically designated as HA and their conjugate base as A^-.

Both CH_3COOH and H_2O have K_a values and are weak acids, so how do we know which one acts as the acid? In a reaction like this, the stronger weak acid (greater K_a) will act as the acid. Since CH_3COOH has a larger K_a than H_2O, it reacts as the acid, forming a conjugate base. Because H_2O is the weaker acid, it is in turn the stronger base of the reactants and will accept the H^+ when CH_3COOH dissociates.

A weak base such as an amine will accept a proton from water to form a protonated amine, in which case the water acts as an acid. The protonated amine and amine are conjugate acid–base pairs of the same functional group.

In this case, the H_2O is the stronger weak acid, donating a proton, and the amine is the stronger weak base, accepting a proton. Conjugate acid–base pairs differ by one H^+.

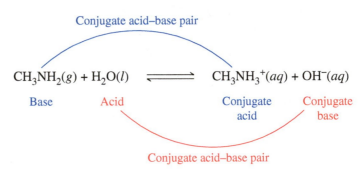

Sample Problem 9.8 Identifying Conjugate Acid–Base Pairs

Label the acid, base, conjugate acid, and conjugate base in the following reaction:

$$HF(aq) + H_2O(l) \rightleftharpoons F^-(aq) + H_3O^+(aq)$$

Solution

Compare HF on the reactant side to F^- on the product side. A proton is donated from the HF, leaving F^- as a product. So HF is acting as an acid. The product (F^-) is the conjugate base. Similarly, H_2O is accepting a proton to form the hydronium ion, so water is acting as a base, producing a conjugate acid, the hydronium ion.

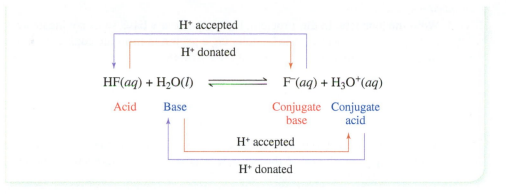

Sample Problem 9.9 Determining Formulas of Conjugates

Provide the following:

  a. conjugate acid of OH^- b. conjugate acid of HPO_4^{2-}

  c. conjugate base of H_2S d. conjugate base of HCO_3^-

Solution

Keep in mind the definitions of Brønsted–Lowry acids and bases when trying to answer this type of problem.

  a. To find the conjugate acid of OH^-, OH^- would be acting as a base by accepting a proton. The addition of H^+ to OH^- produces the neutral molecule H_2O as its conjugate acid because the $1+$ and $1-$ charges cancel out.

  b. If HPO_4^{2-} acts as a base, it would be accepting a proton (H^+), thereby forming $H_2PO_4^-$ as its conjugate acid. (Note that when H^+ and HPO_4^{2-} combine, the charge changes from $2-$ to $1-$.)

  c. To find the conjugate base of H_2S, H_2S would be acting as an acid by donating a proton, H^+. When a proton is donated, what is left over from the H_2S is HS^-. The $1-$ charge is left behind when H^+ is donated.

  d. If HCO_3^- acts as an acid, it would be donating a proton, forming CO_3^{2-} as its conjugate base. The donation of H^+ leaves behind a negative charge, making the total charge on the conjugate base $2-$.

Writing Weak Acid–Base Equations

Solving a Problem

Complete the following reaction if $H_2PO_4^-$ acts as a base. Label the acid, base, conjugate acid, and conjugate base for the following reaction:

$$H_2PO_4^- (aq) + H_2O(l) \rightleftharpoons$$

Mastering Video
Practicing the Concepts
Identifying Conjugate Acids and Bases

Solution

Keep in mind the following rules to correctly write and balance a weak acid–base equilibrium equation.

    Rules for writing products of a weak acid–base equation.

  • The conjugates will always appear on the product side of a chemical equilibrium.

  • When a proton is donated from the acid, the conjugate base formed will have one more negative charge than the acid from which it was formed.

  • When a proton is accepted by the base, the conjugate acid formed will have one more positive charge than the base from which it was formed.

  • The total charge of the reactants equals the total charge of the products.

—Continued next page

Continued—

STEP 1 Write the products. In this problem, $H_2PO_4^-$ acts as a base, so its conjugate acid will be H_3PO_4. The water acts as an acid, donating a proton to form the conjugate base OH^-. Using the rules outlined, the final equation is written

$$H_2PO_4^-(aq) \ + \ H_2O(l) \ \rightleftharpoons \ H_3PO_4(aq) \ + \ OH^-(aq)$$

Base   Acid   Conjugate   Conjugate

acid   base

STEP 2 Check your work. If you have assigned charges correctly, the total charge on either side of the equation should be the same. Here, the reactants have a total charge of $1-$, and the products have a total charge of $1-$.

Sample Problem 9.10   Writing Weak Acid–Base Equations

Complete the following reaction if HSO_4^- acts as an acid. Label the acid, base, conjugate acid, and conjugate base for the following reaction:

$$HSO_4^-(aq) \ + \ H_2O(l) \ \rightleftharpoons$$

Solution

Keep in mind the rules for correctly writing and balancing a weak acid–base equilibrium equation.

a. **Write the products.** In this problem, the hydrogen sulfate ion (HSO_4^-) is acting as an acid and will donate its final proton, yielding sulfate, SO_4^{2-}. The water will act as a base, accepting the donated proton, forming hydronium, H_3O^+.

$$HSO_4^-(aq) \ + \ H_2O(l) \ \rightleftharpoons \ SO_4^{2-}(aq) \ + \ H_3O^+(aq)$$

Acid   Base   Conjugate   Conjugate

base   acid

b. **Check your work.** The total charge on either side of the equation should be the same. In this case, both the reactants and products have a charge of $1-$.

INTEGRATING Chemistry

Find out how blood ▶ acidity affects oxygen binding and transport.

Weak Acids, Oxygen Transport, and Le Châtelier's Principle

The binding, transport, and release of oxygen through the body are, in part, controlled by weak acid–base equilibria. The protein hemoglobin (abbreviated Hb) carries oxygen from the lungs to the tissues via the bloodstream. Both protons (H^+) and oxygen (O_2) bind to hemoglobin, but with opposite affinity, so one or the other is transported. This reaction can be represented by the equilibrium equation shown in the figure below.

If the concentrations of either acid or oxygen change, the equilibrium will shift. For example, at the lungs, the oxygen concentration is higher, so the equilibrium shifts to the right of the equation shown, binding more oxygen and releasing H^+. As the oxygenated hemoglobin (HbO_2) reaches the tissues where levels of H^+ are slightly higher, oxygen is released, shifting the equilibrium to the left and forming more of the deoxygenated hemoglobin (HHb^+) for transport back to the lungs where the process repeats.

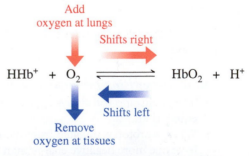

The equilibrium of deoxygenated hemoglobin (HHb^+) and oxygenated hemoglobin (HbO_2) shifts depending on the amount of oxygen and protons present.

Practice Problems

9.31 Using Tables 9.1 and 9.6, identify the stronger acid in each pair:

a. $H_2PO_4^-$ or HPO_4^{2-} b. HF or HCOOH

c. HBr or HNO_2

9.32 Using Tables 9.1 and 9.6, identify the stronger acid in each pair:

a. HCN or H_2CO_3 b. H_3PO_4 or $H_2PO_4^-$

c. HCl or HSO_4^-

9.33 Identify the acid and base on the reactant side of the following equations and identify the conjugate acid and base on the product side:

a. $HSO_4^-(aq) + H_2O(l) \rightleftharpoons H_3O^+(aq) + SO_4^{2-}(aq)$

b. $NH_4^+(aq) + H_2O(l) \rightleftharpoons H_3O^+(aq) + NH_3(aq)$

c. $HCN(aq) + NO_2^-(aq) \rightleftharpoons CN^-(aq) + HNO_2(aq)$

9.34 Identify the acid and base on the reactant side of the following equations and identify the conjugate acid and base on the product side:

a. $H_3PO_4(aq) + H_2O(l) \rightleftharpoons H_3O^+(aq) + H_2PO_4^-(aq)$

b. $CO_3^{2-}(aq) + H_2O(l) \rightleftharpoons OH^-(aq) + HCO_3^-(aq)$

c. $H_3PO_4(aq) + NH_3(aq) \rightleftharpoons NH_4^+(aq) + H_2PO_4^-(aq)$

9.35 Write the formula and name of the conjugate base formed from each of the following acids:

a. HF   b. H_2O

c. H_2CO_3 d. HSO_4^-

9.36 Write the formula and name of the conjugate base formed from each of the following acids:

a. HCO_3^- b. H_3O^+

c. HPO_4^{2-} d. HNO_2

9.37 Write the formula and name of the conjugate acid formed from each of the following bases:

a. CO_3^{2-} b. H_2O

c. $H_2PO_4^-$ d. Br^-

9.38 Write the formula and name of the conjugate acid formed from each of the following bases:

a. SO_4^{2-} b. CN^-

c. OH^- d. ClO_2^-, chlorite ion

9.39 Complete the following reactions and identify the conjugate acid–base pairs if the named reactant acts as an acid.

a. $HCOOH(aq) + H_2O(l) \rightleftharpoons$

Formic acid, found in ants and bee stings

b. $H_2PO_4^-(aq) + H_2O(l) \rightleftharpoons$

Dihydrogen phosphate, a component of cell buffering

c. $HCO_3^-(aq) + H_2O(l) \rightleftharpoons$

Bicarbonate, a component in blood buffering

9.40 Complete the following reactions and identify the conjugate acid–base pairs if the named reactant acts as a base.

a. $\overset{\displaystyle O}{\overset{\|}{CH_3C}}COO^-(aq) + H_2O(l) \rightleftharpoons$

Pyruvate, a product of glucose metabolism

b. $H_2PO_4^-(aq) + H_2O(l) \rightleftharpoons$

Dihydrogen phosphate, a component in cell buffering

c. $HCO_3^-(aq) + H_2O(l) \rightleftharpoons$

Bicarbonate, a component in blood buffering

9.5 pH and the pH Scale

The Autoionization of Water, K_w

We have seen that water can act as a weak acid or base depending on whether a base or an acid is present in the solution. In pure water, the molecules set up an equilibrium, reacting with each other by donating and accepting protons.

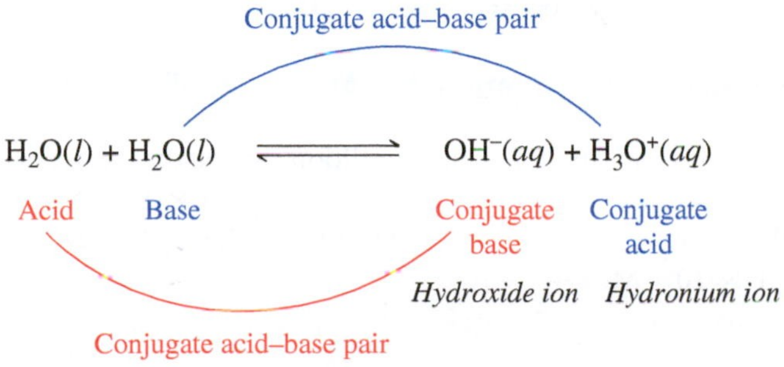

9.5 Inquiry Question

How is the acidity of a solution determined?

Conjugate acid–base pair

$$H_2O(l) + H_2O(l) \rightleftharpoons OH^-(aq) + H_3O^+(aq)$$

Acid   Base   Conjugate   Conjugate

                      base   acid

Hydroxide ion   Hydronium ion

Conjugate acid–base pair

Water can act as an acid or base, forming hydroxide and hydronium ions at equilibrium.

This reaction, which is always present in water, is called the **autoionization of water.** If we were to write the equilibrium constant expression for water, K_w (keeping in mind that pure liquids do not appear), we would write

$$K_w = [OH^-][H_3O^+]$$

Wait—didn't we say in Chapter 8 that pure water does not conduct electricity because there are no dissolved ions present? Now we are saying that ions *are* found in pure water? Let's consider the measured value for K_w:

$$K_w = 1 \times 10^{-14} \text{ at } 25\,°C$$

Remembering the definition of K ([products]/[reactants]) and recognizing that K_w is an extremely small value tell us that the ionized products are found in very small amounts in pure water. In fact, K_w is 100 times smaller than the weakest acid found in Table 9.6. Pure water has some H_3O^+ and OH^- ions present, but the amounts are so small that there are not enough ions present in pure water to conduct electricity.

Because the autoionization of water always occurs in aqueous solution, all aqueous solutions have small amounts of H_3O^+ and OH^- present.

$[H_3O^+]$, $[OH^-]$, and pH

In pure water, both the hydroxide and hydronium ion are being formed equally from the transfer of protons between water molecules, so $[H_3O^+] = [OH^-]$. At 25 °C, both of these these values are 1×10^{-7} M. Recall that "M" is the concentration unit molarity, or moles per liter. When these concentrations are equal, the solution is **neutral.** Keeping in mind that K_w is constant, if an acid is added to water, there is an increase in $[H_3O^+]$ and a decrease in $[OH^-]$, making the solution acidic. If base is added, $[OH^-]$ increases and $[H_3O^+]$ decreases, making a basic solution (see **FIGURE 9.3**). Most aqueous solutions are *not* neutral and have unequal concentrations of H_3O^+ and OH^-.

The amount of H_3O^+ in an aqueous solution defines the acidity of the solution. Concentrations of H_3O^+ usually range from about 1 M to 1×10^{-14} M (really close to zero). This range of numbers is extremely wide. For this reason, the mathematical function "log" is used to compare $[H_3O^+]$ because it gives a set of numbers that usually falls between 0 and 14. This set of numbers describes the **pH** scale (see **FIGURE 9.4**). This scale was developed for the simple comparison of $[H_3O^+]$ values. The letter "p" literally means the mathematical function negative log ($-\log$), so a pH can be determined from the $[H_3O^+]$ as

$$pH = -\log[H_3O^+]$$

TABLE 9.7 shows the relationship between pH and $[H_3O^+]$.

FIGURE 9.3 Relationships between hydronium and hydroxide in neutral, acidic, and basic solutions. In a neutral solution, the $[H_3O^+]$ and the $[OH^-]$ are equal. In acidic solutions, the $[H_3O^+]$ is greater than the $[OH^-]$. In a basic solution, the $[OH^-]$ is greater than the $[H_3O^+]$.

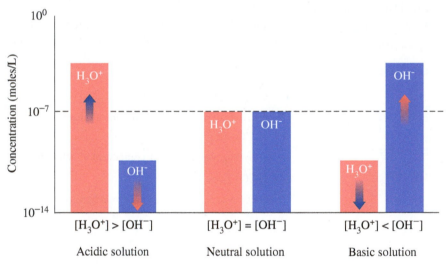

TABLE 9.7 Relationship Between pH and $[H_3O^+]$

Solution Acidity	pH	$[H_3O^+]$
Basic	> 7	< 1.0×10^{-7} M
Neutral	= 7	= 1.0×10^{-7} M
Acidic	< 7	> 1.0×10^{-7} M

Sample Problem 9.11   The Relationship Among Acidity, pH, and $[H_3O^+]$

State whether solutions with the following conditions are acidic, basic, or neutral.
a. pH = 5.0
b. $[H_3O^+] = 2.3 \times 10^{-9}$ M
c. pH = 12.0
d. $[H_3O^+] = 4.3 \times 10^{-4}$ M
e. $[H_3O^+] = 1 \times 10^{-7}$ M

Solution
Refer to Table 9.7 or Figure 9.4.
a. acidic   b. basic   c. basic   d. acidic   e. neutral

Measuring pH

Living things prefer environments that are kept at a constant pH. For example, normal blood pH is strictly regulated between 7.35 and 7.45. The pH of a solution is commonly measured either electronically by using an instrument called a pH meter or by using pH indicator paper. The pH paper is embedded with indicators that change color depending on the pH of the test solution (see **FIGURE 9.5**).

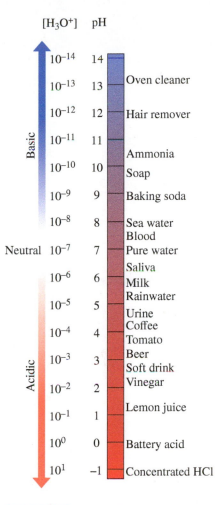

FIGURE 9.4 The relationship between $[H_3O^+]$ and pH of some common substances. Notice that the higher the pH, the smaller the $[H_3O^+]$ and vice versa.

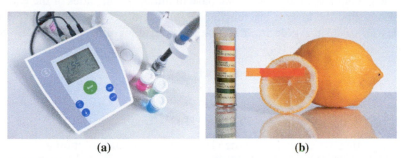

(a)   (b)

FIGURE 9.5 pH measurement. The pH of a solution can be determined using (a) a pH meter or (b) pH indicator paper.

Calculating pH

Suppose that we have a 0.050 M HCl solution and want to calculate its pH. Because strong acids fully ionize in solution, $[HCl] = [H_3O^+]$. This means that we can calculate the pH directly for a strong acid as

$$pH = -\log[H_3O^+]$$
$$pH = -\log(0.050)$$
$$pH = -(-1.30) = 1.30$$

Most scientific calculators have a $\boxed{\log}$ button. Be sure that you can locate this button on your calculator and can perform the example calculation correctly. (There is another logarithm function used in mathematics called the natural log, or *ln*, which is a completely different function and will not be discussed in this chapter.)

Let's take a look at the significant figures in the preceding calculation. In pH calculations with logarithms, the number of significant figures in the $[H_3O^+]$ will be the number of decimal places in the pH value.

$$[H_3O^+] = 0.050 \qquad pH = 1.30$$
Two significant figures   *Two decimal places*

Sample Problem 9.12 **Calculating pH from $[H_3O^+]$**

Determine the pH for the following solutions:

$[H_3O^+] = 1 \times 10^{-9}$ M

$[H_3O^+] = 4.5 \times 10^{-7}$ M

Solution

To do these problems, you will be using the log function, $\boxed{\log}$, on your calculator. Depending on the type of scientific calculator that you have, you may have to input numbers differently. Check with your instructor if you have difficulty producing the correct answers using your calculator.

a. $pH = -\log[H_3O^+] = -\log(1 \times 10^{-9}) = -(-9.0) = 9.0$

b. $pH = -\log[H_3O^+] = -\log(4.5 \times 10^{-7}) = -(-6.35) = 6.35$

Remember, the number of significant figures in $[H_3O^+]$ is the number of decimal places in the pH.

Calculating $[H_3O^+]$

Because we can easily measure the pH of most solutions, we are often more interested in finding the $[H_3O^+]$ from a measured pH value. Suppose we measured the pH of a solution to be 3.00 and we want to find the corresponding $[H_3O^+]$:

$$\text{pH} = -\log[H_3O^+]$$

To solve the pH equation for $[H_3O^+]$, we have to multiply both sides of the equation by negative 1 and find the inverse log function, INV log. In case you don't have an $\boxed{\text{INV}}$ button on your calculator, keep in mind that the inverse log function is 10^x. So to solve the equation for $[H_3O^+]$,

$$\boxed{\text{INV}}\log(-\text{pH}) = [H_3O^+]$$

or alternatively,

$$10^{-\text{pH}} = [H_3O^+]$$

From this equation, we can solve for the $[H_3O^+]$ of a pH 3.00 solution as

$$\text{INV log}(-3.00) \text{ or } 10^{-3.00} = [H_3O^+]$$
$$1.0 \times 10^{-3}\,\text{M} = [H_3O^+]$$

Many programmable calculators have an $\boxed{\text{INV}}$ button that can be used to solve the preceding equation. Most other scientific calculators have a $\boxed{10^x}$ button typically found as a second function above the log button. The $\boxed{y^x}$ button can also be used if 10 is entered as y. Be sure that you can perform the example calculation correctly before continuing.

Regarding significant figures, the number of decimal places given in the pH measurement tells us the number of significant figures in the $[H_3O^+]$:

$$\text{pH} = 3.00 \qquad\qquad [H_3O^+] = 1.0 \times 10^{-3}$$
Two decimal places   Two significant figures

Sample Problem 9.13 Calculating $[H_3O^+]$ from a Measured pH

Determine the $[H_3O^+]$ for solutions having the following measured pH values:

a. pH = 7.35   b. pH = 11.0

Solution

To do these problems, you will use the inverse log function, either $\boxed{\text{INV}}\,\boxed{\log}$ or $\boxed{10^x}$, on your calculator. Depending on the type of scientific calculator that you have, you will input numbers differently. Check with your instructor if you have difficulty producing the correct answers using your calculator.

a. $[H_3O^-] = \text{INV log}(-\text{pH})$ or $10^{-\text{pH}} = \text{INV log}(-7.35)$ or $10^{-7.35} = 4.5 \times 10^{-8}\,\text{M}$

b. $[H_3O^+] = \text{INV log}(-\text{pH})$ or $10^{-\text{pH}} = \text{INV log}(-11.0)$ or $10^{-11.0} = 1 \times 10^{-11}\,\text{M}$

Practice Problems

9.41 State if each of the following solutions is acidic, basic, or neutral:

a. blood, pH 7.38

b. vinegar, pH 2.8

c. drain cleaner, pH 11.2

d. coffee, pH 5.5

9.42 State if each of the following solutions is acidic, basic, or neutral:

a. soda, pH 3.2

b. shampoo, pH 5.7

c. laundry detergent, pH 9.4

d. rain, pH 5.8

9.43 State if each of the following solutions is acidic, basic, or neutral:

   a. $[H_3O^+] = 1.2 \times 10^{-8}$ M

   b. $[H_3O^+] = 7.0 \times 10^{-3}$ M

   c. $[H_3O^+] = 4.7 \times 10^{-11}$ M

   d. $[H_3O^+] = 1.0 \times 10^{-7}$ M

9.44 State if each of the following solutions is acidic, basic, or neutral:

   a. $[H_3O^+] = 5.6 \times 10^{-10}$ M

   b. $[H_3O^+] = 6.2 \times 10^{-8}$ M

   c. $[H_3O^+] = 5 \times 10^{-2}$ M

   d. $[H_3O^+] = 1.8 \times 10^{-6}$ M

9.45 Calculate the pH of each of the solutions in Problem 9.43.

9.46 Calculate the pH of each of the solutions in Problem 9.44.

9.47 Calculate the $[H_3O^+]$ for each of the following measurements of pH:

   a. drain cleaner, pH 12.10

   b. coffee, pH 5.5

   c. gastric juice, pH 2.00

   d. toothpaste, pH 8.3

9.48 Calculate the $[H_3O^+]$ for each of the following measurements of pH:

   a. hand soap, pH 9.5

   b. sea water, pH 8.2

   c. apple juice, pH 3.5

   d. beer, pH 4.5

9.6 pKa

Acetic acid and the ammonium ion are both weak acids, but do they have the same strength? That is, do they dissociate by the same amount? We can determine dissociation by comparing their K_a values as we did in Section 9.4. Notice from **TABLE 9.8** that the K_a value for acetic acid, 1.75×10^{-5}, is a number closer to 1 than the K_a for ammonium ion, 6.3×10^{-10}, which tells us that acetic acid is the stronger of these two weak acids.

In the previous section, we saw that the mathematical function "p" means to take the *negative log* of a number. As seen with the pH scale, it is easier to compare whole numbers than those in scientific notation. To make this comparison for acid strength simpler, we can compare **pKa** values, which range from 0 to about 60.

 9.6 Inquiry Question

How is the pKa used to determine the strength of a weak acid?

TABLE 9.8 pKa and Ka Values for Substances Acting as Weak Acids (25 °C)

Name	Formula	pKa	Ka
Hydrogen sulfate ion	HSO_4^-	2.00	1.0×10^{-2}
Phosphoric acid	H_3PO_4	2.12	7.5×10^{-3}
Hydrofluoric acid	HF	3.19	6.5×10^{-4}
Nitrous acid	HNO_2	3.35	4.5×10^{-4}
Formic acid	HCOOH	3.74	1.8×10^{-4}
Lactic acid	$CH_3CH(OH)COOH$	3.85	1.4×10^{-4}
Acetic acid	CH_3COOH	4.76	1.75×10^{-5}
Carbonic acid	H_2CO_3	6.35	4.5×10^{-7}
Dihydrogen phosphate ion	$H_2PO_4^-$	7.18	6.6×10^{-8}
Ammonium ion	NH_4^+	9.20	6.3×10^{-10}
Hydrocyanic acid	HCN	9.21	6.2×10^{-10}
Bicarbonate ion	HCO_3^-	10.32	4.8×10^{-11}
Hydrogen phosphate ion	HPO_4^{2-}	12.00	1.0×10^{-12}
Water	H_2O	14.00	1.0×10^{-14}

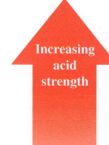

Increasing acid strength

Notice that the pKa value for acetic acid is a smaller number, 4.76, than that of ammonium ion, 9.20. This comparison illustrates the following rule for determining the strength of a weak acid using pKa values: *The smaller the pKa value, the stronger the acid.*

The reactions for these two weak acids in water are shown in **FIGURES 9.6** and **9.7**. Table 9.8 compares the pK_a and K_a values for some common weak acids.

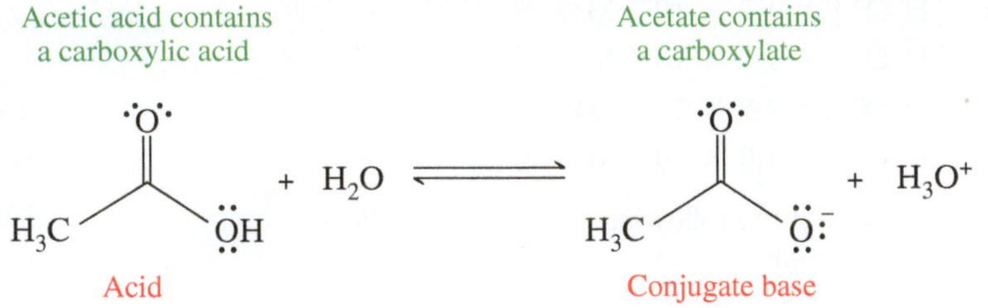

FIGURE 9.6 Equilibrium reaction for the weak acid acetic acid. When acetic acid donates a proton to water, the carboxylate named acetate is formed. Recall that carboxylic acid and carboxylate are different forms of the same functional group.

Acetic acid contains a carboxylic acid

Acetate contains a carboxylate

+ H_2O ⇌ + H_3O^+

H_3C — OH
Acid

H_3C — O
Conjugate base

$$K_a = 1.75 \times 10^{-5}$$
$$pK_a = -\log K_a = 4.76$$

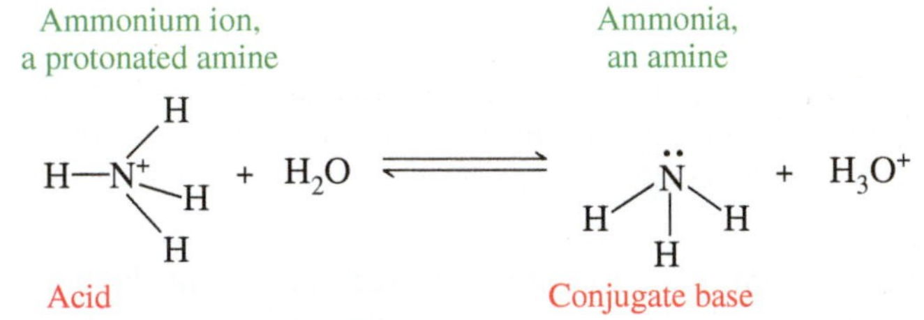

FIGURE 9.7 Equilibrium reaction for the weak acid ammonium ion. When ammonium ion donates a proton to water, ammonia, a weak base, is formed. Recall that a protonated amine and an amine are different forms of the same functional group.

Ammonium ion, a protonated amine

Ammonia, an amine

H—N$^+$—H + H_2O ⇌ N + H_3O^+
Acid

Conjugate base

$$K_a = 6.3 \times 10^{-10}$$
$$pK_a = -\log K_a = 9.20$$

Sample Problem 9.14 Determining Strength of Weak Acids Using pK_a Values

Use Table 9.8 to determine the stronger acid of the following pairs of weak acids:
a. acetic acid or lactic acid
b. carbonic acid or bicarbonate ion

Solution

Using the rule the smaller the pK_a value, the stronger the weak acid, and Table 9.8:
a. Lactic acid is the stronger acid.
b. Carbonic acid is the stronger acid.

Practice Problems

9.49 Using Table 9.8, determine the stronger acid from the following pairs of acids:
a. $H_2PO_4^-$ or HPO_4^{2-}
b. H_2SO_4 or acetic acid
c. formic acid or carbonic acid
d. ammonium ion or hydrocyanic acid

9.50 Using Table 9.8, determine the stronger acid from the following pairs of weak acids:
a. HCl or HCN
b. acetic acid or ammonium ion
c. carbonic acid or bicarbonate ion
d. ammonium ion or hydrogen phosphate ion

9.51 Which is a stronger acid, lactic acid, $pK_a = 3.9$, or pyruvic acid, $pK_a = 2.5$?

9.52 Which is the stronger acid, niacin, $pK_a = 4.85$, or acetylsalicylic acid, $pK_a = 3.5$?

9.7 The Relationship Between pH, pK_a, Drug Solubility, and Diffusion

We have seen two numbers associated with weak acid solutions, the pH and the pK_a. Comparing these two numbers for a weak acid can tell us whether the weak acid or its conjugate base form is present in higher amounts in aqueous solution. Why would we want to know something like this?

To fully answer this, we must consider two topics discussed in previous chapters: solubility, and diffusion across the cell membrane. In Section 7.4, when we discussed solubility of pharmaceuticals, we noted that if molecules contain charges, then they are more readily soluble in aqueous solution. This is advantageous for drugs that are administered orally. However, if substances are to move into the bloodstream or to be excreted to the urine, they will also have to diffuse across the cell membrane, a hydrophobic barrier (see Section 8.6). A neutral, uncharged form can more easily diffuse across the cell membrane.

Keeping this in mind, let's look at a common pharmaceutical, propranolol, a medication prescribed for high blood pressure. Even though this is a larger molecule than many we have seen, focus on the amine functional group, highlighted in yellow in the figure below, which is the part of the molecule that will be charged and uncharged at different pH values. In propranolol, the acid form (protonated amine) is charged and the conjugate base form (amine) is uncharged. This is just like the other examples we have shown for conjugate acid–base pairs in that one of the two forms is ionic (contains a charge) and the other is neutral (uncharged).

9.7 Inquiry Question

How does the charge of a pharmaceutical change with pH?

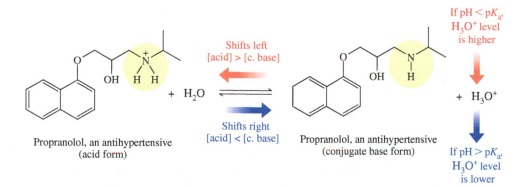

Propranolol, an antihypertensive (acid form)

Propranolol, an antihypertensive (conjugate base form)

Shifts left [acid] > [c. base]

Shifts right [acid] < [c. base]

If pH < pK_a, H$_3$O$^+$ level is higher

If pH > pK_a, H$_3$O$^+$ level is lower

The pK_a for propranolol is 9.42. If this drug were in the stomach, where there is much more acid (H$_3$O$^+$) present (the pH is around 2), much of the substance would be in its charged, acid form so it would be soluble in the stomach. As the drug moves into the small intestine, the pH rises dramatically, up to about 8, so more of the uncharged, conjugate base form would be present. This uncharged form is better able to diffuse across the intestinal lining and into the bloodstream.

This follows from Le Châtelier's principle. If the pH is lower than the pK_a, there is more H$_3$O$^+$ present, shifting the equilibrium to produce more of the acid. If the pH is higher than the pK_a, there is less H$_3$O$^+$ present and more conjugate base is present.

The pK_a (and the K_a) tells us the strength of the weak acid, that is, how much dissociates to form the conjugate base and hydronium ion (H$_3$O$^+$). The pH tells us how much hydronium ion is present in the solution. pK_a values are constant at a given temperature; that is, they do not change if acid or base is added to an aqueous solution. In contrast, the pH does change if acid or base is added to the solution. Knowing both helps us determine whether the weak acid or its conjugate base is the predominant form in a solution.

Many pharmaceuticals contain a weak acid or base group so that both the weak acid form and conjugate base form (one with a charge and one without a charge) of the drug will be present at equilibrium. This allows both solubility and diffusion through a membrane to occur when the drug is administered. The amount of each form can change in a particular environment based on pH levels. Next, we look at how much the amounts change based on pH.

The Henderson–Hasselbalch Equation

Suppose we want to know the actual ratio of conjugate base to acid present. We can use a modified form of the equilibrium expression (introduced in Section 9.3) called the **Henderson–Hasselbalch equation** to calculate a ratio of conjugate base to acid present for a weak acid at a particular pH. This equation is written as

$$pH = pK_a + \log \frac{[c. base]}{[acid]}$$

To solve the Henderson–Hasselbalch equation for the ratio of conjugate base to acid, a little rearranging must be done to the equation:

$$pH - pK_a = \log \frac{[c. base]}{[acid]}$$

Remember that, as with the pH calculations, to move the log from the right side to the left side of the equation, we take the inverse of the log function, which is 10^x.

$$10^{pH-pK_a} = \frac{[c. base]}{[acid]}$$

So for our earlier example of propranolol, where the pK_a was 9.42, if the pH were 2 as it is in the stomach,

$$10^{2-9.42} = \frac{[c. base]}{[acid]}$$

$$10^{-7.42} = 3.8 \times 10^{-8} = \frac{[c. base]}{[acid]}$$

In this example, the number 3.8×10^{-8} is less than 1. Because the ratio is conjugate base to acid, the amount of acid is more than the amount of conjugate base. For propranolol, remember, the acid form is charged. The relationships regarding pH, pK_a, and conjugate base to acid ratios from the Henderson–Hasselbalch equation are summarized in **TABLE 9.9**.

TABLE 9.9 Relationship Between pH, pK_a, and [c. base]/[acid] and Form Predominating

pH vs. pK_a Value	Value of [c. base]/[acid]	Predominant Form
pH < pK_a	$\frac{[c. base]}{[acid]} < 1$	Acid predominates
pH = pK_a	$\frac{[c. base]}{[acid]} = 1$	Acid = conjugate base
pH > pK_a	$\frac{[c. base]}{[acid]} > 1$	Conjugate base predominates

Sample Problem 9.15   **Relative Amounts of Acid and Conjugate Base**

The pK_a for acetylsalicylic acid, aspirin, is 3.5. Draw the form (acid or conjugate base) that predominates at normal physiological pH of 7.4. Is that form charged or uncharged?

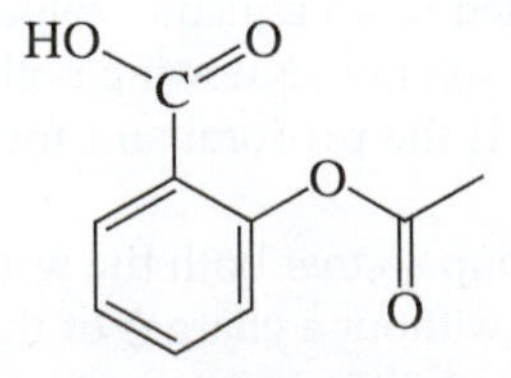

$pK_a = 3.5$

Acetylsalicylic acid, aspirin

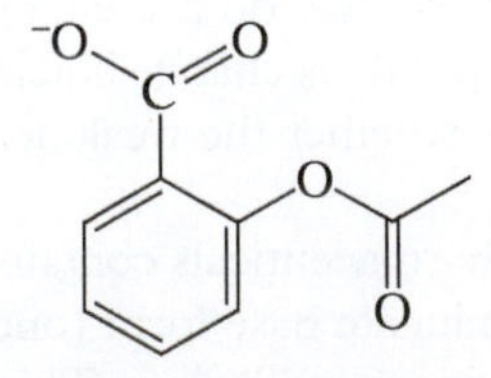

Acetylsalicylate, the conjugate base of acetylsalicylic acid

Solution

To answer this problem, use Table 9.9 to compare the values of pK_a and pH and determine which form, acid or conjugate base, is present in larger amounts. In this case, pH > pK_a so the conjugate base form is present in larger amounts. Examining the structure of acetylsalicylic acid shows that the acid is a carboxylic acid and the carboxylate form, which is charged, would therefore be the predominant form. The conjugate base (charged) has increased solubility over the acid (uncharged).

Sample Problem 9.16   **Using the Henderson–Hasselbalch Equation**

If the pK_a for acetylsalicylic acid, aspirin, is 3.5, use the Henderson–Hasselbalch equation to determine the ratio of conjugate base to acid present and indicate which form predominates at the following pH values:

  a. 1.5   b. 3.5   c. 7.4

Solution

To calculate the ratio of conjugate base to acid, use the rearranged form of the Henderson–Hasselbalch equation:

a.
$$10^{pH - pK_a} = \frac{[c.\,base]}{[acid]}$$

$$10^{1.5 - 3.5} = \frac{[c.\,base]}{[acid]}$$

$$10^{-2} = 0.01 = \frac{[c.\,base]}{[acid]}$$

Because the ratio is less than 1, more acid form is present than conjugate base. For aspirin, the acid form is uncharged, so much of the aspirin gets absorbed in the stomach.

b.
$$10^{3.5 - 3.5} = \frac{[c.\,base]}{[acid]}$$

$$10^{0} = 1 = \frac{[c.\,base]}{[acid]}$$

The ratio of conjugate base to acid is equal to 1 when the pH and pK_a are equal.

c.
$$10^{7.4 - 3.5} = \frac{[c.\,base]}{[acid]}$$

$$10^{3.9} = 8000 = \frac{[c.\,base]}{[acid]}$$

Because the ratio is much greater than 1, the conjugate base form predominates. For aspirin, the conjugate base form is charged, allowing for greater solubility of the drug at the pH of the bloodstream.

Regarding significant figures, as in Section 9.5, the number of decimal places in the pH − pK_a difference is the number of significant figures in the ratio. Because the pH values have one decimal place, the ratios have one significant figure.

INTEGRATING Chemistry

Find out how ▶
pK_a and pH affect
drug delivery.

Tissue pH and Drug Delivery

Delivering a drug through the mostly polar bloodstream and across a nonpolar cell membrane requires a compound that is both reasonably soluble in water and nonpolar enough to cross a cell membrane. One set of drugs that take advantage of pH and pK_a are local anesthetics.

Amine group

Lidocaine, a local anesthetic in
protonated form (RNH$^+$)

pK_a = 7.9

All local anesthetics contain an amine functional group, which in its acid form has a positive charge (protonated amine) and in its conjugate base form is uncharged. To lessen pain, local anesthetics must cross the nerve-cell membrane, which contains a nonpolar barrier created by the tails of phospholipids. However, the active form of the anesthetic is the acid form, which cannot cross the nerve-cell membrane because of its positive charge. Only the uncharged conjugate base form is able to cross the nerve-cell membrane. Therefore, the effectiveness of a local anesthetic relies on the amounts of both forms of the drug. The uncharged form penetrates the nerve-cell membrane, and the charged form relieves pain. The structure of lidocaine is shown as a representative of this class of drugs.

Lidocaine is used topically to relieve itchiness and can also be applied to teeth and gums during oral procedures. The acid and conjugate base forms are in equilibrium in solution. The uncharged form is nonpolar enough to pass through the nerve-cell membrane, but the charged or protonated form is the active form inside the nerve cell, reducing pain. The pK_a of the amino group in lidocaine is 7.9. At physiological pH, about one-quarter of the lidocaine is in the uncharged, conjugate base form and is able to get into the nerve cells. Once inside, the drug reestablishes equilibrium, forming some of the acid, which is active and can relieve pain. Both forms are necessary for drug delivery and activation.

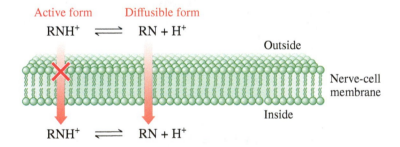

RNH$^+$ represents the acid form of the local anesthetic, and RN represents the uncharged, conjugate base form. The amount of each form is dictated by the pK_a of the drug and the pH of the solution.

Diseased tissues tend to be acidic, which lowers the pH in the tissues. In the case of lidocaine, the pH would then be further from its pK_a of 7.9 and even less of the uncharged form would be present to diffuse through the nerve cells, so more medication would be required to achieve the same effect.

Practice Problems

9.53 The antihypertensive medication alprenolol is shown as an acid below. The pK$_a$ for the acid is 9.6.

 a. Which form is charged: acid, conjugate base, or both?

 b. Which form (acid, conjugate base, or both) will predominate in a stomach with a pH between 1 and 3?

 c. Which form, acid or conjugate base, will be able to more easily diffuse through a cell membrane?

Alprenolol, an antihypertensive,
pK$_a$ = 9.6

9.54 The anticonvulsant medication valproic acid is shown as an acid below. The pK$_a$ for the acid is 4.8.

 a. Which form is charged: acid, conjugate base, or both?

 b. Which form (acid, conjugate base, or both) will predominate in the first part of the small intestine (jejunum), where the pH is between 6 and 7?

 c. Which form, acid or conjugate base, will be able to more easily diffuse through a cell membrane?

Valproic acid, an anticonvulsant,
pK$_a$ = 4.8

9.55 Consider the vitamin niacin, which has a pK$_a$ value of 4.85.

 a. Will the acid or the conjugate base predominate at the following pH values: 3.00, 4.85, and 7.40?

 b. At each pH, is that form charged or uncharged?

 c. Calculate the ratio of [c. base]/[acid] present at each pH.

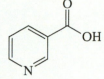

Niacin, pK$_a$ = 4.85

9.56 Consider the neurotransmitter dopamine, which has a pK$_a$ value of 8.90.

 a. Will the acid or the conjugate base predominate at the following pH values: 7.45, 8.90, 13.40?

 b. At each pH, is that form charged or uncharged?

 c. Calculate the ratio of [c. base]/[acid] present at each pH.

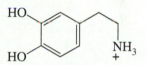

Dopamine, acid form, pK$_a$ = 8.90

9.57 Procaine and lidocaine are used to numb the gums during dental procedures. Procaine has a pK$_a$ of 9.1, whereas lidocaine has a pK$_a$ of 7.9. Which do you think will relieve pain faster in the gums? Explain.

9.58 Both mepivacaine and bupivacaine are local anesthetics used in epidural blocks. Mepivacaine has a pK$_a$ of 7.6, whereas bupivacaine has a pK$_a$ of 8.1. Based on their pK$_a$ values, which will have more of the uncharged form available at physiological pH, thereby blocking pain quicker? Explain.

DISCOVERING THE CONCEPTS

 INQUIRY ACTIVITY—The Bicarbonate Buffer System

Information

The bicarbonate buffer system is an effective physiological buffer in the blood, helping maintain a normal physiological pH of 7.4. CO_2 produced at the tissues rapidly equilibrates (comes to equilibrium) through carbonic acid to bicarbonate ion in the blood.

—*Continued next page*

Continued—

$$2H_2O(l) + CO_2(g) \underset{equilibrates}{\overset{Rapidly}{\rightleftharpoons}} H_2CO_3(aq) + H_2O(l) \rightleftharpoons H_3O^+(aq) + HCO_3^-(aq)$$

Carbonic acid
(Acid)

Bicarbonate ion
(Conjugate base)

The ventilation rate (rate of breathing) controls the amount of CO_2 present and, therefore, the amount of acid (H_3O^+) in the blood. When normal ventilation rates are altered, the normal pH balance becomes disrupted, causing distress. An increase in ventilation rate would expel more CO_2 than normal.

Questions

1. Suppose the lungs fail to expel normal amounts of CO_2 due to shallow exhaling that can occur during an asthma attack. According to Le Châtelier's principle, if more CO_2 is added to the system illustrated in the equilibrium (not enough expelled), what happens to the acidity of the blood? How might a patient with this condition be treated?
2. If the lungs expel CO_2 faster than normal (ventilation rate becomes high—hyperventilation), which can occur during heavy exercise, and more CO_2 is removed from the system than normal, what will happen to the pH of the blood? How might a patient with this condition be treated?
3. People with diabetes may produce excess acid in their tissues when they metabolize fats for energy (metabolic acidosis). How would a diabetic's ventilation rate adjust to correct this condition?
4. Suppose a person drank too much of a baking soda solution (sodium bicarbonate) to combat heartburn (which is excess stomach acid). What would happen if too much acid were removed from the bloodstream?

9.8 Buffers and Blood: The Bicarbonate Buffer System

9.8 Inquiry Question

How do the properties of buffers apply to the bicarbonate buffer in the blood?

How do we keep the pH levels constant in our blood when we eat a variety of foods at varying pH levels? In our bodies, solutions of weak acids containing both acid and conjugate base help neutralize incoming bases and acids, thereby maintaining pH levels. A solution that contains relatively equal amounts of both a weak acid and its conjugate base or a weak base and its conjugate acid is called a **buffer.** As you saw in Section 9.7, this occurs when the pK_a and the pH are the same. A buffer solution will resist a change in pH if small amounts of acid or base are added. Strong acids and bases are not components of buffers.

Our blood is buffered mainly by the bicarbonate buffer system. Dissolved CO_2 produced during cellular respiration travels through the bloodstream for exhalation at the lungs. Although dissolved, this CO_2 rapidly equilibrates with H_2O, forming carbonic acid and further dissociating to form bicarbonate ions. Often, the intermediates H_2CO_3 and H_2O are not shown in this reaction (see **FIGURE 9.8**). We show them here for completeness, so that you can see how the bicarbonate is generated from CO_2. Technically, CO_2 cannot act as an acid because it has no protons, but here we can see that the protons come from H_2O. If the H_2CO_3 is omitted from the equilibrium, the acid is usually identified as CO_2.

FIGURE 9.8 The bicarbonate buffer equilibrium. CO_2 produced at the tissues rapidly returns to equilibrium through carbonic acid to bicarbonate ion. Because this first equilibrium is so quick, CO_2 in this equilibrium is referred to as the acid.

$$2H_2O(l) + CO_2(g) \underset{equilibrates}{\overset{Rapidly}{\rightleftharpoons}} H_2CO_3(aq) + H_2O(l) \rightleftharpoons H_3O^+(aq) + HCO_3^-(aq)$$

Carbonic acid
(Acid)

Bicarbonate ion
(Conjugate base)

A buffer system like the bicarbonate buffer can be denoted as H_2CO_3/HCO_3^-, which shows the acid and its conjugate base. Sometimes, the conjugate base is represented as an ionic compound, a salt, like this: $H_2CO_3/NaHCO_3$.

Maintaining Physiological pH with Bicarbonate Buffer: Homeostasis

The bicarbonate buffer system helps to maintain optimal physiological pH in our bodies. The ability of an organism to regulate its internal environment by adjusting factors such as pH, temperature, and solute concentration is called **homeostasis.** Let's examine how the bicarbonate buffer system is able to regulate blood pH under adverse conditions.

Changes in Ventilation Rate

During normal breathing, appropriate amounts of CO_2 gas are removed from the bloodstream upon exhalation in the lungs, and blood pH is maintained. A person who **hypoventilates** may fail to exhale enough CO_2 from the lungs due to shallow breathing, causing CO_2 gas to build up in the bloodstream. Hypoventilation can occur, for example, when a person is suffering from emphysema, a condition that blocks gas diffusion in the lungs.

According to Le Châtelier's principle (Section 9.3), a buildup of CO_2 ultimately produces more H_3O^+ in the bicarbonate equilibrium, making the blood more acidic. This condition is known as **respiratory acidosis** and can occur in a number of health conditions (**TABLE 9.10**). Persons suffering from this condition must be treated to raise their blood pH back into the normal range. Treatment is done by administering a bicarbonate solution intravenously, which will have the effect of shifting the equilibrium to the left.

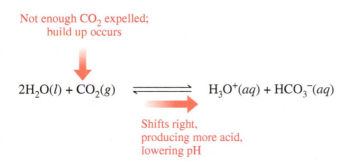

Not enough CO_2 expelled; build up occurs

$$2H_2O(l) + CO_2(g) \rightleftharpoons H_3O^+(aq) + HCO_3^-(aq)$$

Shifts right, producing more acid, lowering pH

TABLE 9.10 Acidosis and Alkalosis: Symptoms, Causes, and Treatments

Respiratory Acidosis: CO_2 ↑, pH ↓		Metabolic Acidosis: H_3O^+ ↑, pH ↓	
Symptoms:	Failure to ventilate, suppression of breathing, disorientation, weakness, coma	Symptoms:	Increased ventilation, fatigue, confusion
Causes:	Lung disease blocking gas diffusion (e.g., emphysema, pneumonia, bronchitis, and asthma); depression of respiratory center by drugs, cardiopulmonary arrest, stroke, poliomyelitis, or nervous system disorders	Causes:	Renal disease, including hepatitis and cirrhosis; increased acid production in diabetes mellitus, hyperthyroidism, alcoholism, and starvation; loss of alkali in diarrhea; acid retention in renal failure
Treatment:	Correction of disorder, infusion of bicarbonate	Treatment:	Sodium bicarbonate given orally, dialysis for renal failure, insulin treatment for diabetic ketosis
Respiratory Alkalosis: CO_2 ↓, pH ↑		**Metabolic Alkalosis: H_3O^+ ↓, pH ↑**	
Symptoms:	Increased rate and depth of breathing, numbness, light-headedness, tetany	Symptoms:	Depressed breathing, apathy, confusion
Causes:	Hyperventilation because of anxiety, fever, exercise; reaction to drugs such as salicylate, quinine, and antihistamines; conditions causing hypoxia (e.g., pneumonia, pulmonary edema, and heart disease)	Causes:	Vomiting, diseases of the adrenal glands, ingestion of excess alkali
Treatment:	Elimination of anxiety-producing state, rebreathing into a paper bag	Treatment:	Infusion of saline solution, treatment of underlying diseases, in extreme cases administer NH_4Cl

Breathing into a paper bag can help a person hyperventilating inhale more CO_2.

Conversely, a person who **hyperventilates** (exhaling excessively versus inhaling), which can occur under conditions of anxiety, is exhaling too much CO_2 from the lungs. Hyperventilation therefore draws H_3O^+ from the bloodstream, making the blood more basic. This condition is known as **respiratory alkalosis.** A person suffering from this condition needs to get more CO_2 back into the bloodstream by breathing a CO_2-rich atmosphere. This can easily be done by having the person breathe into a paper bag, which keeps some of the previously exhaled CO_2 available for the next breath. This increases the CO_2 present in the bloodstream, shifting the equilibrium back to the right, thereby producing more H_3O^+ and reestablishing the equilibrium.

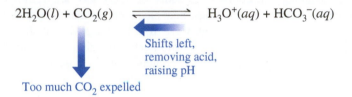

$$2H_2O(l) + CO_2(g) \rightleftharpoons H_3O^+(aq) + HCO_3^-(aq)$$

Shifts left, removing acid, raising pH

Too much CO_2 expelled

Changes in Metabolic Acid Production

The chemical reactions that occur in our bodies can also directly change the pH of our blood by producing too much or too little H_3O^+. People with diabetes use much less glucose for energy than people without diabetes because glucose is unable to get inside their cells for use. Because of this, a person with diabetes often uses fatty acids as a carbon source for energy production. A by-product of fatty acid chemical breakdown is acid production. In this case, the imbalance is not caused by breathing but by chemical reactions in the body, so it is termed **metabolic acidosis.** As in the case of respiratory acidosis, administering bicarbonate neutralizes the excess acid, forming more CO_2 that can then be exhaled.

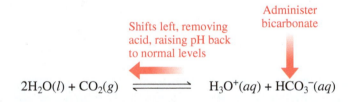

Administer bicarbonate

Shifts left, removing acid, raising pH back to normal levels

$$2H_2O(l) + CO_2(g) \rightleftharpoons H_3O^+(aq) + HCO_3^-(aq)$$

The body can also lose too much acid, thereby making the blood basic. This condition, known as **metabolic alkalosis**, can occur under conditions of excessive vomiting, when the body is trying to replace the acid lost from the stomach, thereby making the blood basic. To lower the pH back to normal, in some cases, ammonium chloride (NH_4Cl, a weak acid) can be administered to neutralize the excess base buildup.

INTEGRATING
Chemistry

Find out how the ▶ body compensates for acidosis and alkalosis.

Regulating CO_2 and Bicarbonate in the Body

In this section we saw that the bicarbonate buffer equilibrium can shift based on the levels of CO_2 or bicarbonate (HCO_3^-) present in the bloodstream. How does the body compensate for that initial imbalance to the equilibrium?

Suppose a person has an asthma attack and is unable to expel enough CO_2. The person's level of CO_2 increases, which can result in respiratory acidosis, where the buildup of CO_2 leads to an increase in H_3O^+, making the blood more acidic. The body automatically responds by changing the concentration of bicarbonate. This secondary response of restoring a balance in pH is called *compensation*, since the body compensates for the change in CO_2 by keeping more bicarbonate in the bloodstream. These higher CO_2 levels trigger the kidneys to decrease the level of bicarbonate excreted in the urine, offsetting the higher CO_2 levels.

The body also regulates CO_2 levels by changing ventilation or breathing rates through a set of chemical sensors in the lower brain stem (medulla) that detect levels of CO_2. These sensors are connected to respiratory centers in the brain stem that control the rate and depth of breathing. For example, if the levels of CO_2 rise, these sensors compensate by triggering the autonomous nervous system to increase the rate of breathing, which then lowers CO_2 levels.

So the body has two systems compensating for changes in the bicarbonate equilibrium: The lungs regulate pH through CO_2, and the kidneys regulate bicarbonate by changing the level excreted in the urine.

If CO_2 levels rise, the lungs compensate by increasing the ventilation rate.

If HCO_3^- levels rise, the kidneys compensate by increasing urine output.

$$2H_2O(l) + CO_2(g) \rightleftharpoons H_3O^+(aq) + HCO_3^-(aq)$$

If CO_2 levels drop, the lungs compensate by decreasing the ventilation rate.

If HCO_3^- levels drop, the kidneys compensate by decreasing urine output.

The body compensates for initial imbalances to the bicarbonate equilibrium. The lungs (left) regulate CO_2 levels through ventilation, and the kidney (right) regulates bicarbonate levels through excretion.

Practice Problems

9.59 During stress or trauma, a person can start to hyperventilate. The person may then be instructed to breathe into a paper bag to avoid fainting.

$$CO_2(g) + 2H_2O(l) \rightleftharpoons H_3O^+(aq) + HCO_3^-(aq)$$

a. How does blood pH change during hyperventilation?

b. In which direction will the bicarbonate equilibrium shift during hyperventilation?

c. What is this condition called?

d. How does breathing into a paper bag help return the blood pH to normal?

e. In which direction will the bicarbonate equilibrium shift as the pH returns to normal?

9.60 A person who overdoses on antacids may neutralize too much stomach acid, which causes an imbalance in the bicarbonate equilibrium in the blood.

$$CO_2(g) + 2H_2O(l) \rightleftharpoons H_3O^+(aq) + HCO_3^-(aq)$$

a. How does blood pH change when this occurs?

b. In which direction will the bicarbonate equilibrium shift during an Alka-Seltzer overdose?

c. What is this condition called?

d. To treat severe forms of this condition, patients are administered NH_4Cl, which acts as a weak acid. How does this restore the bicarbonate equilibrium?

e. In which direction will the bicarbonate equilibrium shift as the pH returns to normal?

CHAPTER REVIEW

The study guide will help you check your understanding of the main concepts in Chapter 9. You should be able to answer problems for each learning outcome in this list. To check your mastery, try the problems listed after each.

STUDY GUIDE	CHAPTER GUIDE

9.1 Acids and Bases—Definitions

Contrast the properties of an acid with those of a base.

- Describe the physical characteristics of an acid and a base. (Try 9.1)
- Identify an acid and a base in a chemical equation using the Arrhenius or Brønsted–Lowry definition. (Try 9.5)

Inquiry Question

How do acids and bases differ?

The Arrhenius definition states that acids produce H^+ and bases produce OH^- when dissolved in water. The Brønsted–Lowry definition builds on the Arrhenius definition by stating that an acid donates H^+ and a base accepts H^+. The H^+ or proton in solution contacts a water molecule, thereby forming the hydronium ion, H_3O^+.

9.2 Strong Acids and Bases

Predict the products of a neutralization reaction.

- Name the six strong acids. (Try 9.7, 9.61)
- Characterize strong bases. (Try 9.9)
- Compare a strong acid to a weak acid. (Try 9.11, 9.63)
- Name acids from their anion and vice versa. (Try 9.13, 9.65)
- Write and balance a neutralization reaction. (Try 9.15, 9.17, 9.65)

Inquiry Question

What happens when an acid reacts with a base?

$$HCl(aq) + NaOH(aq) \longrightarrow NaCl(aq) + H_2O(l)$$

Strong acid   Strong base   Salt   Water

When a strong acid combines with a strong base, a neutralization reaction occurs. The products of neutralization are a salt (ionic compound) and water. Neutralization reactions are typically exothermic. The use of an antacid provides a common example of a neutralization reaction. Antacids are basic compounds that neutralize excess stomach acid. Strong acids and bases completely ionize or dissociate in solution. There are six common strong acids: $HClO_4$, H_2SO_4, HI, HBr, HCl, and HNO_3. Water can act as either an acid or a base in a chemical reaction.

9.3 Chemical Equilibrium

Apply the properties of equilibrium to Le Châtelier's principle.

- Define chemical equilibrium. (Try 9.19, 9.69)
- Write an equilibrium expression for K. (Try 9.21, 9.71)
- Apply Le Châtelier's principle to chemical equilibrium. (Try 9.25, 9.27, 9.73)

Inquiry Question

How do the properties of equilibrium apply to Le Châtelier's principle?

Add N_2 — Shifts right

$$N_2(g) + 3H_2(g) \rightleftharpoons 2NH_3(g)$$

Remove H_2 — Shifts left

Many chemical reactions are reversible, operating in the forward direction up to a point when some of the products revert back to reactants. A chemical equilibrium is established when forward and reverse reactions occur at the same rate. The equilibrium constant, K, defines the extent of a chemical reaction at equilibrium as a ratio of the concentration of the products to the concentration of the reactants. If a chemical reaction at equilibrium is disturbed, the reaction can regain its equilibrium according to Le Châtelier's principle by shifting to offset the disturbance. For example, if more reactants are added to a chemical reaction that is currently at equilibrium, the equilibrium will shift to the right, forming more products to offset the addition. Equilibrium is also affected by changes in temperature.

9.4 Weak Acids and Bases

Write weak acid–base equilibrium equations.

- Apply the principles of chemical equilibrium to weak acids and bases. (Try 9.81c and d)
- Determine strengths of weak acids based on their pK_a values. (Try 9.31)
- Identify conjugate acid–base pairs. (Try 9.33, 9.35, 9.75)
- Complete a chemical equation for a conjugate acid–base in water. (Try 9.39)

Inquiry Question

How are weak acid–base equations written?

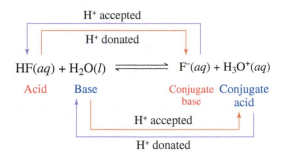

Weak acids can establish equilibrium because they only partially dissociate ($<$5%) in solution. The equilibrium constant for a weak acid is called the acid dissociation constant, K_a. When a weak acid dissociates (loses H^+), it produces a conjugate base. When a weak base accepts H^+, it produces a conjugate acid. These are referred to as conjugate acid–base pairs and differ by one H^+. The more a weak acid dissociates, the higher its K_a value, and the stronger the acid.

9.5 pH and the pH Scale

Calculate pH.

- Determine if a solution is acidic, basic, or neutral if given its pH or $[H_3O^+]$. (Try 9.41, 9.43, 9.77)
- Calculate the pH if given the $[H_3O^+]$. (Try 9.45, 9.77)
- Calculate the $[H_3O^+]$ if given the pH. (Try 9.47, 9.79)

Inquiry Question

How is the acidity of a solution determined?

The amount of H_3O^+ in an aqueous solution determines its acidity. Water ionizes slightly, producing hydronium (H_3O^+) and hydroxide (OH^-) ions. An excess of H_3O^+ in a solution makes a solution acidic. An excess of OH^- makes a solution basic. When the concentrations of hydronium and hydroxide ions are equal, a solution is neutral. The pH scale is a measure of acidity, with values typically falling between 0 and 14. Neutral solutions have a pH value of 7, acidic solutions have values less than 7, and basic solutions have pH values higher than 7. The pH is mathematically related to the concentration of H_3O^+ by the following equation: $pH = -\log[H_3O^+]$.

9.6 pK_a

Predict the relative strengths of weak acids.

- Predict the relative strength of a weak acid from its pK_a value. (Try 9.49, 9.51)

Inquiry Question

How is the pK_a used to determine the strength of a weak acid?

The pH value changes with $[H_3O^+]$, yet the pK_a value ($pK_a = -\log K_a$) is constant for a specific weak acid at a certain temperature. The smaller the pK_a value for a weak acid, the stronger the acid. The pH has an effect on which form of the weak acid predominates, the acid form or the conjugate base form. When the pH of a solution is the same as the pK_a, the acid and conjugate base forms are present in equal amounts. When the pH of a solution is higher than the pK_a, the conjugate base form is present in greater amounts than the acid. When the pH of a solution is lower than the pK_a, for a weak acid, the acid form is present in greater amounts than the conjugate base.

—Continued next page

Continued—

9.7 The Relationship Between pH, pK_a, Drug Solubility, and Diffusion

Predict the charge and relative amounts of a pharmaceutical based on its pH and pK_a.

- Predict the relative amounts of uncharged versus charged compounds present by comparing pK_a values. (Try 9.55a, 9.85a, 9.87b)
- Based on pH and pK_a values, predict whether acid or base predominates. (Try 9.55b, 9.85b, 9.87a)
- Determine the ratio of [c. base]/[acid] present for a given weak acid using the Henderson–Hasselbalch equation. (Try 9.55c, 9.85c, 9.87c)
- Determine which form of a weak acid is charged and how that affects solubility and diffusion of pharmaceuticals. (Try 9.53a and c, 9.85a and c)

Inquiry Question

How does the charge of a pharmaceutical change with pH?

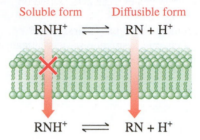

Most pharmaceuticals contain a weak acid or base in their structure. The charge of this group, usually an amine or a carboxylic acid, changes as the pH changes. This affects the drug's solubility and ability to diffuse across a cell membrane. The pK_a is specific for a given acid, while the pH can change in an aqueous solution. When the pH is lower than the pK_a of the weak acid, the acid form predominates. When the pH equals the pK_a, there are equal amounts of acid and conjugate base. When the pH is higher than the pK_a, the conjugate base form predominates. The ratio of conjugate base to acid present at any given pH can be determined using the Henderson-Hasselbalch equation.

9.8 Buffers and Blood—The Bicarbonate Buffer System

Apply the properties of a buffer to the bicarbonate buffer in the blood.

- Apply the properties of a buffer by predicting the direction in which the bicarbonate buffer equilibrium will shift with changes in ventilation rate. (Try 9.59, 9.83)

Inquiry Question

How do the properties of buffers apply to the bicarbonate buffer in the blood?

Buffer solutions consist of approximately equal amounts of a weak acid and its conjugate base. Buffers resist changes in pH when acid or base is added to a solution. The bicarbonate buffer is an important buffer system in the blood. Blood pH is maintained in a narrow range of 7.35–7.45. If the blood pH drops below this range, a condition called acidosis occurs. If the blood pH becomes elevated, a condition called alkalosis exists.

 Mastering Videos

PRACTICING THE CONCEPTS

The following videos can be accessed through the Pearson eText or your Mastering Chemistry course.

- Le Châtelier's Principle
- Identifying Conjugate Acids and Bases

Additional Problems

9.61 Identify the following as strong or weak acids:
  a. H_2SO_4 **b.** H_2CO_3 **c.** HCl   **d.** HF

9.62 Identify the following as strong or weak acids:
  a. HNO_3 **b.** HNO_2 **c.** H_3PO_4 **d.** HBr

9.63 What are some similarities and differences between strong and weak acids?

9.64 What are some ingredients found in antacids? What do they do?

9.65 Name and write the anion formed from the following acids.
  a. hydrofluoric acid, HF
  b. formic acid, HCOOH
  c. sulfurous acid, H_2SO_3

9.66 Name and write the anion formed from the following acids.
  a. hydrosulfuric acid, H_2S
  b. lactic acid, $CH_3CH(OH)COOH$
  c. carbonic acid, H_2CO_3

9.67 Complete and balance the following neutralization reactions:

a. $HCl(aq) + NH_4OH(aq) \rightarrow$

b. $H_2SO_4(aq) + KHCO_3(s) \rightarrow$

9.68 Complete and balance the following neutralization reactions:

a. $HNO_3(aq) + Ba(OH)_2(aq) \rightarrow$

b. $HBr(aq) + KOH(aq) \rightarrow$

9.69 A reaction that has reached equilibrium has which of the following properties?

a. The amount of the products is greater than the amount of reactants.

b. The amount of products is equal to the amount of reactants.

c. The rate of the forward reaction is greater than the rate of the reverse reaction.

d. The rate of the forward reaction is equal to the rate of the reverse reaction.

9.70 If the value of K for a given reaction is less than 1, which of the following is true?

a. The concentration of the products is greater than the concentration of the reactants.

b. The concentration of the products is less than the concentration of the reactants.

c. The concentration of the products is equal to the concentration of the reactants.

9.71 For the following reaction,

$$2HI(g) \rightleftharpoons H_2(g) + I_2(g)$$

a. Write an equilibrium constant expression.

b. If the value for $K = 2.1 \times 10^{-2}$ at 720 K, which substance predominates, HI or H_2 and I_2?

9.72 For the following reaction,

$$N_2O_4(g) \rightleftharpoons 2NO_2(g)$$

a. Write an equilibrium constant expression.

b. If the value for $K = 46$ at 500 K, which gas predominates, N_2O_4 or NO_2?

9.73 Consider the following reaction:

$$CO(g) + 2H_2(g) \rightleftharpoons CH_3OH(g) + Heat$$

a. Is the reaction exothermic or endothermic as written?

b. Provide an equilibrium expression (K) for the reaction.

c. How will the equilibrium shift if hydrogen gas is added to the reaction?

d. How will the equilibrium shift if carbon monoxide is removed from the reaction?

e. How will the equilibrium shift if the temperature is lowered?

9.74 Consider the following reaction:

$$\text{Glycogen}(aq) \text{ (stored glucose)} + H_2O(l)$$

$$\text{Blood glucose } (aq) + Heat$$

a. Is the reaction exothermic or endothermic as written?

b. Provide an equilibrium expression (K) for the reaction.

c. How will the equilibrium shift if blood glucose is removed from the reaction?

d. How will the equilibrium shift if the temperature is raised?

9.75 Write the formula of the conjugate base formed from each of the following weak acids.

a. nitrous acid, HNO_2

b. methylammonium, $CH_3NH_3^+$

c. formic acid, $HCOOH$

9.76 Write the formula of the conjugate acid formed from each of the following weak bases.

a. lactate, $CH_3CH(OH)COO^-$

b. aminomethane, CH_3NH_2

c. phosphate, PO_4^{3-}

9.77 Determine the pH for the following solutions. State whether each solution is acidic, basic, or neutral.

a. $[H_3O^+] = 2.5 \times 10^{-8}$ M

b. $[H_3O^+] = 7.0 \times 10^{-2}$ M

c. $[H_3O^+] = 4 \times 10^{-11}$ M

9.78 Determine the pH for the following solutions. State whether each solution is acidic, basic, or neutral.

a. $[H_3O^+] = 1 \times 10^{-4}$ M

b. $[H_3O^+] = 6 \times 10^{-9}$ M

c. $[H_3O^+] = 5.2 \times 10^{-2}$ M

9.79 Calculate the H_3O^+ concentration for a solution with the following pH:

a. 3.52 **b.** 5.0 **c.** 9.25

9.80 Calculate the H_3O^+ concentration for a solution with the following pH:

a. 8.2 **b.** 6.75 **c.** 12.1

9.81 Consider the acetic acid buffer system with acetic acid, CH_3COOH, and its salt sodium acetate, $NaCH_3COO$:

$$CH_3COOH(aq) + H_2O(l) \rightleftharpoons H_3O^+(aq) + CH_3COO^-(aq)$$

a. What is the purpose of a buffer system?

b. What is the purpose of $NaCH_3COO$ in the buffer?

c. How does the buffer react when acid (H_3O^+) is added?

d. How does the buffer react when base (OH^-) is added?

9.82 Consider the lactic acid buffer with lactic acid, $CH_3CH(OH)COOH$, and its salt sodium lactate, $NaCH_3CH(OH)COO$:

a. What is the purpose of a buffer system?

b. Why is a salt of the acid needed?

c. How does the buffer react when acid (H_3O^+) is added?

d. How does the buffer react when base (OH^-) is added?

9.83 In blood plasma, pH is maintained by the carbonic acid–bicarbonate buffer equilibrium as follows:

$$CO_2(g) + 2H_2O(l) \rightleftharpoons H_3O^+(aq) + HCO_3{}^-(aq)$$

a. If excess acid is added to the bloodstream by a physiological process other than breathing, what is this condition called?

b. Is the pH of the blood below or above normal?

c. How could such a condition be treated?

9.84 Adding a few drops of a strong acid to water will lower the pH appreciably. However, adding the same number of drops to a buffer does not appreciably alter the pH. Why?

pH = 7.0   pH = 3.0   pH = 7.0   pH = 6.9

Water   *Buffer*

9.85 Consider the compounds below, shown in their acid form, and answer the following questions for each compound.

a. Which form is charged: acid, conjugate base, or both?

b. Which form (acid, conjugate base, or both) will predominate in the first part of the small intestine (jejunum), where the pH is between 6 and 7?

c. Which form, acid or conjugate base, will be able to more easily diffuse through a cell membrane?

I. Acetoacetic acid, a ketone body, $pK_a = 3.59$

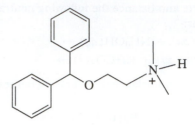

II. Diphenhydramine, an antihistamine, $pK_a = 8.98$

9.86 Consider the compounds below, shown in their conjugate base form, and answer the following questions for each compound.

a. Which form is uncharged: acid, conjugate base, or both?

b. Which form (acid, conjugate base, or both) will predominate in the first part of the small intestine (jejunum), where the pH is between 6 and 7?

c. Which form, acid or conjugate base, will be able to more easily diffuse through a cell membrane?

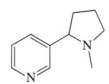

I. Epinephrine, used to treat allergic reactions, $pK_a = 8.59$

II. Nicotine, a stimulant found in tobacco smoke, $pK_a = 8.50$

9.87 Consider the sedative diazepam, which has a pK_a value of 3.40.

a. Will the acid or conjugate base form predominate at the following pH values: 2.75, 3.40, and 6.75?

b. For each pH value, is the given form charged or uncharged?

c. Calculate the ratio of [c. base]/[acid] at each pH.

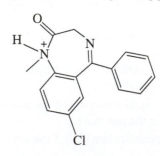

Diazepam, a sedative, $pK_a = 3.40$

9.88 Consider the stimulant ephedrine, which has a pK_a value of 9.65.

    a. Will the acid or conjugate base form predominate at the following pH values: 3.50, 9.65, and 13.85?

    b. For each pH value, is the given form charged or uncharged?

    c. Calculate the ratio of [c. base]/[acid] at each pH.

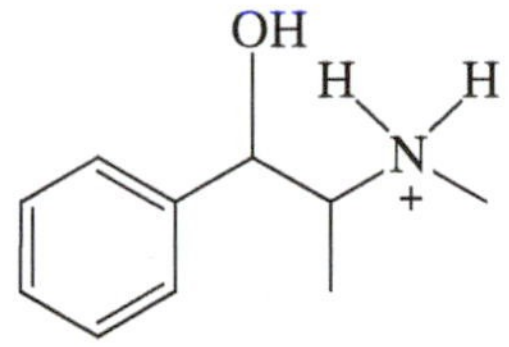

Ephedrine, a stimulant, pK_a = 9.65

9.89 The acid reflux drug Nexium® (esomeprazole) has a pK_a of 4.78. The conjugate base form (shown below) is found in the oral pill. Once ingested in the stomach (pH between 1 and 3), the acid form predominates and acts by blocking the H^+ channels found in the stomach lining so that acid production into the stomach slows.

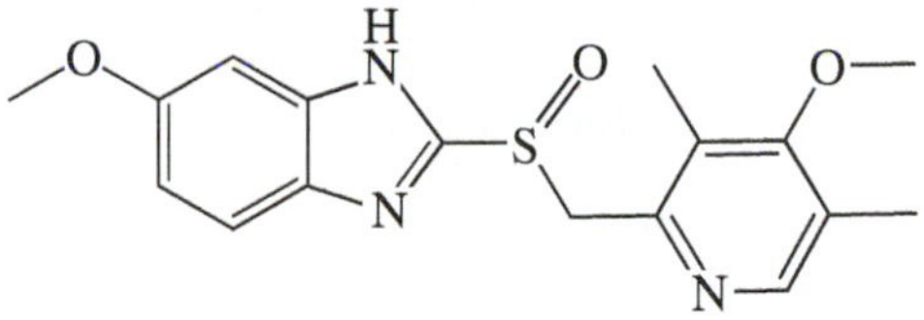

Nexium, used for acid reflux, pK_a = 4.78

    a. Which form of the drug is charged (acid or conjugate base)?

    b. Draw the acid form of Nexium. (*Hint:* The base is the secondary amine.)

Challenge Problems

9.90 Phenol is a very weak acid. Write the conjugate acid–base reaction for phenol in water.

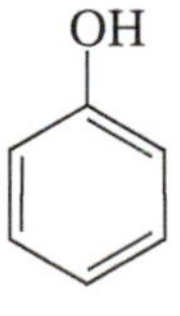

9.91 Le Châtelier's principle applies to oxygen binding to hemoglobin. Hemoglobin is a protein that binds and transports oxygen. It also has locations where it binds other molecules. These molecules include CO_2, H^+, and 2,3-bisphosphoglycerate (2,3-BPG). The equilibrium is shown below.

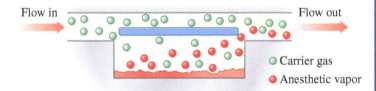

    a. If the concentration of oxygen (O_2) is high, which way does the equilibrium shift in the above equation?

    b. If the H^+ is high in the blood (acidosis condition), in which direction will the equilibrium shift?

    c. If CO_2 builds up in the bloodstream, which way will the equilibrium shift?

9.92 In a vaporizer, a carrier gas flows through a vaporizing chamber, allowing mixing with a liquid anesthetic.

$$\text{Inhaled anesthetic}_{\text{liquid}} \rightleftharpoons \text{Inhaled anesthetic}_{\text{vapor}}$$

Flow in   Flow out

○ Carrier gas
● Anesthetic vapor

Vaporizing chamber

Applying Le Châtelier's principle, as the vapor is removed from the vaporizing chamber with the carrier gas, in which direction will the equilibrium shown shift?

9.93 Amino acids, the building blocks of proteins, are small organic molecules that contain both an amine group and a carboxylic acid. Both of these groups are weak acids. The amino acid valine is shown. The pK_a for the protonated amine is 2.32 and the pK_a for the carboxylic acid is 9.62. Draw the structure of the dominant form of this amino acid in the stomach when the pH is 1.

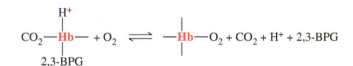

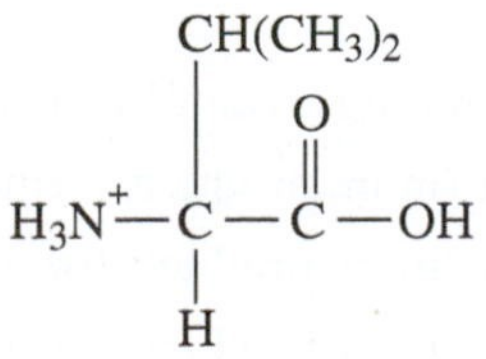

9.94 Naproxen, the active ingredient in Aleve®, has the structure shown. The carboxylic acid group has a pK_a of 4.2.

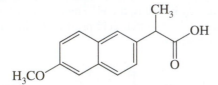

Naproxen, pK_a = 4.2

a. Draw the carboxylate form of naproxen.

b. If a person has diseased tissue with a lower pH, will more or less of the carboxylate form be present?

9.95 To determine the concentration of an unknown weak acid solution, a known amount of strong base can be titrated (added) to it until complete neutralization occurs (an endpoint is reached). If a 10.0-mL sample of vinegar (aqueous acetic acid, CH_3COOH) requires 16.5 mL of 0.500 M NaOH to reach the endpoint, what is the molarity of the acetic acid solution? (*Hint:* 1 mole of NaOH will neutralize 1 mole of acetic acid.)

Answers to Odd-Numbered Problems

Practice Problems

9.1 **a.** acid   **b.** base

c. acid   **d.** base

9.3 Most hydrogen atoms contain one proton, one electron, and no neutrons. Therefore, a positively charged hydrogen ion (H^+) is one proton.

9.5 **a.** HBr is the acid (proton donor) and H_2O is the base (proton acceptor).

b. H_2O is the acid (proton donor) and HCO_3^- is the base (proton acceptor).

9.7 The strong acids are **a.** H_2SO_4 and **b.** HCl.

9.9 The strong bases are **b.** KOH and **c.** LiOH.

9.11 The compound that completely ionizes in water is **a.** HNO_3. NH_4OH is a weak base and HF is a weak acid.

9.13 **a.** hydrobromic acid, HBr

b. chloric acid, $HClO_3$

c. nitrous acid, HNO_2

9.15 **a.** $HNO_3(aq) + LiOH(s) \rightarrow H_2O(l) + LiNO_3(aq)$

b. $H_2SO_4(aq) + Ca(OH)_2(s) \rightarrow 2H_2O(l) + CaSO_4(aq)$

9.17 **a.** $3HBr(aq) + Al(OH)_3(s) \rightarrow 3H_2O(l) + AlBr_3(aq)$

b. $2HI(aq) + CaCO_3(s) \rightarrow CaI_2(aq) + CO_2(g) + H_2O(l)$

9.19 Reversible reactions are reactions that can proceed in both the forward and reverse directions.

9.21 **a.** $K = \dfrac{[CO_2][H_2]}{[CO][H_2O]}$ **b.** $K = \dfrac{[CH_3COO^-][H_3O^+]}{[CH_3COOH]}$

9.23 **a.** reactants   **b.** products

c. both reactants and products present in equal amounts

9.25 **a.** shifts left (more reactants formed)

b. shifts right (more products formed)

c. shifts right (more products formed)

d. shifts right (more products formed)

9.27 **a.** shifts left (more reactants formed)

b. shifts right (more products formed)

c. shifts right (more products formed)

d. shifts left (more reactants formed)

9.29 The escape of CO_2 gas removes CO_2; equilibrium favors the formation of the products. The soda no longer bubbles.

9.31 **a.** $H_2PO_4^-$ **b.** HF   **c.** HBr

9.33 **a.** acid HSO_4^-; conjugate base SO_4^{2-}; base H_2O; conjugate acid H_3O^+

b. acid NH_4^+; conjugate base NH_3; base H_2O; conjugate acid H_3O^+

c. acid HCN; conjugate base CN^-; base NO_2^-; conjugate acid HNO_2

9.35 **a.** F^-, fluoride   **b.** OH^-, hydroxide

c. HCO_3^-, bicarbonate   **d.** SO_4^{2-}, sulfate

9.37 **a.** HCO_3^-, bicarbonate   **b.** H_3O^+, hydronium

c. H_3PO_4, phosphoric acid   **d.** HBr, hydrobromic acid

9.39 **a.** $HCOOH(aq) + H_2O(l) \rightleftharpoons HCOO^-(aq) + H_3O^+(aq)$
Formic acid   Base   Conjugate base   Conjugate acid

b. $H_2PO_4^-(aq) + H_2O(l) \rightleftharpoons HPO_4^{2-}(aq) + H_3O^+(aq)$
Acid   Base   Conjugate base   Conjugate acid

c. $HCO_3^-(aq) + H_2O(l) \rightleftharpoons CO_3^{2-}(aq) + H_3O^+(aq)$
Acid   Base   Conjugate base   Conjugate acid

9.41 **a.** basic   **b.** acidic

c. basic   **d.** acidic

9.43 **a.** basic   **b.** acidic

c. basic   **d.** neutral

9.45 **a.** 7.92   **b.** 2.15

c. 10.33   **d.** 7.00

9.47 **a.** 7.9×10^{-13} M   **b.** 3×10^{-6} M

c. 1.0×10^{-2} M   **d.** 5×10^{-9} M

9.49 a. $H_2PO_4^-$ **b.** H_2SO_4
   c. formic acid **d.** ammonium ion

9.51 pyruvic acid

9.53 a. acid **b.** acid **c.** conjugate base

9.55 At pH 3.00: **a.** The acid form predominates.
   b. uncharged **c.** [c. base]/[acid] = 0.014

   At pH 4.85: **a.** There are equal amounts of both forms present. **b.** not applicable **c.** [c. base]/[acid] = 1

   At pH 7.40: **a.** The conjugate base form predominates. **b.** charged **c.** [c. base]/[acid] = 350

9.57 Lidocaine. In both cases, the pH < pK_a, yet because lidocaine's pK_a is closer to physiological pH, there will be more of the conjugate base (the uncharged form) present and more will cross into the nerve cells, becoming the active form.

9.59 a. Hyperventilation will lower the CO_2 level in the blood, which decreases the H_3O^+ and increases the blood pH.

   b. The equilibrium will shift to the left.

   c. This condition is called respiratory alkalosis.

   d. Breathing into a bag will increase the CO_2 level, which increases H_3O^+ and lowers the blood pH.

   e. The equilibrium will shift to the right.

Additional Problems

9.61 a. strong **b.** weak
   c. strong **d.** weak

9.63 a. Both strong and weak acids produce H_3O^+ in water. Weak acids are only slightly ionized, whereas a strong acid exists as ions in solution (fully ionizes).

9.65 a. F^-, fluoride **b.** $HCOO^-$, formate
   c. HSO_3^-, hydrogen sulfite, and SO_3^{2-}, sulfite

9.67 a. $HCl(aq) + NH_4OH(aq) \rightarrow H_2O(l) + NH_4Cl(aq)$
   b. $H_2SO_4(aq) + 2KHCO_3(s) \rightarrow 2H_2O(l) + K_2SO_4(aq) + 2CO_2(g)$

9.69 d.

9.71 a. $K = \dfrac{[H_2][I_2]}{[HI]^2}$ **b.** The HI predominates.

9.73 a. exothermic **b.** $K = \dfrac{[CH_3OH]}{[CO][H_2]^2}$

   c. It will shift to the right. **d.** It will shift to the left.
   e. It will shift to the right.

9.75 a. NO_2^- **b.** CH_3NH_2 **c.** $HCOO^-$

9.77 a. 7.60, basic **b.** 1.15, acidic **c.** 10.4, basic

9.79 a. 3.0×10^{-4} M **b.** 1×10^{-5} M **c.** 5.6×10^{-10} M

9.81 a. A buffer system keeps the pH constant when small amounts of acid or base are added.

   b. The conjugate base, CH_3COO^-, from the salt $NaCH_3COO$ neutralizes any acid added.

   c. Added H_3O^+ reacts with CH_3COO^-.

   d. Added OH^- reacts with CH_3COOH.

9.83 a. metabolic acidosis **b.** below normal
   c. administer bicarbonate (HCO_3^-)

9.85 I. a. conjugate base **b.** conjugate base **c.** acid
   II. a. acid **b.** acid **c.** conjugate base

9.87 At pH 2.75: **a.** The acid form predominates.
   b. charged **c.** [c. base]/[acid] = 0.22

   At pH 3.40: **a.** There are equal amounts of acid and conjugate base. **b.** not applicable
   c. [c. base]/[acid] = 1

   At pH 6.75: **a.** The conjugate base form predominates.
   b. uncharged **c.** [c. base]/[acid] = 2200

9.89 a. The acid is charged.
   b.

Weak acid form

Challenge Problems

9.91 a. right, forming more of the oxygenated form of hemoglobin (oxyhemoglobin)

   b. left, forming more of the deoxygenated form of hemoglobin (deoxyhemoglobin)

   c. left, forming more of the deoxygenated form of hemoglobin (deoxyhemoglobin)

9.93

9.95

$$\frac{0.500 \text{ mole NaOH}}{1 \text{ L}} \times \frac{1 \text{ L}}{1000 \text{ mL}} \times 16.5 \text{ mL} = 0.00825 \text{ mole NaOH}$$

$$0.00825 \text{ mole NaOH} \times \frac{1 \text{ mole acetic acid}}{1 \text{ mole NaOH}} = 0.00825 \text{ mole acetic acid}$$

$$\frac{0.00825 \text{ mole acetic acid}}{10.0 \text{ mL}} \times \frac{1000 \text{ mL}}{1 \text{ L}} = 0.825 \text{ M acetic acid}$$

10 Proteins— Workers of the Cell

Our skin forms wrinkles as we age because two proteins in the skin, collagen and elastin, are produced in lower quantities. Our cells form collagen through a series of reactions catalyzed by enzymes, which are also proteins. To learn more about structural proteins like collagen and functional proteins like enzymes, read Chapter 10.

LEARNING TIP Make Word Associations

When learning the amino acid abbreviations, you can better retain information by making word and picture associations (also called mnemonic devices). For example, W is the abbreviation for the amino acid tryptophan, which doesn't make much sense until you say it like this: t**W**iptophan.

WHAT DO COLLAGEN, hemoglobin, lactase, ion channels, insulin, and gluten have in common? All are proteins. Much of the structure and function of our bodies is built on these large biomolecules. They transport oxygen in blood, serve as the main components of skin and muscle, defend our bodies against infection, direct cellular chemistry as enzymes, and control metabolism as hormones.

Proteins are polymers (*poly*—"many," *meros*—"part") composed entirely of amino acids covalently bonded in specific sequences. There are 20 different commonly occurring amino acids, and proteins can contain as few as 50 or even thousands of amino acids. The ordering of the amino acids in a protein determines both its structure and biological function.

Chapter 10 begins with amino acid structure. Next, we look at how strings of amino acids twist and fold to form unique three-dimensional structures with specific functions. We end with a specific type of protein called an *enzyme*, which reinforces the relationship between protein structure and function.

DISCOVERING THE CONCEPTS

? **INQUIRY ACTIVITY**—Exploring Amino Acids

Information

In plants and animals, a set of 20 amino acids is required to make proteins. Human beings can manufacture only 10 of them. The other 10 must be obtained through the diet. These latter 10 are termed *essential* amino acids. Dietary proteins are complete or incomplete depending on whether they contain all essential amino acids. Plant sources are more likely to contain incomplete proteins, but when combined, they can provide complete protein sources. For example, protein from rice is deficient in the amino acid lysine, but rich in the amino acid methionine. Beans are deficient in methionine and rich in lysine. So a meal of rice and beans can provide complete protein.

All 20 amino acids have some features in common:

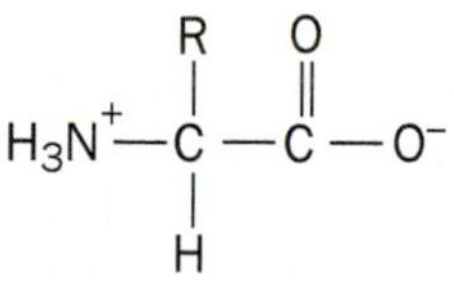

The general formula of an amino acid.

The R is the part of the 20 amino acids that differs. In amino acids, R is often referred to as the "side chain." All the naturally occurring amino acids are L-*enantiomers*. The 20 amino acids can be classified by the polarity of their side chain: polar neutral, polar charged (acidic or basic), or nonpolar.

At physiological pH, the amine group is protonated, and the carboxylic acid is a conjugate base (carboxylate). (See Section 9.7 for a review of pH and pK_a.) This neutral ionic form, with a net charge of zero, is called a *zwitterion*.

Questions

1. What is an *essential* amino acid?
2. Gelatin is produced from an animal protein called *collagen* that does not contain all 20 amino acids and lacks the essential amino acid tryptophan. Is gelatin considered a complete protein or an incomplete protein?
3. Identify the chiral center (using an asterisk) in the general formula of an amino acid shown in the preceding figure.
4. Name the two functional groups common to all 20 amino acids.
5. Locate the side chain (R) on each amino acid shown below. Identify the functional group(s) found in the side chain (R). Is the R group polar (neutral), polar (acidic), polar (basic), or nonpolar?

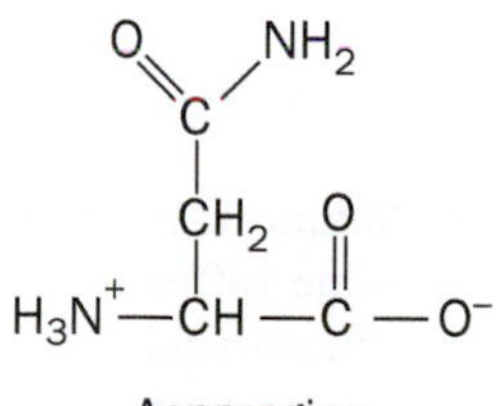

| Asparagine | Methionine | Threonine |
| Asn, N | Met, M | Thr, T |

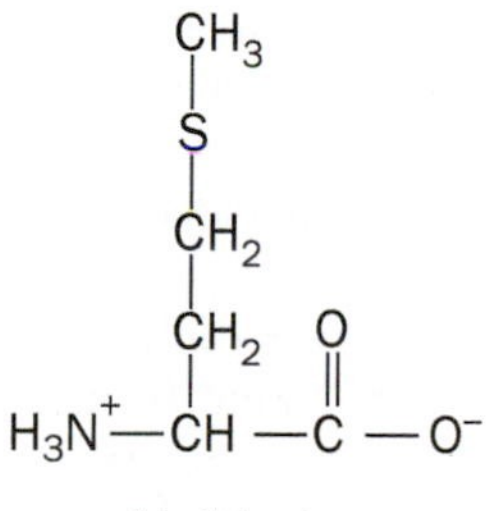

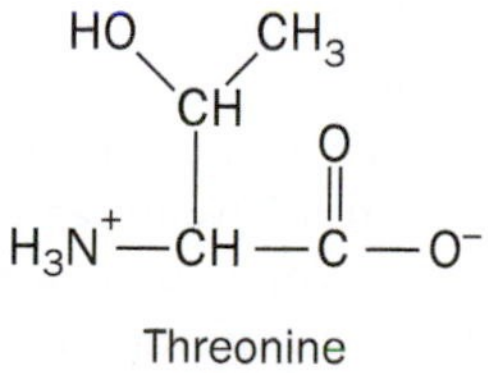

6. The amino acids shown represent L-amino acids. Draw the general structure for the corresponding D-amino acids.
7. The amine and carboxylic acid functional groups on an amino acid are predominantly charged at physiological pH, 7.4.
   a. Is the pK_a for the amine functional group above or below 7.4?
   b. Is the pK_a for the carboxylic acid functional group above or below 7.4?

? 10.1 Inquiry Question

What are the characteristics of an amino acid?

10.1 Amino Acids—The Building Blocks

The name *amino acid* gives us a clue to its structure. "Amino" reminds us that the structure contains a protonated amine ($-NH_3^+$), and the "acid" reminds us that it contains a carboxylic acid in its conjugate base form of carboxylate ($-COO^-$). These two functional groups are bonded to a central carbon atom called the **alpha (α) carbon.** The protonated amine bonded to this carbon is often referred to as the **alpha (α) amino group** and the carboxylate ion as the **alpha (α) carboxylate group.** The α carbon is also bonded to a hydrogen atom and a larger side chain designated by R, which is responsible for the unique identity and characteristics of each amino acid.

The α carbon is a chiral center because it has four different groups bonded to it in all the amino acids except for one, glycine, where the R group is H. As we saw in Chapter 4, compounds that contain a single chiral center can have two forms called *enantiomers,* which are nonsuperimposable mirror images. Only one enantiomer of each amino acid is found in proteins.

We can draw Fischer projections for amino acids like those drawn for carbohydrates. If we draw the carboxylate at the top and the R group at the bottom, the enantiomer with the protonated amine on the left is called the L-*stereoisomer* and its mirror image is the D-*stereoisomer.*

Each amino acid has two enantiomers. Only the L-amino acids are found in proteins.

The L-amino acids are the building blocks of proteins. Some D-amino acids do occur in nature but rarely in proteins.

The side chain, or R group, gives each amino acid its unique identity and characteristics. Twenty amino acids are found in most proteins. Nine different families of organic compounds are represented in the structures of the amino acid side chains— alkanes (hydrocarbon), aromatics, sulfides, alcohols, phenols, thiols, amides, carboxylic acids (present as carboxylate ions), and amines (present as protonated amines). You may want to refer back to Table 4.4 to get reacquainted with the structures of these functional groups.

The chemical properties and behaviors of the functional groups are present in the amino acids and are used to classify the amino acids into a few simple categories (nonpolar and polar—neutral, acidic, and basic). The structures of the amino acids— including their side chains, names, functional groups, and abbreviations—are shown in **TABLE 10.1**.

TABLE 10.1 The 20 Common Amino Acids in Proteins (Common Functional Groups Noted in Side Chains)

Nonpolar Amino Acids

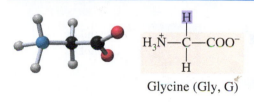

$$H_3\overset{+}{N}-\underset{\underset{H}{|}}{\overset{\overset{H}{|}}{C}}-COO^-$$

Glycine (Gly, G)

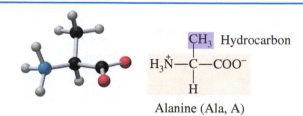

CH_3 Hydrocarbon

$$H_3\overset{+}{N}-\underset{\underset{H}{|}}{\overset{\overset{CH_3}{|}}{C}}-COO^-$$

Alanine (Ala, A)

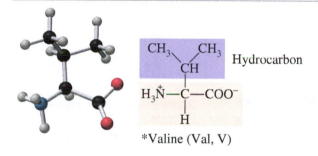

$CH_3 \quad CH_3$
$\underset{}{CH}$ Hydrocarbon

$$H_3\overset{+}{N}-\underset{\underset{H}{|}}{\overset{\overset{CH}{|}}{C}}-COO^-$$

*Valine (Val, V)

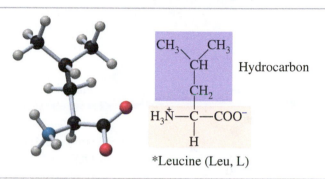

$CH_3 \quad CH_3$
CH Hydrocarbon
CH_2

$$H_3\overset{+}{N}-\underset{\underset{H}{|}}{\overset{\overset{CH_2}{|}}{C}}-COO^-$$

*Leucine (Leu, L)

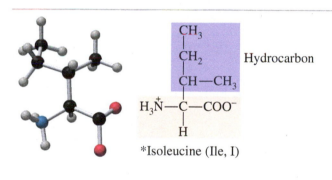

CH_3
CH_2 Hydrocarbon
$CH-CH_3$

$$H_3\overset{+}{N}-\underset{\underset{H}{|}}{\overset{\overset{CH-CH_3}{|}}{C}}-COO^-$$

*Isoleucine (Ile, I)

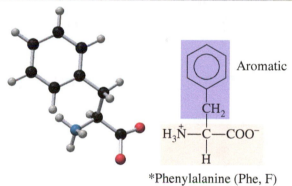

Aromatic

CH_2

$$H_3\overset{+}{N}-\underset{\underset{H}{|}}{\overset{\overset{CH_2}{|}}{C}}-COO^-$$

*Phenylalanine (Phe, F)

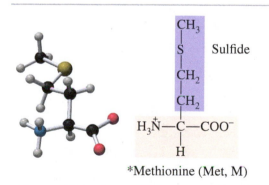

CH_3
S Sulfide
CH_2
CH_2

$$H_3\overset{+}{N}-\underset{\underset{H}{|}}{\overset{\overset{CH_2}{|}}{C}}-COO^-$$

*Methionine (Met, M)

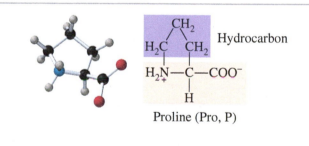

CH_2
$H_2C \quad CH_2$ Hydrocarbon

$$H_2\overset{+}{N}-\underset{\underset{H}{|}}{C}-COO^-$$

Proline (Pro, P)

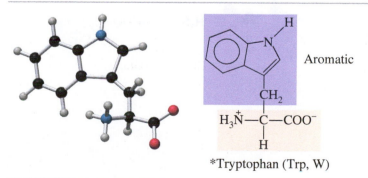

Aromatic

CH_2

$$H_3\overset{+}{N}-\underset{\underset{H}{|}}{\overset{\overset{CH_2}{|}}{C}}-COO^-$$

*Tryptophan (Trp, W)

—Continued next page

TABLE 10.1 The 20 Common Amino Acids in Proteins (Common Functional Groups Noted in Side Chains) (*continued*)

Polar Amino Acids (Neutral)

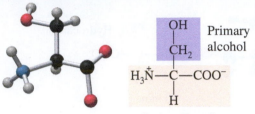

OH Primary
| alcohol
CH₂

$H_3\overset{+}{N}$—C—COO⁻
|
H

Serine (Ser, S)

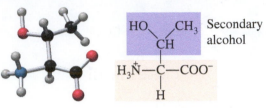

HO   CH₃ Secondary
 \   /   alcohol
  CH

$H_3\overset{+}{N}$—C—COO⁻
|
H

*Threonine (Thr, T)

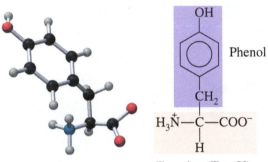

OH
|
⬡   Phenol
|
CH₂

$H_3\overset{+}{N}$—C—COO⁻
|
H

Tyrosine (Tyr, Y)

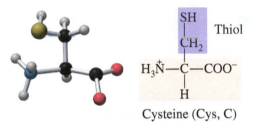

SH   Thiol
|
CH₂

$H_3\overset{+}{N}$—C—COO⁻
|
H

Cysteine (Cys, C)

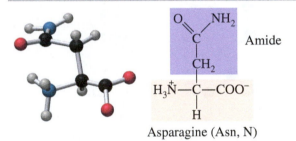

O   NH₂
\\ /
 C   Amide
 |
CH₂

$H_3\overset{+}{N}$—C—COO⁻
|
H

Asparagine (Asn, N)

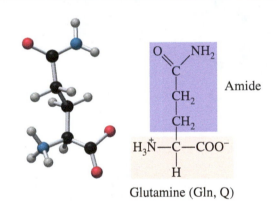

O   NH₂
\\ /
 C
 | Amide
CH₂
|
CH₂

$H_3\overset{+}{N}$—C—COO⁻
|
H

Glutamine (Gln, Q)

Polar Amino Acids (Acidic)

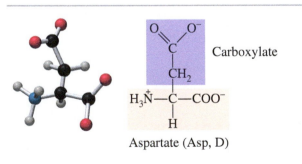

O   O⁻
\\ /
 C   Carboxylate
 |
CH₂

$H_3\overset{+}{N}$—C—COO⁻
|
H

Aspartate (Asp, D)

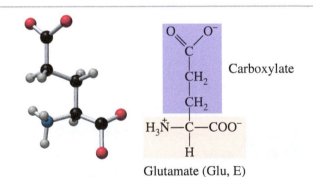

O   O⁻
\\ /
 C
 | Carboxylate
CH₂
|
CH₂

$H_3\overset{+}{N}$—C—COO⁻
|
H

Glutamate (Glu, E)

TABLE 10.1 The 20 Common Amino Acids in Proteins (Common Functional Groups Noted in Side Chains) *(continued)*

Polar Amino Acids (Basic)

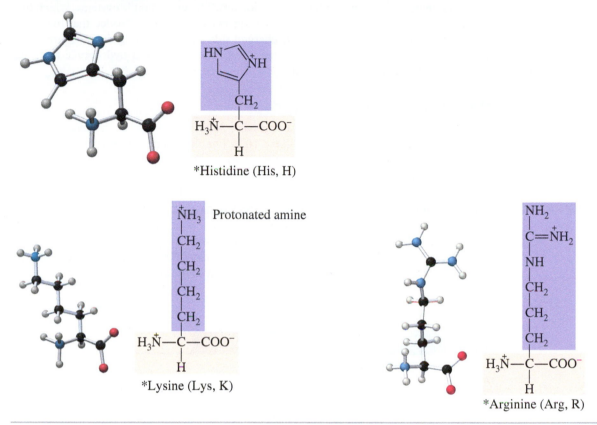

*Histidine (His, H)

Protonated amine

*Lysine (Lys, K)

*Arginine (Arg, R)

*Essential amino acids

The 10 amino acids designated with an asterisk (*) in Table 10.1 are called essential amino acids because they cannot be synthesized in the human body and must be obtained in the diet. Two of these amino acids, arginine and histidine, are essential in the diets of children but not adults. The nonessential amino acids can be converted to the essential amino acids through enzymatic reactions in the body.

Proteins that contain all of the essential amino acids are known as *complete proteins*. Soybeans and most proteins found in animal products such as eggs, milk, fish, poultry, and meat are complete proteins. Some proteins from plant sources, such as nut milks and beans, are *incomplete proteins* because they lack one or more of the essential amino acids. A complete protein meal can be obtained by combining foods like rice and beans or peanut butter on whole-grain bread.

Classification of Amino Acids

The amino acids in Table 10.1 are separated into two broad classifications based on the characteristics of the side chains: **nonpolar amino acids** and **polar amino acids.** The polar amino acids are further divided into neutral, acidic, and basic.

Now let's examine the side chains of the amino acids that are classified as "nonpolar." We see that, with few exceptions, the side chains of these amino acids are composed entirely of carbon and hydrogen. Recall from Chapter 7 that the carbon–hydrogen bond is nonpolar, and compounds composed of only carbon and hydrogen are nonpolar and hydrophobic (water-fearing). So, the side chains of this group of amino acids are nonpolar and hydrophobic. In the case of tryptophan, the nitrogen atom makes up such a small part of the side chain that it contributes very little to its polarity. In the case of methionine, the C—S bonds present are nonpolar because C and S have the same electronegativity—see Chapter 3, Figure 3.14.

Continue by examining the side chains of each of the amino acids classified as "polar." These side chains contain functional groups such as hydroxyl (—OH) and amide (—CONH$_2$) that create an uneven distribution of electrons in the side chain, making them polar. All but one of the polar side chains can form hydrogen bonds with water and, therefore, are hydrophilic (water-loving). The exception is cysteine, which has a polar thiol (—SH) group but does not form hydrogen bonds. Notice that the polar acidic and polar basic amino acids have charged side chains, allowing them to form ion–dipole interactions with water, making them even more polar and hydrophilic than the amino acids classified as polar neutral.

Abbreviations for Amino Acids

Amino acids represent the vocabulary of proteins. Amino acids have both three-letter and one-letter abbreviations, as shown in Table 10.1. The three-letter abbreviations were the first used, but as scientists began to sequence longer chains of amino acids, they reduced the abbreviations to a single letter to transfer data more easily. The one-letter amino acid abbreviations were developed by Dr. Margaret Oakley Dayhoff, a pioneer in protein sequencing. Of the 20 amino acids, 11 abbreviations start with the first letter of their name, six because they are the only amino acid that starts with that letter and five because they are the more common amino acid that starts with the letter. The remaining nine amino acids have different letters. **TABLE 10.2** provides information about the origin of the one-letter abbreviations that can help you remember the abbreviations.

TABLE 10.2 Origins of One-Letter Amino Acid Codes

Amino Acid	Three-Letter Code	One-Letter Code	Rationale for One-Letter Code
Cysteine	Cys	C	Only amino acid that starts with **the letter**
Histidine	His	H	
Isoleucine	Ile	I	
Methionine	Met	M	
Serine	Ser	S	
Valine	Val	V	
Alanine	Ala	A	Most common amino acid that starts with **the letter**
Glycine	Gly	G	
Leucine	Leu	L	
Proline	Pro	P	
Threonine	Thr	T	
Arginine	Arg	R	Name starts with the R sound, a**R**ginine
Phenylalanine	Phe	F	Name starts with the F sound, **F**enylalanine
Tyrosine	Tyr	Y	Name contains a **Y**
Tryptophan	Trp	W	t**W**iptophan
Aspartate	Asp	D	Aspar**D**ate
Asparagine	Asn	N	Contains an N, asparagi**N**e
Glutamate	Glu	E	glu**E**tamate; or glutamate is longer than aspartate and E comes after D in the alphabet
Glutamine	Gln	Q	**Q**-tamine
Lysine	Lys	K	**K** is near L in the alphabet

| Sample Problem 10.1 | **Identifying Amino Acids** |

For the following amino acids:
1. Provide the name and the three-letter and one-letter abbreviations.
2. State whether the side chain is polar (neutral), polar (acidic), polar (basic), or nonpolar.
3. State whether the side chain is hydrophilic or hydrophobic.
4. Identify the functional group in the side chain.

a.

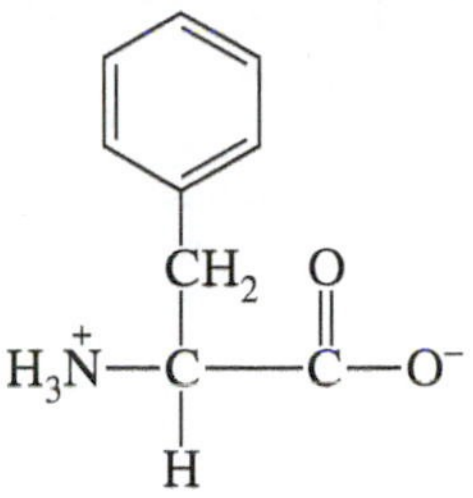

b.

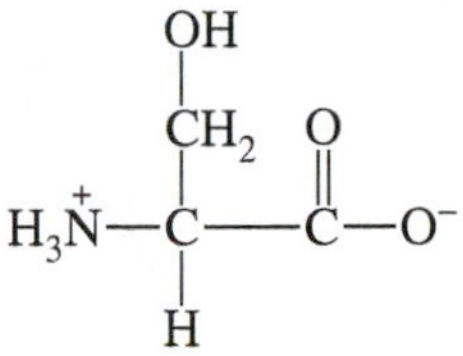

OH
|
CH_2 O
| ‖
$H_3\overset{+}{N}$—C—C—O⁻
|
H

$H_3\overset{+}{N}$—C—C—O⁻ (a. structure with phenyl-CH_2 side chain)

Solution

a. Referring to Table 10.1,
  1. phenylalanine, abbreviated Phe or F   2. nonpolar
  3. hydrophobic   4. aromatic hydrocarbon

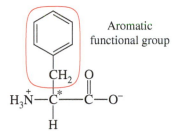

  Aromatic functional group

b. Referring to Table 10.1,
  1. serine, abbreviated Ser or S   2. polar (neutral)
  3. hydrophilic   4. alcohol (primary)

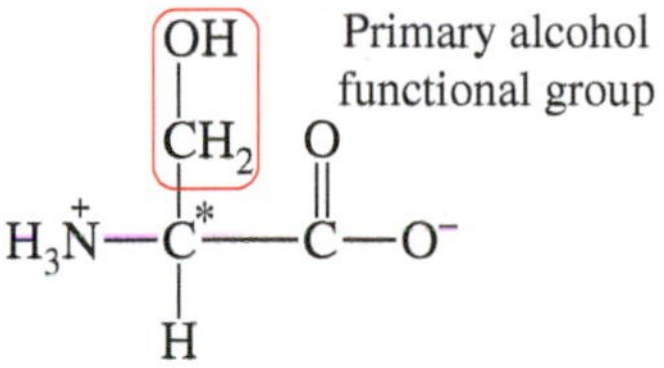

OH — Primary alcohol functional group
|
CH_2 O
| ‖
$H_3\overset{+}{N}$—C*—C—O⁻
|
H

Amino Acids as Weak Acids

Because all amino acids have amine and carboxylic acid functional groups, they have charged and uncharged forms. Amino acids and many other biological molecules contain more than one weak acid and have more than one pK_a value, so they exist in different acid/base forms depending on the pH of the solution. The amino acid alanine shown in the margin is in the form that predominates at physiological pH, 7.4. The amine group is protonated (acid form, charged) and the carboxylic acid group is deprotonated (conjugate base, charged). This ionic form containing no net charge (+ and − cancel each other out) is called a **zwitterion.**

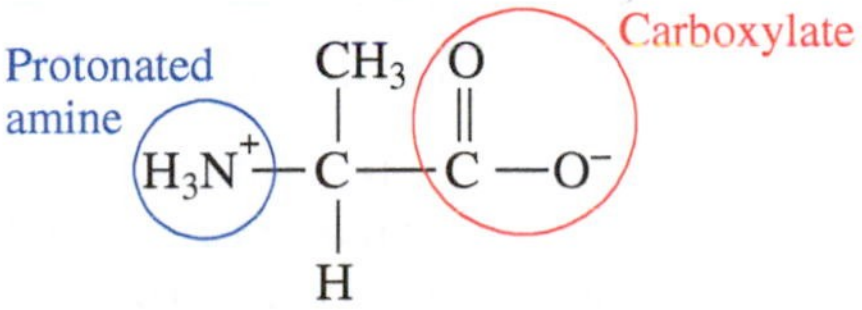

Each amino acid has a unique pH value at which only the zwitterion form is present. This point is called the **isoelectric point (pI).** At the pI of alanine, the negative charge on the carboxylate is balanced by the positive charge on the protonated amine, and the net charge of the amino acid is zero. The pI for the amino acid alanine is 6.0. Its pI is half-way between the pK_a values for the protonated amine and the carboxylic acid, 2.3 and

Protonated amine
Carboxylate

CH_3 O
| ‖
H_3N^+—C—C—O⁻
|
H

Alanine

9.7, respectively. In general, for amino acids, the pK_a for the carboxylic acid is around 2 and the pK_a for the amine is over 9.

In the figure below, the amino acid alanine is shown at different pH values. At pH values below the pI, the acidic forms of the functional groups (protonated) predominate. At pH values above the pI, the basic forms of the functional groups (deprotonated) predominate.

$$H_3N^+\!\!-\!\!\overset{\overset{\displaystyle CH_3}{|}}{\underset{\underset{\displaystyle H}{|}}{C}}\!\!-\!\!\overset{\overset{\displaystyle O}{\|}}{C}\!\!-\!\!OH \quad \underset{\text{Acid added}}{\overset{\text{Base added}}{\rightleftharpoons}} \quad H_3N^+\!\!-\!\!\overset{\overset{\displaystyle CH_3}{|}}{\underset{\underset{\displaystyle H}{|}}{C}}\!\!-\!\!\overset{\overset{\displaystyle O}{\|}}{C}\!\!-\!\!O^- \quad \underset{\text{Acid added}}{\overset{\text{Base added}}{\rightleftharpoons}} \quad H_2N\!\!-\!\!\overset{\overset{\displaystyle CH_3}{|}}{\underset{\underset{\displaystyle H}{|}}{C}}\!\!-\!\!\overset{\overset{\displaystyle O}{\|}}{C}\!\!-\!\!O^-$$

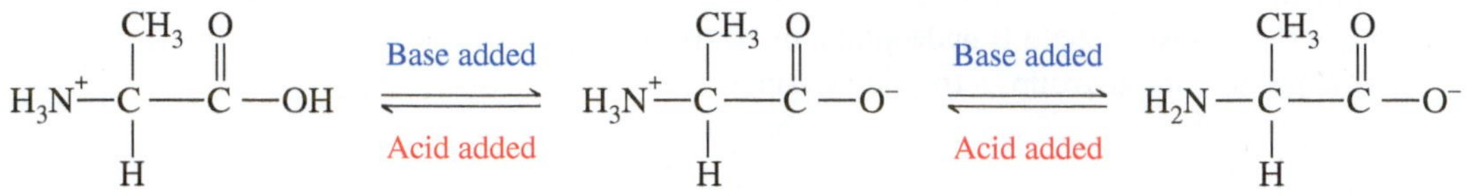

Alanine ion
pH lower (more acidic) than 6
(net charge = 1+)

Alanine zwitterion
pH = pI = 6.0
(net charge = 0)

Alanine ion
pH higher (more basic) than 6
(net charge = 1–)

$pK_{a-COOH} = 2.3$

$pK_{a-\overset{+}{N}H_3} = 9.7$

At acidic pH, both amine and carboxylic acid functional groups are acids (protonated). At the isoelectric point (when pH = pI), the amine group is still an acid and the carboxylic acid is a conjugate base (carboxylate). At basic pH, both functional groups are conjugate bases (deprotonated). The net charge on alanine changes as the pH changes.

Mastering Video
Practicing the Concepts
pH and Amino Acid Structure

Sample Problem 10.2   Amino Acids in Acid or Base

Write the zwitterionic form of the amino acid cysteine shown. If the pI for cysteine is 5.1, what will the net charge of the amino acid be at the following pH values?

a. 2.0   b. 5.1   c. 10.5

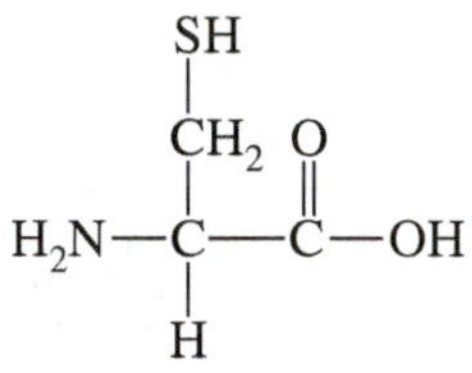

$$H_2N\!\!-\!\!\overset{\overset{\displaystyle CH_2}{\overset{|}{\overset{\displaystyle SH}{|}}}}{\underset{\underset{\displaystyle H}{|}}{C}}\!\!-\!\!\overset{\overset{\displaystyle O}{\|}}{C}\!\!-\!\!OH$$

Cysteine

Solution

The zwitterionic form contains a protonated amine and carboxylate group.

a. At a pH of 2.0, the amine and the carboxylic acid are in their acidic forms. The net charge is 1+.
b. When pH = pI, the zwitterion form is present and the net charge is 0.
c. At a pH of 10.5, the amine and carboxylic acid are in their basic forms. The net charge is 1–.

$$H_3N^+\!\!-\!\!\overset{\overset{\displaystyle CH_2}{\overset{|}{\overset{\displaystyle SH}{|}}}}{\underset{\underset{\displaystyle H}{|}}{C}}\!\!-\!\!\overset{\overset{\displaystyle O}{\|}}{C}\!\!-\!\!O^-$$

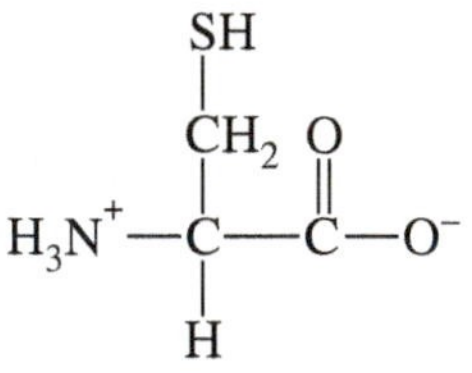

Cysteine zwitterion

Practice Problems

Find the answers to the odd-numbered Practice Problems at the end of the chapter.

10.1 Draw the structure for each of the following amino acids and put an asterisk (*) next to any chiral carbon centers in your structure:

a. alanine   b. lysine

c. tryptophan   d. aspartate

10.2 Classify each of the amino acids in Problem 10.1 as polar (neutral, acidic, or basic) or nonpolar and as hydrophobic or hydrophilic.

10.3 Give the three-letter and one-letter abbreviations and identify the functional group in the side chain for each amino acid in Problem 10.1.

10.4 Draw the structure for each of the following amino acids and put an asterisk (*) next to any chiral carbon centers in your structure:

a. leucine   b. glutamate

c. methionine   d. threonine

10.5 Classify each amino acid in Problem 10.4 as polar (neutral, acidic, or basic) or nonpolar and as hydrophobic or hydrophilic.

10.6 Give the three-letter and one-letter abbreviations and identify the functional group in the side chain for each amino acid in Problem 10.4.

10.7 Draw the structure for the amino acid represented by each of the following abbreviations:

    a. G   b. His

    c. Q   d. Ile

10.8 Draw the structure for the amino acid represented by each of the following abbreviations:

    a. Pro   b. N

    c. Val   d. Y

10.9 Give the names of each amino acid you drew in Problem 10.7.

10.10 Give the names of each amino acid you drew in Problem 10.8.

10.11 Isoleucine has the zwitterion structure shown. Draw the structure and give the net charge of isoleucine that will predominate at the indicated pH values (pI = 6.0).

    a. pH 1.8

    b. pH 6.0

    c. pH 11.5

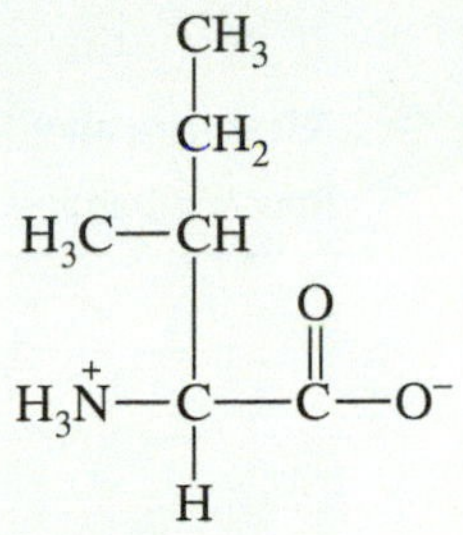

Isoleucine structure:

$$CH_3 - CH_2 - H_3C-CH$$
$$H_3\overset{+}{N}-C-C-O^- \quad (H)$$
Isoleucine

10.12 Glycine has the zwitterion structure shown. Draw the structure and give the net charge of glycine that will predominate at the indicated pH values (pI = 6.0).

    a. pH 1.5

    b. pH 12.0

    c. pH 6.0

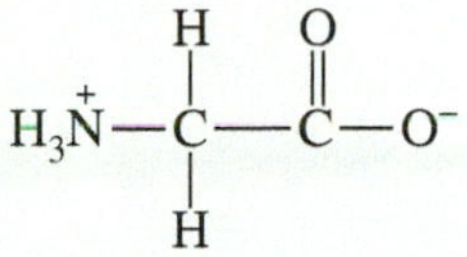

Glycine structure:

$$H \quad O$$
$$H_3\overset{+}{N}-C-C-O^-$$
$$H$$
Glycine

DISCOVERING THE CONCEPTS

❓ INQUIRY ACTIVITY—Condensation and the Peptide Bond

Information

Amino acids combine to form peptides through a *condensation* (also known as *dehydration*) reaction. One water molecule is removed when a carboxylate and protonated amine react. The condensation of two amino acids is shown.

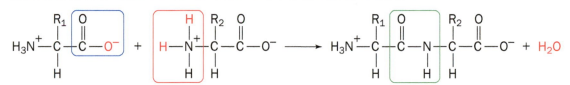

Amino acid 1   Amino acid 2   Peptide bond

When two amino acids combine as shown, a *dipeptide* is formed. A string of many amino acids bonded through peptide bonds is a *polypeptide*. A *protein* is a polypeptide that has a biological function.

Questions

1. Name the organic functional group in a peptide bond.
2. In the preceding condensation reaction, a dipeptide is formed. How many amino acids would be present in a tetrapeptide? _____ How many peptide bonds would be formed in a tetrapeptide? _____
3. The breaking of a peptide bond occurs through a hydrolysis reaction. In the body, these reactions are catalyzed by enzymes called *proteases*. Draw the products of the following reaction if the dipeptide undergoes hydrolysis.

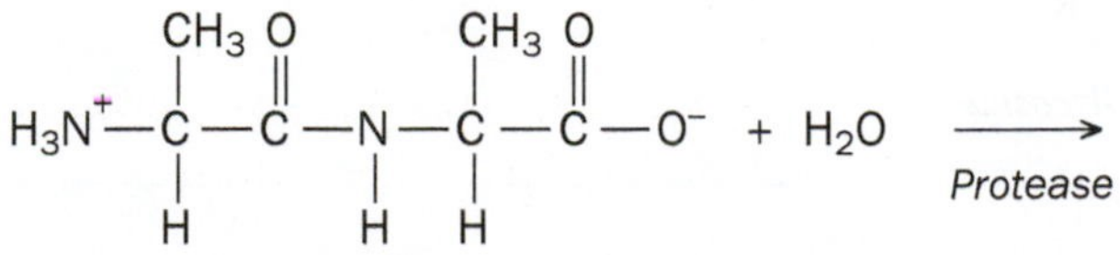

$$CH_3 \quad O \quad\quad CH_3 \quad O$$
$$H_2\overset{+}{N}-C-C-N-C-C-O^- \;+\; H_2O \xrightarrow{\text{Protease}}$$
$$H \quad\quad H \; H$$

The dipeptide Ala—Ala

10.2 Protein Formation

? 10.2 Inquiry Question

How is a peptide bond formed?

In Chapter 6, we showed how two monosaccharides combine through a condensation reaction to form a disaccharide. In Chapter 7, we saw that a triglyceride contains a glycerol and three fatty acid molecules. These also combine through a condensation reaction.

Condensation reactions can also occur between amino acids, and the product is a dipeptide. In this case, the carboxylate ($-COO^-$) of one amino acid molecule reacts with the protonated amine ($-NH_3^+$) of a second amino acid. A water molecule is removed, and an **amide** functional group is formed.

Amides are a family of organic compounds characterized by a functional group containing a carbonyl ($C=O$) bonded to a nitrogen atom. The nitrogen of the amide is bonded to two other atoms, which may be two hydrogen atoms, two carbon atoms, or one carbon atom and one hydrogen atom.

Biological Condensation Reactions

What do most condensation reactions have in common? The reacting functional groups have an $-H$ and an $-OH$ that can be removed during the condensation, resulting in different functional groups, as outlined in **TABLE 10.3**. Remember from Chapter 5 that a condensation reaction operating in reverse is a hydrolysis reaction and the $-H$ and $-OH$ are added to the atoms whose bond was hydrolyzed.

Before we continue discussing protein formation, let's review condensation and hydrolysis reactions as they apply to the biomolecules we have seen.

TABLE 10.3 Common Condensation Reactions Between Functional Groups

Condensation Reaction	Representative Reaction in Biomolecules
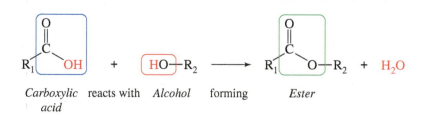 *Carboxylic acid* reacts with *Alcohol* forming *Ester*	Triglyceride and phospholipid formation between fatty acids and glycerol
(see below) *Carboxylate* reacts with *Protonated amine* forming *Amide*	Peptide bond formation between amino acids
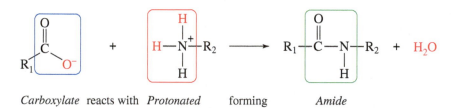 *Hemiacetal* reacts with *Alcohol* forming *Glycoside*	Glycosidic bond formation between monosaccharides

Writing Condensation and Hydrolysis Products

Write the products for the reaction shown.

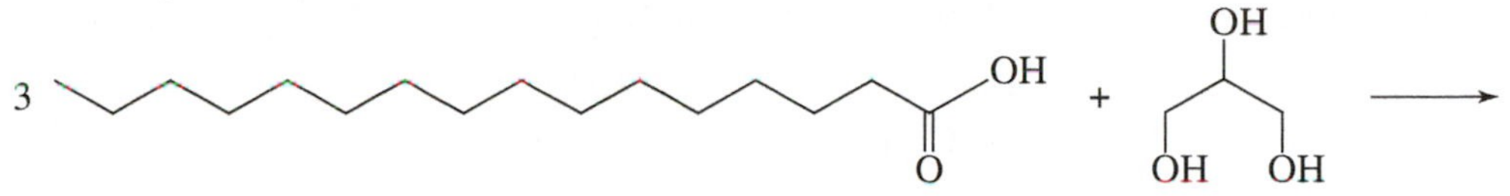

3 [16:0 fatty acid structure] OH + [glycerol structure with OH, OH, OH] ⟶

Solution

STEP 1 Decide whether a condensation or hydrolysis is occurring. Molecules are combining, so this is a condensation.

STEP 2 Identify the functional groups present. In this problem, three carboxylic acids (orange circle; one in each of the three [16:0] fatty acid molecules) and three alcohol functional groups (purple ovals) are present. These will combine to form esters.

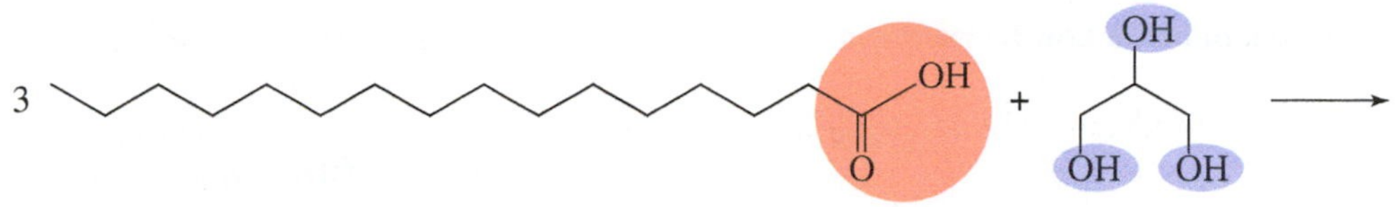

3 [16:0 fatty acid structure] OH + [glycerol structure with OH, OH, OH] ⟶

STEP 3 Write the products. In a condensation, an —H and —OH will be removed from the functional groups, forming three molecules of water. The resulting triglyceride molecule contains three ester functional groups. The —H and —OH are highlighted in color to show the functional group they came from.

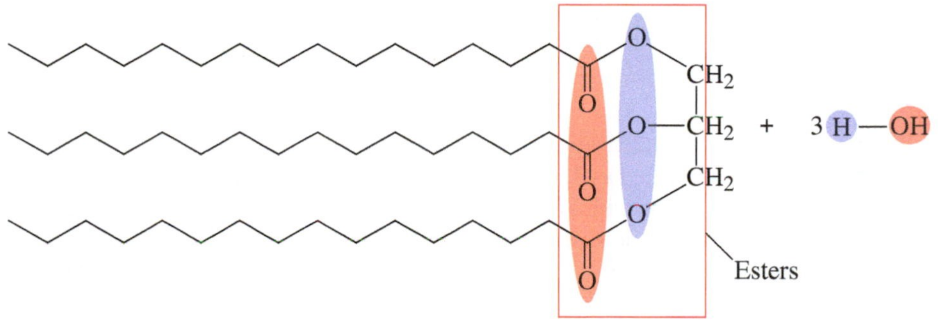

[triglyceride structure] + 3 H—OH

Esters

Sample Problem 10.3   Writing Condensation and Hydrolysis Products

Write the products for the reaction shown.

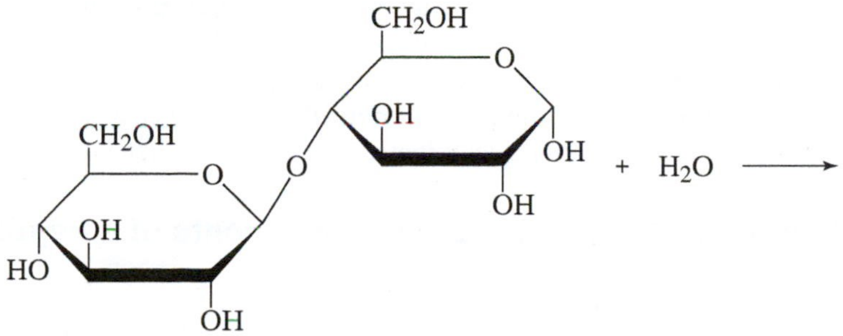

[disaccharide structure with CH₂OH groups] + H₂O ⟶

Solution

STEP 1 Decide whether a condensation or hydrolysis is occurring. A disaccharide is reacting with water, so a hydrolysis is occurring.

STEP 2 Identify the functional groups present. In this problem, there are several alcohol groups and some ether groups.

—Continued next page

Continued—

STEP 3 Write the products. A hydrolysis is going to break the glycosidic bond between the two monosaccharides. The atoms from water are added as an —H and —OH (shown in red). The positioning of the —OH groups formed at the glycosidic bond remains as it was prior to hydrolysis.

Amides and the Peptide Bond

When amino acids combine in a condensation reaction, the amide bond that is formed is called a **peptide bond.**

Alanine (Ala, A)   Valine (Val, V)   (Ala—Val, AV)

The product of the condensation reaction of alanine and valine, as shown, is a **dipeptide,** which is represented as Ala—Val, or AV. In this dipeptide, alanine is located at the **N-terminus** (or amino terminus) because it has an unreacted (free) α-amino group. Similarly, valine is located at the **C-terminus** (or carboxy terminus) because it has an unreacted carboxylate group. By convention, peptides (and larger peptide-containing structures) are always written from the N-terminus to the C-terminus (left to right) whether using full structures or the one- or three-letter abbreviations.

Dipeptides can be formed from any two amino acids. Each pair of different amino acids can combine in two ways, producing two different dipeptides. The two dipeptides, Ala—Val and Val—Ala, are structural isomers, different compounds, and have different properties. The amino acid sequence is critical to both its structure and the function.

Dipeptides are the smallest peptides. Any compound containing amino acids joined by a peptide bond can be called a peptide. A compound with three amino acids is a tripeptide, one with four amino acids is a tetrapeptide, and so on. As the number of amino acids increases, the compound is referred to as a **polypeptide.** A biologically active polypeptide typically containing 50 or more amino acids joined together by peptide bonds is a **protein.**

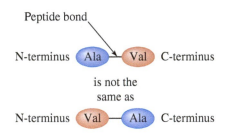

Sample Problem 10.4 Identifying the Components of a Peptide

Consider the following dipeptide:

a. Circle the amino acid at the N-terminus, and give its name.

b. Put a square around the amino acid at the C-terminus, and give its name.

c. Locate the peptide bond by drawing an arrow to it.

d. Give the three-letter and one-letter abbreviations for this dipeptide.

Solution

a–c.

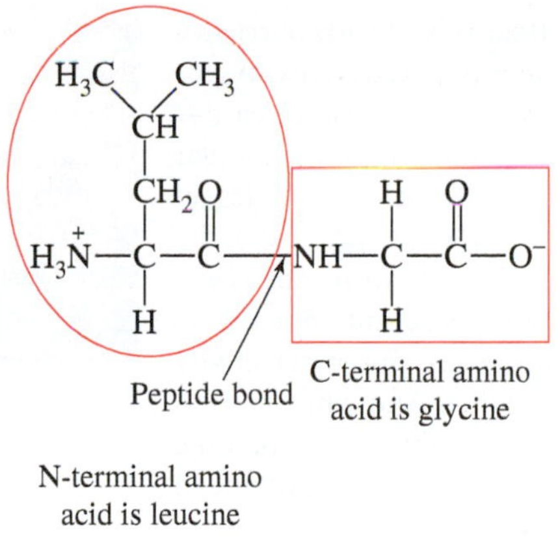

Peptide bond

N-terminal amino
acid is leucine

C-terminal amino
acid is glycine

d. The dipeptide is Leu—Gly, or LG.

Peptides Cause Problems in Celiac Disease

The substance *gluten* is a mixture of proteins found in the grains of wheat, rye, barley, and oats. Wheat flour is the main ingredient in most breads and pastas and is prevalent in the Western diet. The two main proteins in gluten, glutenin and gliadin, contain regions rich in the amino acids proline and glutamine—a unique amino acid sequence.

When gluten enters the digestive system, some of the peptide bonds are hydrolyzed by enzymes called proteases, which break down the gluten, leaving behind proline- and glutamine-rich peptides in the digestive system. As shown in the figure below, some of the glutamines in these fragments are modified to glutamate by an enzyme classified as a glutaminase, which changes the side chain in glutamine from an amide ($-CONH_2$) to a carboxylate group ($-COO^-$). This transition changes a polar neutral residue to a polar acidic residue. The addition of the negative charge to these peptide sequences causes people with the autoimmune disorder *celiac disease* to launch an immune response against these peptides. Symptoms include inflammation of the small intestine, low intestinal absorption of nutrients, and malnutrition.

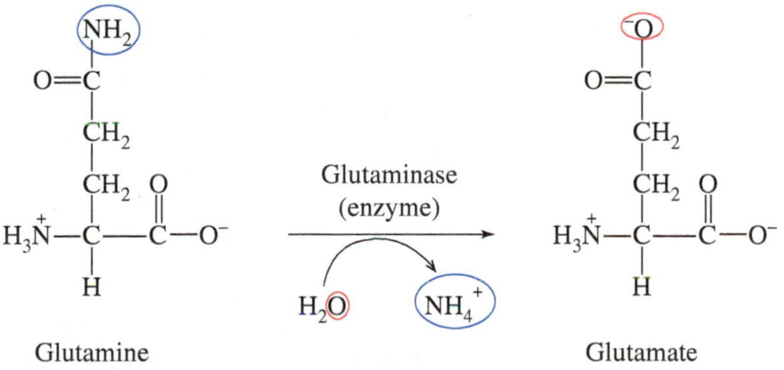

Glutamine

Glutaminase
(enzyme)

Glutamate

The change of glutamine to glutamate introduces additional negative charges to peptide sequences in the gut, which causes inflammation in people with celiac disease. The enzyme glutaminase substitutes the oxygen from a water molecule (orange circles) for the NH_2 on the side chain of glutamine, producing ammonium ions, NH_4^+ (blue circles). If not needed by the body, the ammonium is excreted.

People with celiac disease inherit a certain set of proteins as part of their immune system that attack the modified gluten peptides as if they were foreign substances that would harm the body. Only about 1% of the population has celiac disease, but such people must avoid gluten at all costs and need to supplement their diet with vitamins

—*Continued next page*

INTEGRATING Chemistry

◀ Find out why the wheat proteins in gluten cause illness in some people.

Continued—

due to their inability to effectively absorb nutrients. Up to 6% of the population report sensitivity to gluten, which—while less serious than celiac disease—causes digestive problems, with symptoms similar to irritable bowel syndrome. Such sensitivities in the general population have led to an increase in the availability of gluten-free products in recent years. While the cause of this non-celiac gluten sensitivity is still uncertain, people with gluten sensitivity should avoid gluten in their diet. And while gluten-free foods may be considered healthy, a person on a gluten-free diet must be sure to consume a balanced diet to maintain proper nutrition.

Practice Problems

10.13 Write the products for the following condensation or hydrolysis reactions:

a.

$$H_3\overset{+}{N}-\underset{\underset{H}{|}}{\overset{\overset{OH}{|}{\overset{|}{CH_2}}}{C}}-\overset{O}{\overset{||}{C}}-O^- \ + \ H_3\overset{+}{N}-\underset{\underset{H}{|}}{\overset{\overset{H}{|}}{C}}-\overset{O}{\overset{||}{C}}-O^- \longrightarrow$$

b.

$$H_3C-\overset{O}{\overset{||}{C}}-O-(CH_2)_7CH_3 \ + \ H_2O \longrightarrow$$

Octylacetate, found in citrus oil

10.14 Write the products for the following condensation or hydrolysis reactions:

a.

Isomaltose

$$+ \ H_2O \longrightarrow$$

b.

$$H_3\overset{+}{N}-\underset{\underset{H}{|}}{\overset{\overset{CH_2}{|}}{C}}-\overset{O}{\overset{||}{C}}-\underset{\underset{H}{|}}{N}-\underset{\underset{H}{|}}{\overset{\overset{H}{|}}{C}}-\overset{O}{\overset{||}{C}}-\underset{\underset{H}{|}}{N}-\underset{\underset{H}{|}}{\overset{\overset{CH_2}{|}{\overset{|}{OH}}}{C}}-\overset{O}{\overset{||}{C}}-O^- \ + \ 2H_2O \longrightarrow$$

10.15 Consider the following tripeptide:

$$H_3\overset{+}{N}-\underset{\underset{H}{|}}{\overset{\overset{CH_2}{|}{\overset{CH_2}{|}{\overset{O}{\underset{NH_2}{\overset{\diagup}{C}}}}}}{C}}-\overset{O}{\overset{||}{C}}-\underset{\underset{H}{|}}{N}-\underset{\underset{H}{|}}{\overset{\overset{CH_2}{|}}{C}}-\overset{O}{\overset{||}{C}}-\underset{\underset{H}{|}}{N}-\underset{\underset{H}{|}}{\overset{\overset{CH_3}{\underset{\diagup}{\overset{HO}{\overset{|}{CH}}}}}{C}}-\overset{O}{\overset{||}{C}}-O^-$$

a. Circle the N-terminal amino acid, and give its name. Draw a square around the C-terminal amino acid, and give its name.

b. Give the one-letter and three-letter abbreviations of this tripeptide.

10.16 Consider the following tripeptide:

$$H_3\overset{+}{N}-\underset{\underset{H}{|}}{\overset{\overset{CH_3}{|}}{C}}-\overset{O}{\overset{||}{C}}-\underset{\underset{H}{|}}{N}-\underset{\underset{H}{|}}{\overset{\overset{CH_2}{|}{\overset{CH_2}{|}{\overset{O}{\underset{O^-}{\overset{\diagup}{C}}}}}}{C}}-\overset{O}{\overset{||}{C}}-\underset{\underset{H}{|}}{N}-\underset{\underset{H}{|}}{\overset{\overset{CH_2}{\overset{/\ \ \backslash}{H_2C\ \ \ \ CH_2}}}{C}}-\overset{O}{\overset{||}{C}}-O^-$$

a. Circle the N-terminal amino acid, and give its name. Draw a square around the C-terminal amino acid, and give its name.

b. Give the one-letter and three-letter abbreviations of this tripeptide.

10.17 Draw the structural formula for each of the following peptides:

    a. Phe—Glu   b. KCG

    c. His—Met—Gln   d. WY

10.18 Draw the structural formula for each of the following peptides:

    a. Ala—Asn—Thr   b. DS

    c. Val—Arg   d. IYP

10.19 Identify the N-terminus and the C-terminus for each of the peptides in Problem 10.17.

10.20 Identify the N-terminus and the C-terminus for each of the peptides in Problem 10.18.

10.3 The Three-Dimensional Structure of Proteins

In Chapter 3, we saw that molecules are three-dimensional and that the shapes of simple molecules are determined by the charge clouds around a given atom using the VSEPR model. Peptides and proteins also have three-dimensional shapes or structures, but even the simplest dipeptide composed of two glycines has 17 atoms. Determining its overall three-dimensional shape using the VSEPR model is nearly impossible.

Because proteins are more complex than simple organic molecules, we discuss protein structure at multiple levels. The structure of a protein organizes into four levels: into primary (1°), secondary (2°), tertiary (3°), and quaternary (4°) levels. Each level describes a different set of interactions between amino acids in the protein. Together, the four levels provide a complete description of the three-dimensional structure of a protein.

? 10.3 Inquiry Question

How is a protein's structure held together?

Primary Structure

Like different colored beads in a necklace, the **primary (1°) structure** of a protein is the order in which the amino acids (beads) are joined together by peptide bonds to form the **protein backbone** that lines up from the N-terminus to the C-terminus. The side chains of the amino acids can be thought of as pendants extending from the protein backbone "necklace."

The sequence of amino acids matters. To illustrate, consider the word *protein*. Rearranging the letters in the word, we can make the word *pointer*, which has no meaning in common with the word *protein*. The same seven letters are used to make each word, but the meaning of each word is different. Similarly, arranging the amino acids in a different order creates a different peptide that no longer has the same function.

FIGURE 10.1 shows the primary structure of the eight-amino-acid peptide *angiotensin II*. This small peptide is involved in blood-pressure regulation in humans. Its amino acid sequence is Asp—Arg—Val—Tyr—Ile—His—Pro—Phe (DRVYIHPF). Any other ordering of these same eight amino acids would not function as angiotensin II. The order of the amino acids determines the function of a protein.

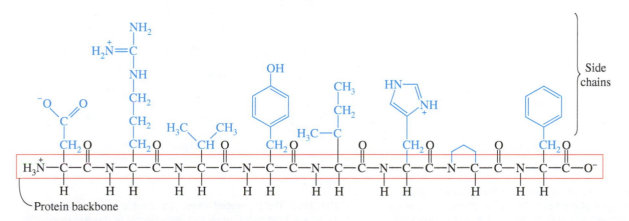

FIGURE 10.1 Primary structure of angiotensin II. Angiotensin II is a peptide involved in the regulation of blood pressure in humans. Notice that the side chains (blue) are not part of the backbone. Can you locate the N-terminus and C-terminus?

Secondary Structure

Have you ever stretched a Slinky® toy or a spring and noticed the spiral shape of its coils? The **secondary (2°) structure** of a protein can form coils or other repeating patterns of structure within the three-dimensional shape of a protein.

The two most common secondary structures are the **alpha helix (α helix)** and the **beta-pleated sheet (β-pleated sheet).** The secondary structures in a protein are stabilized by hydrogen bonding along the protein backbone and involve amino acids that are near each other in the primary structure. Recall from Chapter 7 that hydrogen bonding is not a covalent bond but a strong attractive force.

The α helix is a coiled structure in which the protein backbone adopts the form of a right-handed coiled spring. As shown in **FIGURE 10.2**, the helical structure is stabilized by hydrogen bonds formed between the carbonyl ($C{=}O$) oxygen atom (δ^-) of one amino acid and the N—H hydrogen atom (δ^+) of the amino acid located on the fourth amino acid away from it in the primary structure. The positioning of the hydrogen bonds along the axis of the helix allows it to stretch and recoil. Multiple hydrogen-bonding interactions along the backbone make the helix a strong structure. In the α helix, the side chains of the amino acids project outward away from the axis of the helix.

The β-pleated sheet is an extended structure in which segments of the protein chain align to form a zigzag structure much like a folded paper fan. As shown in **FIGURE 10.3**, strands called beta strands are held together side by side by hydrogen-bonding interactions between their backbones. In the β-pleated sheet, the side chains of the amino acids project above and below the sheet.

In summary, secondary structures involve hydrogen bonding along the backbone of the protein chain. Interactions of the amino acid side chains within the structure of a protein lead to the next level of structure.

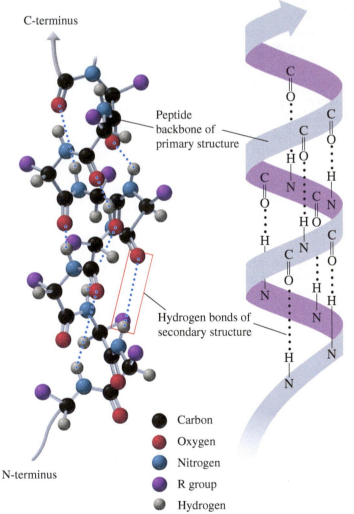

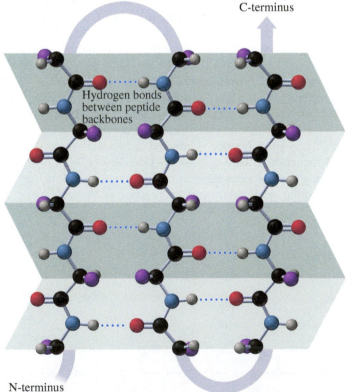

FIGURE 10.2 The α helix. The α helix has the hydrogen bonds positioned along the axis of the helix. There are three amino acids between the carbonyl and amino hydrogen that form the hydrogen-bonding interaction. Often, the backbone is represented by a ribbon, as shown on the right.

FIGURE 10.3 The β-pleated sheet. Adjacent strands of the protein backbone are held together by hydrogen bonds in the β-pleated sheet. Alternating R groups (front and back) give the backbones a zigzag or pleated shape.

Protein Secondary Structures and Alzheimer's Disease

INTEGRATING
Chemistry

◀ Find out how
changes in protein
secondary structure
can cause diseases
like Alzheimer's.

Alzheimer's disease is a common form of dementia that causes memory loss and the inability to think clearly, becoming severe enough to interfere with daily tasks. The disease is progressive and most often occurs in adults over the age of 65. Although researchers are far from understanding the causes of Alzheimer's disease, people with this disease have distinctly different brain tissue than people without the disease. The brain tissue of people with Alzheimer's disease contains plaques between the brain's nerve cells. The buildup of plaques between neurons and neurofibrillary tangles inside the neurons are thought to interfere with nerve-impulse transmission and contribute to neuronal cell death.

The protein plaques are made up of a protein fragment 42 amino acids long called beta-amyloid protein. This protein sequence has been identified in both healthy and diseased individuals. In a person without Alzheimer's, the structure of beta-amyloid is alpha-helical, but in a person with Alzheimer's, the beta-amyloid proteins form beta strands. This change makes the beta-amyloid proteins fold into sticky beta-pleated sheets that aggregate, forming plaques at the neurons.

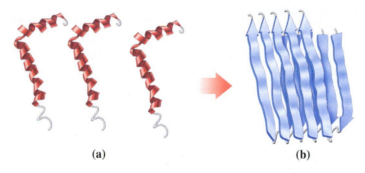

(a)   (b)

In people with Alzheimer's disease, beta-amyloid proteins change
from (a) a normal alpha-helical shape to (b) beta strands that stick
together as a beta-pleated sheet, forming plaques in the brain.

There is no cure for Alzheimer's disease, but medication can slow its progression or lessen symptoms for a limited time. Medications like donepezil and rivastigmine help keep the levels of the brain's nerve transmitters high, which can improve learning and memory.

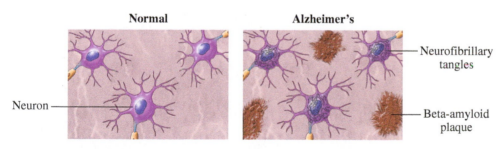

Normal   **Alzheimer's**

Neurofibrillary
tangles

Neuron —

Beta-amyloid
plaque

In the brain of a person with Alzheimer's disease, beta-amyloid plaques between nerve cells
and neurofibrillary tangles inside neurons interfere with nerve signals.

Tertiary Structure

Imagine an office telephone cord connecting the receiver to the console lying on a desk, with short lengths of the coils arranged next to or piled on top of each other. Like the coils of a cord, the α helices and β-pleated sheets of the polypeptide chain interact with each other and their environment to create a three-dimensional shape that is described as a **tertiary (3°) structure**.

The folding and twisting of the polypeptide chain are caused by both hydrophobic and hydrophilic interactions between the side chains of the amino acids and the surrounding aqueous environment. Like the soap micelle in Chapter 7, the nonpolar side chains in proteins are repelled by an aqueous environment and turn toward the interior of the protein, even as the polar side chains are attracted to the aqueous surroundings and appear on the surface. The tertiary structure is stabilized by the attractive forces between the side chains and the aqueous environment of the protein as well as by the attractive forces between the side chains themselves. All of these competing interactions cause the protein to fold into a unique three-dimensional shape.

The interactions involved when a protein folds into its tertiary structure (see **FIGURE 10.4**) include the following:

1. **Nonpolar interactions** The nonpolar amino acid side chains are *unlike* the aqueous (polar) environment. They are repelled by the aqueous environment and aggregate in the interior of the protein. The ability of water to exclude the nonpolar side chains is referred to as the **hydrophobic effect** and drives the three-dimensional folding of proteins. The nonpolar side chains, gathered together in the interior of the protein, interact with each other through London forces, helping to stabilize the tertiary structure. Figure 10.4 shows an example of two aromatic side chains that stack and two methyl groups in hydrophobic interactions.

2. **Polar interactions** Polar amino acid side chains are *like* the surrounding aqueous environment and easily interact with water and each other through the following hydrophilic interactions: dipole–dipole, ion–dipole, and hydrogen-bonding interactions. Because they are charged, polar acidic and polar basic side chains form ion–dipole interactions with water. The side chains of most of the polar amino acids (cysteine being an exception) can interact with each other and with water in the surrounding environment through hydrogen bonding (see Figure 10.4).

3. **Salt bridges (ionic interactions)** Acidic and basic amino acid side chains exist in their ionized form in an aqueous environment. In other words, the acids are carboxylate ions ($-COO^-$), and the bases exist as the protonated amines ($-NH_3^+$). When a protein folds and a carboxylate ion on one amino acid side chain folds up near a protonated amine on a second amino acid side chain, the opposite charges attract, thereby forming a stabilizing ionic interaction also called a *salt bridge* (see Figure 10.4 for an example).

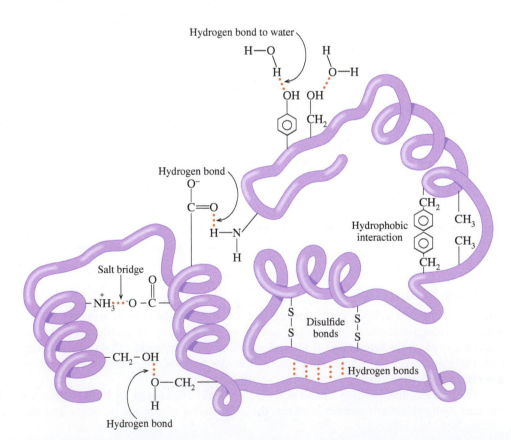

FIGURE 10.4 Protein tertiary structure. The interactions of amino acid side chains with their environment and each other stabilize the tertiary structure of a protein.

4. **Disulfide bonds** If two cysteine side chains are brought close together during the folding of a protein, the two —SH groups (thiols) can react with each other through an oxidation reaction (losing hydrogens) to form a disulfide bond (—S—S—). Examples are shown in Figure 10.4 and below. Note that the disulfide bond is a covalent bond (electrons being shared between two atoms), in contrast to the other stabilizing attractive forces.

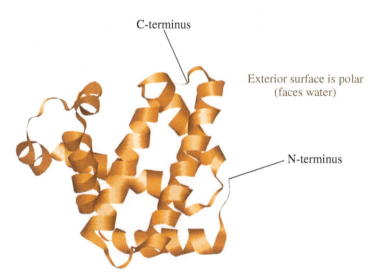

A disulfide bond can form when two cysteine amino acids close to each other in space undergo oxidation.

Proteins can be loosely classified into groups based on how they fold into their resulting three-dimensional shape. **Globular proteins** fold into a compact, spherical shape. Such proteins have the tertiary structure just described; that is, they have polar amino acid side chains on their surface, enhancing their solubility in water, while the nonpolar amino acid side chains gather in the center of the spherical structure, forming a nonpolar core. Myoglobin is an example of a globular protein (see **FIGURE 10.5**). It stores oxygen in skeletal muscle. Although myoglobin is present in human muscle, it exists in much higher concentration in the muscles of sea mammals, such as dolphins and whales, and is the reason these animals can stay underwater for long periods of time without surfacing to breathe. Enzymes and many cellular proteins are globular proteins (see Section 10.5).

Fibrous proteins, on the other hand, have long, threadlike structures. Together, the aligned helices form strong, durable structures. Two familiar examples of fibrous proteins are keratins and collagen (see **FIGURE 10.6**).

C-terminus

Exterior surface is polar (faces water)

N-terminus

FIGURE 10.5 Globular protein. Globular proteins like myoglobin fold into spherical structures, with polar amino acid side chains on their surface and nonpolar amino acid side chains in their interior. The ribbon shown represents the protein backbone.

(a) Alpha-keratin coiled coil

(b) Collagen triple helix

FIGURE 10.6 Fibrous proteins. (a) Human hair contains the coiled coils of alpha-keratin. (b) Skin contains the protein collagen, which has three alpha helices twisted around each other in a triple helix, providing flexibility and ropelike strength.

Keratins are found in hair, nails, and the scales of reptiles. The fibrous protein alpha-keratin gets its strength from the wrapping of two α helices into a coiled structure of coils called a protofibril, which further wraps into fibers forming hair and wool in animals.

Collagen is found in connective tissue, skin, tendons, ligaments, and cartilage. It contains a modified amino acid called hydroxyproline, the amino acid proline with an extra —OH found on the R group. Collagen achieves a similar strength by wrapping three α helices into a triple helix. A network of collagen triple helices forms flexible fibers in the skin that degrade as we age, causing wrinkles. The additional hydroxyl group on the amino acid hydroxyproline allows the polypeptide chains to form hydrogen bonds between the chains, providing extra strength to the triple helix formed in collagen. Fibrous proteins tend to be insoluble in water.

Quaternary Structure

Imagine two or more of the piled-up cords mentioned in the description of tertiary structure neatly arranged next to each other. The **quaternary (4°) structure** describes the interactions of two or more polypeptide chains to form a single biologically active protein. The individual polypeptide chains or subunits are held together by the same interactions that stabilize the tertiary structure of a single protein.

The best-studied example of a protein with a quaternary structure is the oxygen transporter hemoglobin. Hemoglobin consists of four polypeptide chains or subunits—two identical alpha subunits and two identical beta subunits (as in Figure 10.7d). In adult hemoglobin, all four subunits must be present for the protein to properly function as an oxygen carrier.

It is important to note that not all biologically active proteins have a quaternary (4°) structure. Some proteins, myoglobin for example, are made up of a single polypeptide chain. **FIGURE 10.7** and **TABLE 10.4** summarize the structural levels of proteins.

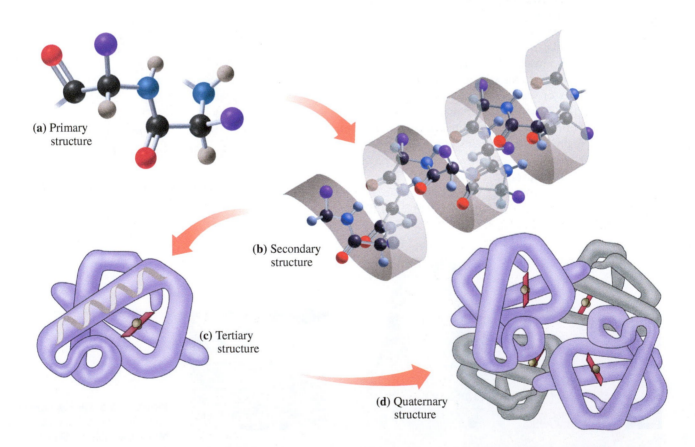

(a) Primary structure

(b) Secondary structure

(c) Tertiary structure

(d) Quaternary structure

FIGURE 10.7 The levels of protein structure. Primary, secondary, and tertiary protein structures build upon each other as shown. Some proteins also have a quaternary structure.

TABLE 10.4 Levels of Protein Structure and Stabilizing Forces Present	
Level of Structure	**Forces Stabilizing Structure**
Primary (1°)	Peptide bonds
Secondary (2°)	Hydrogen bonding along the protein backbone atoms, between amino acids close to each other in sequence
Tertiary (3°)	London forces, hydrogen bonding, dipole–dipole and ion–dipole interactions, ionic salt bridges, and disulfide bonds between the R groups of amino acids far away from each other in sequence
Quaternary (4°)	Same as tertiary structure but between subunits

Mastering Video
Practicing the Concepts
*The Role of Attractive Forces
in Protein Structure*

Sample Problem 10.5 Side Chains in Tertiary Structure

Predict whether each of the following amino acid side chains would more likely be on the surface or in the interior of a globular protein after it has folded into its tertiary structure. Assume the protein is in an aqueous environment.

a. phenylalanine   b. glutamate

Solution

a. In Table 10.1, phenylalanine is classified as a nonpolar amino acid. This means that it would be forced to the interior of the protein due to the hydrophobic effect.

b. In Table 10.1, glutamate is classified as a polar (acidic) amino acid. The negative charge on the carboxylate group would be able to form ion–dipole interactions with water, so it is likely to be on the surface.

Practice Problems

10.21 What type of bonding is present in the primary structure of a protein?

10.22 How many different tripeptides that contain one leucine, one glutamate, and one tryptophan are possible?

10.23 Name the stabilizing attractive force found in secondary structures of proteins.

10.24 When a protein folds into its tertiary structure, does the primary structure change? Explain.

10.25 Describe the differences in the shape of an α helix and a β-pleated sheet.

10.26 What type of interaction would you expect between the side chains of each of the following pairs of amino acids in the tertiary structure of a protein?

a. histidine and aspartate   b. alanine and valine

c. two cysteines   d. serine and asparagine

10.27 What type of interaction would you expect between the side chains of each of the following pairs of amino acids in the tertiary structure of a protein?

a. lysine and glutamate   b. leucine and isoleucine

c. threonine and tyrosine   d. glutamine and arginine

10.28 Determine whether each of the following statements describes the primary, secondary, tertiary, or quaternary structure of a protein.

a. Side chains interact to form disulfide bonds.

b. Peptide bonds join amino acids in a polypeptide chain.

c. Two polypeptide chains are held together by hydrogen bonds.

d. Hydrogen bonding between amino acids in the same polypeptide gives a coiled shape to the protein.

10.29 Determine whether each of the following statements describes the primary, secondary, tertiary, or quaternary structure of a protein.

a. Three polypeptide chains interact to form a biologically active protein.

b. Hydrogen bonds form between adjacent segments of the backbone of the same protein to form a "folded-fan" structure.

c. Nonpolar side chains are repelled by water and move to the interior of the protein.

d. Amino acids react in a condensation reaction to form a peptide bond.

—*Continued next page*

Continued—

10.30 Myoglobin is a protein containing 153 amino acids. Approximately half of the amino acids in myoglobin have polar side chains.

    a. Where would you expect these amino acid side chains to be located in the tertiary structure of the protein?

    b. Where would you expect the nonpolar side chains to be?

    c. Would you expect myoglobin to be more or less soluble in water than a protein composed mainly of nonpolar amino acids? Why?

10.31 What is the difference between the secondary structure present in the beta-amyloid protein in a person without Alzheimer's disease and that in a person with Alzheimer's disease?

10.32 How do beta-amyloid plaques form in a person with Alzheimer's disease?

10.4 Denaturation of Proteins

10.4 Inquiry Question

What causes proteins to denature?

Have you ever cracked an egg and dropped its contents into a hot pan? When you do so, the clear part of the egg, called the egg white, quickly turns from clear to white. You are observing the **denaturation** of the proteins in the egg white.

Denaturation of a protein is a process that disrupts the stabilizing attractive forces in the secondary, tertiary, or quaternary structure. When a protein is denatured, its primary structure is not changed.

In the example of the egg white, the denaturing agent is heat. An increase in temperature disrupts attractive forces such as hydrogen bonding, London forces, and other interactions. A change in pH can also denature a protein. A change in the pH of a protein's environment can alter the charges on the acidic and basic amino acid side chains, which in turn affects their ability to form salt bridges. Certain organic compounds or heavy metal ions can also denature a protein by disrupting disulfide bridges. **TABLE 10.5** gives several examples of denaturing agents and their effects on the various stabilizing forces of protein structure.

When the attractive forces in a protein are disrupted, the protein is denatured.

Heat, acid, base, heavy metal salts, agitation

Active protein   Denatured protein

TABLE 10.5 Examples of Protein Denaturation

Denaturing Agent	Forces or Bonds Disrupted	Examples
Heat above 50 °C	Hydrogen bonds, hydrophobic interactions	Cooking food
Acids, bases, ionic compounds	Salt bridges, hydrogen bonds	Lactic acid from bacteria, which denatures milk proteins in the preparation of yogurt and cheese
Reducing agents	Disulfide bonds	Thiols, which are used in hair straightening and permanent waves
Detergents	Hydrophobic interactions	Membrane proteins
Heavy metal ions Ag^+, Pb^{2+}, Hg^{2+}	Disulfide bonds, salt bridges	Mercury and lead poisoning
Mechanical agitation	Hydrogen bonds, London forces	Whipped cream, meringue made from egg whites

Without the stabilizing interactions necessary to maintain its three-dimensional structure, a protein will unfold into a shapeless string of amino acids. When a protein loses its three-dimensional shape, it also loses its biological activity. The conditions to denature a protein are mild when compared to protein hydrolysis, which requires extremely high temperature and harsh chemical conditions to break peptide bonds.

Hair relaxing and perming involve protein denaturation. If you have curly hair and want it to be straight, you get it relaxed, or straightened. If you have straight hair and you want it to be curly, you get a permanent wave. Both of these processes involve denaturing the proteins in your hair by disrupting the disulfide bonds found in the hair protein keratin, reshaping the hair and forcing the disulfide bonds to re-form at different points. Ammonium thioglycolate is one of the chemicals used to reduce the disulfides, and hydrogen peroxide is often used to oxidize the resulting thiols back to disulfide bonds.

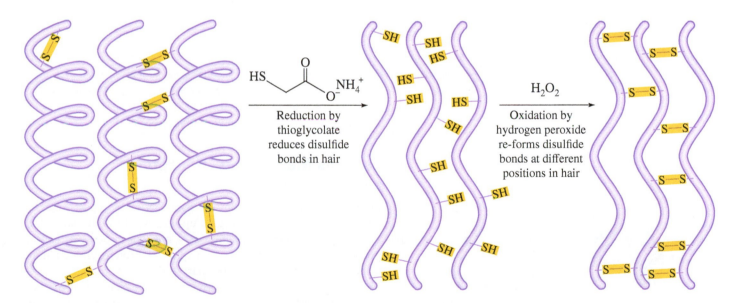

Hair relaxation involves denaturing keratin proteins in hair through reduction and oxidation of disulfide bonds.

Denaturing proteins can be used as a treatment for lead or mercury poisoning. A person who has accidentally ingested a heavy metal like lead or mercury is given egg whites to drink. The proteins in the egg whites are denatured by the mercury or lead, and the combination forms a precipitate. Typically, an emetic is then administered to induce vomiting to discharge the metal–protein precipitate from the body.

Practice Problems

10.33 List the type of attractive force disrupted and the level of protein structure changed by the following denaturing treatments:

    a. adding salt to soy milk to make tofu

    b. whipping egg whites into stiff peaks to fold into an angel food cake

10.34 List the type of attractive force disrupted and the level of protein structure changed by the following denaturing treatments:

    a. adding a reducing agent to hair to relax (straighten) it

    b. baking the proteins in dough to make bread

10.35 Identify each of the following statements as characteristic of protein denaturation or protein hydrolysis.

    a. Milk curdles when lemon juice is added to it.

    b. A protein breaks up into amino acid fragments.

10.36 Identify each of the following statements as characteristic of protein denaturation or protein hydrolysis.

    a. A protein unfolds due to stretching.

    b. A person with straight hair gets a permanent to add curls.

10.5 Protein Functions

? 10.5 Inquiry Question

What functions do proteins play in the body?

Next, we consider some important roles of proteins and examples. Even though all proteins are made of the same 20 amino acids, the sequence of amino acids dictates the structure that the protein adopts, which, in turn, dictates protein function. Proteins have many functions in the body. **TABLE 10.6** shows a variety of protein functions and gives some common examples.

TABLE 10.6 Protein Functions and Examples

Function	Description	Protein Examples
Messenger	Control activities between cells	*Insulin* released from pancreas binds to receptors on other cell types.
Receptor	Control activities in the cell	*Insulin receptor* binds insulin on the cell surface, stimulating glucose uptake into the cell.
Transport	Transport nutrients	*Glucose transporter* transports glucose from the bloodstream into the cell.
		Hemoglobin transports O_2 through the bloodstream to the tissues.
Storage	Store nutrients	*Ferritin* stores iron in cells.
		Casein is a storage protein in milk.
		Gluten is a storage protein in grains.
Contraction	Contract muscles	*Myosin* and *actin* make up the thick and thin filaments in muscle tissue.
Protection	Protect cells	*Antibodies* bind to foreign substances in the body, identifying them for elimination.
Structure	Maintain shape	*Collagen* supports the structure in ligaments, tendons, skin, nails, and horns. *Actin* is also found in the supportive structure of the cell (cytoskeleton).
Catalysis	Catalyze biochemical reactions	*Pepsin* breaks peptide bonds in proteins in the stomach during digestion.

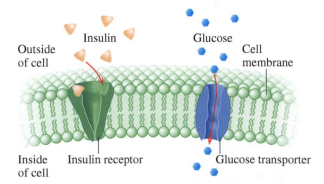

FIGURE 10.8 In the presence of insulin, the glucose transporter moves glucose inside the cell. When the glucose level increases in the bloodstream, insulin is released and travels through the bloodstream, binding to insulin receptor proteins on cells. Insulin binding signals the cell to activate glucose transporter proteins at the cell membrane, allowing glucose to flow into the cell.

Proteins as Messengers, Receptors, and Transporters

Insulin is a protein released from the pancreas as we digest sugar. Its purpose is to signal the cells to take up glucose. In doing so, insulin plays the role of a hormone, acting as a messenger between cells. A **hormone** is a chemical, sometimes a peptide or protein, created in one part of the body that affects another part of the body.

How does insulin get its message to a cell? Sending a message leads us to another type of protein, a receptor. **Receptors** are proteins facing the outer surface of a cell that bind to a hormone or other messenger, triggering a signal inside the cell. In the case of insulin, it signals the synthesis of another protein, the glucose transporter (GLUT), a protein facilitating transport of glucose across the cell membrane. For a discussion of transport across the cell membrane, refer to Section 8.7. The glucose transporter is an **integral membrane protein** spanning a phospholipid bilayer. **FIGURE 10.8** shows how insulin binds to the insulin receptor and affects glucose concentrations in the cell.

Hemoglobin—Your Body's Oxygen Transporter

Another well-studied transport protein that we could not live without is hemoglobin. As mentioned in Section 10.3, the globular protein hemoglobin transports oxygen in the blood. **FIGURE 10.9** shows hemoglobin's four polypeptide subunits: two subunits in red and two subunits in blue. Note that each of the subunits in hemoglobin contains many helices in its secondary structure (represented by the curly ribbons) folded and packed together. These four folded subunits of hemoglobin are attracted to each other through attractive forces, including hydrogen bonds, London forces, and salt bridges, to form a protein quaternary structure that is biologically active.

Each subunit contains a nonprotein part called a **prosthetic group** that is vital to the protein's function. Hemoglobin's prosthetic group is called a heme (shown in green in Figure 10.9). Each heme group binds Fe^{2+}, which, in turn, binds oxygen (O_2). Each Fe^{2+} can bind one oxygen molecule for transport. Therefore, one molecule of hemoglobin can transport up to four molecules of oxygen.

The binding of O_2 to the Fe^{2+} of hemoglobin in the oxygen-rich environment of the lungs induces a change in the shape of the hemoglobin molecule. Biochemists call reshaping in which bond angles are only rotated and not broken a **conformational change.** The conformational change allows hemoglobin to hold the oxygen molecules long enough to deliver the oxygen to tissues where oxygen levels are low. At the tissues, the oxygen dissociates from the hemoglobin, and the shape of the hemoglobin changes back to its deoxygenated form. Because of hemoglobin's unique structure, it is able to interact with its environment and function as an oxygen transport and delivery system.

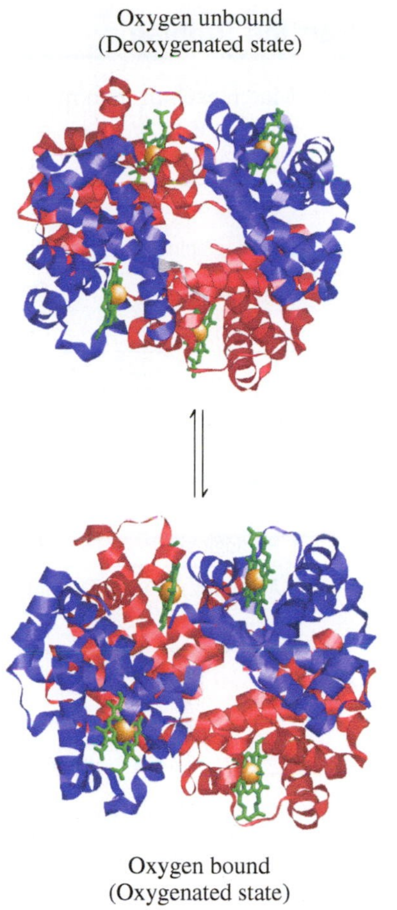

Antibodies—Your Body's Defense Protein

When foreign substances like bacteria enter your body, your immune system produces proteins called antibodies (also called *immunoglobulins)* to recognize and destroy the foreign substances in the bacteria. The foreign substance recognized by an antibody is called an **antigen.** As an example, the influenza virus responsible for the "flu" has proteins on the surface of the virus that are recognized as antigens.

FIGURE 10.10 shows the structure of an antibody. It consists of four polypeptide subunits, two heavy chains (higher mass, shown in blue and red), and two light chains (lighter mass, shown in yellow and green). The secondary structure contains β-pleated sheets (represented by the flat ribbons) that are stacked tightly together. As shown in the figure, the quaternary structure is held together through disulfide bridges between the polypeptide chains, forming a unique "Y" shape, which is the biologically active form of the protein.

The stem of the Y is similar in all antibodies and can bind to receptors on a variety of cells in the body. Antibodies bind antigens at the top of each arm of the Y, at the antigen-binding sites. Each antibody has a unique primary sequence at the top of the Y that recognizes a single antigen on a foreign substance. This ability to recognize a single antigen and the distinctive Y structure of an antibody are well suited to bind antigens and present them for future destruction by the immune system.

Oxygen unbound
(Deoxygenated state)

Oxygen bound
(Oxygenated state)

FIGURE 10.9 Hemoglobin. Hemoglobin consists of four polypeptide subunits (two α chains shown in red and two β chains shown in blue) held together mainly by salt bridges to form the biologically active protein. The heme is shown in green. When oxygen binds to Fe^{2+} (orange ball), a conformational change occurs as salt bridges between the subunits collapse, closing the hole formed in the center of the protein. Can you see the difference?

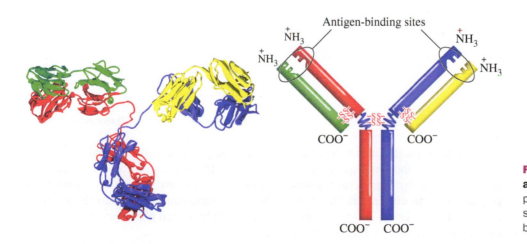

FIGURE 10.10 Structure of an antibody. Antibodies contain four polypeptide chains held together in a "Y" shape by attractive forces and disulfide bridges.

Practice Problems

10.37 Match each protein in column A with its function in column B:

Column A	Column B
Collagen	Storage protein
Hemoglobin	Structural connector
Antibody	Oxygen transporter
Casein	Bind foreign substances in body

10.38 Match each protein in column A with its shape in column B:

Column A	Column B
Collagen	Y-shaped
Hemoglobin	Ropelike
Antibody	Integral membrane protein
Glucose transporter	Roughly spherical

10.39 The glucose transporter provides facilitated transport across the cell membrane. Do you think this transporter is active (a) after you eat a meal? (b) after you wake up in the morning? Support your answers.

10.40 Hemoglobin is found in blood, which is composed mostly of water. Would you expect the amino acid side chains on the surface of hemoglobin to be polar or nonpolar? Support your answer.

10.41 If you come in contact with a virus like the influenza virus, proteins on the surface of the virus act as antigens, and the immune system produces antibodies to mark the infecting viruses for removal. Which part of an antibody binds to these surface proteins?

10.6 Enzymes—Life's Catalysts

? 10.6 Inquiry Question

How does an enzyme catalyze a biochemical reaction?

Have you ever wondered how the food you eat is so quickly transformed into energy that moves muscles and maintains your body? The process involves remarkable proteins called enzymes. Enzymes are typically large globular proteins and are present in every cell of your body. Enzymes act as *catalysts,* compounds that accelerate the reactions of *metabolism* (chemical reactions in the body) but are not consumed or changed by those reactions. Like all other catalysts, an enzyme cannot force a reaction to occur that would not normally occur. Instead, an enzyme simply makes a reaction occur faster.

Like most other proteins, enzymes are large molecules with complex three-dimensional shapes. Because many enzymes in your body function in an aqueous environment, they fold so that polar amino acids are on their surface. The final folded (tertiary) structure of an enzyme plays an important role in its function.

Let's begin our examination of enzyme structure and function by looking at a well-characterized enzyme, hexokinase. This enzyme's job is to catalyze the transfer of a phosphate group from the energy molecule adenosine triphosphate (ATP) to a six-carbon sugar (hexose) like D-glucose. It functions in the first step of glycolysis (the process by which glucose is broken down in the body) by adding a phosphate to the sixth carbon of glucose, forming the product glucose-6-phosphate.

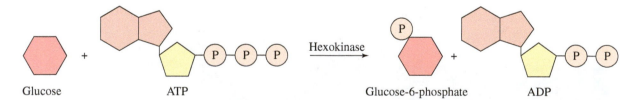

Glucose   +   ATP $\xrightarrow{\text{Hexokinase}}$ Glucose-6-phosphate   +   ADP

The enzyme hexokinase catalyzes the transfer of a phosphate group from ATP to D-glucose.

In the chemical equation, the enzyme name usually appears above or below the reaction arrow. The phosphate group is represented by a P in a circle. It contains a phosphorus surrounded by four oxygen atoms. In Section 3.2, we saw phosphate as one of the polyatomic ions, and in Chapter 4, phosphate was reintroduced as a functional group. At the pH of the body, phosphate occurs as the anion hydrogen phosphate, HPO_4^{2-}, also called inorganic phosphate, and abbreviated P_i.

The Active Site

FIGURE 10.11 shows the folded structure of the enzyme hexokinase. The protein backbone of hexokinase is represented as a ribbon. As hexokinase folds into its tertiary structure, an indentation forms on one part of its surface. This pocket is known as the **active site,** and it is lined with amino acid side chains. As its name implies, the active site is the functional part of the enzyme where catalysis occurs. The active site of hexokinase is labeled in Figure 10.11.

Like a left foot fitting into a left shoe, glucose, the reactant for hexokinase, fits snugly into its active site. In an enzyme-catalyzed reaction, the reactant is called the **substrate.** The active site of hexokinase fits D-glucose. Not even L-glucose, its enantiomer, or even D-galactose, its epimer (one chiral center difference), will fit into the active site.

Because the shape of an enzyme's active site is complementary to the shape of its substrate, most enzymes have only a few substrates, and many have only one. This property of enzymes is called **substrate specificity.** Just as we know which shoe fits on which foot, many enzymes are specific to one enantiomer of the same compound. The opposite enantiomer will not fit perfectly in the active site and cannot undergo catalysis.

Hexokinase also has a nonprotein "helper" in the form of a magnesium ion (Mg^{2+}) in its active site. (Mg^{2+}) assists with catalysis by stabilizing the ATP molecule. There are two categories of nonprotein helpers—**cofactors** and **coenzymes.** Cofactors are inorganic substances like the magnesium ion that can be obtained in a normal diet from minerals. Coenzymes are small organic molecules, many of them derived from vitamins. For example, the coenzyme flavin adenine dinucleotide (FAD) is synthesized in the body from the vitamin riboflavin (vitamin B_2).

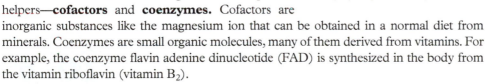

FIGURE 10.11 The active site of hexokinase. Hexokinase's protein structure is shown as a ribbon. During catalysis, glucose fits in the pocket of the active site between the upper and lower lobes of the protein. It is held in place by many hydrogen-bonding interactions (red dashes) with amino acid side chains.

Sample Problem 10.6   Enzymes and Substrates

Describe the function of the active site of an enzyme.

Solution

The active site of an enzyme is a pocket uniquely fitted to the substrate and contains the amino acid side chains that catalyze the reaction.

Enzyme–Substrate Models

How does a substrate like glucose move into the active site of an enzyme like hexokinase and situate itself for catalysis? Glucose is drawn to the active site of hexokinase by attractive forces like hydrogen bonding. Five polar amino acid side chains—asparagine (Asn), aspartate (Asp), glutamate (Glu), lysine (Lys), and threonine (Thr)—form hydrogen bonds to glucose, orienting it correctly within hexokinase's active site for catalysis (see Figure 10.11).

(a) Lock-and-key model

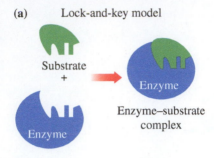

Substrate
+

Enzyme

Enzyme–substrate
complex

(b) Induced-fit model

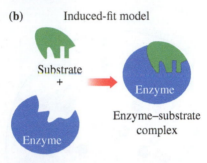

Substrate
+

Enzyme

Enzyme–substrate
complex

FIGURE 10.12 Models for enzyme–substrate interaction. (a) In the lock-and-key model, the enzyme and substrate have rigid complementary shapes that fit together to form ES. (b) In the induced-fit model, the enzyme and substrate have flexible but similar shapes that adjust when the substrate gets closer to the enzyme, thereby forming a unique enzyme–substrate complex. Did you notice that the enzyme active site before substrate binding (left side) is different for the two models?

This initial interaction of an enzyme with a substrate is called the **enzyme–substrate complex (ES).** The formation of this complex occurs before catalysis can begin. This complex is held in place by attractive forces. Only through formation of the enzyme–substrate complex can catalysis occur and products be formed and released. We can represent a simple enzyme catalyzed reaction as

$$E + S \rightleftharpoons ES \longrightarrow E + P$$

Several models describe how the enzyme interacts with the substrate to form the ES. Two will be described here. The first model introduced was the **lock-and-key model.** In this model, the active site is thought of as a rigid, inflexible shape that is an exact complement to the shape of its substrate. According to the lock-and-key model, each enzyme accommodates one and only one substrate, much as a lock works with only one key (see **FIGURE 10.12a**).

Today, we know that many enzymes can react with two or more similar substrates, so a new model called the **induced-fit model** was developed. In this model, the enzyme's active site is flexible and has a shape that is roughly complementary to the shape of its substrate. As the substrate interacts with the enzyme, the active site undergoes a conformational change, adjusting to fit the shape of the substrate (see **FIGURE 10.12b**). The shape of the substrate may change slightly as well.

When hexokinase and glucose combine, its enzyme–substrate complex follows the induced-fit model. In **FIGURE 10.13**, the shape of hexokinase is shown in a ball-and-stick representation. Notice how the shape changes when the enzyme interacts with glucose. The two lobes of the enzyme close snugly around the substrate to form ES.

Rates of Reaction

Recall from Chapter 5 that a catalyst speeds up a reaction by *lowering* the activation energy. Because enzymes are catalysts, they do not affect the thermodynamics of the reaction (whether the reaction is exergonic or endergonic); they only speed up the formation of the products. An enzyme-catalyzed reaction does this by first forming ES and

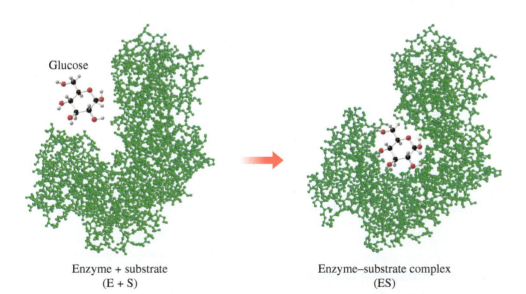

Glucose

Enzyme + substrate
(E + S)

Enzyme–substrate complex
(ES)

FIGURE 10.13 Interaction of hexokinase and glucose through induced fit. When glucose enters the active site of hexokinase (left), the enzyme's right and left sides, called lobes, come together, wrapping glucose snugly into its active site, forming the enzyme–substrate complex (right).

only then proceeding to form products. **FIGURE 10.14** shows how a reaction energy diagram first introduced in Chapter 5 would be modified for an enzyme-catalyzed reaction.

Going back to the hexokinase reaction, glucose and ATP react in the active site of the enzyme and a phosphate is transferred from ATP to glucose. The products of the reaction are glucose-6-phosphate and adenosine diphosphate, ADP. If hexokinase is not present, the reaction between glucose and ATP occurs more slowly. What is it about the environment of the active site of hexokinase that causes these two molecules to react more quickly to form products?

Thus far, we have seen that the first step in an enzyme-catalyzed reaction is the formation of ES. We have also seen that catalysts lower the activation energy of a reaction. Putting these two pieces of information together, we find that it makes sense that the formation of ES is key to lowering the activation energy and allowing a biochemical reaction to proceed and quickly form products.

Lowering activation energy is accomplished during the formation of ES through interactions between the enzyme and the substrate. Each interaction releases a small amount of energy, stabilizing the complex. These small interactions combine to substantially lower the activation energy for an enzyme-catalyzed reaction compared to its uncatalyzed reaction. Let's look at some interactions that help to lower the activation energy.

FIGURE 10.14 Effect of an enzyme on a chemical reaction. Like other catalysts, an enzyme lowers the activation energy of a reaction, thereby increasing the rate of the reaction. Enzymes accomplish this by first forming ES.

Proximity

The active site of an enzyme has a small volume. When ES forms, the active site is "filled" with substrate. This means that the reacting molecules (glucose and ATP, in the case of hexokinase) are in close proximity to each other and to the catalytic side chains of the amino acids lining the active site. The closer they are, the more likely they will react. The activation energy is lowered by this effect. The substrates don't have to find each other as they would in solution—they are already close together in the active site.

The amino acid side chains in the active site of an enzyme are like tools that the enzyme uses to help facilitate the reaction. These side chains are often the functional groups of the acidic and basic amino acids.

Orientation

In the active site of an enzyme, the substrate molecules are held at the appropriate distance and in correct alignment to each other to allow the reaction to occur. The arrangement of the amino acid side chains in the active site creates interactions that orient the substrates so they will react. Without an enzyme to orient the substrates, substrates bumping into each other in solution may not react because of incorrect positioning. The enzyme guarantees that the substrates line up correctly. Correct orientation helps lower the activation energy required.

Bond Energy

Think of a bond as a rubber band. Which is easier to cut with scissors: a rubber band that is stretched or one that is slack? The stretched one is easier to cut because the rubber molecules are weakened or strained, and it takes less energy to break the molecules with scissors. Likewise, when an enzyme interacts with its substrate to form ES, the bonds of the substrate molecule(s) are weakened (strained). The weakening of the bonds means that they will more readily react. In other words, weaker bonds break more easily, and the activation energy is lowered by this effect.

Looking back at our hexokinase reaction, the coenzyme Mg^{2+} holds the ATP molecule in one area of the active site and the glucose interacts with another part of the site. The amino acid side chains in the active site of hexokinase form multiple hydrogen bonds with the glucose (refer to Figure 10.11). Each hydrogen-bonding interaction formed stabilizes ES and lowers the activation energy.

When glucose enters the active site, the enzyme undergoes a significant conformational change, causing the two lobes of the enzyme to close around the substrate (following the induced-fit model). With the two lobes of the enzyme closer together, the ATP is moved closer to the glucose and is in proper alignment for the reaction.

In the absence of an enzyme, it is more difficult for substrates to get close enough and orient themselves to react.

In the presence of an enzyme, it is easier for substrates to get close enough and orient themselves to react.

When in the active site of hexokinase, glucose is positioned so that the hydroxyl group on C6, activated by an aspartate side chain, is able to remove a phosphate from ATP to form glucose − 6-phosphate + ADP (see **FIGURE 10.15**). Because the enzyme is less attracted to the products than the substrates, the lobes of the enzyme move apart and the products are released.

FIGURE 10.15 Catalysis by hexokinase. The coenzyme Mg^{2+} holds ATP in place (blue lines), and hydrogen bonds (red lines) hold glucose in the active site. The O on C6 of glucose is perfectly positioned to form a bond to the end phosphate of ATP. The products—ADP and glucose-6-phosphate—do not fit the active site as snugly as the substrates and are released.

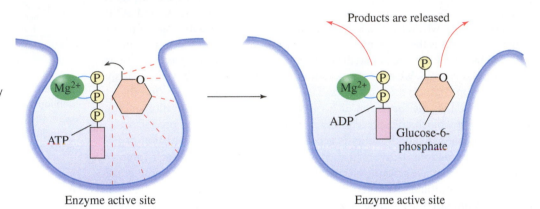

Enzyme active site

Enzyme active site

Products are released

Sample Problem 10.7   Rate of Catalysis

List three factors that contribute to lowering the activation energy when ES is formed.

Solution

Three factors are (1) close proximity of reactants to each other and to interactions with amino acid side chains, (2) favorable orientation of reactants, and (3) weakening of bond energy in the reactants.

Practice Problems

10.42 What is the name given to the reactant of an enzyme-catalyzed reaction?

10.43 What is the name given to the area on an enzyme where catalysis occurs?

10.44 What is the name given to a small, organic nonprotein part of an enzyme that is involved in catalysis?

10.45 What level of protein structure is involved in the formation of an enzyme's active site?

10.46 Describe how a substrate is drawn to an enzyme to form ES.

10.47 Which model for enzyme–substrate interaction describes the action of hexokinase? Describe how the enzyme changes when the enzyme and glucose form ES.

10.48 Describe the key difference in the lock-and-key and induced-fit models.

10.49 What kind of interaction attracts the cofactor Mg^{2+} and ATP to each other? (*Hint:* Look at the structure of the phosphate group.)

10.50 In this section, we saw that glucose is held in the active site of hexokinase by hydrogen bonding. The outline of the active site of a hypothetical enzyme and its substrate glyceraldehyde is represented in the following figure. Draw dotted lines to show the possible hydrogen bonds between glyceraldehyde and the hypothetical enzyme's active site.

DISCOVERING THE CONCEPTS

? INQUIRY ACTIVITY—Factors That Affect Enzyme Activity

Information

The activity of an enzyme (a measure of how fast it catalyzes a chemical reaction) can be affected by changes in the enzyme's environment. Enzymes are most efficient at a pH and temperature optimum.

Questions

1. Pepsin and trypsin are enzymes that catalyze the digestion of proteins. Pepsin has a pH optimum of 2 and trypsin has a pH optimum of 7 to 8. Which of these two enzymes is likely to digest proteins in the stomach?
2. Lactase, the enzyme that hydrolyzes the disaccharide lactose, operates at a pH optimum of about 6.5 and a temperature optimum of 37 °C. How would the following changes affect the rate of catalysis—increase, decrease, or stay the same?
   a. lowering the pH to 2
   b. raising the temperature to 50 °C
   c. increasing the amount of lactose available

Enzyme Inhibition

An enzyme's activity can be reversibly decreased or inhibited if (a) a molecule structurally similar to the normal substrate competes for the active site (similar functional groups or charges are present) or (b) if a molecule binds to a second site on the enzyme, changing the shape of the active site.

Questions

1. Which of the two scenarios, (a) or (b), would be called *competitive inhibition*?
2. Which of the two scenarios, (a) or (b), would be called *noncompetitive inhibition*?
3. Succinate is the substrate for the enzyme succinate dehydrogenase, one of the eight enzymes found in the citric acid cycle. Malonate is a reversible inhibitor of succinate dehydrogenase. What type of inhibitor is malonate likely to be? Explain your reasoning.

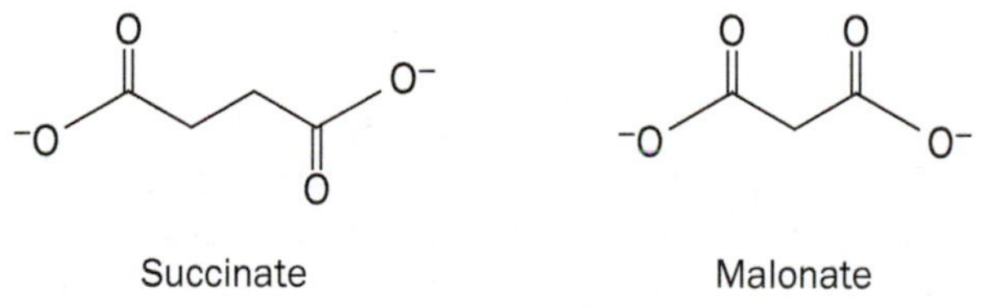

Succinate   Malonate

4. The enzyme hexokinase catalyzes the first step in glycolysis: glucose → glucose-6-phosphate. The activity of this enzyme can be inhibited by the buildup of the product glucose-6-phosphate. Binding glucose-6-phosphate to a second site on the enzyme changes the shape of the active site. Which type of inhibition is this an example of?
5. Which type of inhibition do you think could be overcome by adding more substrate? Explain your answer.
6. Which type of inhibition does not allow the enzyme to reach top speed, even if more substrate is added? Explain your answer.
7. The analgesic aspirin acts as a type of inhibitor called a "suicide" inhibitor. Aspirin forms a covalent bond to the amino acid serine, which has its side chain in the active site of the enzyme cyclooxygenase. Provide an explanation for the name of this inhibition.

10.7 Factors That Affect Enzyme Activity

 10.7 Inquiry Question

What factors affect enzyme activity?

If you slice an apple in half and allow it to sit untouched, the flesh of the apple will turn brown. The brown color is caused by an oxidation reaction (using oxygen from the air) catalyzed by the enzyme polyphenol oxidase. The product of this reaction is responsible for the brown color. One method often used to slow down browning is to sprinkle apple slices with lemon juice. How does lemon juice slow down the browning of apples?

First, lemon juice contains vitamin C, which can react with the product, temporarily removing the brownish color. Second, lemon juice contains citric acid, which slows down, or *inhibits*, the enzymatic reaction by changing the pH environment of the enzyme. Enzyme-catalyzed reactions, like most other chemical reactions, are affected by the reaction conditions. These include changes in substrate concentration, pH, temperature, and the presence of inhibitors. Changing reaction conditions affects how fast an enzyme converts its substrate to product. Measuring how fast an enzymatic reaction occurs is a measure of an enzyme's **activity.**

Substrate Concentration

To understand how the concentration of the substrate affects enzyme activity, let's consider this example. Imagine a factory that assembles bicycles. The supplier brings in enough parts to produce 10 bicycles, and the workers with their tools build 10 bicycles in one hour. The manager of the factory asks the supplier to bring in enough parts to make 15 bicycles, and the workers also complete that task in one hour.

Impressed, the manager asks the supplier to bring in enough parts to make 50 bicycles. In one hour, the workers assemble only 25 bicycles. The manager then asks the supplier to bring in enough parts to make 100 bicycles, but the workers still assemble only 25 in one hour. The workers are working at their maximum capacity by assembling 25 bicycles in one hour. Regardless of the number of extra parts the manager orders, if the resources at the factory do not increase (for example, the number of workers, number of tools), only 25 bicycles can be built in one hour. Much like workers and their tools, enzymes also operate at a particular rate, no matter how many substrate molecules (or bicycle parts) are present.

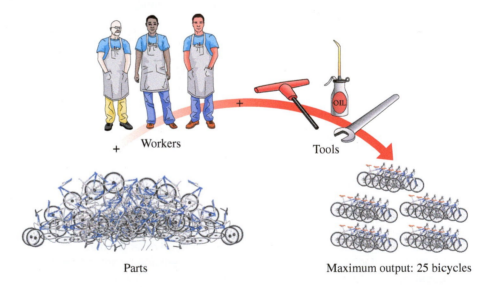

Parts   Maximum output: 25 bicycles

When workers (enzymes) are operating at a maximum rate, a steady state exists and will not increase even if an excess of bicycle parts (substrate) is present.

As we saw earlier, the first step in an enzyme-catalyzed reaction is the formation of ES. If the amount of enzyme remains unchanged, an increase in the substrate concentration increases the enzyme's activity up to the point where the enzyme becomes saturated with its substrate. That is, all of the enzyme becomes saturated with substrate, leaving no more free active sites. The enzyme is working at maximum activity. Increasing the amount of substrate will not increase the activity further. At maximum activity, the conditions under which the enzyme is operating are considered to be in a **steady state.** Under steady-state conditions, substrate is being converted to product as efficiently as possible (see **FIGURE 10.16**).

pH

In addition to substrate concentration, another condition that affects enzyme activity is the pH of an enzyme's environment. Let's go back to the workers at the bicycle factory for an explanation. Suppose that it is flu season and several workers come to work with early stages of the flu, breathe on their coworkers, and infect most of the workforce. With fewer able-bodied workers at the factory, fewer bicycles can be assembled in an hour. Like the environment at the factory, when an enzyme's environment is changed (in this case, the pH), its tertiary structure is disrupted, altering the active site and causing the enzyme's activity to decrease.

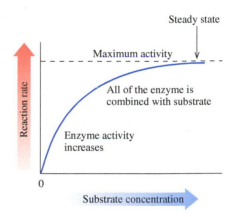

FIGURE 10.16 Effect of substrate concentration on enzyme activity. Increasing the amount of substrate increases the enzyme's activity until the enzyme becomes saturated with substrate and a steady state is reached.

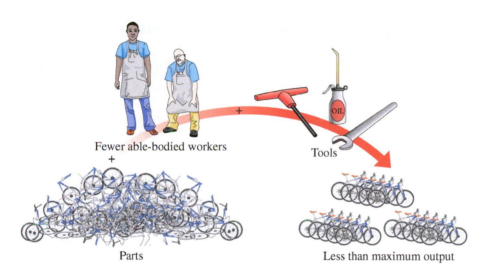

When workers are ill (fewer active enzymes are present), the rate of output is decreased.

Enzymes are most active at a pH known as their **pH optimum.** At this pH, the enzyme maintains its tertiary structure and, therefore, its active site. Changes in the pH not only affect the structure of the enzyme but also may change the nature of an amino acid side chain. For example, if an enzyme requires a carboxylate ion ($-COO^-$) to function properly, lowering the pH could change some of the carboxylate ion to its carboxylic acid form ($-COOH$). This change would cause the enzyme's activity to decrease.

In the body, an enzyme's pH optimum is based on the location of the enzyme. For example, enzymes in the stomach function at a much lower pH because of the acidity. pH optimum values for several common enzymes are shown in **TABLE 10.7**.

TABLE 10.7 pH Optimum for Selected Enzymes

Enzyme	Location	Substrate	pH Optimum
Pepsin	Stomach	Peptide bonds	2
Sucrase	Small intestine	Sucrose	6.2
Urease	Liver	Urea	7.4
Hexokinase	All tissues	Glucose	7.5
Trypsin	Small intestine	Peptide bonds	8
Arginase	Liver	Arginine	9.7

Temperature

As with pH, enzymes have a **temperature optimum** at which they are most active. Let's consider the bicycle factory workers if the air conditioning broke down on a hot day in the middle of August. It is likely that the workers would not be able to work as efficiently to assemble 25 bicycles in one hour.

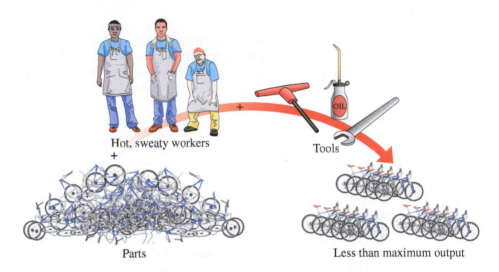

Hot, sweaty workers + Parts + Tools → Less than maximum output

When all the workers are present, but the temperature of the factory makes them hot (some enzymes are denatured by temperature), the rate of output is decreased.

The temperature optimum for most human enzymes is normal body temperature, 37 °C. Above optimum temperature, enzymes lose activity due to the disruption of the attractive forces stabilizing the tertiary structure. At high temperatures, an enzyme denatures, which in turn modifies the structure of the active site. At low temperatures, enzyme activity is reduced due to the lack of energy present for the reaction to take place at all.

Because enzymes are the major culprits in food spoilage, we store foods in a refrigerator or freezer to slow the spoilage process. The enzymes present in bacteria can also be destroyed by high temperatures, in processes like boiling contaminated drinking water or sterilizing instruments and other equipment in hospitals and laboratories.

Inhibitors

Certain molecules known as **inhibitors** can also cause enzymes to lose activity. If we consider the bicycle factory, suppose some of the workers' tools became rusted and cannot be used to assemble the bicycles. This would definitely inhibit bicycle production.

Enzyme inhibitors function in different ways, but they all prevent the active site from interacting with the substrate to form ES. Some inhibitors cause enzymes to lose catalytic activity only temporarily, whereas others function to cause enzymes to lose catalytic activity permanently.

In **reversible inhibition,** the inhibitor causes the enzyme to lose catalytic activity. However, if the inhibitor is removed, the enzyme becomes functional again. Reversible inhibitors can be competitive or noncompetitive.

Competitive inhibitors are molecules that compete with the substrate for the active site. A competitive inhibitor has a structure that resembles the substrate of the enzyme. The competitive inhibitor will interact with the active site to form an enzyme–inhibitor complex, but usually no reaction will take place. As long as the inhibitor remains in the active site, the enzyme cannot interact with its substrate to form products. This lowers the enzyme's activity. **FIGURE 10.17** diagrams how a competitive inhibitor works.

A medical therapy based on competition involves the enzyme liver alcohol dehydrogenase (LAD). LAD oxidizes ethanol, the alcohol found in wine and spirits, as well

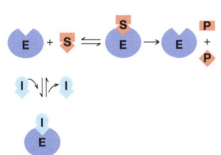

FIGURE 10.17 Competitive inhibition. The inhibitor (I) is similar to the substrate, allowing it to compete for the active site and prevents binding of the substrate. This inhibition is reversible if more substrate is added.

as the compounds ethylene glycol and methanol, both found in antifreeze. All three are substrates of LAD and compete for the active site. Notice that the structures of the three are very similar.

$$CH_3CH_2OH \qquad HOCH_2CH_2OH \qquad CH_3OH$$
Ethanol   Ethylene glycol   Methanol

In fact, one remedy for ethylene glycol poisoning in pets and methanol poisoning in humans is the slow intravenous infusion of ethanol, maintaining a controlled concentration in the bloodstream over several hours. This slows the production of the toxic metabolic products of ethylene glycol or methanol, giving the kidneys time to filter out the excess substrates in the urine.

Noncompetitive inhibitors typically do not resemble the substrate, so they do not compete for the enzyme's active site. Instead, these inhibitors bind to another site on the enzyme that is usually remote from the active site. When a noncompetitive inhibitor binds to the enzyme, it changes the shape of the enzyme. Therefore, the active site loses its shape or is distorted, so it is no longer able to interact effectively with the substrate. Again, as long as the inhibitor remains bound to the enzyme, the enzyme cannot function. **FIGURE 10.18** diagrams how a noncompetitive inhibitor works.

As suggested in Figures 10.17 and 10.18, the inhibition caused by competitive and noncompetitive inhibitors can be reversed. Inhibition caused by a competitive inhibitor can be reversed by adding more substrate to the reaction. Because the inhibitor and the substrate compete for the active site, the higher the concentration of the substrate, the more likely that it will win the competition for the active site. In the case of the noncompetitive inhibitor, adding more substrate has no effect. Regardless of the amount of enzyme, a certain portion of the enzyme is inactivated by the inhibitor. Reversing the effect of a noncompetitive inhibitor typically requires a special chemical reagent to remove the inhibitor and restore the catalytic activity of the enzyme. Going back to the bicycle factory, if a technician removed the rust from all the workers' tools so that they were again in working order, the number of bicycles produced by the factory would increase.

In **irreversible inhibition,** the inhibitor forms a covalent bond with an amino acid side chain on the enzyme. With the inhibitor covalently bonded in the active site, the substrate is excluded or the catalytic reaction is blocked. Regardless of the method, irreversible inhibitors permanently inactivate enzymes (see **FIGURE 10.19**). Likewise, if a fire destroyed a major portion of the bicycle factory, bicycle production would cease.

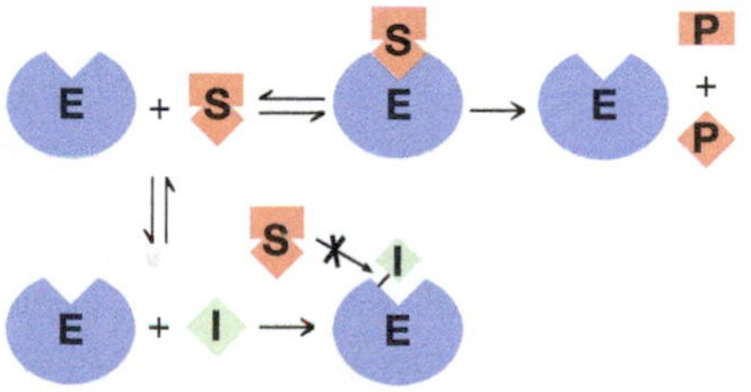

Heavy transition metals like silver, mercury, and lead can inactivate enzymes irreversibly by binding to the thiol ($-SH$) functional groups of the amino acid cysteine if found in the active site. As we saw in Section 10.4, this can also denature the proteins.

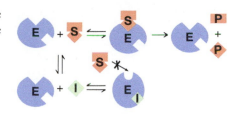

FIGURE 10.18 Noncompetitive inhibition. Normal enzyme–substrate functioning is shown at the top of the figure. When a noncompetitive inhibitor binds to a site other than the active site (notch at the bottom right of the enzyme), the shape of the active site becomes distorted, preventing proper binding and catalysis of the substrate at the active site (bottom part of the figure). This inhibition is reversible if more substrate is added.

FIGURE 10.19 Irreversible inhibition. Normal enzyme–substrate functioning is shown at the top of the figure. When an irreversible inhibitor forms a covalent bond with the enzyme, it renders the enyzme permanently inactive (bottom part of the figure).

Antibiotics Inhibit Bacterial Enzymes

Enzyme inhibitors have long been used in the battle against diseases. The well-known antibiotic penicillin is an irreversible inhibitor. Penicillin binds to the active site of an enzyme that bacteria use in the synthesis of their cell walls. The cell wall is a structure present in bacterial cells but not human cells. When the bacterial enzyme bonds with penicillin, the enzyme loses its catalytic activity, and the growth of the bacterial cell wall slows. Without a proper cell wall for protection, bacteria cannot survive, and the infection stops.

(R is a variable group)

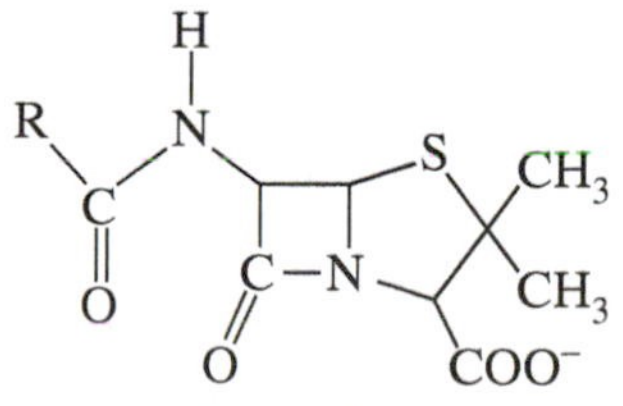

Penicillin

INTEGRATING Chemistry

◀ Find out how penicillin acts as an antibiotic.

Sample Problem 10.8 **Factors Affecting Enzyme Activity**

Lactase catalyzes the hydrolysis of the disaccharide lactose into the monosaccharides galactose and glucose. It is active in the small intestine, where the pH is around 6.8. Supplements of lactase can help people who are lactose-intolerant to digest lactose.

$$\text{Lactose} \xrightarrow{\text{Lactase}} \text{galactose} + \text{glucose}$$

a. Would the enzyme be just as active in the stomach with a pH of 1–3? Support your answer.
b. Would a lactase supplement act just as well on lactose if it were first stirred into cold milk before drinking? Support your answer.

Solution

a. No. The enzyme would be less active in the stomach because it operates at a pH optimum of 6.8.
b. No. The enzyme has a temperature optimum of 37 °C, body temperature. Its activity would be much lower in cold milk.

Practice Problems

10.51 How would the following changes affect enzyme activity for an enzyme whose optimal conditions are normal body temperature and physiological pH?

a. raising the temperature from 37 °C to 60 °C
b. lowering the pH from 7.5 to 4.0

10.52 Chymotrypsin is an enzyme located in the small intestine that catalyzes peptide bond hydrolysis in proteins. Based on the examples given in Table 10.1, would you expect chymotrypsin to be most active at a pH of 4.5, 7.8, or 10.0?

10.53 The enzyme urease functions in the body to catalyze the formation of ammonia and carbon dioxide from urea as shown:

$$\underset{\substack{| \\ \text{H}_2\text{N}-\overset{\displaystyle \overset{\text{O}}{\|}}{\text{C}}-\text{NH}_2}}{} + \text{H}_2\text{O} \xrightarrow{\text{Urease}} 2\text{NH}_3 + \text{CO}_2$$

Describe what effect the following changes would have on the rate of this reaction assuming a steady state has been reached:

a. adding excess urea
b. lowering the temperature to 0 °C

10.54 Indicate whether each of the following describes a competitive or a noncompetitive inhibitor.

a. The structure of the inhibitor is similar to that of the substrate.
b. Adding more substrate to the reaction has no effect on the enzyme activity.
c. The inhibitor competes with the substrate for the active site.
d. The structure of the inhibitor has no resemblance to the structure of the substrate.
e. Adding more substrate to the reaction restores the enzyme activity.

The study guide will help you check your understanding of the main concepts in Chapter 10. You should be able to answer problems for each learning outcome in this list. To check your mastery, try the problems listed after each.

STUDY GUIDE	CHAPTER GUIDE

10.1 Amino Acids—The Building Blocks

Characterize the structure of an amino acid.

- Draw the general structure of an amino acid. (Try 10.1, 10.7)
- Identify amino acids by their one-letter and three-letter abbreviations (Try 10.3, 10.57)
- Identify amino acids based on their polarity. (Try 10.5, 10.57)
- Apply the definition of isoelectric point to predict the charge of a neutral amino acid at points below, at, and above the pI value. (Try 10.11, 10.63)

Inquiry Question

What are the characteristics of an amino acid?

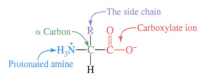

Amino acids contain a central carbon atom, called the α carbon, bonded to four different groups—a protonated amine (amino) group, a carboxylate group, a hydrogen atom, and a side chain (R). Because of this structure, amino acids, with the exception of glycine, are chiral compounds. The L-enantiomers of amino acids are the building blocks of proteins. The 20 amino acids are found in most proteins. They are characterized by various side chains. The properties of the side chains determine whether the amino acids are classified as nonpolar or polar. Amino acids act as weak acids and bases. At physiological pH, the neutral amino acids exist as zwitterions with a net charge of zero. The isoelectric point (pI) exists when there is no net charge present.

10.2 Protein Formation

Form peptide bonds from amino acids.

- Predict the products of a biological condensation or hydrolysis reaction, including peptide bond formation. (Try 10.13, 10.107)
- Name a short peptide by one-letter and three-letter abbreviations given its structure. (Try 10.15b)
- Draw a short peptide sequence given its one-letter or three-letter abbreviation. (Try 10.17, 10.67a)
- Identify the N-terminus and C-terminus of a short peptide sequence. (Try 10.15a, 10.65c)

Inquiry Question

How is a peptide bond formed?

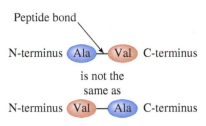

Several biomolecules join through condensation reactions when functional groups combine. Amino acids join through a condensation reaction of the protonated amine group of one and the carboxylate group of the other. The bond that forms between the two amino acids is called a peptide bond, and the new structure is a dipeptide. The newly formed dipeptide has an N-terminus with a free protonated amine group and a C-terminus with a free carboxylate group. A compound containing 50 or more amino acids linked by peptide bond is a polypeptide and, if it has biological activity, is called a protein. Proteins are polymers of amino acids.

10.3 The Three-Dimensional Structure of Proteins

Describe how the levels of protein structure maintain protein shape.

- Describe the attractive forces present as a protein folds into its three-dimensional shape. (Try 10.27, 10.67b)
- Distinguish the levels of protein structure. (Try 10.29, 10.69, 10.71)

Inquiry Question

How is a protein's structure held together?

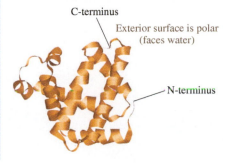

Four levels are used to describe protein structure. Each level is held together through covalent bonding or attractive forces. The primary structure (1°) is the sequence of the amino acids that form the protein backbone.

—Continued next page

Continued—

The bonding interaction is the peptide bond. The secondary structure (2°) involves the interactions of amino acids near each other in the primary structure and describes patterns of regular or repeating structure. The most common secondary structures are the α helix and the β-pleated sheet. The secondary structure is stabilized by hydrogen bonding between atoms in the backbone. The tertiary structure (3°) is formed by folding the secondary structure onto itself and is driven by the hydrophobic interactions of amino acid side chains with their aqueous environment. This level is stabilized by the attractive forces between side chains and disulfide bonds. Proteins that fold into a roughly spherical shape are called globular proteins. Proteins that maintain elongated structures are referred to as fibrous proteins. Some proteins have a quaternary structure (4°), which involves the association of two or more peptides to form a biologically active protein. The quaternary structure is stabilized by the same forces as the tertiary structure.

10.4 Denaturation of Proteins

Define protein denaturation and discuss its causes.

- List the causes of protein denaturation and the attractive forces affected. (Try 10.33, 10.75)
- Distinguish protein denaturation and protein hydrolysis. (Try 10.35, 10.76)

Inquiry Question

What causes proteins to denature?

Denaturation of a protein disrupts the stabilizing attractive forces in the secondary, tertiary, or quaternary structure, often unfolding the protein. When a protein is denatured, its primary structure is not changed. Proteins can be denatured by heat, a change in the pH of their environment, reaction with small organic compounds and heavy metals such as lead or mercury, or mechanical agitation. A denatured protein is no longer biologically active.

10.5 Protein Functions

Compare the various functions of proteins.

- Identify various functions of proteins. (Try 10.37)
- Provide examples of protein structure dictating protein function. (Try 10.39, 10.81)

Inquiry Question

What functions do proteins play in the body?

Proteins have a variety of functions in the body. They act as messengers between cells, receptors on the surface of cells, and transporters through the body or across the cell. Proteins are used to store nutrients, contract muscles, protect the cell, and support its structure. Proteins catalyze biochemical reactions as enzymes.

10.6 Enzymes—Life's Catalysts

Describe how an enzyme catalyzes a biochemical reaction.

- Define active site and substrate. (Try 10.43, 10.83)
- Distinguish cofactor, coenzyme, and prosthetic group. (Try 10.44, 10.83, 10.89)
- Distinguish the lock-and-key model from the induced-fit model. (Try 10.48, 10.85, 10.87)
- Discuss factors that lower the activation energy and speed reaction for an enzyme-catalyzed reaction. (See Sample Problem 10.6)

Inquiry Question

How does an enzyme catalyze a biochemical reaction?

Enzyme–substrate complex

Enzymes are large, globular proteins that serve as catalysts in biological systems. The functional part of an enzyme is the active site, which is a small groove or cleft on the surface of the molecule where catalysis occurs. Substrates are the reactants in the reactions catalyzed by enzymes. Because of the three-dimensional shape of the active site, enzymes have few substrates (many have only one) that will bind and react. The lock-and-key and induced-fit models explain how an enzyme interacts with its substrate to form an enzyme-substrate complex, ES. The formation of ES lowers the activation energy for the catalyzed reaction in several ways. These include increasing proximity, optimizing orientation, and modifying bond energy.

10.7 Factors That Affect Enzyme Activity

Discuss factors that enhance and inhibit enzyme activity.

- Describe how substrate concentration, pH, temperature, and inhibition affect enzyme activity. (Try 10.51, 10.95)
- Distinguish competitive, noncompetitive, and irreversible inhibition. (Try 10.54, 10.99, 10.103)

Inquiry Question

What factors affect enzyme activity?

Enzyme activity is measured by how fast an enzyme catalyzes a reaction. Environmental factors such as substrate concentration, pH, temperature, and the presence of inhibitors can affect enzyme activity. Enzymes have a pH optimum and temperature optimum. Inhibitors decrease or eliminate an enzyme's catalytic abilities. The effect of an inhibitor can be reversible or irreversible. Reversible inhibitors can be competitive inhibitors, which compete with the substrate for the enzyme's active site, or noncompetitive inhibitors, which bind to the enzyme at a site other than the active site, changing the shape of the active site.

 Mastering Videos

PRACTICING THE CONCEPTS

The following videos can be accessed through the Pearson eText or your Mastering Chemistry course.

- pH and Amino Acid Structure
- The Role of Attractive Forces in Protein Structure

Additional Problems

10.55 What functional groups are common to all α-amino acids?

10.56 Give the name and three-letter abbreviation for the amino acid described by each of the following:
- **a.** the nonpolar amino acid with a sulfur atom in its side chain
- **b.** a polar amino acid with a single nitrogen atom in its side chain
- **c.** the nonpolar amino acid with only one carbon in its side chain

10.57 Give the name and three-letter abbreviation for the amino acid described by each of the following:
- **a.** the polar amino acid with a benzene ring in its side chain
- **b.** the nonpolar amino acid whose side chain forms a ring with its α-amino group
- **c.** the polar amino acid with a sulfur atom in its side chain

10.58 Give the name and three-letter abbreviation for the amino acid that does not have a chiral center.

10.59 Two of the amino acids have two chiral centers. Give the name and three-letter abbreviation of each.

10.60 Aspartame, which is commonly known as NutraSweet™, contains the following dipeptide:

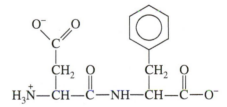

- **a.** What are the amino acids in aspartame?
- **b.** Give the three-letter abbreviation of the N-terminal amino acid.
- **c.** Circle the peptide bond.
- **d.** Draw the structure of the isomer of this dipeptide where the C-terminal and N-terminal amino acids are switched.

10.61 Vegetables, nuts, and seeds are often deficient in one or more of the essential amino acids. From the protein source column in the following table, choose combinations of two or more foods you could eat to get a complete protein.

Amino Acid Deficiency in Some Foods

Protein Source	Amino Acid Missing
Rice	Lys
Oatmeal	Lys
Peas	Met
Beans	Trp, Met
Almonds	Lys, Trp
Corn	Lys, Trp

10.62. Explain why the following amino acid in the form shown cannot exist in aqueous solution:

$$CH_2CH(CH_3)_2$$
$$H_2N-\underset{\underset{H}{|}}{\overset{|}{C}}-\overset{\overset{O}{\|}}{C}-OH$$

10.63 Consider the amino acid valine, shown in its zwitterion form. The pI of valine is 6.0.

  a. Draw the structure of the predominant form of valine at pH = 10.0.

  b. Draw the structure of the predominant form of valine at pH = 3.0.

$$CH(CH_3)_2$$
$$H_3N^+-\underset{\underset{H}{|}}{\overset{|}{C}}-\overset{\overset{O}{\|}}{C}-O^- \qquad pI = 6.0$$

10.64 Draw the structure of the possible dipeptides formed from one alanine combining with one cysteine.

10.65 Consider the amino acids glycine, proline, and lysine.

  a. How many tripeptides can be formed from these three amino acids if each is used only once in the structure?

  b. Using three-letter abbreviations for the amino acids, give the sequence for each of the possible tripeptides.

  c. Draw the structure of the tripeptide that has proline as its N-terminal amino acid and glycine as its C-terminal amino acid. Circle each peptide bond.

10.66 **a.** Draw the structure of Val—Ala—Leu.

  b. Would you expect to find this segment at the center or on the surface of a globular protein? Why?

10.67 **a.** Draw the structure of Ser—Lys—Asp.

  b. Would you expect to find this segment at the center or on the surface of a globular protein? Why?

10.68 Name the covalent bond that helps to stabilize the tertiary structure of a protein.

10.69 Identify some differences between the following pairs:

  a. primary and secondary protein structures

  b. complete and incomplete proteins

  c. fibrous and globular proteins

10.70 Identify some differences between the following pairs:

  a. α helix and β-pleated sheet

  b. ionic interactions (salt bridge) and polar interactions

  c. polar and nonpolar amino acids

10.71 Identify the level of protein structure associated with each of the following:

  a. more than one polypeptide

  b. hydrogen bonding between backbone atoms

  c. the sequence of amino acids

  d. intermolecular forces between R groups

10.72 Identify the level of protein structure associated with each of the following:

  a. α helix

  b. disulfide bridge

  c. peptide bond

  d. salt bridges between polypeptides

10.73 For each of the following proteins, note whether the main secondary structure feature is α helix, β-pleated sheet, or both.

  a. collagen   **b.** alpha-keratin

  c. hemoglobin   **d.** immunoglobulin

  e. hexokinase

10.74 A piece of a polypeptide is folded so that a serine side chain is located on the surface of the polypeptide. Describe the effect on the protein of changing that amino acid to isoleucine.

10.75 Indicate what type(s) of intermolecular forces are disrupted and what level of protein structure is changed by the following denaturing treatments:

  a. an egg placed in water at 100 °C and boiled for 10 minutes

  b. acid added to milk during the preparation of cheese

  c. egg whites whipped in a mixing bowl to make meringue

10.76 Describe the changes that occur in the primary structure when a protein is denatured versus when a protein is hydrolyzed.

10.77 What types of covalent bonds can be disrupted when a protein is denatured? Name a denaturing agent that could accomplish this.

10.78 Describe what happens to the protein structure of hair when it is curled with a heated curling iron.

10.79 Collagen contains an amino acid that is a modified form of the naturally occurring amino acid.
  a. Name the natural amino acid.
  b. How is the structure of the side chain of this amino acid modified?
  c. What is the result of the modification in terms of the structure and function of the collagen?

10.80 To transport oxygen, hemoglobin requires a prosthetic group. Explain what a prosthetic group is and name the one found in hemoglobin.

10.81 Transport proteins are integral membrane proteins and have nonpolar amino acids on their surface and polar amino acids in the interior. This is in contrast to proteins like hemoglobin that are found in the bloodstream. Explain why the surface of the integral membrane protein is nonpolar.

10.82 Describe the role enzymes serve for chemical reactions in the body.

10.83 What occurs at the active site of an enzyme?

10.84 Match the terms (1) ES, (2) enzyme, and (3) substrate with the following descriptions:
  a. has a tertiary structure that recognizes the substrate
  b. is the combination of an enzyme with the substrate
  c. has a structure that fits the active site of an enzyme

10.85 Match the terms (1) active site, (2) lock-and-key model, and (3) induced-fit model with the following descriptions:
  a. the portion of an enzyme where catalytic activity occurs
  b. the active site adapts to the shape of a substrate
  c. the active site has a rigid shape

10.86 Do the amino acids that are in the active site of an enzyme have to be near each other in the enzyme's primary structure? If no, explain.

10.87 The enzyme trypsin catalyzes the breakdown of many structurally diverse proteins in foods. Does the induced-fit or lock-and-key model explain the action of trypsin better? Explain.

10.88 The enzyme sucrase catalyzes the hydrolysis of the disaccharide sucrose but not the disaccharide lactose. Does the induced-fit or lock-and-key model explain the action of sucrase better? Explain.

10.89 A diet deficient in thiamine results in a condition called beriberi, which is characterized by fatigue, weight loss, and poor appetite, among other symptoms. Patients with beriberi do not properly metabolize glucose and other sugars. If thiamine is not an enzyme, what kind of compound do you suppose it is? Explain.

10.90 What type of interactions between an enzyme and its substrate help to stabilize ES?

10.91 Does each of the following statements describe a simple enzyme (no cofactor or coenzyme necessary), an enzyme that requires a cofactor, or an enzyme that requires a coenzyme?
  a. contains Ca^{2+} in the active site
  b. consists of one polypeptide chain in its active form
  c. contains vitamin B_6 in its active site

10.92 A substrate is held in the active site of an enzyme by attractive forces between the substrate and the amino acid side chains. For the outlined regions A, B, and C on the following substrate molecule:
  a. Name one possible attractive force it could form in the active site.
  b. Could the amino acids serine, lysine, or glutamate be present in the active site? Support your answer.

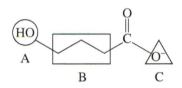

10.93 If each of the following amino acid side chains is present in the active site of an enzyme, indicate whether it would (a) serve a catalytic function, (b) serve to hold the substrate, or (c) both.
  a. aspartate   **b.** phenylalanine

$$-CH_2-\overset{\overset{\displaystyle O}{\|}}{C}-O^-$$

$$-CH_2-\langle\!\!\!\bigcirc\!\!\!\rangle$$

  c. valine   **d.** lysine

$$-CH\overset{\displaystyle CH_3}{\underset{\displaystyle CH_3}{\big\langle}}$$

$$-CH_2CH_2CH_2CH_2NH_3^+$$

10.94 Chymotrypsin, an enzyme that hydrolyzes peptide bonds in proteins, functions in the small intestine at a pH optimum of 7.7 to 8.0. How is the rate of the chymotrypsin-catalyzed reaction affected by each of the following conditions?
  a. decreasing the concentration of proteins
  b. changing the pH to 3.0
  c. running the reaction at 75 °C

10.95 Pepsin, an enzyme that hydrolyzes peptide bonds in proteins, functions in the stomach at a pH optimum of 1.5 to 2.0. How is the rate of a pepsin-catalyzed reaction affected by each of the following conditions?
  a. increasing the concentration of proteins
  b. changing the pH to 5.0
  c. running the reaction at 0 °C

10.96 Problems 10.94 and 10.95 both mention enzymes that hydrolyze peptide bonds. How do you account for the fact that pepsin has a high catalytic activity at pH 1.5 but chymotrypsin has very little activity at pH 1.5?

10.97 After the reaction catalyzed by an enzyme is complete, why do the products of the reaction migrate away from the active site?

10.98 Describe the key difference between a competitive inhibitor and a noncompetitive inhibitor.

10.99 How does an irreversible inhibitor function differently than a reversible inhibitor?

10.100 When lead acts as a poison, it can do so either by replacing another ion (such as zinc) in the active site of an enzyme or by reacting with cysteine side chains to form covalent bonds. Which of these is irreversible and why?

10.101 Increasing the substrate concentration of an enzyme-catalyzed reaction increases the rate of the reaction until the enzyme becomes saturated and reaches its maximum activity. If the amount of *enzyme* in the reaction is increased under saturating conditions, would the rate of the reaction increase, decrease, or remain unchanged? Explain.

10.102 A hypothetical enzyme has an optimal temperature of 40 °C. If we wish to double the rate of the reaction catalyzed by this enzyme, can we do so by increasing the temperature to 80 °C? Explain your answer.

10.103 Cadmium is a poisonous metal used in industries that produce batteries and plastics. Cadmium ions (Cd^{2+}) are inhibitors of hexokinase. Increasing the concentration of glucose or ATP, the substrates of hexokinase, or Mg^{2+}, the cofactor of hexokinase, does not change the rate of the cadmium-inhibited reaction. Is cadmium a competitive or noncompetitive inhibitor? Explain.

10.104 Meats spoil due to the action of enzymes that degrade the proteins. Fresh meats can be preserved for long periods of time by freezing them. Explain how freezing meats works to prevent spoilage.

10.105 Many drugs are competitive inhibitors of enzymes. When scientists design inhibitors to serve as drugs, why do you suppose they choose to design competitive inhibitors instead of noncompetitive inhibitors?

10.106 Fresh pineapple contains the enzyme bromelain, which degrades proteins.

a. The directions on a package of Jell-O® (a protein-containing food product) say to add canned pineapple (which is heated to high temperatures to preserve it), not fresh pineapple. Why?

b. Fresh pineapple is used in a marinade to tenderize tough meat. Why?

Challenge Problems

10.107 Lipoproteins transport dietary fats through the bloodstream and deliver them to the cells. Cholesterol is mainly transported as a cholesterol ester. The formation of a cholesterol ester is shown. Considering the structure of cholesterol and the fatty acid shown, draw the cholesterol ester produced.

Fatty acid   Cholesterol

10.108 The synthesis of aspirin, acetylsalicylic acid, proceeds through a condensation reaction where an ester is formed (esterification). When salicylic acid and acetic anhydride are reacted, the products produced are acetylsalicylic acid and acetic acid. Draw the products.

Salicylic acid   Acetic anhydride

10.109 How is the structure of a soap micelle (Chapter 7) similar to the structure of a globular protein? (*Hint:* Both of these are found in aqueous solution.)

10.110 Insulin is a protein hormone that functions as two polypeptide chains whose amino acid sequences are as follows:

- A chain: GIVEQCCTSICSLTQLENYCN
- B chain: FVNQHLCGDHLVEALYLV CGERGFFYTPKT

a. The highlighted region in chain B forms an α helix. In an α helix, the R groups are found on the outside of the helix, protruding out from the center, and every fourth amino acid appears on the same side of the helix. Considering the polarity of the R groups, is this helix amphipathic (having a polar part and nonpolar part)? Which side do you think faces the exterior?

b. Considering the amino acid sequences, suggest how these two polypeptide chains might be held together in an active insulin molecule.

10.111 The active site of a hypothetical enzyme contains a pocket of nonpolar amino acid side chains next to a cleft containing acidic and basic amino acid side chains.

    a. Which amino acids are likely responsible for catalysis? Explain.

    b. Which amino acids are present mainly to hold the substrate in the active site?

    c. Would you predict that the substrate is polar or nonpolar based on your answers to part a and part b? Why?

10.112 Contact lens wearers often soak their lenses in a solution containing enzymes to remove protein deposits from the lenses. Why is it necessary to rinse the lenses thoroughly before placing them on the eyes?

Answers to Odd-Numbered Problems

Practice Problems

10.1 **a.**

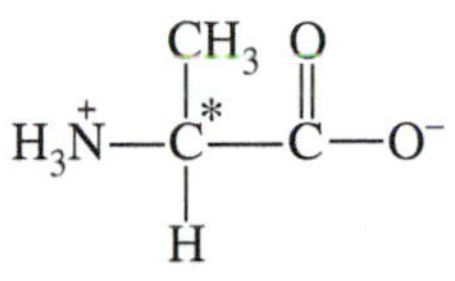

b.

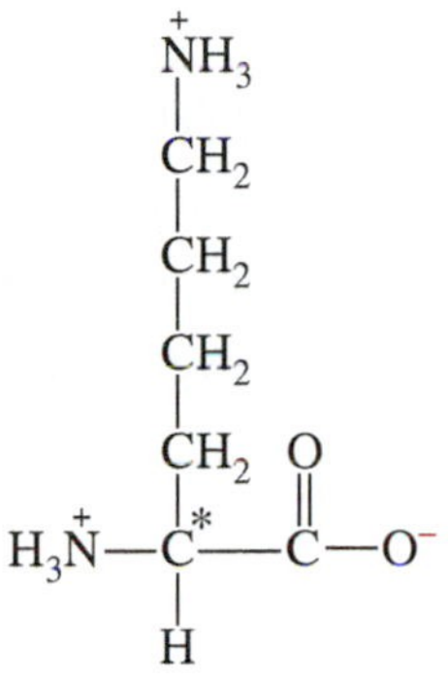

c.

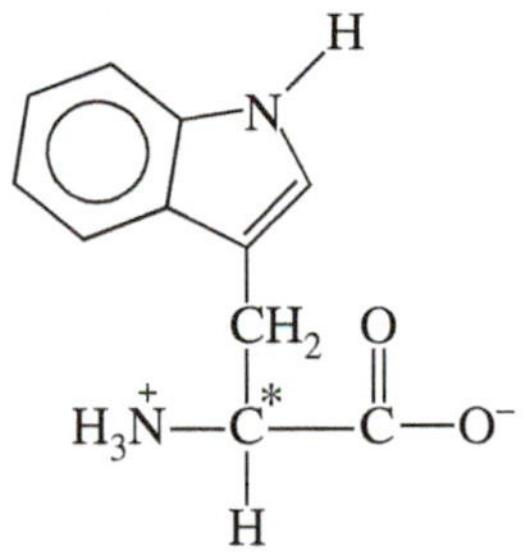

d.

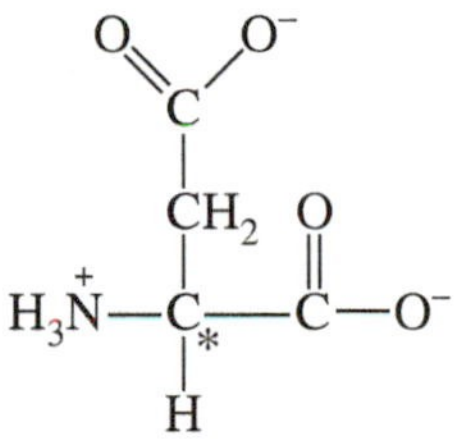

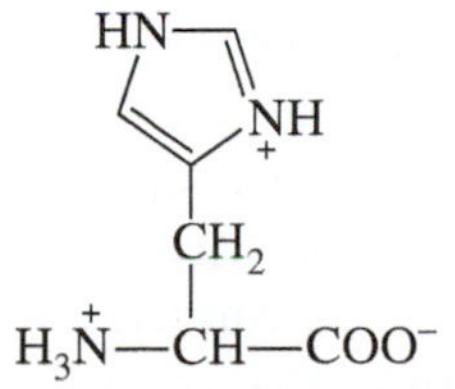

c.

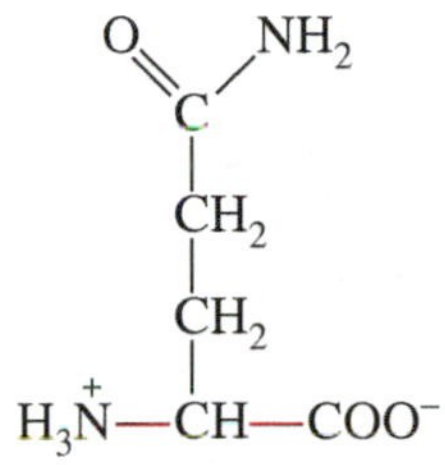

d.

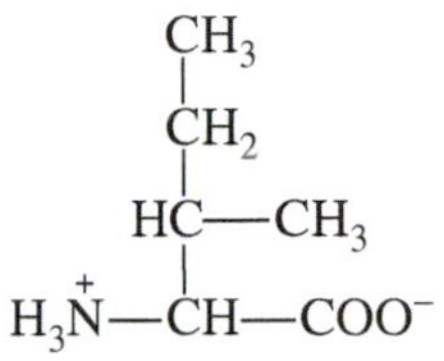

10.9   **a.** glycine   **b.** histidine

      c. glutamine   **d.** isoleucine

10.11 **a.** Net charge: 1+   **b.** Net charge: 0

       pH = 1.8   pH = 6.0

10.3   **a.** Ala, A, hydrocarbon

      b. Lys, K, protonated amine

      c. Trp, W, aromatic

      d. Asp, D, carboxylate

10.5   **a.** nonpolar; hydrophobic

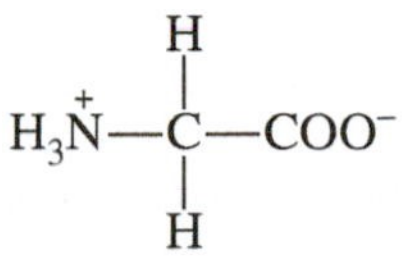

      b. polar (acidic); hydrophilic

      c. nonpolar; hydrophobic

      d. polar (neutral); hydrophilic

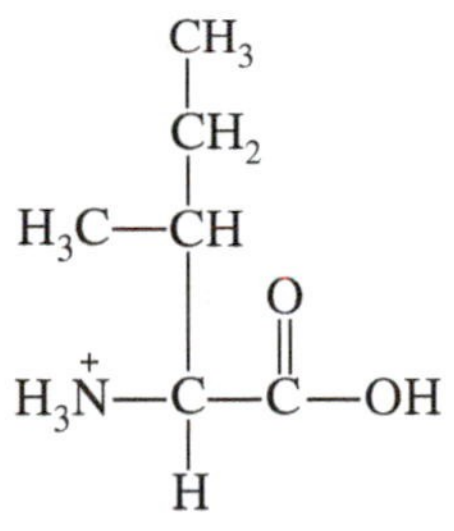

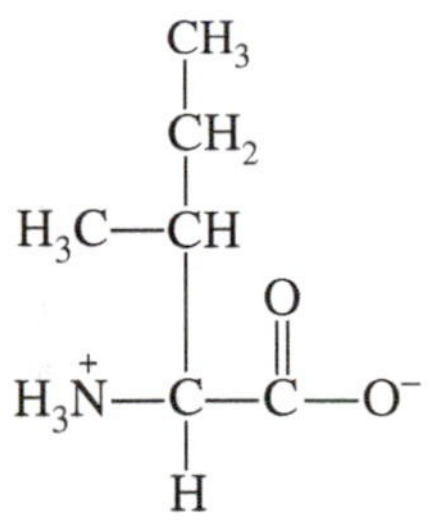

    c. Net charge: 1−

       pH = 11.5

10.7 **a.**

b.

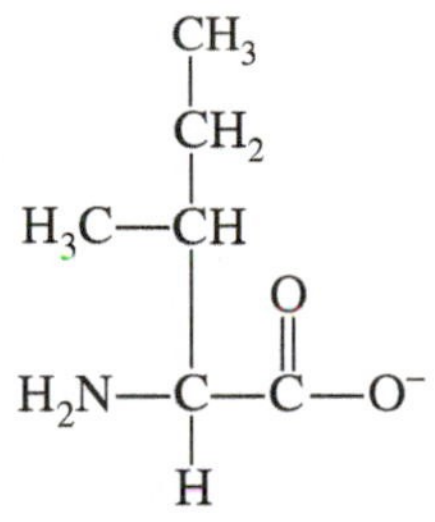

10.13 a.

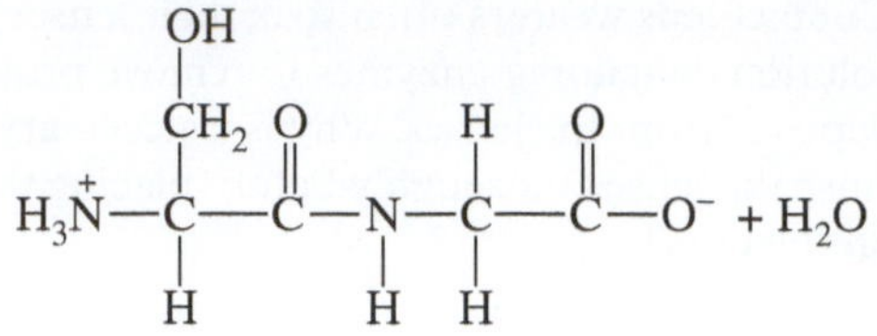

$$H_3\overset{+}{N}-C-C-N-C-C-O^- + H_2O$$

b. $CH_3COOH + CH_3(CH_2)_7OH$

10.15 a.

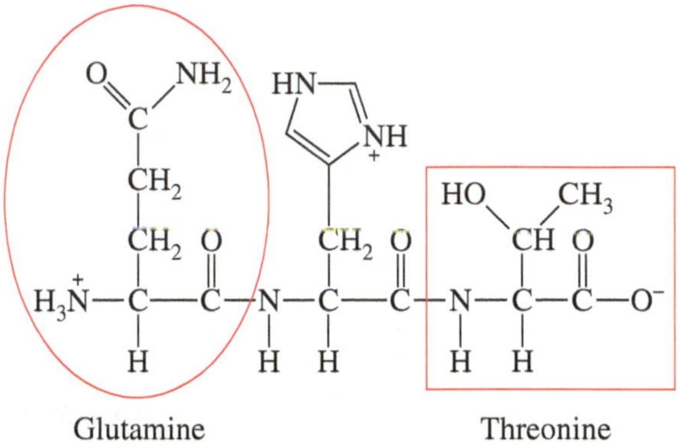

Glutamine   Threonine

b. QHT, Gln—His—Thr

10.17 and 10.19

a.

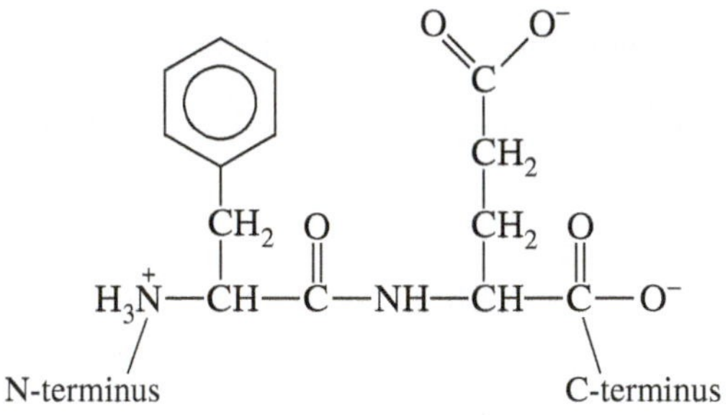

N-terminus   C-terminus

b.

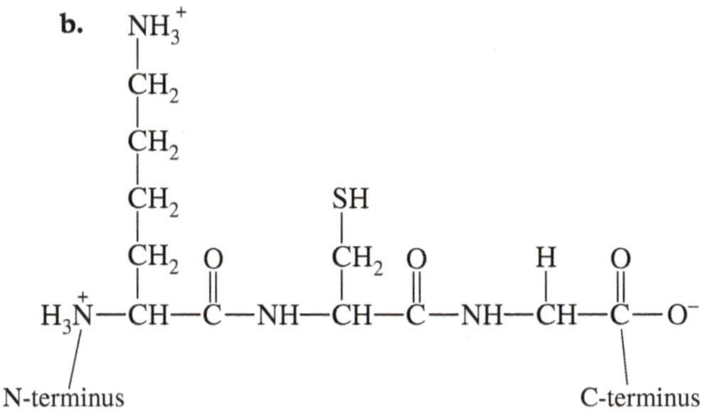

N-terminus   C-terminus

c.

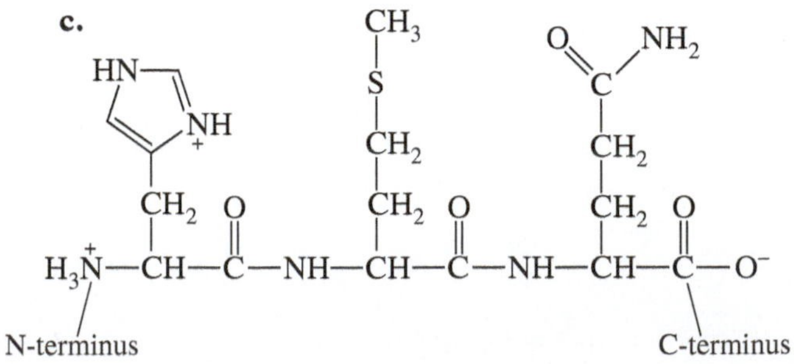

N-terminus   C-terminus

d.

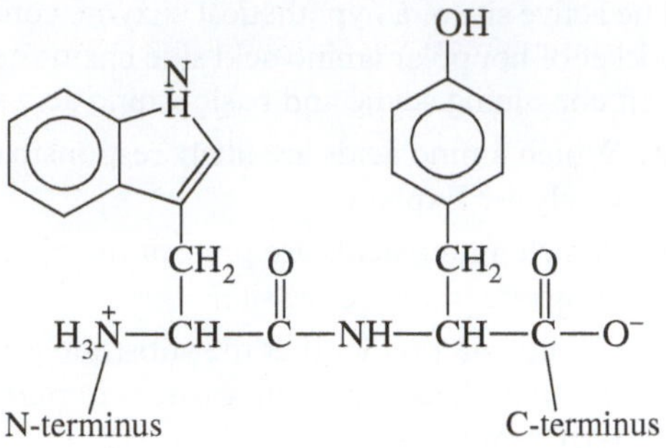

N-terminus   C-terminus

10.21 covalent bonding through peptide bonds

10.23 hydrogen bonding

10.25 An α helix is a coiled structure with the amino acid side chains protruding outward from the helix. The β-pleated sheet is an open, extended, zigzag structure with the side chains of the amino acids oriented above and below the sheet.

10.27 a. salt bridge

b. London force (nonpolar interaction)

c. hydrogen bonding

d. hydrogen bonding and ion–dipole

10.29 a. quaternary   **b.** secondary

c. tertiary   **d.** primary

10.31 Beta-amyloid proteins in a person without Alzheimer's fold into an alpha helix, whereas those in a person with Alzheimer's disease fold into beta-pleated strands.

10.33 a. ionic and hydrogen bonding; secondary, tertiary, and quaternary structure

b. hydrogen bonding and London forces; secondary, tertiary, and quaternary structure

10.35 a. protein denaturation

b. protein hydrolysis

10.37 collagen, structural connector

hemoglobin, oxygen transporter

antibody, bind foreign substances in body

casein, storage protein

10.39 After a meal, the glucose transporter is active because glucose levels are high in the bloodstream and insulin is released. Upon waking in the morning, the glucose transporter is less active because glucose and insulin levels in the bloodstream are low.

10.41 the antigen-binding site

10.43 active site

10.45 tertiary

10.47 The induced-fit model. When glucose is fit into the active site, the enzyme undergoes a conformational change, closing around the substrate.

10.49 Because the phosphates on ATP have negative charges and Mg^{2+} has positive charges, the attraction is ionic.

10.51 a. The enzyme activity will decrease if the temperature is raised above the temperature optimum.

b. The activity will be lowered if the pH is changed.

10.53 a. no effect   **b.** rate decreases

Additional Problems

10.55 carboxylate and protonated amine

10.57 a. tyrosine, Tyr   **b.** proline, Pro

c. cysteine, Cys

10.59 isoleucine, Ile, and threonine, Thr

10.61 several possibilities, including rice and beans, and peas and corn

10.63

$$CH(CH_3)_2$$

$H_2N - C - C - O^-$ (with O double bond, H below), pH = 10

$H_3N^+ - C - C - OH$ (with O double bond, H below), pH = 3

10.65 a. 6

b. Gly—Pro—Lys, Gly—Lys—Pro, Pro—Gly—Lys, Pro—Lys—Gly, Lys—Pro—Gly, Lys—Gly—Pro

c.

10.67 a.

b. This segment would be on the surface in the aqueous environment. The side chains are hydrophilic and would interact with the water in the surrounding environment.

10.69 a. Primary structures are held together by covalent bonding, and secondary structures are held together by hydrogen bonding.

b. Complete proteins contain all essential amino acids and incomplete proteins do not.

c. Fibrous proteins have elongated structures; globular proteins have roughly spherical structures. Globular proteins are soluble in water, and fibrous proteins are not.

10.71 a. quaternary structure

b. secondary structure

c. primary structure

d. tertiary and quaternary structures

10.73 a. α helix   **b.** α helix   **c.** α helix

d. β-pleated sheet   **e.** both

10.75 a. hydrogen bonds and nonpolar attractions; secondary, tertiary, and quaternary

b. hydrogen bonds and salt bridges; secondary, tertiary, and quaternary

c. hydrogen bonds and nonpolar attractions; secondary, tertiary, and quaternary

10.77 disulfide bonds; thioglycolate

10.79 a. proline

b. An OH group is added to the side chain to form hydroxyproline.

c. The OH on the side chain allows for the formation of more hydrogen bonds between side chains, which increases the strength of the collagen.

10.81 The middle portion of these proteins spans the nonpolar portion of the membrane. This surface interacts with the nonpolar environment and therefore must be nonpolar.

10.83 The active site is the location on an enzyme where the substrate binds and catalysis occurs.

10.85 a. (1) active site

b. (3) induced-fit model

c. (2) lock-and-key model

10.87 Because it can have more than one substrate, trypsin's action is better described by the induced-fit model.

10.89 The explanation is that thiamine is involved in enzymatic reactions but is not an enzyme. It is a coenzyme.

10.91 a. requires a cofactor

b. describes a simple enzyme

c. requires a coenzyme

10.93 a. (c)   **b.** (b)

c. (b)   **d.** (c)

10.95 a. The rate would increase due to the increased probability of collision until a steady state is reached.

b. The rate would decrease due to nonoptimal pH and possible protein denaturation.

c. The rate would decrease due to slower movement of molecules.

10.97 The products do not fit snugly into the active site, so they are released.

10.99 An irreversible inhibitor forms covalent bonds to the enzyme and permanently inactivates it. The effects of a reversible inhibitor can be overcome by adding more substrate or by removing the inhibitor.

10.101 If the amount of enzyme were increased in a reaction occurring at a steady state, the maximum rate of the reaction would increase since more enzyme is present to convert substrate into product.

10.103 Cadmium is a noncompetitive inhibitor because increasing the substrate and cofactor has no effect on the rate. This implies that the cadmium is binding to another site on the enzyme.

10.105 When designing an inhibitor, the substrate of an enzyme usually is known even if the structure of the enzyme is not. It is easier to design a molecule that resembles the substrate (competitive) than to find an inhibitor that binds to a second site on an enzyme (noncompetitive).

Challenge Problems

10.107

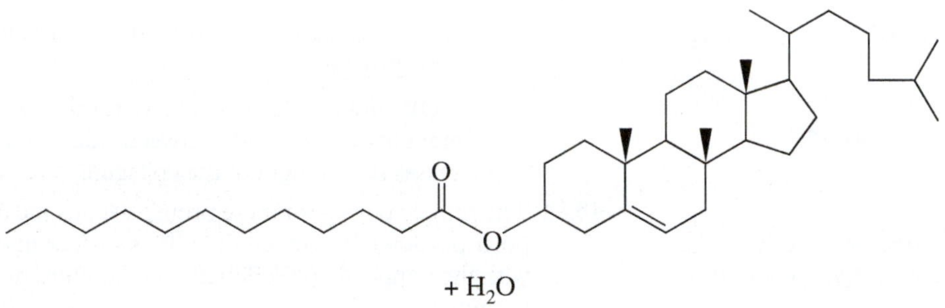

$$+ H_2O$$

10.109 Both have nonpolar interiors and polar surfaces. On a micelle, the polar heads of the fatty acid salts face outward into the aqueous environment like the polar amino acid side chains on the globular protein. The nonpolar tails of the fatty acid salts gather together in the interior of the micelle, just as the nonpolar amino acid side chains gather in the interior of the globular protein.

10.111 **a.** The polar amino acids are likely involved in catalysis since a few charged amino acids are present in the active site.

   b. The nonpolar amino acids are likely involved in aligning the substrate correctly.

   c. The substrate is likely nonpolar with a small polar portion.

Nucleic Acids—Big Molecules with a Big Role

The nucleic acid DNA defines inherited characteristics like the shape of your nose and the color of your eyes, hair, and skin. These large molecules are made up of smaller component parts. Chapter 11 describes the structure of these large molecules and how DNA is used to make proteins in a cell.

> **LEARNING TIP   Group Studying Can Be Effective**
>
> If you study with a group, set a goal, make an agenda, ensure that everyone is prepared, and keep your goal in mind. Everyone should be able to ask or answer questions and come away with a greater understanding of the material you're studying.

Whether working in a hospital, at a crime scene, or at an ancient burial site, we can isolate the nucleic acid deoxyribonucleic acid, or DNA, to positively identify humans and their characteristics. What is unique about this molecule? What does it look like? **DNA** is the molecule in our cells that stores and directs information responsible for cell growth and reproduction. DNA, found in the nucleus of the cell, contains genetic information in what is called the **genome.** One part of the genome, the **gene,** contains information to make a particular protein for the cell. The human genome contains about 20,000 genes.

Every cell in your body contains the same DNA, so why is a skin cell different from a liver cell? Different cell types express different parts of their genome. Together, your DNA in different cells governs your physical traits such as your eye and hair color, your growth rate, and your metabolic rate.

Nucleic acids like DNA and its counterpart ribonucleic acid (**RNA**) are big molecules made up of repeating building blocks called *nucleotides.* Exploring the structure of the nucleotide building blocks is a starting point for understanding the structure of the nucleic acids and genomics, the study of genes and their interactions.

DISCOVERING THE CONCEPTS

? INQUIRY ACTIVITY—Components of Nucleotides

Information

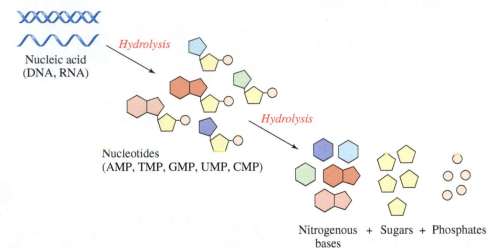

Nucleic acid
(DNA, RNA)

Hydrolysis

Nucleotides
(AMP, TMP, GMP, UMP, CMP)

Hydrolysis

Nitrogenous + Sugars + Phosphates
bases

Scheme 1

Component molecules found in nucleic acids.

Questions

1. Using scheme 1, name the three component molecules found in a nucleotide.
2. What type of chemical reaction occurs when a nucleotide is *formed* from its component molecules?

Nitrogenous Bases

There are five different *nitrogenous bases* found in nucleotides, and these are classified into two families: purines and pyrimidines. Their general structure is as follows:

Purine   Pyrimidine

The five nitrogenous bases found in nucleotides and nucleic acids are shown below:

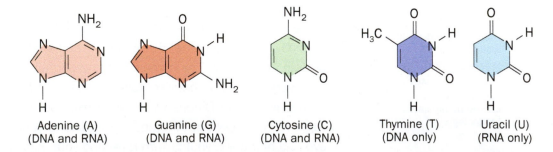

Adenine (A)
(DNA and RNA)

Guanine (G)
(DNA and RNA)

Cytosine (C)
(DNA and RNA)

Thymine (T)
(DNA only)

Uracil (U)
(RNA only)

Questions

3. Which of the nitrogenous bases are purines? Which are pyrimidines?
4. Which are found in DNA? Which in RNA?
5. Compare the structures of uracil and thymine. How are they different?

Pentose Sugars

There are two aldopentose sugars found in nucleic acids: ribose and deoxyribose. They are shown here in the Haworth projection. The sugars in nucleic acids are numbered with a prime (′) to distinguish their carbons from the carbons of the nitrogenous bases.

Questions

6. a. Compare the structures of ribose and deoxyribose. How are they different?
   b. Which one do you think is found in DNA? RNA?
7. Notice that the carbons in the pentose ring are numbered with a prime (′) symbol, whereas the carbons in the purine and pyrimidine rings are not. Why do you think this would be important?

Phosphates

A phosphate is a functional group that contains phosphorus surrounded by four oxygen atoms. Phosphate was first introduced as a polyatomic ion in Chapter 3 and then in the functional group table in Chapter 4 (Table 4.4). Under physiological conditions, the anion, hydrogen phosphate, HPO_4^{2-}, is the dominant form.

Putting It All Together

As implied in scheme 1, condensation reactions between phosphate, sugar, and nitrogenous base produce a nucleotide. The phosphate —OH condenses with —OH on carbon 5′ of the sugar. The purine nitrogen numbered 9 condenses with the —OH on carbon 1′ of a sugar. The pyrimidine nitrogen numbered 1 condenses with the —OH on carbon 1′ of a sugar.

Questions

8. Draw the structure of the nucleotide adenosine monophosphate (AMP).

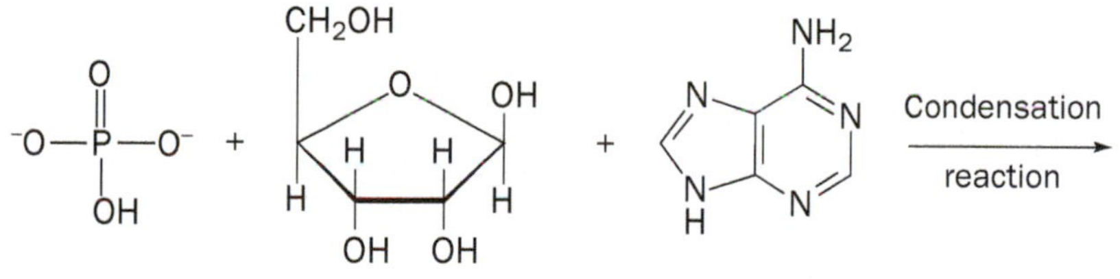

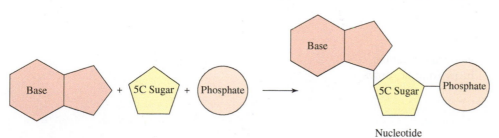

9. If AMP is adenosine monophosphate, how would the structure of ATP, adenosine triphosphate, differ?

Pentose sugars in RNA and DNA

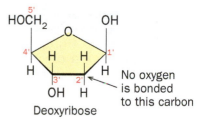

Ribose

No oxygen is bonded to this carbon

Deoxyribose

11.1 Components of Nucleic Acids

Just as proteins are strings of amino acids, nucleic acids are strings of molecules called **nucleotides.** Most proteins contain 20 different amino acids in a given sequence, whereas most nucleic acids are created from a set of 4 nucleotides in a given sequence. Each nucleotide has three basic components: a nitrogenous base (purine or pyrimidine), a five-carbon sugar (pentose), and a phosphate functional group. The differences between DNA and RNA lie in slight variations of the components found in these nucleotides.

What's an Inquiry Question?

Inquiry Questions are designed to focus your reading on the main concepts by section. As you read each section, see if you can answer each Inquiry Question in your own words.

11.1 Inquiry Question

What are the components of nucleotides and nucleic acids?

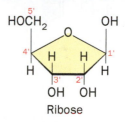

The component parts of a nucleotide are a nitrogenous base, a pentose, and a phosphate.

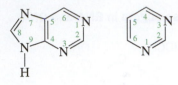

Purine   *Pyrimidine*

Nitrogenous Bases

There are four different nitrogenous bases found in a nucleic acid. Each of the bases has one of two nitrogen-containing aromatic rings, either the purine ring or the pyrimidine ring, shown at left.

DNA contains two purines, adenine (A) and guanine (G), as well as two pyrimidines, thymine (T) and cytosine (C). RNA contains the same bases, except that thymine (5-methyluracil) is replaced with uracil (U) (see **FIGURE 11.1**).

FIGURE 11.1 Purine and pyrimidine bases. DNA contains the nitrogenous bases A, G, C, and T. RNA contains A, G, C, and U. Notice that the structural difference between T and U is a single methyl group.

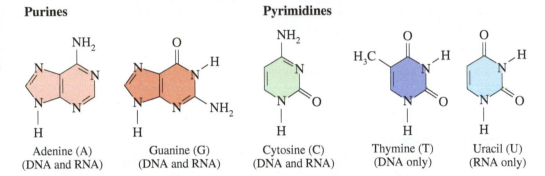

Purines

Adenine (A)
(DNA and RNA)

Guanine (G)
(DNA and RNA)

Pyrimidines

Cytosine (C)
(DNA and RNA)

Thymine (T)
(DNA only)

Uracil (U)
(RNA only)

Ribose and Deoxyribose

Nucleotides also contain five-carbon pentose sugars. To distinguish the carbons in the nitrogenous bases from the carbons in the sugar rings, a prime symbol (′) is added to the carbon numbering of the sugars (1′, 2′, 3′, 4′, 5′).

RNA contains the pentose *ribose* (the "R" in RNA), and DNA contains the pentose *deoxyribose* (the "D" in DNA). Deoxyribose lacks oxygen on carbon 2′ of the pentose; otherwise, its structure is identical to that of ribose (see **FIGURE 11.2**).

Pentose sugars in RNA and DNA

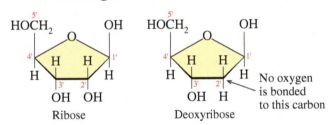

Ribose

Deoxyribose

No oxygen is bonded to this carbon

FIGURE 11.2 Five-carbon sugars. The five-carbon pentose sugar found in RNA is ribose and that in DNA is deoxyribose. Notice that the carbon numbering includes the prime (′) symbol.

Sample Problem 11.1   Components of Nucleic Acids

Identify each of the following bases as a purine or a pyrimidine and name it.

a.

b.

Solution

a. purine; guanine
b. pyrimidine; uracil

Condensation of the Components

In Chapter 6, we saw that two monosaccharides could join through a condensation reaction. In that condensation reaction, a molecule of water is removed and a glycosidic bond forms between two sugars. Similarly, a pentose and a nitrogenous base can join by this reaction when a nitrogen in the base (N1 of pyrimidines or N9 of purines) bonds to C1′ of the pentose, forming a carbon-to-nitrogen glycosidic bond. The —H and —OH removed form water. The product of this reaction is called a nucleoside. The formation of a **nucleoside** is the first step in the synthesis of a nucleotide.

Joining a pentose and a nitrogenous base produces a nucleoside and water.

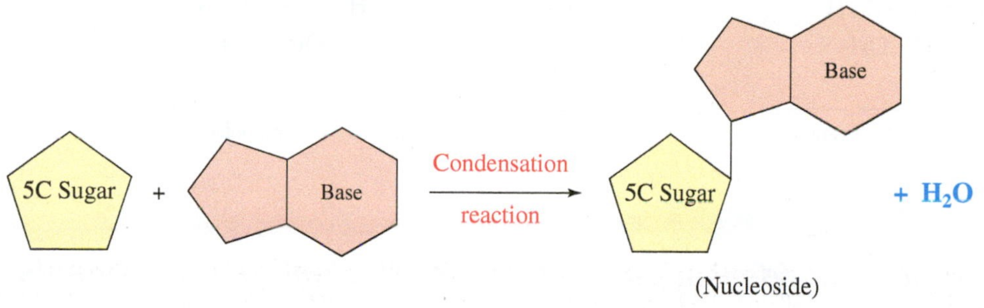

The condensation reaction between ribose and the base adenine is shown, forming the nucleoside called *adenosine*.

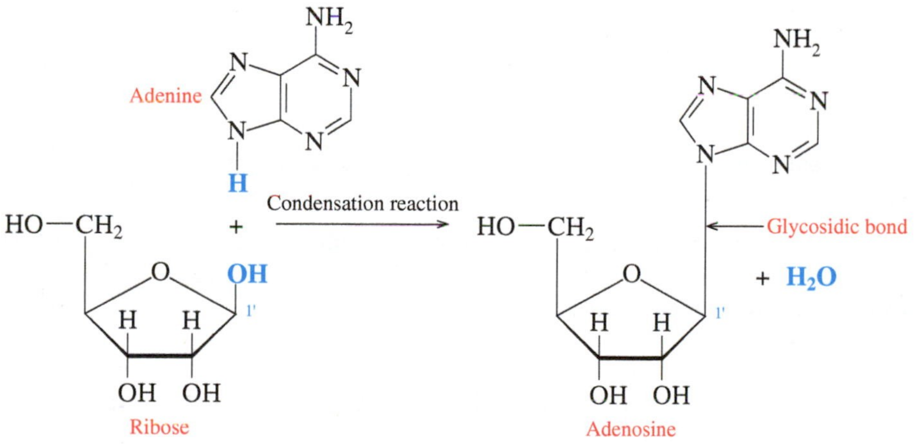

Suppose a hydrogen phosphate group (HPO_4^{2-}), referred to as phosphate here for simplicity, reacts with the —OH group on C5′ of a nucleoside. The molecule formed in this condensation reaction is a nucleo*tide*.

Joining a phosphate to a nucleoside produces a nucleotide and water.

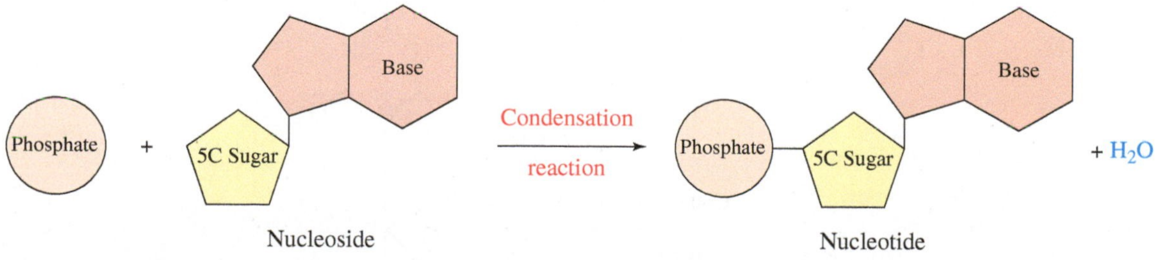

In the case of adenosine, the nucleotide *adenosine* mono*phosphate* (AMP) is produced.

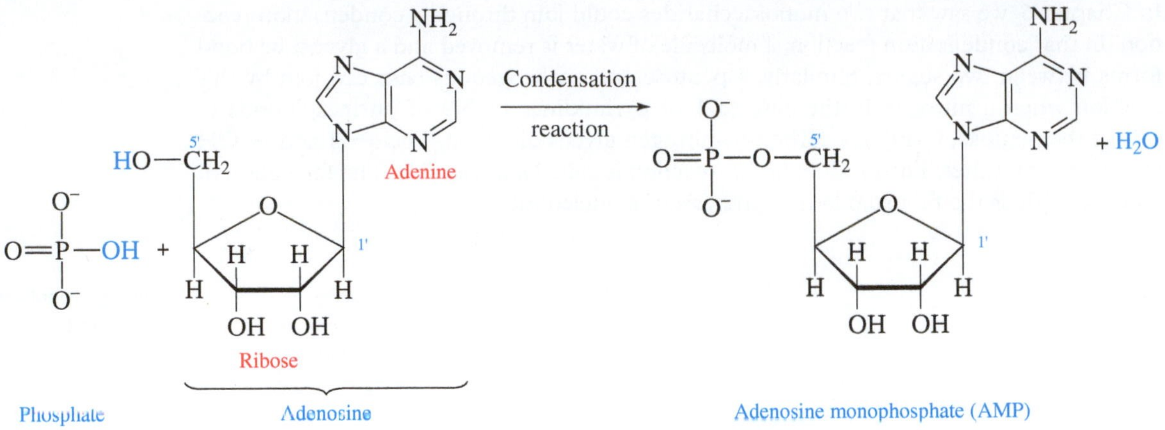

Phosphate   Adenosine   Adenosine monophosphate (AMP)

Solving a Problem

Writing Condensation Products for Nucleotide Components

Write the products for the condensation reaction shown.

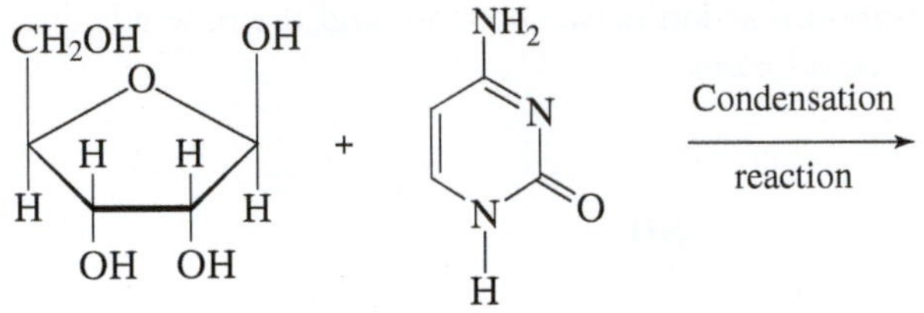

Solution

STEP 1 Identify the components. The components are a pentose (ribose) and a pyrimidine (cytosine).

STEP 2 Locate the functional groups undergoing condensation. When a pentose combines with a pyrimidine, the —OH (blue) of C1′ condenses with the —H (blue) of N1, forming a glycosidic bond and H_2O.

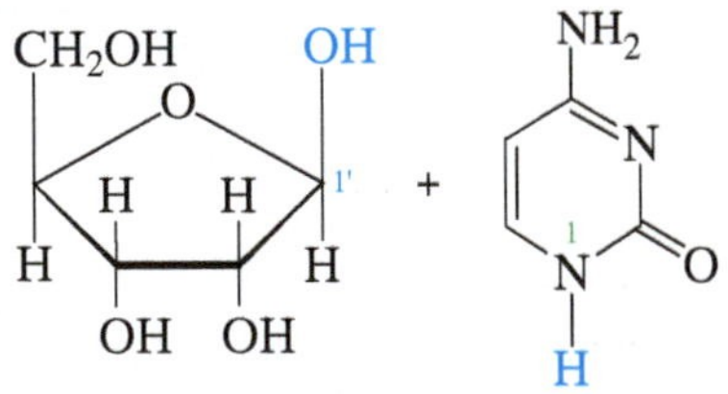

STEP 3 Write the product. The product will be the nucleoside cytidine and a molecule of water.

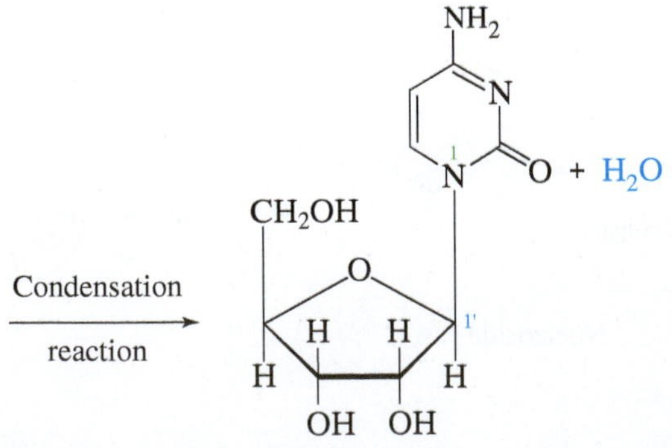

Sample Problem 11.2 Writing Condensation Products

Write the products for the condensation reaction shown.

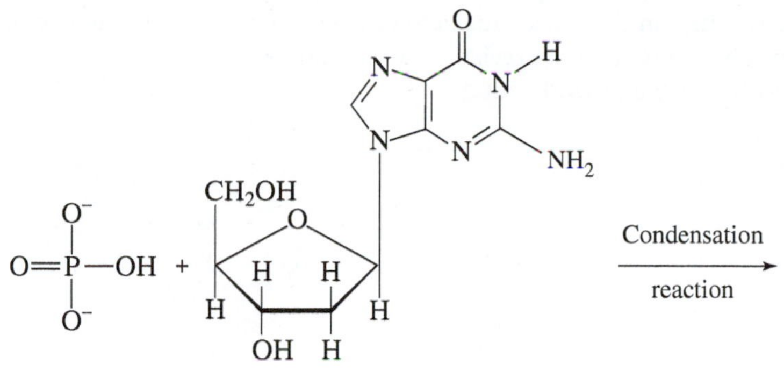

Solution

STEP 1 Identify the components. The components are a nucleoside (deoxyguanosine) and a phosphate group.

STEP 2 Locate the functional groups undergoing condensation. When a nucleoside combines with a phosphate, the —H (blue) on C5′ condenses with the phosphate —OH (blue), forming a nucleotide and H_2O.

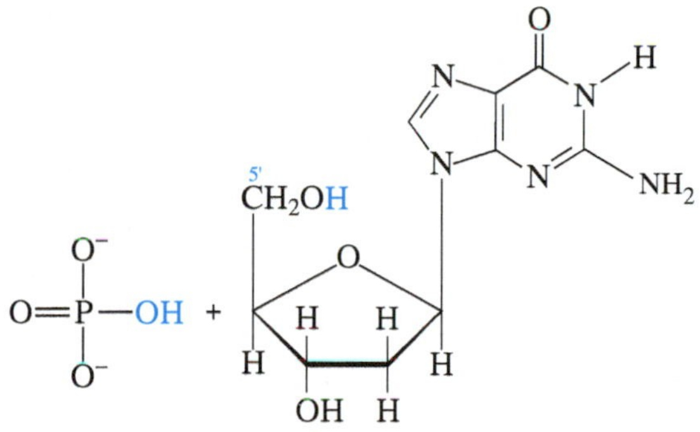

STEP 3 Write the product. The product will be the nucleotide shown and a molecule of water.

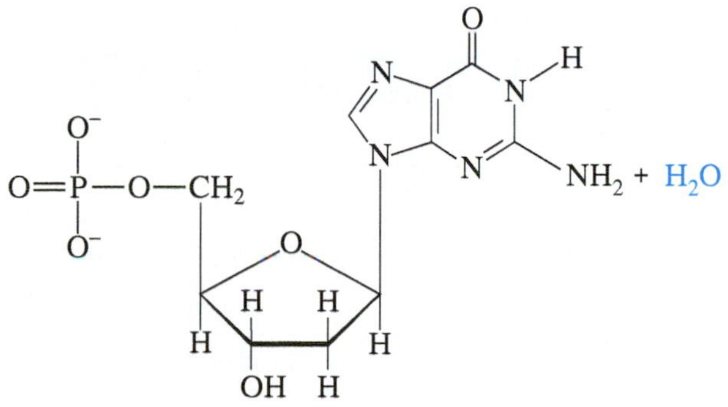

Naming Nucleotides

Nucleotides include the nucleoside name and the number of phosphates present. Because the names can become quite long, nucleotide names often are abbreviated. The abbreviation indicates the type of sugar (ribose or deoxyribose) and the nitrogenous base. If deoxyribose is found in the nucleotide, a lowercase d is inserted at the beginning of the abbreviation. The names and abbreviations of the nucleotides found in nucleic acids are shown in **FIGURE 11.3** and **TABLE 11.1**.

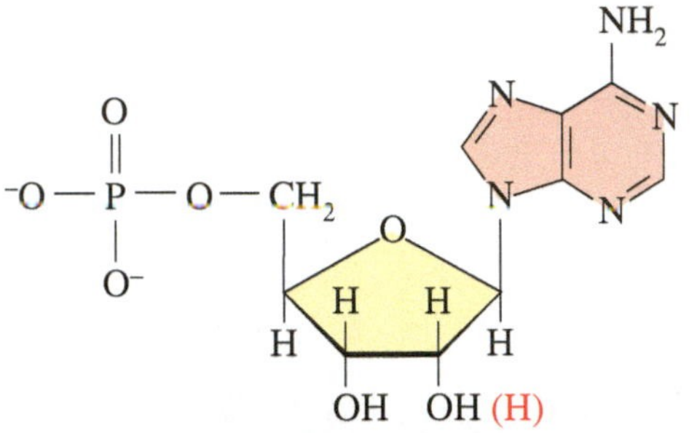

Adenosine monophosphate (AMP)
Deoxyadenosine monophosphate (dAMP)

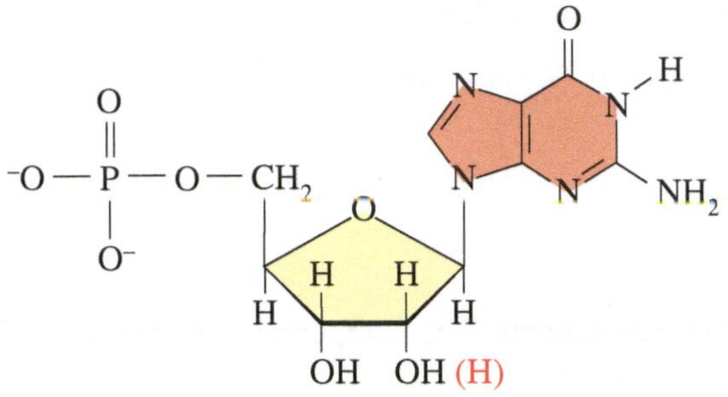

Guanosine monophosphate (GMP)
Deoxyguanosine monophosphate (dGMP)

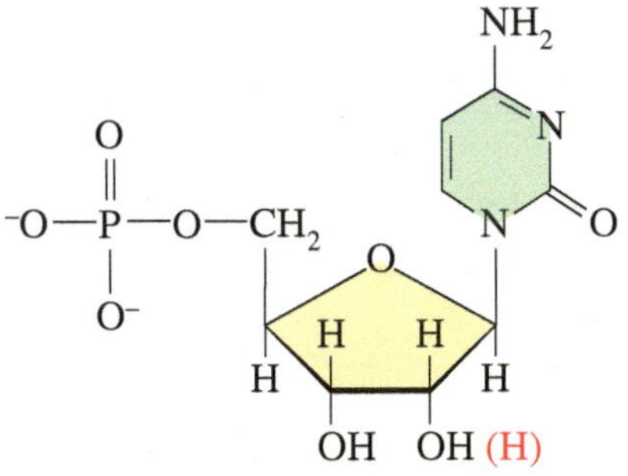

Cytidine monophosphate (CMP)
Deoxycytidine monophosphate (dCMP)

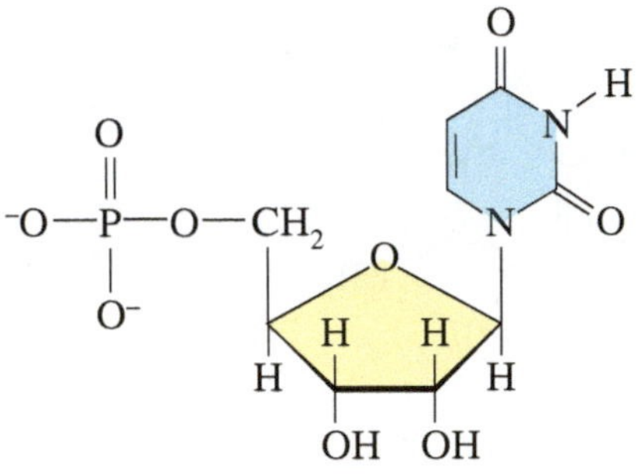

Uridine monophosphate (UMP)

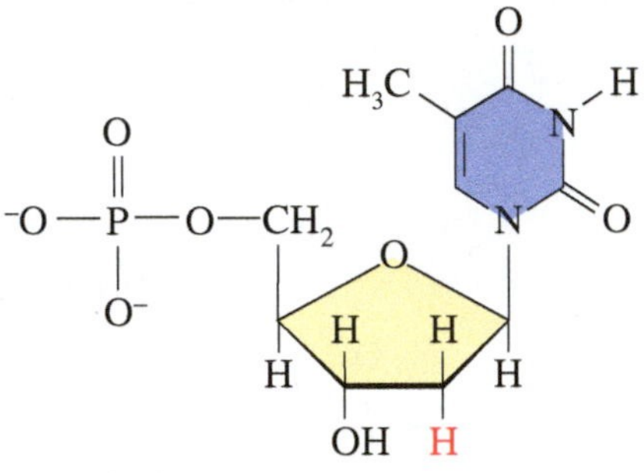

Deoxythymidine monophosphate (dTMP)

FIGURE 11.3 Naming nucleotides. The nucleotides of RNA (black names) are identical to those of DNA (red names) except that in DNA the sugar is deoxyribose (H instead of OH on carbon 2′) and deoxythymidine replaces uridine.

TABLE 11.1 Nucleoside and Nucleotide Names in DNA and RNA

Base	Nucleoside	Nucleotide
RNA		
Adenine (A)	Adenosine (A)	Adenosine monophosphate (AMP)
Guanine (G)	Guanosine (G)	Guanosine monophosphate (GMP)
Cytosine (C)	Cytidine (C)	Cytidine monophosphate (CMP)
Uracil (U)	Uridine (U)	Uridine monophosphate (UMP)
DNA		
Adenine (A)	Deoxyadenosine (A)	Deoxyadenosine monophosphate (dAMP)
Guanine (G)	Deoxyguanosine (G)	Deoxyguanosine monophosphate (dGMP)
Cytosine (C)	Deoxycytidine (C)	Deoxycytidine monophosphate (dCMP)
Thymine (T)	Deoxythymidine (T)	Deoxythymidine monophosphate (dTMP)

Sample Problem 11.3   Nucleotides

Identify which nucleic acid (DNA or RNA) contains each of the following nucleotides. State the components of each nucleotide:
  a. deoxyguanosine monophosphate (dGMP)
  b. adenosine monophosphate (AMP)

Solution

  a. This is a DNA nucleotide because it contains deoxyribose. It also contains guanine and one phosphate.
  b. This is an RNA nucleotide because it contains ribose (does not have the deoxy- prefix). It also contains adenine and one phosphate.

Practice Problems

Find the answers to the odd-numbered Practice Problems at the end of the chapter.

11.1 Identify each of the following as a purine or a pyrimidine and name them.

  a.   b.

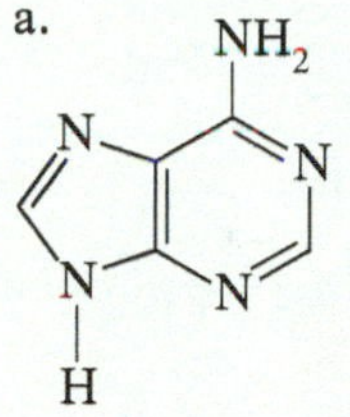

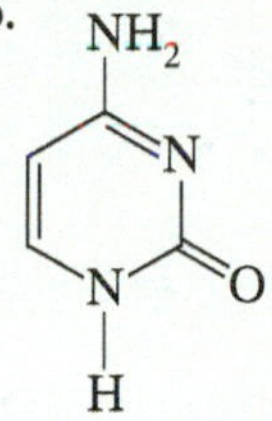

11.2 Identify each of the following as a purine or a pyrimidine:
  a. guanine   b. cytosine

11.3 Identify the bases in Problem 11.1 as present in RNA, DNA, or both.

11.4 Identify the bases in Problem 11.2 as present in RNA, DNA, or both.

11.5 What is the difference between ribose and deoxyribose?

11.6 Which of the pentose sugars is found in DNA? Which is found in RNA?

11.7 List the names and abbreviations of the four nucleotides in DNA.

11.8 List the names and abbreviations of the four nucleotides in RNA.

11.9 Provide the products for each of the following condensation reactions:

  a.

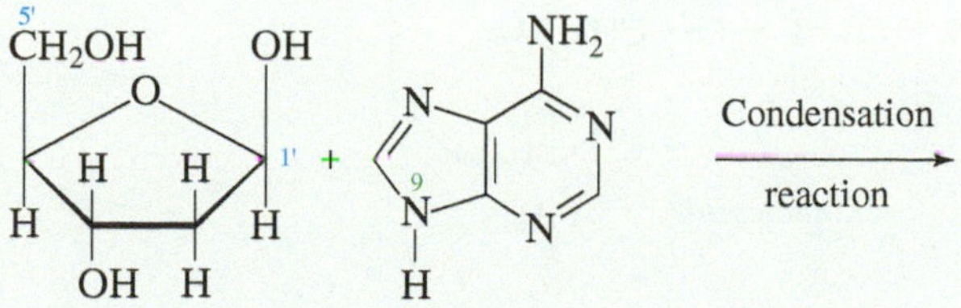

b.

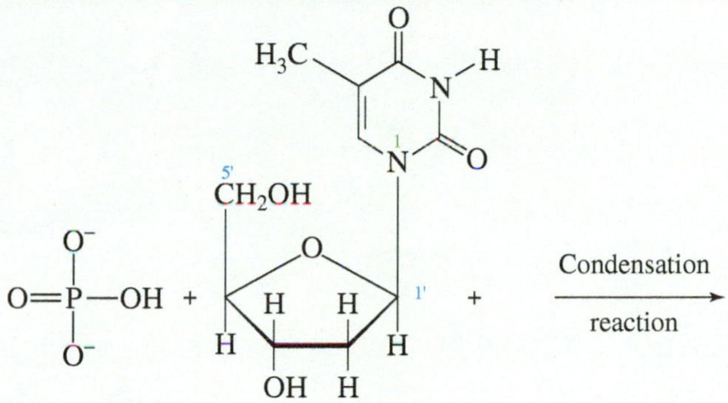

11.10 Provide the products for each of the following condensation reactions:

a.

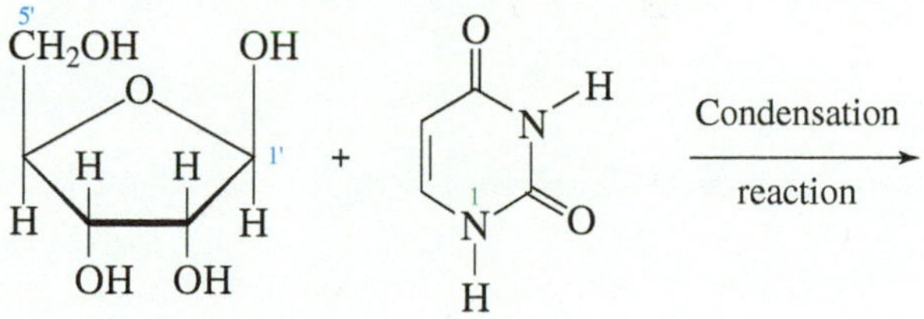

b.

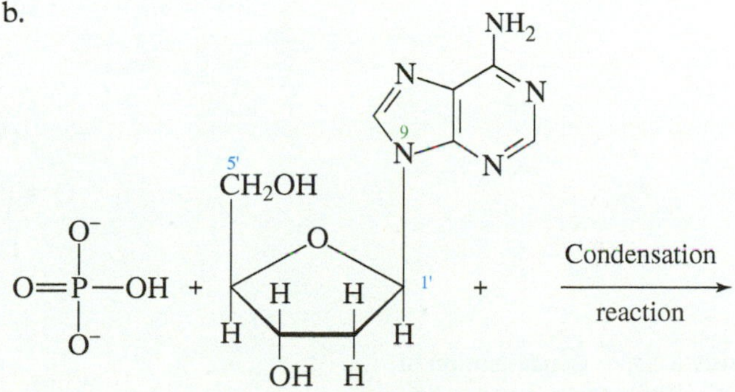

11.2 Nucleic Acid Formation

11.2 Inquiry Question

How are nucleic acids formed?

Many nucleotides linked together form **nucleic acids.** Much as amino acids in proteins are linked together by peptide bonds, nucleotides are linked together through **phosphodiester bonds**, where the oxygens in the phosphate are connected between the 3′ and 5′ carbons of adjacent sugar molecules (see **FIGURE 11.4**).

Primary Structure: Nucleic Acid Sequence

If we think of a nucleic acid as beads on a string, there would be four different colored beads in our necklace, representing four nitrogenous bases. The string consists of an alternating sugar–phosphate backbone, with the bases extending from the sugar. Recall that a protein's primary structure is the sequence of amino acids. A nucleic acid's primary structure is indicated by its nucleotide sequence. The backbone of a nucleic acid runs directionally, with the phosphates connected between the 3′ carbon of one sugar and the 5′ carbon of the neighboring sugar. This is analogous to a protein sequence running from N-terminus to C-terminus.

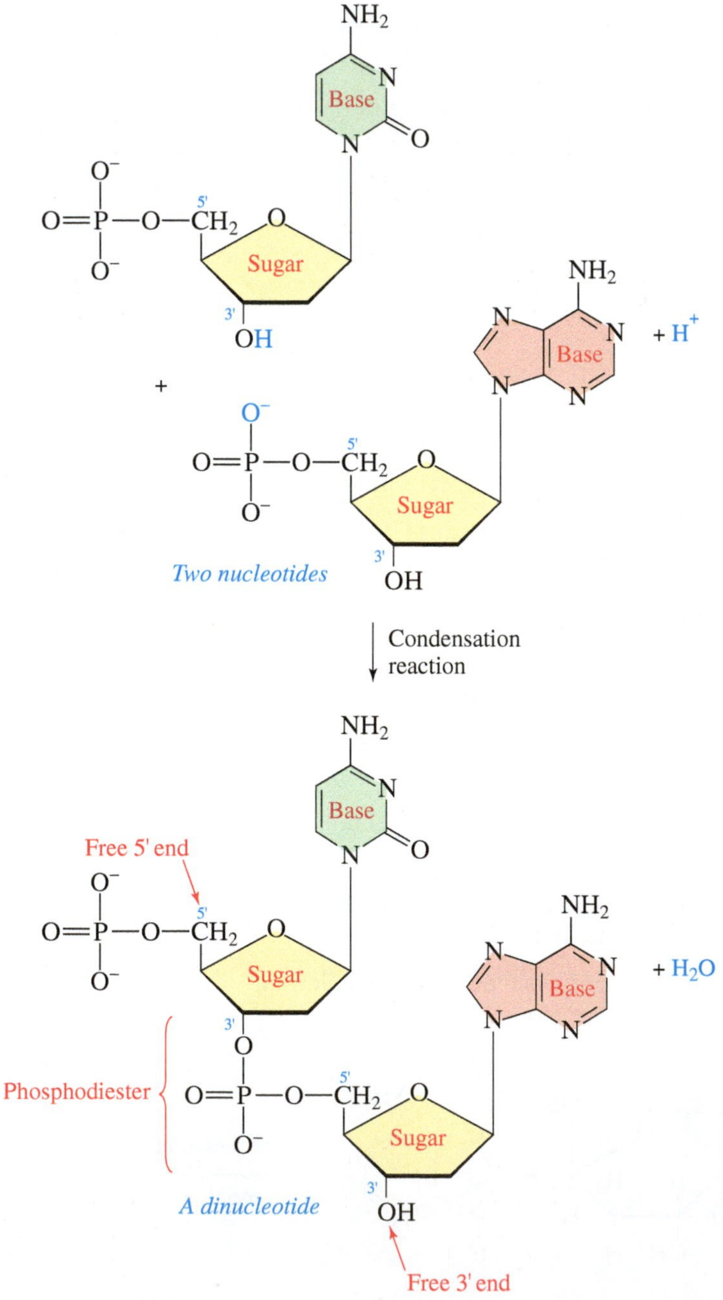

FIGURE 11.4 Condensation of nucleotides. Nucleotides bond to each other through a condensation reaction between the 3′ —OH of one nucleotide's sugar and the phosphate group (bonded to the 5′ —OH of the sugar) of a second nucleotide, forming a dinucleotide.

Nucleic acid sequences can be designated by one-letter base abbreviations. The convention is that if a single nucleic acid strand is shown as being drawn horizontally, the 5' end of a nucleic acid is at the left end and the 3' end is at the right end.

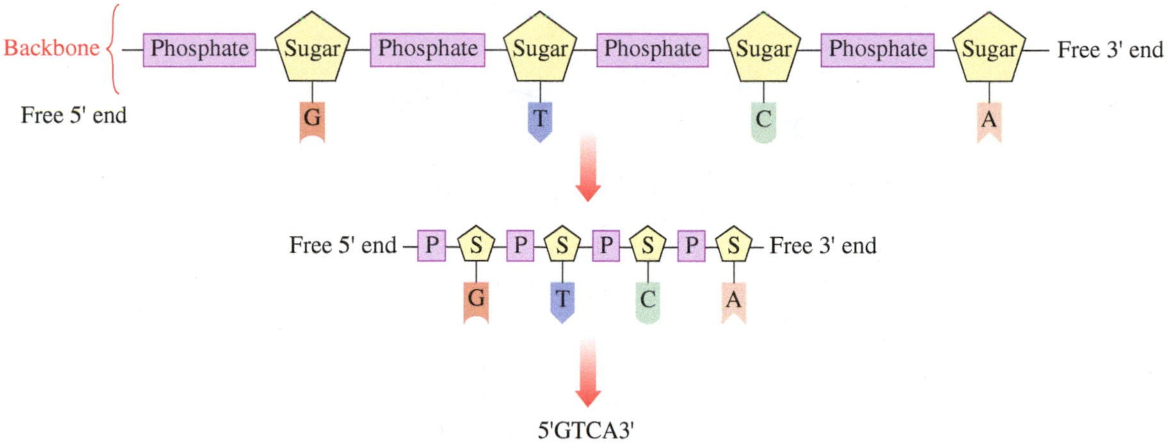

Backbone { — Phosphate — Sugar — Phosphate — Sugar — Phosphate — Sugar — Phosphate — Sugar — Free 3' end

Free 5' end   G   T   C   A

Free 5' end — P — S — P — S — P — S — P — S — Free 3' end

G   T   C   A

5'GTCA3'

A nucleic acid's sequence can be more simply represented by its one-letter base codes.

Sample Problem 11.4   Nucleic Acid Formation

Draw the RNA dinucleotide (two nucleotides joined by a phosphodiester) formed if two adenosine monophosphates undergo a condensation reaction. Label the 5' and 3' ends of your structure. Identify the phosphodiester bond.

Mastering Video
Practicing the Concepts
Condensation Reactions in Nucleic Acid Chemistry

Solution

The adenosines are joined between the 3' —OH of ribose from one adenosine and the phosphate —OH of the second adenosine, forming a phosphodiester.

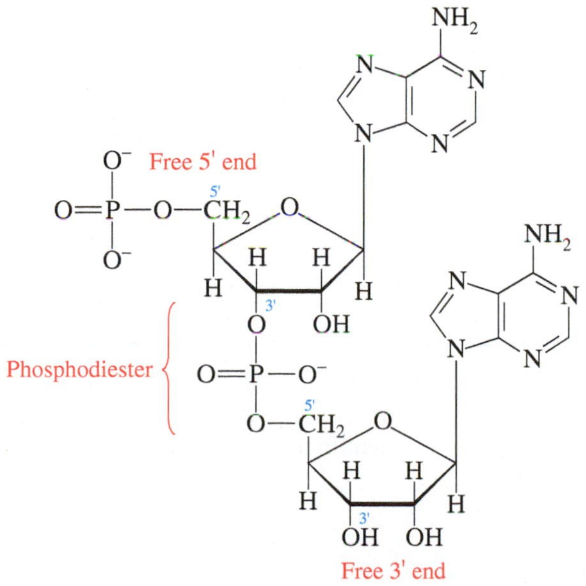

Sample Problem 11.5 Nucleic Acid Abbreviations

Provide an abbreviation for the following nucleic acid sequence:

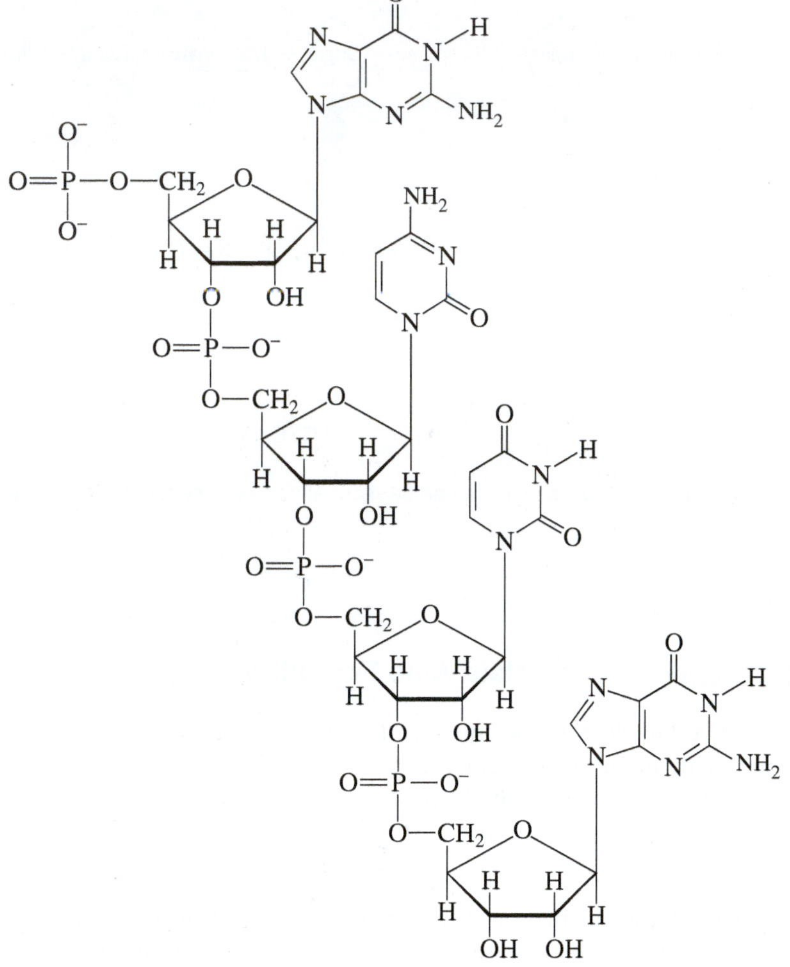

Solution

To solve this problem, you must locate the 5′ end of the nucleic acid and identify each nitrogenous base. The 5′ end of this molecule is at the top left. The sequence is abbreviated beginning with base letters at the 5′ end. This nucleic acid is abbreviated 5′GCUG3′, or simply GCUG.

Practice Problems

11.11 What is the name of the bond that joins nucleotides in a nucleic acid?

11.12 Describe the differences in the two ends of a nucleic acid.

11.13 Draw the dinucleotide GC that would be found in RNA. Label the 5′ and 3′ ends of your structure. Identify the phosphodiester bond.

11.14 Draw the dinucleotide AT that would be found in DNA. Label the 5′ and 3′ ends of your structure. Identify the phosphodiester bond.

DISCOVERING THE CONCEPTS

? INQUIRY ACTIVITY—The Unique Structure of DNA

Characteristics of DNA

- Double stranded—Two strands of nucleic acid line up in opposite directions (antiparallel) with the bases in the center.
- Hydrogen bonded—The bases hydrogen bond between the strands like rungs in a ladder: G≡C, A=T. These pairings are referred to as *complementary base pairs.*
- Double helix—Because of the chirality of the sugars, the two antiparallel strands have a natural twist to them, forming a helical shape called a double helix.

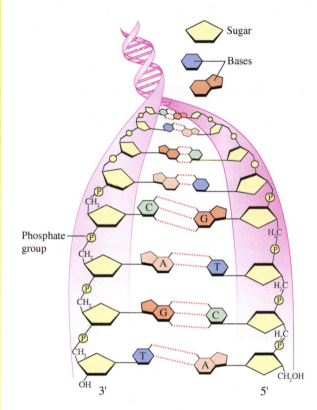

A section of double-stranded DNA can be abbreviated as

5′ACTGTCAG3′
3′TGACAGTC5′

Questions

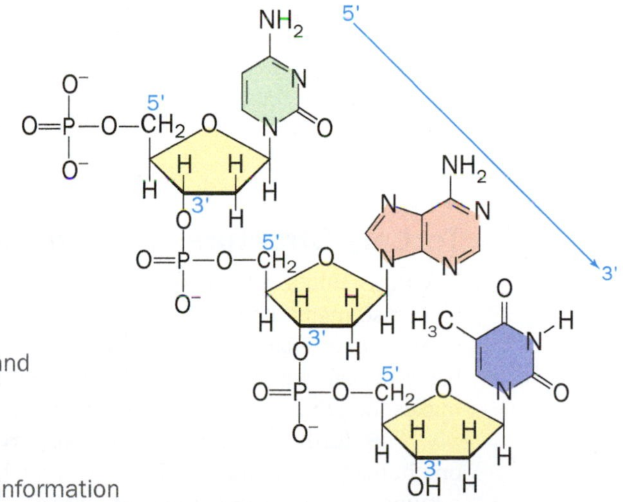

1. a. How would you designate the nucleic acid shown at right in one-letter abbreviation?
    b. Is this a strand of DNA or RNA? How do you know?
    c. Provide an abbreviation for the strand of DNA with complementary bases to the one shown in part a. Indicate the 5′ and 3′ ends.

2. Provide the complementary DNA strand to the following strand:

   5′AATTCCGCTAACG3′

3. From your own experiences and the information presented, list information on both the structure and function of DNA.

11.3 DNA

? **11.3 Inquiry Question**

What are the unique structural features of DNA?

The base sequences in nucleic acids stored as DNA in the cell's nucleus hold the code for cellular protein production. A few key discoveries beginning in the late 1940s led to what we now understand as the structure of DNA. Erwin Chargaff noted that the amount of adenine (A) is always equal to the amount of thymine (T) (A = T), and the amount of guanine (G) is always equal to the amount of cytosine (C) (G = C). This also implies that the number of purines equals the number of pyrimidines in DNA.

In DNA the number of purines equals the number of pyrimidines. A = T and G = C.

Adenine = Thymine   Guanine = Cytosine

Secondary Structure: Complementary Base Pairing

DNA's secondary structure is described by the interaction of two nucleic acids to form a **double helix.** This was first proposed in 1953 by James Watson and Francis Crick. A double helix can be envisioned as a twisted ladder (see **FIGURE 11.5**). Imagine that the sugar–phosphate backbones make up the rails of the ladder and the bases extending inward from the backbone interact as the rungs in the center of the ladder.

The two strands both have the bases in the center. Their backbones run in opposite directions, or as biochemists describe them, the backbones run *antiparallel* to each other. One strand goes in the 5′ to 3′ direction, and the other strand goes in the 3′ to 5′ direction.

Each of the rungs in DNA's helical ladder contains one base from each nucleic acid strand. The two bases in each ladder rung associate through hydrogen bonding. Watson and Crick realized that all the rungs are the same length, so they must contain one purine and one pyrimidine. The pairs A–T and G–C are called **complementary base pairs** (see **FIGURE 11.6**). Adenine and thymine form two hydrogen bonds, while guanine and cytosine form three hydrogen bonds. The DNA in one human cell contains about 3 billion of these base pairs!

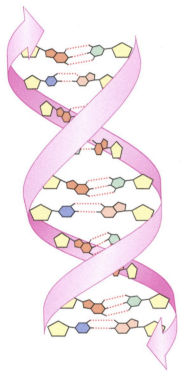

FIGURE 11.5 A DNA double helix. The two backbones run antiparallel to each other (noted by arrows on ends), and the bases line up in the interior space.

Sample Problem 11.6 **Complementary Base Pairs**

Write the base sequence and label the 3′ and 5′ ends of the complementary strand for a segment of DNA with the base sequence 5′ACGATCT3′.

Solution

The complementary strand contains the complementary bases to those of the given strand. The two strands run opposite each other so the answer is as follows:

Given strand: 5′ACGATCT3′

Complementary strand: 3′TGCTAGA5′

Tertiary Structure: Chromosomes

Try this. Hold a rubber band vertically at the top and bottom and twist it. You can only twist it so many times before it starts to double up on itself. As in proteins, tertiary structure refers to a large biomolecule's overall shape when it folds onto itself. Because DNA has a helical twist (the double helix), any further twisting (doubling it up on itself) that makes the DNA more compact constitutes its tertiary structure.

This further twisting is called **supercoiling.** The 3 billion base pairs of DNA in one human cell would stretch out as a double helix to about 6 feet in length. The DNA in a human cell is separated into 46 pieces (23 from mother, 23 from father) that are supercoiled around proteins called *histones.* These pieces of DNA wound around histones pack into **chromosomes** (see **FIGURE 11.7**). Chromosomes are an efficient package for lots of DNA information.

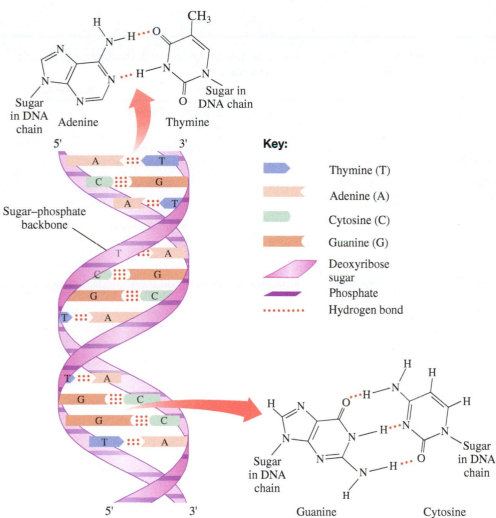

Key:

- Thymine (T)
- Adenine (A)
- Cytosine (C)
- Guanine (G)
- Deoxyribose sugar
- Phosphate
- ········· Hydrogen bond

FIGURE 11.6 Complementary base pairs. Hydrogen bonds between the complementary base pairs hold the nucleic acid strands together in DNA's double helix.

Mastering Video
Practicing the Concepts
Nucleic Acid Structure

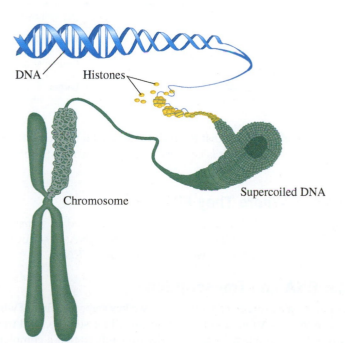

FIGURE 11.7 Chromosome structure. The DNA is wound around proteins called *histones*, which then wind into supercoiled structures. These fibers are organized into the chromosome structure.

Practice Problems

11.15 Describe the orientation of antiparallel strands in DNA. Use the terms 3′ and 5′ in your description.

11.16 Describe the double helical structure of DNA.

11.17 How are the two strands of nucleic acid in DNA held together?

11.18 Define complementary base pairing.

11.19 Write the base sequence and label the 3′ and 5′ ends of the complementary strand for a segment of DNA with the following base sequences:

    a. 5′AAAA3′   b. 5′CCCCTTTT3′
    c. 5′ACATTGG3′   d. 5′TGTGAACC3′

11.20 Write the base sequence and label the 3′ and 5′ ends of the complementary strand for a segment of DNA with the following base sequences:

    a. 5′AAAAAACC3′   b. 5′GGGGGAT3′
    c. 5′AAAATTTT3′   d. 5′CGCGATATTA3′

11.21 Fill in the following table with the analogous nucleic acid structures:

	Protein	Nucleic Acid
Primary structure	Sequence of amino acids	
Secondary structure	Local hydrogen bonding between backbone atoms	
Tertiary structure	Folding of backbone onto itself associates amino acids far away in sequence	

11.22 List the similarities and differences in the secondary structure of a protein and the secondary structure of DNA.

11.4 RNA and Protein Synthesis

? 11.4 Inquiry Question

What are the roles of RNA in protein synthesis?

RNA can be considered the "middleman" in the process of creating a protein from a gene in DNA. Like DNA, RNA is a string of nucleotides. Yet some important differences exist. These differences are listed in **TABLE 11.2**.

TABLE 11.2 Differences Between RNA and DNA

Differences	RNA	DNA
Sugar	Ribose	Deoxyribose
Base	Uracil	Thymine
Stranding	Single strand of nucleic acid	Double strand of nucleic acid
Size	Much smaller	Larger

One of the main differences is that RNA does not contain the base thymine. In RNA, the base uracil is substituted, and it is complementary to adenine, forming two hydrogen bonds (A═U).

RNA Types and Where They Fit In

The three main types of RNA found in the cell are involved in transforming a DNA sequence into a protein sequence. The names of the three RNAs describe their role in this transformation: *messenger RNA*, *ribosomal RNA*, and *transfer RNA*.

Messenger RNA and Transcription

The process of making a protein from DNA has two key steps, the first of which is to make a gene copy from the DNA found in the cell nucleus. This step is called transcription. In **transcription**, DNA's double helix temporarily unwinds so that a complementary copy

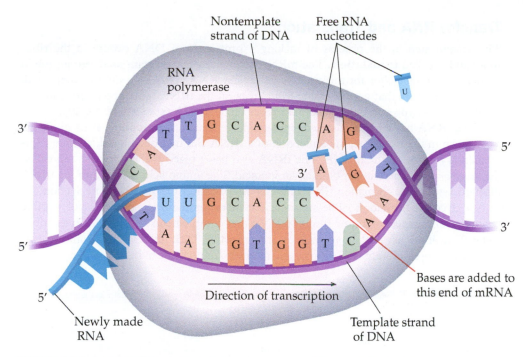

FIGURE 11.8 Transcription. In transcription, a messenger RNA is transcribed from one strand of genetic DNA (the template) as a complementary copy. Notice that the thymine (T) that pairs with adenine (A) in DNA is replaced with uracil (U) in RNA.

can be made from one of the strands, referred to as the template strand. Free RNA nucleotides are linked, forming a complementary copy of the DNA template called messenger RNA (**mRNA**). Messenger RNA is a single-stranded nucleic acid containing bases complementary to those on the original DNA strand (see **FIGURE 11.8**). Gene copying is catalyzed by **RNA polymerase**, an enzyme complex containing several subunits that binds to the DNA. Once completed, the mRNA travels from the nucleus to an organelle called the ribosome, where the mRNA sequence is processed into protein.

Ribosomal RNA and the Ribosome

The **ribosome** is one of many small structures found in cells called **organelles.** The ribosome can be thought of as a protein factory. It is composed of ribosomal RNA (**rRNA**) and protein. It is the place where the nucleotide sequence of mRNA is interpreted into an amino acid sequence. The ribosome has two rRNA/protein subunits called the small subunit and the large subunit. The general shape of each is shown in **FIGURE 11.9**. The mRNA strand fits into a groove on the small subunit, with the bases pointing toward the large subunit.

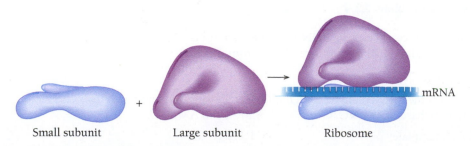

FIGURE 11.9 Ribosome structure. A typical ribosome consists of a small subunit and a large subunit. Both subunits contain rRNA and protein. The mRNA fits in a groove between them.

Transfer RNA and Translation

The second step in the process of making a protein from DNA occurs in the ribosome and is called **translation.** The mRNA sequence must be translated into a protein sequence. The facilitator for this process is transfer RNA (**tRNA**), a tightly compacted, T-shaped structure that gets its shape from areas on the tRNA sequence where complementary bases hydrogen bond with other bases on the tRNA (see **FIGURE 11.10**).

The tRNA has a three-base sequence (triplet) called an **anticodon** at its *anticodon loop.* When in the ribosome, the anticodon of a tRNA can hydrogen bond to three complementary bases on the mRNA. The tRNA also has a place at the opposite end called the *acceptor stem* where it can bind an amino acid. The amino acid is joined to the tRNA through ester bond formation. (We saw this esterification reaction back in Section 7.3 with triglyceride formation.) The amino acid is incorporated into a growing polypeptide chain one amino acid at a time through tRNA that complements the mRNA in the ribosome. We will discuss protein synthesis more in Section 11.5. Each of the 20 amino acids has one or more different tRNAs available to bring amino acids to the ribosome.

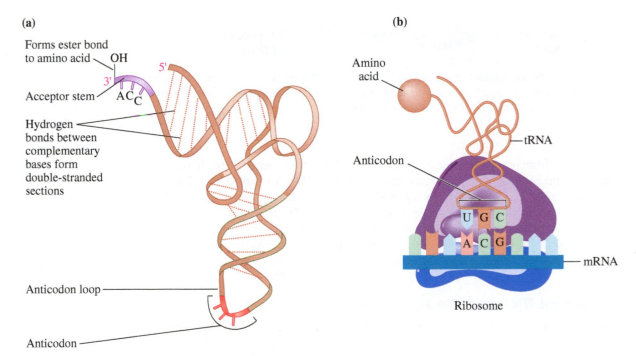

(a)

Forms ester bond to amino acid

Acceptor stem

Hydrogen bonds between complementary bases form double-stranded sections

Anticodon loop

Anticodon

(b)

Amino acid

tRNA

Anticodon

mRNA

Ribosome

FIGURE 11.10 tRNA structure. (a) The ribbon structure of tRNA showing complementary base pairing. (b) The tRNA with amino acid and mRNA in the ribosome. A typical tRNA molecule has an anticodon loop that complements three bases on mRNA, and has an acceptor stem at the 3′ end of the nucleic acid where an amino acid attaches.

Sample Problem 11.7 **Transcription**

The sequence of bases in a DNA template strand is 5′TTAACG3′. What is the corresponding mRNA that is produced from this DNA?

Solution

To form the mRNA, the bases in the DNA template are paired with their complementary bases: G with C, C with G, T with A, and A with U.

| DNA template: | 5′TTAACG3′ |
| Complementary mRNA: | 3′AAUUGC5′ |

Practice Problems

11.23 Name the three types of RNA and their functions.

11.24 List the mRNA bases that complement the bases A, T, G, and C in DNA.

11.25 In your own words, define the term *transcription*.

11.26 In your own words, define the term *translation*.

11.27 The sequence of bases in a DNA template strand is 5′ACCGCCTTAAGG3′. What is the corresponding mRNA produced?

11.28 The sequence of bases in a DNA template strand is 5′GGCTTATTGCCA3′. What is the corresponding mRNA produced?

DISCOVERING THE CONCEPTS

? INQUIRY ACTIVITY—The Genetic Code

Exploring Translation

The genetic code wheel for amino acids found on the next page in **FIGURE 11.11** can be used to translate a triplet of three nucleotides from a nucleic acid into one amino acid.

Codons in mRNA	5′GGU | GAC | CUA | AUC3′
Translation	↓ ↓ ↓ ↓
Amino acid sequence	Gly – Asp–Leu – Ile (three-letter abbreviation)
	G D L I (one-letter abbreviation)

Sample 1. Four codons of mRNA are translated to a tetrapeptide sequence.

Questions

1. How many nucleotide bases (A, G, C, U) are found in one codon?
2. Based on Figure 11.11 and sample 1 above, the mRNA sequence 5′UUU3′ codes for the amino acid phenylalanine. For what amino acid would the codon 5′GCA3′ code?
3. Name the amino acid corresponding to each of the following codons (use Figure 11.11):
   a. 5′AUU3′
   b. 5′CAC3′
   c. 5′GGA3′
4. What is the sequence of amino acids coded by the following codons in mRNA? 5′CUA \| GUC \| UGC3′
5. The process of forming an amino acid sequence from mRNA as we did in the previous question is called translation. Is this term appropriate considering the process that is taking place? Describe how you can remember this term.

Protein Synthesis

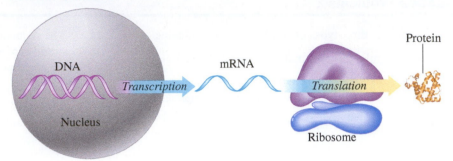

Scheme 1

The central dogma of molecular biology shown above describes the transfer of sequence-specific information, in this case, protein synthesis. Protein synthesis begins with transcription of gene DNA to a complementary copy of mRNA, which moves to the ribosome for translation. Amino acids are linked together at the ribosome, forming protein until a stop codon on the mRNA is reached.

Questions

6. In what part of the cell does transcription take place?
7. In what organelle does translation take place?
8. Of the three molecules, DNA, mRNA, or protein, which moves from the nucleus to the ribosome during the process of protein synthesis?
9. During transcription, a messenger RNA (mRNA) molecule is made from a genetic piece of DNA. The mRNA is complementary to the DNA molecule from which it was copied. If a piece of the DNA has the sequence 5′ACGTAGTCACGT3′, what would the complementary mRNA sequence be? Indicate the 5′ and 3′ ends on your sequence.
10. Using the genetic code wheel in Figure 11.11, what would the corresponding amino acid sequence be for the mRNA sequence in question 9?

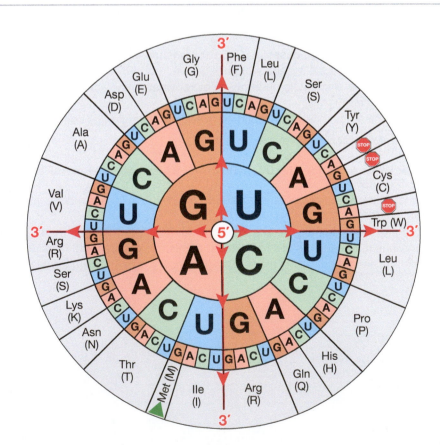

FIGURE 11.11 The genetic code wheel for amino acids. To use the wheel, start from the 5′ center and work your way outward to 3′ to identify the amino acid corresponding to the mRNA codon. For example, for the mRNA codon 5′UCA3′, you would move outward from the center U to the C and then to the A, indicating the amino acid serine (Ser, S). Notice that more than one codon codes for the same amino acid. Stop codons are indicated with stop signs and signal the end of a peptide chain. The green start arrow signifies the start codon, 5′AUG3′, that initiates protein synthesis if found at the beginning of a nucleic acid sequence. If this codon is found inside a gene, it codes for methionine.

11.5 Putting It Together: The Genetic Code and Protein Synthesis

If you were on a beach at night and saw a light in the ocean give three short flashes, three long flashes, and then three more short flashes, what would you do? Many of us, not recognizing the Morse code distress signal SOS, would do nothing. People able to translate Morse code would immediately get help if they saw this signal. Similarly, the genetic code from our gene sequences of DNA is a code transcribed into mRNA and decoded as a protein sequence. This *translation* involves tRNA and takes place at the ribosome.

? **11.5 Inquiry Question**

How is the genetic code used to synthesize proteins?

The Genetic Code

The mRNA transcribed from the DNA contains a sequence of bases specifying the protein to be made. A given mRNA triplet called a **codon** translates to a specific amino acid. For example, the sequence UUU in the mRNA specifies the amino acid phenylalanine.

The **genetic code** wheel (see Figure 11.11) shows the codons of mRNA, from the center moving outward from 5′ to 3′, for the 20 amino acids. Sixty-four codon combinations are possible from the four bases A, G, C, and U.

Further examination of the gray portion of the wheel shows a green start arrow and some stop signs. These mRNA codons start and stop protein synthesis, respectively. The triplet AUG is the start codon. If found at the 5′ end of mRNA, AUG initiates the synthesis of a protein from the mRNA. If AUG is found within the mRNA, it codes for the amino acid methionine. The three codons UAA, UAG, and UGA are stop codons. When they appear in the mRNA sequence, they are a signal to stop adding amino acids to the growing protein chain. See if you can translate the three mRNA codons below into the three amino acids shown using the genetic code wheel (Figure 11.11).

Codons in mRNA	5′UUU|GGG|CGC3′		
Translation	↓	↓	↓
Amino acid sequence	Phe – Gly – Arg		(three-letter abbreviation)
	F G R		(one-letter abbreviation)

Sample Problem 11.8   Codons

What is the sequence of amino acids coded by the following codons in mRNA?

5′CAA\|AUC\|GUA3′

Solution

Using the genetic code wheel in Figure 11.11, working from the center outward, CAA codes for glutamine, AUC for isoleucine, and GUA for valine. The amino acid sequence is Gln–Ile–Val, or QIV.

Protein Synthesis

Let's pause to review the entire process of protein synthesis, which begins with DNA and the process of transcription (see **FIGURE 11.12**).

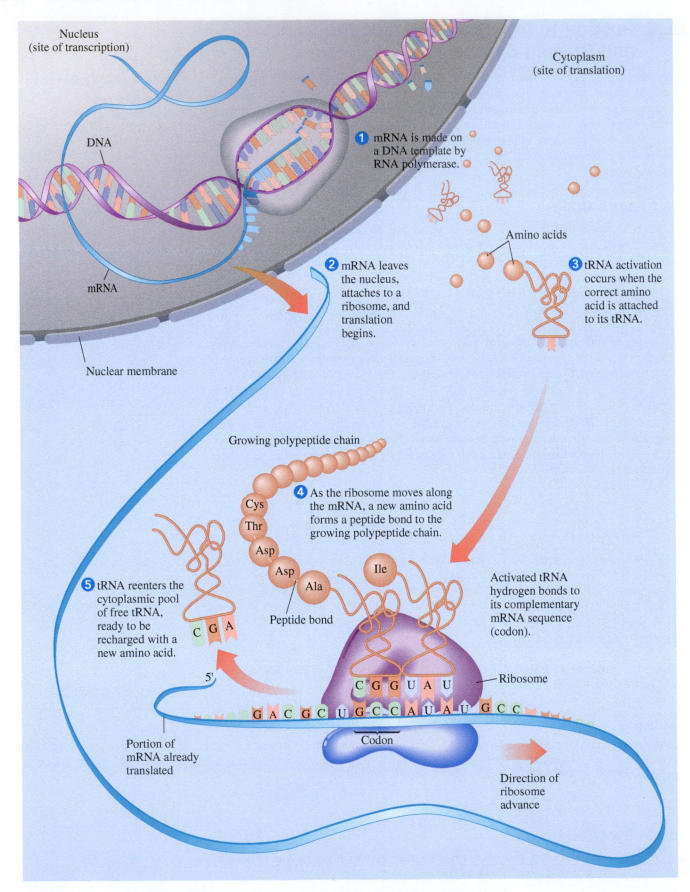

Nucleus (site of transcription)

Cytoplasm (site of translation)

DNA

1 mRNA is made on a DNA template by RNA polymerase.

Amino acids

mRNA

2 mRNA leaves the nucleus, attaches to a ribosome, and translation begins.

3 tRNA activation occurs when the correct amino acid is attached to its tRNA.

Nuclear membrane

Growing polypeptide chain

4 As the ribosome moves along the mRNA, a new amino acid forms a peptide bond to the growing polypeptide chain.

Cys
Thr
Asp
Asp
Ala
Ile

Activated tRNA hydrogen bonds to its complementary mRNA sequence (codon).

5 tRNA reenters the cytoplasmic pool of free tRNA, ready to be recharged with a new amino acid.

Peptide bond

C G A

5'

Ribosome

C G G U A U

G A C G C U G C C A U A U G C C

Codon

Portion of mRNA already translated

Direction of ribosome advance

FIGURE 11.12 **Protein synthesis.** Protein synthesis begins with transcription of gene DNA to the mRNA (upper left), which moves to the ribosome for translation via activated tRNA molecules. Amino acids are linked until a stop codon on the mRNA is reached.

Transcription

DNA found in the nucleus unwinds at the site of a gene. A complementary copy of this DNA template, called mRNA, is created. See Figure 11.12 ❶. *RNA polymerase* is the name of the enzyme complex that binds to the DNA and links nucleotides to create mRNA. The mRNA then travels out of the nucleus to a ribosome. See Figure 11.12 ❷.

tRNA Activation

Before the tRNA can be used in the ribosome, an amino acid must be attached to its acceptor stem. An enzyme called tRNA synthetase attaches the correct amino acid to the acceptor stem of the tRNA specific for that amino acid. The activated tRNA is now ready to hydrogen bond to the complementary mRNA codon in the ribosome. See Figure 11.12 ❸.

Translation

Protein synthesis begins when mRNA positions itself at the ribosome. The first codon in mRNA is the start codon, AUG. An activated tRNA with an anticodon of UAC and the amino acid methionine attached enters the ribosome and hydrogen bonds to the mRNA. A second activated tRNA matching the next codon on the mRNA enters an adjacent position on the ribosome. The two amino acids then join, forming a peptide bond, and the methionine detaches from the first tRNA. The deactivated tRNA in the first position leaves the ribosome, and the second tRNA shifts into the first position with the dipeptide attached.

This shifting is called *translocation*. As the ribosome moves along the mRNA, a peptide bond joins each subsequent amino acid that enters the ribosome. In this way, a growing polypeptide chain emerges. See Figure 11.12 ❹. The tRNAs dislodge from the ribosome and return to the free tRNA pool in the cytoplasm to become recharged with another amino acid. See Figure 11.12 ❺.

Termination

Eventually, the ribosome encounters a stop codon, and protein synthesis ends. The polypeptide chain is released from the ribosome. The initial amino acid methionine is often removed from the beginning of the polypeptide chain. The growing polypeptide chain folds into its tertiary structure, forming any disulfide links, salt bridges, or other interactions that make the polypeptide a biologically active protein.

Sample Problem 11.9 **Transcription and Translation**

Provide the mRNA and protein sequence that would result from the following DNA template strand:

$$3'\text{TAA} | \text{CCG} | \text{GCG} | \text{AAT} 5'$$

Solution

Each of the triplets in the DNA sequence has a complementary mRNA sequence that runs antiparallel to the template strand, so it starts with the 5′ end:

mRNA codons	5′AUU|GGC|CGC|UUA3′
Three-letter amino acid sequence	Ile – Gly – Arg – Leu
One-letter sequence	I G R L

Practice Problems

11.29 Why are there at least 20 tRNAs?

11.30 List the difference between a codon and an anticodon.

11.31 Provide the three-letter amino acid sequence expected from each of the following mRNA segments:
a. 5′AAA|CCC|UUG|GCC3′
b. 5′CCU|CGA|AGC|CCA|UGA3′
c. 5′AUG|CAC|AAA|GAA|GUA|CUU3′

11.32 Provide the three-letter amino acid sequence expected from each of the following mRNA segments:
a. 5′AAA|AAA|AAA3′
b. 5′UUU|CCC|UUU|CCC3′
c. 5′UAC|GGG|AGA|UGU3′

11.33 In your own words, define translocation.

11.34 In your own words, describe how a peptide chain is extended.

11.35 The following portion of DNA is in the template DNA strand:

3′TGT|GGG|GTT|ATT5′

a. Write the corresponding mRNA section. Show the nucleic acid sequence as triplets and label the 5′ and the 3′ ends.
b. Write the anticodons corresponding to the codons on the mRNA.
c. Write the three-letter and one-letter amino-acid sequences that will be placed in a peptide chain.

11.36 The following portion of DNA is in the template DNA strand:

3′GCT|TTT|CAA|AAA5′

a. Write the corresponding mRNA section. Show the nucleic acid sequence as triplets and label the 5′ and the 3′ ends.
b. Write the anticodons corresponding to the codons on the mRNA.
c. Write the three-letter and one-letter amino acid sequences that will be placed in a peptide chain.

11.6 Genetic Mutations

What will happen to protein synthesis if the DNA sequence is changed? Any change in a DNA nucleotide sequence is called a **mutation.** Let's examine the possibilities.

- *No change in protein sequence.*   Sometimes, a change in a DNA base will have no effect. Only about 2% of the DNA in your chromosomes encodes for proteins. The rest of your DNA is noncoding DNA, originally called "junk" DNA because it took scientists time to determine its purpose. This DNA is not junk; it contains recognition sites for protein binding. Also, recall that there is more than one codon that codes for each amino acid. For example, if the codon UUU were changed to UUC, the amino acid phenylalanine would still be placed in the growing polypeptide chain (see the genetic code wheel, Figure 11.11). These types of mutations are called **silent mutations.**
- *A change in protein sequence occurs, but it has no effect on protein function.*   If, for example, the codon AUU were mutated to GUU, the amino acid isoleucine would be changed to valine. These two amino acids are similar in polarity and size, and it is likely that such a substitution would not have much of an effect on the protein function. This is another type of silent mutation.
- *A change in protein sequence occurs and affects protein function.*   If, for example, the codon AUU were mutated to AAU, isoleucine would be changed to asparagine. In this case, a nonpolar amino acid is replaced with a polar amino acid, which could affect the structure and the function of the protein. Other mutations that can have a negative effect on protein synthesis include mutating a codon into a stop codon or inserting or deleting a base. The latter has the effect of shifting the triplets that are read in the mRNA and would change the identity of all subsequent amino acids after the insertion or deletion.

Sample Problem 11.10 Mutations

An mRNA has the codon sequence 5′CCC|AGA|GCC3′. If a base substitution in the DNA changes the mRNA codon of AGA to GGA, how is the amino acid sequence affected in the resulting protein? Can you predict if this might affect the protein function? Refer to the genetic code wheel in Figure 11.11 and the amino acid table in Chapter 10 (Table 10.1) to help you answer this question.

Solution

The initial mRNA sequence of 5′CCC|AGA|GCC3′ codes for the peptide proline–arginine–alanine. When the mutation occurs, the new sequence of mRNA codons, 5′CCC|GGA|GCC3′, codes for proline–glycine–alanine. Arginine is replaced by glycine, which changes a polar, charged amino acid into a nonpolar, uncharged amino acid. This change could have an effect on protein structure and function.

Sources of Mutations

How do mutations occur? Sometimes when DNA replicates itself, errors occur at random. This is considered a **spontaneous mutation.** Environmental agents such as chemicals or radiation that produce mutations in DNA are called **mutagens.** Many mutagens can cause cancer and are designated **carcinogens.** Viruses can also cause mutations (Section 11.7).

One common chemical mutagen is sodium nitrite ($NaNO_2$), which is used as a preservative in processed meats like hot dogs, bacon, and bologna. In the presence of secondary amines (found in proteins), sodium nitrite forms a compound called a nitrosamine, a known carcinogen in animals. Nitrosamines assist in the conversion of cytosine into uracil, which effectively converts a C–G base pair into a U–A base pair.

If it is a known mutagen, why is sodium nitrite still used to preserve meat? Sodium nitrite effectively kills the bacteria *Clostridium botulinum,* responsible for the foodborne illness botulism, which one can get from improperly cured meats. The amount of nitrites used in the preservation process is small, and the benefits to meat preservation outweigh the risk of mutation.

$$NaNO_2 \quad + \quad \underset{\underset{H}{|}}{R_2 \diagdown \underset{N}{} \diagup R_1} \quad \longrightarrow \quad \underset{R_2}{\overset{R_1}{}} N{-}N{=}O$$

| Sodium nitrite, food preservative used in meats | Amine from protein | Nitrosamine |

If a mutation occurs in a **somatic cell** (any cell type other than egg or sperm), it affects only the individual organism and can cause conditions like cancer. Mutations that occur in **germ cells** (sperm or egg cells), however, can be passed on to future generations. A germ cell mutation can lead to malformation (and subsequent malfunction) of a particular protein. Germ cell mutations cause **genetic diseases.** More than 6000 genetic diseases caused by a mutation in a single gene have been identified. See **TABLE 11.3** for some examples.

$$\text{DNA} \xrightarrow{\text{Mutation}} \underset{\text{of DNA}}{\text{Alteration}} \longrightarrow \underset{\text{protein}}{\text{Defective}} \longrightarrow \underset{\text{or cancer (somatic cells)}}{\text{Genetic disease (germ cells)}}$$

INTEGRATING Chemistry

◀ Find out how DNA mutations can lead to disease.

TABLE 11.3 Some Genetic Diseases

Genetic Disease	Protein or Chromosomal Defect	Disease Symptoms
Galactosemia	Transferase enzyme required for the metabolism of galactose-1-phosphate is absent	Cataracts, mental retardation
Cystic fibrosis	Mutation in a gene producing a protein that regulates salt transport into and out of cells	Thick secretions of mucus, difficulty breathing, blocked pancreatic function
Down syndrome	Formation of three chromosomes instead of a pair of chromosomes, usually on chromosome 21	Heart and eye defects, cognitive and physical issues
Familial hypercholesterolemia	Mutation in a gene on chromosome 19, which regulates cholesterol levels	High cholesterol levels, early coronary heart disease age 30–40)
Muscular dystrophy (Duchenne)	One of 10 forms of MD; a mutation in the X chromosome results in the low or abnormal production of the protein dystrophin.	Muscle-atrophy disease appears around age 5, with death by age 20; occurs in about 1 of 10,000 males
Huntington's disease (HD)	Mutation in a gene on chromosome 4, which can now be mapped to test people in families with HD	Nervous tremors leading to total physical impairment
Sickle-cell anemia	Defective hemoglobin from a mutation in a gene on chromosome 11	Anemia from decreased oxygen-carrying ability of red blood cells
Hemophilia	One or more defective blood-clotting factors	Poor blood coagulation, excessive bleeding, and internal hemorrhages
Tay–Sachs disease	Defective hexosaminidase A	Accumulation of lipids in the brain, resulting in mental retardation, loss of motor control, and early death

Practice Problems

11.37 In your own words, define mutation.

11.38 In your own words, define a silent mutation.

11.39 Consider the following portion of mRNA produced by the normal order of DNA nucleotides:

$$5'CUU|AAA|CGA|GUU3'$$

a. Write the amino acid sequence that would be produced from this mRNA.

b. Write the amino acid sequence if a mutation changes CUU to AUU. Is this likely to affect protein function?

c. Write the amino acid sequence if a mutation changes CGA to AGA. Is this likely to affect protein function?

d. What happens to protein synthesis if a mutation changes AAA to UAA?

e. What happens to the protein sequence if an A is added to the beginning of the chain and the sequence changes to 5'ACU|UAA|ACG|AGU3'?

f. What happens to the protein sequence if the C is removed from the beginning of the chain and the sequence changes to 5'UUA|AAC|GAG3'?

11.40 Consider the following portion of mRNA produced by the normal order of DNA nucleotides:

$$5'ACA|UCA|CGG|GUA3'$$

a. Write the amino acid sequence that would be produced from this mRNA.

b. Write the amino acid sequence if a mutation changes UCA to ACA. Is this likely to affect protein function?

c. Write the amino acid sequence if a mutation changes CGG to GGG. Is this likely to affect protein function?

d. What happens to protein synthesis if a mutation changes UCA to UAA?

e. What happens to the protein sequence if a G is added to the beginning of the chain and the sequence changes to 5'GAC|AUC|ACG|GGU3'?

f. What happens to the protein sequence if the A is removed from the beginning of the chain and the sequence changes to 5'CAU|CAC|GGG3'?

11.7 Viruses

Viruses are small particles containing from 3 to 200 genes that can infect any cell type. Viruses are not considered cells because they cannot make their own proteins or their own energy. They contain only parts needed to infect a cell. Viruses have their own nucleic acid but use the ribosomes and RNA of the infected cell—also called the **host cell**—to make their proteins. Some infections caused by viruses invading human cells are listed in **TABLE 11.4**.

Viruses are very simple. Even though they have a variety of shapes, they all contain a nucleic acid (either DNA or RNA) enclosed in a protein coat called a **capsid.** Many viruses also contain an additional protective coat called an **envelope** surrounding the capsid (see **FIGURE 11.13**). Many animal viruses also contain exterior spikes made of protein. The function of all viruses is the same: to monopolize the functions of the host cell for the benefit of the virus.

11.7 Inquiry Question
How does a virus use nucleic acids?

TABLE 11.4 Some Diseases Caused by Viral Infection

Disease	Virus
Common cold	Coronavirus (more than 100 types)
Flu	Influenza virus
Warts	Human papillomavirus (HPV)
Herpes	Herpes simplex virus
Leukemia, AIDS	Retroviruses
Hepatitis	Hepatitis virus (types A–C are most common)
Mumps	Rubulavirus
Epstein–Barr	Epstein–Barr virus (EBV)

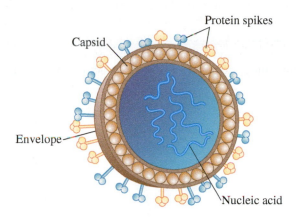

FIGURE 11.13 Virus structure. All viruses contain nucleic acid protected by a protein coat called a capsid.

A virus infects a cell when an enzyme in the protein coat makes a hole in the host cell, allowing the viral nucleic acid to enter and mix with the host cell's material. If the virus contains DNA, the host cell will begin to replicate the viral DNA in the same way it would replicate its own DNA. Viral DNA produces viral RNA, which proceeds to make the proteins for the virus. The completed virus particles are then assembled and released from the cell to infect more cells. This release often occurs by a process called *budding* (see **FIGURE 11.14**).

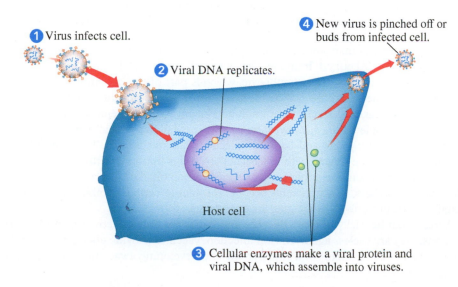

FIGURE 11.14 The life cycle of a virus. After a virus attaches to the host cell, it injects its viral DNA and uses the host cell's nucleic acids, enzymes, amino acids, and ribosomes to make viral mRNA, new viral DNA, and viral proteins. The newly assembled viruses are released to infect other cells.

INTEGRATING
Chemistry

Find out how we ▶
control the flu
using a vaccine

Common Viral Diseases

Before a virus causes a disease, it must enter a host cell. Most viruses gain entry through mucous membranes or body openings like the mouth, nose, eyes, ears, and genital areas. Three common illnesses caused by viruses are the common cold, the flu, and HIV/AIDS.

The common cold is a viral infection. About 200 different cold viruses can cause the common cold, making it difficult to cure. Cold viruses can mutate quickly from one person to the next. Usually, a cold will last about a week before the body's immune system effectively fights it off, so the most common treatment is to rest and drink plenty of fluids.

The influenza, or "flu," virus types A and B cause seasonal epidemics during the winter months. An epidemic is an outbreak of an infection that spreads quickly in a particular location. Because flu viruses mutate quickly, scientists reformulate flu vaccines each year. **Vaccines** are inactive or weakened forms of viruses that boost the immune response by causing the body to produce antibodies that will fight off the virus when actually infected. Different strains of flu virus have different protein sequences on the exterior that spike the surface of the virus. The annual flu vaccine contains a mixture of inactive flu viruses expected to emerge during the flu season and are prepared as shown in **FIGURE 11.15**. When you get a flu shot, you expose yourself to a low dose of flu antigen so that your immune system can prepare for an actual viral attack.

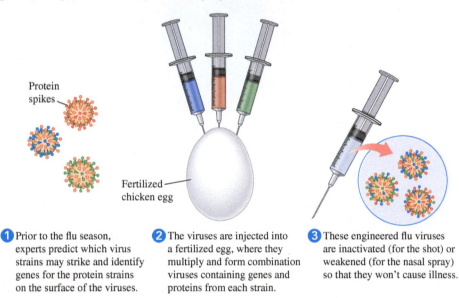

Protein
spikes

Fertilized
chicken egg

1 Prior to the flu season, experts predict which virus strains may strike and identify genes for the protein strains on the surface of the viruses.

2 The viruses are injected into a fertilized egg, where they multiply and form combination viruses containing genes and proteins from each strain.

3 These engineered flu viruses are inactivated (for the shot) or weakened (for the nasal spray) so that they won't cause illness.

FIGURE 11.15 Preparation of the influenza vaccine. Scientists reformulate the flu virus each year by researching which flu viruses are making people sick now and their ability to spread during the next flu season. These top picks are combined to produce an inactive virus vaccine that will build up antibodies in the immune system to prepare for a future exposure to the flu.

Another virus, human immunodeficiency virus type 1, or HIV-1, is responsible for the disease AIDS (Acquired Immune Deficiency Syndrome). HIV-1 infects a type of white blood cell known as a T4 lymphocyte, which is part of the human immune system. The depletion of these immune cells reduces a person's ability to fight off other infections. Because of this weakened state of the immune system, AIDS patients are susceptible to infections that people with a healthy immune system wouldn't typically get, such as skin sarcomas (a type of cancer) and bacterial infections like pneumonia. Today, people with HIV and AIDS can be treated with a "cocktail" of drugs that inhibit the life cycle of the virus during replication and protein synthesis. Such combination therapies can keep the HIV infection in check. When patients become resistant to certain combinations, different cocktails can be prescribed, prolonging life expectancy.

Viruses can be tricky to treat. Because antibiotics are only effective against bacterial infections, they are useless against a viral infection. Vaccines provide good protection, but isolating the key components of a viral particle and developing a vaccine is costly.

Sample Problem 11.11 Viruses

Why are viruses unable to replicate on their own?

Solution

Viruses contain only a protective protein coat and DNA or RNA, and lack the necessary replication machinery, including enzymes, nucleosides, and ribosomes.

Practice Problems

11.41 For each of the following, note whether the component can be found in a virus, a cell, or both.
  a. DNA   b. DNA polymerase
  c. capsid

11.42 For each of the following, note whether the component can be found in a virus, a cell, or both.
  a. RNA   b. protein coat
  c. ribosome

11.43 Why do viruses need to enter a host cell?

11.44 Name two components common to all viruses.

11.45 How does a vaccine protect against a viral disease?

11.46 Why won't antibiotic medications treat the flu?

11.8 Recombinant DNA Technology

Since the discovery of DNA as the genetic material in all cells, scientists have been finding ways to manipulate DNA to produce the proteins needed as medicines as well as proteins helpful in crop resistance to pests. More recently, scientists have found a way to manipulate DNA to repair defective genes, which can then be inserted directly into cells. In this section we discuss how these techniques work.

Just as the term sounds, **recombinant DNA** involves recombining DNA from different sources. In this process, often called *genetic engineering* or *gene cloning*, the genome of one organism is altered by splicing in a section of DNA containing a gene from a second organism, or more recently, a new corrected gene sequence. Why would anyone want to do that? Inserting a more complex organism's gene into a simpler organism (like bacteria) with a shorter life cycle allows production of large amounts of a desired protein more quickly.

The idea of combining genetic material is not new—humans have been crossbreeding plants and animals for centuries, exchanging DNA to produce desired traits. The recombinant DNA techniques developed in the mid-1970s and the advances in the last decade both work more quickly than crossbreeding and are more predictable.

? 11.8 Inquiry Question

How can DNA be manipulated outside of the cell?

Recombining DNA to Produce Proteins

One of the first applications of recombinant DNA technology incorporated the human insulin gene into the common bacteria *Escherichia coli* (*E. coli*), thereby allowing these bacteria to produce the human insulin protein. Before 1982, pig or cow insulin was harvested for use by people with diabetes. The amino acid sequences of the pig and cow insulin differ in two and three amino acids, respectively, from human insulin. Although pig and cow insulin sequences are similar to human insulin, impurities in the extraction process resulted in rejection in some people. Almost all insulin used today is expressed human insulin, causing fewer side effects.

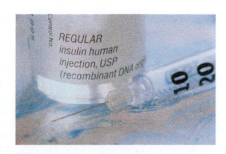

Human insulin was one of the first recombinant DNA proteins developed.

Recombinant DNA technology has also allowed the insertion of genes into food plants, affording crops advantages during growth. In the late 1990s, Monsanto introduced Roundup Ready® cotton to farmers. This variety of cotton has a bacteria gene inserted into it that produces an enzyme resistant to Roundup®, a weed killer. When farmers apply Roundup to a Roundup Ready cottonfield, weeds are killed and the cotton plant survives. Today, many food crops have been genetically modified and are identified as GM foods. China, Russia, and the European Union identify foods as genetically modified. In 2016, the United States passed a GM labeling law; however, implementation has been delayed by questions regarding the type of labeling and the foods covered. The most common GM crops are soybean, corn, cotton, canola, and sugar beet. Because ingredients produced from these crops are found in processed foods, it is likely that you have encountered a GM food at your local grocery store.

Let's look at the basic steps to recombining DNA and producing a protein in another organism (see **FIGURE 11.16**).

- *Identify and isolate a gene of interest.* A gene must be located and cut out of its chromosome. This gene of interest (like insulin) is referred to as the *donor DNA.* The removal is done with enzymes called **restriction enzymes** that recognize specific DNA sequences four to eight bases in length. One of these enzymes, *Eco*RI (pronounced "echo–R–one," from the bacteria *E. coli*) recognizes the following DNA sequence, and cuts the DNA as shown.

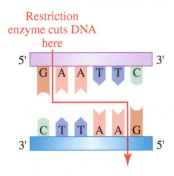

- *Insert donor DNA into the organism DNA using a vector.* A **vector** is a transporter for the donor DNA. It can incorporate donor DNA into the genome of the organism. One common vector found in bacteria is a piece of circular DNA called a **plasmid.** Using the same restriction enzyme, the plasmid is opened and the isolated gene DNA is inserted into the plasmid, which is then introduced into a bacterium. The vector DNA is incorporated into the bacterial DNA, and the bacteria reproduce, grow, and divide.
- *Express the incorporated gene in the new organism.* Bacteria that have the donor DNA incorporated are induced to produce the protein from the gene of interest. This protein production of a nonnative gene is called **expression.** The protein can be subsequently isolated and purified for its intended use.

CRISPR—Twenty-First-Century Gene Editor

Recently, scientists have developed a tool that can insert a DNA sequence directly into a cell's DNA, much like you would insert extra words into a sentence by typing. This is different than the previously discussed recombinant DNA technology because now gene insertion can be done without the use of a vector. Instead, the DNA is directly delivered to the cell that needs the new gene. **CRISPR** (pronounced "crisper"), which stands for *C*lusters of *R*egularly *I*nterspaced *S*hort *P*alindromic *R*epeats, can edit the base sequence of a gene without inserting DNA from another source.

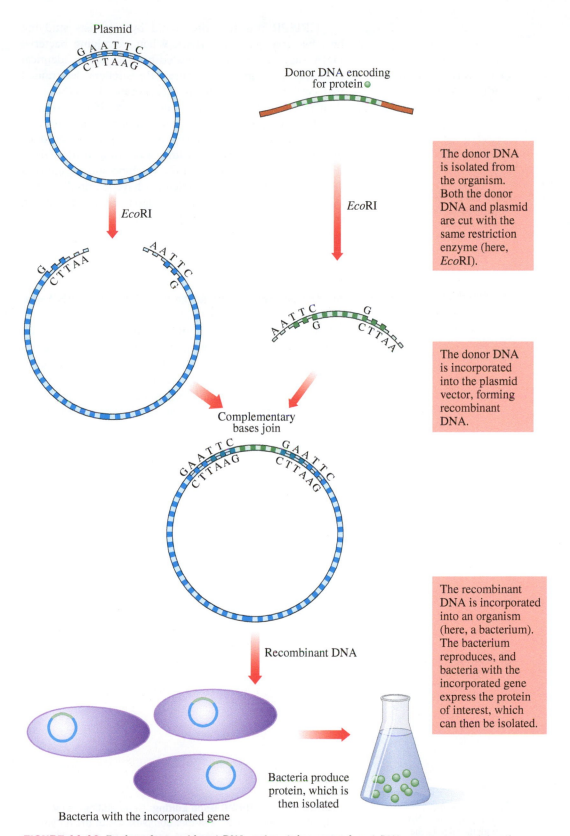

FIGURE 11.16 Basics of recombinant DNA and protein expression. A DNA segment containing the gene to be cloned is isolated and incorporated into a plasmid vector, and the protein is then expressed and isolated from bacteria.

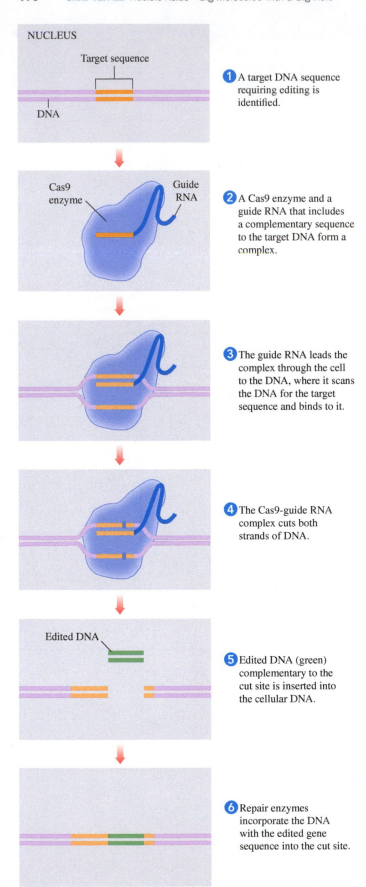

NUCLEUS

Target sequence

DNA

1 A target DNA sequence requiring editing is identified.

Cas9 enzyme

Guide RNA

2 A Cas9 enzyme and a guide RNA that includes a complementary sequence to the target DNA form a complex.

3 The guide RNA leads the complex through the cell to the DNA, where it scans the DNA for the target sequence and binds to it.

4 The Cas9-guide RNA complex cuts both strands of DNA.

Edited DNA

5 Edited DNA (green) complementary to the cut site is inserted into the cellular DNA.

6 Repair enzymes incorporate the DNA with the edited gene sequence into the cut site.

CRISPR was first discovered by scientists studying how bacteria fight off viruses. When examining bacterial DNA, they found short, repeated base sequences identical to those in the infecting viruses. Researchers determined that certain bacteria actually incorporate viral DNA into their own DNA as a defense mechanism. If a virus were to enter one of these bacteria a second time, the bacterium would use the previously incorporated repeats to produce viral RNA that matches up with the invading virus's DNA. This signals an enzyme called *Cas9* to bind and cut the viral DNA, rendering it useless. Scientists have manipulated this bacterial defense mechanism in the lab to allow gene editing in cells so that we can now delete, modify, or replace any genes in any organism. Because the technique is so precise, many genes can be manipulated in the same cell.

Suppose you wanted to repair a defective gene in a cell. You would design a piece of RNA, called *guide RNA*, that recognizes a unique site at the gene. The guide RNA also binds to the Cas9 enzyme and directs it to the targeted gene DNA. Once introduced into the cell, this Cas9–guide RNA complex finds its complementary DNA sequence on the target DNA. When the complex is bound to the target DNA, the Cas9 enzyme acts as a molecular scissors and cuts the DNA. The cell then begins to repair the cut site using its own enzymes. If DNA complementary to the site is introduced, this new DNA with the corrected gene sequence will be incorporated. This process is summarized in **FIGURE 11.17**.

CRISPR technology holds promise for changing the DNA sequences of disease-causing genes and engineering genes to perform new functions. Clinical trials on diseases such as sickle-cell anemia and certain melanoma cancers are expected to begin before 2020.

FIGURE 11.17 CRISPR Gene Editing. Gene editing using CRISPR involves a guide RNA that complexes with the enzyme Cas9, guiding it to a targeted DNA sequence for gene editing. Cas9 acts as a molecular scissors that cuts the DNA. Once cut, a new DNA sequence can be introduced and will be incorporated by the cell's own DNA repair enzymes.

Recombinant DNA and Fluorescent Proteins

Have you ever seen the brightly colored fluorescent fish at the local pet store and wondered where they came from? Such fluorescent Glofish® are the result of recombining the DNA for a fluorescent protein into the fish's DNA. Of course, breeding fluorescent fish is only a side effect of a Nobel prize–winning set of discoveries that are being used to "light up" proteins and cells of interest.

How does a scientist know if a gene from one organism is incorporated into the cell, as described in Figure 11.16? One way this problem is solved is to attach a *reporter* gene to the gene of interest. The first fluorescent reporter gene was called green fluorescent protein, or GFP, found in the Pacific Northwest jellyfish *Aequorea victoria*. When the gene for GFP is linked to a gene of interest, scientists can simply shine UV light onto the organism or cell culture to see whether the gene has been incorporated. Today, several colors of fluorescent protein reporter genes are available and are being used successfully in cancer research. In this case, the fluorescent reporter genes are linked to genes directing blood vessel growth because in cancer cells, these genes are more active. The reporter genes light up, showing cancer cell growth. Fluorescent reporter genes like GFP are a powerful tool for monitoring gene expression in a variety of cell types. They have been incorporated into many organisms from bacteria to mice to the commercially available Glofish. Fluorescent reporter genes allow scientists to *see* changes in cells previously invisible to us.

INTEGRATING Chemistry

◀ Find out how fluorescent proteins are lighting the way in cancer research.

Fluorescent proteins have been expressed in Glofish, seen here in a variety of colors.

Sample Problem 11.12 CRISPR Technology

Describe how a gene from one organism can be incorporated into a second organism using CRISPR.

Solution

The steps in CRISPR include using a guide RNA that is complementary to the target DNA to be edited. This guide RNA also binds to the enzyme Cas9 and guides it to the target site. Once the Cas9–guide RNA complex finds the target DNA, the Cas9 enzyme cuts the target DNA, and a new DNA piece can be inserted. The new DNA is linked into the cut site by the cell's own DNA repair enzymes.

Practice Problems

11.47 In your own words, define gene expression.

11.48 In your own words, define recombinant DNA.

11.49 Describe the function of a vector.

11.50 Describe the structure of a plasmid.

11.51 Describe how restriction enzymes are used in recombinant DNA technology.

11.52 CRISPR was first discovered as a defense mechanism in what organism?

11.53 Describe the function of the CRISPR enzyme Cas9.

11.54 Name the gene that tags other genes through fluorescent coloring.

11

CHAPTER REVIEW

The study guide will help you check your understanding of the main concepts in Chapter 11. You should be able to answer problems for each learning outcome in this list. To check your mastery, try the problems listed after each.

STUDY GUIDE	CHAPTER GUIDE

11.1 Components of Nucleic Acids

Identify the parts of nucleotides and nucleic acids.

- Distinguish the five nitrogenous bases found in nucleic acids as purine or pyrimidine. (Try 11.1, 11.55)
- Identify the five nitrogenous bases found in nucleic acids by name. (Try 11.1, 11.57)
- Distinguish ribose and deoxyribose. (Try 11.5, 11.60)
- Write condensation products for nucleoside and nucleotide formation. (Try 11.9)

Inquiry Question

What are the components of nucleotides and nucleic acids?

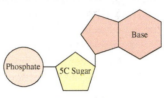

Nucleotide

Deoxyribonucleic acid (DNA) and ribonucleic acid (RNA) are nucleic acids. They consist of strings of nucleotides. A nucleotide has three components: a nitrogenous base, a five-carbon sugar, and a phosphate. A nucleoside consists of the nitrogenous base and the five-carbon sugar. The sugar deoxyribose is found in DNA, and the sugar ribose is found in RNA. The bases adenine (A), guanine (G), and cytosine (C) are found in both DNA and RNA. Thymine (T) is a fourth base found in DNA, and uracil (U) is a fourth base found in RNA. Nucleotides are named as the nucleoside + the number of phosphates (up to three) bonded to it. The components of nucleosides, as well as those of nucleotides, link together through condensation reactions.

11.2 Nucleic Acid Formation

Form a nucleic acid from nucleotides.

- Write the product of a condensation of nucleotides. (See Sample Problem 11.4, Try 11.13)
- Abbreviate a nucleic acid using one-letter base coding. (See Sample Problem 11.5)
- Characterize the structural features of nucleic acids. (Try 11.11, 11.61)

Inquiry Question

How are nucleic acids formed?

Each nucleic acid has a unique sequence that is its primary structure. Nucleic acids form when nucleotides undergo condensation, linking sugar to phosphate. The 3′ —OH on the sugar of one nucleotide bonds with the phosphate on the 5′ end of a neighboring nucleotide. The backbone of a nucleic acid is formed from alternating sugar–phosphate groups. The nitrogenous bases extend from the backbone sugar. Each nucleic acid has single free 5′ and 3′ ends.

Free 5' end—P–S–P–S–P–S–P–S—Free 3' end / G T C A

11.3 DNA

Describe the unique structural features of DNA.

- Characterize the structural features of DNA. (Try 11.15, 11.17)
- Write the complementary base pairs for a single strand of DNA. (Try 11.19, 11.63)

Inquiry Question

What are the unique structural features of DNA?

A DNA molecule resembles a twisted ladder. It consists of two antiparallel strands of nucleic acid with bases facing inward. The two strands are held together through the bases (rungs of the ladder), which hydrogen bond to complementary bases on the other strand, giving DNA its secondary structure. The base A forms two hydrogen bonds to T, and G forms three hydrogen bonds to C. In most plants and animals, DNA is found in the cell nucleus and is compacted into a tertiary structure called a chromosome. Chromosomes also contain a protein component called a histone, around which the DNA is supercoiled. Humans have 23 pairs of chromosomes in their cells.

11.4 RNA and Protein Synthesis

Specify the roles of RNA in protein synthesis.

- List three types of RNA and their role in protein synthesis. (Try 11.23, 11.65)
- Distinguish transcription from translation. (Try 11.25, 11.26)
- Translate a DNA strand into its complementary mRNA. (Try 11.27, 11.63)

Inquiry Question

What are the roles of RNA in protein synthesis?

RNA differs from DNA in that it contains a ribose sugar instead of a deoxyribose. Instead of thymine, RNA contains uracil, which can hydrogen bond to adenine. RNA is smaller than DNA and is single stranded. The three types of RNA that are involved in transforming the DNA sequence into a protein sequence in the cell are messenger RNA (mRNA), ribosomal RNA (rRNA), and transfer RNA (tRNA). The messenger RNA is involved in transcribing a complementary copy from gene DNA in the cell nucleus and taking that copy to the ribosome. Ribosomes are cell organelles where protein synthesis occurs. They consist of rRNA and protein. The tRNA is a compact RNA structure that acts as a conduit between the code on the messenger RNA and an amino acid sequence. tRNA has two features: the anticodon at one end that is complementary to the codon sequence on the mRNA and the acceptor stem where an amino acid can attach.

11.5 Putting It Together: The Genetic Code and Protein Synthesis

Translate the genetic code into a protein sequence.

- Translate a mRNA sequence into a protein sequence using the genetic code. (Try 11.31, 11.71, 11.73)
- Translate a template DNA sequence into a protein sequence using the genetic code. (Try 11.35, 11.85)

Inquiry Question

How is the genetic code used to synthesize proteins?

A genetic code contains a series of base triplet sequences on mRNA specifying the order of amino acids in a protein. The codon AUG signals the start of transcription, and the codons UAG, UGA, and UAA signal it to stop. Protein synthesis begins with transcription where an mRNA creates a complementary copy of a DNA gene. tRNA activation involves binding an amino acid to the tRNA, catalyzed by the enzyme tRNA synthetase. During translation, tRNAs bring the appropriate amino acids to the ribosome, and peptide bonds form until termination when a stop codon is reached. The polypeptide becomes a functional protein upon release.

11.6 Genetic Mutations

Discuss various nucleic acid sequence mutations and possible effects.

- Define mutation. (Try 11.37)
- Determine changes in protein sequence if an mRNA sequence is mutated. (Try 11.39, 11.77)

Inquiry Question

How do genetic mutations occur?

Base alterations from normal cell DNA sequences are called mutations. Some mutations are silent and have no effect on protein synthesis, and some may change the sequence but not alter protein function. Others can affect protein sequence, structure, and function. Mutations can be random during DNA replication, or they can be caused by mutagens such as chemicals or radiation. A mutation in a germ cell can be inherited. If such a mutation results in a defective protein, a genetic disease results.

—Continued next page

Continued—

11.7 Viruses

Apply functions of nucleic acids to viruses.

- List the differences between a virus and a cell. (Try 11.41, 11.43)
- List the structural components of a virus. (Try 11.41, 11.44)
- Describe how a virus infects a cell. (Try 11.43)
- Describe in basic terms how a vaccine protects against a virus. (Try 11.45, 11.78)

Inquiry Question

How does a virus use nucleic acids?

 Viruses are particles containing DNA or RNA and a protein coat called a capsid. They invade a host cell and use the host cell's machinery to replicate more virus particles. Viral infections can be prevented by vaccines in some cases, like influenza, or can be treated by blocking certain steps in viral replication specific to the virus, not the host.

11.8 Recombinant DNA Technology

Explain how nucleic acids can be manipulated in DNA technology.

- Apply knowledge of nucleic acid structure to DNA technology. (Try 11.47, 11.49, 11.51, 11.53)
- Distinguish recombinant DNA and CRISPR. (Try 11.80)

Inquiry Question

How can DNA be manipulated outside of the cell?

Recombinant DNA technology involves the expression of a protein from one organism into a second organism. This can be accomplished after the gene of interest is isolated and clipped from a genome using a restriction enzyme.

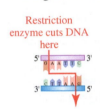
Restriction enzyme cuts DNA here

The gene of interest is incorporated into a vector that transports and incorporates the gene into the second organism. The host organism can then produce the protein of interest during transcription and translation. In CRISPR, genes can be directly edited within a cell using an enzyme, Cas9, that cuts DNA at a specific site where DNA is then introduced. Gene cloning or making an exact copy of a gene is different from organism cloning, in which an exact copy of an organism is produced.

Mastering Videos

PRACTICING THE CONCEPTS

The following videos can be accessed through the Pearson eText or your Mastering Chemistry course.

- Condensation Reactions in Nucleic Acid Chemistry
- Nucleic Acid Structure

Additional Problems

11.55 Identify each of the following bases as a pyrimidine or a purine:

  a. cytosine   **b.** adenine

  c. uracil   **d.** thymine

  e. guanine

11.56 Indicate if each of the bases in Problem 11.55 is found in DNA only, RNA only, or both DNA and RNA.

11.57 Identify the base and sugar in each of the following nucleotides:

  a. CMP   **b.** dAMP

  c. dGMP   **d.** UMP

11.58 Identify the base and sugar in each of the following nucleotides:

  a. dTMP   **b.** AMP

  c. dCMP   **d.** GMP

11.59 How do the bases thymine and uracil differ?

11.60 How do the sugars ribose and deoxyribose differ?

11.61 Fill in the following table comparing structural similarities between proteins and nucleic acids:

	Protein	Nucleic Acid
Repeating unit	Amino acid	
Backbone repeat	$N{-}C_\alpha{-}C{-}N{-}C_\alpha{-}C$, etc.	
One-letter abbreviation	Name of amino acid	
Free left end	Amino or N-terminus	
Free right end	Carboxy or C-terminus	

11.62 Write the complementary base sequence for each of the following DNA segments. Indicate the 5′ and the 3′ ends.

  a. 5′GACTTAGGC3′

  b. 5′TGCAAACTAGCT3′

  c. 5′ATCGATCGATCG3′

11.63 Write the complementary base sequence for each of the following DNA segments. Indicate the 5′ and the 3′ ends.

  a. 5′TTACGGACCGC3′

  b. 5′ATAGCCCTTACTGG3′

  c. 5′GGCCTACCTTAACGACG3′

11.64 Match the following statements with mRNA, rRNA, or tRNA:

  a. combines with proteins to form ribosomes

  b. carries the genetic information from the nucleus to the ribosome

  c. carries amino acids to ribosome for protein synthesis

11.65 Match the following statements with mRNA, rRNA, or tRNA:

  a. contains codons for protein synthesis

  b. contains anticodons

  c. found in the small subunit and the large subunit of the ribosome

11.66 List the possible codons for each of the following amino acids:

  a. threonine   **b.** serine   **c.** cysteine

11.67 List the possible codons for each of the following amino acids:

  a. valine   **b.** proline   **c.** histidine

11.68 Provide the amino acid corresponding to each of the following codons:

  a. ACG   **b.** CCA   **c.** GCA

11.69 Provide the amino acid corresponding to each of the following codons:

  a. UUG   **b.** CGG   **c.** AUC

11.70 Provide the three-letter amino acid sequence expected from each of the following mRNA segments:

  a. 5′CAA|GUU|AGA|CUG3′

  b. 5′CAU|AAA|UUU|CGA|UGA3′

  c. 5′AUG|CCC|GGG|GAG|CGC|AUU3′

11.71 Provide the three-letter amino acid sequence expected from each of the following mRNA segments:

  a. 5′GGC|AUC|UAC|CGA3′

  b. 5′CUA|AGC|UUC|AAC|UGG3′

  c. 5′GUA|CGU|CGG|UUA|CCA|ACC3′

11.72 Endorphins are polypeptides that reduce pain. What is the one-letter amino acid sequence formed from the following mRNA that codes for a pentapeptide that is an endorphin called Leu-enkephalin?

  5′AUG|UAC|GGU|GGA|UUU|CUA|UAA3′

11.73 What is the one-letter amino acid sequence formed from the following mRNA that codes for a pentapeptide that is an endorphin called Met-enkephalin?

  5′AUG|UAC|GGU|GGA|UUU|AUG|UAA3′

11.74 What is the anticodon on tRNA for each of the following codons in an mRNA?

  a. AGC   **b.** UAU   **c.** CCA

11.75 What is the anticodon on tRNA for each of the following codons in an mRNA?

  a. GUG   **b.** CCC   **c.** GAA

11.76 a. A base substitution changes a codon for an enzyme from GCC to GCA. Why is there no change in the amino acid order in the protein?

  b. In sickle-cell anemia, a base substitution in the hemoglobin gene replaces glutamate (a polar amino acid) with valine. Why does the replacement of one amino acid cause such a drastic change in biological function?

11.77 a. A base substitution for an enzyme replaces leucine (a nonpolar amino acid) with alanine. Why does this change in amino acids have little effect on the biological activity of the enzyme?

  b. A base substitution replaces cytosine in the codon UCA with adenine. How might this substitution affect the amino acids in the protein?

11.78 Provide a reason why there is no vaccine for the common cold.

11.79 Name two common viruses that infect humans.

11.80 How is CRISPR different from earlier recombinant DNA techniques?

11.81 What is the function of the guide RNA in CRISPR?

11.82 List two societal benefits to recombinant DNA technology.

11.83 Describe one way GFP is being used to benefit science.

Challenge Problems

11.84 Oxytocin is a small peptide hormone involved in the process of birthing and lactation. It contains the nine amino acids shown.

CYIQNCPLG

a. How many nucleotides would be found in the mRNA for this protein?

b. Suggest an mRNA sequence for the peptide. Show the 5′ and 3′ ends.

c. Suggest a complementary template DNA sequence based on the mRNA sequence suggested in part b.

11.85 Alpha-melanocyte stimulating hormone (α-MSH) is a 13-amino-acid peptide hormone responsible for pigmentations in hair and skin. Its peptide sequence is shown.

SYSMQHFRWGKPV

a. How many nucleotides would be found in the mRNA for this protein?

b. Suggest an mRNA sequence for the peptide. Show the 5′ and 3′ ends.

c. Suggest a complementary template DNA sequence based on the mRNA sequence suggested in part b.

11.86 a. If the DNA chromosomes of salmon contain 28% adenine, what is the percent of thymine, guanine, and cytosine?

b. If the DNA chromosomes of humans contain 20% cytosine, what is the percent of guanine, adenine, and thymine?

11.87 A protein contains 35 amino acids. How many nucleotides would be found in the mRNA code for this protein?

11.88 The DNA double helix can unwind, or denature, at temperatures between 90 °C and 99 °C. Denaturing occurs when H bonds are broken. Which of the following strands of DNA would be expected to denature at a higher temperature? Provide an explanation.

DNA strand 1	5′ATTTTCCAAAGTATA3′
	3′TAAAAGGTTTCATAT5′
DNA strand 2	5′CGCGAGGGTCCACGC3′
	3′GCGCTCCCAGGTGCG5′

Answers to Odd-Numbered Problems

Practice Problems

11.1 a. purine; adenine

b. pyrimidine; cytosine

11.3 a. both DNA and RNA

b. both DNA and RNA

11.5 Ribose contains an —OH on C2′, whereas deoxyribose contains only —H at C2′.

11.7 deoxyadenosine monophosphate (dAMP), deoxythymidine monophosphate (dTMP), deoxycytidine monophosphate (dCMP), deoxyguanosine monophosphate (dGMP)

11.9 a.

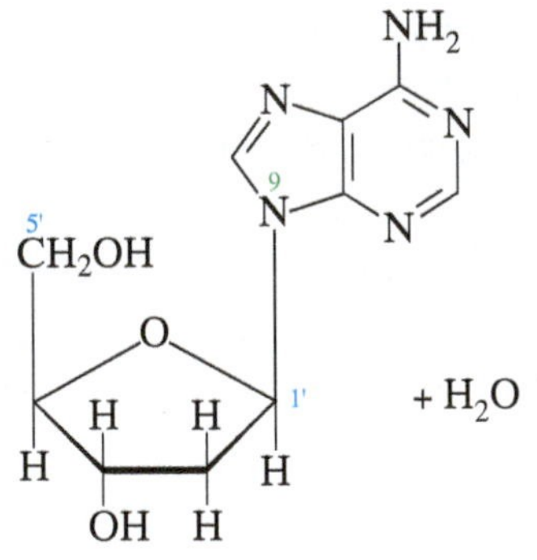

b.

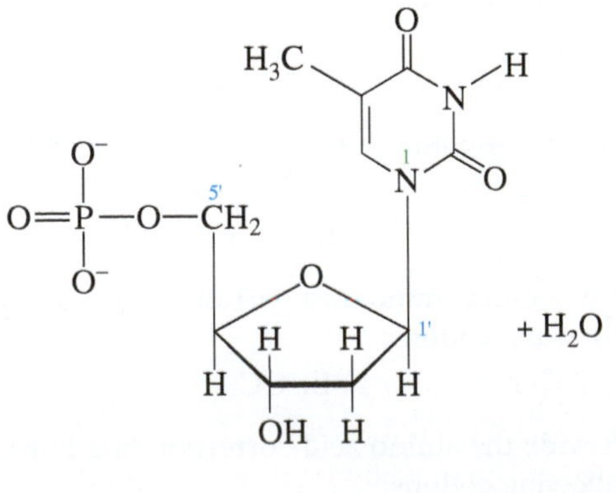

11.11 The nucleotides in nucleic acids are held together by phosphodiester bonds. These bonds are formed between the 3′ —OH of a sugar (ribose or deoxyribose) and a phosphate group on the 5′-carbon of another sugar.

11.13

Phosphodiester

11.15 The two strands of nucleic acid that make up DNA are oriented in opposite directions (antiparallel) to each other. In one of the strands, the sugars are connected from the 3′C to the 5′C, and in the other strand they run 5′ to 3′.

11.17 The two DNA strands are held together by hydrogen bonds between the bases in each strand.

11.19 **a.** 3′TTTT5′   **b.** 3′GGGGAAAA5′
   c. 3′TGTAACC5′   **d.** 3′ACACTTGG5′

11.21

	Protein	Nucleic Acid
Primary structure	Sequence of amino acids	Sequence of nucleotides
Secondary structure	Local hydrogen bonding between backbone atoms	Hydrogen bonding between bases in nucleic acid strands
Tertiary structure	Folding of backbone onto itself associates amino acids far away in sequence	Supercoiling of DNA into tight, compact structures like chromosomes

11.23 messenger RNA—copies DNA and then moves from nucleus to ribosome

ribosomal RNA—structural component of the ribosome

transfer RNA—places a specific amino acid in a growing protein chain at the ribosome

11.25 In transcription, the sequence of nucleotides on one strand of a DNA template (the double helix is temporarily unwound) is used to produce a complementary mRNA copy using the enzyme RNA polymerase.

11.27 3′UGGCGGAAUUCC5′

11.29 Twenty amino acids are brought into the ribosome, each bonded to a different tRNA.

11.31 **a.** Lys–Pro–Leu–Ala
   b. Pro–Arg–Ser–Pro–Stop
   c. Met–His–Lys–Glu–Val–Leu

11.33 Translocation is the movement of the second tRNA to the spot vacated by the first tRNA, allowing for the next tRNA to bind to the ribosome. The amino acid is added to the growing peptide.

11.35 **a.** 5′ACA|CCC|CAA|UAA3′
   b. 3′UGU|GGG|GUU|AUU5′
   c. three-letter code: Thr–Pro–Gln–Stop; one-letter code: TPQ-Stop

11.37 A mutation is any change to the DNA sequence of a gene.

11.39 **a.** Three-letter code: Leu–Lys–Arg–Val, one-letter code: LKRV
   b. IKRV. Changing leucine to isoleucine will not have an effect on the protein function because both are nonpolar amino acids.
   c. LKRV. No, because both codons code for arginine.
   d. Leucine will be the last amino acid incorporated into the growing protein chain because UAA is a stop codon.
   e. If A is added to the beginning of the chain, the codons would be shifted, and the new sequence would be Thr–Stop–Thr–Ser. The growing chain would stop after incorporating the first threonine.
   f. If C is removed from the beginning of the chain, the codons would be shifted, and the new sequence would be Leu–Asn–Glu (LNE).

11.41 **a.** both   **b.** cell   **c.** virus

11.43 Viruses cannot replicate themselves without a host cell. They do not have the cell machinery.

11.45 A vaccine allows the body to mount an immune response by producing antibodies to a less active form of a virus. If an active form is encountered later, the body can fight against it more effectively.

11.47 Gene expression is producing a protein from a DNA gene sequence (usually of a nonnative gene).

11.49 A vector is a transporter for donor DNA. It enables incorporation of the donor DNA into the organism DNA.

11.51 Restriction enzymes are used to cut the gene to be cloned out of the original DNA.

11.53 Cas9 acts as a molecular scissors to cut double-stranded DNA.

Additional Problems

11.55 **a.** pyrimidine   **b.** purine
   c. pyrimidine   **d.** pyrimidine
   e. purine

11.57 a. cytosine, ribose   **b.** adenine, deoxyribose

     c. guanine, deoxyribose   **d.** uracil, ribose

11.59 Thymine contains a methyl group at carbon-5 ($-CH_3$) that uracil lacks.

11.61

	Protein	Nucleic Acid
Repeating unit	Amino acid	Nucleotide
Backbone repeat	$N-C_\alpha-C-N-C_\alpha-C$, etc.	sugar–phosphate–sugar–phosphate, etc.
One-letter abbreviation	Name of amino acid	Name of base
Free left end	Amino or N-terminus	5′ end
Free right end	Carboxy or C-terminus	3′ end

11.63 a. 3′AATGCCTGGCG5′

     b. 3′TATCGGGAATGACC5′

     c. 3′CCGGATGGAATTGCTGC5′

11.65 a. mRNA   **b.** tRNA   **c.** rRNA

11.67 a. GUU, GUC, GUA, GUG

     b. CCU, CCC, CCA, CCG

     c. CAU, CAC

11.69 a. leucine   **b.** arginine   **c.** isoleucine

11.71 a. Gly–Ile–Tyr–Arg

     b. Leu–Ser–Phe–Asn–Trp

     c. Val–Arg–Arg–Leu–Pro–Thr

11.73 (start) YGGFM (stop)

11.75 a. CAC   **b.** GGG   **c.** CUU

11.77 a. Both alanine and leucine have small, nonpolar R groups. Both amino acids are found in similar environments in proteins, so this substitution will not likely affect protein structure or function.

     b. If a serine codon (UCA) is replaced with a stop codon (UAA), any remaining amino acids will not be linked to the growing polypeptide chain. This mutation can affect both the structure and function of the protein.

11.79 Three possible answers discussed in the chapter are the common cold virus, influenza, and HIV-1/AIDS.

11.81 The guide RNA recognizes a unique site on a gene and signals Cas9 to cut the DNA at the guide site.

11.83 When the gene sequence for GFP is linked to the recombined gene, it can be used to visually confirm the presence of recombined DNA.

11.85 a. 3 nucleotides/amino acid + 1 start codon + 1 stop codon = (13 amino acids × 3) + 3 + 3 = 45 total nucleotides

     b. Answers will vary slightly depending on the codon selected.

     5′AGU|UAU|AGU|AUG|CAA|CAU|UUU| CGU|UGG|GGU|AAA|CCU|GUU3′

     c. Answers should complement the answer in part b.

     3′TCA|ATA|TCA|TAC|GTT|GTA|AAA|GCA| ACC|CCA|TTT|GGA|CAA5′

11.87 3 nucleotides/amino acid + 1 start codon + 1 stop codon = (35 amino acids × 3) + 3 + 3 = 111 nucleotides

12

Food as Fuel—An Overview of Metabolism

Our bodies perform chemical reactions constantly to produce energy from food and create the molecules for survival. Such metabolic reactions require proper nutrition and health for our bodies to perform at their best. We are now ready to apply the concepts of chemistry to better understand how the body produces and stores energy.

LEARNING TIP   Study in Regular Intervals

Your retention increases if you review the information you just learned today a day or two from now. A single, initial exposure to content is not enough for you to remember most of it. If you space out your study time into one- or two-day intervals during the first week after your initial exposure, your ability to recall the information on a test two or three weeks later will increase.

DO YOU KNOW people who claim to have a high metabolism? When they say this, they usually mean they can eat more than the average person and still maintain their body weight. Likewise, many nutrition products and weight-loss plans promise to "boost your metabolism"—if your metabolism is "boosted," you will lose weight. **Metabolism** refers to the chemical reactions occurring in the body that break down or build up molecules. In biological systems, the chemical reactions of metabolism often occur in a series of steps called a **metabolic pathway.**

In Chapter 5, we saw that chemical reactions can absorb or give off energy. In the body, this chemical energy is captured in the nucleotide **a**denosine **tri**phosphate, or ATP, which then serves as the fuel for bodily functions.

In this chapter, we explore the absorption of biomolecules from food, metabolic pathways involved in glucose metabolism, and the way other biomolecules, such as proteins and lipids, are part of metabolic pathways.

DISCOVERING THE CONCEPTS

? INQUIRY ACTIVITY—Anabolism and Catabolism

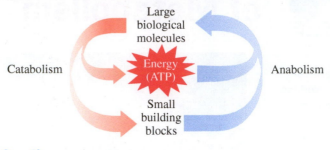

FIGURE 1. Overview of metabolism. Each arrow represents a series of chemical reactions.

Questions

1. Based on Figure 1, name the reactants and products in the table below.

Type of reaction	Reactants	Products
Catabolism		
Anabolism		

2. Are reactions of catabolism or anabolism endergonic? Exergonic?
3. In your own words, provide a definition for catabolism.
4. In your own words, provide a definition for anabolism.
5. Provide a definition for metabolism.
6. Of the four different reaction types, oxidation, reduction, condensation, and hydrolysis, which are more likely seen in catabolism, and which in anabolism? Support your answer.
7. The chemical reactions that break down glucose are called glycolysis, while the reactions that create glucose by the body are called gluconeogenesis.
   a. Is glycolysis or gluconeogenesis considered catabolism? Anabolism?
   b. The reactions in glycolysis are not identical to the reactions in gluconeogenesis, although some intermediates are the same. Why do you think this would be so?
8. Chemical reactions outside the body often give off or take in large amounts of energy as heat. Do you think this is true of metabolic reactions? Explain.

What's an Inquiry Question?

? Inquiry Questions are designed to focus your reading on the main concepts by section. As you read each section, see if you can answer each Inquiry Question in your own words.

? 12.1 Inquiry Question

What is metabolism?

12.1 How Metabolism Works

All animals, including humans, must ingest food for fuel to carry out daily functions such as circulation, respiration, and movement. How does eating food produce energy? How do our bodies convert food into useful energy? Animals extract energy (measured as Calories) from the covalent bonds contained in carbohydrates, fats, and proteins. As an example, let's look at what happens when someone eats a bean and cheese burrito.

In the first stage of metabolism, the large biomolecules in food are digested or broken down into their smaller units through hydrolysis reactions. Polysaccharides, like starch in the tortilla, are hydrolyzed into monosaccharide units, the triglycerides in the cheese to glycerol and fatty acids, and the proteins in the beans and cheese into their amino acid units. These hydrolysis products, the smaller molecules produced by the breakdown of the large biomolecules, are absorbed through the intestinal wall into the bloodstream, and are eventually transported to different tissues for use by the cells.

Once in the cells, the hydrolysis products are broken down through oxidative processes into a few common metabolites containing two or three carbons (see **FIGURE 12.1**). **Metabolites** are chemical intermediates formed by enzyme-catalyzed reactions in the body. At this point, as long as the cells have oxygen and are producing energy, two-carbon acetyl groups can be broken down further to carbon dioxide (one carbon) through a series of chemical reactions called the citric acid cycle. This cycle works in conjunction

O   O
‖   ‖
$H_3C — C — C — O^-$

(a) Pyruvate

O
‖
$H_3C — C$ ┤ Coenzyme A

(b) Acetyl group

FIGURE 12.1 Common metabolites. (a) Pyruvate (a three-carbon molecule) and (b) acetyl groups (containing two carbons) found in acetyl coenzyme A are common intermediates in many biochemical pathways where carbon is metabolized.

A bean and cheese burrito contains carbohydrates, proteins, and triglycerides that can be digested and metabolized for energy.

with the electron transport and oxidative phosphorylation pathways to produce energy that is transferred through the cells by the molecules ATP, nicotinamide adenine dinucleotide (NADH), and flavin adenine dinucleotide (FADH$_2$). We will learn more about these cycles and pathways later in the chapter. One burrito contains a lot of carbon atoms, but we have a lot of cells that require fuel. **FIGURE 12.2** provides an overview of catabolism showing how nutrients from foods are broken down in the body.

In our cells, when molecules from food oxidize, the energy is repackaged into molecules with high chemical potential energy instead of dissipating as heat. The chemical potential energy in ATP is transformed into kinetic energy or used to make other molecules in the body.

Chemical reactions that occur in living systems are called *biochemical reactions.* Most of the time in biological systems, chemical reactions occur in a series called a metabolic pathway. For example, the sugar molecule glucose (containing six carbons) is broken down to two molecules of pyruvate (three carbons each) through a series of chemical reactions collectively referred to as *glycolysis.*

Metabolism can be considered in two parts, catabolism and anabolism. **Catabolism** refers to chemical reactions in which larger molecules are broken down into a few common metabolites. Because catabolism breaks down molecules, we see many hydrolysis and oxidation reactions in catabolic pathways. These reactions tend to be exergonic ($-\Delta G$).

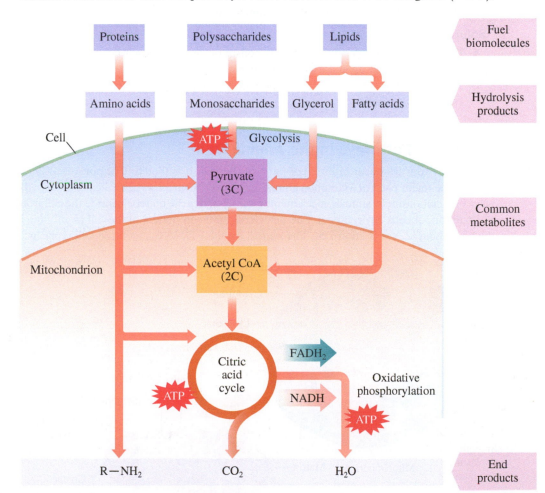

FIGURE 12.2 Overview of catabolism. Food fuels are hydrolyzed and further oxidized to a few common metabolites as they move from the intestine to the interior regions of the cell. These fuels ultimately produce CO_2, H_2O, and waste amine products ($R—NH_2$), and store energy produced as ATP.

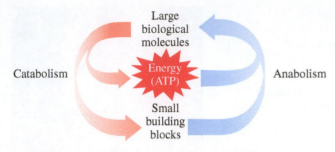

FIGURE 12.3 Catabolism and anabolism. Catabolic and anabolic reactions together define metabolism. Catabolism produces the energy-rich molecules and hydrolysis products that can be used as building blocks for anabolism (production of large biological molecules).

Anabolism refers to chemical reactions in which metabolites combine to form larger molecules. In anabolic pathways, many of the reactions are condensation or reduction reactions. These reactions tend to be endergonic $(+\Delta G)$.

The energy released during catabolic reactions is captured in molecules such as ATP, and is used to drive the anabolic reactions. ATP and other energy-rich molecules couple the reactions of catabolism and anabolism. In this way, the energy given off in catabolism is transferred to the energy-requiring reactions in anabolism (see **FIGURE 12.3**). Reactions in metabolism are catalyzed by enzymes that work to control the rates of biochemical reactions. The properties of catabolism and anabolism are summarized in **TABLE 12.1**.

TABLE 12.1 Common Properties of Biochemical Reactions in Metabolism

Property	Catabolism	Anabolism
Chemical reactions	Hydrolysis	Condensation
	Oxidation	Reduction
Energy	Exergonic (produce energy)	Endergonic (require energy)
Reaction rate	Catalyzed by enzymes	Catalyzed by enzymes

Sample Problem 12.1   **Distinguishing Anabolism from Catabolism**

Indicate whether the following processes represent anabolism or catabolism.
a. burning fat molecules during an intense workout
b. making proteins from component amino acids

Solution

a. Burning fat molecules represents catabolism because the larger fat molecules are broken up and used to produce energy for movement.
b. Making proteins from smaller component building blocks like amino acids represents anabolism.

Metabolic Pathways in the Animal Cell

Metabolic pathways occur in different parts of the cell. It is therefore important to know some of the major parts of a common animal cell (see **FIGURE 12.4**) to locate the main pathways in metabolism. In animals, a cell membrane separates the materials inside the cell from the exterior aqueous environment. The main structural component of the cell membrane is the phospholipid (see Section 7.5). The *nucleus* contains DNA that controls cell replication

FIGURE 12.4 The cell. The major components of an animal cell showing the major organelles. The mitochondrion shown in detail on the right is where most of the ATP is produced in the cell.

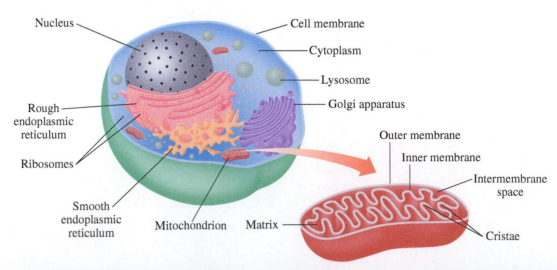

and protein synthesis. The **cytoplasm** consists of all the material between the nucleus and the cell membrane. The **cytosol** is the fluid part of the cytoplasm. It is the aqueous solution of electrolytes and enzymes that catalyzes many of the cell's chemical reactions.

Within the cytoplasm are specialized structures called *organelles* that carry out specific functions in the cell. We have already seen that the ribosomes are the sites of protein synthesis (Chapter 11). The **mitochondria** are the energy-producing factories of the cells. A mitochondrion consists of an outer membrane and an inner membrane, with an intermembrane space between them. The fluid section encased by the inner membrane is called the *matrix*. Enzymes located in the matrix and along the inner membrane catalyze the oxidation of carbohydrates, fats, and amino acids.

All of the oxidation pathways produce CO_2, H_2O, and energy, which are used to form energy-rich compounds. **TABLE 12.2** provides a summary of the functions of the cellular components in animal cells. Keep in mind that a cell can have more than one of the same organelle.

TABLE 12.2 Locations and Functions of Components in Animal Cells

Component	Description and Function
Cell membrane	Separates the contents of a cell from the external environment and contains structures that communicate with other cells
Cytoplasm	Consists of all the cellular contents between the cell membrane and nucleus
Cytosol	The fluid part of the cytoplasm that contains enzymes for many of the cell's chemical reactions
Endoplasmic reticulum (ER)	Processes proteins for secretion and synthesizes phospholipids (rough ER); synthesizes fats and steroids (smooth ER)
Golgi apparatus	Modifies and secretes proteins from the endoplasmic reticulum and synthesizes glycoproteins and cell membranes
Lysosomes	Contain hydrolytic enzymes that digest and recycle old cell structures
Mitochondria	Contain the structures for the synthesis of ATP from energy-producing reactions
Nucleus	Contains genetic information for the replication of DNA and the synthesis of protein
Ribosomes	Sites of protein synthesis using mRNA templates

Sample Problem 12.2   Metabolism and Cell Structure

Identify the cell component associated with each of the following descriptions.
a. organelle in the cell where energy is produced
b. location of genetic information and DNA replication
c. contains all the cellular components outside the nucleus of the cell

Solution

a. mitochondrion   b. nucleus   c. cytoplasm

Practice Problems

Find the answers to the odd-numbered Practice Problems at the end of the chapter.

12.1 How can you identify a catabolic reaction?

12.2 How can you identify an anabolic reaction?

12.3 Indicate whether the following processes represent anabolism or catabolism.

a. glycolysis   b. synthesis of fats

12.4 Indicate whether the following processes represent anabolism or catabolism.

a. conversion of glycolysis products to fats
b. generation of ATP from breakdown of fructose

12.5 Name the types of chemical reactions that tend to be found in catabolic pathways.

12.6 Name the types of chemical reactions that tend to be found in anabolic pathways.

12.7 Name a molecule that transfers energy within the cell.

12.8 Pyruvate and acetyl groups are common chemical intermediates called _____.

12.2 Metabolically Relevant Nucleotides

 12.2 Inquiry Question

What do the nucleotides that transfer energy look like?

In Chapter 11, we saw that nucleotides are the building blocks of the nucleic acids DNA and RNA. Nucleotides also have important metabolic functions in cells. They act as energy exchangers and can also be coenzymes.

Metabolic nucleotides have two forms: a high-energy form and a low-energy form. They consist of some basic components introduced in Chapter 11: the nucleoside adenosine, a phosphate, and a five-carbon sugar. Many of these molecules also incorporate a vitamin within their structure.

Because the structures of the basic components were shown in Chapter 11, they are illustrated here using their component names. **TABLE 12.3** shows the structures of the metabolic nucleotides described in the following sections in both their high- and low-energy forms. The nucleotide portion (base, sugar, and at least one phosphate) and the vitamin portion are also indicated.

TABLE 12.3 Metabolically Relevant Nucleotides

High-Energy Form	Low-Energy Form

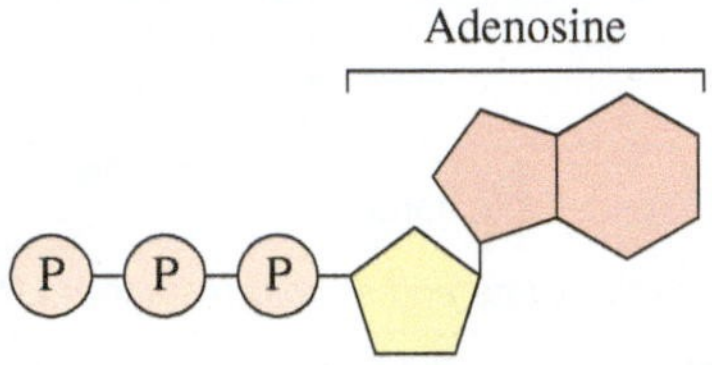

Adenosine

P–P–P

Adenosine triphosphate
ATP

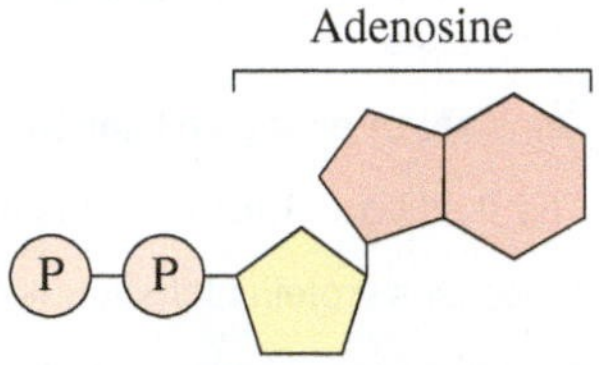

Adenosine

P–P

Adenosine diphosphate
ADP

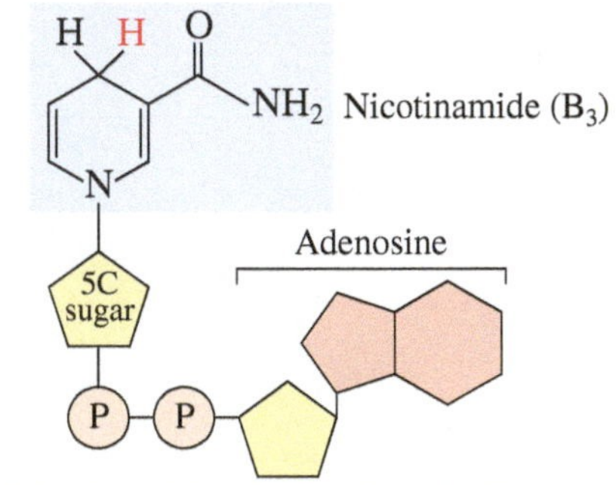

H H O
NH₂ Nicotinamide (B₃)

5C sugar
Adenosine
P–P

Nicotinamide adenine dinucleotide
NADH
(reduced form)

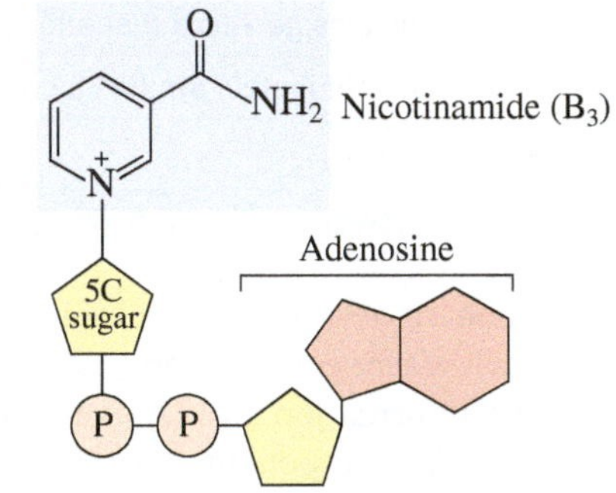

O
NH₂ Nicotinamide (B₃)

5C sugar
Adenosine
P–P

Nicotinamide adenine dinucleotide
NAD⁺
(oxidized form)

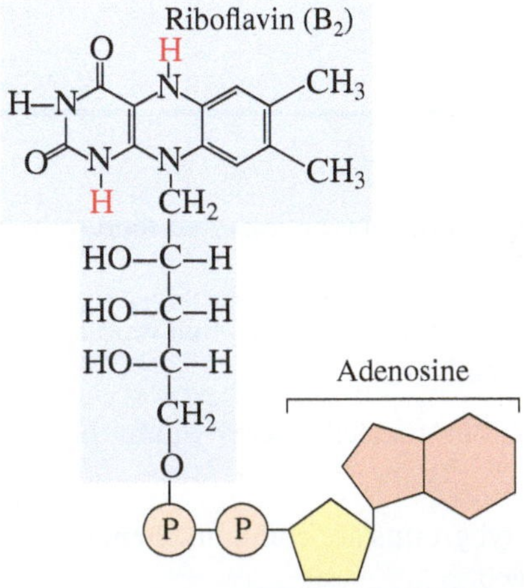

Riboflavin (B₂)

O H N CH₃
H–N
O N N CH₃
H CH₂
HO–C–H
HO–C–H
HO–C–H
CH₂
O
P–P
Adenosine

Flavin adenine dinucleotide
FADH₂
(reduced form)

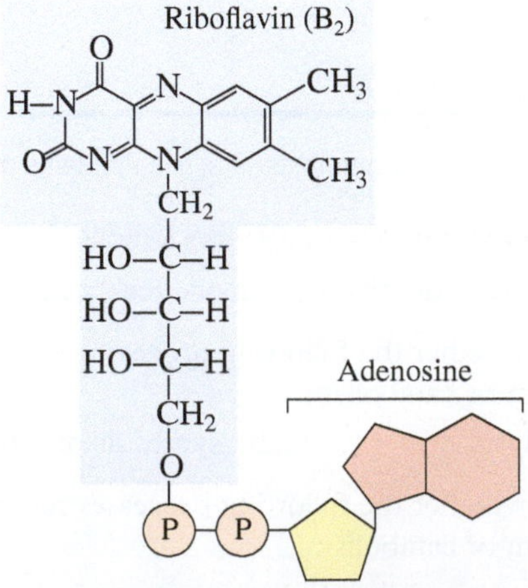

Riboflavin (B₂)

O N CH₃
H–N
O N N CH₃
CH₂
HO–C–H
HO–C–H
HO–C–H
CH₂
O
P–P
Adenosine

Flavin adenine dinucleotide
FAD
(oxidized form)

High-Energy Form

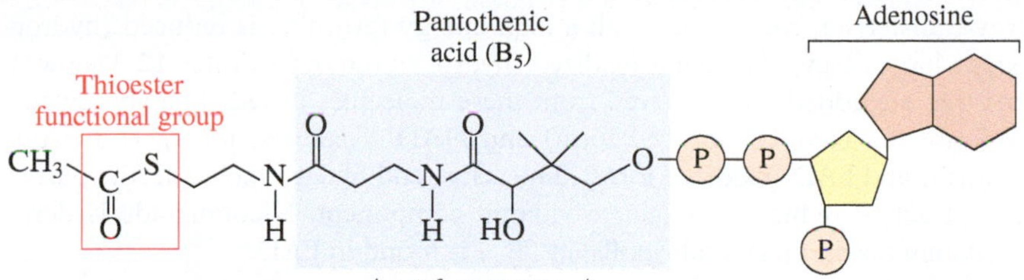

Pantothenic acid (B₅)

Adenosine

Thioester functional group

Acetyl coenzyme A

Low-Energy Form

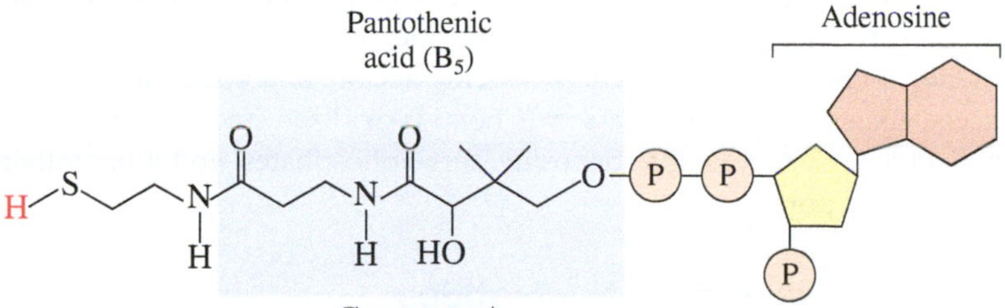

Pantothenic acid (B₅)

Adenosine

Coenzyme A

ATP/ADP

The nucleotide ATP is often referred to as the energy currency of the cell. ATP can undergo hydrolysis to adenosine diphosphate (ADP), as shown in the following equation. The nucleotide components are shown beneath the chemical structure for simplicity. During hydrolysis, energy is released as a product (ΔG is negative) so in this case, ATP is the high-energy form and ADP is the low-energy form. The energy given off during the hydrolysis of ATP can be coupled to drive a chemical reaction that requires energy during anabolism.

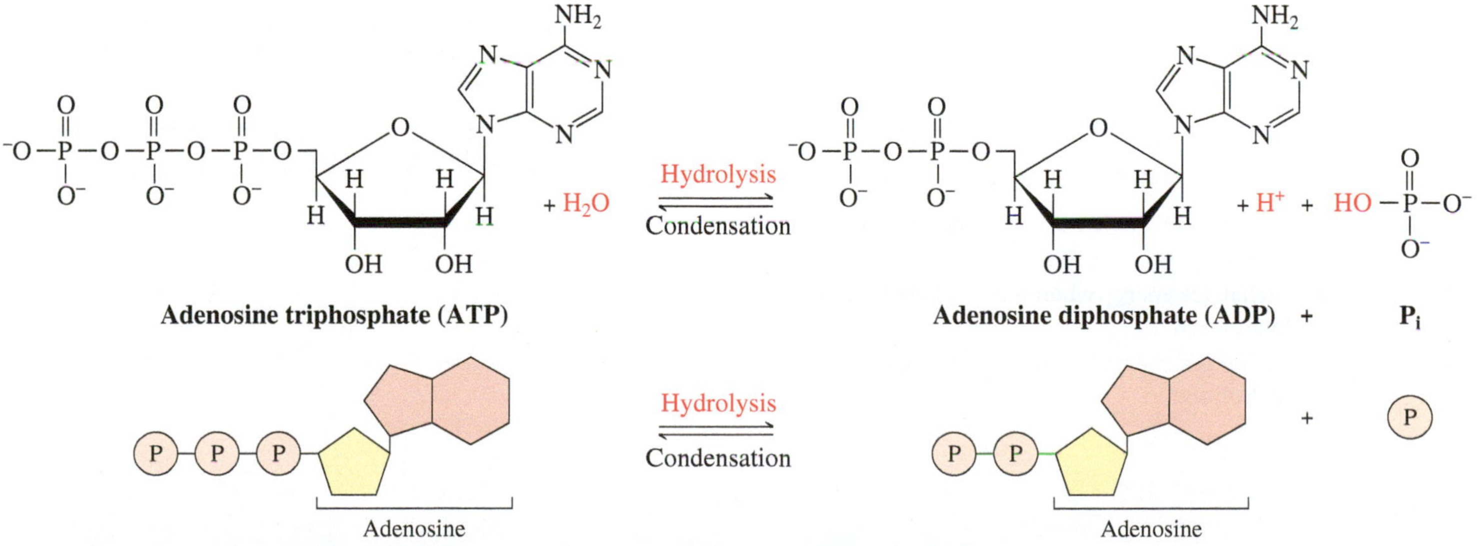

Adenosine triphosphate (ATP)   **Adenosine diphosphate (ADP)** +   **P$_i$**

The hydrolysis of ATP is an exergonic reaction ($-\Delta G$). The reaction is commonly abbreviated ATP $\longrightarrow$ ADP + P$_i$

NADH/NAD$^+$ and FADH$_2$/FAD

Nicotinamide adenine dinucleotide (NAD$^+$) and flavin adenine dinucleotide (FAD) are two energy-transferring compounds with a high-energy form that is reduced (hydrogen added) and a low-energy form that is oxidized (hydrogen removed). Table 12.3 shows the hydrogens that are added and removed from these molecules in red. The abbreviations for these forms are NADH (reduced form) and NAD$^+$ (oxidized form), and FADH$_2$ (reduced form) and FAD (oxidized form). The active end of each molecule, the part that is being oxidized or reduced, contains a vitamin component. Nicotinamide is derived from the vitamin niacin (B$_3$), and riboflavin (B$_2$) is found in FAD.

Acetyl Coenzyme A and Coenzyme A

Another important energy exchanger containing a nucleotide is coenzyme A (CoA). The two forms of this compound are acetyl coenzyme A (high energy) and coenzyme A (low energy). Energy is released from acetyl coenzyme A when the C—S bond in the thioester functional group is hydrolyzed, producing acetate and coenzyme A. Coenzyme A forms acetyl coenzyme A when the S—H bond is oxidized (red H removed) and an acetyl group is added. CoA contains adenosine, three phosphates, and a pantothenic acid (vitamin B$_5$)–derived portion.

Sample Problem 12.3 Metabolic Nucleotides

Provide the abbreviation for the high-energy form of the following:
a. nicotinamide adenine dinucleotide
b. flavin adenine dinucleotide

Solution

The high-energy forms of these nucleotides are the reduced forms (contain H).
a. NADH
b. FADH$_2$

Practice Problems

12.9 Identify the metabolic nucleotide described by the following:

a. contains the vitamin riboflavin

b. contains a thioester functional group

12.10 Identify the metabolic nucleotide described by the following:

a. contains a form of the vitamin niacin

b. the main energy currency in the body

12.11 Identify the metabolic nucleotide described by the following:

a. exchanges energy when a phosphate bond is hydrolyzed

b. exchanges energy when a C—S bond is hydrolyzed

12.12 Using abbreviations (not structures), write the reaction of nicotinamide adenine dinucleotide that gives off energy $(-\Delta G)$.

12.13 Using abbreviations (not structures), write the reaction of flavin adenine dinucleotide that gives off energy $(-\Delta G)$.

12.14 Using abbreviations (not structures), write the reaction of coenzyme A that gives off energy $(-\Delta G)$.

12.3 Digestion—From Food Molecules to Hydrolysis Products

When food enters the body, it begins to break down in a process called **digestion.** In digestion, large fuel biomolecules are hydrolyzed into smaller hydrolysis products. These hydrolysis products are then able to be absorbed by the body and delivered to the cells, where they can be further catabolized.

Carbohydrates

Have you ever noticed a sweet taste in your mouth after eating a saltine cracker? This happens because the enzyme alpha-amylase, which is secreted in saliva, begins to digest the starch (amylose and amylopectin) in the cracker. This salivary amylase hydrolyzes some of the α-glycosidic bonds in the starch molecules, producing glucose, the disaccharide maltose, and oligosaccharides. The slightly sweet taste is due to the presence of glucose produced during starch hydrolysis.

Only monosaccharides are small enough to be transported through the intestinal wall and into the bloodstream. To complete the digestion of starch, enzymes in the small intestine hydrolyze starch into monosaccharides. The disaccharides lactose, sucrose, and maltose are also hydrolyzed in the small intestine. These hydrolysis reactions were first discussed in Section 5.5, and the digestion process is outlined in **FIGURE 12.5**.

We eat other carbohydrates like cellulose that cannot be digested because we lack the enzyme cellulase that hydrolyzes its β-glycosidic bonds. These indigestible fibers—found in whole grains, beans, and vegetables—are referred to as insoluble fibers. Although not useful as fuel for the body, they are important for a healthy digestive tract. Insoluble fiber stimulates the large intestine to help the body excrete waste.

The first step in breaking down carbohydrates, proteins, and dietary fats is hydrolysis.

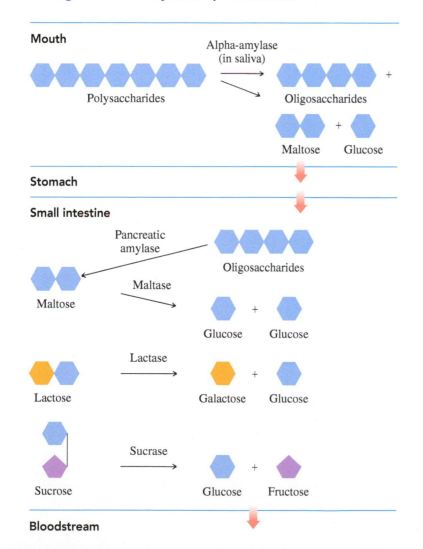

FIGURE 12.5 Digestion of carbohydrates. The digestion of carbohydrates begins in the mouth and is completed in the small intestine.

Fats

Dietary fats, or lipids, such as the triglycerides and cholesterol are nonpolar molecules, so their digestion in aqueous digestive juices is a little tricky. To assist in the digestion of dietary fats, a substance called *bile* is excreted from the gallbladder into the duodenum, the upper portion of the small intestine, during digestion. Bile contains amphipathic molecules called *bile salts*, which behave like soaps. They contain a nonpolar part and a polar part. The amphipathic bile salts place their nonpolar face toward the dietary fats and their polar face toward the water, forming micelles much like the soap micelles that break up a greasy dirt stain on clothing.

This process of breaking up larger nonpolar globules into smaller droplets (micelles) is called **emulsification.** The smaller micelles are able to move the dietary fats closer to the intestinal cell wall so that nonhydrolyzable lipids like cholesterol can be absorbed and hydrolyzable lipids like triglycerides can be hydrolyzed by the enzyme pancreatic lipase into free fatty acids and monoglycerides, which are then absorbed in the small intestine.

Once across the intestinal wall, free fatty acids and monoglycerides are reassembled as triglycerides. Cholesterol is linked to another free fatty acid, forming a cholesterol ester. The triglycerides and cholesterol esters are then repackaged with protein as a lipoprotein called a **chylomicron** (pronounced ky-lo-MY-kron). Chylomicrons ultimately transport triglycerides through the bloodstream to the tissues where they are used for energy production or are stored in the cells (see **FIGURE 12.6**).

FIGURE 12.6 Digestion of dietary fats. Hydrolyzable lipids like triglycerides react with pancreatic lipase in the small intestine, forming monoglycerides and free fatty acids for transport across the intestinal wall. They are reassembled into triglycerides, which combine with cholesterol ester and protein into droplets called chylomicrons that are delivered to the tissues.

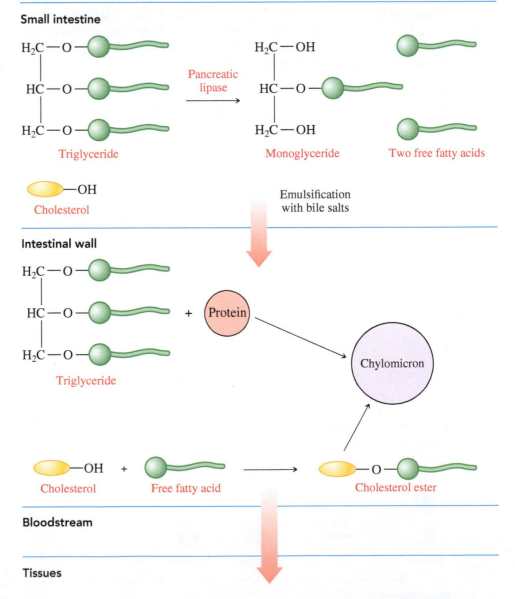

Proteins

Protein digestion begins in the stomach, where proteins are denatured (unfolded) by the acidic digestive juices. Digestive enzymes like pepsin, trypsin, and chymotrypsin hydrolyze peptide bonds as the denatured proteins begin their journey through the stomach and into the intestine. The resulting amino acids are absorbed through the intestinal wall into the bloodstream for delivery to the tissues. The digestion of proteins is depicted in **FIGURE 12.7**.

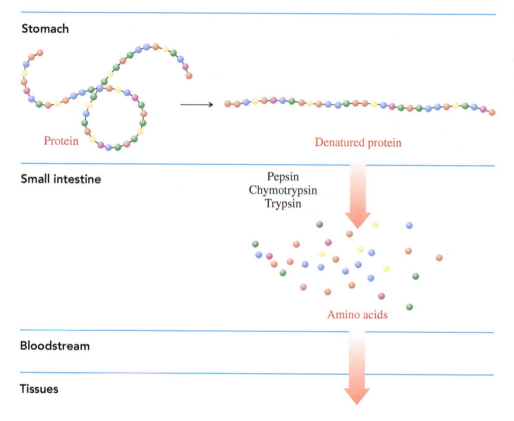

FIGURE 12.7 Digestion of proteins. Proteins are denatured in the stomach and hydrolyzed to amino acids in the small intestine, where they are absorbed into the bloodstream and delivered to the tissues.

Sample Problem 12.4 Digestion

Indicate the enzyme(s) involved in the digestion of each of the following:
   a. lactose
   b. triglyceride
   c. protein

Solution

   a. The disaccharide lactose is broken into the monosaccharides galactose and glucose by the enzyme lactase (see Figure 12.5).
   b. A triglyceride is broken down into two free fatty acids and a monoglyceride by an enzyme called pancreatic lipase (see Figure 12.6).
   c. A protein is broken up into component amino acids by digestive enzymes, some of which are pepsin, trypsin, and chymotrypsin (see Figure 12.7).

Practice Problems

12.15 Name a carbohydrate (if any) that undergoes digestion in each of the following sites:

a. mouth

b. stomach

c. small intestine

12.16 α-Amylase is produced in the _____ and it catalyzes _____.

12.17 Describe how cholesterol is packaged after absorption in the intestine.

12.18 Explain the role of bile salts in the digestion of fats.

12.19 Name the end products for digestion of proteins.

12.20 Name the end products for digestion of starch.

12.4 Glycolysis—From Hydrolysis Products to Common Metabolites

? 12.4 Inquiry Question

How is glucose catabolized through glycolysis?

The main source of fuel for the body is glucose, which is catabolized through a chemical pathway called glycolysis. Glucose is such an important fuel source that when there is not enough glucose entering the body during periods of sleeping or fasting, the body makes its own glucose through the anabolic process called **gluconeogenesis.**

As discussed in Section 10.5, glucose enters the cell from the bloodstream through a glucose transporter protein in the cell membrane. Inside the cell, glycolysis occurs in the liquid portion of the cytoplasm (cytosol) when the six-carbon monosaccharide glucose is broken down into two three-carbon molecules of pyruvate. Other hydrolysis products such as fructose, amino acids, and free fatty acids can also be used as fuel by the cells by entering glycolysis later in the pathway or by chemical conversion to one of the common metabolites.

The Chemical Reactions in Glycolysis

In the body, energy must be transferred in small amounts, minimizing the heat released in the process. Reactions that produce energy are coupled with reactions that require energy, thereby helping to maintain a constant body temperature. In glycolysis, energy is transferred through phosphate groups undergoing condensation and hydrolysis reactions.

There are ten chemical reactions in glycolysis that result in the formation of two molecules of pyruvate from one molecule of glucose. As shown in **FIGURE 12.8**, the first five reactions require an energy investment of two molecules of ATP, which are used to add two phosphate groups to the sugar molecule. This newly formed molecule is then split into two sugar phosphates. Reactions 6 through 10 generate two high-energy NADH molecules during the addition of two more phosphates and four ATP molecules when the four phosphates are removed from the sugar phosphates.

Through the ten reactions in glycolysis, one molecule of glucose is converted into two molecules of pyruvate. Because two molecules of ATP were originally required in the energy-investment phase and four are produced in the energy-generation phase, the net energy output for one molecule of glucose is two NADH and two ATP. The net chemical reaction is shown.

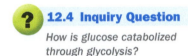

D-Glucose

$+ 2NAD^+ + 2ADP + 2P_i \longrightarrow$

$$2H_3C - \underset{\underset{O}{\|}}{C} - \underset{\underset{O}{\|}}{C} - O^- + 2NADH + 2H^+ + 2ATP$$

Pyruvate

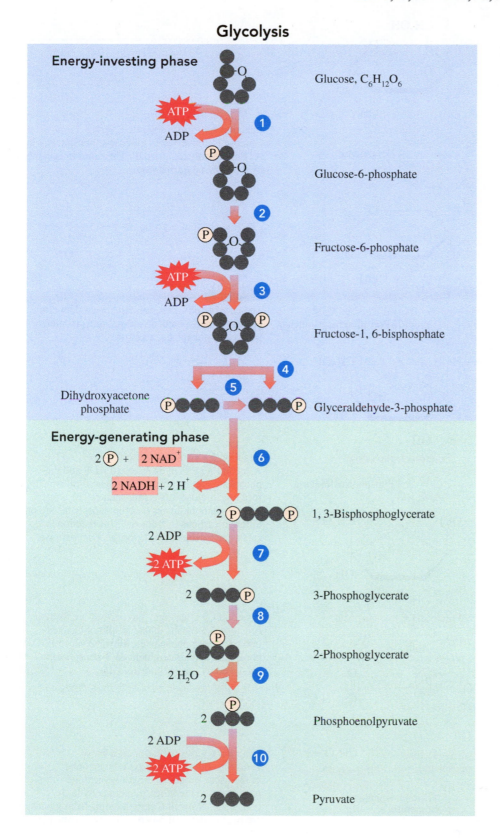

FIGURE 12.8 Schematic of glycolysis. In glycolysis, a six-carbon (gray circles) glucose is catabolized into two three-carbon pyruvate molecules. One molecule of glucose produces two ATP (net) and two NADH molecules.

The ten reactions in the pathway of glycolysis are detailed on the next two pages. Note that some of the reactions are reversible (reach an equilibrium), whereas others are irreversible. This is important for regulating the pathway.

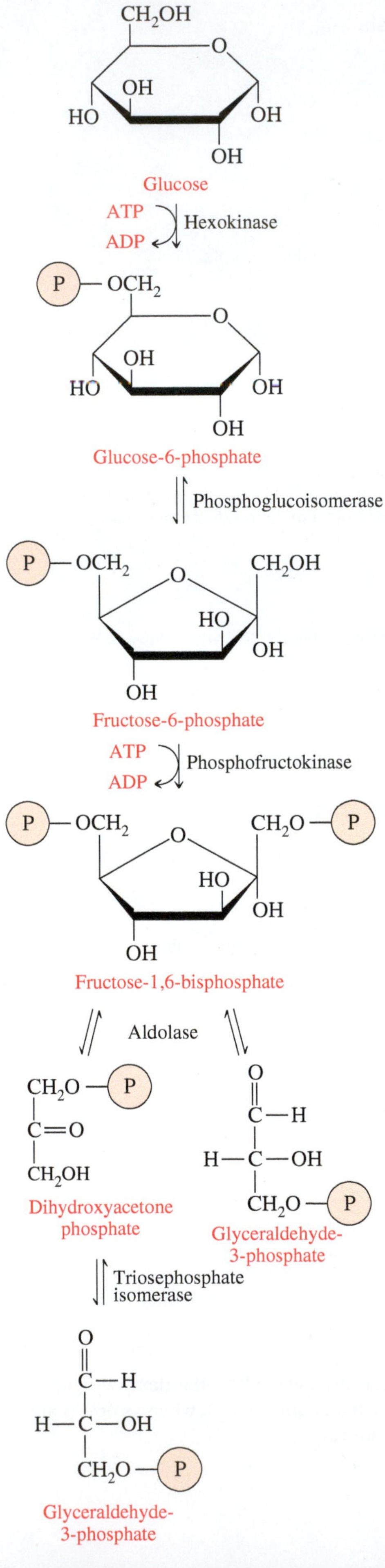

CH_2OH

Glucose

ATP ⟍ Hexokinase
ADP ↙

Reaction ❶ Phosphorylation: First ATP invested
Glucose is converted to glucose-6-phosphate when
ATP is hydrolyzed to ADP. The reaction is catalyzed
by the enzyme *hexokinase*.

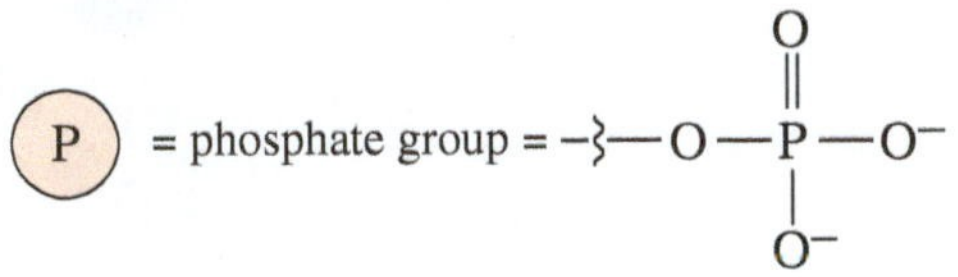

P — OCH_2

Glucose-6-phosphate

$$P = \text{phosphate group} = -\{-O-\overset{\displaystyle O}{\underset{\displaystyle O^-}{\overset{\|}{\underset{|}{P}}}}-O^-$$

‖ Phosphoglucoisomerase

Reaction ❷ Isomerization
The enzyme *phosphoglucoisomerase* converts
glucose-6-phosphate, an aldose, to its isomer,
fructose-6-phosphate, a ketose.

P — OCH_2 CH_2OH

Fructose-6-phosphate

ATP ⟍ Phosphofructokinase
ADP ↙

Reaction ❸ Phosphorylation: Second ATP invested
A second ATP is hydrolyzed to ADP and the
phosphate is transferred to fructose-6-phosphate,
forming fructose-1,6-bisphosphate. The word
bisphosphate indicates that the phosphates are on
different carbons in fructose. This reaction is
catalyzed by the enzyme *phosphofructokinase*.

P — OCH_2 CH_2O — P

Fructose-1,6-bisphosphate

‖ Aldolase ‖

Reaction ❹ Cleavage: Two trioses are formed
Fructose-1,6-bisphosphate is split or cleaved
into two triose phosphates, dihydroxyacetone
phosphate and glyceraldehyde-3-phosphate,
catalyzed by the enzyme *aldolase*.

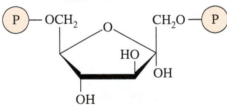

CH_2O — P O
$|$ $\|$
$C=O$ C — H
$|$ $|$
CH_2OH H — C — OH
                       $|$
Dihydroxyacetone CH_2O — P
phosphate
                  Glyceraldehyde-
                  3-phosphate

‖ Triosephosphate
isomerase

Reaction ❺ Isomerization of a triose
Triosephosphate isomerase converts one
of the triose products, dihydroxyacetone
phosphate, to the other, glyceraldehyde-3-
phosphate. Now all 6 carbon atoms from glucose
are in two identical 3-carbon triose phosphates.

O
$\|$
C — H
$|$
H — C — OH
$|$
CH_2O — P

Glyceraldehyde-
3-phosphate

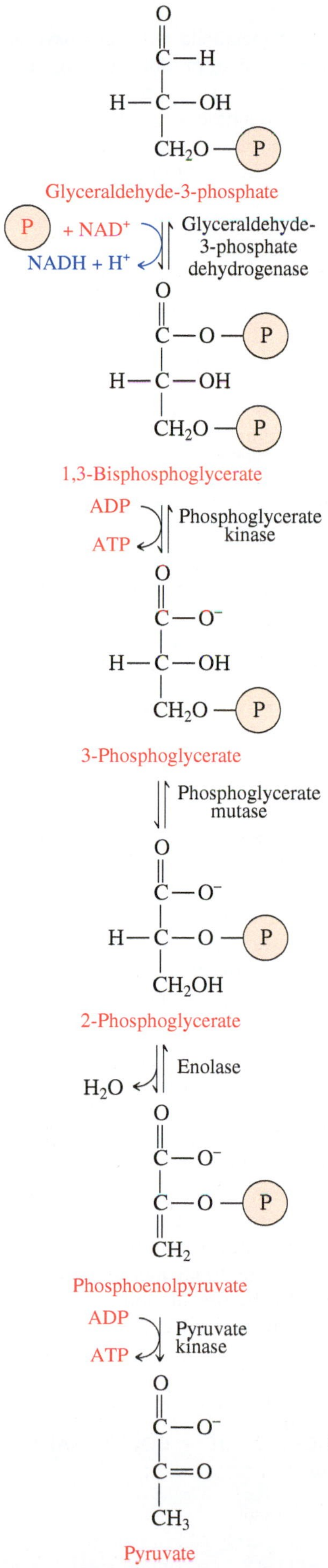

Glyceraldehyde-3-phosphate

Reaction 6 First energy production yields NADH
The aldehyde group of glyceraldehyde-3-phosphate is oxidized and phosphorylated by *glyceraldehyde-3-phosphate dehydrogenase*. The coenzyme NAD$^+$ is reduced to the high-energy compound NADH and H$^+$ in the process.

1,3-Bisphosphoglycerate

Reaction 7 Second energy production yields ATP
The energy-rich 1,3-bisphosphoglycerate now drives the formation of ATP when *phosphoglycerate kinase* transfers one phosphate from 1,3-bisphosphoglycerate to ADP.

3-Phosphoglycerate

Reaction 8 Formation of 2-phosphoglycerate
A *phosphoglycerate mutase* transfers the phosphate group from carbon 3 to carbon 2 to yield 2-phosphoglycerate.

2-Phosphoglycerate

Reaction 9 Removal of water makes a high-energy enol (alkene-alcohol)
An *enolase* catalyzes the removal of water to yield phosphoenolpyruvate, a high-energy compound that can transfer its phosphate in the next step.

Phosphoenolpyruvate

Reaction 10 Third energy production yields a second ATP
ATP is generated in this final reaction when the phosphate from phosphoenolpyruvate is transferred. The reaction is catalyzed by *pyruvate kinase*.

Pyruvate

Mastering Video
Practicing the Concepts
Glycolysis

Regulation of Glycolysis

As a dam regulates water output from a reservoir, most metabolic pathways have at least one reaction step that regulates the flow of reactants to final products through the pathway. Regulation steps are irreversible and are exergonic $(-\Delta G)$. The main step of regulation in glycolysis is step 3. The enzyme phosphofructokinase, which catalyzes the phosphorylation of fructose-6-phosphate to fructose-1,6-bisphosphate, is heavily regulated by the cells. ATP acts as an inhibitor of phosphofructokinase. If cells have plenty of ATP, glycolysis slows down.

Sample Problem 12.5 **Glycolysis**

If one NADH is generated in step 6 of glycolysis, how are a total of two NADH generated from one molecule of glucose?

Solution

Two molecules of glyceraldhyde-3-phosphate (three-carbon molecules) are produced from one molecule of glucose, so this reaction occurs two times for every molecule of glucose going through the pathway.

The Fates of Pyruvate

The metabolite pyruvate generated from glycolysis can be catabolized further, producing more energy for the body. The fate of pyruvate depends on the availability of oxygen in the cell. When ample oxygen is available (referred to as **aerobic** conditions), pyruvate is oxidized further to acetyl coenzyme A. When oxygen is in short supply (referred to as **anaerobic** conditions), pyruvate is reduced to lactate.

During times of strenuous exercise, oxygen is in short supply in the muscles.

Aerobic Conditions

When oxygen is readily available in the cell, pyruvate produces more energy for the cell. When pyruvate breaks down further, the carboxylate functional group of pyruvate is liberated as CO_2 during a process called **oxidative decarboxylation.** "Oxidative" tells us that this reaction is an oxidation–reduction reaction. In this case, the metabolite coenzyme A is oxidized (H removed) to its high-energy form acetyl CoA, and the NAD^+ is reduced to NADH. "Decarboxylation" tells us that a carboxylate (COO^-) is removed as CO_2. Here the pyruvate is decarboxylated, producing the two-carbon acetyl group. The acetyl group binds to coenzyme A through a sulfur atom, creating a thioester functional group during the oxidation, forming the metabolite acetyl CoA. This reaction occurs in the mitochondria.

Anaerobic Conditions

Under anaerobic conditions, the middle carbonyl in pyruvate is reduced (hydrogen added) to an alcohol functional group, and lactate is formed. The hydrogen (and energy) required for this reaction is supplied by NADH and H^+, producing NAD^+. Because no ATP is generated by pyruvate under anaerobic conditions, the NAD^+ produced funnels back into glycolysis to oxidize more glyceraldehyde-3-phosphate (step 6), providing a small but necessary amount of ATP. This reaction occurs in the cytoplasm.

A familiar single-celled organism, yeast, converts pyruvate to ethanol under anaerobic conditions. This process is called **fermentation.** In the preparation of alcoholic beverages like wine, yeast produces pyruvate from glucose in grape juices and, under low-oxygen conditions, transforms pyruvate into ethanol.

As wine ferments in the cask, yeast forms a layer at the top, producing ethanol in the liquid mixture.

$$H_3C-\overset{\overset{\displaystyle O}{\|}}{C}-\overset{\overset{\displaystyle O}{\|}}{C}-O^- + NADH + 2H^+ \longrightarrow H_3C-\overset{\overset{\displaystyle OH}{|}}{\underset{\underset{\displaystyle H}{|}}{C}}-H + CO_2 + NAD^+$$

Pyruvate   Ethanol

FIGURE 12.9 provides a summary of the aerobic and anaerobic fates of pyruvate.

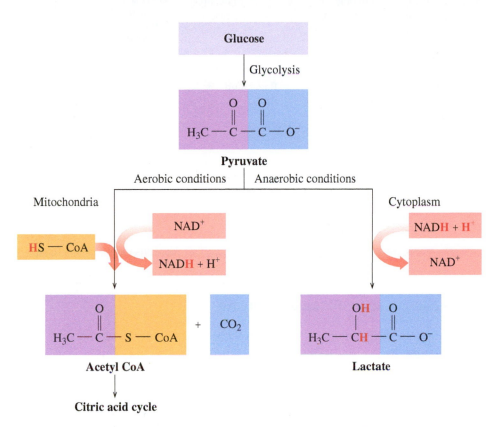

FIGURE 12.9 The fates of pyruvate. Pyruvate is converted to acetyl CoA under aerobic conditions and to lactate under anaerobic conditions.

Fructose and Glycolysis

Can fructose enter glycolysis to produce energy? As we saw in Chapter 6, fructose tastes sweeter than an equivalent amount of glucose, so less can be used to produce the desired sweetness. In fact, refined sugars like high-fructose corn syrup use more fructose and less glucose to reach a given level of sweetness.

Fructose is readily taken up in the muscle and liver. In the muscles, it is converted to fructose-6-phosphate, entering glycolysis at step 3 (refer to glycolysis reactions earlier in this section). In the liver, it is enyzmatically converted to the trioses dihydroxyacetone phosphate and glyceraldehyde-3-phosphate used in step 5 of glycolysis. Any fructose that enters a cell must flow from reaction 5 through 10. Because fructose uptake by the cells is not regulated by insulin like glucose uptake is, all fructose in the bloodstream enters the cells and is forced into catabolism.

Remember that glycolysis is regulated earlier in the pathway at step 3. The triose products created by fructose in the liver provide an excess of reactants that funnel into step 6, creating excess pyruvate and acetyl CoA that, if not required for energy by the cells, is converted to fat.

Food products containing excessive amounts of fructose, like sugary drinks, are harmful because the fructose enters the bloodstream (and then the cells) more quickly than fructose found in fruits does. Fructose in fruits is in combination with fiber, which slows sugar absorption. Ingesting high levels of fructose can lead to excess end products in glycolysis and, eventually, fat synthesis. Maintaining a healthy balance of sugars in combination with fiber is necessary for preventing catabolic overload of sugar in the cells.

INTEGRATING
Chemistry

◀ Find out why eating fructose is not the same as eating glucose.

—Continued next page

Continued—

In the liver, the breakdown of glucose is regulated, while the breakdown of fructose is not. All fructose flows into glycolysis after the bottleneck (the regulation step), allowing all of it to be converted to end products. The result can be excess products, which, when unnecessary, are converted to fat.

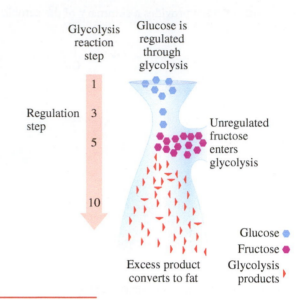

Sample Problem 12.6   Fates of Pyruvate

Is each of the following products from pyruvate produced under anaerobic or aerobic conditions?

    a. acetyl CoA   b. lactate   c. ethanol

Solution

    a. aerobic conditions   b. anaerobic conditions   c. anaerobic conditions

Practice Problems

12.21 Name the starting reactant of glycolysis.

12.22 Name the end product of glycolysis.

12.23 How many ATP molecules are produced when one molecule of glucose undergoes glycolysis?

12.24 How many ATP molecules are invested in the energy-investing stage of glycolysis for one molecule of glucose?

12.25 In terms of high-energy molecules, what is the net output for one molecule of glucose undergoing glycolysis?

12.26 How many NADH molecules are produced when one molecule of glucose undergoes glycolysis?

12.27 Name the coenzyme produced during the conversion of pyruvate to lactate.

12.28 Name the coenzyme produced during the conversion of pyruvate to acetyl CoA.

12.29 Name the products of fermentation.

12.30 The formation of lactate permits glycolysis to continue under anaerobic conditions. Explain.

12.31 What is the main regulation point of glycolysis? How does this explain why fructose is readily converted into fat?

12.32 Explain how the catabolism of fructose differs from that of glucose.

12.5 The Citric Acid Cycle—Central Processing

? 12.5 Inquiry Question

What types of reactions occur during the citric acid cycle?

We have just seen how glucose begins its catabolism in the cell by breaking into two pyruvate molecules. What about amino acids and fatty acids? During aerobic catabolism when O_2 is present, all three of these biomolecules funnel into and out of the citric acid cycle. The **citric acid cycle** is a series of reactions that degrades the two-carbon acetyl groups from the metabolite acetyl CoA into CO_2 and generates the high-energy reduced molecules NADH and $FADH_2$. This cycle is also known as the Krebs cycle or the tricarboxylic acid cycle. We will use the term *citric acid cycle* in this book.

The name *citric acid cycle* originated because the first step in the cycle involves the formation of the six-carbon molecule citrate. Citrate is the conjugate base of citric acid and the predominant form of this weak acid at physiological pH. This formation of citrate that starts the cycle is a condensation reaction between an entering acetyl CoA molecule and the four-carbon molecule oxaloacetate. The six-carbon molecule citrate loses two carbons sequentially as CO_2, first forming the five-carbon molecule α-ketoglutarate and then forming the four-carbon molecule succinyl CoA. These carbon–carbon bond-breaking reactions transfer energy and produce NADH from the coenzyme NAD^+. Succinyl CoA then runs through a set of reactions regenerating oxaloacetate, and the cycle begins again. The citric acid cycle is summarized in **FIGURE 12.10**.

Reactions of the Citric Acid Cycle

There are eight reactions in the citric acid cycle. Each is catalyzed by an enzyme. These reactions occur in the mitochondrial matrix, deep within the mitochondria (see Figure 12.4). The cycle starts with the addition of the two-carbon metabolite acetyl CoA. Recall from the earlier section that acetyl CoA can be produced from pyruvate, the end product of glycolysis. In Section 12.8, we will see that acetyl CoA can also be produced from the catabolism of fatty acids. The eight reactions are shown in **FIGURE 12.11** and described here, beginning with the formation of citrate.

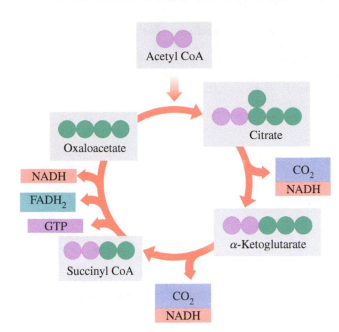

FIGURE 12.10 Fundamentals of the citric acid cycle. The two carbons of acetyl CoA (purple circles) condense with the four carbons of oxaloacetate (green circles) to form citrate. Two carbons are sequentially removed from citrate as CO_2, ultimately re-forming oxaloacetate. During the cycle the high-energy nucleotides NADH, $FADH_2$, and GTP are produced.

Reaction ❶ Formation of Citrate
In reaction 1, the acetyl group from acetyl CoA (two carbons) combines with oxaloacetate (four carbons), forming citrate (six carbons) and CoA.

Reaction ❷ Isomerization to Isocitrate
If you examine the reactant and product of reaction 2, you will see that the two molecules are structural isomers. The —OH and one of the H atoms have been swapped in citrate to form isocitrate. Citrate contains a tertiary alcohol, whereas isocitrate contains a secondary alcohol (see Section 6.2). A secondary alcohol can be further oxidized to a carbonyl, whereas a tertiary alcohol cannot. This rearrangement is necessary because isocitrate is oxidized in the next reaction.

Mastering Video
Practicing the Concepts
The Citric Acid Cycle

Reaction ❸ First Oxidative Decarboxylation (Release of CO_2)
Similar to the decarboxylation of pyruvate to acetyl CoA discussed earlier, reaction 3 is an oxidative decarboxylation. In this reaction, the secondary alcohol isocitrate undergoes oxidation (two hydrogens removed) to a ketone called α-ketoglutarate. A corresponding reduction reaction also takes place. (Remember that oxidation and reduction always occur together.) The coenzyme NAD^+ is reduced to NADH, accepting the proton and electrons removed during the oxidation. The six-carbon isocitrate is decarboxylated to the five-carbon α-ketoglutarate.

Reaction ❹ Second Oxidative Decarboxylation
A second decarboxylation occurs in this step, and a four-carbon molecule is produced. In this reaction, a second oxidation–reduction reaction also takes place. The thiol group of CoA is oxidized (in this case, loses a hydrogen), and a second NAD^+ is reduced to NADH. Alpha-ketoglutarate (five carbons) is decarboxylated into a succinyl group (four carbons). The CoA is bonded to the succinyl group, thus producing succinyl CoA.

Reaction ❺ Hydrolysis of Succinyl CoA
In this step, succinyl CoA undergoes hydrolysis to succinate and coenzyme A. The energy produced during the hydrolysis of the thioester is transferred to produce the high-energy nucleotide guanosine triphosphate or GTP from GDP and P_i. In the cell, GTP is readily converted to ATP through the transfer of a phosphate group.

$$GTP + ADP \longrightarrow GDP + ATP$$

FIGURE 12.11 The reactions of the citric acid cycle. In one turn of the citric acid cycle, oxidative catabolism produces two CO_2 molecules, the reduced coenzymes NADH and $FADH_2$, and a GTP molecule.

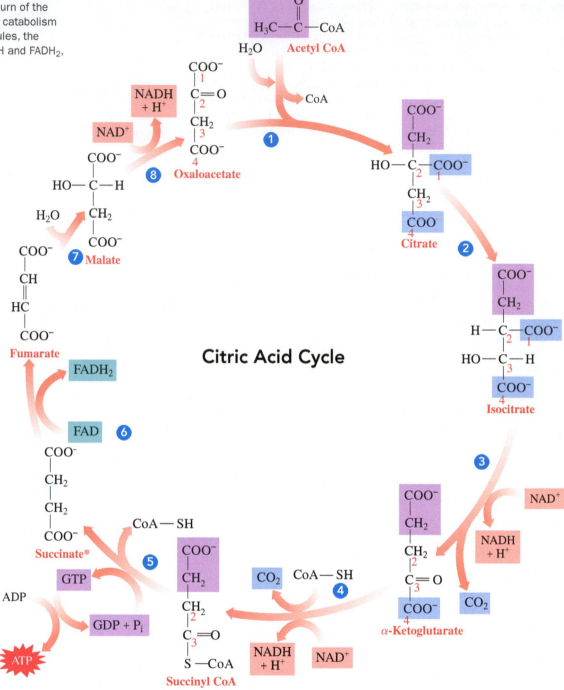

* Succinate is a symmetrical compound, so we can no longer track the acetyl group that entered the cycle as acetyl CoA.

Reaction ⑥ Dehydrogenation of Succinate
In this oxidation, one hydrogen is eliminated from each of the two central carbons of succinate, forming a trans C=C bond, thus producing fumarate. These two hydrogens reduce the coenzyme FAD to $FADH_2$.

Reaction ⑦ Hydration of Fumarate
Water adds to the trans double bond of fumarate as —H and —OH in a hydration reaction, forming malate.

Reaction ⑧ Oxidation of Malate
As in reaction 3, the secondary alcohol on malate is oxidized to a ketone, forming oxaloacetate, providing protons and electrons to reduce a third NAD^+ to NADH.

Citric Acid Cycle: Summary

One turn of the citric acid cycle produces a net energy output of three NADH, one $FADH_2$, and one GTP (which forms ATP). Two CO_2 and one CoA are also produced. Because the reactants in the cycle are regenerated, the net reaction for one turn of this eight-step cycle is

$$\text{Acetyl CoA} + 3NAD^+ + FAD + GDP + P_i + 2H_2O \longrightarrow$$
$$2CO_2 + 3NADH + 3H^+ + FADH_2 + CoA + GTP$$

Sample Problem 12.7 **Citric Acid Cycle**

When one acetyl CoA completes the citric acid cycle, how many of each of the following are produced?
   a. NADH   b. CO_2 c. $FADH_2$

Solution
   a. One turn of the citric acid cycle produces three molecules of NADH.
   b. Two molecules of CO_2 are produced by one turn of the citric acid cycle.
   c. One turn of the citric acid cycle produces one molecule of $FADH_2$.

Practice Problems

12.33 Name the compounds required to start the citric acid cycle.

12.34 Name the products of one turn of the citric acid cycle.

12.35 Name the reactions in the citric acid cycle that involve oxidative decarboxylation.

12.36 Name the reaction in the citric acid cycle that is a condensation.

12.37 Name the reactions of the citric acid cycle that reduce NAD^+.

12.38 Name the reaction of the citric acid cycle that reduces FAD.

12.39 If one molecule of glucose were to proceed through the citric acid cycle, how many ATP, NADH, and $FADH_2$ would be produced?

12.40 If one molecule of glucose were to proceed through the citric acid cycle, how many molecules of CO_2 would be produced?

12.6 Electron Transport and Oxidative Phosphorylation

Let's review the number of ATP molecules produced from the metabolism of a single molecule of glucose running through glycolysis and entering two turns of the citric acid cycle as acetyl CoA. Two ATP are produced in glycolysis, and two ATP are produced in the two turns of the citric acid cycle (as GTP).

    Organisms undergoing aerobic catabolism can produce more energy, but so far, we have not seen a substantial number of ATP produced. Just two ATP are produced in glycolysis and two ATP in the citric acid cycle. Where is all the energy? Remember that high-energy reduced forms of the nucleotides NADH and $FADH_2$ are also produced in glycolysis (two NADH per glucose), from pyruvate oxidation to acetyl CoA (two NADH per glucose), and in the citric acid cycle (six NADH and two $FADH_2$ per glucose).

 12.6 Inquiry Question

How is the energy in NADH and $FADH_2$ transformed into ATP?

Glycolysis	2 NADH and 2 ATP
Oxidation of 2 pyruvate	2 NADH
Citric acid cycle (2 acetyl CoA)	6 NADH, 2 $FADH_2$, and 2 ATP

How do these reduced nucleotides transfer energy? These high-energy reduced nucleotides (NADH and FADH$_2$) transfer their electrons and hydrogens through the inner mitochondrial membrane and combine with oxygen to form H$_2$O. The energy generated as a result of this process is captured and used to drive the reaction of ADP + P$_i$ to form ATP. The process of producing ATP using the energy from the oxidation of reduced nucleotides is called **oxidative phosphorylation.**

Electron Transport

Mitochondria are the ATP factories of the cell. Reduced nucleotides from the citric acid cycle are produced here, and their energy upon oxidation is used to generate ATP. The reactions of the citric acid cycle occur in the matrix of the mitochondria (refer to Figure 12.4), and NADH and FADH$_2$ begin their journey through the inner membrane here.

A set of enzyme complexes, commonly called complexes I through IV, are embedded in the inner membrane of the mitochondria, separating the matrix from the intermembrane space. This membrane also contains a set of electron carriers that transport the electrons and protons of NADH and FADH$_2$ through the inner mitochondrial membrane. Two of the electron carriers, coenzyme Q and cytochrome *c*, are not firmly attached to any one complex and serve to shuttle electrons between the complexes (see **FIGURE 12.12**). We examine each enzyme complex individually starting with complex I to see how electrons shuttle through the complexes.

FIGURE 12.12 The electron transport chain. Electrons flow to molecular oxygen (O$_2$), ultimately producing water through a series of electron carriers embedded in protein complexes in the inner mitochondrial membrane. The complete structure of a mitochondrion is shown in Figure 12.4.

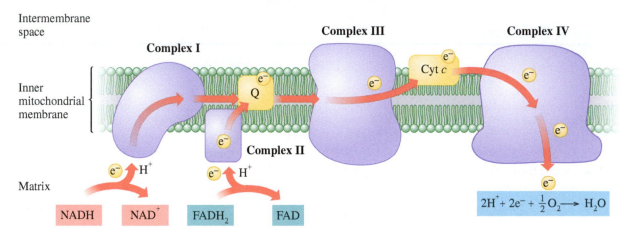

$$Q + 2H^+ + 2e^- \rightleftharpoons QH_2$$

Quinone

Oxidized coenzyme Q (Q)

Hydroquinone

Reduced coenzyme Q (QH$_2$)

Oxidation–reduction reactions for coenzyme Q

Complex I NADH Dehydrogenase
At complex I, NADH enters electron transport. During its oxidation at this complex, also called NADH dehydrogenase, two electrons and two protons are transferred to the electron transporter coenzyme Q, reducing its two ketone groups to alcohols (see figure at left). NAD$^+$ is regenerated and returns to a catabolic pathway as in the citric acid cycle. The overall reaction at complex I is

$$NADH + H^+ + Q \longrightarrow NAD^+ + QH_2$$

Complex II Succinate Dehydrogenase
FADH$_2$ enters electron transport at complex II, also known as succinate dehydrogenase. This enzyme is also part of the citric acid cycle. FADH$_2$ is produced when succinate is converted to fumarate, providing this entry point to electron transport for FADH$_2$. Two electrons and two protons from FADH$_2$ are also transferred to a coenzyme Q to yield QH$_2$, The overall reaction at complex II is

$$FADH_2 + Q \longrightarrow FAD + QH_2$$

Complex III Coenzyme Q Cytochrome *c* Reductase
At complex III, the reduced coenzyme Q (QH$_2$) molecules formed in complex I or II are reoxidized to coenzyme Q (Q) via the enzyme coenzyme Q cytochrome *c* reductase,

and the electrons pass through a series of electron acceptors until they arrive as single electrons in the mobile protein cytochrome c, which moves the electron from complex III to complex IV.

Complex IV Cytochrome c Oxidase

At complex IV, also called cytochrome c oxidase, single electrons are transferred from cytochrome c through another set of electron acceptors to combine with hydrogen ions and oxygen (O_2), forming water. This is the final stop for the electrons.

$$2H^+ + 2e^- + \frac{1}{2}O_2 \longrightarrow H_2O$$

Sample Problem 12.8 **Electron Transport**

Give the abbreviation for each of the following electron transporters:
  a. the reduced form of flavin adenine dinucleotide
  b. the reduced form of coenzyme Q

Solution
  a. $FADH_2$ b. QH_2

Oxidative Phosphorylation

How does the movement of electrons through a set of enzyme complexes in the inner membrane generate ATP for the cell? Biochemist Peter Mitchell first proposed the **chemiosmotic model** (**FIGURE 12.13**), linking electron transport to the generation of a proton (H^+) difference, or *gradient*, on either side of the inner membrane and the resulting production of ATP. In this model, three of the complexes (I, III, and IV) span the inner membrane pumping (relocating) protons out of the matrix and into the intermembrane space while electrons shuttle through the complexes. These protons are not generated directly from the oxidation of NADH and $FADH_2$ but are simply present in the matrix.

A difference in both charge (electrical) and concentration (chemical) of protons on either side of the membrane results in an **electrochemical gradient.** The formation of the proton gradient across the inner mitochondrial membrane provides the energy for ATP synthesis.

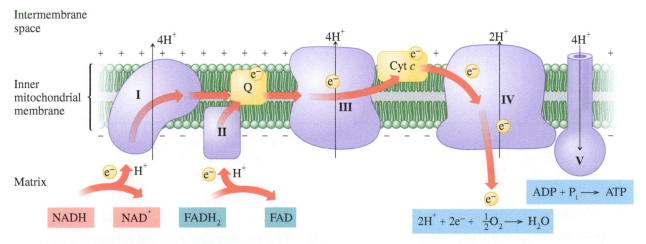

FIGURE 12.13 The chemiosmotic model. Electron flow is accompanied by proton (H^+) transfer across the inner mitochondrial membrane, producing an electrochemical gradient. Protons can only reenter the matrix through a proton-specific channel, complex V. This movement of protons through complex V provides the energy for ATP synthesis.

The inner mitochondrial membrane is a very tight membrane that is impermeable even to small particles like protons. The only way for protons to move back into the matrix is through protein complex V, called ATP synthase. The movement of protons from an area with many protons to an area with fewer protons releases energy. As protons flow back into the matrix through complex V, the resulting release of energy drives the synthesis of ATP.

$$ADP + P_i + Energy \longrightarrow ATP$$

Can you see why this ATP synthesis is called oxidative phosphorylation? It is *oxidative* because the energy driving the reaction is produced by oxidation reactions. *Phosphorylation* comes from the reaction type, since a phosphate group is added to ADP, forming ATP.

INTEGRATING Chemistry

Find out why brown ▶ fat is brown.

Thermogenesis—Uncoupling ATP Synthase

If the protons pumped during electron transport and their return to the matrix during ATP production are disrupted, ATP is not produced. If electron transport and ATP synthesis are uncoupled, the energy that would have been harnessed as ATP is released as heat. This generation of heat in the body is called **thermogenesis**.

Some animals adapted to cold climates produce *thermogenin*, a protein that uncouples electron transport and oxidative phosphorylation, allowing protons to leak back into the matrix, generating heat. Mammals like bears can regulate their body temperature during hibernation in part through thermogenesis. Such animals have a higher amount of a tissue called *brown fat*, which appears brown due to the high concentrations of mitochondria present. The cytochrome molecules present in mitochondria contain iron ions responsible for the brown color. Newborn babies have higher levels of brown fat than do adults because newborns do not have much body mass to maintain body temperature. Brown fat deposits are located near major blood vessels that carry the warmed blood through the body, allowing a newborn to generate more heat to warm its body surface.

Sample Problem 12.9 Chemiosmotic Model

Provide another name for complex V and describe its function.

Solution

Complex V is also known as ATP synthase. It provides a channel for H^+ to flow back into the matrix (after being pumped out by complexes I, III, and IV) and is the site for ATP synthesis, which occurs in the matrix.

Practice Problems

12.41 Name the electron carrier that transports electrons from complex I to complex III.

12.42 Name the electron carrier that transports electrons from complex III to complex IV.

12.43 Name the complex where water is produced.

12.44 Name the complex where $FADH_2$ enters electron transport.

12.45 How many electrons can be transported by the mobile electron carrier coenzyme Q? By cytochrome *c*?

12.46 How do the structures of the complexes I–V differ from the structure of coenzyme Q?

12.47 According to the chemiosmotic theory, how does the proton gradient provide energy to synthesize ATP?

12.48 How is the proton gradient established?

12.49 What reaction occurs during oxidative phosphorylation?

12.50 Where in the mitochondria is ATP synthesized?

DISCOVERING THE CONCEPTS

❓ INQUIRY ACTIVITY—ATP Production

The reduced nucleotides NADH and $FADH_2$ are converted to ATP via the electron transport chain and oxidative phosphorylation. The most widely accepted values among biochemists are shown.

Nucleotide Input	ATP Output
NADH	2.5
$FADH_2$	1.5

Questions

1. The metabolites for one molecule of glucose through glycolysis are shown below. Taking into consideration the conversion of reduced nucleotides to ATP, what is the total number of ATP produced by one glucose completing glycolysis?

$$Glucose \longrightarrow 2pyruvate + 2ATP + 2NADH$$

2. The conversion of pyruvate to acetyl CoA is shown below for one molecule of glucose. Taking into consideration the conversion of reduced nucleotides to ATP, what is the total number of ATP produced by one glucose (two pyruvate) completing this conversion?

$$2Pyruvate \longrightarrow 2acetyl\ CoA + 2NADH$$

3. Each acetyl CoA entering the citric acid cycle produces two CO_2, three NADH, one $FADH_2$, and one GTP (converted directly to one ATP in the cell). Taking into consideration the conversion of reduced nucleotides to ATP, what is the total number of ATP produced by one glucose (two acetyl CoA) going through the citric acid cycle?

4. Taking your answers to questions 1–3 into account, what is the total number of ATP produced during the complete oxidative catabolism of one molecule of glucose?

12.7 ATP Production

Oxidative phosphorylation couples the energy of electron transport to proton pumping and finally to ATP synthesis. How many ATP are synthesized for each reduced nucleotide entering electron transport? Because NADH enters the electron transport chain at complex I and $FADH_2$ enters at complex II, the number of protons pumped and the number of ATP produced for these two molecules differ. Ten H^+ are pumped into the inner membrane space for every NADH entering electron transport, and six H^+ are pumped into the inner membrane space for each $FADH_2$ transported (refer to Figure 12.13). The most widely accepted value for the number of H^+ flowing back into the matrix to synthesize one ATP is four. So, the number of ATP synthesized per NADH is 2.5, and the number of ATP synthesized per $FADH_2$ is 1.5.

❓ 12.7 Inquiry Question

How many ATP can be produced during oxidative catabolism?

Nucleotide Input	Protons (H^+) Pumped	ATP Output
NADH	10	2.5
$FADH_2$	6	1.5

Sample Problem 12.10 **ATP Production**

Why does the oxidation of NADH provide energy for the formation of 2.5 ATP molecules, whereas $FADH_2$ produces 1.5 ATP?

Solution

Electrons from the oxidation of NADH enter electron transport earlier through complex I than do those from $FADH_2$. For every four protons pumped through the inner membrane, one ATP can be synthesized. One NADH pumps 10 protons into the intermembrane, which can synthesize 2.5 ATP, whereas one $FADH_2$ pumps six protons, forming 1.5 ATP.

Counting ATP from One Glucose

How many ATP molecules can be generated from one molecule of glucose undergoing complete catabolism? As noted in Section 12.1, the end products for the oxidative catabolism for any fuel (foodstuff) are CO_2 and H_2O. For glucose ($C_6H_{12}O_6$), the overall balanced oxidation equation is

$$C_6H_{12}O_6 + 6O_2 \longrightarrow 6CO_2 + 6H_2O + Energy$$

This equation may look the same as one seen previously for the combustion of glucose. While the net reaction is the same, in the body there are many more steps that occur. The energy is not burned away as in combustion, but stored in nucleotides and ATP for future use. Let's review the pathways of the biochemical oxidation of glucose and count the total number of ATP produced.

Glycolysis

In glycolysis, the oxidation of glucose produces two NADH molecules and two ATP molecules. Recall that glycolysis occurs in the *cytoplasm*, and electron transport draws NADH from the matrix inside the *mitochondria*. The two NADH from glycolysis must be shuttled into the matrix to enter electron transport. The direct shuttling of two NADH into the matrix results in the production of five ATP.

$$Glucose \longrightarrow 2pyruvate + 2ATP + 2NADH \ (5ATP)$$

Oxidation of Pyruvate

After glycolysis, the two pyruvates enter the mitochondria, where they are further oxidized to produce a total of two acetyl CoA, two CO_2, and two NADH. The oxidation of two pyruvates thus leads to the production of five ATP.

$$2Pyruvate \longrightarrow 2acetyl \ CoA + 2NADH \ (5ATP)$$

Citric Acid Cycle

The two acetyl CoA produced from the two pyruvate next enter the citric acid cycle. Each acetyl CoA entering the cycle yields two CO_2, three NADH, one $FADH_2$, and one GTP (converted directly to ATP). Because two acetyl CoA enter the cycle from one glucose, a total of six NADH, two $FADH_2$, and two ATP is produced.

6 NADH	$\longrightarrow$	15 ATP
2 $FADH_2$	$\longrightarrow$	3 ATP
Directly in pathway	$\longrightarrow$	2 ATP

Total ATP for two acetyl CoA: 20 ATP

Total ATP from Glucose Oxidation

By summing the ATP produced from glycolysis, oxidation of pyruvate, and the citric acid cycle, a net number of ATP produced from one glucose can be estimated. A summary is shown in **TABLE 12.4** and diagrammed in **FIGURE 12.14**.

TABLE 12.4 ATP Produced by the Complete Oxidation of Glucose

Pathway	Reduced Nucleotides Produced	ATP Yield
Glycolysis (produced directly in pathway)	2 NADH$_{cytoplasm}$ $\longrightarrow$ 2 NADH$_{matrix}$	5 ATP
		2 ATP
2 Pyruvate $\rightarrow$ 2 acetyl CoA	2 NADH	5 ATP
Citric acid cycle	6 NADH	15 ATP
Two turns of the cycle accommodate two acetyl CoA (produced as GTP in pathway)	2 FADH$_2$	3 ATP
		2 ATP
	TOTAL ATP:	**32 ATP**

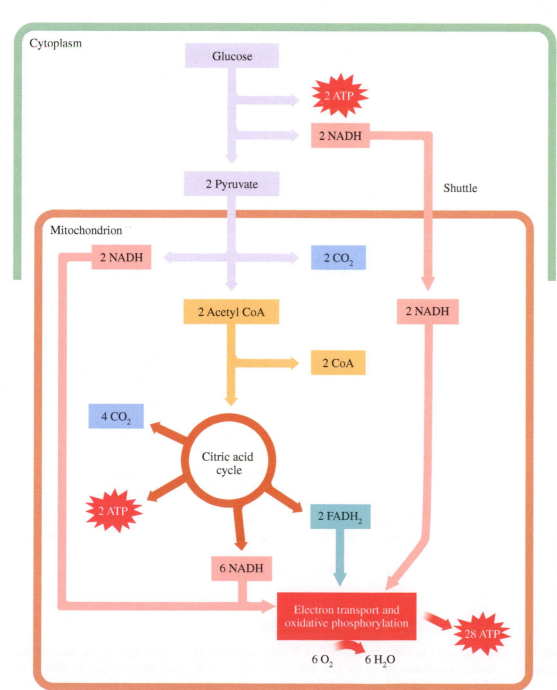

FIGURE 12.14 Complete glucose oxidation. The complete oxidation of glucose to CO_2 and H_2O yields a total of 32 ATP. This accounts for 2 ATP from glycolysis, 2 ATP from the citric acid cycle, and 28 ATP from the oxidation of NADH and FADH$_2$ during oxidative phosphorylation. The arrow coloration shows the fate of the major metabolites.

Sample Problem 12.11 ATP Production

Indicate the total number of ATP produced by the following oxidations:

a. one pyruvate to one acetyl CoA
b. one acetyl CoA turning through the citric acid cycle

Solution

a. The oxidation of pyruvate to acetyl CoA produces one NADH, which yields 2.5 ATP.
b. One turn of the citric acid cycle produces

$$\begin{array}{ccccc} 1\text{ GTP} & + & 3\text{ NADH} & + & 1\text{ FADH} \\ \downarrow & & \downarrow & & \downarrow \\ 1\text{ ATP} & + & 7.5\text{ ATP} & + & 1.5\text{ ATP} \\ & & \downarrow & & \\ & & 10\text{ ATP} & & \end{array}$$

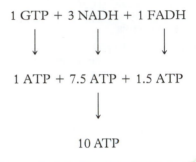

Practice Problems

12.51 List the energy yield in ATP molecules for each of the following:

a. $\text{NADH} \longrightarrow \text{NAD}^+$
b. $\text{Glucose} \longrightarrow \text{2pyruvate}$
c. $\text{2Pyruvate} \longrightarrow \text{2acetyl CoA} + 2CO_2$
d. $\text{Acetyl CoA} \longrightarrow 2CO_2$

12.52 List the energy yield in ATP molecules for each of the following:

a. $FADH_2 \longrightarrow \text{FAD}$
b. $\text{Glucose} + 6O_2 \longrightarrow 6CO_2 + 6H_2O$
c. $\text{Glucose} \longrightarrow \text{2lactate}$
d. $\text{Pyruvate} \longrightarrow \text{lactate}$

DISCOVERING THE CONCEPTS

INQUIRY ACTIVITY—Fatty Acid Catabolism (Beta Oxidation)

Information

Examine Figure 12.15 to answer the questions below.

Questions

1. How is the fatty acyl CoA at the top of the cycle different from the fatty acyl CoA at the bottom of the cycle? (*Acyl* is used instead of *acid* because the fatty acid is attached to another molecule, in this case a CoA.)
2. How many reduced nucleotides are produced in one cycle? Name them.
3. How many acetyl CoA molecules are produced in one cycle of oxidation?
4. a. How many cycles of beta oxidation would it take to catabolize a stearic acid molecule (a fatty acid, [18:0]) into all acetyl CoA units?
   b. How many acetyl CoA molecules would be produced?
   c. How many reduced nucleotides would be produced?
5. If a molecule of glucose produces a net 32 ATP when completely catabolized, which do you think will produce more energy, one molecule of glucose or one molecule of stearic acid? Justify your answer.

12.8 Other Fuel Choices

12.8 Inquiry Question

How do fatty acids and amino acids produce ATP?

Glucose is our main source of fuel. When there is more than enough to take care of the energy requirements of the cell, glucose is stored in the liver and muscles as glycogen. Hydrolysis of glycogen (called *glycogenolysis*) produces glucose when glucose concentrations are low (during sleeping and fasting). Once glycogen stores are depleted, glucose can be synthesized from noncarbohydrate sources (originating from proteins and triglycerides) that feed into *gluconeogenesis*. Amino acids and fatty acids can also feed into oxidative catabolism at different points, allowing ATP production to continue in the absence of glucose. In this section, we briefly discuss the catabolism of fatty acids and amino acids.

Energy from Fatty Acids

If glycogen and glucose are not available, cells that still require ATP can oxidize fatty acids to acetyl CoA through a degradation pathway known as **beta oxidation (β oxidation).** In β oxidation, carbons are removed two at a time from an activated fatty acid. An activated fatty acid is a fatty acid bonded to coenzyme A and is called **fatty acyl CoA.** The removal of two carbons during β oxidation produces an acetyl CoA and a fatty acyl CoA shortened by two carbons. The four-step cycle repeats until the original fatty acid is completely degraded to two-carbon acetyl CoA units. The acetyl CoA generated can then enter the citric acid cycle. This reaction cycle occurs in the mitochondrial matrix.

Fatty acyl CoA containing 18 carbons

The Beta Oxidation Cycle

β oxidation (see **FIGURE 12.15**) includes a cycle of four reactions that convert the $-CH_2-$ of the β carbon to a β ketone. Once this ketone is formed, the two-carbon acetyl group splits from the fatty acyl carbon chain, shortening the fatty acid. One cycle of β oxidation yields one acetyl CoA, one FADH$_2$, and one NADH.

Reaction ❶ Oxidation (Dehydrogenation)
The first reaction removes one hydrogen from the alpha and beta carbons, and a double bond is formed. These hydrogens are transferred to FAD to form FADH$_2$.

Reaction ❷ Hydration
In reaction 2, water is added to the α and β carbon double bond as $-H$ and $-OH$, respectively.

Reaction ❸ Oxidation (Dehydrogenation)
The alcohol formed on the β carbon is oxidized to a ketone. As we have seen before in the citric acid cycle, the hydrogen from the alcohol reduces NAD$^+$ to NADH.

Reaction ❹ Removal of Acetyl CoA
In the fourth reaction of the cycle, the bond between the α and β carbon is broken and a second CoA is added, forming an acetyl CoA and a fatty acyl CoA shortened by two carbons. The fatty acyl CoA can be run through the cycle again.

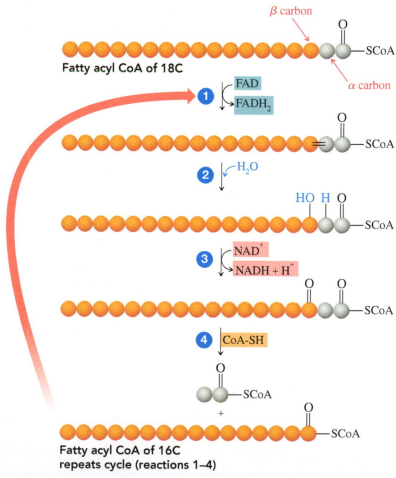

FIGURE 12.15 β **oxidation.** A fatty acyl CoA is broken up two carbons at a time into acetyl CoA units through a four-step cycle known as β oxidation. One FADH$_2$, one NADH, and one acetyl CoA are produced during each turn of the cycle.

The net reaction for one cycle of β oxidation is summarized as

Fatty acyl CoA$_{n \text{ carbons}}$ + NAD$^+$ + FAD + H$_2$O + CoA $\longrightarrow$

fatty acyl CoA$_{n-2 \text{ carbons}}$ + acetyl CoA + NADH + H$^+$ + FADH$_2$

Cycle Repeats and ATP Production

How many cycles of β oxidation does one fatty acid go through, and how many acetyl CoA are produced? Let's consider the saturated fatty acid stearic acid (18 carbons) as it undergoes β oxidation. Every two carbons will produce an acetyl CoA, so nine acetyl CoA are produced. This will take eight turns of the β oxidation cycle because the last turn produces two acetyl CoA. How many ATP will one stearic acid molecule produce?

Turns of β oxidation

8 7 6 5 4 3 2 1

—SCoA

Fatty acyl CoA of 18C

Solving a Problem

Producing ATP from β Oxidation

Calculate the total number of ATP produced from the β oxidation of one stearic acid [18:0] entering the β oxidation cycle as an acyl CoA.

Solution

STEP 1 Determine the total number of acetyl CoA. Each acetyl CoA contains 2 carbons, so for a fatty acid with 18 carbons, a total of 9 acetyl CoA will be produced.

STEP 2 Determine the number of cycles of β oxidation. If 9 acetyl CoA are produced, then the stearic acid went through eight cycles because the last cycle produces 2 acetyl CoA from the final 4 carbons.

STEP 3 Calculate the total number of ATP produced from NADH, FADH$_2$, and the acetyl CoA generated. In Section 12.7, we calculated that each turn of the citric acid cycle ultimately produces 10 ATP for each acetyl CoA that enters. Also, 1 NADH and 1 FADH$_2$ are produced per cycle of β oxidation, giving a total of 8 NADH and 8 FADH$_2$ per stearic acid molecule.

ATP Production from β Oxidation for a Stearic Acid [18:0] Molecule	
9 Acetyl CoA	
9 Acetyl CoA $\times$ 10 ATP/acetyl CoA	90 ATP
8 turns of β oxidation	
8 NADH $\times$ 2.5 ATP/NADH	20 ATP
8 FADH$_2$ $\times$ 1.5 ATP/FADH$_2$	12 ATP
	Total: **122 ATP**

Sample Problem 12.12 Producing ATP from β Oxidation

Calculate the number of ATP produced from the β oxidation of one myristic acid [14:0] entering the β oxidation cycle as an acyl CoA.

Solution

STEP 1 Determine the total number of acetyl CoA. Each acetyl CoA contains 2 carbons, so for a fatty acid with 14 carbons, a total of 7 acetyl CoA will be produced.

STEP 2 **Determine the number of cycles of β oxidation.** If 7 acetyl CoA are produced, then the myristic acid went through six cycles. The last cycle produces 2 acetyl CoA from the final 4 carbons.

STEP 3 **Calculate the total number of ATP produced from NADH, FADH$_2$, and the acetyl CoA generated.** In Section 12.7, we calculated that each turn of the citric acid cycle ultimately produces 10 ATP. In addition, 1 NADH and 1 FADH$_2$ are produced per cycle of β oxidation, producing a total of 6 NADH and 6 FADH$_2$ per stearic acid molecule.

7 Acetyl CoA	
7 Acetyl CoA × 10 ATP/acetyl CoA	70 ATP
6 turns of β oxidation	
6 NADH × 2.5 ATP/NADH	15 ATP
6 FADH$_2$ × 1.5 ATP/FADH$_2$	9 ATP
Total:	**94 ATP**

Too Much Acetyl CoA: Ketosis

In the absence of carbohydrates, the body breaks down its body fat through β oxidation to continue ATP production. This breaking down may seem efficient to the dieter considering a low-carb diet, yet some tissue types like the brain need glucose for energy. Even in the absence of carbohydrates, the liver will produce glucose from pyruvate through gluconeogenesis for tissues like the brain.

The oxidation of large amounts of fatty acids can cause acetyl CoA molecules to accumulate in the liver. When accumulation occurs, the two-carbon acetyl units condense in the liver, forming the four-carbon ketone molecules β-hydroxybutyrate and acetoacetate, and the molecule acetone. These are collectively referred to as **ketone bodies.** Their formation, termed *ketogenesis*, is outlined in **FIGURE 12.16**. The condition known as **ketosis** occurs when an excessive amount of ketone bodies is present in the body (and cannot be efficiently metabolized). This condition is often seen in diabetics who have difficulty getting glucose into the cells for energy, individuals on low-carbohydrate or high-fat diets, and individuals undergoing starvation.

Because two of the ketones are carboxylic acids, the excessive formation of ketone bodies can lower the blood pH and cause *metabolic acidosis*. Acetone vaporizes easily, giving a person suffering from ketosis an odd, sweet-smelling breath upon exhalation, similar to that of someone who has been drinking alcohol.

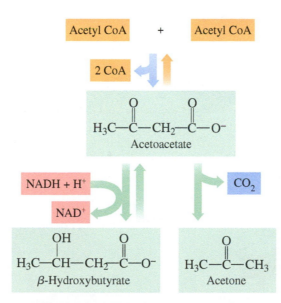

FIGURE 12.16 Ketogenesis. Excess acetyl CoA molecules not necessary for ATP production combine in the anabolic pathway ketogenesis (blue arrow) to produce the ketone bodies acetoacetate, β-hydroxybutyrate, and acetone (green boxes and arrows).

INTEGRATING Chemistry

◀ Find out why low-carb diets are not a good idea metabolically.

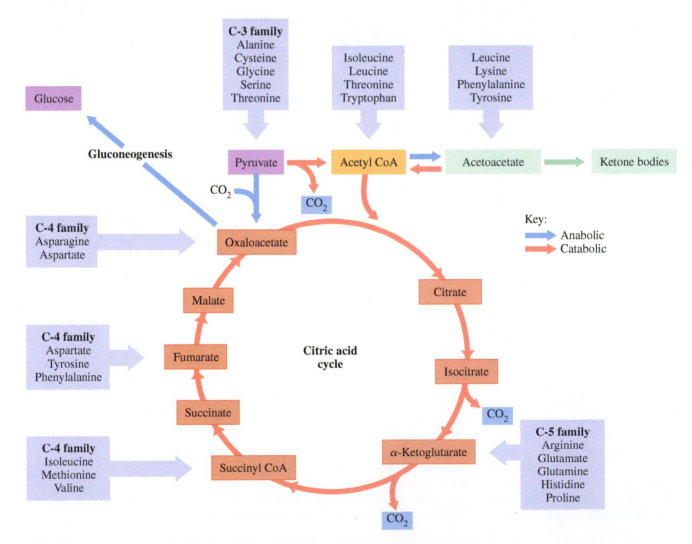

Urea

Excess ammonium is converted to urea for excretion through the urea cycle.

Energy from Amino Acids

While we can extract energy from proteins, it is not our body's preferred energy source because of the by-products that must be processed. The biggest one of these is nitrogen, which is removed as the first step in amino acid degradation. Through a process called **transamination**, the α-amino group is removed from an amino acid, producing an α-keto acid that can then be converted into intermediates for other metabolic pathways.

The process of transamination produces excess ammonium ions (NH_4^+) however, which must be excreted because a buildup of ammonium can be toxic. A series of reactions called the **urea cycle** converts ammonium ions into urea, which can then be excreted in the urine. Excess processing of ammonium can put a strain on the liver, where the urea cycle takes place, and the kidneys, where urea is ultimately excreted.

Various amino acids offer a way to replenish the intermediates in the citric acid cycle and therefore have the ability to generate ATP. Which intermediate they replenish depends on the catabolic pathway for the individual amino acid. For example, alanine degrades to a three carbon (C-3) metabolite and can enter the pathways as pyruvate. Amino acids that degrade to four carbons (C-4) are converted to oxaloacetate, and those that degrade to five carbons (C-5) are converted to α-ketoglutarate. Some amino acids can enter at more than one point depending on cellular requirements (see **FIGURE 12.17**).

In this way, we get some energy from amino acids, but it is only about 10% of the required energy under normal conditions. More energy is extracted from amino acids in conditions like fasting or starvation when carbohydrate and fat stores are depleted. Under these conditions, proteins in body tissues are degraded for fuel.

FIGURE 12.17 Amino acids feed into oxidative catabolism. Carbon atoms from degraded amino acids are converted to the intermediates of the citric acid cycle and other pathways.

Sample Problem 12.13   Carbon Atoms from Amino Acids

Which amino acids provide carbon atoms that enter the citric acid cycle as
α-ketoglutarate?

Solution

Those amino acids with five carbons feed into α-ketoglutarate. These are arginine,
glutamate, glutamine, histidine, and proline.

Putting It Together: Linking the Pathways

This chapter introduced several catabolic pathways and the high-energy molecules they
produce. **FIGURE 12.18** gives a visual summary of the material discussed in this chapter.
Degradation of food biomolecules begins with digestion, and when the cell requires
energy and oxygen is plentiful, larger molecules are metabolized into smaller metabolites

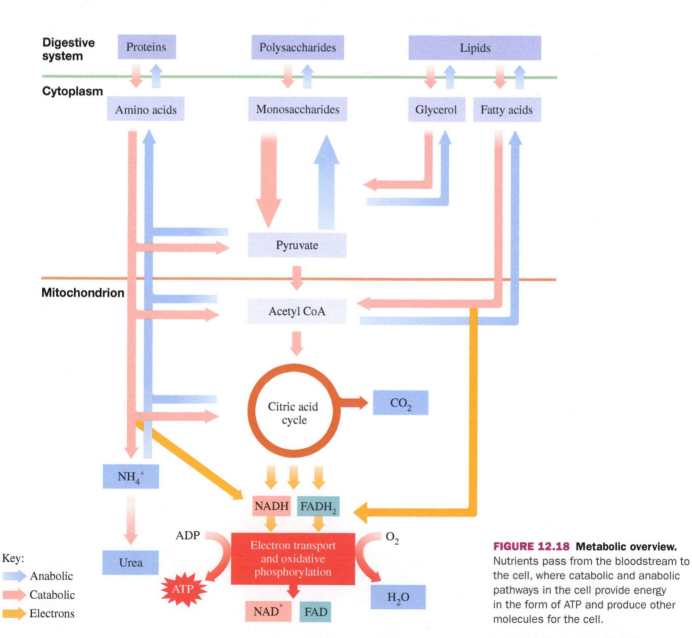

FIGURE 12.18 Metabolic overview.
Nutrients pass from the bloodstream to
the cell, where catabolic and anabolic
pathways in the cell provide energy
in the form of ATP and produce other
molecules for the cell.

that ultimately funnel into the citric acid cycle, electron transport, and oxidative phosphorylation. Through anabolic pathways, larger molecules can be synthesized from the smaller metabolites when necessary. The biological hydrolysis products can be shifted into anabolic or catabolic pathways depending on the requirements of the cell. Glucose can be degraded to acetyl CoA entering the citric acid cycle to produce energy or be converted to glycogen for storage in the cells. Amino acids provide nitrogen for anabolism of nitrogen compounds, but their carbons can enter the citric acid cycle as α-keto acids if necessary.

Practice Problems

12.53 Name the location in the cell where β oxidation takes place.

12.54 Name the metabolic nucleotides necessary for β oxidation.

12.55 Capric acid is a saturated fatty acid, [10:0].
 a. Draw fatty acyl capric acid activated for β oxidation.
 b. Identify the α and β carbons in fatty acyl capric acid.
 c. Write the overall equation for the complete β oxidation of capric acid.

12.56 Arachidic acid is a saturated fatty acid, [20:0].
 a. Draw fatty acyl arachidic acid activated for β oxidation.
 b. Identify the α and β carbons in fatty acyl arachidic acid.
 c. Write the overall equation for the complete β oxidation of arachidic acid.

12.57 Under what conditions are ketone bodies produced in the body?

12.58 Explain why diabetics produce high levels of ketone bodies.

12.59 Why does the body convert NH_4^+ into urea?

12.60 Draw the structure of urea.

12.61 Name the metabolic substrate(s) that can be produced from the carbon atoms of each of the following amino acids:
 a. alanine   b. aspartate
 c. valine   d. glutamine

12.62 Name the metabolic substrate(s) that can be produced from the carbon atoms of each of the following amino acids:
 a. leucine   b. asparagine
 c. cysteine   d. arginine

The study guide will help you check your understanding of the main concepts in Chapter 12. You should be able to answer problems for each learning outcome in this list. To check your mastery, try the problems listed after each.

STUDY GUIDE	CHAPTER GUIDE

12.1 How Metabolism Works

Describe features of metabolic reactions.

- Distinguish catabolism from anabolism. (Try 12.1)
- Identify reactions as catabolic or anabolic. (Try 12.3, 12.65)
- Name the parts of a cell associated with metabolism. (Try 12.7, 12.63)

Inquiry Question

What is metabolism?

Metabolism refers to biochemical reactions occurring in the body. Catabolism refers to reactions that break down larger molecules into smaller ones. In the body, biochemical reactions are catalyzed by enzymes and are usually grouped into pathways. Biochemical reactions that produce energy tend to be coupled to reactions requiring energy. Metabolic pathways tend to be compartmentalized in different parts of the cell. The mitochondria are the energy-producing factories of the cells.

TABLE 12.1 Properties of Biochemical Reactions in Metabolism

Property	Catabolism	Anabolism
Chemical reactions	Hydrolysis	Condensation
	Oxidation	Reduction
Energy	Exergonic (produce energy)	Endergonic (require energy)
Reaction rate	Catalyzed by enzymes	Catalyzed by enzymes

12.2 Metabolically Relevant Nucleotides

Distinguish the high- and low-energy forms of metabolically relevant nucleotides.

- Identify the metabolically relevant nucleotides. (Try 12.9, 12.11)
- Distinguish the low-energy/high-energy oxidized/reduced forms of the relevant nucleotides. (Try 12.13, 12.83)

Inquiry Question

What do the nucleotides that transfer energy look like?

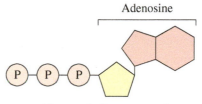
Adenosine

Nucleotides are used in metabolism to transfer energy throughout the cell. ATP is considered the main energy currency of the cell and produces energy when hydrolyzed to $ADP + P_i$. NADH and $FADH_2$ contain a nucleotide portion and a vitamin portion. They are important coenzymes that transport hydrogens and electrons in the cell. Coenzyme A (CoA) also contains a nucleotide and vitamin portion. Each of these nucleotides has a high-energy and low-energy form.

12.3 Digestion—From Food Molecules to Hydrolysis Products

Describe the products of food molecule hydrolysis.

- Compare digestion of carbohydrates, lipids, and proteins. (Try 12.15, 12.19, 12.67)

Inquiry Question

How are food molecules digested?

Food molecules are broken down into their component parts through hydrolysis prior to absorption into the bloodstream. Monosaccharides and amino acids travel directly through the bloodstream to the cells for absorption, whereas triglycerides are packaged into lipoproteins called chylomicrons for delivery.

—Continued next page

Continued—

12.4 Glycolysis—From Hydrolysis Products to Common Metabolites

Demonstrate the reactions of glycolysis.

- Follow a molecule of glucose through the ten reactions of glycolysis. (Try 12.21, 12.71)
- Discuss the anaerobic and aerobic fates of pyruvate. (Try 12.27, 12.75)
- Contrast glycolysis for glucose and for fructose. (Try 12.31, 12.77)

Inquiry Question

How is glucose catabolized through glycolysis?

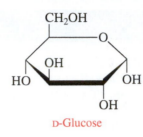

D-Glucose

Glycolysis is a series of ten reactions that catabolize a six-carbon glucose molecule to two three-carbon pyruvate molecules. These ten reactions yield two ATP and two NADH molecules per glucose. Under aerobic conditions, pyruvate is further oxidized in the mitochondria to acetyl CoA. In the absence of oxygen, pyruvate is reduced to lactate and regenerates NAD^+ so that glycolysis can continue. Glucose can be stored as glycogen in the liver and muscle for later use. Fructose can undergo glycolysis. In the liver, its catabolism is unregulated and can produce excess products that become stored in the body as fat.

12.5 The Citric Acid Cycle—Central Processing

Classify the reactions of the citric acid cycle.

- Identify the reactions in the citric acid cycle. (Try 12.35, 12.37, 12.79)
- List the energy output of the citric acid cycle. (Try 12.39)

Inquiry Question

What types of reactions occur during the citric acid cycle?

The citric acid cycle occurs in the mitochondrial matrix and combines acetyl CoA (two carbons) with oxaloacetate (four carbons), producing citric acid. Citric acid undergoes oxidation, decarboxylation, dehydrogenation, and a hydration, yielding two CO_2, GTP, three NADH, and one $FADH_2$. The cycle regenerates oxaloacetate to begin again. GTP readily converts to ATP in the cell.

12.6 Electron Transport and Oxidative Phosphorylation

Discuss the oxidation of reduced nucleotides and production of ATP.

- Describe the function of each enzyme complex (I–IV) during electron transport. (Try 12.41, 12.43)
- Discuss the function of coenzyme Q and cytochrome *c*. (Try 12.45)
- Describe the production of ATP at complex V using the chemiosmotic model. (Try 12.47, 12.85)

Inquiry Question

How is the energy in NADH and $FADH_2$ transformed into ATP?

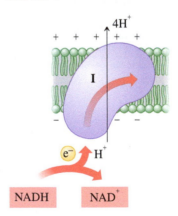

The reduced nucleotides NADH and $FADH_2$ become oxidized, transporting H^+ and electrons through a series of enzyme complexes in the inner mitochondrial membrane. The final electron acceptor in this process is O_2, which combines with H^+ to form H_2O. The complexes act as proton pumps, moving protons from the matrix to the inner membrane space during electron transport. This produces an electrochemical gradient. The protons can return to the matrix through complex V, ATP synthase, which generates ATP. This process is known as oxidative phosphorylation. When ATP and proton transport are uncoupled, the energy in the proton gradient is released as heat, which assists in maintaining body temperature in a process called thermogenesis.

12.7 ATP Production

Calculate the number of ATP produced during oxidative catabolism.

- Convert the number of reduced nucleotides produced (NADH, $FADH_2$) to a corresponding number of ATP. (Try 12.51)
- Calculate the number of ATP produced during the oxidative catabolism of a molecule of glucose. (See Table 12.4; Try 12.89)

Inquiry Question

How many ATP can be produced during oxidative catabolism?

Nucleotide Input	Protons (H^+) Pumped	ATP Output
NADH	10	2.5
$FADH_2$	6	1.5

For every four H^+ pumped into the inner membrane space and returned to the matrix, one ATP can be synthesized. The oxidation of one NADH provides enough energy to synthesize 2.5 ATP. One $FADH_2$ provides energy to synthesize 1.5 ATP. Under aerobic conditions, the complete oxidation of one molecule of glucose produces a total of 32 ATP.

12.8 Other Fuel Choices

Examine how fatty acids and amino acids contribute to ATP production

- Calculate the number of ATP produced from a saturated fatty acid undergoing β oxidation. (Try 12.55, 12.93, 12.97)
- Describe the metabolic pathways of β oxidation, production of ketone bodies, and the urea cycle. (Try 12.59, 12.61, 12.98)

Inquiry Question

How do fatty acids and amino acids produce ATP?

Fatty acids produce ATP when glucose supplies are low. Fatty acids link to coenzyme A, forming activated fatty acyl CoA, which is transported to the mitochondria for catabolism in a reaction called β oxidation. Fatty acyl CoA is oxidized, producing a new fatty acyl CoA that is two carbons shorter and one molecule of acetyl CoA. Each turn of β oxidation produces one NADH and one $FADH_2$. High levels of acetyl CoA in the cell activate the ketogenesis pathway, forming ketone bodies that can lead to ketosis and acidosis. Amino acids can produce ATP when other fuel supplies are low and the cell does not require other nitrogen-containing compounds. When amines are removed from amino acids as ammonium ions, they are converted to urea for excretion. The carbons from amino acids can feed into oxidative catabolism as different intermediates depending on the amino acid.

 Mastering Videos

PRACTICING THE CONCEPTS

The following videos can be accessed through the Pearson eText or your Mastering Chemistry course.

- Glycolysis
- The Citric Acid Cycle

Summary of Reactions

Hydrolysis of ATP

$$ATP \longrightarrow ADP + P_i + Energy$$

Formation of ATP

$$ADP + P_i + Energy \longrightarrow ATP$$

Digestion of Proteins

$$Protein + H_2O \xrightarrow[Enzyme]{H^+} amino\ acids$$

Hydrolysis of Disaccharides

$$Lactose + H_2O \xrightarrow{Lactase} galactose + glucose$$

$$Sucrose + H_2O \xrightarrow{Sucrase} glucose + fructose$$

$$Maltose + H_2O \xrightarrow{Maltase} glucose + glucose$$

Digestion of Triglycerides

$$Triglycerides + 3H_2O \xrightarrow{Lipases} glycerol + 3\ fatty\ acids$$

Glycolysis

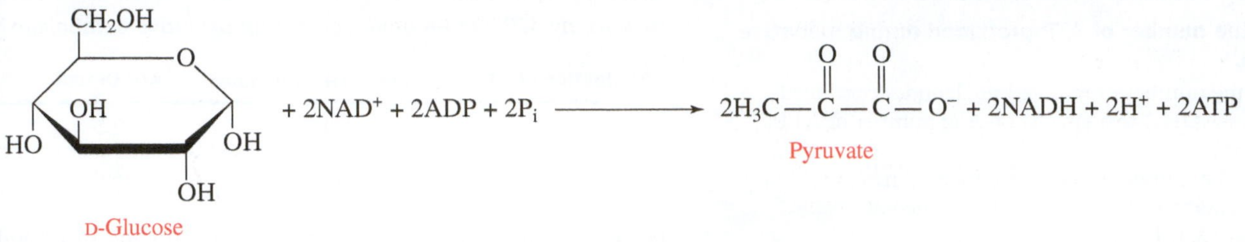

$$\text{D-Glucose} + 2NAD^+ + 2ADP + 2P_i \longrightarrow 2H_3C-\overset{\overset{\displaystyle O}{\|}}{C}-\overset{\overset{\displaystyle O}{\|}}{C}-O^- + 2NADH + 2H^+ + 2ATP$$

Pyruvate

Oxidation of Pyruvate to Acetyl CoA

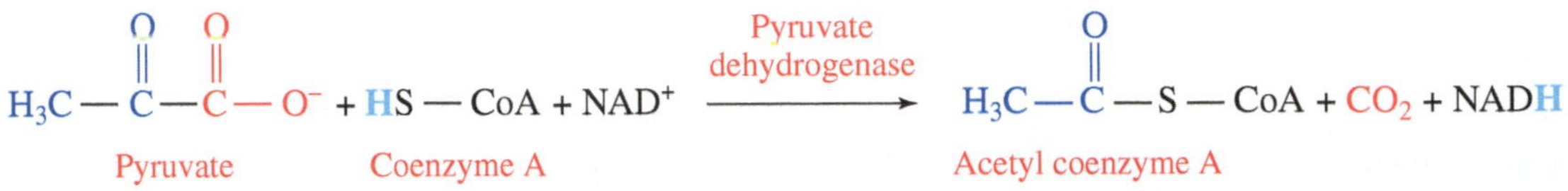

$$H_3C-\overset{\overset{\displaystyle O}{\|}}{C}-\overset{\overset{\displaystyle O}{\|}}{C}-O^- + HS-CoA + NAD^+ \xrightarrow{\text{Pyruvate}\atop\text{dehydrogenase}} H_3C-\overset{\overset{\displaystyle O}{\|}}{C}-S-CoA + CO_2 + NADH$$

Pyruvate   Coenzyme A   Acetyl coenzyme A

Reduction of Pyruvate to Lactate

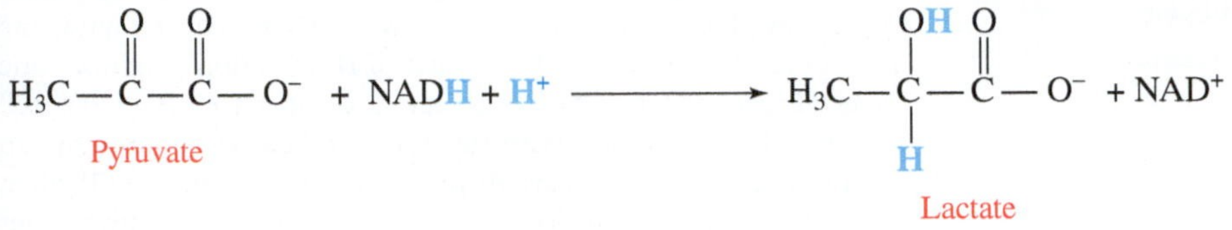

$$H_3C-\overset{\overset{\displaystyle O}{\|}}{C}-\overset{\overset{\displaystyle O}{\|}}{C}-O^- + NADH + H^+ \longrightarrow H_3C-\overset{\overset{\displaystyle OH}{|}}{\underset{\underset{\displaystyle H}{|}}{C}}-\overset{\overset{\displaystyle O}{\|}}{C}-O^- + NAD^+$$

Pyruvate   Lactate

Citric Acid Cycle

$$\text{Acetyl CoA} + 3NAD^+ + FAD + GDP + P_i + 2H_2O \longrightarrow 2CO_2 + 3NADH + 3H^+ + FADH_2 + CoA + GTP$$

Complete Oxidation of Glucose

$$C_6H_{12}O_6 + 6O_2 \longrightarrow 6CO_2 + 6H_2O$$

β Oxidation of Fatty Acid

$$\text{Fatty acyl CoA}_{n\text{ carbons}} + NAD^+ + FAD + H_2O + CoA \longrightarrow \text{fatty acyl CoA}_{n-2\text{ carbons}} + \text{acetyl CoA} + NADH + H^+ + FADH_2$$

Additional Problems

12.63 Name the location in the cell where the following catabolic processes take place:

    a. glycolysis   **c.** ATP synthesis

    b. citric acid cycle

12.64 Name the location in the cell where the following catabolic processes take place:

    a. pyruvate oxidation

    b. electron transport

    c. β oxidation

12.65 Identify each pathway as anabolic or catabolic.

    a. gluconeogenesis

    b. protein synthesis

12.66 Identify each pathway as anabolic or catabolic

    a. beta oxidation   **b.** citric acid cycle

12.67 Identify the type of food—carbohydrate, fat, or protein—that gives each of the following digestion products:

    a. glucose   **d.** glycerol

    b. fatty acid   **e.** amino acids

    c. maltose

12.68 How and where does sucrose undergo digestion in the body? Name the products.

12.69 How and where does lactose undergo digestion in the body? Name the products.

12.70 If glycolysis occurs in the cytoplasm and the citric acid cycle occurs in the mitochondrial matrix, how do the products of glycolysis get inside the mitochondrial matrix?

12.71 Which of the following reactions in glycolysis produce ATP or NADH?

a. 1,3-bisphosphoglycerate to 3-phosphoglycerate

b. glucose-6-phosphate to fructose-6-phosphate

c. phosphoenolpyruvate to pyruvate

12.72 Which of the following reactions in glycolysis produce ATP or NADH?

a. glucose to glucose-6-phosphate

b. glyceraldehyde-3-phosphate to 1,3-bisphosphoglycerate

c. dihydroxyacetone phosphate to glyceraldehyde-3-phosphate

12.73 Which of the reactions given in Problems 12.71 represent isomerizations where the reactants and products are structural isomers?

12.74 Which of the reactions given in Problems 12.72 represent isomerizations where the reactants and products are structural isomers?

12.75 After running a marathon, a runner has muscle pain and cramping. What might have occurred in the muscle cells to cause this?

12.76 After eating a large, starchy meal, what does the body do with all the glucose from starch?

12.77 Fructose can be catabolized through glycolysis. How is the metabolism different for a person who is inactive versus a person who is metabolizing fructose during exercise?

12.78 When pyruvate is used to form acetyl CoA, the product has only two carbon atoms. What happened to the third carbon?

12.79 Refer to the diagram of the citric acid cycle on page 498 to answer each of the following:

a. Name the six-carbon compounds.

b. Name the five-carbon compounds.

c. Name the oxidation reactions.

d. Name the reactions where secondary alcohols are oxidized to ketones.

12.80 Refer to the diagram of the citric acid cycle on page 498 to answer each of the following:

a. Name the four-carbon compounds.

b. Name the reactant that undergoes a hydration reaction.

c. Name the reaction that is coupled to GTP formation.

d. Provide the number of CO_2 molecules produced per turn of the citric acid cycle.

12.81 If there are no reactions in the citric acid cycle that use oxygen, O_2, why does the cycle operate only in aerobic conditions?

12.82 What products of the citric acid cycle are used in electron transport?

12.83 Identify the following as the reduced or oxidized form:

a. NAD^+ **b.** $FADH_2$ **c.** QH_2

12.84 Identify the following as the reduced or oxidized form:

a. NADH   **b.** FAD   **c.** Q

12.85 During electron transport, H^+ are pumped out of the _____, across the inner membrane, and into the _____.

12.86 During ATP synthesis, H^+ move from the _____, across the inner membrane, and into the _____.

12.87 Mammals can regulate their body heat through a process called thermogenesis. What part of metabolism changes to allow for the production of heat?

12.88 What is the effect of proton accumulation in the intermembrane space?

12.89 How many ATP are produced when glucose is oxidized to pyruvate compared to when glucose is oxidized to CO_2 and H_2O?

12.90 Name the reaction that removes nitrogen from an α-amino acid, forming an α-keto acid.

12.91 In what organ does the urea cycle take place?

12.92 What metabolic substrate(s) can be produced from the carbon atoms of each of the following amino acids?

a. histidine   **c.** methionine

b. isoleucine   **d.** phenylalanine

12.93 Consider the complete oxidation of capric acid, a saturated fatty acid, [10:0].

a. How many acetyl CoA units are produced?

b. How many cycles of β oxidation occur?

c. How many ATP are generated from the complete oxidation of capric acid?

12.94 Consider the complete oxidation of arachidic acid, a saturated fatty acid, [20:0].

a. How many acetyl CoA units are produced?

b. How many cycles of β oxidation occur?

c. How many ATP are generated from the complete oxidation of arachidic acid?

Challenge Problems

12.95 Identify all steps in the oxidative catabolism of glucose that can be considered oxidative decarboxylations.

12.96 How many turns of the citric acid cycle does it take for the two carbons from an acetyl group entering the citric acid cycle to be liberated as CO_2?

12.97 Lauric acid, a saturated fatty acid, [12:0], is found in coconut oil.

    a. Draw fatty acyl lauric acid activated for β oxidation.

    b. Identify the α and β carbons in fatty acyl lauric acid.

    c. Write the overall equation for the complete β oxidation of lauric acid.

    d. How many acetyl CoA units are produced?

    e. How many cycles of β oxidation occur?

    f. Account for the total ATP yield from β oxidation of lauric acid [12:0] by completing the following table:

        _____Acetyl CoA × 10 ATP/acetyl CoA   _____ATP

        _____NADH × 2.5 ATP/NADH   _____ATP

        _____$FADH_2$ × 1.5 ATP/$FADH_2$ _____ATP

                               Total: _____**ATP**

12.98 A person is brought to the emergency room in what appears to be a drunken stupor. On closer examination, she seems to be breathing very rapidly and you notice a sweet-smelling odor on her breath. You find out this person has not been drinking alcohol. What is her likely condition?

12.99 Can acetyl CoA feed into gluconeogenesis, producing glucose for the body? Can fatty acids be used to produce glucose?

12.100 A friend has been on a low-carb diet to lose weight. Initially, she lost 10 pounds, but after a few weeks she is having a hard time losing any more weight, even though her dieting patterns are the same. She is also becoming confused at times and can't focus. What is happening to her metabolically, and how would you explain it to her?

12.101 Excess nitrogen in the body can form uric acid crystals responsible for the painful condition called gout (see the Integrating Chemistry feature in Section 8.2). Besides proteins, name a biomolecule synthesized in the body that uses nitrogen.

Answers to Odd-Numbered Problems

Practice Problems

12.1 In metabolism, a catabolic reaction breaks apart molecules, releases energy, and often involves oxidation.

12.3 **a.** catabolism   **b.** anabolism

12.5 hydrolysis and oxidation reactions

12.7 ATP (adenosine triphosphate)

12.9 **a.** FAD/$FADH_2$ **b.** acetyl CoA

12.11 **a.** ATP   **b.** acetyl CoA

12.13 $FADH_2 \longrightarrow FAD + H_2$

12.15 **a.** starch   **b.** none

    c. starch, oligosaccharides, disaccharides

12.17 Cholesterol is esterified and packaged into lipoproteins.

12.19 amino acids

12.21 D-glucose

12.23 4 ATP (net 2 ATP)

12.25 2 NADH and 2 ATP

12.27 NAD^+

12.29 ethanol, CO_2, and NAD^+

12.31 The main regulation point of glycolysis is phosphofructokinase, the enzyme involved in step 3. Because the products of the metabolism of fructose enter glycolysis at step 5, they bypass step 3 regulation, which leads to the production of excess pyruvate and acetyl CoA not required by the cells. The excess is ultimately converted to fat.

12.33 acetyl CoA and oxaloacetate

12.35 isocitrate $\longrightarrow$ α-ketoglutarate (reaction 3) and α-ketoglutarate $\longrightarrow$ succinyl CoA (reaction 4)

12.37 isocitrate $\longrightarrow$ α-ketoglutarate (reaction 3), α-ketoglutarate $\longrightarrow$ succinyl CoA (reaction 4), and malate $\longrightarrow$ oxaloacetate (reaction 8)

12.39 2 ATP, 6 NADH, and 2 $FADH_2$

12.41 coenzyme Q

12.43 complex IV

12.45 up to two; only one

12.47 As protons flow through ATP synthase, energy is released to produce ATP.

12.49 ATP is synthesized from ADP.
     $ADP + P_i + Energy \longrightarrow ATP$

12.51 **a.** 2.5 ATP   **b.** 7 ATP

    c. 5 ATP   **d.** 10 ATP

12.53 mitochondrial matrix

12.55 **a.** and **b.**

β carbon

O

$S—CoA$

α carbon

c.

O

$S—CoA + 4NAD^+ + 4FAD + 4H_2O + 4CoA$

$\downarrow$

$5\,Acetyl\ CoA + 4NADH + 4H^+ + 4FADH_2$

12.57 Ketone bodies form when excess acetyl CoA results from the breakdown of large amounts of fat.

12.59 NH_4^+ is toxic if it accumulates in the body.

12.61 **a.** pyruvate   **b.** oxaloacetate, fumarate
   c. succinyl CoA   **d.** α-ketoglutarate

Additional Problems

12.63 **a.** cytoplasm   **b.** mitochondrial matrix
   c. mitochondrial matrix

12.65 **a.** anabolic   **b.** anabolic

12.67 **a.** carbohydrate   **b.** fat
   c. carbohydrate   **d.** fat
   e. protein

12.69 Lactose undergoes digestion in the small intestine to yield galactose and glucose.

12.71 **a.** produces ATP   **b.** neither
   c. produces ATP

12.73 **b.** glucose-6-phosphate to fructose-6-phosphate

12.75 The runner's muscles may be switching over to anaerobic catabolism to keep ATP production going. The muscles produce lactate and H^+ during this process, causing soreness.

12.77 Fructose that is metabolized in the muscle enters glycolysis at reaction 3 as fructose-6-phosphate and is enzymatically regulated. This would be the case for someone who is exercising and using fructose for muscle movement. Fructose that is not metabolized in the muscle is metabolized in the liver and enters glycolysis at step 5. All fructose entering the liver is metabolized to glycolysis products (pyruvate/acetyl CoA), an excess of which would be stored as fat.

12.79 **a.** citrate and isocitrate

   b. α-ketoglutarate

   c. isocitrate $\longrightarrow$ α-ketoglutarate (reaction 3), α-ketoglutarate $\longrightarrow$ succinyl CoA (reaction 4), succinate $\longrightarrow$ fumarate (reaction 6), and malate $\longrightarrow$ oxaloacetate (reaction 8)

   d. isocitrate $\longrightarrow$ α-ketoglutarate (reaction 3) and malate $\longrightarrow$ oxaloacetate (reaction 8)

12.81 O_2 is used during electron transport. The coenzymes NADH and $FADH_2$ produced in the citric acid cycle are oxidized to NAD^+ and FAD during electron transport.

12.83 **a.** oxidized   **b.** reduced   **c.** reduced

12.85 matrix; intermembrane space

12.87 Protons that were pumped out of the matrix during electron transport leak back in, producing heat instead of synthesizing ATP.

12.89 The oxidation of glucose to pyruvate produces 7 ATP, 5 from NADH, whereas 32 ATP are produced from the complete oxidation of glucose to CO_2 and H_2O.

12.91 the liver

12.93 **a.** 5   **b.** 4   **c.** 66 ATP

12.95 oxidation of pyruvate: pyruvate $\longrightarrow$ acetyl CoA; reaction 3 of citric acid cycle: isocitrate $\longrightarrow$ α-ketoglutarate; reaction 4 of the citric acid cycle: α-ketoglutarate $\longrightarrow$ succinyl CoA

12.97 a. and **b.**

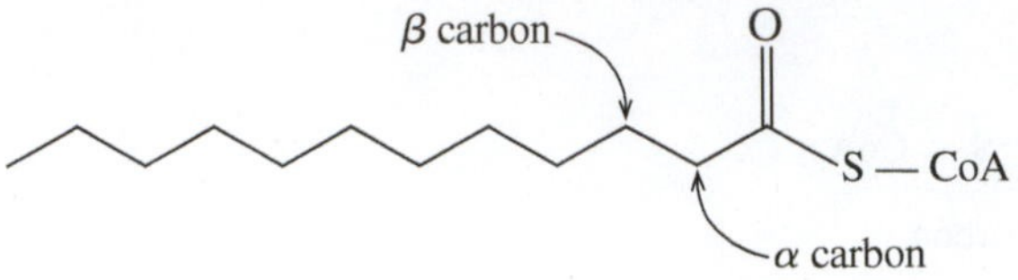

c.

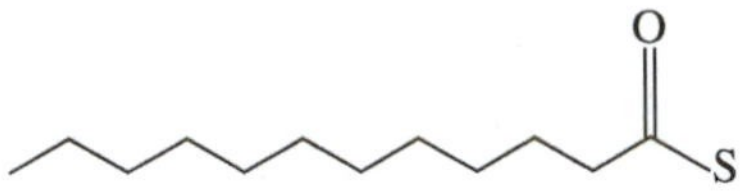

$$\text{S—CoA} + 5\text{NAD}^+ + 5\text{FAD} + 5\text{H}_2\text{O} + 5\text{CoA} \longrightarrow 6\text{Acetyl CoA} + 5\text{NADH} + 5\text{H}^+ + 5\text{FADH}_2$$

d. 6

e. 5

f.

____6____	Acetyl CoA × 10 ATP/acetyl CoA	____60____ ATP
____5____	NADH × 2.5 ATP/NADH	____12.5____ ATP
____5____	$FADH_2$ × 1.5 ATP/$FADH_2$	____7.5____ ATP

Total: 80 ATP

12.99 Only oxaloacetate or pyruvate is a starting point for gluconeogenesis (see the arrows in Figure 12.18). Acetyl CoA cannot be converted to pyruvate, so fatty acids cannot generate glucose.

12.101 The nitrogenous bases found in nucleotides and nucleic acids are one example.

Credits

Frontmatter

p. iii: (*top*) Laura Frost/Kendra Carboneau; p. iii: (*bottom*) Todd Deal/Jeremy Wilburn; p. xii: Freeman, S., Eddy, S.L. McDonough, M.,Smith, M.K., Okoroafor, N., Jordt, H.,Wenderoth, M.P. (2014). Active learning increases student performance in science, engineering, and mathematics. Proceedings of the National Academy of Sciences USA. 111(23), 8410–8415; p. xiii:Wilson, K.J., Brickmann, P., Brame, C.J. (2018). Evidence-based teaching guides: Group work. CBE-Life Sceicnes Education. 17 (1). DOI: https://doi.org/10.1187/cbe.17-12-0258; p. xiii: Frost, L. D. (2010). *Making Chemistry Relevant: Strategies for Including All Students in a Learner-Sensitive Classroom Environment.* S. Basu-Dutt, Ed. New York: John Wiley & Sons; p. xiv: Callendar, A.A. & McDaniel, M.A. (2007). The benefits of embedded question adjuncts for low and high structure builders. Journal of Educational Psychology, 99(2), 339–338

Chapter 1

p. 1: Nguyen and McDaniel *in* Benassi,V.A., Overson, C.E., & Hakala, C.M. (2014). *Applying Science of Learning in Education: Infusing psychological science into the curriculum.* Retrieved from the Society for the Teaching of Psychology website: http://teachpsych. org/ebooks/asle2014/index.php, p. 112; p. 1: CandyBox Images/Shutterstock; p. 3: (*left*) Josef Bosak/Shutterstock; p. 3: (*middle left and right*) Laura D. Frost; p. 3: (*middle right*) titelio/Shutterstock; p. 4: (*top*) mevans/E+/Getty Images; p. 4: (*middle*) Li Wa/Shutterstock; p. 4: (*bottom*) Ruslan Kerimov/Shutterstock; p. 7: Richard Megna/Fundamental Photographs, NYC; p. 11: Michael C. Gray/Shutterstock; p. 12: Jan Kaliciak/Shutterstock; p. 16: (*left*) Big Pants Production/Shutterstock; p. 16: (*middle*) Garsya/Shutterstock; p. 16: (*right*) pikselstock/Shutterstock; p. 18: (*left*) Laura D. Frost; p. 18: (*middle*) frantic00/Shutterstock; p. 21: STOCK4B GmbH/Alamy Stock Photo; p. 24: Cathy Yeulet/123RF; p. 25: (*top*) Ron Kloberdanz/Shutterstock; p. 25: (*bottom*) Science Photo Library/Alamy Stock Photo; p. 26: Eric Schrader; p. 27: Javier Larrea/age fotostock/Alamy Stock Photo; p. 30: Stephen Mcsweeny/Shutterstock; p. 31: ESB Professional/Shutterstock; p. 32: Kichigin/Shutterstock; p. 38: grebeshkovmaxim/Shutterstock

Chapter 2

p. 49: Dunlosky, J. Rawson, K., Marsh, E. Nathan M. Willingham, D. (2013). *Improving Students' Learning with Effective Learning Techniques: Promising Directions from Cognitive and Educational Psychology.* Psychological Science in the Public Interest 14(1) 4–58; p. 49: APHP St-Louis-Garo/Phanie/Alamy Stock Photo; p. 50: (*left*) topseller/Shutterstock; p. 50: (*bottom left*) elen_studio/Shutterstock; p. 50: (*bottom middle*) Alexander Dashewsky/Shutterstock; p. 50: (*bottom right*) Evgeniy Goncharenko/123RF; p. 51: Dennis S.K. Collection via Wikimedia Commons; p. 52: (*left*) Dennis S.K. Collection via Wikimedia Commons; p. 52: (*right*) Fablok/Shutterstock; p. 56: Franco Volpato/Shutterstock; p. 57: Data from National Isotope Development Center; p. 57: Science History Images/Alamy Stock Photo; p. 59: BSIP SA/Alamy Stock Photo; p. 64: BSIP SA/Alamy Stock Photo; p. 66: Gustoimages/Science Source; p. 68: (*top left*) BSIP SA/Alamy Stock Photo; p. 68: (*bottom*) Westgate EJ, FitzGerald GA (2005) Pulmonary Embolism in a Woman Taking Oral Contraceptives and Valdecoxib. *PLoS Med* 2(7): e197; p. 69: Editorial Image, LLC/Alamy Stock Photo; p. 70: Robert Friedland/Science Source; p. 73: BSIP SA/Alamy Stock Photo

Chapter 3

p. 77: Roediger, H.L., & Karpicke, J.D. (2006). Test-enhanced learning: Taking memory tests improves long-term retention. *Psychological Science,* 17, 249–255. (Retrieval practice works better than re-reading); p. 77: Cultura Creative (RF)/Alamy Stock Photo; p. 84: Mike Casper/Alamy Stock Photo; p. 88: Food Collection/Alamy Stock Photo; p. 92: (*top*) Anastasia Bulanova/Shutterstock; p. 92: (*bottom*) huayang/Moment/Getty Images; p. 99: Laura D. Frost; p. 101: Turtle Rock Scientific/Science Source

Chapter 4

p. 125: Mayer, R. E. (2009). *Multimedia Learning.* New York, NY: Cambridge University Press; p. 125: JohnKwan/Shutterstock; p. 126: Photo by Abigail Deal; p. 129: Liliboas/iStock Editorial/Getty Images Plus; p. 132: AsiaPix/Big Cheese Photo/Getty Images; p. 136: Gary Moon/AGE Fotostock; p. 143: (*left*) Dmitry Kalinovsky/Shutterstock; p. 143: (*top right and bottom right*) Eric Schrader; p. 144: Steffen Foerster/Shutterstock; p. 145: Korionov/Shutterstock; p. 149: ckellyphoto/Fotolia; p. 156: (*left*) Bochkarev/Shutterstock; p. 156: (*middle*) Kurhan/Shutterstock; p. 156: (*top right*) Monticello/Shutterstock; p. 156: (*bottom right*) Wiktory/Shutterstock; p. 160: (*all*) Eric Schader; p. 164: (*left*) kaband/Shutterstock; p. 164: (*right*) Coprid/Shutterstock; p. 165: Taigi/Shutterstock

Chapter 5

p. 181: Nate Kornell & Robert A. Bjork (2008) Optimizing self-regulated study: The benefits—and costs—of dropping flashcards, Memory, 16:2, 125–136, DOI: 10.1080/09658210701763899 Karpicke, J. D. (2009). Metacognitive control and strategy selection: Deciding to practice retrieval during learning. *Journal of Experimental Psychology: General,* 138(4), 469–486. doi:http://dx.doi.org/10.1037/a0017341 CH06; p. 181: karenfoleyphotography/Alamy Stock Photo; p. 182: (*top*) Laura D. Frost; p. 182: (*bottom*) Eric Schrader; p. 184: (*left*) Yanlev/Fotolia; p. 184: (*right*) Halfpoint/Fotolia; p. 185: (*left*) Laura D. Frost; p. 185: (*right*) Charles D. Winters/Science Source; p. 187: Eric Schrader; p. 188: (*left*) Karin Hildebrand Lau/Shutterstock; p. 188: (*right*) Shawn Hempel/Shutterstock; p. 190: Greg Stanfield/Shutterstock; p. 194: Mark Goldman/Icon SMI 749/Mark Goldman/Icon SMI/Newscom; p. 195: Jaroslaw Grudzinski/Shutterstock; p. 204: BELMONTE/BSIP/AGE Fotostock; p. 207: Eric Schrader

Chapter 6

p. 219: Scholey, A.B., Moss, M.C., Neave, N.,Wesnes, K. (1999). Cognitive performance, hyperoxia, and heart rate following oxygen administration in healthy young adults. *Physiology & Behavior,* 67(5), 783–789; p. 219: Andriy Popov/123RF; p. 221: Liljam/Shutterstock; p. 223: (*left*) Avalon Studio/E+/Getty Images; p. 223: (*right*) Keni/Shutterstock; p. 223: (*bottom*) FotografiaBasica/Vetta/Getty Images; p. 233: (*left*) kcline/E+/Getty Images; p. 233: (*right*) Elenathewise/Getty Images; p. 234: FreezeFrameStudio/E+/Getty Images; p. 242: (*top*) Martyn F. Chillmaid/Science Source; p. 242: (*bottom*) Jürgen Fälchle/Fotolia; p. 243: Eric Schrader; p. 248: indigolotos/Fotolia; p. 249: (*top*) Lidante/Shutterstock; p. 249: (*bottom*) sumire8/Shutterstock; p. 254: Biophoto Associates/Science Source; p. 255: (*top*) Tiger Images/Shutterstock; p. 255: (*top middle*) Sasirin Pamai/Shutterstock; p. 255: (*bottom middle*) David Toase/Stockbyte/Getty Images; p. 255: (*bottom*) Olha Solodenko/123RF

Chapter 7

p. 268: Rohrer, D., & Taylor, K. (2007). The shuffling of mathematics problems improves learning, *Instructional Science, 35(6),* 481–498; p. 268: Belchonock/123RF; p. 273: Eric Schrader; p. 276: Niccolo'Simoncini/Shutterstock; p. 281: Foto-zone/Alamy Stock Photo; p. 288: (*top*) Miiisha/Shutterstock; p. 288: (*bottom*) Mygueart/iStock/Getty Images Plus/Getty Images; p. 293: Filipe B.Varela/Shutterstock; p. 296: (*left*) Andrew Silk/ZUMAPRESS/Newscom; p. 296: (*middle*) JB Reed/Bloomberg/Getty Images; p. 296: (*right*) Studioshots/Alamy Stock Photo; p. 305: Filipe B. Varela/Shutterstock

Chapter 8

p. 313: Mueller, P.A., Oppenheimer, D.M. (2014), *Psychological Science,* 23 April 2014, DOI: 10.1177/0956797614524581; p. 313: Robert Holland/Alamy Stock Photo; p. 314: (*top*) Paul Silverman/Fundamental Photographs, NYC; p. 314: (*lower left*) Alx/Fotolia; p. 314: (*lower middle*) Monia/Fotolia; p. 314: (*lower right*) Eric Schrader; p. 315: George Wypych, Handbook of Solvents. ChemTec Publishing, (2001), 1675 pages.; p. 315: (*top*) Mike Napieralski/Pearson Education, Inc; p. 315: (*middle*) David Crockett/Fotolia; p. 315: (*bottom*)Klaus Guldbrandsen/Science Source; p. 317: Turtle Rock Scientific/Science Source; p. 318: (*top*) Dr P. Marazzi/Science Source; p. 318: (*middle*) remik44992/Alamy Stock Photo; p. 318: (*bottom*) Science Photo Library/Alamy Stock Photo; p. 321: Eric Schrader; p. 322: Turtle Rock Scientific/Science Source; p. 328: (*left*) Perutskyi Petro/123RF; p. 328: (*right*) vipman/Shutterstock; p. 330: Eric Schrader; p. 331: (*top*) Robert Kyllo/Shutterstock; p. 331: (*bottom*) Eric Schrader; p. 332: pkline/Getty Images; p. 334: (*left*) Jeff Lueders/Alamy Stock Photo; p. 334: (*right*) GGRIGOROV/Shutterstock; p. 336: Photo by Abigail Deal; p. 338: Sandia National Laboratories; p. 339: David M. Phillips/Science Source; p. 341: Picsfive/Shutterstock; p. 346: Photo by Abigail Deal

Chapter 9

p. 352: Brown, P.C., Roediger, H.L., McDaniel, M. (2014) Make it Stick: The Science of Successful Learning. Belknap Press, Cambridge, MA. Chapter 3; p. 352: Alena TS/Shutterstock; p. 352: (*inset*) Greenni/Shutterstock; p. 355: ggw/Shutterstock; p. 358: Blue Lemon Photo/Fotolia; p. 371: (*left*) photong/Shutterstock; p. 371: (*right*) Photo Researchers/Science History Images/Alamy Stock Photo; p. 382: JGI/Jamie Gril/Blend Images/Getty Images

Chapter 10

p. 392: Worthen, J. B., & Hunt, R. R. (2011). *Mnemonology: Mnemonics for the 21st Century.* New York: Psychology Press; p. 392: XiXinXing/Getty Images; p. 406: Martin Shields/Alamy Stock Photo; p. 409 rendering by L. Frost from The Protein Data Bank/RCSB; p. 411 (mid-right) rendering by L. Frost from the Protein Data Bank/RCSB; p. 411: (*left*) David M. Phillips/Science Source; p. 411: (*middle*) Leung Cho Pan/123RF; p. 411: (*right*) Steve Gschmeissner/Science Photo Library/Alamy Stock Photo; p. 414: Forgiss/Fotolia p. 417 (right) rendering by L. Frost from the Protein Data Bank/RCSB; p. 419 rendering by L. Frost from the Protein Data Bank/RCSB; p. 420 rendering by L. Frost from the Protein Data Bank/RCSB;

Chapter 11

p. 439: Chew, S. Developing a Mindset for Successful Learning, https://www.samford.edu/departments/academicsuccess-center/how-to-study, retrieved 4/4/18; p. 439: Stockbroker/Alamy Stock Photo; p. 467: Leland Bobbe/The Image Bank/Getty Images; p. 471: CB2/ZOB/WENN/Newscom

Chapter 12

p. 479: Cepeda, N.J., Vul., E. Rohrer, D., Wixted, J.T. Paschler, H. (2008). Spacing Effects in Learning: A temporal Ridgeline of Optimal Retention. *Psychological Science,* 19(11), 1095–1102; p. 479: FatCamera/iStock/Getty Images; p. 481: Niko Endres/Fotolia; p. 487: Duplass/Shutterstock; p. 494: (*top*) JANICE FUHRMAN/KRT/Newscom; p. 494: (*bottom*) Dean Drobot/Shutterstock; p. 502: Laura D. Frost

Glossary/Index

Anabolism Metabolic chemical reactions in which smaller molecules are combined to form larger ones. These reactions tend to be reductive and require energy. 480, 482

Anaerobic In the absence of oxygen. 494–495

Anaerobic conditions, glycolysis, 494–495

Analgesics, 146

Anesthetics, 378

Angiotensin II, 407

Animal cell, 482–483

Animal fat, 147, 299–300

Anions An ion with a net negative charge; formed when an atom gains electrons. 80–82. *See also* Ionic compound

Anomer A sugar diastereomer differing only in the position of the hydroxyl at the anomeric carbon. 238–241

Anomeric carbon In a monosaccharide, the carbon that was the carbonyl in the linear structure; for example, carbon 1 of D-glucose. 238–241

Antibiotics, 427

Antibodies, 417

Anticoagulants, 258

Anticodon The triplet of bases in the center loop of tRNA that is complementary to a codon on mRNA. 456, 461

Anticodon loop, 456

Antigen A molecule capable of eliciting the immune response of producing antibodies in the body. 417

Antiparallel, 452

Aqueous *(aq)* label, 11

Aqueous solution A solution in which water is the solvent. 314

  acids and bases in, 353–354
  cell membranes, transport across, 342–344
  chemical equations for, 322–326
  colloids and suspensions, 315–316
  concentration, 328–333
  diffusion and dialysis, 340–341
  dilutions, 334–336
  electrolytes, overview, 322–324
  equilibrium constant *(K)*, 360–362
  formation of, overview, 316–317
  Henry's law, pressure and, 318–320
  hydration, 317
  ionic solutions and equivalents, 326–327
  iso-, hypo-, and hypertonic solutions, 337
  as mixtures, 314–316
  molarity, calculation of, 329–330
  osmosis, 337–342
  parts per million and billion, 332–333
  percent concentration, 330, 332
  percent mass/mass, 331
  percent mass/volume, 331
  percent volume/volume, 331
  solubility and saturated solutions, 317
  solubility and temperature, 318
  stock solutions, 336
  water, properties of, 314–315

Arachidic acid, 147

Arachidonic acid, 163

Arginase, 425

Arginine (Arg, R), 397, 398, 458, 510–512

Aromatic alcohols, 139

Aromatic aldehyde, 223

Aromatic compounds A family of cyclic organic compounds whose functional group is a benzene ring. 144–145

Aromatic hydrocarbons, overview of, 137, 139

Aromaticity A term used to describe the unexpected stability of aromatic compounds. 144–145

Arrhenius, Svante, 353

Arrhenius acids and bases, 353

Artificial sweeteners, 250

Ascorbic acid, 298

Asparagine (Asn, N), 393, 396, 398, 458, 510–512

Aspartame, 250

Aspartate (Asp, D), 396, 398, 458, 510–512

Aspirin, 126, 423

Asthma, 382–383

-ate, acid suffix, 356

Atmosphere (atm) A pressure unit commonly used to measure atmospheric pressure equal to 101,325 Pa. 269–270

Atmospheric pressure, boiling point and, 280

Atom The smallest unit of matter that can exist and keep its chemically unique characteristics. 4
  atomic mass and isotopes, 54–56
  atomic number and mass number, 51–53
  components of, 50–51
  electron arrangement in, 78–80
  mole, 98–103
  radioactivity and radioisotopes, 56–60

Atomic mass The weighted average mass of all naturally occurring isotopes of an element. 55

Atomic mass unit (amu) The small unit of mass used to quantify the mass of very small particles; 1 amu is equal to one-twelfth of a carbon atom containing six protons and six neutrons. 51, 101

Atomic number The number above the symbol of an element on the periodic table. It is equal to the number of protons in the element. 51–53

ATP (adenosine triphosphate) The main molecule that transfers energy in the cells of living systems. 184
  amino acid metabolism, 510–512
  anabolism and catabolism, overview, 480–483
  cell membranes, transport across, 343–344
  citric acid cycle, 496–499
  condensation and hydrolysis reactions, 204–206
  electron transport chain, 499–502
  fatty acid catabolism, 507–509
  glycolysis, 490–496
  hexokinase and, 418–422, 492
  metabolic energy from, 479, 485, 511–512
  oxidative phosphorylation, 499–502
  production of, summary, 503–506
  structure of, 484
  thermogenesis, 502

ATP synthase, 502

Attractive force An attraction of opposite charges between molecules. 281
  in biomolecules, 290–291
  boiling point and vapor pressure, 280, 283, 287–290
  cell membrane and, 301–303
  determining forces in a compound, 286
  in dietary lipids, 299–300
  dipole-dipole attractions, 282
  in gases, 269–276
  hydrogen bonding, 282–285
  ion-dipole attraction, 285
  in liquids and solids, 277–291
  London forces, 282
  melting point, 289
  soap and, 268
  solubility and, 292–298
  types of, 281–287

Autoionization of water *(K_w)* Spontaneous reaction of two water molecules to form a hydronium ion (H_3O^+) and a hydroxide ion (OH^-). 369–370

Avogadro, Amedeo, 98

Avogadro's number (N) The number of particles present in a mole of a substance (6.02×10^{23}). 98–103, 329

B

Bacteria, antibiotics and, 427

Barley, 248

Base A substance that dissolves in water and produces hydroxide ions (OH^-), according to the Arrhenius theory. All bases are proton acceptors according to the Brønsted–Lowry theory. 353
  antacids, 357–358
  in aqueous solution, 353–354
  bicarbonate buffer system, 379–383
  conjugate acids and bases, 366–368, 376–377
  Henderson-Hasselbalch equation, 376–377
  neutralization reactions, 356–357
  pH and pH scale, 369–372
  strong bases, 355–356
  tissue pH and drug delivery, 375–378
  weak acid-base equations, writing of, 367–368
  weak bases, 356, 363–368

Basic A term that describes an aqueous solution in which the concentration of hydroxide ions (OH^-) is greater than the concentration of hydronium ions (H_3O^+). Basic solutions have a pH higher than 7. 353

Becquerel (Bq) A unit of radiation activity equal to 1 nuclear decay event (disintegration) per second. 65

Becquerel, Henri, 56

Beeswax, 205

Behnic acid, 147

Benedict's test (reagent) A common test for the presence of a reducing sugar. If a reducing sugar is present, a brick–red precipitate results. 241–242

Bent shape, 106

Benzaldehyde, 223

Benzene, 144–145

Beta (β)-amyloid protein, 409

Beta (β) anomer, 238–241

Beta (β)-carotene, 143

Beta (β) decay, 63

Beta (β) oxidation The degradation of fatty acids by removing two carbon segments from a fatty acid at the oxidized (β) carbon. 506–509

Beta (β) particle A form of nuclear radiation consisting of a high energy electron emitted from an unstable neutron. 57–58

Beta (β)-pleated sheet A secondary protein structure with a zigzag structure resembling a folded fan. 408

Bicarbonate
  acid dissociation constant *(K_a)*, 365
  formula and uses, 83
  important ions in body, 84
  ion formation, 82
  in IV replacement solutions, 327

Bicarbonate buffer system, 379–383

Bilayer, phospholipid, 301–303

Bile, 488

Bile salt, 488

Binary compound A compound that is composed of only two elements. 96–97

Biochemical reaction Chemical reactions that occur in living systems. 191, 481. *See also* Metabolism
  condensation and hydrolysis reactions, 203–206
  oxidation in cells, 202
  representation of, 196–197

Biodiesel, 135–136

Biological half-life The time it takes for a radioactive substance to lose half of its radioactivity due to bodily functions. 66